Materials Aspect of Thermoelectricity

Materials Aspect of Thermoelectricity

EDITED BY
CTIRAD UHER
UNIVERSITY OF MICHIGAN, ANN ARBOR, USA

CRC Press
Taylor & Francis Group
Boca Raton London New York

CRC Press is an imprint of the
Taylor & Francis Group, an **informa** business

CRC Press
Taylor & Francis Group
6000 Broken Sound Parkway NW, Suite 300
Boca Raton, FL 33487-2742

Printed on acid-free paper
Version Date: 20160825

International Standard Book Number-13: 978-1-4987-5490-3 (Hardback)

Library of Congress Cataloging-in-Publication Data

Names: Uher, Ctirad, editor.
Title: Materials aspect of thermoelectricity / editor, Ctirad Uher.
Description: Boca Raton : CRC Press, 2017. | Includes index.
Identifiers: LCCN 2016032831 | ISBN 9781498754903 (hardcover : acid-free paper)
Subjects: LCSH: Thermoelectricity. | Thermoelectric materials. | Materials--Thermal properties. | Expansion (Heat)
Classification: LCC TA418.52 .M374 2017 | DDC 621.31/243--dc23
LC record available at https://lccn.loc.gov/2016032831

Visit the Taylor & Francis Web site at
http://www.taylorandfrancis.com

and the CRC Press Web site at
http://www.crcpress.com

Printed and bound in the United States of America by Publishers Graphics, LLC on sustainably sourced paper.

Contents

Preface

Since the publication in 2012 of the most recent volume of *Thermoelectrics and Its Energy Harvesting*, edited by the late David M. Rowe, the research activity in the field of thermoelectricity has continued its rapid pace. Novel families of materials have been discovered and major improvements in the classical thermoelectric materials made. It is thus appropriate to collect, in a new volume, the major advancements made in the material aspects of thermoelectricity and provide their critical assessment in regard to the broadening of application opportunities for thermoelectric energy conversion. While the thermoelectric community has been greatly missing the dedicated work of Professor Rowe in editing the past three very successful volumes published by CRC Press, I have attempted to continue that effort by assembling authoritative reviews by leading scientists working in the field of thermoelectricity. In this new volume, I aim to focus on the material aspects of recent developments, as this area is evolving very rapidly and, potentially, could have an immense impact on the entire field of thermoelectricity.

Material Aspects of Thermoelectricity consists of 21 chapters written by top researchers who were directly involved with the respective areas of research. I have requested them to keep their chapters in the style of review articles and to not be afraid of critically assessing the field.

I have followed the practice of Professor Rowe and played the role of an acquisition editor rather than a copy editor. Although I have attempted to clarify meaning on numerous occasions, I have tried to maintain the international flavor of the book by inviting scientists from different continents. I have also tried to give the opportunity to younger and more mid-career researchers who have been exceptionally successful in carving the path and opening new horizons in the field of thermoelectrics.

I thank all contributing authors for their time, energy, and willingness to prepare their chapters, and I hope this new volume will be welcomed by the thermoelectricity community and beyond. Finally, I gratefully acknowledge the tremendous support of my graduate students, Alex Page and Trevor Bailey, who assisted me with processing the manuscripts.

Ctirad Uher

Editor

Ctirad Uher is a C. Wilbur Peters Professor of Physics at the University of Michigan in Ann Arbor.

He earned his BSc in physics with a University Medal from the University of New South Wales in Sydney, Australia. He carried out his graduate studies at the same institution under Professor H. J. Goldsmid on the topic of "Thermomagnetic effects in bismuth and its dilute alloys," and received his PhD in 1975. Subsequently, Professor Uher was awarded the prestigious Queen Elizabeth II Research Fellowship, which he spent at Commonwealth Scientific and Industrial Research Organization (CSIRO), National Measurement Laboratory (NML), in Sydney. In 1989, the University of New South Wales awarded him the title of DSc for his work on transport properties of semimetals.

Professor Uher started his academic career in 1980 as an assistant professor of physics at the University of Michigan. He progressed through the ranks and became full professor in 1989. He served as an associate chair of the Department of Physics and subsequently as an associate dean for research at the College of Literature, Sciences and Arts. In 1994, he was appointed as chair of physics, the post he held for the next 10 years.

Professor Uher's more than 40 years of research is described in more than 400 refereed publications in the area of transport properties of solids, superconductivity, diluted magnetic semiconductors, and thermoelectricity. In the field of thermoelectricity, to which he returned during the past 20 years, he worked on the development of skutterudites, half-Heusler alloys, modified lead telluride materials, magnesium silicide solid solutions, tetrahedrites, and Molecular Beam Epitaxy (MBE)-grown thin film forms of Bi_2Te_3-based materials. He has written a number of authoritative review articles and has presented his research at numerous national and international conferences and invited talks. In 1996, he was elected fellow of the American Physical Society. Professor Uher was honored with the title of Doctor Honoris Causa from the University of Pardubice in the Czech Republic in 2002, and in 2010 he was awarded a named professorship at the University of Michigan. He received the prestigious China Friendship Award in 2011.

Professor Uher supervised 15 PhD thesis projects and mentored numerous postdoctoral researchers, many of whom are leading scientists in academia and in research institutions all over the world.

Professor Uher served on the board of directors of the International Thermal Conductivity Conferences and as a board member of the International Thermoelectric Society. In 2004–2005, he was elected vice-president of the International Thermoelectric Society and during 2006–2008 served as its president.

Contributors

Abdeljalil Assoud
Department of Chemistry
and
Waterloo Institute for Nanotechnology
University of Waterloo
Waterloo, Ontario, Canada

Umut Aydemir
Department of Materials Science and Engineering
Northwestern University
Evanston, Illinois

Lidong Chen
State Key Laboratory of High Performance Ceramics and Superfine Microstructure
and
CAS Key Laboratory of Materials for Energy Conversion
Shanghai Institute of Ceramics
Chinese Academy of Sciences
Changning, Shanghai, People's Republic of China

Hang Chi
Department of Physics
University of Michigan
Ann Arbor, Michigan
and
Condensed Matter Physics and Materials Science Department
Brookhaven National Laboratory
Upton, New York

Caitlin M. Crawford
Department of Physics
Colorado School of Mines
Golden, Colorado

Titas Dasgupta
Department of Metallurgical Engineering and Materials Science
Indian Institute of Technology Bombay
Mumbai, Maharashtra, India

Johannes de Boor
Institute of Materials Research
German Aerospace Center
Köln, Germany

Franck Gascoin
Laboratoire de Cristallographie et Science des Matèriaux (CRISMAT)
Université de Normandie
Caen, France

Prashun Gorai
National Renewable Energy Laboratory
Colorado School of Mines
Golden, Colorado

Quansheng Guo
Department of Chemistry
and
Waterloo Institute for Nanotechnology
University of Waterloo
Waterloo, Ontario, Canada

Joseph P. Heremans
Department of Mechanical and Aerospace Engineering
and
Department of Physics
and
Department of Materials Science and Engineering
The Ohio State University
Columbus, Ohio

Mercouri G. Kanatzidis
Department of Chemistry
Northwestern University
Evanston, Illinois
and
Materials Science Division
Argonne National Laboratory
Argonne, Illinois

Gun-Ho Kim
School of Mechanical and Nuclear Engineering
Ulsan National Institute of Science and Technology (UNIST)
Ulsan, South Korea

Holger Kleinke
Department of Chemistry
and
Waterloo Institute for Nanotechnology
University of Waterloo
Waterloo, Ontario, Canada

Kunihito Koumoto
Toyota Physical and Chemical Research Institute
Nagakute, Japan

Jing-Feng Li
Key Laboratory of New Ceramics and Fine Processing
School of Materials Science and Engineering
Tsinghua University
Beijing, People's Republic of China

Wei Liu
Department of Physics
University of Michigan
Ann Arbor, Michigan
and
Max Planck Institute for Chemical Physics of Solids
Dresden, Germany

Yuanfeng Liu
Department of Materials Science and Engineering
University of Michigan
Ann Arbor, Michigan

Ruiming Lu
Department of Materials Science and Engineering
University of Michigan
Ann Arbor, Michigan

Xu Lu
College of Physics
Chongqing University
Shapingba, Chongqing, People's Republic of China

Yuzuru Miyazaki
Department of Applied Physics
Graduate School of Engineering
Tohoku University
Sendai, Japan

Donald T. Morelli
Department of Chemical Engineering and Materials Science
Michigan State University
East Lansing, Michigan

Takao Mori
International Center for Materials Nanoarchitectonics (WPI-MANA)
National Institute for Materials Science
and
Graduate School of Pure and Applied Sciences
University of Tsukuba
Tsukuba, Japan

Eckhard Müller
Institute of Materials Research
German Aerospace Center
Koeln, Germany
and
Institute for Inorganic and Analytical Chemistry
Justus Liebig University Giessen
Giessen, Germany

Alexander Page
Department of Physics
University of Michigan
Ann Arbor, Michigan

Kevin P. Pipe
Department of Mechanical Engineering
University of Michigan
Ann Arbor, Michigan

Pierre F.P. Poudeu
Department of Materials Science and Engineering
University of Michigan
Ann Arbor, Michigan

Pengfei Qiu
State Key Laboratory of High Performance Ceramics and Superfine Microstructure
and
CAS Key Laboratory of Materials for Energy Conversion
Shanghai Institute of Ceramics
Chinese Academy of Sciences
Changning, Shanghai, People's Republic of China

Pranati Sahoo
Department of Materials Science and Engineering
University of Michigan
Ann Arbor, Michigan

Xun Shi
State Key Laboratory of High Performance Ceramics and Superfine Microstructure
and
CAS Key Laboratory of Materials for Energy Conversion
Shanghai Institute of Ceramics
Chinese Academy of Sciences
Changning, Shanghai, People's Republic of China

G. Jeffrey Snyder
Department of Applied Physics and Materials Science
California Institute of Technology
Pasadena, California
and
Department of Materials Science and Engineering
Northwestern University
Evanston, Illinois

Vladan Stevanović
Department of Physics
Colorado School of Mines
Golden, Colorado

Xianli Su
State Key Laboratory of Advanced Technology for Materials Synthesis and Processing
Wuhan University of Technology
Wuhan, Hubei, People's Republic of China

Koichiro Suekuni
Graduate School of Advanced Sciences of Matter
and
Institute for Advanced Materials Research
Hiroshima University
Higashi-Hiroshima, Japan

Toshiro Takabatake
Graduate School of Advanced Sciences of Matter
and
Institute for Advanced Materials Research
Hiroshima University
Higashi-Hiroshima, Japan

Gangjian Tan
Department of Chemistry
Northwestern University
Evanston, Illinois

Xinfeng Tang
State Key Laboratory of Advanced Technology for Materials Synthesis and Processing
Wuhan University of Technology
Wuhan, Hubei, People's Republic of China

Yinglu Tang
Laboratory Materials for Energy Conversion
Empa, Swiss Federal Laboratories for Materials Science and Technology
Dübendorf, Switzerland

and

Department of Applied Physics and Materials Science
California Institute of Technology
Pasadena, California

and

Department of Materials Science and Engineering
Northwestern University
Evanston, Illinois

Ruoming Tian
Toyota Physical and Chemical Research Institute
Nagakute, Japan

Eric S. Toberer
Department of Physics
Colorado School of Mines
Golden, Colorado

Chunlei Wan
State Key Laboratory of New Ceramics and Fine Processing
School of Materials Science and Engineering
Tsinghua University
Beijing, People's Republic of China

Bartlomiej Wiendlocha
Faculty of Physics and Applied Computer Science
AGH University of Science and Technology
Krakow, Poland

Chris Wolverton
Department of Materials Science and Engineering
Northwestern University
Evanston, Illinois

Jihui Yang
Department of Materials Science and Engineering
University of Washington
Seattle, Washington

Ronggui Yang
Department of Mechanical Engineering
University of Colorado
Boulder, Colorado

Alex Zevalkink
Max Planck Institute for Chemical Physics of Solids
Dresden, Germany

Qingjie Zhang
State Key Laboratory of Advanced Technology for Materials Synthesis and Processing
Wuhan University of Technology
Wuhan, Hubei, People's Republic of China

Li-Dong Zhao
School of Materials Science and Engineering
Beihang University
Beijing, People's Republic of China

1

Discovery and Design of New Thermoelectric Materials

Eric S. Toberer, Prashun Gorai, and Vladan Stevanović
Colorado School of Mines

CONTENTS

1.1 Introduction

State-of-the-art thermoelectric generator efficiency has risen to over 15% due to the development of new materials [1]. While impressive, significant gains in conversion efficiency are still required for thermoelectric generators to be competitive against traditional mechanical heat engines [2,3]. In this chapter, we focus on the identification of new materials with the potential for even higher thermoelectric performance. Traditional experimentally led searches are compared to new approaches that utilize high-throughput (HT) computations to screen materials.

Experimentally driven exploration of new materials has been the incumbent philosophy since the dawn of humanity. Here, we consider current design principles for material exploration as well as outstanding questions linking the chemical structure and transport properties. These principles are highlighted *via* case examples of experimentally led

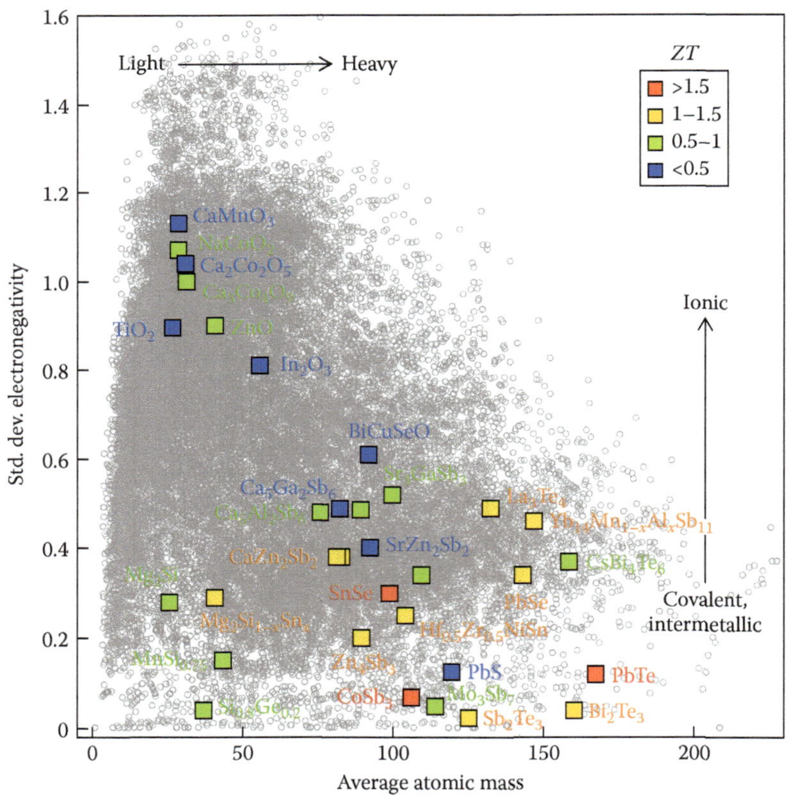

FIGURE 1.1
Only a handful of materials (33 shown) among the 40,000 known stoichiometric and crystalline metal–nonmetal inorganic solids (gray circles) from the Inorganic Crystal Structure Database [4] have so far been explored as thermoelectric materials (colored squares). Known thermoelectric materials span a diverse chemical space represented by the standard deviation in electronegativity and average atomic mass. A majority of the known thermoelectric materials are built from high-mass atoms with covalent or intermetallic bonding. (Reproduced from *Comp. Mater. Sci.*, 112, Gorai, P., Gao, D., Ortiz, B. et al., TE Design Lab: A virtual laboratory for thermoelectric material design, 368, Copyright 2016, with permission from Elsevier.)

searches that relied on serendipity, design principle development, and extrapolation to yield new classes of thermoelectric materials. These include the discovery of $CsBi_4Te_6$, the rise of Zintl compounds for high-temperature power conversion, and the "rattling" compounds with glass-like thermal conductivity.

We live in an age of computational search enabled by recent advances in both hardware and software. Within thermoelectric materials, ground-state electronic structure can be readily calculated on thousands of compounds with minimal human involvement [6–15]. While ground-state properties are not a complete picture of a material, there is a possibility to distill insights from such vast arrays of data. Significant challenges remain, however, for HT material exploration of thermoelectric properties. First, the search space is infinite. Within inorganic materials, tens of thousands of known, stoichiometric compounds (see Figure 1.1) represent a infinitesimal sliver of the possible alloys, dopants, metastable compounds, and hypothetical structure space. Consideration of the vast array of organic compounds and associated blends is even more daunting. Second, the critical terms for thermoelectric transport are nontrivial to calculate even in simple compounds. Specifically, charge carrier and phonon scattering require calculation of the transition probability for all possible interactions. Third, the transformation from chemical structure in real space to the transport properties residing in reciprocal space (e.g., group velocity) obscures the relationship between structure and properties. In other words, the multidimensional function that connects structure to properties is highly nonlinear. As a result, searches for advanced materials cannot rely on structural descriptors alone. As evidence of these challenges, HT thermoelectric searches have not yet yielded a revolutionary class of materials. Nevertheless, the potential for computation to radically change how new materials are discovered is enormous and serves to inspire this chapter.

1.2 Transport and Design Parameters

In this section, we review the design parameters that are critical to thermoelectric transport. While Boltzmann transport has been summarized many times, we review it here, as it serves to both guide understanding of experimental measurements and direct computational searches. We then reformulate the thermoelectric figure of merit ZT using solutions to the Boltzmann transport equation (BTE) to yield a "quality factor" β. To date, virtually all materials with high figure of merit are band conductors where free carriers are weakly interacting. In light of this, we frame the discussion of design parameters in terms of band conductors.

1.2.1 Boltzmann Transport Theory

1.2.1.1 Generalized Description

For both electron and phonon transport, we will utilize the BTE to describe the time evolution of the particle distribution function $f(\mathbf{r},\mathbf{p},t)$. Here, f provides a statistical distribution of particles over the coordinate ($\mathbf{r}$), momentum ($\mathbf{p}$), and time (t). It is important to note that this approach relies on several approximations: (1) it is a single-particle approach that neglects correlation effects; (2) collisions are treated as quantum events, while particles themselves are described semiclassically; and (3) collisions are usually binary and assumed to be instantaneous and localized in space. These approximations are valid for dilute carrier gases (low doping), in slowly varying fields and for systems with relatively high carrier mobility (the carrier mean free path needs to be larger than its de Broglie wavelength). Classic thermoelectric materials are typically band conductors, and we may expect that the quasiclassical treatment of carriers and the assumptions about collisions to hold from the intrinsic materials properties perspective. However, they are typically nearly degenerately doped and used at relatively high temperatures, which may lead to the breaking of some of these assumptions. Also, in strongly correlated systems and structures that conduct *via* carrier hopping, the charge carrier mobilities are usually low, leading to relatively short mean free paths. Additionally, nanostructuring and/or alloying that is used to reduce phonon mean free paths might also lead to the breaking down of some of these assumptions when applied to lattice thermal conductivity (LTC). To our knowledge, a systematic and quantitative analysis of the conditions under which approximations adopted in the BTEs break in thermoelectric materials is currently missing from the literature.

Under these assumptions and in thermodynamic equilibrium, f reduces to the standard Fermi–Dirac (electrons) or Bose–Einstein (phonons) distribution. Under nonequilibrium conditions, the time evolution of f at a point (**r**, **p**) of the phase space and time t is governed by particle drift due to applied fields that introduce external forces (**F**), the diffusion of particles ($\nabla_r f$), and collisions [16]:

$$\frac{\partial f}{\partial t} = -\frac{\boldsymbol{p}}{m}.\nabla_{\mathrm{r}} f - \boldsymbol{F}.\nabla_{\mathrm{p}} f + \left(\frac{\partial f}{\partial t}\right)_{coll} \tag{1.1}$$

where ∇_r and ∇_p stand for the gradients with respect to the coordinate and momentum, respectively. An important consequence is that in the absence of scattering ($\partial f/\partial t = 0$), one can rewrite Equation 1.1 as $df/dt = 0$, which implies that every state that is occupied at time t will remain occupied at any $t + \Delta t$. It is, therefore, only the scattering from and into a given state that changes its occupation as a function of time. Following the symbol conventions from Reference 16, the collision (scattering) term can be written for electrons (analogously for phonons) as

$$\left(\frac{\partial f}{\partial t}\right)_{coll} = \sum_{\mathrm{p}'} f(\boldsymbol{p}')[1 - f(\boldsymbol{p})]S(\boldsymbol{p}',\boldsymbol{p}) - f(\boldsymbol{p})[1 - f(\boldsymbol{p}')]S(\boldsymbol{p},\boldsymbol{p}') \tag{1.2}$$

where the first term in the sum describes scattering into the state with momentum $\boldsymbol{p}$ contributing to the increase of f, while the second describes scattering from $\boldsymbol{p}$ that acts to reduce f. Here, collisions are instantaneous in time and occur at a particular point, thus simplifying f to $f(\boldsymbol{p})$. $S(\boldsymbol{p}',\boldsymbol{p})$ stands for the probability per unit of time for the scattering from $\boldsymbol{p}'$ into $\boldsymbol{p}$ that could, in principle, be obtained using Fermi's golden rule. The product $[1 - f(\boldsymbol{p})]S(\boldsymbol{p}', \boldsymbol{p})$ is then nothing but the rate $1/\tau(\boldsymbol{p}',\boldsymbol{p})$ for the scattering process from $\boldsymbol{p}'$ to $\boldsymbol{p}$. As a result, Equation 1.2 can be rewritten as

$$\left(\frac{\partial f(\boldsymbol{p})}{\partial t}\right)_{coll} = \sum_{\mathrm{p}'} \frac{f(\boldsymbol{p}')}{\tau(\boldsymbol{p}',\boldsymbol{p})} - \frac{f(\boldsymbol{p})}{\tau(\boldsymbol{p},\boldsymbol{p}')} \tag{1.3}$$

1.2.1.2 Relaxation Time Approximation

In the case of low (nondegenerate) doping, $f(\boldsymbol{p})$ and $f(\boldsymbol{p}')$ are usually small. If taken as approximately equal and for distributions $f \approx f_0 + \delta f$ that are only slightly away from the equilibrium distribution, δf can then be taken outside the summation (note that $(\partial f_0/\partial t)_{coll} = 0$). The whole sum, which is now over the scattering rates, can be described using a single effective τ_{eff}. These considerations motivated the relaxation time approximation:

$$\left(\frac{\partial f(\boldsymbol{p})}{\partial t}\right)_{coll} = -\frac{f - f_0}{\tau_{eff}} \tag{1.4}$$

In the absence of external forces, this effective τ represents the time constant for the system to *return* (relax) to the equilibrium distribution.

Physically, relaxation back to f_O requires scattering events to cause transitions in momentum and consequently in energy. Electrons are predominantly scattered due to phonon distortions of the lattice in thermoelectric materials at high temperature. Such distortions alter the energy of the allowed electron wave functions and may induce a local polarization of the lattice. Likewise, defects and impurities serve as an additional source of electron scattering through the breakdown of periodicity and creation of local electric fields. At low temperatures, electron–electron scattering is reasonably uncommon due to the challenges of simultaneously satisfying the Pauli exclusion principle and energy and momentum conservation. As the Fermi–Dirac distribution smears with temperature, electron–electron scattering can become significant. Identifying materials with weak electron scattering and resolving the energy and momentum dependence of scattering are critical aspects of HT material exploration for thermoelectrics.

For phonon scattering, a material will intrinsically scatter phonons through phonon–phonon and phonon–electron interactions. Additionally, atomic and microstructural features can further break down the intrinsic periodicity and lead to further scattering. Much like electron scattering, a critical goal for HT thermoelectric searches is identifying materials with strong phonon scattering.

Note that τ is a function of the momentum and consequently of energy; different carriers may relax at different rates. For example, charge carriers at the conduction band edge (E_C) are rarely scattered by acoustic phonons due to the low density of available final states. As such, perturbations to the distribution function at E_C are slowly resolved through scattering events. In contrast, perturbations to f at higher energy may rapidly relax to f_O due to increased scattering rates. This energy dependence of τ is typically described as a power law (Equation 1.5). The constant relaxation time approximation (CRTA; $r = 0$) is often used in the interpretation of first-principles calculations (Equation 1.5). Here, the time constant for f to return to f_0 is assumed to be independent of carrier energy or momentum. Real systems, on the other hand, do not show a constant relaxation time and the functional form of $\tau(E)$ may be complex. In practice, Matthiessen's approach (Equation 1.6) is often used to estimate an effective τ from individual scattering sources (i). Likewise, phonon scattering is strongly energy dependent. For example, alloy mass contrast scattering of phonons can be viewed as Rayleigh scattering and the scattering rate scales with ω^{-4}.

$$\tau(E) \sim \tau_0 \left(\frac{E}{E_0}\right)^r \tag{1.5}$$

$$\frac{1}{\tau} = \sum_i \frac{1}{\tau_i} \tag{1.6}$$

1.2.1.3 Core Equations

For electron transport, the BTE within the relaxation time approximation can be evaluated in the presence of external electric fields and thermal gradients to yield the electrical conductivity (σ), the Seebeck coefficient (α), and the Lorenz number (L). Approximating transport as isotropic allows the following simplified integrals. In the isotropic limit, a simple, energy-dependent transport distribution function $\Sigma(E) = \sum_{n,\mathbf{k}} v_{n,i}(\mathbf{k})^2 \tau_n(\mathbf{k})\delta(E - E_n(\mathbf{k}))$ occurs in all of these terms. Here, $v_{n,j}(\mathbf{k})$ is the velocity (group) of the carrier in the state with the band index n and crystal momentum $\hbar\mathbf{k}$ assuming that the external field is applied along one of the crystal axes (j), $\tau_n(\mathbf{k})$ is its relaxation time, and the sum across $\delta(E-E_n(\mathbf{k}))$ represents the density of states. As first-principles calculations consider reciprocal space as a discrete grid, the expressions (Equation 1.7) are often written as sums across the first Brillouin zone rather than in integral form. Such a format regains the anisotropic character but requires a dense **k**-mesh to be accurate.

$$\sigma = -q^2 \int \Sigma(E) \frac{\partial f_0}{\partial E} dE \tag{1.7}$$

$$\alpha = \frac{-q}{T\sigma} \int (E - E_F)\Sigma(E) \frac{\partial f_0}{\partial E} dE \tag{1.8}$$

$$L = \frac{-1}{T^2\sigma} \int (E - E_F)^2 \Sigma(E) \frac{\partial f_0}{\partial E} dE \tag{1.9}$$

The visual interpretation of these equations can be seen in Figure 1.2 for a degenerately doped material with the Fermi level (chemical potential) inside of the band. A parabolic dispersion is assumed, leading to linear behavior in v^2 with energy and a density of states that scales with $E^{0.5}$. The charge carrier relaxation time decays with increasing energy due to the assumption of electron–phonon scattering. Here, the low density of states near the band edge leads to few available electron transitions and thus long relaxation times. Weighing the transport distribution function, $\Sigma(E)$, with $\partial f/\partial E$ and $(E - E_F)^n$ yields the integrands of the earlier expressions. Consideration of Figure 1.2 highlights the importance of asymmetry in $\Sigma(E)$ to induce a nonzero Seebeck coefficient upon integration.

A similar expression can be developed for the lattice component of thermal conductivity, κ_L. An isotropic approximation yields Equation 1.10, where each branch of the phonon dispersion, i, has a frequency-dependent heat capacity, C; phonon group velocity, v; and relaxation time, τ. Each branch has vibrational modes across a frequency range spanned from ω_{min} to ω_{max}. Rather than transporting charge (as in Equation 1.7), heat is transported and the quantity of heat transported depends on the heat capacity per mode, the total number of modes at some frequency ω, and the Bose–Einstein distribution. All of these terms are embedded in $C_i(\omega)$.

$$\kappa_L = \frac{1}{3} \sum_i \int_{\omega_{min}}^{\omega_{max}} C_i v_i^2 \tau_i d\omega \tag{1.10}$$

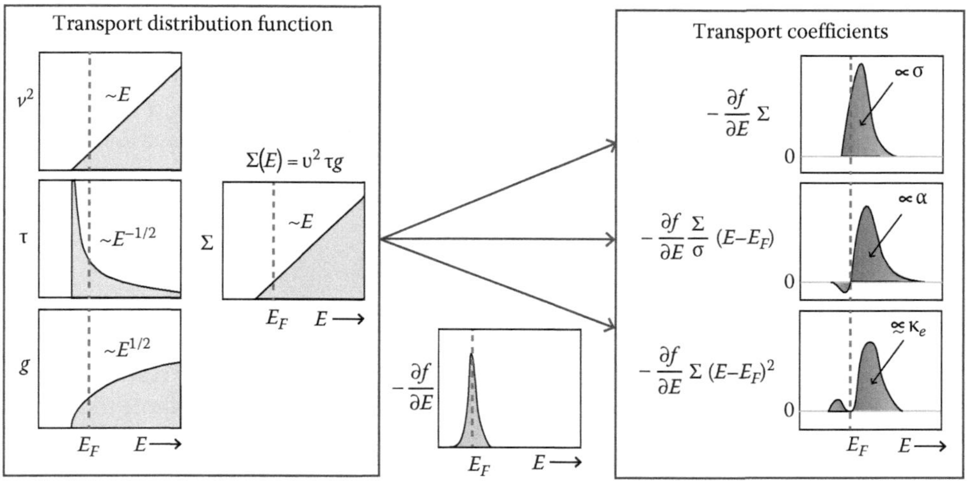

FIGURE 1.2
The transport distribution function (Σ), chemical potential (dashed line), and temperature (*via* the Fermi–Dirac distribution, f) define the critical transport integrals for thermoelectric performance. On the left, the components of Σ are shown for a degenerately doped parabolic band with acoustic phonon scattering. On the right is a visual representation of the integrands that determine σ, α, and L. Changes to $v(E)$, $\tau(E)$, or $g(E)$ will affect all macroscopic properties.

While optimizing the electronic transport, distribution function (Σ) is nontrivial for enhanced thermoelectric performance; interpreting Equation 1.10 for optimal phonon transport is comparatively simple. One seeks materials that simultaneously possess low heat capacity, small group velocity, and short relaxation times.

1.2.2 Performance Metrics

1.2.2.1 Figure of Merit ZT

The core figure of merit for thermoelectric performance (ZT) emerges both from macroscopic and microscopic considerations of generator efficiency. At the macroscopic scale, ZT emerges from an analysis of system performance within the constant property model (where σ, α, and κ are constants as a function of position or temperature). Similarly, analysis using the compatibility factor approach reveals that a fully self-compatible ($u = s$) generator has an efficiency strictly determined by ZT and the Carnot efficiency [17].

$$\eta = \frac{T_h - T_c}{T_h} \frac{\sqrt{1+ZT_{avg}} - 1}{\sqrt{1+ZT_{avg}} + (T_c/T_h)} \tag{1.11}$$

$$\eta = 1 - \exp\left(-\int_{T_c}^{T_h} \frac{\eta_r}{T} dT\right) \quad \eta_r = \frac{\sqrt{1+ZT} - 1}{\sqrt{1+ZT} + 1} \tag{1.12}$$

While ZT is the relevant metric, the components of ZT are challenging to directly calculate from first principles. Ideally, one would calculate σ, α, κ_e, and κ_L as a function of temperature and electron chemical potential. As discussed in the following, these terms are challenging to calculate for complex materials or in an HT fashion. Further, one would include the breakdown in periodicity inherent in thermoelectric materials (i.e., dopants, defects, and alloying species). As such, there is the need for approximations and reformulations of ZT that are compatible with the data available from HT calculations.

1.2.2.2 Quality Factor

The terms that constitute ZT are all nontrivial and have a strong dependence on carrier concentration; as such, it can be advantageous to separate out the carrier density from the nominally "intrinsic" material properties. When one approximates transport in terms of single carrier-type transport within an isotropic parabolic band, manipulation of the terms within ZT yields Equation 1.13. Here, the chemical potential is found in two functions u and v, while the nominally intrinsic terms can be found strictly in β. If the band edge has no degeneracy beyond spin, the density of states parameterization, m^*_{DOS}, and the band effective mass, m^*_b, are equivalent. For the case of N equivalent bands at the band edge, $m^*_{DOS} = N^{2/3} m^*_b$.

$$ZT = \frac{u\beta}{v\beta + 1} \tag{1.13}$$

$$\beta = \left(\frac{k_B}{e}\right)^2 \frac{2e(k_B T)^{3/2} T}{(2\pi)^{3/2} \hbar^3} \frac{\mu\left(m^*_{DOS}\right)^{3/2}}{\kappa_L} \tag{1.14}$$

This approximation allows the separation of "intrinsic" material terms (i.e., m^*_{DOS}, μ, κ_L) from the electron chemical potential, η. In concert, these material properties are termed the quality factor, β (Equation 1.14). Note that β is often associated with *Chasmar* and *Stratton* due to their excellent discussion but is previously discussed by *Goldsmid* and *Douglas* in 1954 [18,19]. Figure 1.3 shows that ZT scales monotonically with β and further depends on optimizing η. Good thermoelectric performance requires both large β and the ability to optimize η for maximum ZT.

From the perspective of material search, one can thus remove materials with low β without having to consider dopability. However, high β does not guarantee good thermoelectric performance as the optimization of the electron chemical potential may not be experimentally realizable. For example, many Zintl compounds have emerged as promising thermoelectric materials if doped n-type through several material searches [6,8]. However, optimizing the carrier concentration to the appropriate level has been hindered by Fermi level pinning due to compensating native defects [20,21].

We emphasize that β is strictly a reformulation of ZT to separate doping from the largely intrinsic material properties. It does not circumvent the need to measure or calculate electron and phonon scattering. Nor does it appropriately treat convoluted Fermi surfaces, as it assumes N equivalent spherical surfaces. Its appeal is the ability to rank materials with a simple scalar value and the fact that β's constituent terms highlight the key strengths and opportunities of a material.

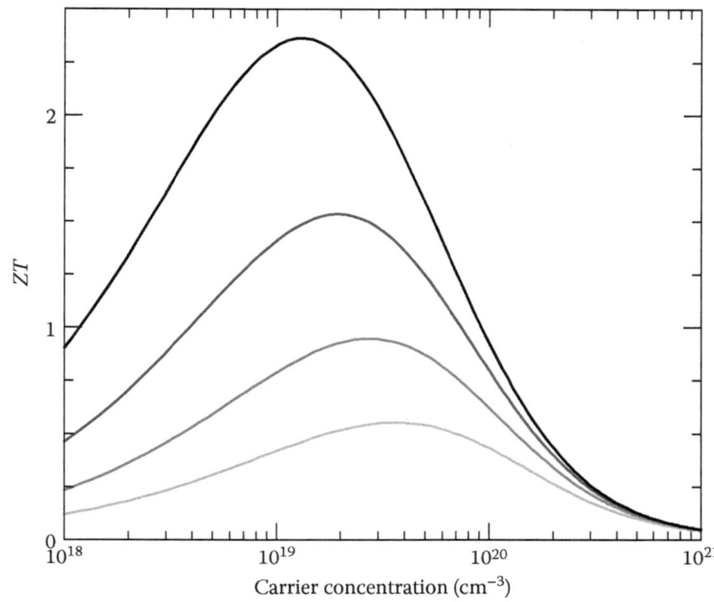

FIGURE 1.3
Thermoelectric performance can be estimated using a single parabolic band model for transport from solutions to the Boltzmann transport equation within the relaxation time approximation. Here, the quality factor $\beta \sim (\mu_0 (m^*_{DOS})^{3/2})/\kappa_L$ doubles between each line; curves generated using μ_0 = 100 (lowest curve)–800 (highest) $cm^2\ V^{-1}\ s^{-1}$, $\kappa_L = 1\ W\ m^{-1}\ K^{-1}$, $m^*_{DOS} = 1\ m_e$, and T = 300 K. The rise in optimized ZT is sublinear to the increase in β.

1.3 Experimental Approaches to Thermoelectric Material Discovery

Historically, thermoelectric (TE) material development has relied heavily on extrapolation from known materials to suggest new compounds for exploration. These searches have been experimentally driven with computation providing *ex post facto* insight into the reasons for the observed performance (if present at all). In this section, current structure–property relationships that can guide future searches are reviewed. Case studies of thermoelectric searches that have yielded new classes of materials are then discussed in the context of these structure–property guidelines.

1.3.1 Search Strategies for Experimentally Led Expeditions

Exploratory investigation of new materials for thermoelectric performance is time consuming and thus requires balancing the risk versus the reward. The farther away in chemical space that a material is from known thermoelectrics, the greater the risk. On the other hand, developing a new class of materials adds a new continent to our map of thermoelectric materials. To provide guidance and potentially reduce the risk when exploring new chemical spaces, the following proposes some simple heuristics.

1.3.1.1 Thermal Conductivity

In the search for materials with low thermal conductivity, one is searching for materials with strong phonon scattering, low group velocity phonons, and potentially low volumetric heat capacity [22]. The heat capacity is often ignored; however, there is the potential for low thermal conductivity in microporous materials. Thermoelectric materials typically operate well above the Debye temperature, and thus, the volumetric heat capacity simply scales linearly with the number density of atoms. In light of this, porous materials such as MOFs or zeolites could potentially show much lower thermal conductivities simply due to their open structure. For cryogenic applications, one may seek materials with extremely high Debye temperatures (i.e., stiff systems) so that the volumetric heat capacity remains low [22].

For phonons to scatter, a breakdown in the periodic, harmonic approximation must occur. This can occur through phonons interacting with other particles (e.g., phonons, electrons, magnons) or loss of registry in the lattice. Concerning loss of registry, high defect concentrations (e.g., Zn_4Sb_3) can be expected to induce strong point defect scattering. Such information may be part of the original structural study of the material, provided care was taken in the refinement. On a slightly larger length scale, some materials naturally form high densities of twins or other extended defects (e.g., $SrZnSb_2$, perovskite superstructures) [24]. However, predicting these natural nanostructures *a priori* is challenging. At high temperatures, phonon–phonon scattering is typically the most critical source of scattering. Phonon–phonon scattering is mediated *via* anharmonic interactions in the crystal and requires states for the allowed transition. For an experimentally led investigation, both the phonon band structure and the strength of the anharmonicity are not obvious in advance. Structural descriptors that suggest the potential for high anharmonicity include soft bonding, high coordination numbers, low symmetry to the coordination sphere, and the presence of multiple equivalent atomic positions. For example, many of the Zintl compounds show anion lone pairs that may exhibit anharmonic interactions with surrounding atoms. Further complication arises from the variation in anharmonicity associated with each phonon mode. For example, Figure 1.4 shows the variation in mode Grüneisen ($\gamma_{mode} = -(V/\omega)(\partial\omega/\partial V)$) of d-Si for the acoustic and optical branches across the Brillouin zone. As the square of γ_{mode} is the critical term in the scattering matrix element, the magnitude, rather than the sign, determines the scattering strength. Note that the rise in γ_{mode} near Γ is due to the ω^{-1} in the calculation of γ_{mode}.

Complementary to scattering, low phonon group velocity is a critical metric in the exploratory search for new thermoelectric materials. For the acoustic branches, one is simply looking for a material with soft bonding, high mass density, and a propensity to shear. Soft bonding is often found in ionic materials, compounds with significant nuclear screening, and layered compounds with weak bonding (e.g., SnSe). Achieving low group velocity in the optical branches emerges from high mass contrast within the unit cell. Mass and local elastic contrast creates gaps in the dispersion, which lowers the average velocity. As the number of vibrational branches grows with the number of atoms in the unit cell, the impact of these gaps on the group velocity increases.

The use of nanostructuring has developed over the last two decades as a critical tool to lower thermal conductivity. If nanostructuring is employed, one primarily looks at the speed of sound as an indicator of the lower bound that nanostructuring can achieve [25]. Conveniently, the speed of sound is easier to predict than scattering rates from simple inspection of the structure and bonding. To achieve exceptionally high ZT, it may be that we need materials where high mobility is preserved and routes other than nanostructuring

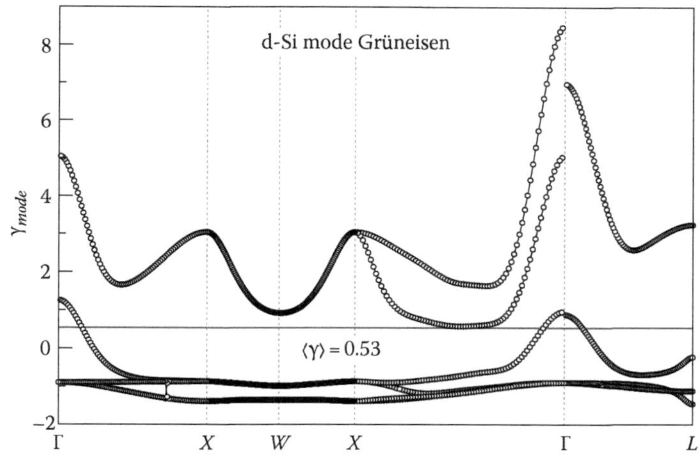

FIGURE 1.4
The mode Grüneisen within d-Si varies significantly in both magnitude and sign across the six phonon branches. The measured Grüneisen, $\langle\gamma\rangle$, from thermal expansion experiments [23] partially cancels the positive and negative values. For assessing phonon–phonon scattering and ultimately κ_L, the more relevant term is $\langle\gamma^2\rangle$.

are employed to achieve low κ_L. In this case, exploratory searches face the challenge of identifying materials with large anharmonicity to enhance phonon–phonon scattering.

1.3.1.2 Electronic Transport

The development of heuristics for electron transport is fundamentally more challenging than phonon transport due to the difference in (quasi-)particle type. As bosons, all phonons participate in thermal transport at high temperature, leading to an averaging across the dispersion. In contrast, charge carrier transport is unaffected by states far away from the electron chemical potential. As such, chemical modifications that alter the electron group velocity, density of states, or scattering of states near the band edge can profoundly alter the macroscopic charge transport properties. For example, Figure 1.5 shows how different defects in $ZnSiP_2$ alter the band edge density of states [26]. Unfortunately, this chemical sensitivity makes the development of robust heuristics difficult for electronic transport. Further, the transformation from a real space potential to reciprocal space electronic structure *via* the Bloch formalism requires far more than intuition for complex materials. Figure 1.6 shows the spread in valence band effective mass across 2000 compounds as a function of number of atoms in the primitive cell and electronegativity contrast. The absence of a clear trend that exceeds the scatter highlights the challenge associated with the development of simple heuristics for predicting electronic structure.

Thermoelectric materials are typically semiconductors and the first-round screening when exploring new materials is to consider valence balance. Careful examination of bonding is required to assess if the system is electron deficient or rich. In some cases, the stoichiometric electronic state is metallic but close to a band edge. For example, Mo_3Sb_7 is electron deficient by one electron per formula unit; substitution with Te brings the chemical potential into the bandgap [27,28]. Similar behavior is seen in $Yb_{14}MnSb_{11}$ upon substitution with Al^{3+} or La^{3+} for Yb^{2+} or Mn^{2+} [29]. For intermetallic materials, even electron counting may be challenging. While intermetallic compounds like the Heusler systems

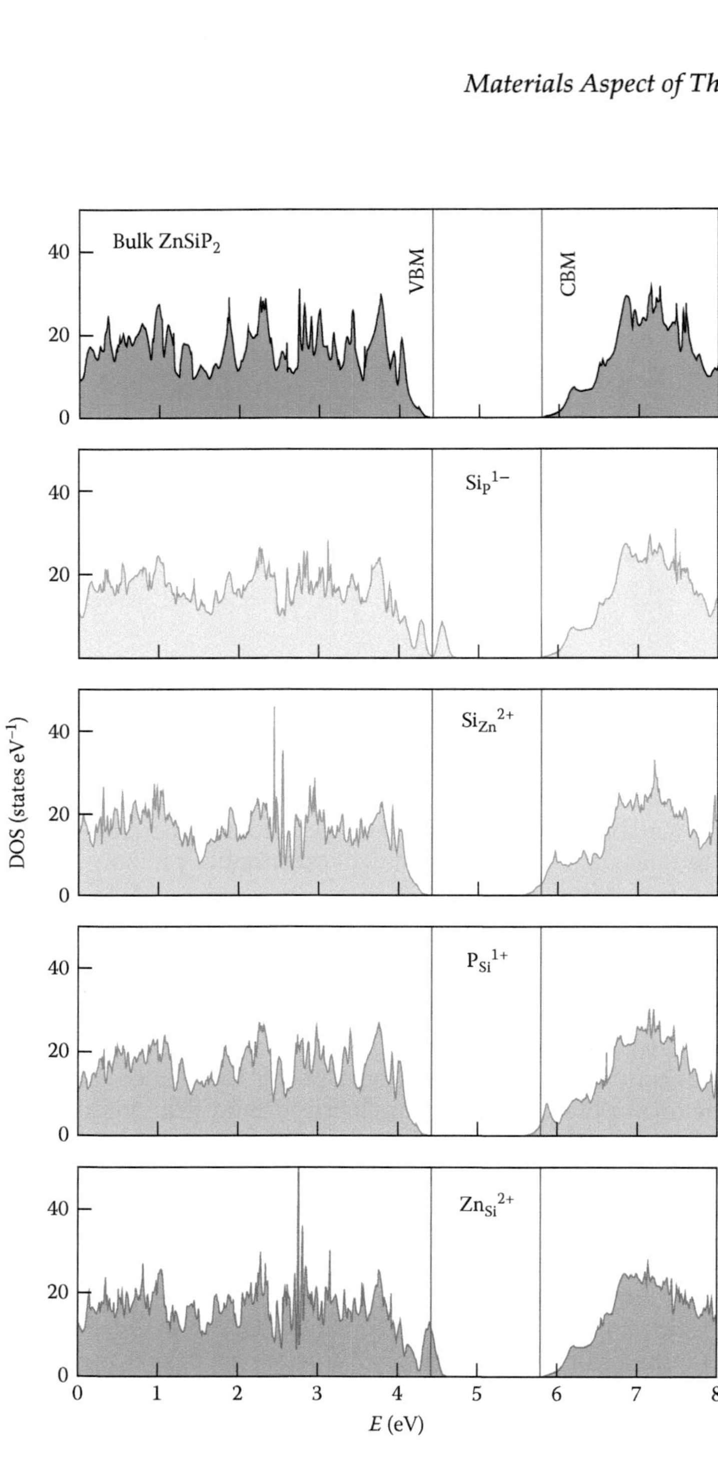

FIGURE 1.5
To illustrate the profound effect defects can have on electronic properties, the density of states of $ZnSiP_2$ near the band edge is calculated. The upper panel is the perfect chalcopyrite structure, while the lower four panels show the impact of the lowest-energy native defects. As such, comparing ground-state calculations with transport measurements can necessitate more sophisticated calculations than the standard band structure calculations for the primitive cell. (From Martinez, A.D., Warren, E.L., Gorai, P. et al., 2016, Solar energy conversion properties and defect physics of $ZnSiP_2$, *Energy Environ. Sci.*, 9: 1031. Reproduced by permission of The Royal Society of Chemistry.)

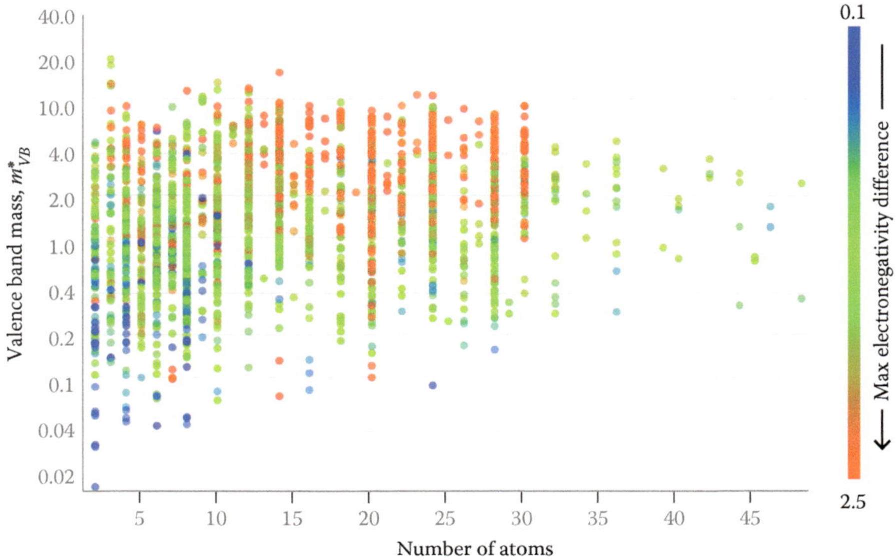

FIGURE 1.6
Extracting relationships between real space structure and bonding and the electronic structure is highly nontrivial; here, the dependence of the valence band mass on the number of atoms in the primitive cell and the maximum contrast in constituent electronegativity (heat) are considered for 2000 semiconductors/insulators [5].

are known to exhibit a bandgap associated with filling the bands to certain electron counts per formula unit, such rules are absent in exotic intermetallic systems [30].

Electron–phonon scattering rates are similarly quite challenging to predict. For acoustic deformation potential scattering, $\tau \propto C_{11}(m^*_{DOS})^{-3/2}\Xi^{-2}$, where C_{11} is the elastic constant for longitudinal vibrations and Ξ is the deformation potential. While Ξ is unknown *a priori*, one can see that stiffer compounds will be less affected by electron–phonon scattering due to the decreased deformation of the lattice. Likewise, compounds with low m^*_{DOS} offer fewer electronic states to transition into and thereby a larger τ. Note that increased band degeneracy will increase m^*_{DOS}, potentially offsetting gains in $\alpha(n)$ from band degeneracy. Nevertheless, simple initial electronic structure calculations, even in an experimentally led search, can provide some insight into mobility *via* effective mass and bulk modulus.

1.3.2 Case Examples of Experimentally Driven Searches

Most experimental searches for new thermoelectric materials have extrapolated from the local chemistry and properties of known thermoelectric materials to suggest other compositions to explore. A minority of searches have led with hypotheses associated with experimental or theoretical physics phenomena (e.g., model Hamiltonian, properties of nonthermoelectric materials).

1.3.2.1 Extrapolation and Intuition

Classic thermoelectric materials searches have been serial efforts that develop an intuition through observation of prior results. We consider two successful examples of extrapolation that yielded new thermoelectric materials and structure–property relationships.

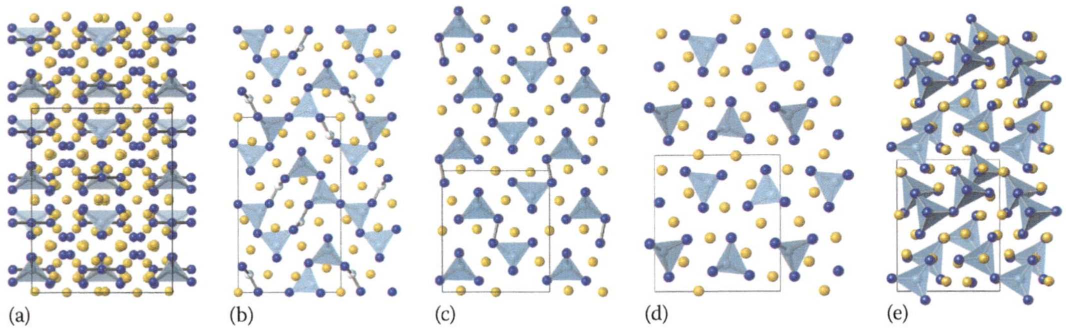

FIGURE 1.7
Following the discovery of high ZT in (a) $Yb_{14}MnSb_{11}$, structure–property trends in Zintl compounds based on MSb_4 were extensively investigated. Here, (b) $Yb_9Zn_{4.2}Sb_9$, (c) $Ca_5Al_2Sb_6$, (d) Ca_3AlSb_3, and (e) Sr_3GaSb_3 highlight the structural diversity available with this building block.

Since their discovery and classification in the 1930s, Zintl compounds had been considered to be intrinsic semiconductors with poor conductivity. Beginning in the 1990s, the thermoelectric community began to appreciate that such compounds could, in fact, be doped and early examples showed moderate ZT [31]. The chemical complexity found across Zintl compounds creates an enormous search space; early efforts by the *Kauzlarich*, *Kanatzidis*, and *Nolas* groups helped build a foundation of understanding. Notable early successes include $Ba_8Ga_{16}Ge_{30}$, $CsBi_4Te_6$ (discussed in Section 1.3.2.2), and $YbZn_2Sb_2$ [32]. Particularly notable is the discovery of $Yb_{14}MnSb_{11}$, which showed the promise of Zintl compounds for high temperature power generation. Subsequent studies have extrapolated from this success, studying other Zintl compounds with polyanionic MSb_4 tetrahedral structures (see Figure 1.7) [33–36]. Much of this work is inspired by manually searching through the Inorganic Crystal Structure Database (ICSD) to "rediscover" Zintl compounds that have never been studied for their transport properties.

The search for thermoelectric materials for cooling applications has been particularly challenging over the last half century, with virtually all thermoelectric materials having ZT at room temperature (RT) less than 0.5. Alloys based on Bi_2Te_3 and $Bi_{1-x}Sb_x$ are two notable exceptions. Bi_2Te_3's excellent thermoelectric performance arises from a layered structure with low shear modulus, heavy constituent atoms, high band degeneracy, and high mobility. Inspired by this excellent performance, the *Kanatzidis* group undertook a search for new structures from alkali metals, bismuth, and chalcogen atoms [37]. These efforts yielded many new compounds, including $Rb_{0.5}Bi_{1.83}Te_3$, $RbPb_2Bi_3Te_7$, $BaBiTe_3$, and $K_2Bi_8Q_{13}$ (Q = S, Se). In an effort to extend Q to the telluride analog, $Cs_2Bi_8Te_{13}$ was prepared but surprisingly yielded $CsBi_4Te_6$ instead [38]. $CsBi_4Te_6$ (Figure 1.8) proved to be an excellent low-temperature thermoelectric material with mobility even higher than Bi_2Te_3. This historical perspective highlights how extrapolation in concert with embracing serendipity can yield advanced materials.

1.3.2.2 Novel Physics

The preceding two examples extrapolate from known thermoelectric materials to drive experimental searches. In contrast, the development of "rattling" thermoelectric materials was grounded in observations of simple halide salts. Thermal transport in KCl was

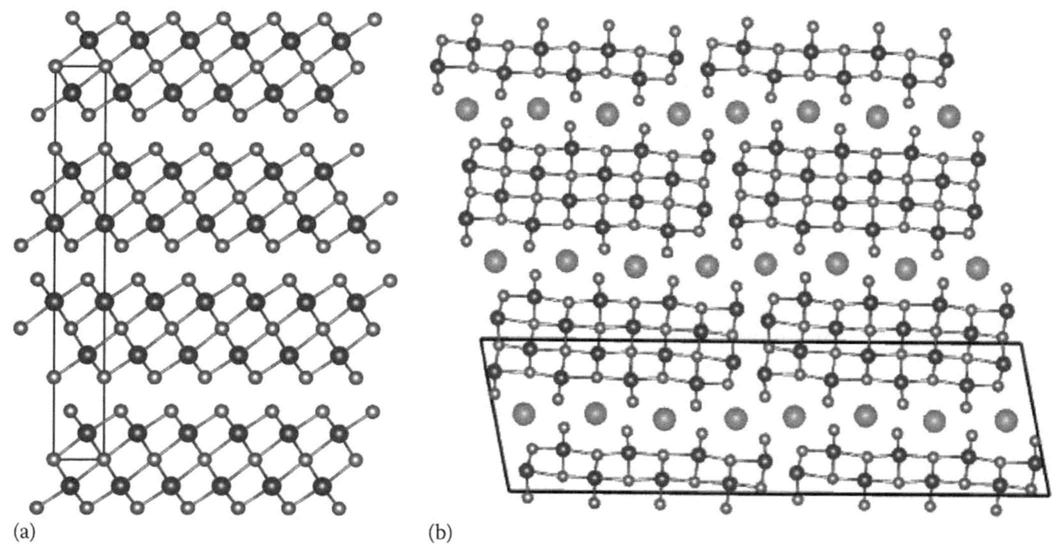

FIGURE 1.8
Bi_2Te_3 (a) served as an inspiration for a search of complex tellurides; this ultimately yielded $CsBi_4Te_6$ (b). These compounds share a high degree of anisotropy and distorted rock salt slabs of BiTe that terminate with Te surfaces. $CsBi_4Te_6$ has similar transport properties to Bi_2Te_3, yielding an excellent ZT at and below room temperature.

known to be radically suppressed by substitution with NO_2^- ions. The rotational modes of NO_2^- served as localized states for scattering phonons [39,40]. The fundamental physics of this phenomenon had been well studied. As noted in the first edition of the *Thermoelectrics Handbook*, the potential to apply this physics to thermoelectric materials was a great opportunity and required the search for materials with specific structural properties. Initial efforts focused on searching for materials with large atomic displacement parameters as a proxy for atoms that are only weakly coordinated with the framework. Early successes were clathrate and skutterudite compounds with rattling interstitial atoms. Compounds such as Cu_2SbSe_3 and tetrahedrite $Cu_{12}Sb_4S_{13}$ suggest that other local structures can also produce low-frequency vibrational modes that inhibit phonon transport [41,42].

1.3.2.3 Prospects for Experimentally Driven Search

Historically, the experimental search for new thermoelectric materials has been a largely serial process that faces several significant challenges. First, each new compound must be prepared at a series of dopant levels to assess the performance near an optimized carrier concentration. This is challenging from a synthetic perspective due to the need to discover appropriate dopants and the serial process of controlling and assessing the impact of dopant concentration. Second, the measured transport properties provide minimal insight into the critical terms that live in reciprocal space (e.g., $E_i(\vec{k})$, $\tau_i(\vec{k})$ for both electrons and phonons). As such, causal relationships between structure and properties are elusive. Third, doping may fundamentally change the band edge electronic structure and mask performance that may be realizable with other dopants. Finally, measurement uncertainty, including systematic errors between groups, provides an additional level of noise in the development of chemical intuition. In contrast, calculations using a given code and associated functional are repeatable between groups to high precision and has the potential to accelerate discovery.

Over the last decade, the role of computation in experimentally driven efforts has expanded and is beginning to mirror approaches found in other disciplines. For example, the photovoltaic community's "rapid development" approach initially employs computation to consider a large initial ensemble of compounds and rapidly eliminate materials with critical flaws (e.g., optical absorption strength, bandgap). HT experiment then rapidly provides properties as a function of growth conditions for the remaining compounds. Concurrently, higher cost calculations are run on a smaller set to eliminate other critical flaws (e.g., defects to determine Fermi level pinning). Within thermoelectric materials, an increasingly tight feedback loop between experiment and theory is beginning to emerge and can be expected to accelerate material discovery and optimization.

1.4 Computationally Driven High-Throughput Searches

The core strength of computations is the ability to provide access to relevant materials properties. For many physical quantities (e.g., lattice constants, bulk modulus, band masses, and density of states), computational results can be quantitatively accurate and can be obtained with relatively inexpensive calculations. More challenging quantities (e.g., details of the electronic structure, defect, and dopant energies) can be determined accurately using more expensive computational approaches. However, despite the decades' worth of effort in advancing first-principles calculations, predicting thermoelectric performance remains extremely challenging, primarily due to the difficulties associated with the direct calculations of transport properties with quantitative accuracy. To date, the majority of computational approaches for predicting thermoelectric performance of materials, in particular those that aim at HT searches, have tried to circumvent this problem by relying on approximations and assumptions that may not hold uniformly across different materials. These approximations include those inherent to the relaxation time approximation, assumptions concerning scattering, and the rigid band approach, which neglects the effects of temperature, dopants, and defects on the electronic structure. Further, the state-of-the-art computational methods are not always employed due to their high cost, and even these methods are not infallible.

From the perspective of HT searches, the ability to parallelize and automate both calculations and the subsequent data analysis is crucial. Modern high-performance computing architectures in combination with software solutions that facilitate management of a large number of calculations are critical enablers of HT searches. Still, the desire to search for promising candidate materials over relatively large chemical spaces introduces additional challenges. Namely, there is always a trade-off between the accuracy of predictions and the speed of calculations. The usual approach is either to adopt a certain set of approximations or to develop proxies or descriptors for the relevant materials properties that are difficult to compute directly. Either strategy needs to allow sufficiently fast calculations.

On the other hand, the ability to perform HT searches leads to a situation where experimental validation is being eclipsed by the rate of new calculations. "Closing the loop" between theory and experiment is crucial to validating and refining theoretical predictions. From an experimental perspective, validation requires success along the entire synthesis–doping–sintering–measurement pathway. Additionally, validation requires that experimental results map to computational predictions. For example, addressing the Fermi surface structure requires measurements well beyond the usual transport studies.

After the HT calculations are finished, the following step is usually the ranking of materials based on the computed quantity(s) and/or descriptor(s). These rankings are immensely useful as they fill in the gaps that exist in the present knowledge, provide insights into the chemical trends of computed properties, and are certainly valuable in offering guidance for further studies, both theoretical and experimental. In our view, the biggest drawback is that by neglecting certain aspects of thermoelectric performance, dopability of materials being an important example, it may happen that in reality the best performing material is not the one maximizing a certain metric/descriptor, but the one that exhibits optimal combination of various, oftentimes even conflicting, materials properties. In this section, we review the notable HT computational efforts and provide a critical analysis of the advantages and limitations in the capability to identify champion thermoelectric materials. At the end, we discuss the dopability of materials in light of HT computational searches.

1.4.1 Search Spaces

Every HT search starts with defining the scientific question and the search space. There are multiple factors that influence the choice of the search space. Namely, narrowing the search space can be governed by physical insights and/or intuition that certain classes of materials, defined by their chemical composition or crystal structure or both, show propensity for good thermoelectric performance. Additional constraints may arise to refine the scientific inquiry under consideration or may be imposed by the limitations of the computational method used to calculate electronic structure and/or phonon dispersions. For example, the limitations of standard approximations to density functional theory (DFT), such as the local density and generalized gradient approximations, in handling *d*- and *f*-electron systems are relatively well known (see, e.g., References 43–45 and references therein).

Generally, most HT searches start with a defined material set with known chemical composition and crystal structure that are readily available in material databases. The most common choice is the ICSD [4], which contains inorganic solids with well-characterized chemical composition and crystal structure. For instance, Madsen focused on Sb-containing compounds from the ICSD [6], Wang et al. considered 2500 compounds from the ICSD [11], Zhu and coworkers searched for new thermoelectric materials in the ICSD with XYZ_2 chemical formula (X, Y, rare earths; Z, group VI elements) [46], and the authors of this chapter explored materials and material classes from the ICSD [5,7,8].

Most of the materials reported in the ICSD have only well-defined composition and crystal structures. Screening for relevant materials in a subspace of the ICSD therefore provides important information about the material property that is currently missing from our knowledge base. From a practical point, searching over the ICSD also bypasses questions related to materials growth and stability since a vast majority of the materials reported in the ICSD have been realized experimentally. In contrast, the search that includes previously unreported (i.e., unknown) materials naturally invokes questions regarding the stability of the materials.

While the methods for predicting stability of hitherto unknown inorganic materials have seen significant advances over the last decade [47–50], few searches for new thermoelectric materials have been conducted in this unknown chemical space. Examples can be found in the work of Zhang et al. [41], who combined the stability assessment with the calculations of the electronic band structure, phonon dispersions, and mode Grüneisen parameter (in the quasi-harmonic approximation). They focused on Cu–Sb–Se chemical systems and identified previously unreported $Cu_{12}Sb_4Se_{13}$ as a promising thermoelectric material. Gorai and coworkers [7] expanded their analysis of the thermoelectric performance of binary materials with A_1B_1 chemical formula based on semiempirical models for electronic and phonon transport to hypothetical A_1B_1 systems with the rock salt structure. While the primary goal was to confirm the observed trends among known (ICSD) A_1B_1 materials, this work offered some insights into the stability and propensity for thermoelectric performance of these hypothetical A_1B_1 rock salts. It is also important to mention that the work of Sparks et al. [51,52], which is purely statistical in nature, uses available experimental knowledge and builds models using data mining and machine learning for estimating the potential for thermoelectric performance of both existing and new materials.

1.4.2 High-Throughput Search Metrics

The HT search for thermoelectric materials inherently requires a trade-off between accuracy and cost. Here, we consider strategies to rank candidate materials for their potential thermoelectric performance. Both experimentally and computationally driven HT searches face similar challenges (i.e., performance varies with doping level and temperature), although the absence of scattering calculations in typical computational searches adds additional complexity to the ranking process.

1.4.2.1 Electronic Properties—Reduced Power Factor

In the experimental search for advanced thermoelectric materials, the power factor ($\alpha^2\sigma$) has emerged as a metric to assess materials in the absence of thermal conductivity measurements. While sometimes misleading due to the four orders of magnitude that κ can vary across, the power factor has served as a minimum metric for screening materials. The power factor has also been used to optimize materials; however, this is risky due to the neglect of thermal transport in the optimization.

For HT computational searches, a similar quantity known as the reduced power factor ($\alpha^2\sigma/\tau$) has emerged. The reduced power factor approach assumes that the relaxation time is constant with energy, allowing τ to be removed from the integrals in Equation 1.8 and canceled. Unfortunately, τ remains associated with σ; this is addressed through the normalization by τ. Given an electronic structure, Equations 1.8 and 1.9 can be used to calculate α and σ/τ as a function of η. The appeal of the reduced power factor is that the anisotropy and nonparabolicity of the electronic structure can be incorporated into the assessment of thermoelectric performance. Further, it thus allows transport to be assessed across a broad range of chemical potential and includes the impact of minority carriers [53].

Like the experimental power factor, the reduced power factor faces several challenges. First, it does not incorporate thermal conductivity. Second, the calculation requires a dense **k**-mesh to appropriately capture the charge carrier group velocities. Further, the impact of minority carriers can be overemphasized due to the reduced bandgaps associated with conventional DFT calculations. The results can also be misleading due to the assumption of a constant relaxation time; τ may decrease significantly as the density of available states increases or may show anisotropic behavior. Finally, for comparing materials based on their reduced power factor, skepticism is required due to the different τ between materials.

While most computational papers provide no justification of the validity of the CRTA, a few point to the agreement between calculated and experimental $\alpha(n)$ as evidence that the CRTA approach is valid [54,55]. However, due to the occurrence of $\tau(E)$ in both the numerator and denominator of α, the energy dependence may induce errors that are smaller than the experimental uncertainty. Instead, assessing the variation in $\tau(E)$ will be more visible through comparison of computational and experimental $\sigma(n)$. One could alternatively use the method of four coefficients to assess the energy dependence of scattering [56,57].

To estimate a ZT from the reduced power factor, the Wiedemann–Franz law can be used to expand κ_e. By assuming a value for the ratio κ_L/τ, ZT can be estimated as a function of η. Here again, the uncertainty of κ_L/τ is several orders of magnitude.

$$ZT = \frac{\alpha^2\sigma/\tau}{LT\sigma/\tau + \kappa_L/\tau} \tag{1.15}$$

As a variation, one can also employ a constant mean free path ($l = v_g\tau$) instead of a constant relaxation time. A similar expression to Equation 1.15 can be formed where κ_L/l is unknown. This approach has been applied in the context of nanostructured materials, where a common mean free path is assumed. Experimentally, achieving boundary-limited electron mean free paths in bulk semiconductors is nontrivial. Further, materials that have large ZT due to high mobility may be missed when searching using this metric.

1.4.2.2 Thermal Conductivity

The power factor focuses on identifying materials with suitable electronic performance and, often, the assumption that nanostructuring can ultimately be used to reduce κ_L; a similar metric can be built around low κ_L. For materials with extremely low κ_L, ZT reduces to $\sim\alpha^2/L$. With some caveats, one could thus search for materials with large α and extremely low κ_L. Metrics for κ_L typically utilize Callaway's approximation within the isotropic limit (Equation 1.10) to determine the critical descriptors. To simply rank the κ_L of different materials, one may simplify this expression into descriptors, such as the bulk modulus, density, Grüneisen parameter, and number of atoms in the primitive cell.

1.4.2.3 Prospects

Approaches to rank materials for potential thermoelectric performance fundamentally must address the dispersion and scattering of electrons and phonons as well as the dopability. The quality factor parameterizes the problem so that doping is decoupled from the intrinsic properties but does not circumvent the fundamental challenge of scattering. The reduced power factor provides additional insight into the electronic structure but, likewise, does not overcome the scattering challenge nor does it address thermal transport. Neither approach speaks to the dopability of the system. Methods to predict the lattice component of thermal conductivity have had good success; however, one cannot ignore the electronic contributions to performance. Challenges extend beyond individual searches as well. In the absence of a single metric that is computationally accessible, the field has struggled to unify the diverse results of computation across the literature. This has led to an abundance of small searches in the literature, which are challenging to aggregate or distill into experimental goals. As such, improved search criteria remains an outstanding opportunity to guide and focus the field.

1.4.3 Overview of High-Throughput Searches for Thermoelectric Performance

1.4.3.1 Constant Relaxation Time Approximation

The constant relaxation time is the simplest, lowest-order approximation and is, by far, the most commonly employed approach in computational searches for new thermoelectric materials [6,12,14,46,58]. In the context of thermoelectric materials, the calculation of phenomenological transport coefficients such as σ, α, and κ_e from first-principles calculations using the CRTA was first introduced by Sofo and coworkers [54]. The authors treated τ as a constant parameter ($\tau_{basal} = 2.2 \times 10^{-14}$ s, $\tau_{\perp,basal} = 2.1 \times 10^{-14}$ s) to fit the experimentally measured anisotropic Seebeck coefficient *versus* electrical conductivity plot for Bi_2Te_3. Given τ, the coefficients (σ, α, κ_e) were obtained by solving the BTE in the transport distribution function formalism [59]. The calculated coefficients together with an experimentally determined κ_L (1.5 W m^{-1} K^{-1} along basal, 0.7 W m^{-1} K^{-1} along trigonal) were used to calculate ZT, which showed good agreement for n-type Bi_2Te_3 but not for p-type (see Figure 1.9). Subsequently, this methodology for calculating transport coefficients within the CRTA was implemented by Madsen and Singh in the widely used BoltzTraP code [60]. The code was validated for the cases of Bi_2Te_3 and $CoSb_3$, by reproducing the results for Bi_2Te_3 in Reference 54, and for $CoSb_3$ by showing agreement between measured $1/R_H$, where R_H is the Hall coefficient, and calculated carrier density in the region of parabolic bands.

Madsen is credited with the first computationally driven HT search for new thermoelectric materials [6], wherein all Sb-containing compounds in the ICSD before 1990 that are ordered and do not contain highly electronegative elements, such as N, O, F, Cl, or lanthanide atoms (570 structures in total), were screened based on ZT using the CRTA

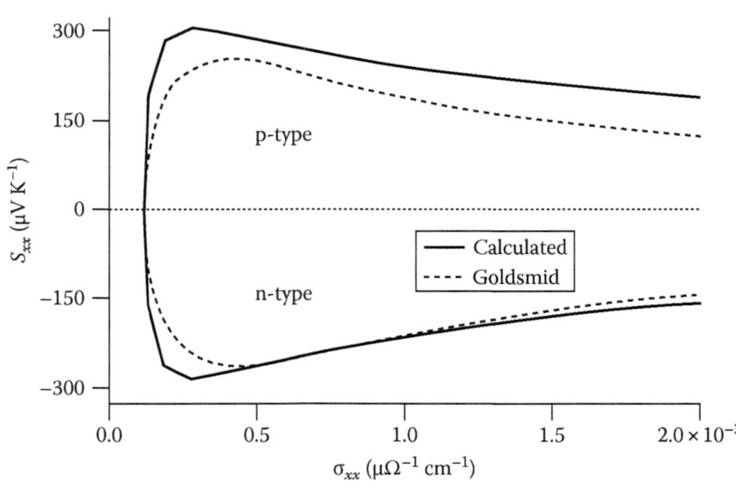

FIGURE 1.9
Calculated Seebeck coefficient S_{xx} versus electrical conductivity σ_{xx} in the basal plane of Bi_2Te_3 compared with experimental data from Reference 62. (Reprinted with permission from Scheidemantel, T.J., Ambrosch-Draxl, C., Thonhauser, T., Badding, J.V., and Sofo, J.O., *Phys. Rev. B*, 68, 125210, 2003. Copyright 2003 by the American Physical Society.)

($\tau = 2 \times 10^{-14}$ s, the same as Bi_2Te_3 in Reference 54) to calculate transport coefficients and an assumed κ_L (= 2 W m^{-1} K^{-1}). Previously known Sb-containing thermoelectrics, such as skutterudites [61], were identified as good TE materials. LiZnSb was proposed as a promising candidate TE material with a predicted ZT from 1.8 to 2.4 depending on the exchange correlation functional used in solving the electronic structure. This initial search also highlights the challenge of recruiting experimentalists for verification; of the fourteen promising compounds identified, six remain unexplored for thermoelectric performance to our knowledge. These include intriguing materials such as Zr_5ZnSb_3, NaZnSb, and $TlZn_2Sb_2$.

Since Madsen's pioneering efforts, several computational HT searches have been undertaken that employ the CRTA to calculate the electronic transport coefficients while assuming a constant ($\kappa_L\tau$) ratio (constant κ_L, τ of the order of 10^{-14} s [13,14,58]) for *all* compounds or using the minimum κ_L [25] to describe the phononic transport [46]. The latter search [46] is notable for combining HT search with experimental assessment; the search revealed $TmAgTe_2$ as a promising thermoelectric material and reported a maximum ZT of 0.35 (limited by p-type dopability). Likewise, the sulfide search of Reference 58 thoroughly investigated experimentally doping SnS to appropriate p-type levels. In both cases, defect calculations proved critical in understanding experimental doping observations.

The approximation that τ is an energy-independent constant is consistent with certain experimental measurements, especially when τ is fitted to the experimental data for a specific material. While it is computationally attractive to assume a constant τ to calculate electronic transport coefficients, employing the CRTA to screen chemically or structurally diverse search spaces, as opposed to standalone materials, is prone to errors because τ can significantly differ between materials. Even within a class of materials, there is no direct evidence that τ determined for one compound is valid for another compound in the same class. Hence, HT computational approaches would greatly benefit from a systematic investigation of the validity of the assumptions used when ranking materials based on constant τ. To date, the positive identification of known thermoelectric materials (e.g., PbTe, Bi_2Te_3) remains missing in CRTA searches when a single τ is used across the search space.

1.4.3.2 Constant Mean Free Path Approximation

Recognizing the drawbacks of the CRTA, Curtarolo and coworkers [11] assumed a constant mean free path (l) (equal to the grain size L) for carriers in sintered nanograined materials to circumvent the calculation of τ and screened 2500 materials from the ICSD for large constant mean free path approximation (CMFPA) power factors (P). A TiO_2 Magnéli phase, Ti_5O_9 (ICSD #31399), was identified as a promising candidate with an average normalized $\langle P\rangle/L$ of ~300 μW cm^{-1} K^{-2} nm^{-1} for n-type and ~1000 μW cm^{-1} K^{-2} nm^{-1} for p-type, assuming a constant l = 1 nm (see Figure 1.10). The CMFPA has also been used to estimate the ZT of 75 thermodynamically stable nanograined half-Heusler (HH) alloys [10], with the electronic transport coefficients calculated with a modified version of BoltzTraP and lattice thermal conductivities from second-order interatomic force constants calculated with phonopy [63]. Using the nanograined ZT as the descriptor, the authors identified 5 most promising candidate HHs for RT and high-temperature operations from a set of 75 thermodynamically stable HHs in the ICSD that are not semimetals. The HT screening procedure was not validated against known HHs that are good thermoelectric materials, and the identified candidates had not been considered for thermoelectric applications so far.

Nanostructuring has been long known to improve the performance of thermoelectric materials [64,65] either because of effects such as quantum effects [66], energy filtering [67],

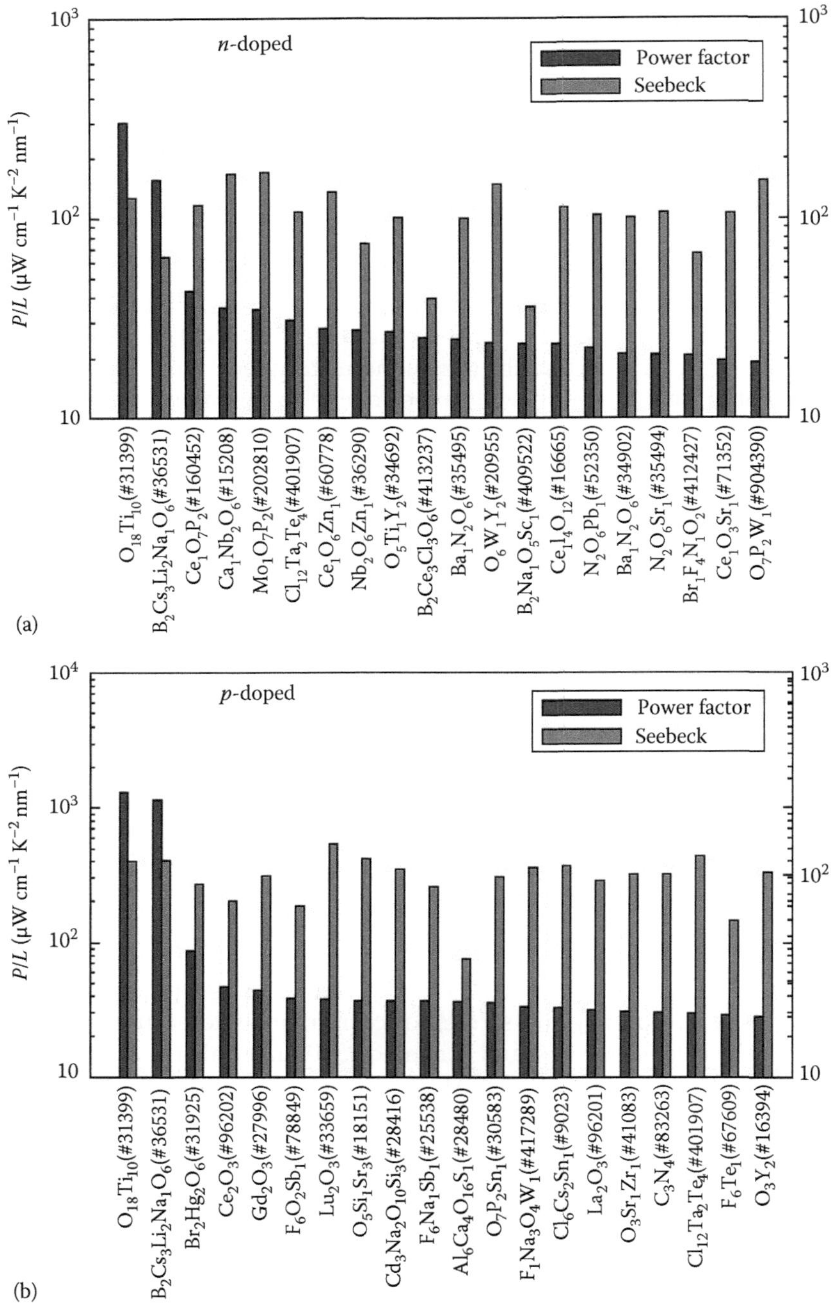

FIGURE 1.10
The average normalized power factor $\langle P\rangle/L$ (where L is grain size of the sintered material) of 20 best candidate thermoelectric materials predicted by Curtarolo et al. using a high-throughput screening approach based on the constant mean free path approximation. Performance is shown separately for (a) n- and (b) p-type materials. (Reprinted with permission Wang, S., Wang, Z., Setyawan, W., Mingo, N., and Curtarolo, S., *Phys. Rev. X*, 1, 021012, 2011. Copyright 2011 by the American Physical Society.)

or most frequently, due to the minimization of κ_L [68]. However, computational screening of materials in nanograined form poses phenomenological challenges as follows: (1) there is a tacit assumption that the candidate materials can be appropriately synthesized in nanostructured form, (2) nanostructuring can introduce detrimental quantum confinement effects, and (3) charge trapping at grain boundaries can overshadow the predicted good thermoelectric properties. From an HT screening perspective, the CMFPA route is promising but still suffers from the drawback that, unless κ_L is determined, the screening of TE materials is still based on power factors rather than *ZT* or descriptors for *ZT*. As in the case of the CRTA approach, large-scale validation of the CMFPA in identifying known thermoelectric materials as candidates is currently missing.

1.4.3.3 Semi-Empirical Models

In an attempt to develop a computationally tractable approach that takes into account both κ_L and the energy dependence of the charge carrier relaxation time, the authors [8] developed a semiempirical model to assess the magnitude of the quality factor β, which contains intrinsic material properties, namely, the intrinsic charge carrier mobility (μ_o), effective mass m^*_{DOS}, and κ_L. For materials that were optimally doped to maximize *ZT*, the value of β calculated from measured RT μ_o, m^*_{DOS}, and κ_L was found to be a good descriptor of ZT_{max} (see Figure 1.11).

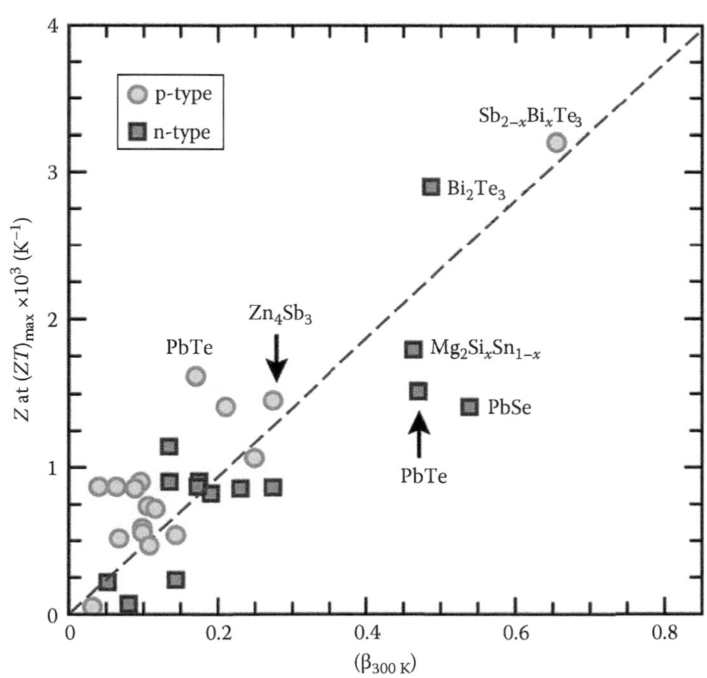

FIGURE 1.11
β determined from room temperature (RT) experimental measurements show a fairly good correlation with *Z* at peak *ZT* for a wide variety of materials, despite differences in peak temperature. $\beta_{300\ K}$ is obtained using Equation 1.14 and RT transport measurements from the literature. (From Yan, J., Gorai, P., Ortiz, B. et al., Material descriptors for predicting thermoelectric performance, *Energy Environ. Sci.*, 8, 983, 2015. Reproduced by permission of The Royal Society of Chemistry.)

The semiempirical models for μ_o and κ_L, which are based on classic semiconductor relations and contain parameters that can be easily calculated from first-principles DFT calculations, were fitted to measured RT μ_o and κ_L values. Motivated by the relationships describing electron–phonon scattering rates of band conductors, μ_o was expressed as

$$\mu_0 = A_0(B)^s \left(m_b^*\right)^{-t} \tag{1.16}$$

where B is the bulk modulus that describes the elastic properties of the material, and m_b^* is the band effective mass.

Fitting parameters A_0, s, and t were fitted to experimental data, while B and m_b^* were computed with first-principles methods. A modified Debye–Callaway model for κ_L is used:

$$\kappa_L = A_1 \frac{\bar{M} v_s^3}{V^{2/3} n^{1/3}} + A_2 \frac{v_s}{V^{2/3}} \left(1 - \frac{1}{n^{2/3}}\right) \tag{1.17}$$

where $\bar{M}$ is the average atomic mass, V is the average volume per atom, v_s is the speed of sound, and n is the number of atoms in the primitive cell.

The model assumes a constant Grüneisen parameter (γ). To a first approximation, v_s depends on B and density d ($v_s = \sqrt{Bd}$). The first term on the right side of Equation 1.17 is the acoustic and the second term the optical component of κ_L. A_1 and A_2 are fitting parameters that were fitted to the RT κ_L of 45 compounds, while $\bar{M}$, V, d, and B were obtained from the first-principles calculations. The authors validated the semiempirical approach by showing that the calculated β (now β_{SE}) can correctly identify known thermoelectric materials [7,8]. Using β_{SE} as the metric to screen a limited search space, potential candidates such as $Cu_2CdSnSe_4$ (*n*-type), KSnSb (*n*-type), $AgAsSe_2$ (*p*-type), and $AgBiTe_2$ (*p*-type) were proposed.

In a subsequent study [7], the β_{SE}-based approach was employed to computationally assess the thermoelectric performance of 287 known A_1B_1 compounds in the ICSD and 231 unknown A_1B_1 compositions in the assumed hypothetical rock salt structure. The study identified an important composition–structure–property relation that is responsible for the predominance of the simple binary A_1B_1 stoichiometry among known thermoelectric materials (see Figure 1.12). The binary A_1B_1 stoichiometry naturally accommodates ions in low oxidation states, which was shown to result in large band degeneracies and small effective masses, especially for p-type transport, in the cubic and orthorhombic structures. The screening of the A_1B_1 chemical space also revealed new candidate p-type materials, such as InI and MnTe, and n-type materials, such as ZnAs and SnO.

While the β_{SE} approach is computationally tractable, several assumptions were made in developing the semiempirical models for μ_o and κ_L, such as isotropic (spherical) hole pockets/electron valleys and a constant Grüneisen parameter. In spite of the underlying assumptions, the fact that known thermoelectrics are correctly identified as promising candidates provides some credence to the approach. As additional experimental data are incorporated into the semiempirical relations, prior rankings can be reassessed without additional DFT calculations. The metric β_{SE} is appropriate for the purposes of screening large chemical spaces and downsizing the candidate pool. Upon downsizing, the most promising candidates may be examined in detail using more computationally expensive methods that overcome some of the assumptions but may not be suitable for HT searches.

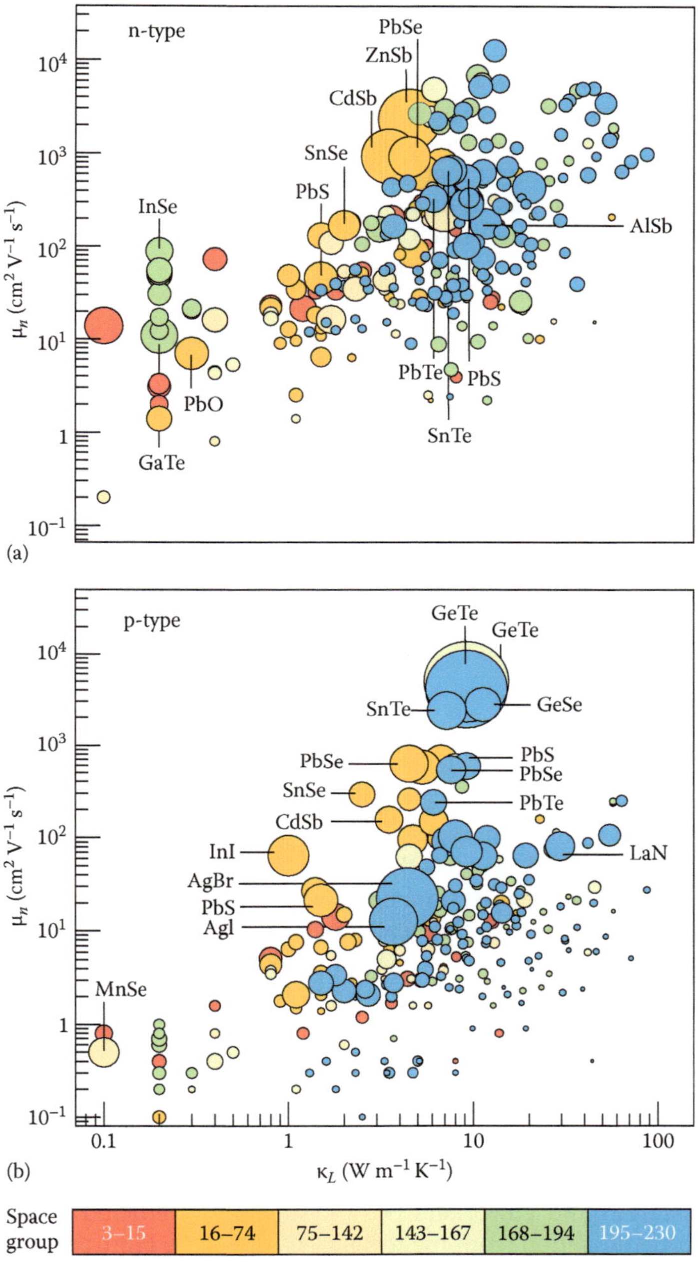

FIGURE 1.12
Calculated β_{SE} of 287 metal–nonmetal A_1B_1 compounds in all of their known structures available in the Inorganic Crystal Structure Database. The circle area scales with the value of β_{SE} for (a) conduction (n-type) and (b) valence (p-type) band transport. An important composition–structure–property relation is observed: large β_{SE} contains ions in a low oxidation state and the structures are dominated by the cubic (blue) and orthorhombic (orange) crystals. (Reprinted with permission from Gorai, P., Parilla, P., Toberer, E.S., and Stevanović, V., Computational exploration of the binary A_1B_1 chemical space for thermoelectric performance, *Chem. Mater.*, 27, 6213. Copyright 2015 American Chemical Society.)

The HT searches based on the CRTA, CMFPA, and semiempirical models are agnostic to the dopability of materials, that is, whether a candidate material can be doped to the appropriate concentration of the desired doping type (n-type vs. p-type). A candidate material predicted to be a good n-type thermoelectric may be heavily p-type doped due to intrinsic defects that cause Fermi level pinning. Since the dopability of materials is not incorporated in these HT search methodologies, it poses enormous practical challenges associated with the *experimental realization* of the predicted candidates. The importance of incorporating dopability in the search methodologies and the associated computational challenges are discussed at length in Section 1.4.4.

1.4.3.4 HT Searches Based on Lattice Thermal Conductivity

Several noteworthy works on the computational modeling of κ_L and HT material searches based on κ_L have recently emerged. We begin with a discussion of strategies to estimate κ_L. In addition to the intrinsic Umklapp scattering process, a wide array of phonon scattering mechanisms, including point defect scattering, grain boundary scattering, and magnetic scattering, give rise to the thermal resistance of a material. The Callaway model [71] (Equation 1.17) is the most widely used and was simplified by Slack and coworkers [72] and Berman [73] to estimate the intrinsic κ_L in a temperature range where only Umklapp processes dominate. These models, which are valid for solids with only one atom per primitive cell, were further extended by Morelli and Slack [74] to more complex crystal structures with n atoms in the primitive cell,

$$\kappa_L = A\frac{\bar{M}\theta^3\delta}{\gamma^2 T n^{2/3}} \tag{1.18}$$

where $\bar{M}$ is the average atomic mass, δ is the volume per atom, γ is the Grüneisen constant, A is a γ-dependent factor [75], and θ is the Debye temperature.

It should be noted that the κ_L estimated using the Slack model assumes that the optical branches do not contribute to the heat transport. Using this model, Morelli and Slack [74] predicted the κ_L of around 60 materials in rock salt, diamond, and zincblende crystal structures and showed fairly good agreement with measured RT κ_L. Toberer and coworkers [22] extended a Debye–Callaway model to include heat transport by optical modes and estimated the κ_L of 39 materials ranging from simple to complex chemistries and crystal structures. The predicted κ_L showed good agreement with measured RT κ_L (see Figure 1.13) to within a factor of half an order of magnitude, for κ_L ranging over three orders of magnitude.

HT computational screening of κ_L, θ, and γ using a quasi-harmonic Debye model was reported by Curtarolo and coworkers [9], although the predicted κ_L showed almost an order of magnitude deviation from the measured RT κ_L (see Figure 1.14a and b). By modifying the definition of the mode-averaged Grüneisen parameter, which enters the quasi-harmonic model as a squared term in the denominator of Equation 1.18, Madsen et al. [76] demonstrated better κ_L predictions using a similar quasi-harmonic Debye model (see Figure 1.15). While simpler to implement, in these phenomenological models, the details of the scattering physics are conveniently clumped and parameterized, which makes them vulnerable to breaking down in more complex systems that are typically not considered in validating the models.

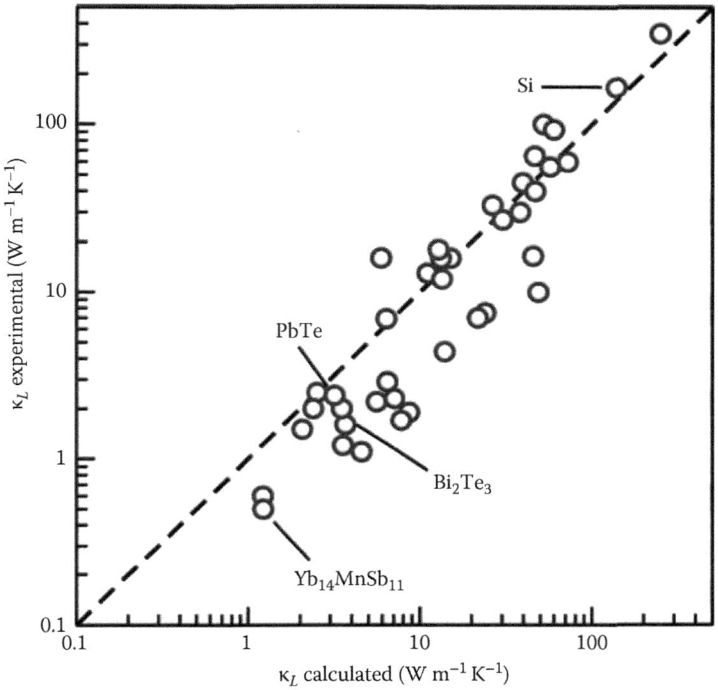

FIGURE 1.13
Experimental versus calculated lattice thermal conductivity (κ_L) of 39 compounds using an extended Debye–Callaway model that accounts for heat transport by optical modes. The acoustic contribution to κ_L contains Umklapp scattering and a boundary scattering contribution given by $\kappa_{L,a} = (v_s d)/(2VN)$, where v_s is the speed of sound, d is the grain size (which is set to 1 μm), V is the volume per atom, and N is the number of atoms in the primitive cell. (From Toberer, E.S., Zevalkink, A., and Snyder, G.J., Phonon engineering through crystal chemistry, *J. Mater. Chem.*, 21, 15843, 2011. Reproduced by permission of The Royal Society of Chemistry.)

HT searches that rely on predicting the κ_L from more fundamental definitions are more well poised to handle complex systems. Using the single-mode relaxation time method [77], the general expression for κ_L at temperature T is

$$\kappa_L = \frac{1}{k_B T^2 V} \sum_{\mathbf{q},i} n_{\mathbf{q},i}(n_{\mathbf{q},i}+1)\left|v^z_{\mathbf{q},i}\right|^2 \hbar^2 \omega^2_{\mathbf{q},i} \tau_{\mathbf{q},i} \tag{1.19}$$

where $\mathbf{q}$ is the phonon wave vector, i is the phonon band index, n is the Bose–Einstein distribution, ω is the frequency, v^z is the phonon group velocity in the z-direction, and τ is the relaxation time.

The determination of τ from third-order anharmonic force constants is computationally intensive. Curtarolo and coworkers [10] screened approximately 79,000 HH materials (from the AFLOWLIB database) to identify 75 thermodynamically stable ordered semiconductor HHs and then used machine learning and *ab initio* calculations to downselect 3 materials with predicted $\kappa_L < 5$ W m^{-1} K^{-1}. In the *ab initio* approach, the authors assumed that the third-order force constants are transferable within compounds sharing the same

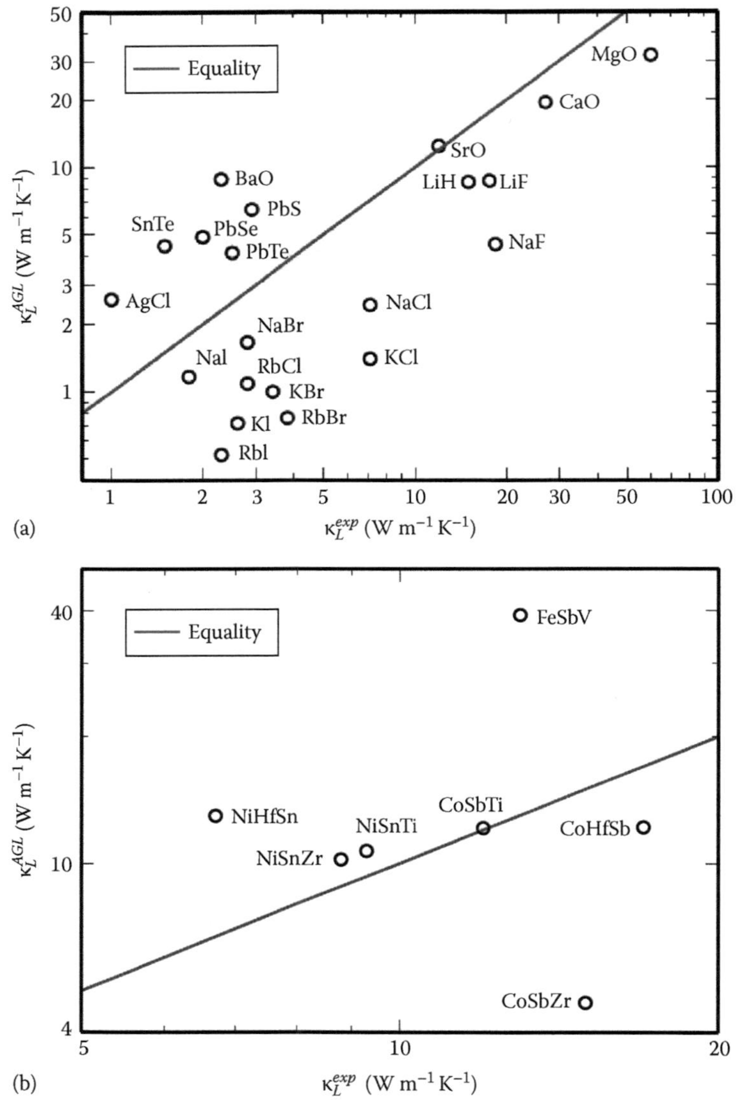

FIGURE 1.14
(a) Estimated versus experimental room temperature κ_L of 21 rock salt and (b) 7 half-Heusler compounds calculated using a quasi-harmonic Debye model. AGL stands for Automatic Gibbs Library, a software module that incorporates this model within AFLOW [69] and Materials Project [70]. (Reprinted with permission from Toher, C., Plata, J.J., Levy, O. et al., *Phys. Rev. B*, 90, 174107, 2014. Copyright 2014 by the American Physical Society.)

crystal structure. In particular, they used the anharmonic force constants calculated for Mg_2Si, which shares crystal structure features similar to HHs. In the machine learning approach, a "random forest regression" method was employed to "learn" from a training set of fully calculated κ_L of 32 compounds and predict κ_L of the screened stable HHs. The transferability of anharmonic force constants from one system to a class of materials is subject to doubt in the same way that applying a constant charge carrier relaxation time (τ) calculated for Bi_2Te_3 applied to all materials is not validated.

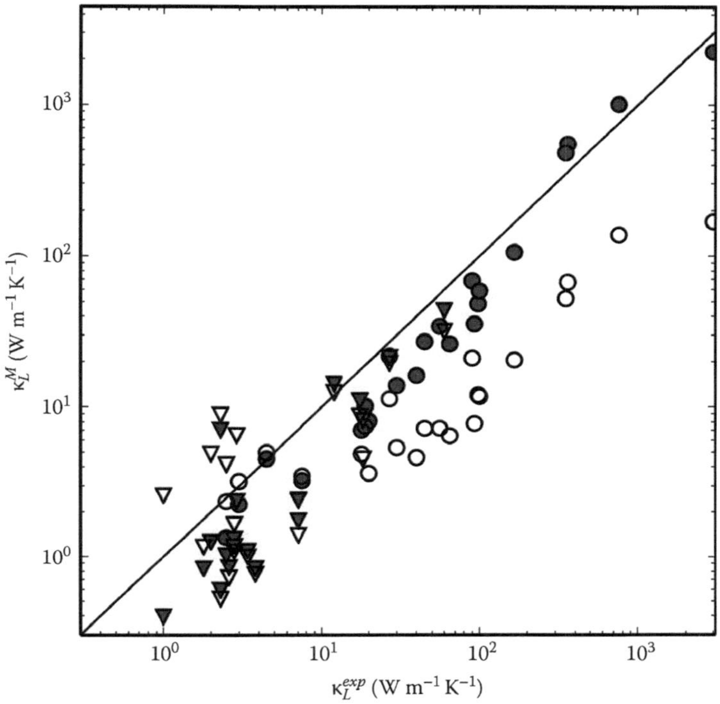

FIGURE 1.15
Calculated versus experimental lattice thermal conductivities κ_L at 300 K of rock salt (triangles) and zincblende (circles) structures. The filled symbols are estimated κ_L modeled in Reference 76, while the open symbols are taken from Reference 9. (Madsen, G.K.H., Katre, A., and Bera, C.: Calculating the thermal conductivity of the silicon clathrates using the quasi-harmonic approximation. *Phys. Stat. Solidi A*. 2016. 213: 802–807. Copyright Wiley-VCH Verlag GmbH & Co. KGaA. Reproduced with permission.)

Seko and coworkers [78] calculated κ_L of 101 compounds (see Figure 1.16) by calculating the third-order force constants for each compound. Bayesian statistics was then used, with the volume of the unit cell (V) and density (ρ) as the predictors, to screen 54,779 compounds from the Materials Project database. The fully calculated κ_L of the 101 compounds shows about half an order of magnitude deviation from the measured κ_L that ranges over three orders of magnitude. The virtual screening yielded 10 compounds with extremely low predicted κ_L, with 9 out of 10 containing Pb and the same number of halides, and 4 out of 10 in the *Pnma* orthorhombic crystal structure. While the fully calculated κ_L is desired for accuracy, calculation of third-order force constants is computationally extremely demanding and, therefore, cannot be implemented in HT fashion. Additionally, the performance of the model (half an order of magnitude deviation) is comparable to that of more simplistic and computationally tractable semiempirical models [8].

1.4.4 Dopability of Candidate Thermoelectric Materials

The dopability of materials is critical for their success as thermoelectrics. All known good thermoelectric materials possess (1) bulk properties that provide a basis for beneficial electronic and phonon transport properties and (2) dopability close to the optimal carrier concentration at which ZT is optimized. Both conditions contribute equally to materials

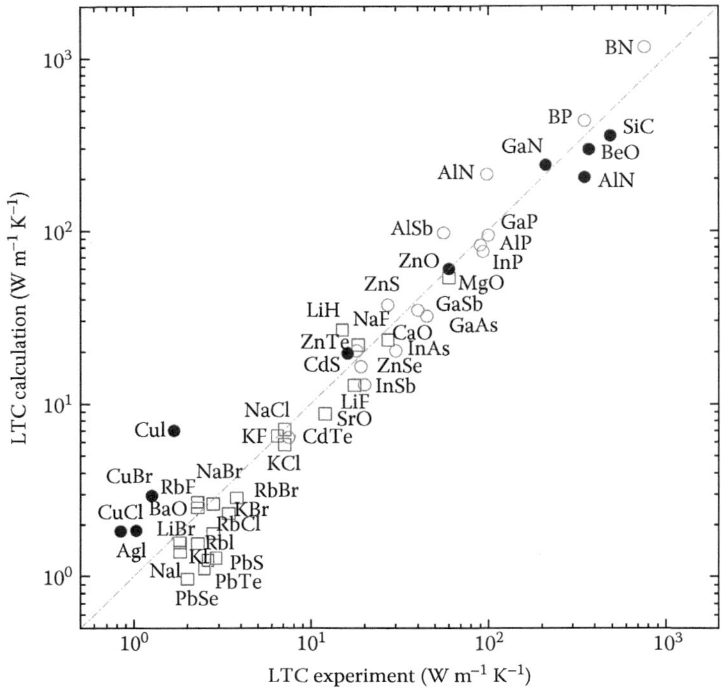

FIGURE 1.16
Lattice thermal conductivity (LTC) of 101 compounds calculated using first-principles calculations of third-order force constants compared with the experimental LTC at 300 K. (Reproduced with permission from Seko, A., Togo, A., Hayashi, H., Tsuda, K., Chaput, L., and Tanaka, I., *Phys. Rev. Lett.*, 115, 205901, 2015. Copyright 2015 by the American Physical Society.)

performance, but so far, only the former has been explicitly treated in the majority of theoretical/computational studies of thermoelectric materials and in virtually all HT searches, despite the decades of work in understanding and modeling defects and doping in semiconductors (see, e.g., References 79–81). Therefore, all previously discussed efforts that aim at assessing the thermoelectric performance of materials on the basis of electronic and phonon band structures alone only shed light on the *potential* for thermoelectric performance. Ultimately, the actual achievable ZT value will depend on the ability to dope candidate materials to the required doping type (p or n) and carrier concentration.

This apparent neglect of dopability among computational efforts first arises from the complexity of assessing the intrinsic potential for thermoelectric performance, as discussed earlier in detail. Second, predicting defects and doping, while certainly possible to do accurately using modern computational techniques, is still considered a laborious task that is relatively difficult to perform in an HT manner. The usual justification is that one wants to construct HT approaches in a funnel-like workflow, that is, to initially reduce the number of systems by ranking materials based on less demanding calculations and then do more detailed studies on a smaller set. But this justification might have a weakness in that it becomes unclear where to stop when choosing materials based on initial ranking. As already noted, it might occur that the new groundbreaking thermoelectric is a material that has an optimal combination of electronic structure properties and dopability, rather than being among the top in rankings that neglect doping.

Several case examples highlight the importance of assessing dopability when searching for new thermoelectrics. Considering the groundbreaking search of Madsen, all of the eight promising compounds that have been studied experimentally were found to be natively p-type [6,86]. Unfortunately, the high thermoelectric performance of each of these eight compounds was for free electrons, not holes. While extrinsic dopants may be able to overcome the degenerate (~10^{19}–10^{20} cm^{-3}) hole doping, this is expected to be nontrivial. Additionally, the work of Gorai et al. [7] suggests that some of the known thermoelectric materials that are intrinsically p-type, such as ZnSb or SnSe in its RT orthorhombic (*Pnma*) structure, have the potential to perform much better if n-type doping can be achieved. However, realizing degenerate doping of a certain type can be challenging. For example, Madsen [6] predicted good n-type thermoelectric performance of LiZnSb but the material is intrinsically p-type.

The problems associated with doping materials opposite to their native doping type are a consequence of the fact that if shallow donors are introduced into the natively p-type system in an attempt to achieve n-type doping, shifting the Fermi energy in the direction of the conduction band results in lowering the formation energy of acceptors, which, in turn, compensates any electron doping. This is graphically demonstrated in Figure 1.17, where a sketch of the defect formation energies of an acceptor defect and a donor is shown as function of the Fermi energy. The horizontal part represents the defect in the charge neutral state, while the oblique part represents the charged (ionized) defect. In equilibrium, the Fermi energy will be pinned at the crossing point of the two defects to balance the charge they are contributing. To achieve the n-type doping in natively p-type material, one needs to find a donor impurity with the formation energy significantly lower that the acceptor. This is challenging because of two things. First, the intrinsic nature of the material suggests that the acceptors are already low in energy; and second, low-energy impurities could also indicate the formation of secondary phases rather than a solid solution.

Modern defect calculations have reached the point of being quantitatively accurate both in revealing the intrinsic nature of materials and in identifying suitable dopants. The approach to directly assess the dopability of materials is well understood [80,87–89] and involves defect calculations for both intrinsic and extrinsic defects/dopants. These calculations are capable of providing the energy required to form a certain defect in the system

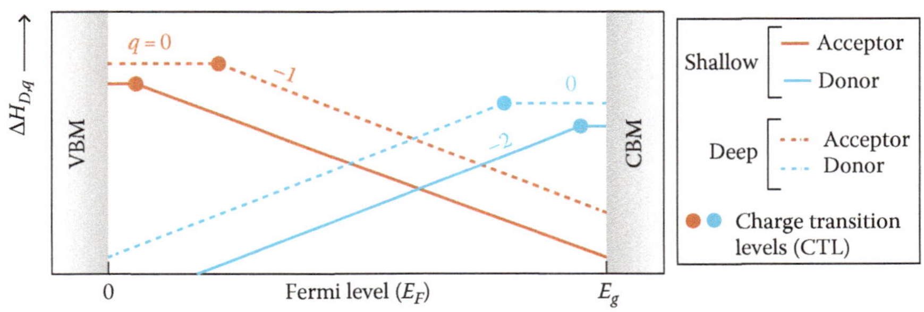

FIGURE 1.17
Sketch of a typical defect formation energy $\Delta H_{D,q}$ versus Fermi level plot. CTL is charge transition level and E_g is bandgap. Acceptors (red) and donors (blue) that have CTLs close to the valence and conduction band edges, respectively, are shallow defects and those with CTLs further away from the corresponding band edges are deep defects. The Fermi level is pinned by the charge balance between the acceptors and donors.

coupled with reservoirs of matter and charge, which are described by the corresponding chemical potentials, and have been used in predicting (a) the solubility of the defect/dopant [90]; (b) the ionization energy of the defect levels, that is, its shallowness [91]; and (c) compensation by either native defects or some other possibly external impurity that is commonly present during growth, such as oxygen or hydrogen [92]. If this is done for all relevant intrinsic defects as well as for a range of external impurities, the picture of the intrinsic conduction type of the material can be created together with the identification of efficient dopants and their solubility limits.

However, the HT approaches would benefit from a shorter route to predicting dopability of materials. The intrinsic nature of semiconductors together with the doping limits has been discussed previously on a more qualitative level as being a consequence of the absolute position of the band edges of materials, that is, their ionization potentials and electron affinities [93,94]. Intuitively, the higher on absolute scale the conduction band is, the more energy is to be gained for the system to develop mechanisms to remove free electrons from the conduction band into some lower-energy electronic states. Analogously, the deeper the valence band, the harder it is to realize free holes. This behavior is illustrated in Figure 1.18 in which absolute band edges from calculations and experiments are shown together with the known intrinsic nature. All values (theory and experiment) are

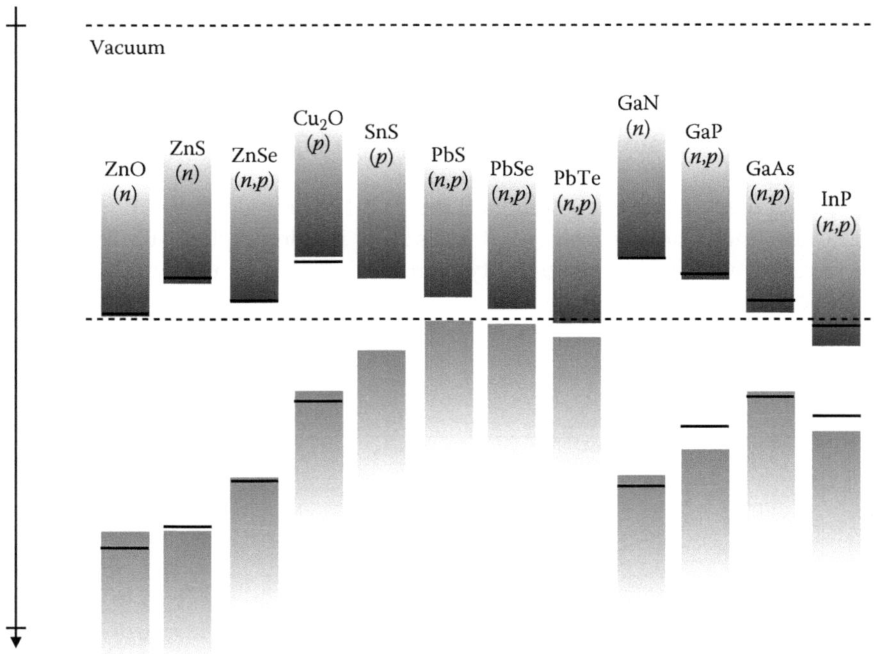

FIGURE 1.18
Computed absolute (relative to vacuum) positions of the band edges of several semiconductor materials together with the known native doping type. The water redox potential at approx. −4.4 eV (dashed line in the middle) is drawn as a reference. Thick black lines represent photoemission data. All values, theory and experiment, are taken from Reference 85 except for the lead chalcogenides, which are computed separately using the same approach.

taken from Reference 85, except for the lead chalcogenides, which are computed separately using the same approach—a combination of DFT for surfaces and many-body GW for the bulk electronic structure. It is clear when moving along the ZnO–ZnS–ZnSe series that the valence band shifts up in energy from ZnO, which is known to be dopable n-type and not p-type, to ZnSe for which both doping types have been achieved. Both cuprous oxide (Cu_2O) and orthorhombic SnS are known to be intrinsically p-type and not dopable otherwise. Only recently some success has been achieved in n-type SnS [95], but only when alloyed with Pb.

Figure 1.18 also shows the position of band edges in the Pb-chalcogenide series, known to be both *p* and *n* dopable. Based on the same arguments, their bipolarity is likely a consequence of the relatively narrow bandgaps. On the other hand, for thermoelectric applications, the magnitude of the bandgap sets the limits on the application temperature, and in larger bandgap systems, the doping limits and doping asymmetry can be expected to be more pronounced. As illustrated by standard pnictide semiconductors (Figure 1.18), while the absolute band positions provides some insight into the dopability of semiconductors, future efforts are needed to obtain more quantitative relationships between absolute position of band edges and material dopability.

A potential alternative approach to describing dopability in an HT manner has emerged recently. Ramprasad and coworkers [96] identified a quantitative correlation between the defect properties of materials, such as formation energies and the position of charge transition levels (CTLs), and the valence band width of the defect-free host. Similarly, Deml et al. [97,98] identified a correlation between the energy to form a neutral oxygen vacancy and the intrinsic properties of host metal oxides and their elemental constituents. Namely, they found using a set of 45 different metal oxides that a linear combination of the enthalpy of formation, valence band width, band gap, and the electronegativity difference between oxygen and its first nearest neighbor directly reproduces computed oxygen vacancy formation energies to within 0.2 eV. Future extension of these kinds of approaches to other defects and other classes of materials could bring the needed combination of accuracy and speed for integration with the existing HT computational methods to improve the search for new thermoelectric materials.

1.5 Conclusions

The potential for computation to cast light on the search for new thermoelectric materials is immense. However, the effective use of this resource and its integration with experimental efforts remains an evolving challenge. At the finest scale, improving the accuracy of individual predictions and validating them experimentally is critical. More broadly, we can search more intelligently by extracting structure–property relationships from the vast available datasets. Together, these developments enable the emergence of new search paradigms (see Figure 1.19). Critical hurdles that remain for HT searches include the following: (1) scattering rates of electrons and phonons, (2) dopability and defects, (3) the failure of the rigid band approximation due to a breakdown of crystallinity (defects, phonons), and (4) addressing transport in correlated systems. The development of accurate, inexpensive, and validated methods for these critical challenges could illuminate the next generation of thermoelectric materials.

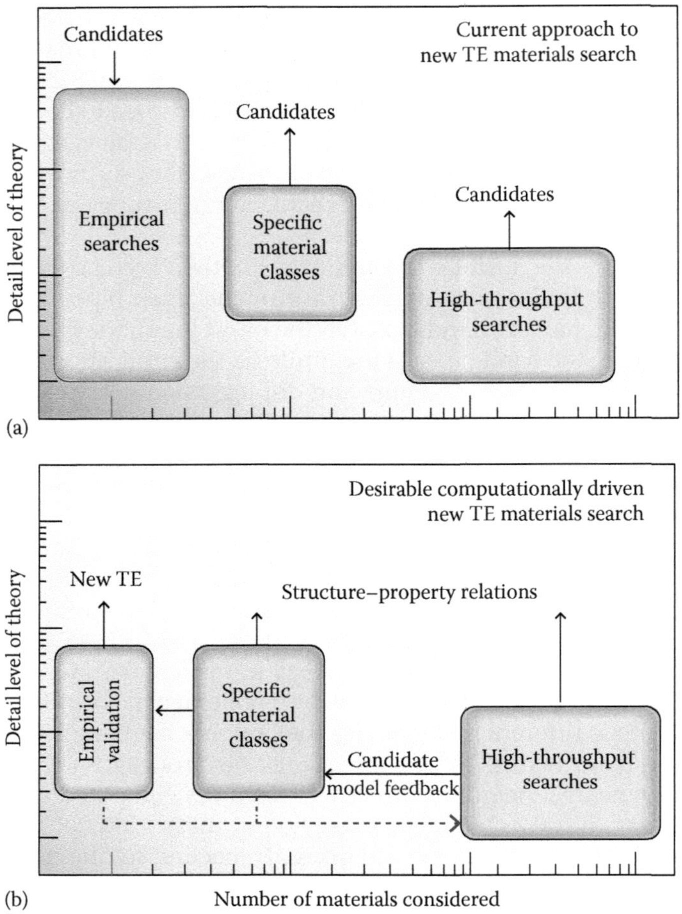

FIGURE 1.19
Schematic representing the current approach to computationally driven search for new TE materials (a) wherein high-throughput searches and searches within specific material classes yield candidate materials, while empirical searches rely on candidates identified based on intuition. To be successful, the computationally driven approach should undergo a paradigm shift to not only converge upon promising candidates but also generate new structure–property relations and establish a more streamlined theory–experiment feedback loop (b). The new learning gained through structure–property relations is useful for designing new TE materials.

References

1. Caillat, T., S. Firdosy, B. C.-Y. Li et al. 1999. Progress status of the development of high-efficiency segmented thermoelectric couples. *AIP Conf. Proc.* 458: 1403.
2. Vining, C. B. 2009. An inconvenient truth about thermoelectrics. *Nat. Mater.* 8: 83.
3. Baranowski, L. L., G. J. Snyder, and E. S. Toberer. 2012. Concentrated solar thermoelectric generator. *Energy Environ. Sci.* 5: 9055.
4. Belsky, A., M. Hellenbrandt, V. L. Karen, and P. Luksch. 2002. New developments in the inorganic crystal structure database (ICSD): Accessibility in support of materials research and design. *Acta Crystallogr. B* 58: 364.

5. Gorai, P., D. Gao, B. Ortiz et al. 2016. TE Design Lab: A virtual laboratory for thermoelectric material design. *Comp. Mater. Sci.* 112: 368.
6. Madsen, G. K. H. 2006. Automated search for new thermoelectric materials: The case of LiZnSb. *J. Am. Chem. Soc.* 128: 12140.
7. Gorai, P., P. Parilla, E. S. Toberer, and V. Stevanović. 2015. Computational exploration of the binary A_1B_1 chemical space for thermoelectric performance. *Chem. Mater.* 27: 6213.
8. Yan, J., P. Gorai, B. Ortiz et al. 2015. Material descriptors for predicting thermoelectric performance. *Energy Environ. Sci.* 8: 983.
9. Toher, C., J. J. Plata, O. Levy et al. 2014. High-throughput computational screening of thermal conductivity, Debye temperature, and Grüneisen parameter using a quasi-harmonic Debye model. *Phys. Rev. B* 90: 174107.
10. Carrete, J., N. Mingo, S.Wang, and S. Curtarolo. 2014. Nanograined half-Heusler semiconductors as advanced thermoelectrics: An ab initio high-throughput statistical study. *Adv. Funct. Mater.* 24: 7427.
11. Wang, S., Z. Wang, W. Setyawan, N. Mingo, and S. Curtarolo. 2011. Assessing the thermoelectric properties of sintered compounds via high-throughput *ab-initio* calculations. *Phys. Rev. X* 1: 021012.
12. Bhattacharya, S. and G. K. H. Madsen. 2015. High-throughput exploration of alloying as design strategy for thermoelectrics. *Phys. Rev. B* 92: 085205.
13. Opahle, I., A. Parma, E. J. McEniry, R. Drautz, and G. K. H. Madsen. 2013. High-throughput study of the structural stability and thermoelectric properties of transition metal silicides. *New J. Phys.* 15: 105010.
14. Opahle, I., G. K. H. Madsen, and R. Drautz. 2012. High throughput density functional investigations of the stability, electronic structure and thermoelectric properties of binary silicides. *Phys. Chem. Chem. Phys.* 14: 16197.
15. Yang, J., H. Li, T. Wu, W. Zhang, L. Chen, and J. Yang. 2008. Evaluation of half-Heusler compounds as thermoelectric materials based on the calculated electrical transport properties. *Adv. Funct. Mater.* 18: 2880.
16. Lundstrom, M. 2009. *Fundamentals of Carrier Transport*. Cambridge, U.K.: Cambridge University Press.
17. Snyder, G. J. and T. S. Ursell. 2003. Thermoelectric efficiency and compatibility. *Phys. Rev. Lett.* 91: 148301.
18. Chasmar, R. P. and R. Stratton. 1959. The thermoelectric figure of merit and its relation to thermoelectric generators. *J. Electron. Control* 7: 52.
19. Goldsmid, H. J. and R. W. Douglas. 1954. The use of semiconductors in thermoelectric refrigeration. *Br. J. Appl. Phys.* 5: 386.
20. Bjerg, L., G. K. H. Madsen, and B. B. Iversen. 2012. Ab initio calculations of intrinsic point defects in ZnSb. *Chem. Mater.* 24: 211.
21. Pomrehn, G. S., A. Zevalkink, W. G. Zeier, A. Walle, and G. J. Snyder. 2014. Defect-controlled electronic properties in AZn_2Sb_2 zintl phases. *Angew. Chem.* 126: 3490.
22. Toberer, E. S., A. Zevalkink, and G. J. Snyder. 2011. Phonon engineering through crystal chemistry. *J. Mater. Chem.* 21: 15843.
23. Gauster, W. B. 1971. Low-temperature Grüneisen parameters for silicon and aluminum. *Phys. Rev. B* 4: 1288.
24. Guiton, B. S. and P. K. Davies. 2007. Nano-chessboard superlattices formed by spontaneous phase separation in oxides. *Nat. Mater.* 6: 586.
25. Cahill, D. G., S. K. Watson, and R. O. Pohl. 1992. Lower limit to the thermal conductivity of disordered crystals. *Phys. Rev. B* 46: 6131.
26. Martinez, A. D., E. L. Warren, P. Gorai et al. 2016. Solar energy conversion properties and defect physics of $ZnSiP_2$. *Energy Environ. Sci.* 9: 1031.
27. Shi, X., Y. Pei, G. J. Snyder, and L. Chen. 2011. Optimized thermoelectric properties of $Mo_3Sb_{7-x}Te_x$ with significant phonon scattering by electrons. *Energy Environ. Sci.* 4: 4086.
28. Toberer, E. S., S. R. Brown, T. Ikeda, S. M. Kauzlarich, and G. J. Snyder. 2008. High thermoelectric efficiency in lanthanum doped $Yb_{14}MnSb_{11}$. *Appl. Phys. Lett.* 93: 062110.

29. Toberer, E. S., C. A. Cox, S. R. Brown et al. 2008. Traversing the metal-insulator transition in a Zintl phase: Rational enhancement of thermoelectric efficiency in $Yb_{14}Al_xMn_{1-x}Sb_{11}$. *Adv. Funct. Mater.* 18: 2795.
30. Kandpal, H. C., C. Felser, and R. Seshadri. 2006. Covalent bonding and the nature of band gaps in some half-Heusler compounds. *J. Phys. D: Appl. Phys.* 39: 776.
31. Kauzlarich, S. M., S. R. Brown, and G. J. Snyder. 2007. Zintl phases for thermoelectric devices. *Dalton Trans.* 21: 2099.
32. Toberer, E. S., A. F. May, and G. J. Snyder. 2010. Zintl chemistry for designing high efficiency thermoelectric materials. *Chem. Mater.* 22: 624.
33. Zevalkink, A., E. S. Toberer, W. Zeier, E. Flage-Larsen, and G. J. Snyder. 2011. Ca_3AlSb_3: An inexpensive, non-toxic thermoelectric material for waste heat recovery. *Energy Environ. Sci.* 4: 510.
34. Zevalkink, A., E. S. Toberer, T. Bleith, E. Flage-Larsen, and G. J. Snyder. 2011. Improved carrier concentration control in Zn-doped $Ca_5Al_2Sb_6$. *J. Appl. Phys.* 110: 013721.
35. Zevalkink, A., W. G. Zeier, G. Pomrehn, E. Schechtel, W. Tremel, and G. J. Snyder. 2012. Thermoelectric properties of Sr_3GaSb_3—A chain-forming Zintl compound. *Energy Environ. Sci.* 5: 9121.
36. Ohno, S., A. Zevalkink, Y. Takagiwa, S. K. Bux, and G. J. Snyder. 2014. Thermoelectric properties of the $Yb_9Mn_{4.2-x}Zn_xSb_9$ solid solutions. *J. Mater. Chem. A* 2: 7478.
37. Chung, D.-Y., T. P. Hogan, M. Rocci-Lane et al. 2004. A new thermoelectric material: $CsBi_4Te_6$. *J. Am. Chem. Soc.* 126: 6414.
38. Chung, D.-Y., T. Hogan, P. Brazis et al. 2000. $CsBi_4Te_6$: A high-performance thermoelectric material for low-temperature applications. *Science* 287: 1024.
39. Pohl, R. O. 1962. Thermal conductivity and phonon resonance scattering. *Phys. Rev. Lett.* 8: 481.
40. Wagner, M. 1963. Influence of localized modes on thermal conductivity. *Phys. Rev.* 131: 1443.
41. Zhang, Y., E. Skoug, J. Cain, V. Ozolins, D. Morelli, and C. Wolverton. 2012. First-principles description of anomalously low lattice thermal conductivity in thermoelectric Cu-Sb-Se ternary semiconductors. *Phys. Rev. B* 85: 054306.
42. Lu, X., D. T. Morelli, Y. Xia et al. 2013. High performance thermoelectricity in earth-abundant compounds based on natural mineral tetrahedrites. *Adv. Energy Mater.* 3: 342.
43. Martin, R. 2004. *Electronic Structure: Basic Theory and Practical Methods.* Cambridge, U.K.: Cambridge University Press.
44. Stevanović, V., S. Lany, X. Zhang, and A. Zunger. 2012. Correcting density functional theory for accurate predictions of compound enthalpies of formation: Fitted elemental-phase reference energies. *Phys. Rev. B* 85: 115104.
45. Lany, S. 2015. Semiconducting transition metal oxides. *J. Phys. Condens. Matter* 27: 283203.
46. Zhu, H., G. Hautier, U. Aydemir et al. 2015. Computational and experimental investigation of $TmAgTe_2$ and XYZ_2 compounds, a new group of thermoelectric materials identified by first-principles high-throughput screening. *J. Mater. Chem. C* 3: 10554.
47. Hautier, G., C. C. Fischer, A. Jain, T. Mueller, and G. Ceder. 2010. Finding nature's missing ternary oxide compounds using machine learning and density functional theory. *Chem. Mater.* 22: 3762.
48. Chen, H., G. Hautier, and G. Ceder. 2012. Synthesis, computed stability, and crystal structure of a new family of inorganic compounds: Carbonophosphates. *J. Am. Chem. Soc.* 134: 19619.
49. Zhang, Z., V. Stevanović, M. d'Avezac, S. Lany, and A. Zunger. 2012. Prediction of A_2BX_4 metal-chalcogenide compounds *via* first-principles thermodynamics. *Phys. Rev. B* 86: 014109.
50. Gautier, R., X. Zhang, L. Hu et al. 2015. Prediction and accelerated laboratory discovery of previously unknown 18-electron ABX compounds. *Nat. Chem.* 7: 308.
51. Gaultois, M. W., T. D. Sparks, C. K. H. Borg, R. Seshadri, W. D. Bonificio, and D. R. Clarke. 2013. Data-driven review of thermoelectric materials: Performance and resource considerations. *Chem. Mater.* 25: 2911.
52. Sparks, T. D., M. W. Gaultois, A. Oliynyk, J. Brgoch, and B. Meredig. 2016. Data mining our way to the next generation of thermoelectrics. *Scripta Mater.* 111: 10.
53. Snyder, G. J. and E. S. Toberer. 2008. Complex thermoelectric materials. *Nat. Mater.* 7: 105.

54. Scheidemantel, T. J., C. Ambrosch-Draxl, T. Thonhauser, J. V. Badding, and J. O. Sofo. 2003. Transport coefficients from first-principles calculations. *Phys. Rev. B* 68: 125210.
55. Bejenari, I. and V. Kantser. 2008. Thermoelectric properties of bismuth telluride nanowires in the constant relaxation-time approximation. *Phys. Rev. B* 78: 115322.
56. Chernik, I. A., V. I. Kaidanov, M. I. Vinogradova, and N. V. Kolomoets. 1968. Investigation of the valence band of lead telluride using transport phenomena. *Sov. Phys. Semicond.* 2: 645–651.
57. Zhitinskaya, M. K., V. I. Kaidanov, and I. A. Chernik. 1966. Nonparabolicity of the conduction band of Lead Telluride. *Sov. Phys.-Solid State* 8: 246.
58. Bera, C., S. Jacob, I. Opahle et al. 2014. Integrated computational materials discovery of silver doped tin sulfide as a thermoelectric material. *Phys. Chem. Chem. Phys.* 16: 19894.
59. Mahan, G. D. and J. O. Sofo 1996. The best thermoelectric. *Proc. Natl. Acad. Sci.* 93: 7436.
60. Madsen, G. K. H. and D. Singh. 2006. BoltzTraP: A code for calculating band-structure dependent quantities. *J. Comput. Phys. Commun.* 175: 67.
61. Sales, B. C., D. Mandrus, B. C. Chakoumakos, V. Keppens, and J. R. Thompson. 1997. Filled skutterudite antimonides: Electron crystals and phonon glasses. *Phys. Rev. B* 56: 15081.
62. Goldsmid, H. J. 1964. *Thermoelectric Refrigeration*. London, U.K.: Temple Press Books Ltd.
63. Togo, A. and I. Tanaka. 2015. First principles phonon calculations in materials science. *Scripta Mater.* 108: 1.
64. Poudel, B., Q. Hao, Y. Ma et al. 2008. High-thermoelectric performance of nanostructured bismuth antimony telluride bulk alloys. *Science* 320: 634.
65. Chen, Z.-G., G. Han, L. Yang, L. Cheng, and J. Zou. 2012. Nanostructured thermoelectric materials: Current research and future challenge. *Prog. Nat. Sci. Mater. Int.* 22: 535.
66. Dresselhaus, M. S., G. Chen, M. A. Tang et al. 2007. New directions for low-dimensional thermoelectric materials. *Adv. Mater.* 19: 1043.
67. Zou, T. H., X. Y. Qin, D. Li et al. 2014. Enhanced thermoelectric performance via carrier energy filtering effect in Zn_4Sb_3 alloy bulk embedded with $(Bi_2Te_3)_{0.2}(Sb_2Te_3)_{0.8}$. *J. Appl. Phys.* 115: 053710.
68. Zhao, L. D., H. J. Wu, S. Q. Hao et al. 2013. All-scale hierarchical thermoelectrics: MgTe in PbTe facilitates valence band convergence and suppresses bipolar thermal transport for high performance. *Energy Environ. Sci.* 6: 3346.
69. Curtarolo, S., W. Setyawan, S. Wang et al. 2012. AFLOWLIB.ORG: A distributed materials properties repository from high-throughput *ab initio* calculations. *Comput. Mater. Sci.* 58: 227.
70. Jain, A., S. P. Ong, G. Hautier et al. 2013. Commentary: The materials project: A materials genome approach to accelerating materials innovation. *APL Mater.* 1: 011002.
71. Callaway, J. 1959. Model for lattice thermal conductivity at low temperatures. *Phys. Rev.* 113: 1046.
72. Morelli, D. T., J. P. Heremans, and G. A. Slack. 2002. Estimation of the isotope effect on the lattice thermal conductivity of group IV and group III-V semiconductors. *Phys. Rev. B* 66: 195304.
73. Berman, R. 1976. *Thermal Conduction in Solids*. Oxford, U.K.: Clarendon Press.
74. Morelli, D. T. and G. A. Slack. 2006. High lattice thermal conductivity solids. In *High Thermal Conductivity Materials*, p. 37, Springer.
75. Julian, C. L. 1965. Theory of heat conduction in rare-gas crystals. *Phys. Rev.* 137: A128–A137.
76. Madsen, G. K. H., A. Katre, and C. Bera. 2016. Calculating the thermal conductivity of the silicon clathrates using the quasi-harmonic approximation. *Phys. Stat. Solidi A*. 213: 802.
77. Srivastava, G. P. 1990. *The Physics of Phonons*. Bristol, England: Adam Hilger.
78. Seko, A., A. Togo, H. Hayashi, K. Tsuda, L. Chaput, and I. Tanaka. 2015. Prediction of low-thermal conductivity compounds with first-principles anharmonic lattice-dynamics calculations and Bayesian optimization. *Phys. Rev. Lett.* 115: 205901.
79. van de Walle, C. G. and J. Neugebauer 2004. First-principles calculations for defects and impurities: Applications to III-nitrides. *J. Appl. Phys.* 95: 3851.
80. Freysoldt, C., B. Grabowski, T. Hickel et al. 2014. First-principles calculations for point defects in solids. *Rev. Mod. Phys.* 86: 253.
81. Alkauskas, A., P. Deák, J. Neugebauer, A. Pasquarello, and C. G. V. de Walle. 2011. *Advanced Calculations for Defects in Materials: Electronic Structure Methods*. Weinheim, Germany: Wiley.

82. Li, W.-F., C.-M. Fang, M. Dijkstra, and M. A. van Huis. 2015. The role of point defects in PbS, PbSe and PbTe: A first principles study. *J. Phys. Condens. Matter* 27: 355801.
83. Peng, H., J.-H. Song, M. G. Kanatzidis, and A. J. Freeman. 2011. Electronic structure and transport properties of doped PbSe. *Phys. Rev. B* 84: 125207.
84. Liu, Y., M. Zhou, and J. He. 2016. Towards higher thermoelectric performance of Bi_2Te_3 *via* defect engineering. *Scripta Mater.* 111: 39.
85. Stevanović, V., S. Lany, D. S. Ginley, W. Tumas, and A. Zunger. 2014. Assessing capability of semiconductors to split water using ionization potentials and electron affinities only. *Phys. Chem. Chem. Phys.* 16: 3706–3714.
86. Toberer, E. S., A. F. May, C. J. Scanlon, and G. J. Snyder. 2009. Thermoelectric properties of p-type LiZnSb: Assessment of *ab initio* calculations. *J. Appl. Phys.* 105: 063701.
87. Lany, S. and A. Zunger. 2008. Assessment of correction methods for the band-gap problem and for finite-size effects in supercell defect calculations: Case studies for ZnO and GaAs. *Phys. Rev. B* 78: 235104.
88. West, D., Y. Y. Sun, and S. B. Zhang. 2012. Importance of the correct fermi energy on the calculation of defect formation energies in semiconductors. *Appl. Phys. Lett.* 101: 082105.
89. Peng, H., D. O. Scanlon, V. Stevanović, J. Vidal, G. W. Watson, and S. Lany. 2013. Convergence of density and hybrid functional defect calculations for compound semiconductors. *Phys. Rev. B* 88: 115201.
90. Zhang, S. B. and S.-H. Wei. 2001. Nitrogen solubility and induced defect complexes in epitaxial GaAs:N. *Phys. Rev. Lett.* 86: 1789.
91. Deák, P., B. Aradi, T. Frauenheim, E. Janzén, and A. Gali. 2010. Accurate defect levels obtained from the HSE06 range-separated hybrid functional. *Phys. Rev. B* 81: 153203.
92. van de Walle, C. G. 2003. Hydrogen as a shallow center in semiconductors and oxides. *Phys. Stat. Solidi (b)* 235: 89.
93. Zhang, S., S.-H. Wei, and A. Zunger. 1999. Overcoming doping bottlenecks in semiconductors and wide-gap materials. *Physica B: Condens. Matter* 273–274: 976.
94. Walukiewicz, W. 2001. Intrinsic limitations to the doping of wide-gap semiconductors. *Physica B: Condens. Matter* 302–303: 123.
95. Ran, F.-Y., Z. Xiao, Y. Toda, H. Hiramatsu, H. Hosono, and T. Kamiya. 2015. N-type conversion of SnS by isovalent ion substitution: Geometrical doping as a new doping route. *Sci. Rep.* 5: 1.
96. Ramprasad, R., H. Zhu, P. Rinke, and M. Scheffler. 2012. New perspective on formation energies and energy levels of point defects in nonmetals. *Phys. Rev. Lett.* 108: 066404.
97. Deml, A. M., V. Stevanović, C. L. Muhich, C. B. Musgrave, and R. O'Hayre. 2014. Oxide enthalpy of formation and band gap energy as accurate descriptors of oxygen vacancy formation energetics. *Energy Environ. Sci.* 7: 1996.
98. Deml, A. M., A. M. Holder, R. P. O'Hayre, C. B. Musgrave, and V. Stevanović. 2015. Intrinsic material properties dictating oxygen vacancy formation energetics in metal oxides. *J. Phys. Chem. Lett.* 6: 1948.

2

Tetradymites: Bi_2Te_3-Related Materials

Joseph P. Heremans* and Bartlomiej Wiendlocha**
**Ohio State University*
***AGH University of Science and Technology*

CONTENTS

2.1 Introduction

As early as 1954, Goldsmid and Douglas [1] identified the materials that are now the most commonly used in the thermoelectric industry for either Peltier cooling or thermoelectric power generation. These materials belong to the class of tetradymite-type semiconductors of general formula $(Bi_{1-x}Sb_x)_2(Te_{1-x}Se_x)_3$ and crystallize in the rhombohedral crystal structure ($R\bar{3}m$). Most Peltier cooling devices operating at approximately room temperature use tetradymites. When high-temperature solders are used, tetradymite-based thermoelectric elements can operate up to 200°C and are therefore also suitable in some thermoelectric power generation applications. There are several reasons for this suitability, which remain as valid today as they were over half a century ago [2]. The naturally high thermoelectric efficiency of this class of materials results from a convergence of favorable properties from both the phonon and the electron spectra.

The thermoelectric efficiency is governed by the figure of merit of the material, $ZT = (\alpha^2\sigma/\kappa)T$, where α is the thermopower and σ and κ are the electrical and thermal conductivities of the material. The thermal conductivity is the sum of its lattice (κ_L) part, due to conduction by phonons, and its electronic (κ_E) part, which is directly related by the Wiedemann–Franz law ($\kappa_E = LT\sigma$) to the electrical conductivity. Here, L is the Lorenz ratio, and for the free electron $L = 2.48 \times 10^{-8}$ V^2 K^{-2}. However, L can deviate from this value to reach perhaps 1.6×10^{-8} V^2 K^{-2} in semiconductors in which the scattering of electrons is very inelastic, that is, when each collision involves a considerable exchange of energy. With this knowledge, the figure of merit can be written as $ZT = (\alpha^2/L)(1/(1 + (\kappa_L/\kappa_E)))$. Since the range over which L can be made to change is limited to ($2 \pm 0.5 \times 10^{-8}$ V^2 K^{-2}), thermoelectrics research consists of maximizing α for a given concentration of charge carriers and minimizing the lattice thermal conductivity relative to the electronic thermal conductivity. This recipe is followed in the quest for a higher ZT in all materials, including the tetradymites, but the latter naturally come with a number of particularly favorable physical properties that are summarized next.

From the point of view of the electronic properties of the tetradymites, the presence of heavy p-block elements (Bi, Sb, Te, Se) promotes strong spin–orbit interactions. These interactions result in electronic energy bands characterized by small energy gaps, high group velocities, and small effective masses. While these properties promote large effective Bohr radii for the electrons and hole wave functions and high electron mobilities, the small effective mass is not usually known to result in a high thermoelectric power factor. However, in many tetradymite compounds, this is remedied by the fact that the density of states can still be high because the band degeneracy (the number of degenerate minima in the conduction band or maxima in the valence band) can be high. Indeed, all practical alloys are designed around this criterion. The intense spin–orbit interactions also result in the presence of topologically protected surface states [3] and record-high spin-Hall angles [4] in some tetradymites, notably Bi_2Se_3. These properties have resulted in a surge of scientific activity around these semiconductors for properties other than thermoelectricity.

The phonon properties favor a naturally low lattice thermal conductivity, again resulting in favorable values of ZT. Bi, Sb, Te, and Se are among the heaviest elements in the periodic table, and the chemical bonds between them are not particularly stiff. As a consequence, the sound velocity in the tetradymites is low. The unit cells are large so that the Brillouin zones and the phonon cutoff wave vectors are small. The atoms are octahedrally coordinated, and the bonds are quite anharmonic. When combined, these properties result in a naturally low intrinsic lattice thermal conductivity, even in perfect crystals, where it is limited only by anharmonic phonon–phonon interactions. The nearly two-dimensional nature of the tetradymites results in strongly anisotropic properties. The ratio of the conductivity along the c-axis and perpendicularly to it is very different for electrical and thermal conductivities. As in all nearly two-dimensional systems, this can be exploited to maximize the thermoelectric figure of merit ZT in materials with some degree of preferential alignment.

Starting from these very favorable properties, modern materials science has resulted in further improvements in *ZT*. Thermoelectric tetradymite alloys with a high figure of merit are prepared using in essence six strategies:

1. Alloying between the binary tetradymites (Bi_2Te_3, Bi_2Se_3, and Sb_2Te_3) is used to reduce the lattice thermal conductivity via alloy scattering of phonons.
2. When possible, alloying is also used to increase the band degeneracy or at least the density of states.
3. Doping is used to reach the optimum concentration of electrons that maximizes *ZT*. Doping can be done either intrinsically, by exploiting the properties of electrically active defects present in the materials, or extrinsically, by adding elements other than from groups V or VI that give or accept free electrons.
4. Aligning the alloys preferentially is desirable, so that the heat and current fluxes are mostly oriented in the plane perpendicular to the *c*-axis. This typically applies more to the *n*-type materials, which have a stronger anisotropy in their electrical than thermal properties. It is done during the preparation process, for example, by zone melting (ZM). Since it is preferable to use polycrystals because of their better mechanical properties, extrusion or hot forging is probably preferable in materials for applications.
5. The alloys can be nanostructured to a grain size that preferentially scatters phonons while affecting electron mobility less than lattice thermal conductivity. This is done by using powder-metallurgical techniques, such as ball milling and sintering, for example, in a spark plasma sintering machine.
6. This review will outline how it has recently been discovered that inducing dislocations can be beneficial to the *ZT*. The combination of scattering on grain boundaries, random atoms in alloys, and dislocations affects phonons across almost their complete frequency spectrum and is called full-spectrum scattering.

Most commercial tetradymite materials used in thermoelectric modules are solid solutions. The *p*-type materials are alloys along the (Bi_2Te_3)–(Sb_2Te_3) pseudobinary line, typically with compositions around $(Bi_{0.20}Sb_{0.80})_2Te_3$ to $(Bi_{0.25}Sb_{0.75})_2Te_3$ [5]. The *n*-type material typically follows the (Bi_2Te_3)–(Bi_2Se_3) pseudobinary line, although quaternaries [6] have also been tried. One often used compositions for *n*-type alloys is around $Bi_2(Te_{0.98}Se_{0.02})_3$ to $Bi_2(Te_{0.95}Se_{0.05})_3$. Alloys of compositions between Bi_2Te_3 and the line compound $Bi_2Te_2S_e$, for example, $Bi_2(Te_{0.8}Se_{0.2})_3$ [2], have also been investigated. There are three main reasons for this, aligned with strategies (1)–(3). First, alloys have a lattice thermal conductivity that is reduced by alloy scattering to below that of pure binaries. Alloy scattering introduces distortions in the lattice at the length scale of the interatomic distance, which is effective at scattering phonons but much less effective at scattering charge carriers. This is because charge carriers have much longer wavelengths than phonons in these solids. Alloy scattering is thus more effective at reducing the lattice thermal conductivity than the electron mobility. Second, the compositions are chosen to have a higher band degeneracy, which helps the thermoelectric power factor. Third, most commercial materials are not extrinsically doped by means of adding chemical elements other than Bi, Sb, Se, or Te to release or bind free electrons, but rely on defect doping that also depends on alloy composition. Defect doping relies on departures of the ratio between the concentration of group V and group VI elements from the exact stoichiometry, making the materials either pnictogen (Bi, Sb) rich or chalcogen (Te, Se) rich. Since the defect structure is highly dependent on the composition, the alloy compositions are designed to generate the optimum electron or hole concentrations.

Thin-film tetradymites for thermoelectric applications will not be reviewed here. Admittedly, early reports exist about nanometer-sized superlattices of Bi_2Te_3, Sb_2Te_3, and Bi_2Se_3 that promised high ZT values, but these results have never been reproduced. There is a possible use for thermoelectric thin films in devices where the thermal response time must be very rapid. Because the response time scales with the length of the legs of a Peltier device, thin-film thermoelectric devices in which the current flows in the direction normal to the film plane could switch from heating to cooling very fast. Unfortunately, this geometry also exacerbates the parasitic losses in contact resistances. The performance of such films is generally inferior to that of bulk devices and depends on criteria that are substantially different from what is described here. Therefore, this review focuses exclusively on bulk tetradymite alloys, and thin-film tetradymite-type thermoelectric structures are discussed separately in Chapter 3.

This review outlines the basic concepts of materials science necessary to understand how the six strategies work. First, the structures and phase diagrams will be described. The following section is dedicated to the electronic band structure. Next, the doping mechanisms are discussed, including defect and extrinsic doping. This is followed by a review of the lattice thermal conductivity of the binary tetradymites (Bi_2Te_3, Bi_2Se_3, and Sb_2Te_3) and their (Bi_2Te_3)–(Sb_2Te_3) and (Bi_2Te_3)–(Bi_2Se_3) pseudobinary alloys. Next discussed is the recent progress in increasing ZT due to the invention and implementation of nanostructuring and dislocation engineering techniques that reduce the lattice thermal conductivity. This is followed by a description of the electronic transport properties and the recent progress made in optimizing that. Here again, the discussion is focused on the binary tetradymites (Bi_2Te_3, Bi_2Se_3, and Sb_2Te_3) and their (Bi_2Te_3)–(Sb_2Te_3) and (Bi_2Te_3)–(Bi_2Se_3) pseudobinary alloys. In this last section, the thermoelectric figure of merit is also discussed, not only in the binaries and the alloys, but then in the recent materials in which figures of merit values as high as $ZT = 1.85$ have been achieved in p-type material and near $ZT = 1.2$ in n-type material.

2.2 Structure

The original, historic tetradymite mineral is a sulfide–telluride, Bi_2STe_2. Generally, the sulfides do not have good thermoelectric properties, even though several were studied [7]. Therefore, this review focuses on the quaternary $(Bi_{1-x}Sb_x)_2(Te_{1-x}Se_x)_3$ alloys that crystallize in the tetradymites' structure and which reside in the main hatched region of compositions shown in Figure 2.1. One edge compound, Sb_2Se_3, crystallizes in an orthorhombic ($D_{2h}^{16} - Pnam$) structure, as do the alloys in the oppositely hatched region in Figure 2.1 near Sb_2Se_3. The open unhatched region in Figure 2.1 indicates the phase space of multiphased alloys. While most of the points inside the rest of the phase space are solid solutions, several ordered intermetallic line compounds exist, for example, Bi_2Te_2Se [8].

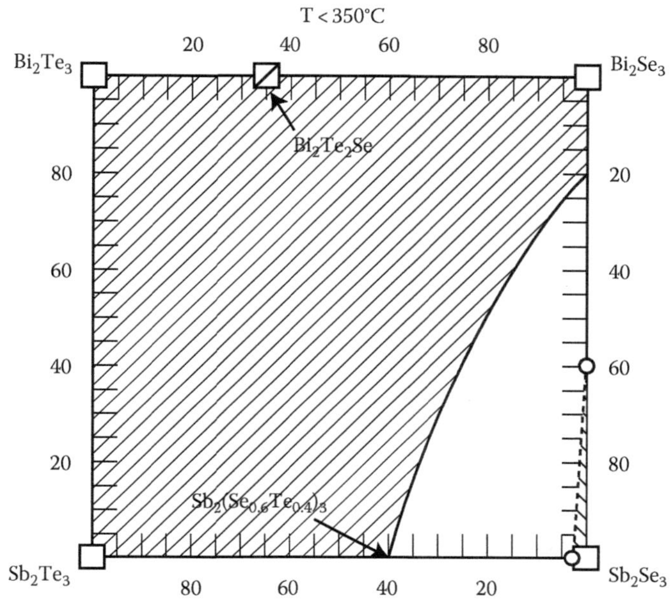

FIGURE 2.1
Phase diagram at temperatures below 350°C of the $(Bi_{1-x}Sb_x)_2(Te_{1-y}Se_y)_3$ system, including binary compounds, alloys, and line compounds. The main hatched area (samples with open squares) indicates the region in which alloys crystallize in the tetradymite ($R\bar{3}m$) structure. Crystals in the oppositely hatched region next to Sb_2Se_3 are orthorhombic (space group *Pnam*); the unhatched region denotes the region where the two phases coexist. (Redrawn from Kuznetsov, V.G. and Palkina, K.K., *Rus. J. Inorg. Chem.*, 8, 624, 1963.)

The tetradymite crystal structure is trigonal (space group $D^5_{3d} - R\bar{3}m$). It is shown in Figure 2.2a for a compound of general formula M_2X_3, where M is a metal atom (Bi, Sb) represented by a full dot and X is a chalcogen atom (Te, Se, and, as discussed earlier, S) represented by an open dot. There are two different sites for the chalcogen atoms, $X^{(1)}$ and $X^{(2)}$. Chalcogen atoms in site $X^{(2)}$ have all metal atoms as nearest neighbors. Chalcogen atoms in site $X^{(1)}$ have also other chalcogen atoms as neighbors. Following the progression of atoms in Figure 2.2b along the trigonal axis (*c*-axis), one can identify a quasi-one-dimensional chain of atoms that follows a progression of atoms in the sequence $-X^{(1)}-M-X^{(2)}-M-X^{(1)}-X^{(1)}-$. The chalcogen atoms in site (2) have $X^{(2)}-M$ bonds that are covalent and ionic, while the atoms in site (1) have $X^{(1)}-X^{(1)}$ van der Waals bonds with their chalcogen neighbors (Figure 2.2b). This conveys to the tetradymites an almost two-dimensional nature, with a strong anisotropy between the properties in the plane and those along the *c*-axis. In *n*-type Bi_2Te_3 at room temperature [5], the ratio of the electrical conductivity in the trigonal plane to that along the trigonal axis is of the order of (5 ± 1):1 (it varies with preparation conditions), while that of the thermal conductivities is of the order of (2.5 ± 0.5):1. The thermopower is relatively isotropic, as is expected from the direct relation between thermopower and the electron entropy, a scalar. Indeed, in solids where the thermopower is anisotropic, this arises usually from multicarrier effects combined with the anisotropy of the electrical conductivity and mobility.

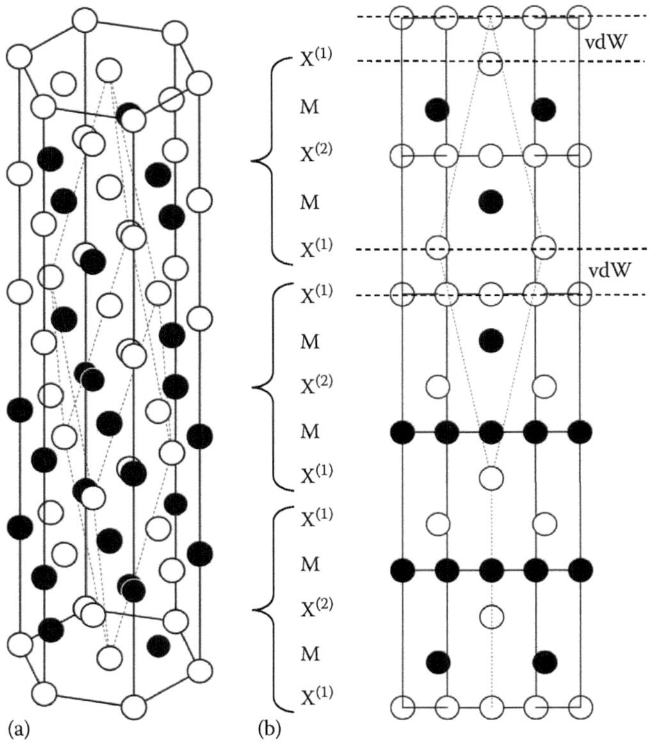

FIGURE 2.2
(a) The trigonal ($R\bar{3}m$) tetradymite crystal structure. Open dots represent chalcogen atoms, Te or Se; filled dots represent metal atoms, Bi or Sb. (b) The van der Waals gap is shown as "vdW." (Redrawn from Wiese, J.R. and Muldawrer, R., *J. Phys. Chem. Solids*, 15, 13, 1960.)

2.3 Electronic Band Structures

2.3.1 Binary Tetradymites: Dispersion Relations and Fermi Surfaces

The electronic band structures of tetradymites were investigated theoretically and experimentally by a large number of authors, making it impossible to mention all of them. Only some of more recent work is referenced here when several sources exist, with older work cited when it stands alone. As far as the band structure computations are concerned, various methods were used and various experimental and theoretical crystal structures were employed. Thus, there are a number of differences between published results. Sometimes, results are even contradictory, for example, whether the gap in Bi_2Se_3 is direct or indirect or what the exact locations are of the valence band maxima in Sb_2Te_3. The reason for this rather unusual situation is the extreme sensitivity of the band structure of tetradymites to the details of the crystal structure parameters, as well as a general problem of density functional methods in predicting the band structure of layered systems where weak van der Waals interactions play an important role [10]. Also, a systematic analysis of Fermi surfaces (FSs) is lacking in the available literature.

TABLE 2.1

Computed Unit Cell Parameters of Bi_2Te_3, Bi_2Se_3, and Sb_2Te_3—Used in the Following Calculations

Parameter	Bi_2Te_3	Bi_2Se_3	Sb_2Te_3
a (Å)	4.348	4.107	4.248
c (Å)	30.271	28.459	29.980
u Bi	0.3998	0.4005	0.3987
v Te(1)/Se(1)	0.2098	0.2110	0.2110

Lattice constants, a and c given in Å, space group no. 166, $R\bar{3}m$, u is the parameter of the (2c) (u, u, u) site, occupied by Bi or Sb, v is the parameter of the (2c) (v, v, v), occupied by Te(1)/Se(1), whereas Te(2)/Se(2) atoms occupy (1a) site, with coordinates (0, 0, 0) in all the cases, and thus are not repeated in the table.

To address this situation, this review presents systematic calculations of the band structures and FSs of the tetradymites, using a consistent computational approach and consistent parameters. The calculations were done using the density functional theory (DFT) and the full-potential linearized augmented plane wave method, as implemented in the WIEN2k code [11]. Equilibrium crystal structures were found, optimizing unit cell volume, angles, and atomic positions. The standard local density exchange-correlation potential local density approximation (LDA) method with the parameterization of Perdew and Wang [12] was chosen because previous work [13] reported better agreement of the calculated physical properties with the experimental results when the relaxation is done using the LDA, instead of the generalized gradient approximations. Unit cell optimization was done within the scalar relativistic approach, on a 2000 k-points mesh, and the results are presented in Table 2.1.

The calculated unit cell parameters are slightly smaller than experimental values, but the differences are not larger than 1.5%. In the next step, spin–orbit interaction was included in calculations for the optimized unit cells, and the modified Becke–Johnson (mBJ) semilocal exchange potential [14] was employed to improve the description of the electronic states near the band gap. A self-consistent cycle was finalized for about 40,000 k-points, and the FS (as well as transport properties as will be reported later) was computed on a 200,000 k-points mesh. The application of the mBJ potential, which was designed to improve the band gap values in s–p semiconductors, turned out to be crucial for the correct shape of the valence band in Bi_2Se_3, since standard potentials predict wrong shape of the valence band near the Γ point, leading to incorrect, indirect band gap (see also discussion in the following). Even though mBJ potential corrected the valence band in Bi_2Se_3, the computed band gap was still not exact, showing that Bi_2Se_3 and tetradymites in general are difficult for the DFT modeling.

Figure 2.3 shows the calculated band structures of Bi_2Te_3 (Figure 2.3a), Bi_2Se_3 (Figure 2.3b), and Sb_2Te_3 (Figure 2.3c) near the Fermi level and mapped out over some selected high-symmetry directions of the Brillouin zone shown in Figure 2.3d. Only the last valence and the first conduction bands are shown. The Fermi level is placed in the middle of the gap, and the band gap is left as calculated using mBJ potential: 0.11 eV (Bi_2Se_3), 0.15 eV (Bi_2Te_3), and 0.14 eV (Sb_2Te_3). The experimental band gaps are [15,16] 0.2–0.33 eV (Bi_2Se_3), 0.13–0.17 eV (Bi_2Te_3), and 0.2–0.3 eV (Sb_2Te_3); thus, the calculated values for Bi_2Se_3 and Sb_2Te_3 are still underestimated. The FSs for degenerately doped *p*- and *n*-type material are shown [17] in Figure 2.4 and 2.5 respectively, for the three selected carrier concentrations, 5×10^{18}, 1.1×10^{19}, and 8×10^{19} cm^{-3}.

Starting with Bi_2Te_3, we see a multivalley structure of the dispersion relations, both for *n*- and *p*-type materials. The valence band maximum (VBM) is not located on high symmetry points of the Brillouin zone, and it is not even aligned with any high-symmetry direction; it lies in the Γ–*Z*–*U*–*F* plane ("mirror plane"). The present calculations show it close

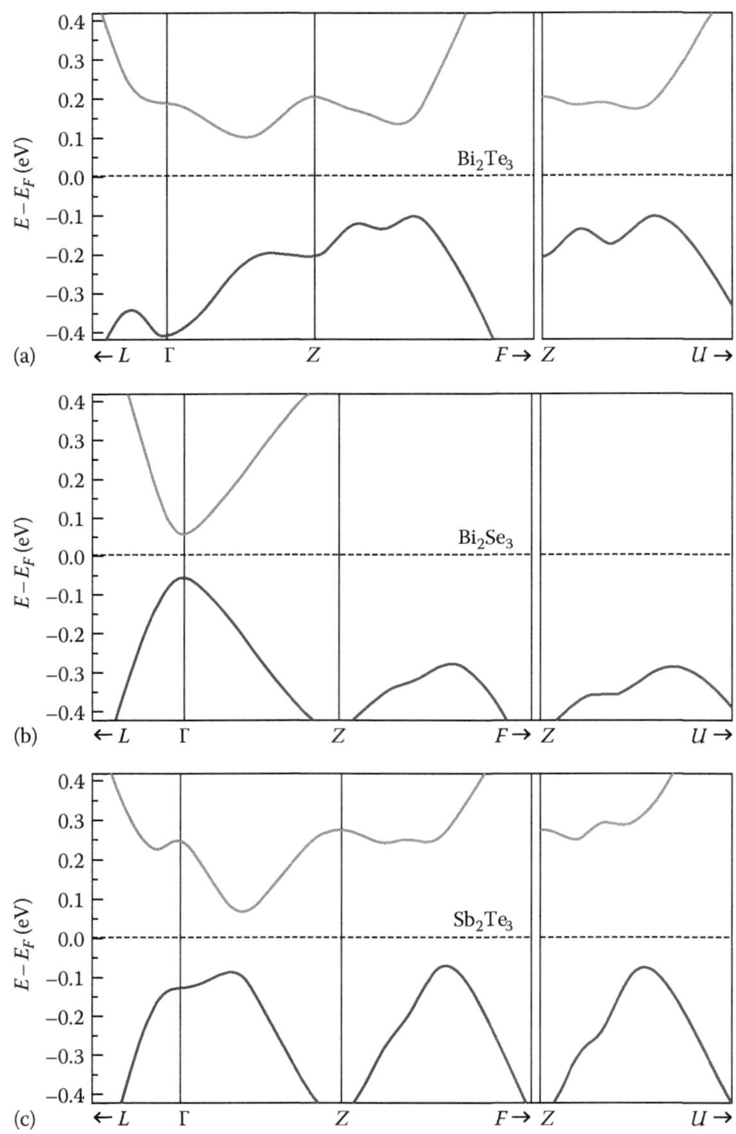

FIGURE 2.3
The dispersion of the lowest conduction and the highest valence bands of the binary tetradymite compounds: (a) Bi_2Te_3, (b) Bi_2Se_3, and (c) Sb_2Te_3. *(Continued)*

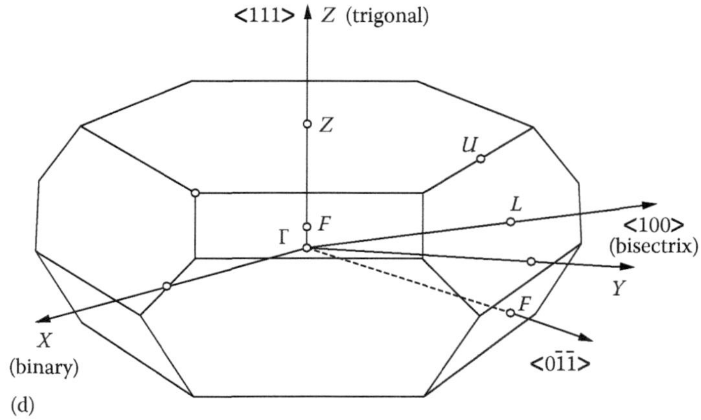

FIGURE 2.3 (*Continued*)
The dispersion of the lowest conduction and the highest valence bands of the binary tetradymite compounds: (d) The Brillouin zone with the high symmetry points. The energies of the valence band maxima along the Γ–*F* directions in the $(Bi_{1-x}Sb_x)_2Te_3$ occur at $x = 0.2$–0.25, resulting in *p*-type material with a very high degeneracy of the valence band maxima.

to the *Z*–*F* direction (see also References 18 and 19 for a discussion of how the locations of VBM and of the conduction band minima [CBM] change when different computational approaches are used). When the hole concentration is small, six hole pockets appear (see Figure 2.4a). When the Fermi level moves deeper into the valence band (VB), contributions to the FS from local maxima seen along the *Z*–*F* and *Z*–*U* directions become visible as additional pockets. For the heaviest doped example of 8×10^{19} cm^{-3}, the *Z*–*F* ellipsoids are connected. This multiple degeneracy of the band structure correlates well with the superior thermoelectric performance of Bi_2Te_3 discussed in Section 2.6.

For the *n*-type material, the conduction band FS (Figure 2.5a) for the lower electron concentrations consists of two electron pockets from the *Z*–Γ–*Z* direction, plus six pockets lying on the Γ–*Z*–*U*–*F* plane, similarly to the *p*-type case. For larger electron concentrations (Figure 2.5b and c), the +*z* and −*z* pockets join, forming large and roughly aspherical FSs.

Antimony telluride, Sb_2Te_3, exhibits many similar features, but with important deviations in the valence band. The VBMs also lie on a Γ–*Z*–*U*–*F* plane and have two maxima, one in the *Z*–*F* direction and the other in the *Z*–*U* direction. In the present calculations, both maxima have nearly the same energy. Thus, for hole concentrations of $p = 5 \times 10^{18}$ cm^{-3}, twelve pipelike FS pieces are seen in Figure 2.4a. As E_F becomes larger (goes deeper), the second VB maximum in the Γ–*Z* direction starts to contribute to the FS (Figure 2.4b). Finally, for $p = 8 \times 10^{19}$ cm^{-3}, the *Z*–*F* pipes and Γ–*Z* pockets join. For the *n*-type Sb_2Te_3 case, heavily electron-populated pockets (Figure 2.5c) have very similar features as the FS of Bi_2Te_3. The FS of materials with lower electron concentrations is shown in Figure 2.5a and b; the CBM in the Γ–*Z* direction lies much deeper than in Bi_2Te_3.

The last binary material covered here, bismuth selenide Bi_2Se_3, possesses qualitatively different FSs. Its CBM and VBM are both centered on the Brillouin zone center (Γ-point), making the band gap direct. Because the material has a strong tendency to be prepared *n*-type, the properties of the conduction band were revealed experimentally first. Transport measurements by Köhler and Fabricius [20] identified the CBM as being at Γ-point, but in many later theoretical papers, the VB shape was wrongly determined. The fact that the VBM is at Γ has been unambiguously confirmed in recent work, both theoretically (GW method) and

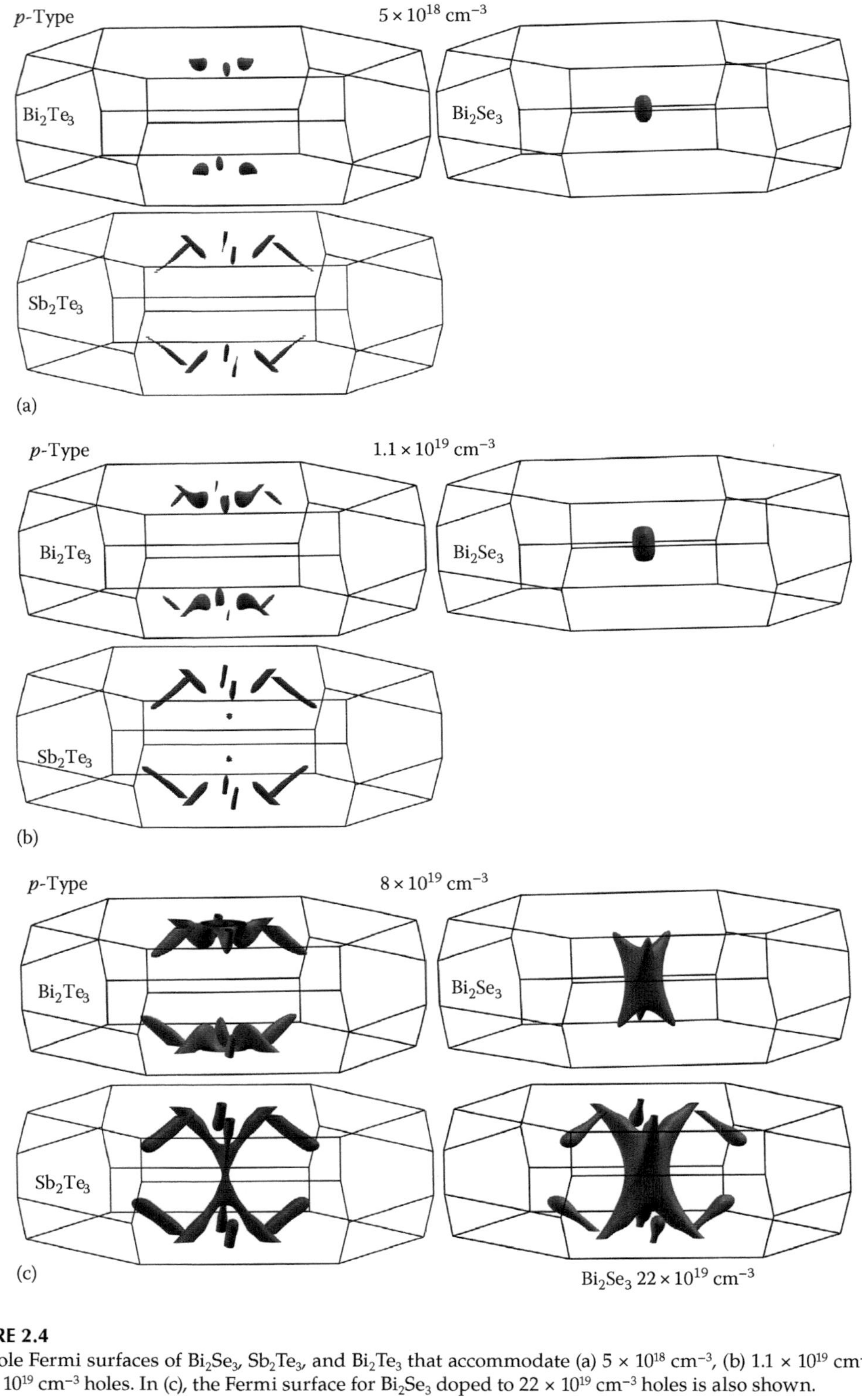

FIGURE 2.4
The hole Fermi surfaces of Bi_2Se_3, Sb_2Te_3, and Bi_2Te_3 that accommodate (a) 5×10^{18} cm^{-3}, (b) 1.1×10^{19} cm^{-3}, and (c) 8×10^{19} cm^{-3} holes. In (c), the Fermi surface for Bi_2Se_3 doped to 22×10^{19} cm^{-3} holes is also shown.

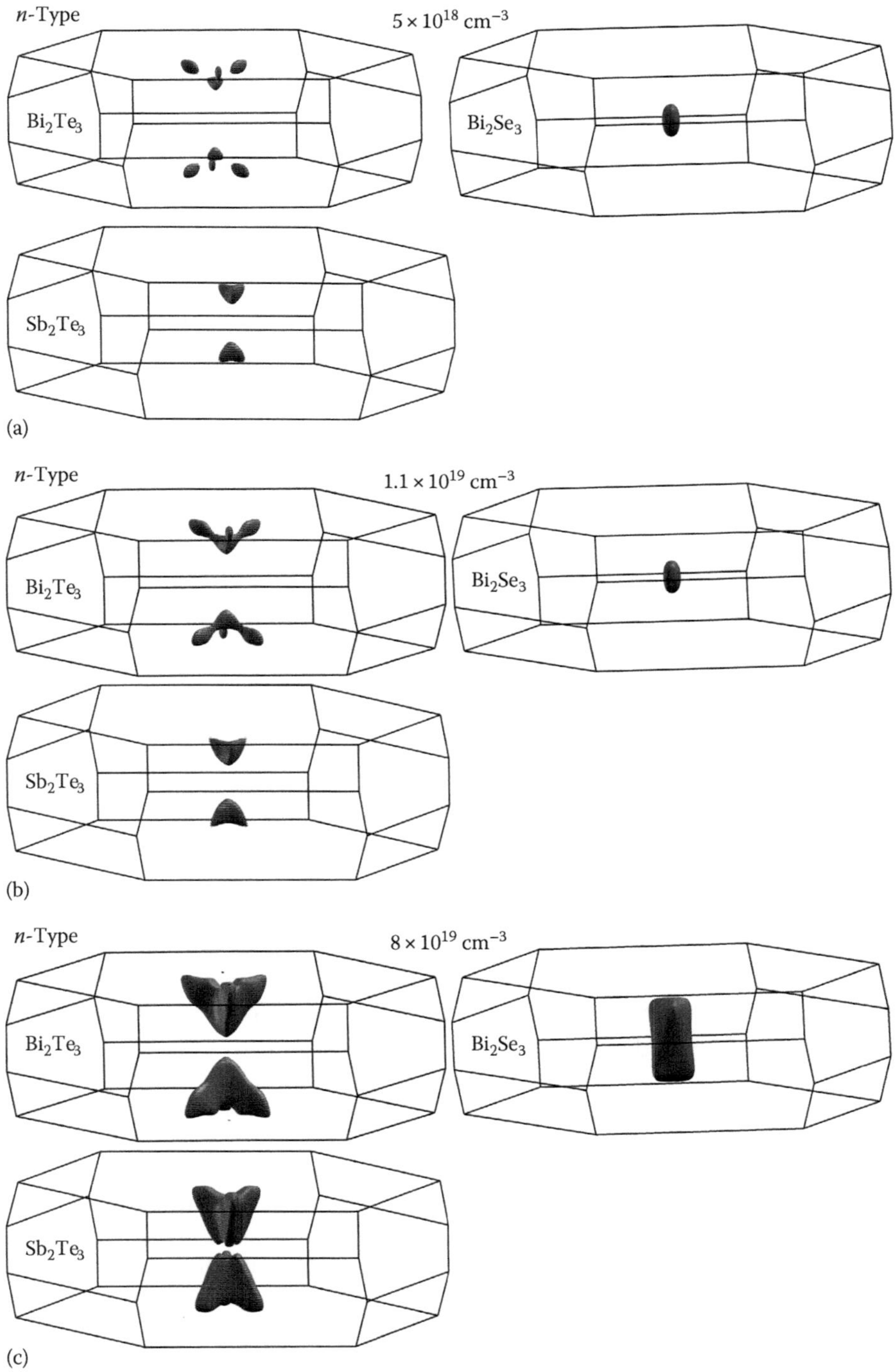

FIGURE 2.5
The electron Fermi surfaces of Bi_2Se_3, Sb_2Te_3, and Bi_2Te_3 that accommodate (a) 5×10^{18} cm^{-3}, (b) 1.1×10^{19} cm^{-3}, and (c) 8×10^{19} cm^{-3} electrons.

experimentally angle-resolved photoemission spectroscopy (ARPES) [21–23]. The method used here employs the mBJ potentials that correctly place the VBM at Γ.

The FS of *p*-type Bi_2Se_3 starts to grow from a Γ-centered spheroidal pocket [10]. At heavier *p*-type concentrations, this center pocket starts to develop pipelike appendices. These additional features, formed from the local VBM in the Γ–*Z*–*U*–*F* plane (visible in *Z*–*F* and *Z*–*U* directions in Figure 2.4) in our computations, show up just above the presented concentration of 8×10^{19} cm^{-3} as additional separated pipes similar to those seen for Bi_2Te_3 and Sb_2Te_3. An additional FS picture for Bi_2Se_3 doped to 2.2×10^{20} cm^{-3} is plotted in Figure 2.4c to demonstrate these clearly. Topologically, the heavy hole FS-doped Bi_2Se_3 (near ~10×10^{19} cm^{-3}) is equivalent to that of Sb_2Te_3, but the latter is squeezed around the Brillouin zone center. The concentration at which the pipelike parts start to develop in Bi_2Se_3 is directly dependent on the energy location of the local "second" VBM [24], as seen in *Z*–*F* and *Z*–*U* directions in Figure 2.3. Its precise position is highly sensitive to the geometry of the unit cell used in computations and to the exchange-correlation potential used. Because of this, different results may be found in the literature. In our case of a relaxed crystal structure with the mBJ potential, the second VB maximum is about 0.2 eV below the VBM at Γ, not far from the recent result calculated using the more accurate GW method [15] (about 0.16 eV), whereas in Reference 25 it falls about 30–50 meV below the first VBM. This higher location of the second VBM in Reference 25 leads to the earlier appearance of the pipes in the FS outlined by Shubnikov–de Haas (SdH) measurements [10]. This resulted in a significant overestimation of the predicted thermopower values for *p*-type Bi_2Se_3 [25], much larger than can be attributed to the effect of scattering on the thermopower. The results computed here predict a *p*-type thermopower of $\alpha = 170\ \mu V\ K^{-1}$ at 300 K and $p = 10^{19}$ cm^{-3}, very close to the experimental values [10].

Figures 2.6 (Bi_2Te_3), 2.7 (Bi_2Se_3), and 2.8 (Sb_2Te_3) show the comparison of the computed FSs with the available ARPES data [26–28]. While those data were acquired mainly to observe the topological metallic surface states, which are clearly in evidence in the figures, the ARPES data show the FSs in the bulk of the material as well. The figures show the rather good agreement between the calculated FS and published experimental results.

2.3.2 Effective Masses in Binary Tetradymites

The effective mass m^* is a parameterization of the second derivative of the energy dispersion relation *vis-à-vis* the momentum **k**-vector. If an energy band in a semiconductor has a simple parabolic and isotropic dispersion, $E = \hbar^2 k^2/2m^*$, the effective mass $m^* = \hbar^2 (d^2E/dk^2)^{-1}$ is a scalar that does not depend on energy or charge carrier concentration. In very anisotropic solids like the tetradymites that have the complex FSs shown in Figures 2.4 through 2.8, it is a tensor and is strongly dependent on the exact value of the **k**-vector and, thus, on the location of the Fermi level and the concentration of carriers. In practice, the mass tensor is not necessarily the most convenient parameterization possible and Figures 2.4 and 2.5 give a better idea of what the FS really looks like. Nevertheless, because the effective mass is conveniently used in transport calculations, values for the mass tensor elements are summarized here. There are several ways to define and measure the effective mass tensor elements in such a manner that they have practical meaning in transport.

First, with a little imagination, it is possible to view the FSs at low carrier concentrations shown in Figures 2.4a and 2.5a as a set of ellipsoids centered either on Γ or on some internal point of the Brillouin zone. If at Γ point, such an ellipsoid has a degeneracy number $N = 1$. The ellipsoids that are at internal points of the Brillouin zone have a degeneracy number of $N = 6$ per point—for example, in the case of *p*-type Sb_2Te_3, there are two such

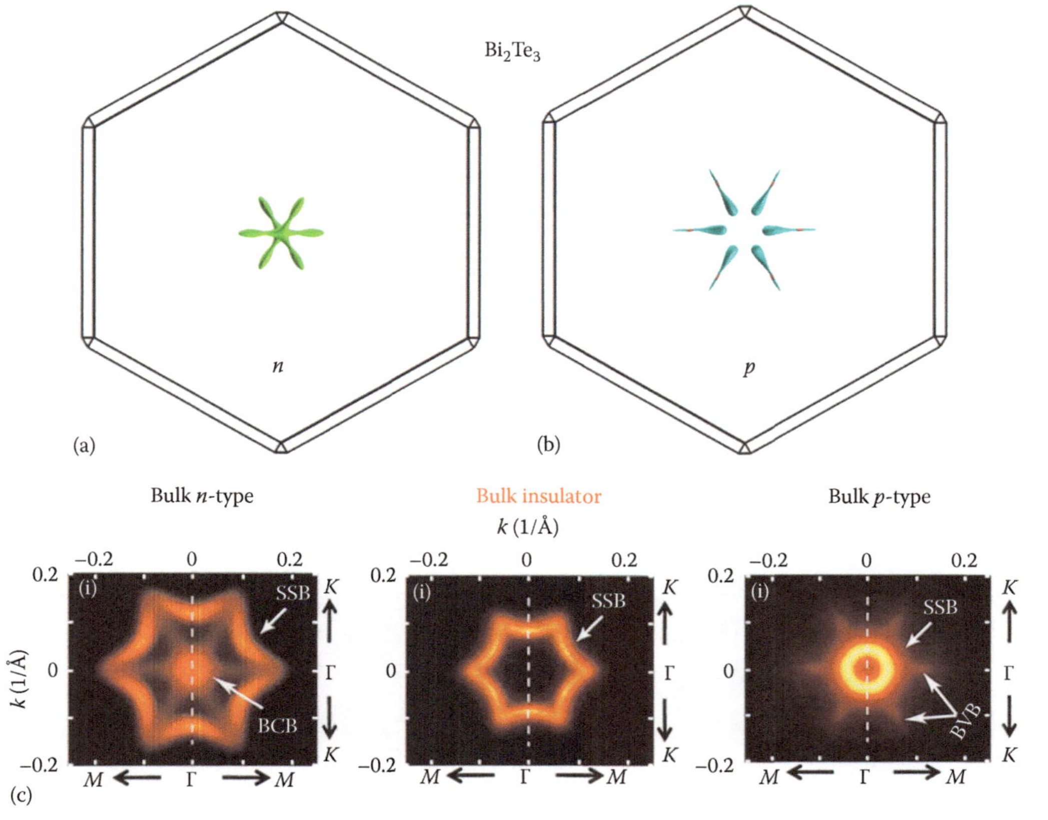

FIGURE 2.6
Projection along the c-axis of (a) the conduction band and (b) the valence band Fermi surfaces of n- and p-type Bi_2Te_3 doped degenerately to 1.1×10^{19} cm^{-3} charge carriers. (c) The calculated Fermi surfaces are compared to the ARPES measurements. In the ARPES pictures, the strong contribution from the surface states, labeled as SSB (surface-state band), is dominating the pictures, but the bulk conduction band as well as bulk valence band contributions are also visible. (From Chen, Y.L., Analytis, J.G., Chu, J.-H. et al., Experimental realization of a three-dimensional topological insulator, Bi_2Te_3, *Science*, 325, 178, 2009. Reprinted with permission of AAAS.)

points and so $N = 12$. This leads to the concepts labeled the 6-valley model or the 12-valley model used in the literature. The ellipsoid model has the advantage that it can be characterized by only three values for the masses, corresponding to the three main axes of each ellipsoid: m_1, m_2, and m_3. The volume of the ellipsoid then is a function of the product of the three principal masses and is characterized by the *density-of-states (DOS) effective mass* $\sqrt[3]{m_1 m_2 m_3}$ for each ellipsoid. A band consisting of N such ellipsoids has a DOS effective mass defined as $m_{DOS} \equiv N^{2/3}\sqrt[3]{m_1 m_2 m_3}$.

Second, there are several ways, both experimental and theoretical, to gain access to these parameters. Experimentally, there are measurements of Landau levels in the tetradymites. These are SdH effects [29] or de Haas–van Alphen (dHvA) effects [29], which give the cross section of the FSs in the direction normal to an external applied magnetic field. From these, and from their temperature dependence, one deduces values for m_1, m_2, and m_3. The

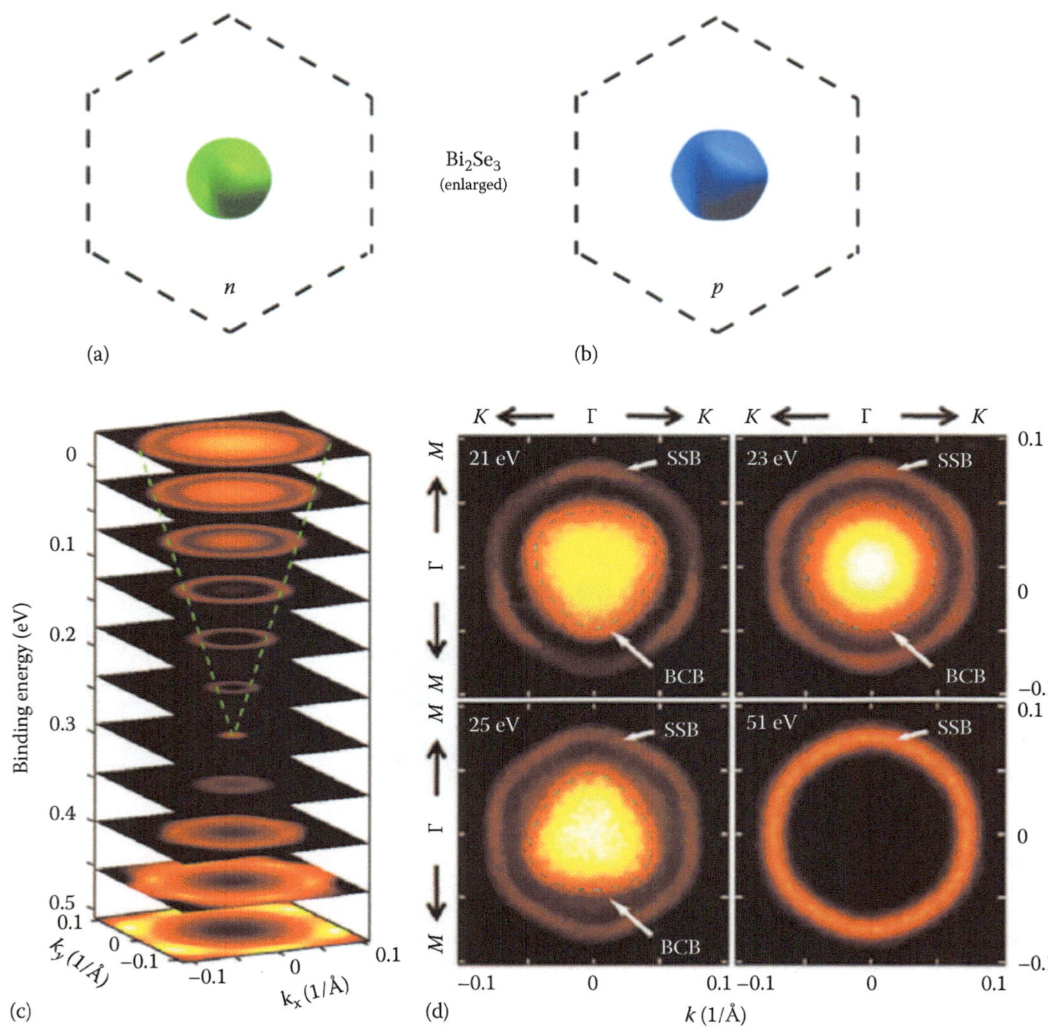

FIGURE 2.7
Projection along the *c*-axis of (a) the conduction band and (b) the valence band Fermi surfaces of (a) *n*- and (b) *p*-type Bi_2Se_3 doped degenerately to 1.1×10^{19} cm^{-3} charge carriers, compared to (c) the ARPES measurements. The theoretical Fermi surfaces are enlarged, relative to the Brillouin zone boundaries; for proper scale, see Figures 2.4 and 2.5. (d) In ARPES pictures, similar to Figure 2.6, the surface-state band contribution is seen. (From Chen, Y.L., Chu, J.-H., Analytis, J.G. et al., Massive Dirac fermion on the surface of a magnetically doped topological insulator, *Science*, 329, 659, 2010. Reprinted with permission of AAAS.)

cyclotron mass for a given direction of the magnetic field is then given by $m_{\omega} = \sqrt{m_{a\perp B} m_{b\perp B}}$; it is the geometric average of the masses along the principal axes (*a* and *b*) of the ellipse that is the cross section of the ellipsoid with the plane normal to the applied magnetic field *B*. Typically, it is given with *B* either parallel or perpendicular to the *c*-axis. The cyclotron mass then represents the masses of the cross sections of the FSs shown in Figures 2.4 and 2.5 with the planes parallel or perpendicular to the *c*-axes. The integration of the volume of

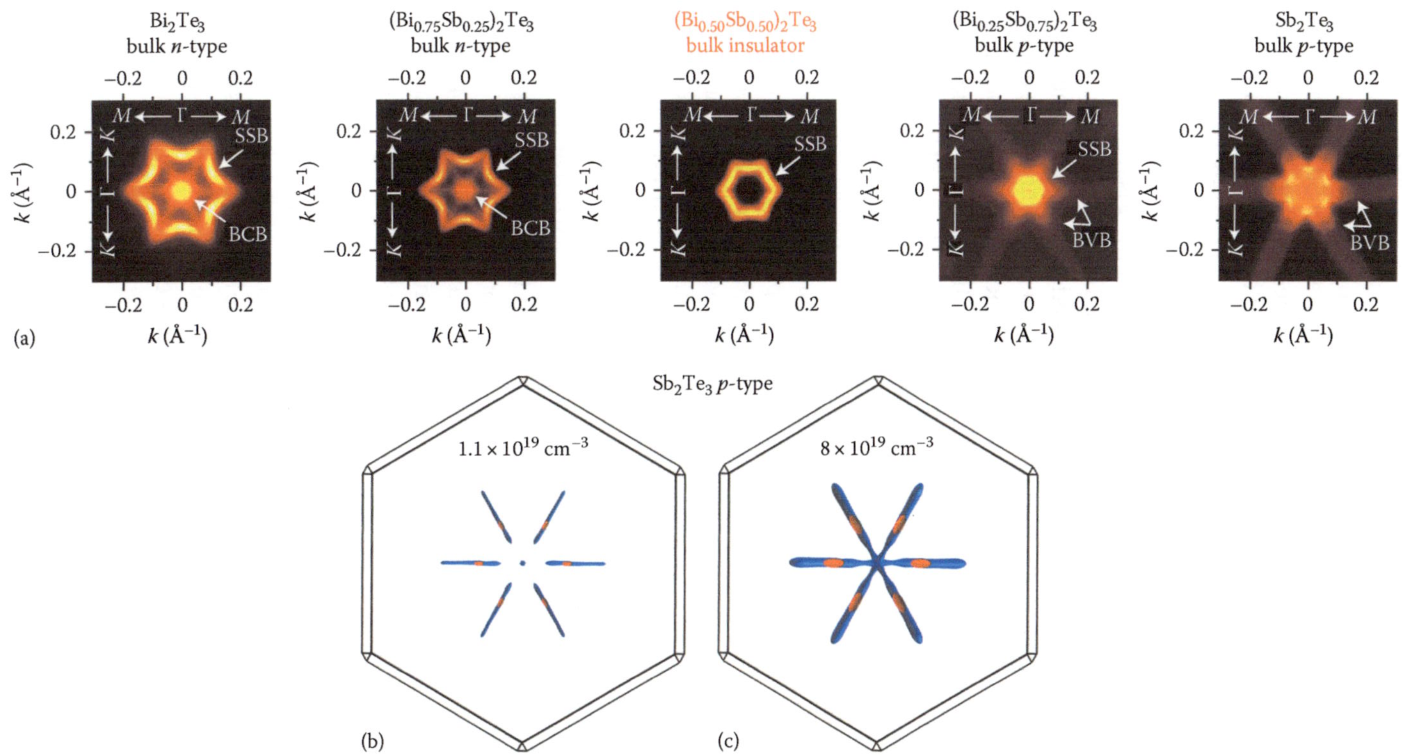

FIGURE 2.8
(a) Evolution of the Fermi surface (FS) for $(Bi_xSb_{1-x})_2Te_3$. For the limiting case of Sb_2Te_3, the ARPES data may be compared to the calculated *p*-type FS, shown in (b) for 1.1×10^{19} cm^{-3} and in (c) for 8×10^{19} cm^{-3} holes (there is no information on the location of the Fermi energy in the experimental pictures). As previously, the surface-state band contribution is labeled SSB. (Reprinted by permission from Macmillan Publishers Ltd. *Nat. Nanotechnol.*, Kong, D., Chen, Y., Cha, J.J. et al., Ambipolar field effect in the ternary topological insulator $(Bi_xSb_{1-x})_2Te_3$ by composition tuning, 6, 705, copyright 2011.)

the ellipsoids gives the DOS mass m_{DOS}. Because the shape of the FS changes with doping level, and also because of the strong spin–orbit interactions in the tetradymites, the effective mass tensor elements and their determinant are strong functions of the electron or hole concentration. One must therefore specify at what concentration the masses are defined. The experimental values of m_1, m_2, m_3, $m_{\omega//}$, $m_{\omega\perp}$, and m_{DOS} obtained from SdH measurements on the binary tetradymites are reported in Table 2.2. Usually, these are the mass values extrapolated to the band edge, that is, in the limit where E_F and n are zero. There are exceptions: the masses reported in Reference 10 cover the hole concentrations from 6 to 18 × 10^{18} cm^{-3}.

The second way to determine m_{DOS} is by measuring both the thermopower α and the Hall effect (R_H) of a sample and using the transport integrals to relate α to the carrier concentration ($n = 1/R_He$) in a relation $\alpha(n)$ that is known as the Pisarenko relation. This procedure will be outlined in the next section, but here the Pisarenko relation is used to determine the effective mass m_{DOS}. Usually, this procedure implies some assumption about the details of the scattering mechanism, which is often assumed to be either energy independent or dependent on $\sqrt{E}$, as is characteristic for acoustic phonon scattering. The resulting values of the "transport mass" m_{DOS} from the literature are also reported in Table 2.2, as they are actually more relevant to thermoelectrics than SdH or dHvA masses. The value for this mass is only valid for a specified carrier concentration, by its very definition. Tichy and Horak [42] measured masses in samples doped from 5 × 10^{18} to 2 × 10^{20} cm^{-3} and gave a mass extrapolated to the band edge. The transport masses from Harman et al. [43] are reported at the concentrations given in the table footnote.

Given the in-plane dielectric constant $\varepsilon_{\perp 0}$ and the in-plane cyclotron mass m_ω $B//c$ (obtained with the field along the c-axis), it is possible to determine the effective in-plane Bohr radius of electrons and holes, $a_B^* = a_B(\varepsilon/\varepsilon_0)(m^*/m_e)$, where ε_0 is the vacuum permittivity and a_B is the Bohr radius of the hydrogen atom, which has the value of 0.053 nm. These values, reported in Table 2.2, measure the spatial extent of the wave function of electrons and holes. The fact that these values are much larger than the in-plane unit cell is the reason behind the high mobility of charge carriers in these materials. The motion of electrons and holes in tetradymites is not very sensitive to perturbations of the atomic potentials at length scales much smaller than a_B^*, which in the tetradymites include most interstitial defects, foreign atoms in alloys, dopant atoms, or antisite defects. In particular, it is instructive to compare a_B^* to the Thomas–Fermi screening length r_{TF} [47]. When the latter becomes longer than a_B^*, the screening of the Coulomb interactions in the degenerate semiconductor extends over a distance longer than the size of any potential well that might arise by a defect in the solid, such as a dopant atom or a vacancy. In metals and degenerately doped semiconductors, Mott's criterion for inducing a metal–insulator transition at low temperature corresponds to the charge carrier concentration where $r_{TF} = a_B^*$ [47]. Expressing r_{TF} in terms of the carrier concentration leads to an expression for a critical charge carrier concentration n_{crit} above which no metal–insulator transition is observed [47] $a_B^* n_{crit}^{1/3} = 1/4(\pi/3)^{1/3}$, also reported in Table 2.2. The very low values for n_{crit} explain why in practice no tetradymite semiconductors become insulators even at very low temperatures.

2.3.3 Alloys

2.3.3.1 Valence Band Structure of $(Bi_{1-x}Sb_x)_2Te_3$

Special attention has been paid to the valence band structure of alloys along the pseudobinary Bi_2Te_3–Sb_2Te_3 line because most useful p-type thermoelectric materials are prepared

TABLE 2.2

Experimental Values for the Dielectric Permittivity in Both DC (ε_0) and in the High-Frequency Limit (ε_∞) and in Both the Direction Parallel to and Perpendicular to the *c*-Axis of the Binary Tetradymite Compounds

Property	Units	Bi_2Se_3	Bi_2Te_3	Sb_2Te_3	
$\varepsilon_{\perp 0}$	—	113 [30]	290 [30]	168 [31]	
$\varepsilon_{//0}$	—		75 [30]	36.5 [31]	
$\varepsilon_{\perp\infty}$	—	25 [32]	85 [30]	50 [33]	
$\varepsilon_{//\infty}$	—	17 [32]	50 [30]	30 [33]	
$\chi_{m//}/4\pi$	$cm^3\ mol^{-1}$	-3.8×10^{-7} [34]	-7.6×10^{-7} [35]	-2.8×10^{-7} [34]	
$\chi_{m\perp}/4\pi$	$cm^3\ mol^{-1}$	-3.8×10^{-7} [34]	-3.4×10^{-7} [35]	—	
E_g—indirect	meV at 293 K	—	130 [36]	210 [36]	
E_g—direct	meV at 77 K	160 [20,37]	—	—	
	meV at 10 K	300 [38]	−0.15 [36]	—	
dE_g/dT	$meV\ K^{-1}$	−0.20 [37]	—	—	
Electrons					
Number of pockets	N	1	6		
$m_\omega\ B//c$	m_e	0.124 [39]	0.06 [40]		
$m_{\omega//}/m_{\omega\perp}$		4 [32]			
Pocket principal axes					
m_1	m_e		0.021 [41]		
m_2	m_e		0.32 [41]		
m_3	m_e		0.081 [41]		
m_{DOS}—transport	m_e	0.16 [42]	0.32(*a*) [43]		
m_{DOS}—SdH	m_e		0.27 [40]		
Bohr radius a_B^*	nm	48	256		
n_{crit}	cm^{-3}	1.4×10^{14}	9×10^{11}		
Holes				UVB	LVB
Number of pockets	*N*	1	6	6	6
$m_\omega\ B//c$	m_e		0.09 [44]	0.049 [45]	0.14 [45]
$m_\omega\ B\perp c$	m_e	0.25 [10]			
Pocket principal axes					
m_1	m_e		0.031 [41]	0.084 [46]	0.034 [45]
m_2	m_e		0.44 [41]	1.24 [46]	0.34 [45]
m_3	m_e		0.086 [41]	0.127 [46]	0.54 [45]
m_{DOS}—transport	m_e	0.25 [10]	0.46(*b*) [43]	0.34(*c*) [43]	
m_{DOS}—SdH	m_e	—	0.35 [44]	0.3 [46]	0.97 [45]
Bohr radius a_B^*	nm	24	171	182	64
n_{crit}	cm^{-3}	1.2×10^{15}	3×10^{12}	3×10^{12}	6×10^{13}

For the same, the diamagnetic susceptibility (χ_m in SI units, $cm^3\ mol^{-1}$), energy gap, SdH and transport effective masses in units of m_e, and the free electron mass. The transport masses from Harman et al. [43] are reported at (a) $5 \times 10^{18}\ cm^{-3}$, (b) $8 \times 10^{18}\ cm^{-3}$, and (c) $7 \times 10^{19}\ cm^{-3}$. For electrons and holes, effective in-plane Bohr radius a_B^* and corresponding critical concentration n_{crit} at the Mott metal–insulator transition.

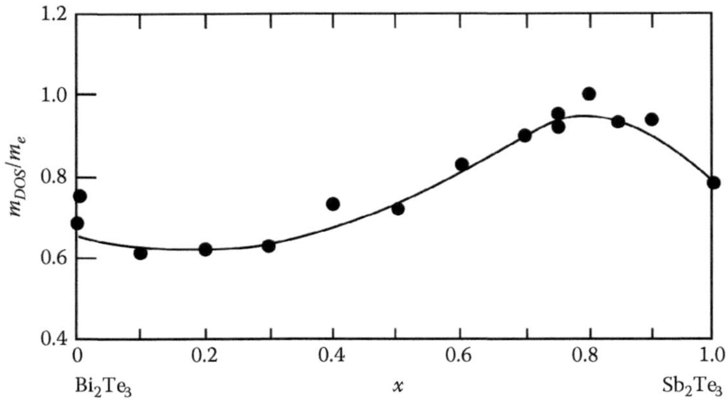

FIGURE 2.9
Evolution of the effective masses in the $(Bi_xSb_{1-x})_2Te_3$ alloys. The total density of states mass m_{DOS} for p-$(Bi_{1-x}Sb_x)_2Te_3$ single crystals in dependence on x. (Adapted from Stordeur, M. et al., *Phys. State Solidi* (*b*), 150, 165, 1988.)

in this system. Figure 2.8 shows the evolution of ARPES measurements on a series of single-crystal alloys. Particularly at $x = 0.75$, an intense DOS is seen centered along the Γ–Z axis in the middle of the Brillouin zone. The structure is likely a superposition of the VB elements of Bi_2Te_3 and Sb_2Te_3 as seen in Figure 2.4c at high hole concentration, with the Sb_2Te_3 contribution ($N = 6 + 6$; see Table 2.2) dominating. A study of the various components of the effective mass tensors in these alloys has been made by Stordeur et al. [46]. The results are shown in Figure 2.9, which reports the various masses of the quasi-ellipsoidal pockets and the final m_{DOS}. In that paper, the masses are derived from transport and are reported to be at the band edge, but the value for Bi_2Te_3 in Figure 2.9 ($0.7m_e$) is double that of the other authors ($0.35m_e$ in Table 2.2). The hole concentration in the samples varies monotonically from ~8×10^{18} cm^{-3} for Bi_2Te_3 to ~8×10^{19} cm^{-3} for Sb_2Te_3. If the masses for Sb_2Te_3 of Stordeur et al. [46] are presumed to be at the measured hole concentration rather than extrapolated to the band edge, their values correspond again quite well to those reported by other authors as shown in Table 2.2. Therefore, while there are some unresolved issues with Reference 46, the m_{DOS} in Figure 2.9 shows a nonmonotonic behavior with x that reaches a maximum at $x = 0.7$–0.8, making that composition the prime choice for high-ZT p-type thermoelectric material.

2.3.3.2 Energy Band Gap of the $Bi_2(Te_{1-x}Se_x)_3$ Alloys

Greenaway and Harbeke [48] measured the optical band gap of alloys along the Bi_2Te_3–Bi_2Se_3 line. They observed a maximum at the composition of the intermetallic Bi_2Te_2Se line compound. Their data can be summarized as follows. For $x < 0.33$, that is, on the Bi_2Te_3-rich side, the gap increases linearly with the Se content, following a law E_g (meV) = 145 + (4.62 × x), ($0 < x < 0.33$). A maximum value for the gap of 299 meV is obtained for $x = 0.33$, beyond which the gap decreases with increasing Se content following E_g (meV) = 361 − (1.94 × x), ($0.33 < x < 1$). It follows that for high-temperature applications of tetradymites, for example, in power generation from waste heat, the line compound Bi_2Te_2Se has the advantage of being less susceptible to bipolar effects, where thermally excited minority charge carriers decrease the total thermopower and induce ambipolar thermal conduction.

2.3.4 Transport Properties Predicted from Band Structure

Theoretical computations also allow the investigation of the changes in the effective mass of the charge carriers as a function of the carrier concentration. The DOS effective mass may be calculated using the formula [49,50] $m_{DOS} = m_e \hbar^3 \pi^2 \sqrt{g(E)g'(E)}$, where $g(E)$ is the density of states per energy and volume and $g'(E)$ is its energy derivative. Results are plotted for *n*- and *p*-type tetradymites in Figure 2.10. For Bi_2Se_3, at the low carrier concentrations, m^*_{DOS} is low, much less than $0.5m_e$, being slightly heavier for *p*- than for *n*-type. For *p*-type material, near 1×10^{19} cm^{-3}, the experimental effective mass is $m^*_{DOS} \approx 0.25m_e$ [10],

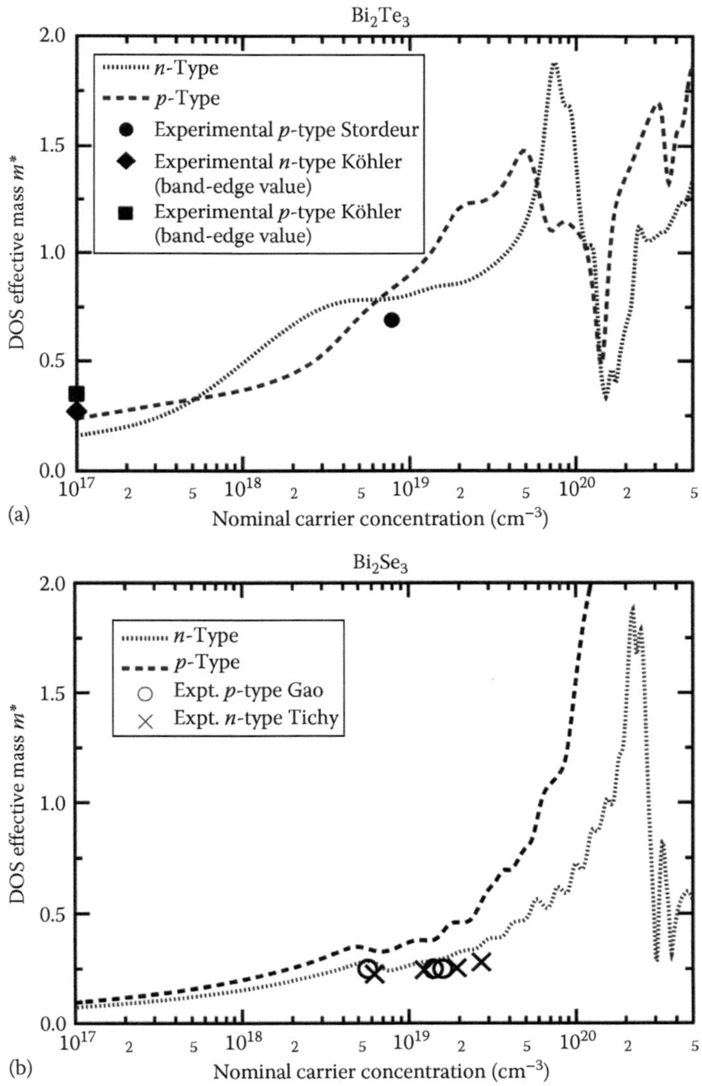

FIGURE 2.10
Theoretically calculated density-of-states effective masses for *n*- and *p*-type tetradymites, as a function of nominal carrier concentration. (a) Bi_2Te_3, (b) Bi_2Se_3. *(Continued)*

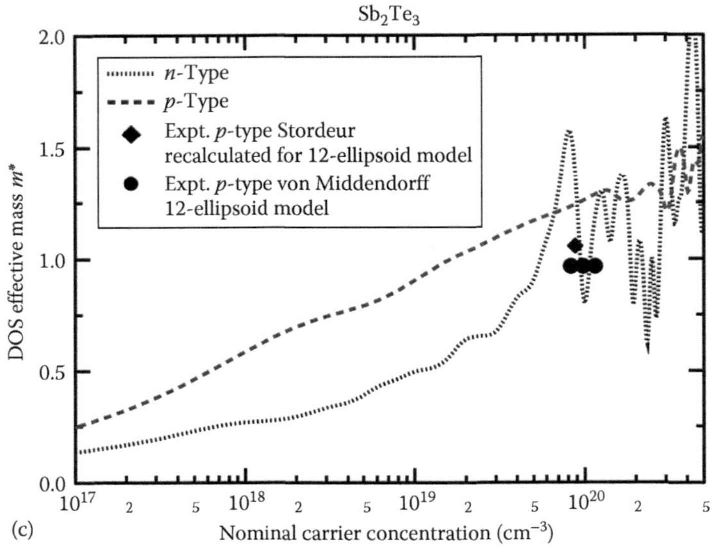

FIGURE 2.10 *(Continued)*
Theoretically calculated density-of-states effective masses for n- and p-type tetradymites, as a function of nominal carrier concentration. (c) Sb_2Te_3. (The experimental data points are taken from Gao, Y.B. et al., *Phys. Rev. B*, 90, 125204, 2014; Köhler, H., *Phys. Stat. Solidi (b)*, 73, 95, 1976; Tichy, L. and Horák, J., *Phys. Rev. B*, 19, 1126, 1979; Köhler, H., *Phys. Stat. Solidi (b)*, 74, 591, 1976; von Middendorff, A. et al., *Solid State Commun.*, 13, 443, 1973; Stordeur, M. et al., *Phys. Stat. Solidi (b)*, 150, 165, 1998.)

whereas the calculated one is larger, but not far from the measured, $m^*_{DOS} \approx 0.37m_e$. When E_F goes deeper and the "pipe region" of the FS starts to develop (5×10^{19}–1×10^{20} cm^{-3} in our study), m^*_{DOS} quickly increases. The experimental [42] n-type mass is also close to the computations and oscillates near $0.25m_e$ for $n \approx 10^{19}$ cm^{-3} as well.

For p- and n-type Bi_2Te_3, the calculated DOS effective mass gradually increases from ~$0.25m_e$ near the band edge to ~$0.75m_e$ around 10^{19} cm^{-3}, with p-type being heavier at the beginning. The concentrations where m^*_{DOS} increases are connected to the appearance of additional electron (hole) pockets, leading to an increase in N. The m^*_{DOS} value reported in Reference 46 is in very good agreement with computations, if, as explained earlier, it is assumed to be at the carrier concentration and not at the band edge. For p-type Sb_2Te_3, due to its many-valley band structure, m^*_{DOS} is large already for the low doping and continuously increases when p increases. In the computations presented here, the large valence band DOS centered at Γ point appears near $p = 10^{20}$ cm^{-3}. The experimental value, extracted from SdH measurements and assuming a 12-valley model for the Sb_2Te_3 valence band structure (see Table 2.2 and references therein), is slightly smaller. As far as n-type is concerned, m^*_{DOS} is smaller, until $n = 10^{20}$ cm^{-3} where a large and nonspheroidal FS is formed. Since, to the best of our knowledge, n-type Sb_2Te_3 has never been prepared at any electron concentration, we cannot compare the theory to measurements for the conduction band of that material.

Transport properties of tetradymites were studied using first-principles computations by several groups. The first studies devoted to Bi_2Te_3 and Sb_2Te_3 are one of the earliest examples of application of Boltzmann transport equations for thermoelectric materials. These, and more recent results, may be found in References 13, 18, 51–53, whereas thermoelectric transport properties of Bi_2Se_3 were studied in Reference 25. To have a direct connection between the thermoelectric properties and the comprehensive band structure results presented in this review, and also to ascertain again the predictive power of such calculations, Boltzmann transport calculations using BoltzTraP code [54] in the constant relaxation time approximation were carried out. The theoretical band gaps for Bi_2Se_3 and Sb_2Te_3 were artificially expanded to 0.3 eV to avoid bipolar effects because the calculated band gaps were underestimated. The results for the conductivity-weighted average thermopower are presented in Figure 2.11 as a function of the charge carrier concentration (the so-called Pisarenko plot). The calculations reproduce the experimentally observed trends for all binary tetradymites quite well; one set of data on a thin film was added [55], but most data are on bulk crystals. Since the thermopower of tetradymites is not very anisotropic [5], we present the computed average α. The theoretical values slightly overestimate the thermopower of p-type Bi_2Te_3, but then again underestimate that of p-type Sb_2Te_3. The deviations are likely to arise from the effect of charge carrier scattering, whereby acoustic phonon scattering decreases α compared to the calculated value, but optical phonon scattering and ionized impurity scattering enhance it.

The analysis of the thermopower α in Figure 2.11 yields the firmest conclusions at low temperature (T = 100 K). Starting with p-type conductivity and low carrier concentrations (10^{18} cm^{-3}), the computed thermopower for Sb_2Te_3 is the largest among the three binary tetradymites because it has the highest valley degeneracy. The thermopower of Bi_2Se_3 is more than two times smaller due to its single FS pocket at Γ-point. Similar observations can be made for n-type materials; however, here Bi_2Te_3 starts with the highest α. When the carrier concentration increases, α gradually decreases for Bi_2Te_3 and Sb_2Te_3, but a plateau is observed between 10^{19} and 10^{20} cm^{-3} for p-type Bi_2Se_3. This is where the Bi_2Se_3 valence band FS starts to form pipelike pockets and E_F approaches the second VBM. When the temperature is higher (300 K), the positive influence of the second VBM on α starts for hole concentrations in the mid-10^{19} cm^{-3}; still, hole concentrations around 10^{20} cm^{-3} would be required to make Bi_2Se_3 competitive against the two other tetradymites. n-Type Bi_2Te_3 and Sb_2Te_3 have similar thermopowers, whereas n-type Bi_2Se_3 exhibits a lower α, again due to its FS that has only a single electron pocket at Γ-point.

Since ZT depends on α^2, the thermopower is the most important single ingredient in optimizing thermoelectric materials. Therefore, it does not appear that a compound with little Te is likely to emerge as a good thermoelectric for Peltier cooling in the tetradymites family. In Figure 2.11, bipolar conduction effects are seen already at 300 K for low-doped Bi_2Te_3 due to its small gap and at 700 K for Bi_2Se_3 and Sb_2Te_3. This reduces the potential of Bi_2Te_3 for higher temperature operation such as for power generation applications. It is here that the intermetallic Bi_2Te_2Se, with its gap reaching 300 meV [48], or perhaps even Bi_2Se_3, shows a potential competitive advantage.

2.4 Doping

2.4.1 Defect Doping

Defect doping relies on slight departures from the strict V_2VI_3 stoichiometry that would result in perfectly charge-neutral crystals. The V stands for a trivalent group V element, Bi or Sb, while the VI stands for a chalcogen at nominal valence −2. The first of the two most

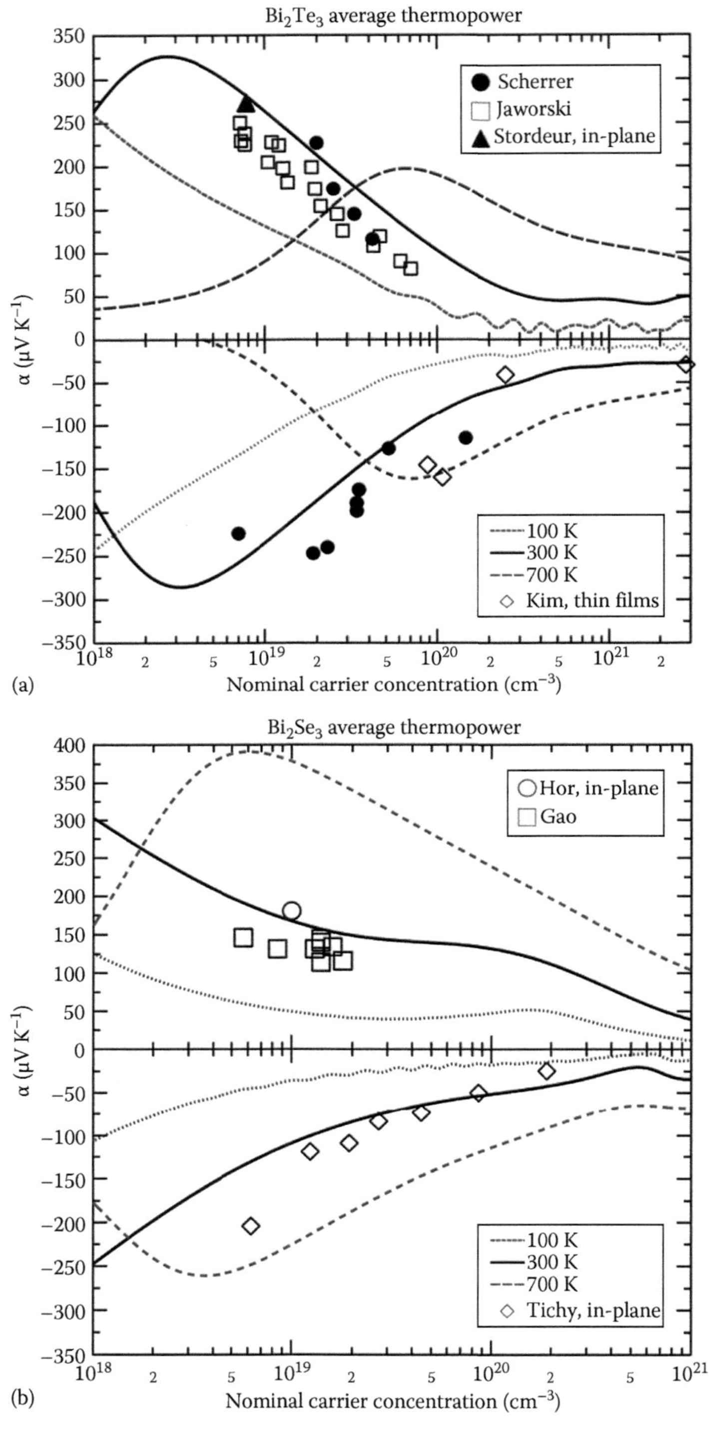

FIGURE 2.11
Pisarenko plots: the thermopower against nominal carrier concentration for (a) Bi_2Te_3, (b) Bi_2Se_3.

(Continued)

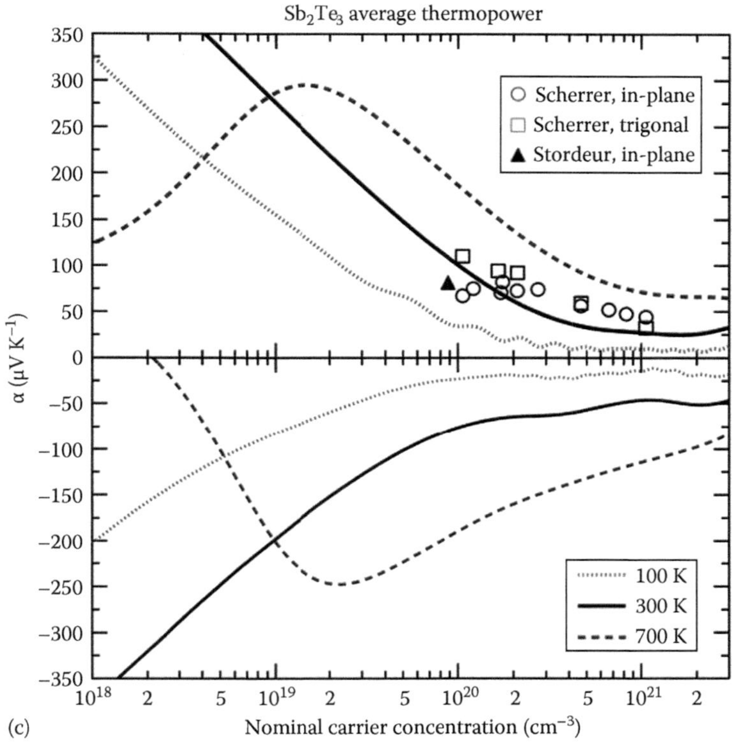

FIGURE 2.11 ***(Continued)***
Pisarenko plots: the thermopower against nominal carrier concentration for (c) Sb_2Te_3. The lines represent the computed conductivity-averaged thermopower. (The experimental data points are taken from Scherrer, H. and Scherrer, S., Bismuth telluride, antimony telluride and their solid solutions, in *CRC Handbook of Thermoelectrics*, ed. D. M. Rowe, CRC Press, Boca Raton, FL, 1995, pp. 211–237; Gao, Y.B. et al., *Phys. Rev. B*, 90, 125204, 2014; Hor, Y.S. et al., *Phys. Rev. B*, 79, 195208, 2009; Tichy, L. and Horák, J., *Phys. Rev. B*, 19, 1126, 1979; Stordeur, M. et al., *Phys. Stat. Solidi (b)*, 150, 165, 1998; Kim, J.-H. et al., *Mater. Trans.*, 54, 618, 2013; Jaworski, C.M. et al., *Phys. Rev. B*, 80, 233201, 2009.)

common defects are group V antisite defects, whereby an Sb atom in Sb_2Te_3 goes to a Te site (Sb_{Te}). Counting both s- and p-electrons, Sb has five electrons on its outer shells where Te would have six. Therefore, the presence of an Sb atom on a Te site leaves the solid shy of one electron: antisite defects dope tetradymites *p*-type. This is also the main defect in unintentionally doped Bi_2Te_3, which tends naturally to grow *p*-type [56], although Bi_2Te_3 can readily be prepared *n*-type in a Te-rich atmosphere, as will be described hereunder. Unlike the two aforementioned binary compounds, Bi_2Se_3 shows a much reduced propensity for Bi–Se mixing. Here, Hor et al. [38] report that the primary defects are doubly charged selenium vacancies, acting as *n*-type dopants. Bi_2Se_3 typically grows *n*-type at an electron concentration near 10^{17} cm^{-3}. There are diverging opinions on the nature of the defect: while Analytis et al. [57] agree with Hor et al. [38], Huang et al. [58] claim that there are Bi_{Se} antisite defects and additional layers. In the *n*-type $Bi_2(Se, Te)_3$ alloys, the native donor defect is identified [59] as due to an interaction between antisite defects formed during the crystal growth and vacancies created by a nonbasal slip motion. Therefore, it involves the simultaneous presence of Te and Bi vacancies. It is important to point out that the doping defect chemistry of tetradymites is rather more complicated than that of most other compound semiconductors, such as PbTe or SnTe. In the latter, excess metal (Pb, Sn) dopes the material *n*-type and excess chalcogen (Se, Te) dopes its *p*-type. In contrast, as explained earlier, excess metal often dopes the tetradymite semiconductors *p*-type.

The concentrations of electrons and holes due to defect doping in materials prepared at thermodynamic equilibrium depend on the exact shapes of the solidus lines in the equilibrium phase diagrams. These have been determined [5] and are shown for Bi_2Te_3 in Figure 2.12a, for Sb_2Te_3 in Figure 2.12b, for *p*-type $(Bi, Sb)_2Te_3$ alloys in Figure 2.12c, and for *n*-type $Bi_2(Se, Te)_3$ alloys in Figure 2.12d. The figures can be thought of as extensions of

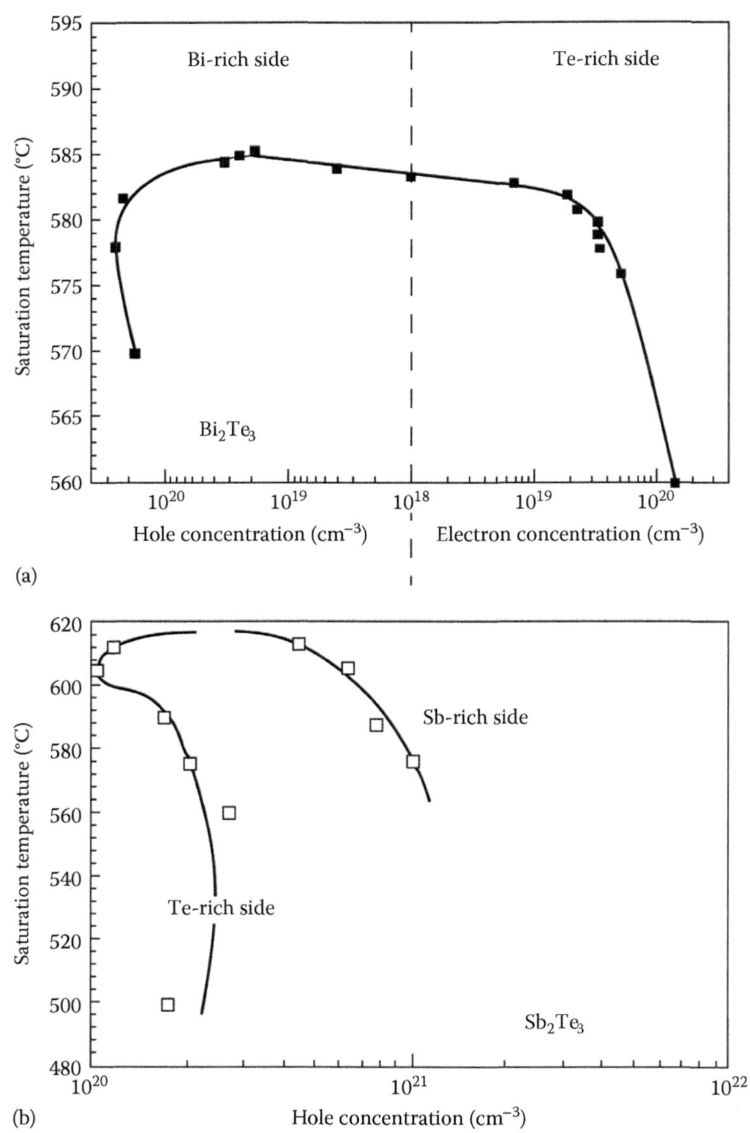

FIGURE 2.12
Solidus of binary (a) Bi_2Te_3 and (b) Sb_2Te_3 and ternary. *(Continued)*

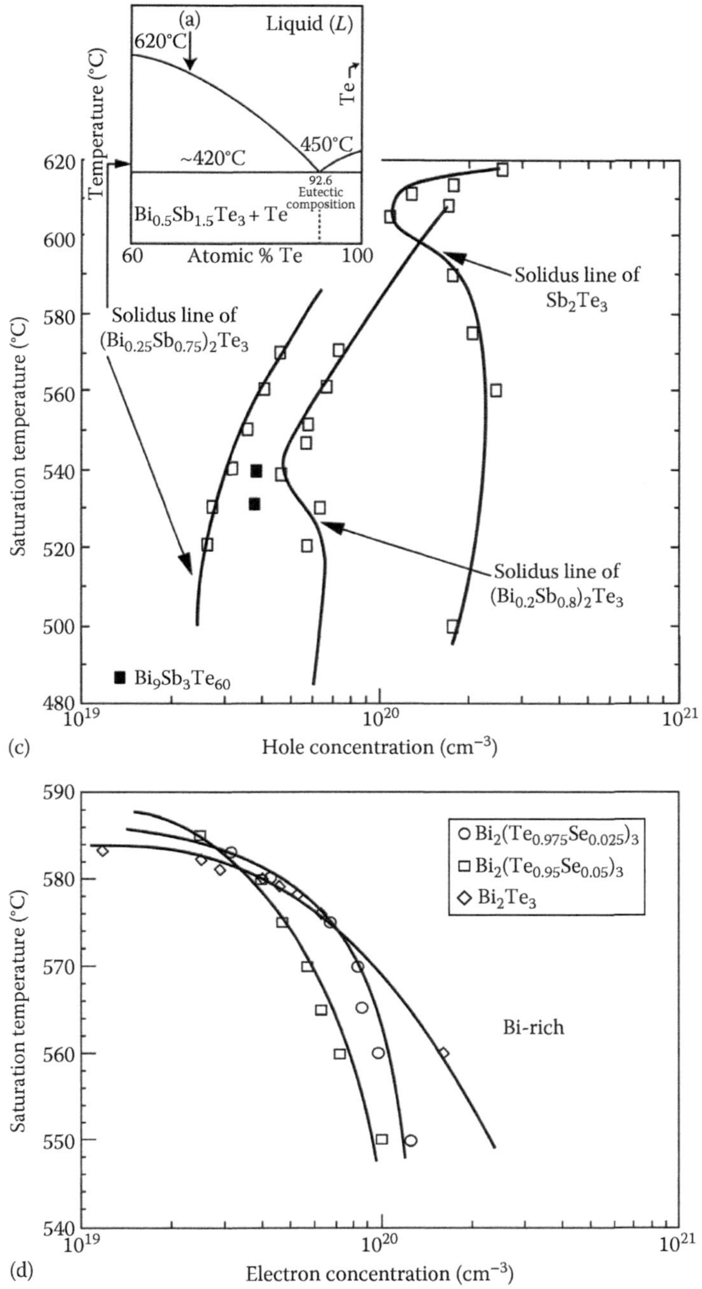

FIGURE 2.12 ***(Continued)***
Solidus of binary (c) $(Bi_{1-x}Sb_x)_2Te_3$ and (d) $Bi_2(Te_{1-x}Se_x)_3$ tetradymites at thermodynamic equilibrium. The inset in (c) shows the equilibrium phase diagram in its more usual form, with as ordinate axis the atomic concentration of Te. Rather than composition or V/VI ratio, the ordinate axes of all other figures give the excess electron or hole concentrations, which, in turn, are a measure of the group V or group VI element excess via defect doping. The ordinate axes are the saturation temperatures for samples prepared either group V or group VI rich. (Redrawn from Scherrer, H. and Scherrer, S., Bismuth telluride, antimony telluride and their solid solutions, in: *CRC Handbook of Thermoelectrics*, D.M. Rowe, ed., CRC Press, Boca Raton, FL, 1995, pp. 211–237.)

classical binary (pnictogen–chalcogen) or alloy phase diagrams, such as the pseudobinary [$(Bi_{0.25}Sb_{0.75})_2Te_3$]–Te phase diagram shown as the inset in Figure 2.12c. In the inset, the *y*-axis is the temperature and the *x*-axis is the concentration of Te, denoted [Te]. The *y*-axis limit of the inset is the solidus of the $(Bi_{0.25}Sb_{0.75})_2Te_3$ alloy. The large Figure 2.12c can be thought of as a blown-out version of the inset, where [Te] does not vary between 60 and 100 at.%, but only deviates from 60 at.% by a few parts per thousand. In other words, the ordinate axis of the inset is not, strictly speaking, a well-defined vertical line, but has a slight bowing smaller than the width of the ink. In this particular example, the material is *p*-type in the main frame of Figure 2.12c, so that this corresponds to a Te-deficient concentration, [Te] < 60 at.%, and the bowing of the solidus line in the inset would be to the left, to an excess Bi and Sb that corresponds to the solidus line indicated in the main frame of Figure 2.12c for $(Bi_{0.25}Sb_{0.75})_2Te_3$. The accuracy of chemical detection methods is not sufficient to represent this bowing in figures as a function of [Te]. However, the hole concentration can be measured by Hall effect measurements and corresponds to the concentration of Sb antisites, itself related to [Te]. In other words, conventional equilibrium phase diagrams where the abscissa axis is labeled as percentage concentration of elements do not have the resolution to identify the width of the solidus. The excess electrons or holes shown in Figure 2.12 at, for example, 10^{19} cm^{-3} would correspond to a deviation from stoichiometry of roughly one part in 10^4. Only electrical measurements can reveal such small deviations.

To create Figure 2.12, the authors grew single crystals by the hot-molten zone technique, a preparation technique at thermodynamic equilibrium. Note that any equilibrium technique can be used, such as an annealing step at the temperature shown as the ordinate axis, the saturation temperature. If the annealing or preparation takes place in a chalcogen-rich environment, then the abscissa in Figure 2.12a and d represents the value of the electron concentration one can reach as a function of the saturation temperature in Bi_2Te_3 or $Bi_2(Se, Te)_3$ alloys. This is due to the equilibrium concentration of chalcogen vacancies at the saturation temperature. Figure 2.12b shows that Sb_2Te_3 prepared even in Te-rich conditions will always come out *p*-type, but with a hole concentration about an order of magnitude lower than if it had been prepared in an Sb-rich atmosphere, so here, excess Te simply reduces the concentration of Sb antisite defects. If the preparation occurs in a pnictogen-rich environment, the abscissa in Figure 2.12a through c shows which value of the hole concentration corresponds to which saturation temperature in Bi_2Te_3, Sb_2Te_3, or $(Bi, Sb)_2Te_3$ alloys. It is important to realize that similar solidus lines exist for all compound semiconductors.

Relying on defect doping can be convenient, but there are several drawbacks. To begin with, the alloy composition determines the degree of band degeneracy, while at the same time being the primary parameter that determines the polarity and concentration of the majority charge carrier. Admittedly, the saturation temperature can be used to adjust the doping level; however, the range that can be achieved this way is quite limited, in particular since low saturation temperatures require very long anneals for kinetic reasons. In thermoelectric applications, where it is important to reach the optimal doping density as well as a high band degeneracy, having one single parameter adjust two important variables can be limiting. Furthermore, when prepared this way, Bi_2Se_3 and alloys near that composition are always *n*-type, while Sb_2Te_3 and alloys near that composition are always *p*-type. These disadvantages can sometimes be overcome by extrinsic doping.

2.4.2 Extrinsic Doping

Substitutional impurities dope the tetradymites following the principles that hold for other compound semiconductors. In the simple rigid-band charge transfer model, one assumes that the dopant impurity is added to the host solid in sufficiently low concentrations to avoid perturbing the band structure of the host material. The role of the dopant impurity is limited to transferring charge from its core electrons to the population of free electrons in the host solid. To the first order, one can expect elements from groups I, II, III, or IV substituting for the group V element to act as acceptors and group VII elements substituting for the group VI element to act as a donor. This works as long as the solubility of the dopant atom in the tetradymite host is sufficient.

The known dopants for the binary tetradymites are listed here. For Sb_2Te_3, Mn [60], Ag [61], Cd [62], and CdS [63] are known to be acceptors and iodine [64], Tl [65], and Ti [66] are donors. Two amphoteric impurities are known: Ge [67] and Pb [67] are mostly acceptors, but Ge can act as a donor [68] at concentrations up to 0.01 at% and Pb [69] up to 0.025 wt%.

Undoped Bi_2Se_3 is typically *n*-type with an electron concentration around $n = 10^{19}$ cm^{-3}. Known donors [70] are the halogens, Ge, Pb, and Cu. Cd [71] and Hg [72] are acceptors. Several more acceptor impurities have been discovered more recently for Bi_2Se_3 that can actually result in *p*-type material, including Ca [38], Mn [27,73], Mg [74,75], and Pb [10]. Yet, it is suspected that the native defects of Bi_2Se_3 still exist even in the presence of these chemical acceptors, and we do not know if the *p*-type material is compensated or if the acceptor impurities alleviate the donor action of the native defects.

In Bi_2Te_3, a technologically relevant range of electron and hole concentrations can be reached with defect doping, as described in Figure 2.12. Extrinsic dopants exist: Cu [76] and halogens (Br, I) [77] are donors, initially introduced as AgI or CuBr, but materials doped with the noble metal halides tended not to be stable over time [6]. Iodine is preferably introduced as SbI_3 or TeI_4 because these compounds exhibit considerable solubility in Bi_2Te_3 and its solutions, and no performance deterioration over long periods near room temperature [6].

2.4.3 Resonant Doping

Among various impurities used to tune the carrier concentration and optimize the thermoelectric performance of Bi_2Te_3, Sn is especially interesting, as it was found to form the so-called resonant level [78–80]. Resonant levels (RLs) are dopants that depart from the usual rigid-band model, in which it is assumed that the presence of the impurity atom does not distort the original band dispersion of the host material. RL impurities have specific electronic states that form a large and narrow DOS peak at the resonance energy, even when the impurity concentration is very low (see References 81–84). This is because they "catalyze" changes in the distribution of the electronic states of the atoms of the host solid. When the resonant impurity concentration reaches a few percent (1%–2% in the case of PbTe:Tl [85] and Bi_2Te_3:Sn [80]) and the orbital character of the resonant state is appropriate (e.g., s-like for Tl and Sn in the aforementioned cases), the RL DOS broadens and hybridizes with the electronic states of the host material. This ensures that the electrons in that state are delocalized and transport charge and that no narrow impurity band forms. In such a case, the RL may lead to a strong increase of the thermopower above the Pisarenko line (thermopower vs. carrier concentration) of other, classical impurities, as was observed in PbTe:Tl and Bi_2Te_3:Sn. This increase in the thermopower, originally explained as a result of the increase of the density of states [81,85], was recently shown for PbTe:Tl to have an effect on transport rather similar to the effect of an increase in band degeneracy, since additional electronic states appear around the highest valence band [82,84]. The measured thermopower at room temperature for Bi_2Te_3:Sn is shown in Figure 2.13e (after Reference 80). Between $p = 3–6 \times 10^{19}$ cm^{-3}, the thermopower of the polycrystalline sample is about

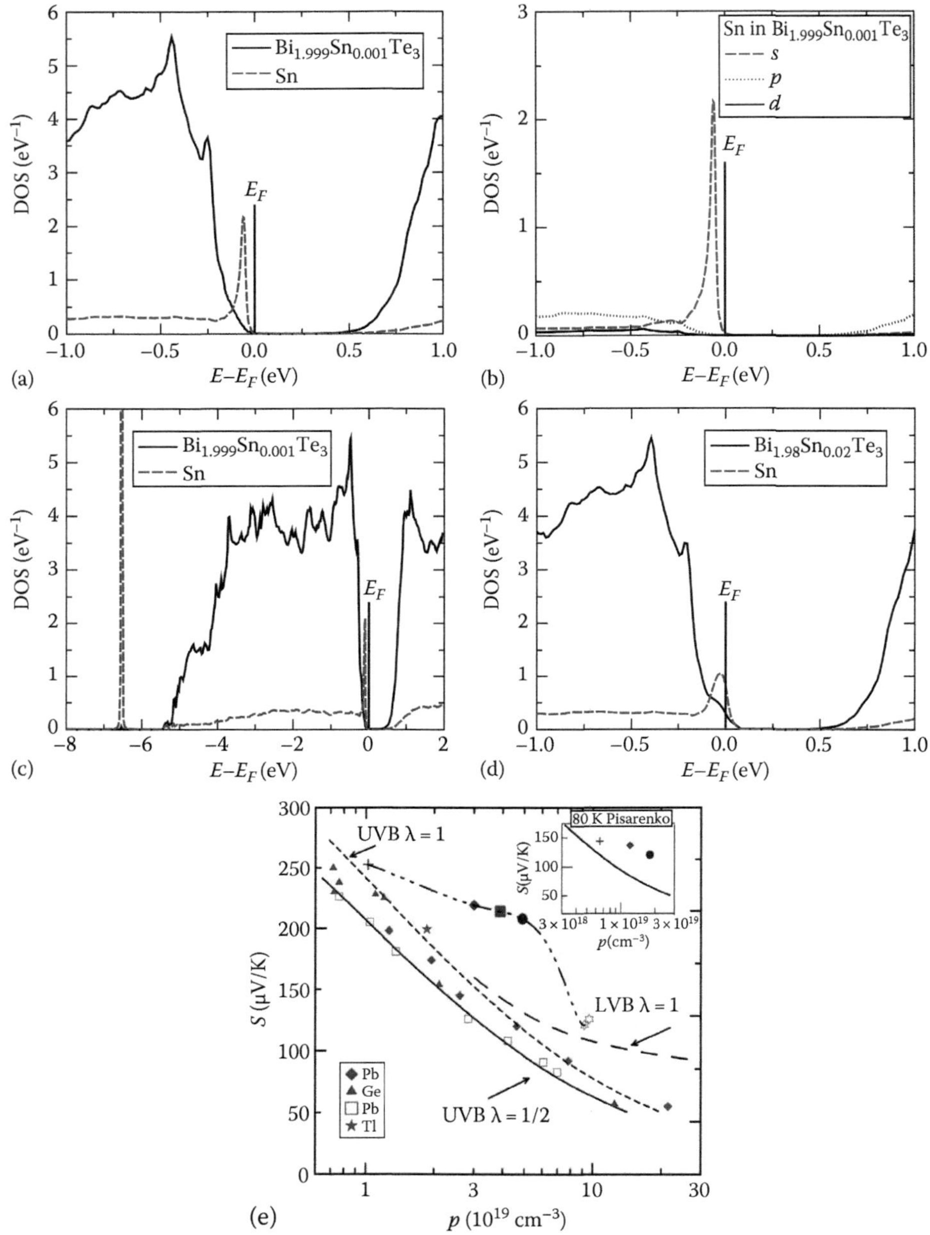

FIGURE 2.13
Density of states (DOS) and thermopower of Bi_2Te_3 doped with the resonant impurity Sn. (a) DOS of Bi_2Te_3 doped with 0.1% Sn near E_F. The atomic contributions plotted separately (atomic contributions here are given per atom, not weighted by their concentration). The resonant peak in DOS on the Sn atom is visible. (b) Angular momentum decomposition of partial Sn DOS. (c) DOS for a broader energy range, showing also second, hyperdeep peak, important for the acceptor behavior of Sn. (d) DOS for 2% doped material. (e) Pisarenko relation (solid and dashed line) for *p*-type Bi_2Te_3 as calculated at 300 K. (Adapted from Jaworski, C.M. et al., *Phys. Rev. B*, 80, 233201, 2009.) The symbols are the following: pluses $Bi_{1.9975}Sn_{0.0025}Te_3$, diamonds $Bi_{1.9925}Sn_{0.0075}Te_3$, dots $Bi_{1.985}Sn_{0.015}Te_3$, squares $Bi_{1.98}Sn_{0.02}Te_3$, and stars $Bi_{1.95}Sn_{0.05}Te_3$. The symbols for Bi_2Te_3 doped with Pb, Ge, and Tl are shown in the figure. The inset contains the calculated Pisarenko relation at 80 K as well as experimental points for $Bi_{2-x}Sn_xTe_3$ as measured in Jaworski et al. [80].

twice as large as for Pb- or Ge-doped material ($\alpha \sim 220\ \mu V\ K^{-1}$ vs. $\alpha \sim 120\ \mu V\ K^{-1}$). The theoretical confirmation of the formation of RL on Sn in Bi_2Te_3 was recently reported [84]. For the purpose of this review, we have performed density functional KKR-CPA calculations [86,87] for Sn-doped Bi_2Te_3, with technical details similar to those presented in [88], and resulting DOS is shown in Figure 2.13a through d. A very sharp and narrow partial DOS peak on Sn atoms is first formed for 0.1% of Sn concentration, as seen in panel (a). This peak originates from the 5s states of tin (panel (b)) and is accompanied by the characteristic [82] hyperdeep peak state located about −5 eV below E_F (panel (c)). The formation of this hyperdeep state is important to achieve the acceptor behavior of the impurity, as it bounds one Sn electron and one Bi electron. Because of this, the Sn atom not only forms the resonant state near E_F but also acts as a monovalent acceptor in a way similar to Sn in pure bismuth semimetal [89]. Finally, when Sn concentration is equal to 2%, a "hump" in DOS near the edge of the valence band is formed (panel (d)) as a result of the broadening of the Sn DOS peak and hybridization of Sn with the host Bi_2Te_3 electronic states. Both theory and experiment confirm that Sn in Bi_2Te_3 does not behave as a regular acceptor in the rigid-band model, which would then only shift the Fermi level, but forms an RL responsible for the observed increase in thermopower.

There were attempts to verify whether Sn forms RLs in other tetradymites, in particular in Sb-based compounds, so far without success. In Bi_2Te_2Se [90–92], the large concentration of native n-type defects (most likely vacancies on Te or Se) creates a compensation effect that can mask any resonant effect (see also [93] for a theoretical analysis). The possibilities of Sn-doping other Bi-based tetradymites require further studies, especially as theory [93] predicts the formation of RL also in Bi_2Te_2Se and Bi_2Se_3.

2.5 Phonons

2.5.1 Binary Tetradymites

Figure 2.14 gives representative in-plane thermal conductivity (κ) data as a function of temperature measured on single crystals of the binary tetradymites. The data on Bi_2Te_3 are reported by Goldsmid [94] on a sample that was lightly n-type, with an electrical resistivity at room temperature of 5 mΩ cm. The lattice thermal conductivity ($\kappa_L = \kappa - \kappa_E$) is separated from the total in that paper, by measuring both the electrical and the thermal conductivity and using the Wiedemann–Franz relation ($\kappa_E = LT\sigma$) to estimate κ_E. This requires a good knowledge of L, often an arduous task. The proper way to do this is to measure other transport properties; develop a set of transport equations that properly reproduce the electrical resistivity, Hall coefficient, and thermopower; and then use that model to derive a calculated value of L. The result [94] of this procedure gives data for the in-plane lattice thermal conductivity of Bi_2Te_3, also shown in Figure 2.14. The data on Bi_2Se_3 were measured by Hor et al. [38] on a not intentionally doped n-type sample with $n = 8 \times 10^{17}\ cm^{-3}$. The data on the in-plane total thermal conductivity of single-crystal Sb_2Te_3 were reported in Lostak et al. [63] on a not intentionally doped p-type sample with a hole concentration of $6.8 \times 10^{19}\ cm^{-3}$. The lattice contribution was not separated from the total thermal conductivity for the latter two samples, but κ_L dominated the conductivity up to at least 200 K. In turn, κ_L at room temperature was likely dominated by anharmonic phonon–phonon scattering.

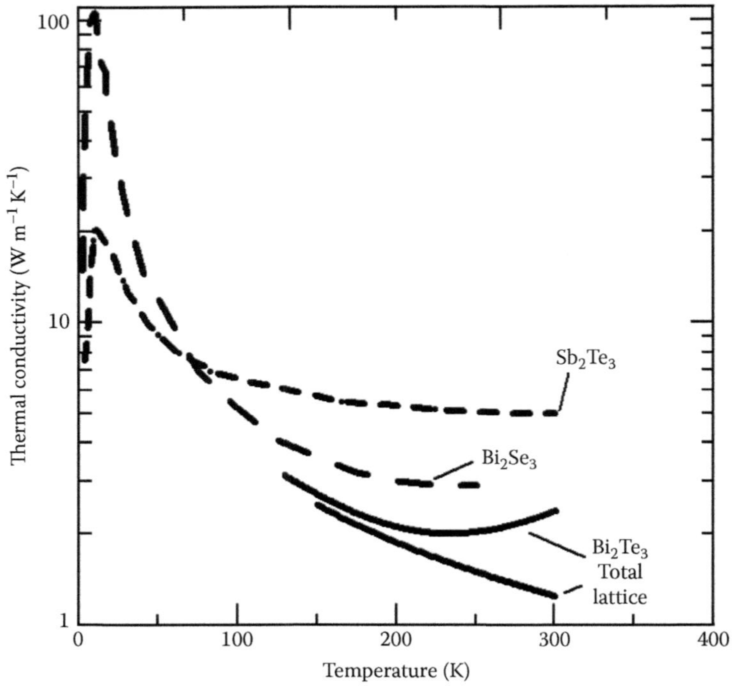

FIGURE 2.14
Temperature dependence of the in-plane thermal conductivity of single crystals of *n*-type Bi_2Te_3 (total and lattice contribution, after Goldsmid [94]), *n*-type Bi_2Se_3 not intentionally doped with $n = 8 \times 10^{17}$ cm^{-3} (after Hor et al. [38]), and *p*-type Sb_2Te_3 with $p = 6.8 \times 10^{19}$ cm^{-3}. (Redrawn from Lostak, P. et al., *J. Cryst. Growth*, 222, 565, 2001.)

Following the kinetic formula initially developed for ideal gases, the lattice thermal conductivity is the sum over all phonon modes (index *i*) and all phonon frequencies (ω) or energies ($E = \hbar\omega$) of the amount of heat that phonon carries: $\kappa_L = 1/3\sum_{i,\omega} C_i(\omega)v_i(\omega)\ell_i(\omega)$, where $C_i(\omega)$ is the specific heat, $v_i(\omega)$ is the group velocity, and $\ell_i(\omega)$ is the mean free path of a particular phonon mode and frequency. Executing this sum requires first the knowledge of the phonon dispersion relations $\omega(q)$ for each mode, where q is the wave vector. The phonon dispersion relations spectra of Bi_2Te_3, Bi_2Se_3, and Sb_2Te_3 are quite similar: the calculated dispersion [95] for Bi_2Te_3 is shown in Figure 2.14. It reproduces the experimental data [96] quite well. The group velocities are calculated as the slopes of the dispersion relations, $v = \nabla_q\omega$. The mean free path is related to the scattering frequency (τ^{-1}) or lifetime (τ) by $\ell_i(\omega) = v_i(\omega)\tau_i(\omega)$. Knowledge of the phonon lifetimes τ or scattering frequencies τ^{-1} can be gathered for intrinsic properties of solids, as described later, and can be engineered to a great degree. Indeed, optimizing *ZT* consists in finding mechanisms that minimize the average phonon lifetime, without affecting the electron scattering time. This is discussed in the following paragraphs.

Several phonon scattering mechanisms exist that limit τ. If they operate independently from each other, it is common to derive τ using Matthiessen's rule. This states that the overall scattering frequency of a phonon undergoing independent scattering processes is obtained by adding the respective scattering frequencies of each process. Intrinsic to all solids is phonon–phonon scattering, though processes called "Umklapp," the German word for *turnaround*. Attributing a frequency τ_U^{-1} to Umklapp scattering, one must further consider the other phonon scattering mechanisms to be alloy scattering τ_A^{-1}, isotope scattering τ_I^{-1}, dislocation scattering τ_D^{-1}, and scattering on nanoparticles τ_N^{-1}. Isotope scattering is mainly effective at low temperature and will be disregarded. Phonons are also scattered by electrons, a mechanism that affects phonons in metals near the Debye temperature [97], but vanishes either below or above it. In copper at room temperature, phonon–electron scattering contributes only a few percent to the phonon scattering, and one can reasonably assume that the effect is smaller in the materials studied here where the charge carrier concentration is three orders of magnitude smaller than in metals. It is therefore also disregarded in this review. The total scattering frequency is then $\tau^{-1} = \tau_U^{-1} + \tau_A^{-1} + \tau_D^{-1} + \tau_N^{-1}$. The different terms are described next.

2.5.2 Intrinsic Phonon–Phonon Scattering

Starting with the intrinsic mechanisms, in good crystals and at temperatures above the Debye temperature (164 K for Bi_2Te_3; 182 K for Bi_2Se_3; 160 K for Sb_2Te_3) [16], phonons are scattered intrinsically by other phonons. This is the case at room temperature for the tetradymites, which are the solids in which the Brillouin zone is small and the Debye temperature consequently low. The probability of intrinsic phonon–phonon scattering events (τ_U^{-1}) is primarily governed by the anharmonicity of the bonds [98], which is quite intense in the tetradymites at room temperature [95]. The physics behind the phonon anharmonicity is directly related to the chemical nature of the bonds in all solids. In the tetradymites [99], the origin of the anharmonicity comes from resonant bonds, a hybridization between different electronic configurations. Here, three valence p-electrons on the pnictogen alternate their occupancy of six available covalent bonds that exist between a given atom and its octahedral neighbors. The conditions that make this happen in the tetradymites include the octahedral coordination of the atoms, the weak s–p hybridization, and the fact that the s-electrons form bands at energies well below those of the bands that come from the p-orbitals. As a result, the bonds are highly polarizable, which is correlated with high anharmonicity, and, as an aside, also with the high static dielectric constants in Table 2.2, which, in turn, lead to the high mobility as explained there.

The anharmonicity is expressed by the Grüneisen parameter, which for each mode and frequency is the logarithmic derivative of the frequency with respect to the volume of the crystal, $\gamma_i(\omega) \equiv -(V/\omega_i)(\partial\omega_i/\partial V)$. This quantity affects each phonon differently and must be mapped over the entire Brillouin zone separately. It can be averaged over the Brillouin zone to produce a single scalar number, which is not that large in Bi_2Te_3, only 1.6 [16]. This single number does not represent the whole effect: the strong anharmonic scattering in tetradymites is best illustrated by a detailed calculation of the linewidth of the phonon dispersion relations, shown in Figure 2.15 at 77 and 450 K. The calculated temperature-induced line shape broadening shown in Figure 2.15 is inversely proportional to the phonon–phonon scattering lifetime. In Figure 2.15, it results from anharmonic phonon–phonon interactions, which are calculated for each phonon mode by integrating its interaction probability with all other phonon modes [95]. At 77 K the phonon dispersion lines are sharp and well defined. The significant broadening of the linewidths shown at 450 K indicates that the phonon lifetimes are reduced by the temperature increase. Some optical modes are diffuse beyond recognition. At 450 K, the lower-energy heat-carrying modes with a significant group velocity maintain some coherence, but nevertheless, anharmonic scattering is very significant.

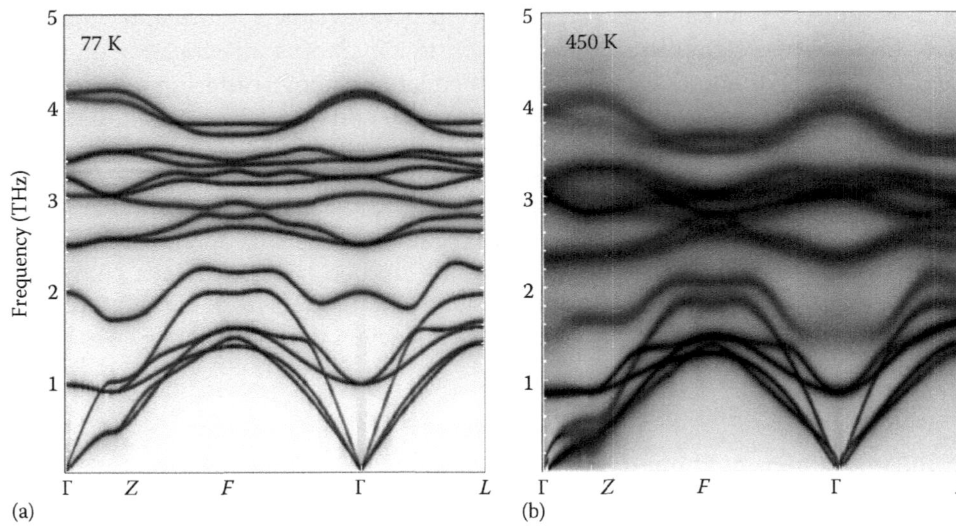

FIGURE 2.15
Phonon dispersions in Bi_2Te_3 at (a) 77 K and (b) 450 K. The phonon lifetime is inversely proportional to the broadening of the lines, illustrating the important contribution of anharmonic scattering in limiting the high-temperature lattice thermal conductivity. (Redrawn from Hellman, O. and Broido, D.A., *Phys. Rev. B*, 90, 134309, 2015; Hellman, O. and Broido, D.A., *Phys. Rev. B*, 92, 219903(E), 2015.)

With τ_U^{-1} now calculated for each mode and frequency and with the knowledge of the phonon spectra, which gives the group velocities, density of states, and heat capacity, the lattice thermal conductivity can be calculated and compared to the experimental data already given in Figure 2.14; the comparison is excellent and given in the original publication [95]. The relative contribution of mode and frequency is quantified by the *cumulative spectral thermal conductivity*, which is defined as the ratio of the contribution to the thermal conductivity of all modes up to a given frequency, that is, $\kappa_{CL}(\omega) = 1/3\sum_{i,\omega'=0}^{\omega} C_i(\omega')v_i(\omega')\ell_i(\omega')$. This quantity, as calculated [95] at 300 K and assuming that only anharmonic phonon–phonon processes scatter phonons, is shown for the in-plane thermal conductivity of binary Bi_2Te_3 in Figure 2.16. It shows that phonons with frequencies below 2 THz (100 K) provide the dominant contribution to the thermal conductivity. The dominant modes are the purely acoustic branches at the lowest frequencies; these are augmented by acoustic phonons intermixed with two low-lying optical phonon branches labeled semiacoustic in Figure 2.16. In very hard solids, the sound velocities are high and only the acoustic modes contribute at 300 K. That is not the case here, because the tetradymites are soft and their Brillouin zones are small, so that the mixed optical/acoustic phonon modes, and even the purely optical modes, contribute much to κ_L near room temperature.

Figure 2.16 conveys how heat is carried by a spectrum of phonons with frequencies ω ranging from 0.1 to over 2 THz. Increasing the figure of merit $ZT = (\alpha^2/L)(1+(\kappa_L/\kappa_E))^{-1}$ requires a reduction of κ_L by scattering phonons across as many frequencies as possible, while not scattering electrons. For a given value of the Grüneisen parameter, anharmonic phonon–phonon interactions at higher temperatures roughly give a scattering rate that scales approximately with the density of states for phonons to scatter in, that is, $\tau_U^{-1} \propto \omega^2$. The next step in minimizing the lattice thermal conductivity is to design

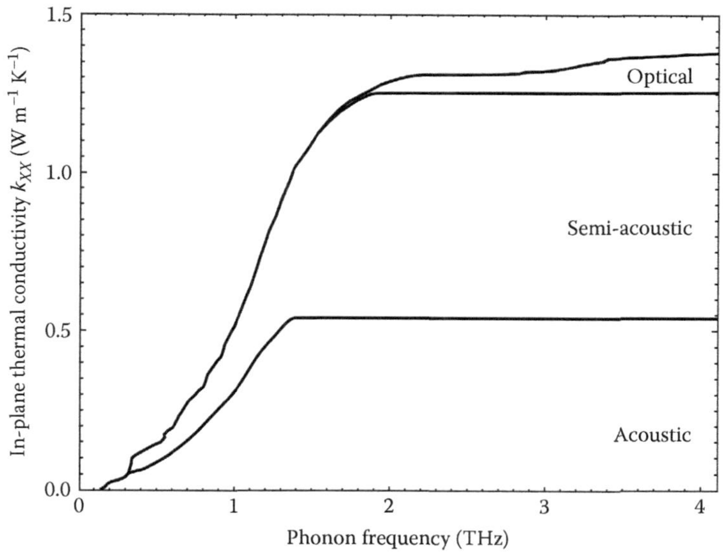

FIGURE 2.16
Calculated spectral contributions to the in-plane lattice thermal conductivity of binary Bi_2Te_3. The ordinate represents the amount of thermal conductivity due to phonons of a frequency equal or lower than the frequency given as abscissa. This is calculated assuming only anharmonic phonon scattering at 300 K. (Reconstructed from Hellman, O. and Broido, D.A., *Phys. Rev. B*, 90, 134309, 2015; Hellman, O. and Broido, D.A., *Phys. Rev. B*, 92, 219903(E), 2015.)

scattering mechanisms for phonons over as large a range of frequencies relevant to heat transport (i.e., shown in Figure 2.16) as possible.

2.5.3 Alloys along the (Bi_2Te_3)–(Sb_2Te_3) and the (Bi_2Te_3)–(Bi_2Se_3) Pseudobinary Lines

Alloy scattering is the first effective mechanism used to reduce the lattice thermal conductivity further. The effect of alloy scattering on κ_L at room temperature is shown in Figure 2.17 for the two alloy systems that are most important to thermoelectric applications. The κ_L of *p*-type alloys is shown along the (Bi_2Te_3)–(Sb_2Te_3) pseudobinary line in the left frame. That of the *n*-type alloys is shown along the (Bi_2Te_3)–(Bi_2Se_3) pseudobinary line in the right frame. The data in the curves are reproduced from the references as follows: curve (I) data are reproduced from Reference 100, curve (II) data from Reference 101, curve (III) data from Reference 102, and curve (IV) data from Reference 103. The discrepancy between the results of the different authors shown in Figure 2.17 for the *p*-type alloys is due to the fact that they used different ways to calculate the Lorentz factor when trying to separate the lattice thermal conductivity from the electronic contribution. The maximum in the middle of the thermal conductivity of the pseudobinary (Bi_2Te_3)–(Bi_2Se_3) alloys is probably due to the formation of the ordered intermetallic compound Bi_2Te_2Se [6], although no superlattice lines were identified in the x-ray spectra taken on those samples for that study [6]. It is interesting to note that the decrease of lattice thermal conductivity with Se content in the $Bi_2(Te_{1-x}Se_x)_3$ alloys at low values of x occurs quite steeply, with the full reduction obtained by $x \approx 10$ at.%.

The substitution of atoms by other atoms in alloys constitutes scattering centers that work only on phonons at the length scale of the interatomic distance. This corresponds to

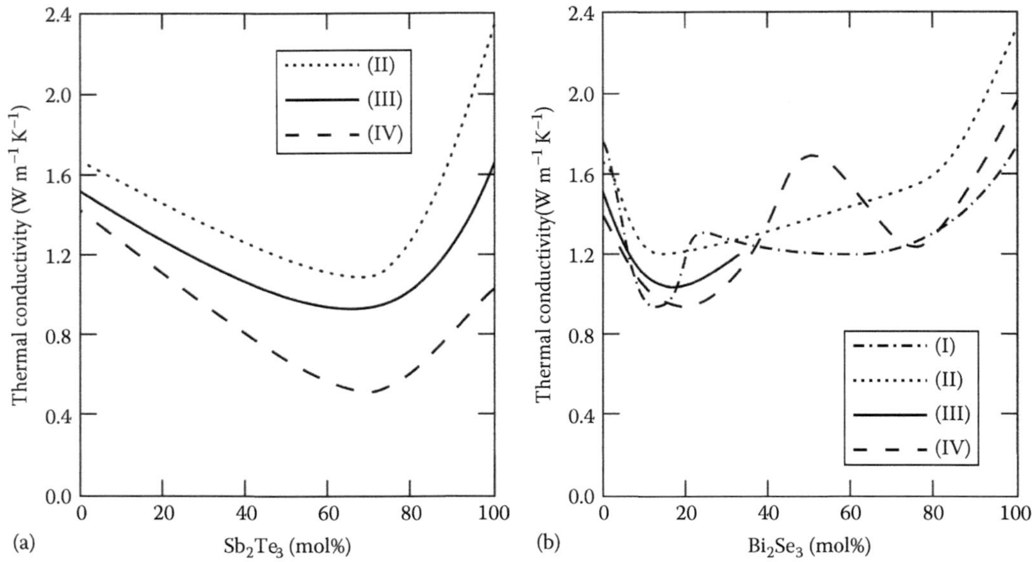

FIGURE 2.17
The lattice thermal conductivity at 300 K as a function of the alloy composition (a) in $(Bi_{1-x}Sb_x)_2Te_3$ alloys and (b) in $Bi_2(Te_{1-x}Se_x)_3$ alloys. (The data in curve (I) are redrawn from Champness et al. [100], in curve (II) from Birkholtz [101], in curve (III) from Goldsmid [102], and in curve (IV) from Rosi et al. [103].)

the wavelength of the most energetic phonons. Alloy scattering and scattering on isotopes are point defect (Rayleigh) scattering with $\tau_A^{-1} \propto \omega^4$. One of the most favorable aspects of the tetradymites in general is their large Bohr radii (see Table 2.2), which makes the electrons relatively insensitive to defects at this length scale. Thus, alloy scattering is particularly effective in increasing the ratio μ/κ_L.

2.5.4 Nanostructures

As outlined earlier, it is desirable to scatter phonons across the entire spectrum of frequencies shown in Figure 2.16 that contribute to thermal conductivity. The mechanisms outlined so far, $\tau_U^{-1} \propto \omega^2$ and $\tau_A^{-1} \propto \omega^4$, are most effective on high-frequency phonons. Boundary scattering is known to limit the low-temperature thermal conductivity of all dielectric solids, but when samples sizes are of the order of mm's, this happens only in the cryogenic temperature range. With the development of nanotechnologies, it was quickly realized that nanometer-sized obstacles in thermoelectric semiconductors could be effective at scattering lower-frequency phonons even at room temperature. Boundaries, grains, and nanometer-scaled obstacles all limit the mean free path of the phonons roughly to a value equal to their size and are frequency and temperature independent. Since the group velocity of acoustic phonons is constant, that implies that boundary or nanostructure scattering is independent of the phonon frequency: $\tau_N^{-1} \propto \omega^0$. What is the grain boundary length scale that generates the most scattering of heat-carrying phonons, yet does not affect the mobility too much?

The calculations in Reference 95 give the mean free path of the different phonons as limited by anharmonic scattering. Therefore, the procedure used to generate Figure 2.16 can be followed to represent the *thermal conductivity accumulation* as a function of the mean free path, that is, $\kappa_{CL}(\ell) = 1/3\sum_{i,\ell'_i(\omega)=0}^{\ell} C_i(\omega)v_i(\omega)\ell'_i(\omega)$, which is given in the original publication [95]. This is the amount of lattice thermal conductivity attributed to phonons with a mean free path *l* or less; it is sometimes also represented normalized to the total thermal conductivity κ_ℓ. This information is useful because it makes it possible to determine the size of nanostructures that will effectively scatter the phonons in a way that is complementary to anharmonic and alloy scattering. The answer for binary tetradymites at room temperature is that about 50% of the conductivity comes from phonons with a mean free path ℓ less than 10 nm and half with one of order $\ell \sim$ 10–100 nm. This suggests that adding nanostructures of order 5–30 nm is the most promising way to reduce κ_ℓ further while remaining at a length scale under the Bohr radii to limit electron scattering.

Only bulk nanostructuring is considered here, since thin-film work falls outside the scope of this review. In alloys, nanostructures can be obtained via either top-down or bottom-up synthesis routes. Here, bottom up means that the nanostructures can be grown intrinsically into the material by appropriate heat treatment, much like pearlitic steels. Top down characterizes the approach where the nanostructured powders of material are prepared first, typically by ball milling or melt spinning or possibly also by wet chemistry preparation techniques. The powders are then sintered into a composite, with a heat treatment that limits grain growth, typically spark plasma sintering.

The top-down approach was used by Poudel et al. [104] on *p*-type alloys. Here, *p*-type $(Bi, Sb)_2Te_3$ ingots were prepared and then ball milled for several hours to transform them into powders with dimensions ranging in the 5–50 nm range. The powders were hot-pressed (HP) in a graphite die to form 100% dense composites. A transmission electron microscope (TEM) microstructure of such composite is shown in Figure 2.18a: the grains are highly crystalline and the grain boundaries are clean, yet the grains vary in size from 5 to 30 nm, the desired length scale. Te precipitates 5–30 nm in diameter are also observed, as well as 2–10 nm diameter Sb-rich nanodots. The thermal conductivity of this material is shown

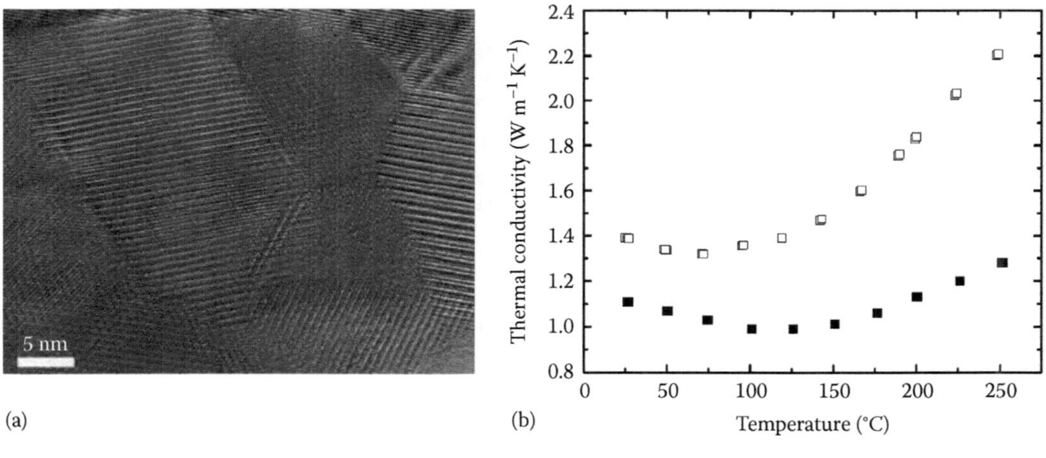

FIGURE 2.18
(a) TEM micrograph of hot-pressed nanocomposite *p*-type $(Bi, Sb)_2Te_3$ alloy showing high-crystallinity grains with clean grain boundaries. (b) Temperature dependence of the resulting thermal conductivity (full squares) is reduced by 20% or more compared to that of an alloy ingot (open squares) of similar composition. (From Poudel, B., Hao, Q., Ma, Y. et al., High-thermoelectric performance of nanostructured bismuth antimony telluride bulk alloys, *Science*, 320, 634–638, 2008. Reprinted with permission of AAAS.)

as a function of temperature in Figure 2.18b, where it is compared to that of a piece of the starting ingot. A 20% reduction in thermal conductivity is observed near room temperature. The reduction above 200°C is larger, but the explanation there is complicated by the increased role of electronic effects. The reduction in thermal conductivity is mostly ascribed to a reduction in the lattice conduction κ_L around 300 K and, at high temperature, to a decrease in ambipolar conduction [105]. Overall, the *ZT* of the nanostructured material is increased by 20% at room temperature. Because the nanostructuring also induced changes in the electronic properties, the *ZT* will be further discussed in Section 2.6.

The bottom-up approach to nanostructuring may already have been observed in the samples of Poudel et al. [104] but will be put more in evidence in the work described in Section 2.5.5. An opportunity to precipitate nanostructures in the (Bi, Sb)$_2$Te$_3$ alloy system arises from the pseudobinary [(Bi$_{0.25}$Sb$_{0.75}$)$_2$Te$_3$]–Te phase diagram already shown as the inset in Figure 2.12c. At about 92.6 atm.% Te, a Te-rich eutectic phase forms with a melting point around 420°C. Te-rich material, for example, of composition indicated by the arrow (a) in Figure 2.12c, will form some precipitates of this phase: the size of the precipitates depends on the cooling history of the sample, but it is possible to make this at the nanoscale. Because most binary or pseudobinary phase diagrams of systems that contain a line compound have a similar shape, with one or two eutectics to the left or to the right of the line compound, using such eutectics to create desirable nanostructures by a top-down metallurgical technique is often attempted. After the initial work of Poudel et al. [104] demonstrated the feasibility of the nanostructuring approach, extensive literature on the subject has appeared that has led to considerable gains in κ_L reduction and *ZT* enhancement in both *n*-type and *p*-type tetradymites. *ZT* values will be reported in Section 2.6, after the electronic properties are discussed.

2.5.5 Full-Spectrum Multiscale Scattering

So far, this review describes strategies to scatter low-frequency phonons with boundary or nanoparticle scattering ($\tau_N^{-1} \propto \omega^0$) and high-frequency phonons with alloy scattering ($\tau_A^{-1} \propto \omega^4$), besides the always present anharmonic scattering. Recently, a full-spectrum approach has been developed [106] that also targets midrange phonons using dislocation scattering. Dislocations are typically lines of scattering centers and scatter mid-energy phonons through most of the 10^9–10^{12} Hz [107]. It must be added that dislocations can be charged and induce electronic effects that are discussed in Section 2.6.

Kim et al. [106] also focus on the *p*-type (Bi$_{0.25}$Sb$_{0.75}$)$_2$Te$_3$ alloy composition. They compare the thermal conductivity and thermoelectric transport properties of samples created with four types of preparation techniques (see Figure 2.19). A conventional stoichiometric ingot is studied as a reference (a chemical analysis revealed it to be Te rich at molar ratios Bi$_{10}$Sb$_{28.9}$Te$_{61.1}$). Stoichiometric ingots were melt-spun and sintered (S-MS). The newest approach was to melt-spin stoichiometric material mixed with a very large (25%) excess Te and sinter that (Te-MS). Te forms a eutectic at 420°C (inset in Figure 2.12c). During the sintering process at 480°C and 70 MPa, the eutectic is expelled as a liquid. This leads to a complicated morphology revealed by TEM in the original publication [106]. The structure has small grains as in [104], but also a salient new feature (Figure 2.19a): the presence of dense arrays of dislocations at the grain boundaries. It is clear (Figure 2.19b) that the thermal conductivity is greatly decreased, and this is attributed to a further reduction of the lattice conductivity, which reaches just above 0.3 W m^{-1} K^{-1} in the Te-MS material from 300 to 350 K, rising to about 0.6 W m^{-1} K^{-1} at 460 K. This happens without prejudice to the electronic properties, leading to an enhanced *ZT* (see Section 2.6).

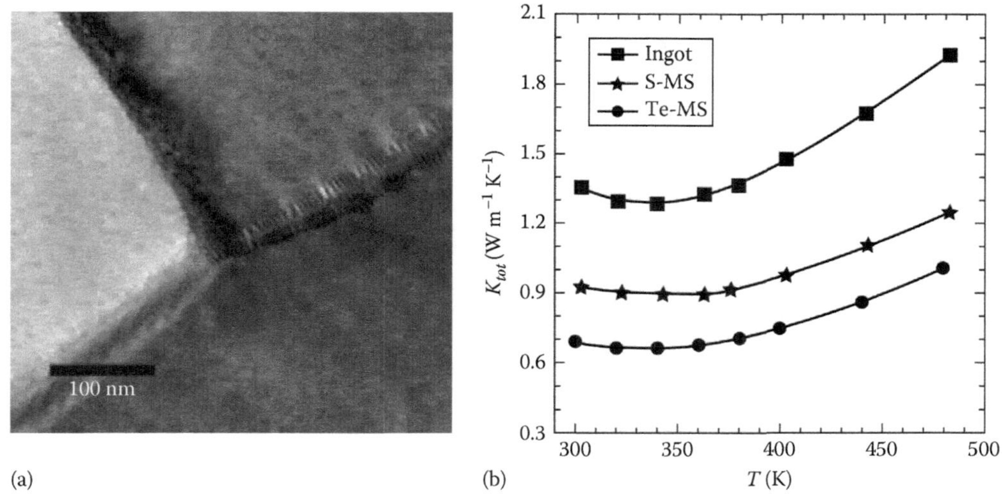

FIGURE 2.19
(a) TEM micrograph of a melt-spun and sintered sample of nanocomposite *p*-type $(Bi, Sb)_2Te_3$ Te-rich alloy showing a regular arrays of dislocations at the grain boundaries. (b) Temperature dependence of the thermal conductivity of a regular ingot and a melt-spun sintered sample of regular $(Bi, Sb)_2Te_3$, compared to a melt-spun sintered sample of very Te-rich $(Bi, Sb)_2Te_3$ (Te–MS). The latter is considerably reduced compared to all former samples. (From Kim, S.I., Lee, K.H., Mun, H.A. et al., Dense dislocation arrays embedded in grain boundaries for high-performance bulk thermoelectrics, *Science*, 348, 109–114, 2015. Reprinted with permission of AAAS.)

The lattice thermal conductivity of *n*-type $Bi_2(Te_{1-x}Se_x)$ alloys has also been reduced [108] by a multiscale, full-spectrum approach similar to that described earlier for *p*-type material. The same principles are used [108]: nanostructuring and dislocation scattering. In this case, the nominal composition of the starting material is $Bi_2(Te_{0.93}Se_{0.07})_3$. The case of *n*-type alloys is complicated by the anisotropy of their properties. It is shown in the next paragraph that for these alloys, the *ZT* along the trigonal axis is about half that in the trigonal plane. Good alloys are preferentially oriented. As a result, simply ball milling and sintering the sample will result in a material that is less oriented and consequently has a lower *ZT* at the onset of the optimization procedure. While nanostructuring in this fashion may limit the lattice thermal conductivity, it denies the benefit of using only the in-plane *ZT* of aligned ingots. This difficulty was circumvented by Hu et al. [108] in the following way: preferentially aligned $Bi_2(Te_{0.93}\ Se_{0.07})_3$ material prepared commercially by ZM was subjected to hot deformation (HD) processes. The HD process consists of cutting cylindrical samples of 10 mm diameter, putting them in a 16 mm diameter die, and subjecting them to a uniaxial pressure at 823 K in the absence of lateral confinement. This allows the samples to deform laterally and creates a very high density of dislocations (see Figure 2.20a). The growth direction of the ingot is parallel to the direction of the pressure and perpendicular to the direction of the deformation. Two levels of HD were applied, labeled HD1 (less deformation) and HD2 (more). In general, the best *ZT* is the in-plane *ZT* of the aligned ZM-HD2 material, where the in-plane thermal conductivity is decreased with respect to that of the pristine ZM material by about 25% (see Figure 2.20b). In that sample, the HD induces highly distorted regions with a spatial extent at the 10–50 nm scale (see Figure 2.20a). Inside these regions, arrays of dislocations are found spaced typically

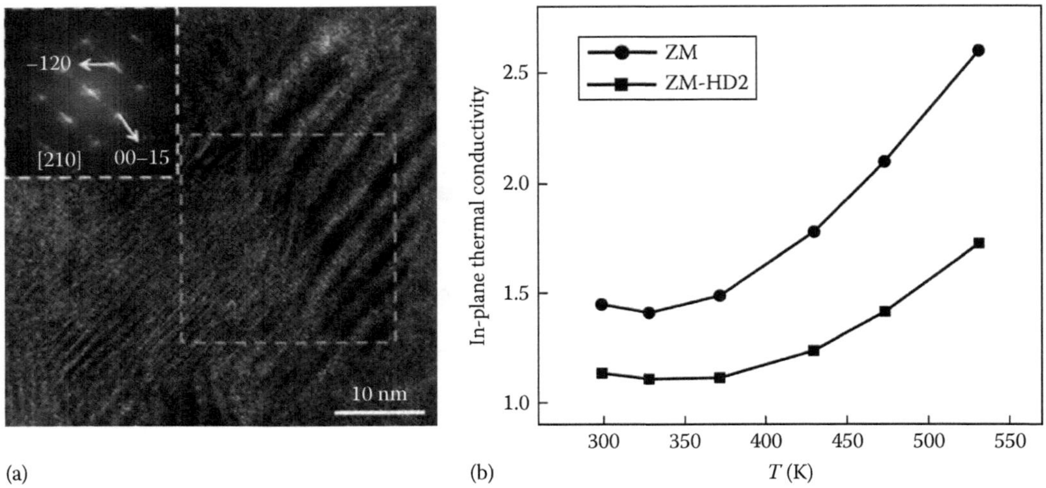

FIGURE 2.20
(a) High-resolution TEM micrograph and (b) in-plane thermal conductivity of aligned *n*-type $Bi_2(Te_{0.93}Se_{0.07})_3$ alloys. ZM stands for zone-melted material, which was subjected to two hot deformations, resulting in materials labeled ZM-HD2. The high-resolution transmission electron microscope (HRTEM) image in (a) is obtained along the [210] zone axis of material ZM-HD2. It shows high distortions, with a high density of dislocations. The inset shows a Fourier transform. The in-plane thermal conductivity of the ZM-HD2 sample is reduced compared to regular ZM material. (From Hu, L., Wu, H., Zhu, T. et al.: Tuning multiscale microstructures to enhance thermoelectric performance of *n*-type bismuth-telluride-based solid solutions. *Adv. Energy Mater.* 2015. 5. 1500411. Copyright Wiley-VCH Verlag GmbH & Co. KGaA. Reproduced with permission.)

3 nm apart, constituting a multiscale set of phonon scattering centers. The thermal conductivity perpendicularly to the preferential alignment, more parallel to the trigonal axis, is actually increased by about 30%, but since the material is intended to be used aligned, this does not harm the in-plane *ZT*.

We address in the next section the question as to whether or not the mechanisms that are used to limit the lattice thermal conductivity also affect the electronic properties and, more precisely, the electron mobility.

2.6 Electronic Transport Properties

2.6.1 Binary Tetradymites

The in-plane electrical conductivity and thermoelectric power of the three binary tetradymites, Bi_2Te_3, Bi_2Se_3, and Sb_2Te_3, are shown in Figures 2.21 through 2.23, respectively. The data on Bi_2Te_3 come from Champness and Kipling [109]. The samples were doped using deviations from stoichiometry. The Hall mobility of the *p*-type sample doped to 8×10^{18} cm^{-3} was 5100 and 40 (cm^2 V^{-1} s^{-1}) at 77 K and 300 K, respectively, while that for the *p*-type sample at 1.6×10^{19} cm^{-3} was 2700 and 35 (cm^2 V^{-1} s^{-1}) at 77 and 300 K, respectively. The temperature dependence of the mobility of both *p*-type samples follows a T^{-1} slope. The *n*-type sample with the lowest electron concentration, 1×10^{18} cm^{-3}, has the mobility of

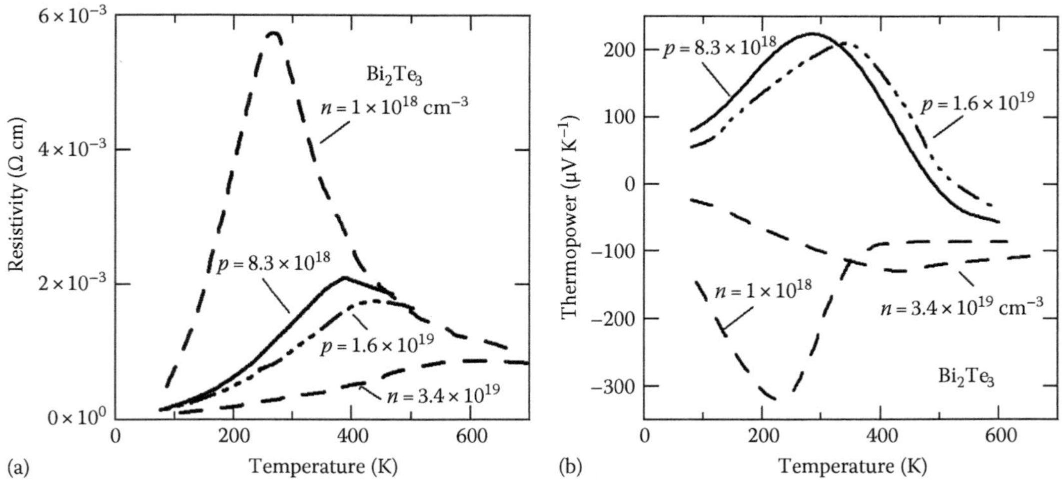

FIGURE 2.21
Temperature dependence of (a) the in-plane resistivity and (b) the thermopower of n-type and p-type Bi_2Te_3 material doped to the carrier concentrations indicated in the figure (in cm^{-3}). (Redrawn from Champness, C.H. and Kipling, A.L., *Can. J. Phys.*, 44, 769, 1965.)

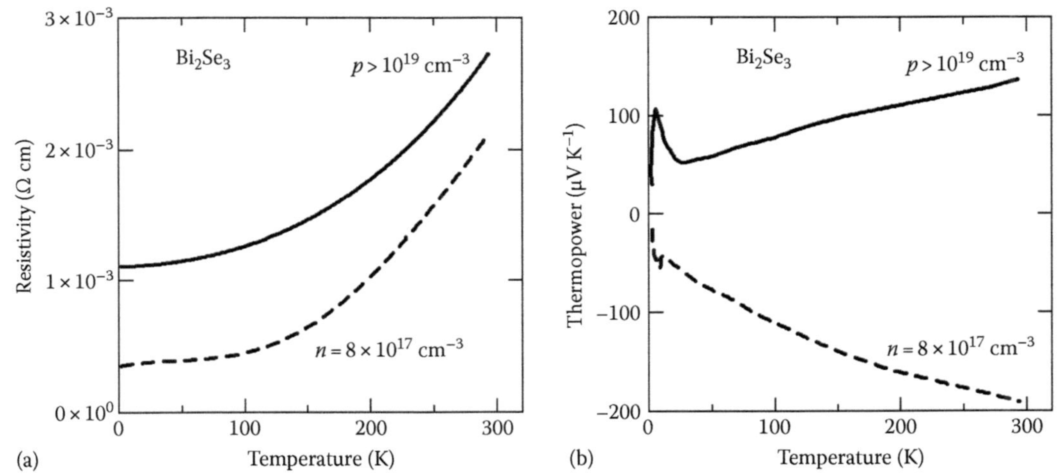

FIGURE 2.22
Temperature dependence of (a) the in-plane resistivity and (b) the thermopower of n-type and p-type Bi_2Se_3 material doped to the carrier concentrations indicated in the figure (in cm^{-3}). (Redrawn from Hor, Y.S. et al., *Phys. Rev. B*, 79, 195208, 2009.)

18,000 and 48 $cm^2\ V^{-1}\ s^{-1}$ at 77 and 300 K, respectively, and a much more rapid T^{-2} slope. At higher electron concentrations, the mobility is strongly reduced; for n = 3.4 × 10^{19} cm^{-3}, it is 2700 and 90 ($cm^2\ V^{-1}\ s^{-1}$) at 77 and 300 K, respectively, again following a T^{-1} slope. From the Hall and thermopower data, a transport effective mass can be calculated as outlined in Section 2.3. In general, this mass increases with increasing carrier concentration in both the n-type and the p-type material. The transport effective mass of the electrons is smaller,

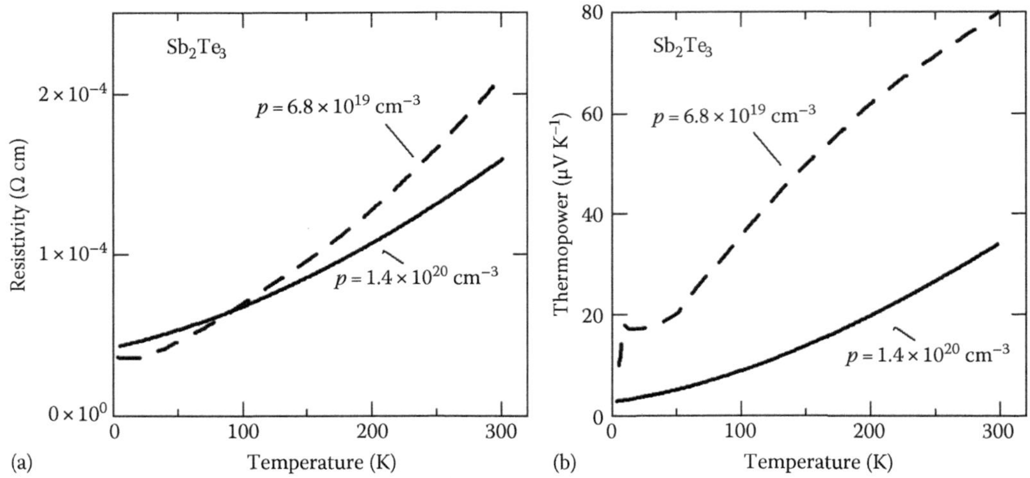

FIGURE 2.23
Temperature dependence of (a) the in-plane resistivity and (b) the thermopower of p-type Sb_2Te_3 material doped to the carrier concentrations indicated in the figure (in cm^{-3}). (Redrawn from Lostak, P. et al., *J. Cryst. Growth*, 222, 565, 2001.)

consistent with the shape of the FSs. Since these values depend on the assumption made for the electron or hole scattering mechanism, their exact values are not very relevant, but they are of the order of 0.15–0.2 free electron masses for the n-type material and 0.2 for the p-type material, assuming that ionized impurity scattering is dominant (if one assumes acoustic phonon scattering, these values roughly double).

The electrical transport properties of alloys are discussed in the following sections. Before entering the detailed discussion, a few general points can be made about the differences between the electronic properties of the p-type materials that are mostly (Bi_2Te_3)–(Sb_2Te_3) pseudobinary alloys and those of the n-type materials that are mostly (Bi_2Te_3)–(Bi_2Se_3) pseudobinary alloys. First, the electrical and thermal conductivities of the p-type alloys generally have the same anisotropy. This results in an isotropic ZT. In contrast, the n-type alloys have an anisotropic ZT. This makes it easier to use processing techniques from powder metallurgy, such as ball milling and sintering, for the p-type alloys than for their n-type counterparts. As already pointed out in Section 2.5.5, it is necessary to preserve the preferential alignment in the n-type alloys. Second, the role of doping the p-type alloys is to reduce the hole concentration from that which results from the native defects; in contrast, the goal of doping in n-type alloys is to increase the electron concentration. Another point concerns the electronic effects that can result from the creation of inhomogeneities, such as nanoprecipitates or dislocations, in thermoelectric materials. These do limit the lattice thermal conductivity, as described in Section 2.5, but there are two additional potential effects that must be considered. There are limitations imposed by the effective medium theory [110,111], which state that a mixture of two noninteracting thermoelectric materials cannot have a ZT higher than the highest of the ZTs of either components. Of course, lattice thermal conductivity reduction by phonon-precipitate scattering is one of the interactions that is not covered by the theory. Inhomogeneous mixtures can create a second effect due to electronic interactions: the thermoelectric properties of the precipitates can be such that they give rise to circulating currents, similar to the ambipolar thermal conductivity, which in turn decrease the thermopower of the mixture and increase its thermal conductivity. Precipitates of the Te-rich eutectic phase in many of the Sb-rich alloys are conducting and can create such problems. Dislocations also can be charged, and while they are very helpful in reducing κ_L, one must pay attention to the role they play on electrical properties and particularly to their potential to generate circulating currents—as revealed by a decrease in thermopower.

2.6.2 *p*-Type Alloys, Typically along the (Bi_2Te_3)–(Sb_2Te_3) Pseudobinary Line

Undoped $(Bi_{1-x}Sb_x)_2Te_3$ alloys are all naturally *p*-type, with a hole concentration that increases with the concentration of Sb_2Te_3 [46]. The thermal conductivity reduction due to alloying has already been described in Figure 2.17 and is mostly due to the lattice contribution. The minimum lattice thermal conductivity occurs near $x = 0.7$, to a value around 0.9 W m^{-1} K^{-1}, which compares to 1.4 W m^{-1} K^{-1} for Bi_2Te_3. By a favorable coincidence, that composition is near the point where the valence bands' FSs of both binaries more or less blend together and where the effective mass has a maximum (see Figure 2.9). The hole concentration of the as-grown *p*-type alloys with $x = 0.75–0.8$ is usually too high compared to the optimum concentration that maximizes *ZT*. This is typically addressed by introducing excess Te, at the risk of forming a precipitate of the Te-rich eutectic shown in Figure 2.12c. Adding Se would result in the preferential formation of selenides [112]. The *ZT* of Te-rich $(Bi_{1-x}Sb_x)_2Te_{3.13}$ alloys is shown in Figure 2.24 [113]. This clearly illustrates that optimum

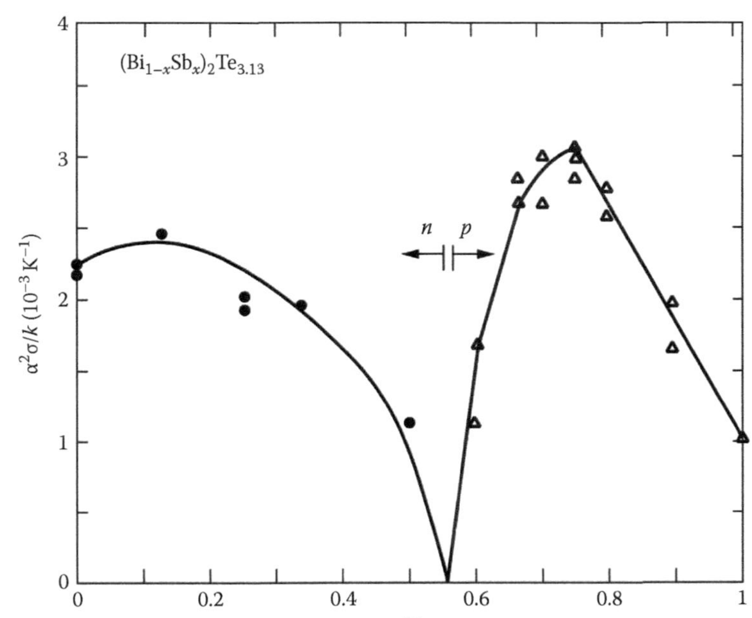

FIGURE 2.24
The composition dependence of the room-temperature figure of merit of pseudobinary Te-rich tetradymite alloys along the (Bi_2Te_3)–(Sb_2Te_3) pseudobinary line at room temperature. Undoped alloys would have all been *p*-type, but because of the 4.3% excess Te, the samples with Sb content at or below 50% are *n*-type. (Redrawn from Champness, C.H. et al., *Can. J. Phys.*, 43, 653, 1965.)

TABLE 2.3

Anisotropy of Single Crystals of the p-Type Alloys (T = 300 K)

	$(Bi_{0.2}Sb_{0.8})_3Te_3$	$(Bi_{0.2}Sb_{0.8})_3Te_3$	$(Bi_{0.25}Sb_{0.75})_3Te_3$	$(Bi_{0.25}Sb_{0.75})_3Te_3$
n (10^{19} cm^{-3})	7.5	5.8	4.9	2.9
ρ_{33}/ρ_{11}	2.95	2.76	3.3	3.22
α_{33}/α_{11}	1.05	1.11	0.98	0.99
κ_{11}/κ_{33}	2.44	2.18	2.33	1.94
ZT_{11}/ZT_{33}	1.07	1.04	1.5	1.64

Source: After Scherrer, H. and Scherrer, S., Bismuth telluride, antimony telluride and their solid solutions, in: *CRC Handbook of Thermoelectrics*, D.M. Rowe, ed., CRC Press, Boca Raton, FL, 1995, pp. 211–237.

Two alloy compositions are studied and the values given for various doping levels.

ZT values are achieved for partially compensated p-type alloys with x = 0.75–0.8. It also shows that this alloy system could also host n-type alloys with a reasonable ZT near x = 0.15, although we are not aware of a commercial application of these. The anisotropy of single crystals of p-type $(Bi_{1-x}Sb_x)_2Te_3$ with x = 0.75 and 0.80 is reported in [5]. The anisotropy of the electrical and thermal conductivity, of the thermopower, and of the ZT of single crystals of $(Bi_{0.8}Sb_{0.2})_3Te_3$ and of $(Bi_{0.75}Sb_{0.25})_3Te_3$ at 300 K is shown in Table 2.3. By another favorable coincidence, the anisotropy of the electrical conductivity and of the thermal conductivity and thermopowers of $(Bi_{0.8}Sb_{0.2})_3Te_3$ scale in such a way that the anisotropy effects on electrical and thermal conductivity cancel each other. The ZT is then approximately isotropic. This makes it possible to use randomly oriented crystals of the p-type tetradymites in nanostructured composites. Figure 2.25 shows the thermoelectric transport properties of p-type thermoelectric materials from an industrial source, whose ZT is about 10% below today's best materials used in modules that give ΔT_{max} around 75 K (hot side at 300 K). The ingot is fine grained and cylindrical in shape. The results are shown with the sample aligned so that the electrical and thermal fluxes are either perpendicular or parallel to the cylinder axis. The anisotropies of the electrical and thermal measurements, if they are significant at all in this p-type material, are compensated. This results in a material with an isotropic ZT. The compensation may not be exactly as suggested by the figure, since different pieces of material were used for the parallel and perpendicular measurements. Hall measurements give a hole concentration of $(2.4 \pm 0.2) \times 10^{19}$ cm^{-3}. The experimental electrical conductivity (σ), Hall effect (R_H), thermopower (α), and isothermal Nernst coefficient (N) were used to determine the hole concentration n and mobility μ, the DOS effective mass m_D^*, and the relaxation time scattering exponent λ, which is defined by the energy dependence of the relaxation time $\tau(E) = \tau_0 E^{\lambda}$, following a process known as the "method of the four coefficients" [114]. For acoustic phonon scattering, $\lambda = -1/2$; for polar optical phonon scattering, $\lambda = +1/2$; and for ionized impurity scattering, $\lambda = +3/2$. Here, λ is calculated from α, N, and $\mu = R_H/\rho$ for degenerate semiconductors with a parabolic band dispersion [114]: $\lambda = -(3N/2\mu)(N/\mu-\alpha)^{-1}$. Between 100 and 300 K, $\lambda \approx -0.6 \pm 0.1$, indicating acoustic phonon scattering; the DOS effective mass at 100 K is $m_D^* \sim 1.2\,m_e$, where m_e is the free electron mass. Sb_2Te_3 at $(2.4 \pm 0.2) \times 10^{19}$ cm^{-3} would have $m_D^* \sim 0.7 m_e^*$ [46]. In contrast, the DOS mass at the valence band edge of alloys with x = 0.8 can reach $m_D^* \sim 1 m_e$. These numbers agree very well with the calculations in Figure 2.10 and the masses of Stordeur et al. [46]. Further, from the mobility and carrier concentration, an estimate can be made of the hole mean free path, which in the parallel direction ranges from 60 nm at 80 K to

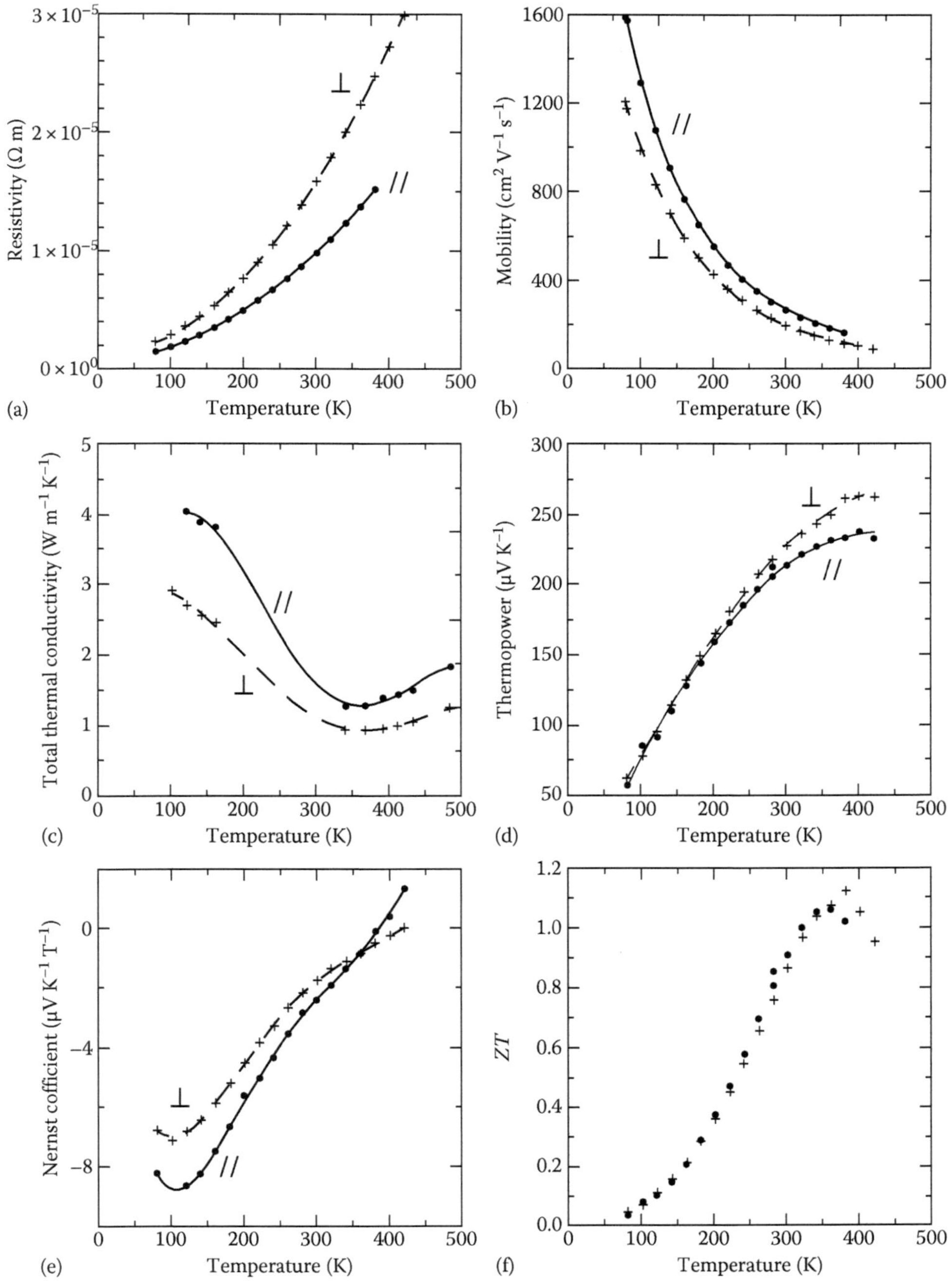

FIGURE 2.25
Thermoelectric transport properties of p-type thermoelectric materials from an industrial source, with a ZT about 10% lower than that of the best materials in use today. Hall measurements on the p-type material give a hole concentration of $(2.4 \pm 0.2) \times 10^{19}$ cm^{-3}. (a) Electrical resistivity, (b) carrier mobility, (c) total thermal conductivity, (d) thermopower, (e) Nernst coefficient, and (f) ZT. The results are shown with the sample aligned so that the electrical and thermal fluxes are either in the direction parallel (closed dots, full lines) to the cylindrical axis of the ingot or perpendicular to it (+ signs, dashed lines).

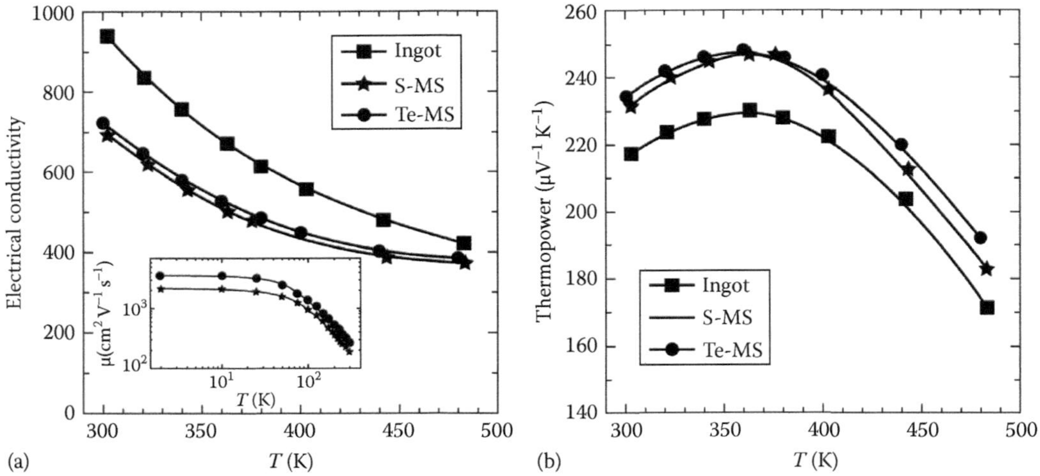

FIGURE 2.26
(a) Electrical conductivity and (b) thermopower of melt-spun sintered sample of nanocomposite *p*-type $(Bi_{0.25}Sb_{0.75})_2Te_3$ Te-rich alloy from Figure 2.19. The temperature dependence of the electrical conductivity of a regular ingot and a melt-spun sintered sample of regular $(Bi_{0.25}Sb_{0.75})_2Te_3$ are compared to a melt-spun sintered (Te-MS) sample of very Te-rich $(Bi, Sb)_2Te_3$. The Te-MS material has the highest *ZT* reported on tetradymites to the best of our knowledge. (Redrawn from Kim, S.I. et al., *Science*, 348, 109, 2015.)

13 nm at room temperature. Both values are smaller than the Bohr radii and limited by phonon scattering.

This heritage material sets the basis for comparison with the results of recent optimization studies. In particular, Figure 2.26 shows the temperature dependence of the electrical conductivity and thermopower of S-MS sample of very Te-rich $(Bi_{0.25}Sb_{0.75})_2Te_3$ (Te-MS), whose thermal conductivity was shown in Figure 2.19 to be reduced compared to that of similar materials. The Te-MS material has the highest *ZT* reported on tetradymites to the best of our knowledge. Its transport properties are compared to those of a regular ingot, a ball-milled (BM) sample, and an S-MS sample of $(Bi, Sb)_2Te_3$. Several Te-MS samples were measured, and their hole concentrations as measured by Hall effect are around $(1.8 \pm 0.2) \times 10^{19}$ cm^{-3} with mobilities around 300 ± 20 cm^2 V^{-1} s^{-1} (the error bars cover sample-to-sample reproducibility) [106]. The Te-MS material has fewer holes than the commercial material shown in Figure 2.25, but a comparable mobility, illustrating that the nanostructures and dislocations have a limited effect on the mobility.

A summary of the *ZT* values that have resulted from other optimization studies is given in Figure 2.27. This figure is updated from an earlier similar figure published by Goldsmid [115]. The references here are (a) [116], (b) [117], (c) [118], (d) [119], (e) [120], (f) [121], (g) [122], (h) [123], (i) [124], (j) [125], and (k) [106].

2.6.3 *n*-Type Alloys, Typically along the (Bi_2Te_3)–(Bi_2Se_3) Pseudobinary Line

As already mentioned, most *n*-type materials are alloys along the (Bi_2Te_3)–(Bi_2Se_3) pseudobinary line. The *ZT* at room temperature of these alloys is shown as a function of composition in Figure 2.28 [100]. The anisotropy of the electrical and thermal conductivity, of the thermopower, and of the *ZT* of single crystals $Bi_2(Te_{1-x}Se_x)_3$ with $x = 0.025$ and 0.05 is

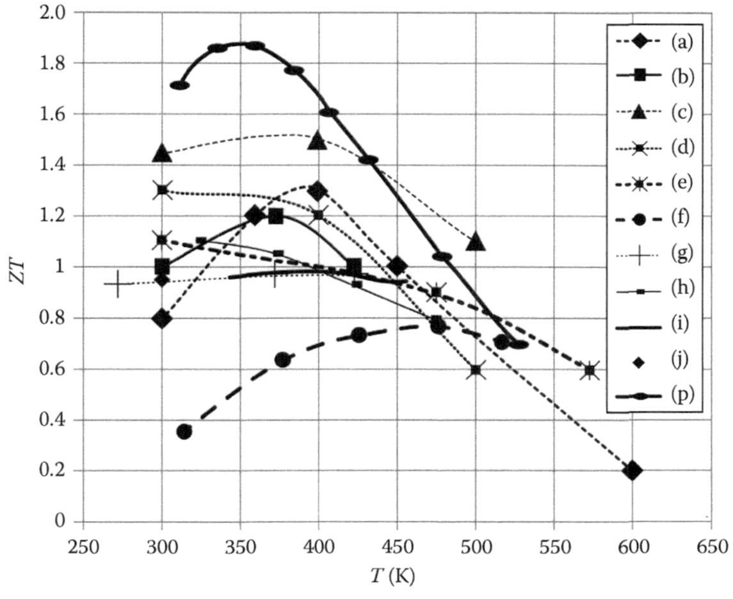

FIGURE 2.27
Compilation of the temperature dependence of the thermoelectric figure of merit of *p*-type tetradymite alloys. The data are updated from Goldsmid [115]. The references for the different curves are (a) Li et al. [116], (b) Bulat et al. [117], (c) Xie et al. [118], (d) Nguyen et al. [119], (e) Son et al. [120], (f) Imamuddin and Dupre [121], (g) Matsuura and Rowe [122], (h) Snyder [123], (i) Goldsmid and Cochrane [124], (j) Bogomolov et al. [125], and (p) the Te-MS material reported in Kim et al. [106].

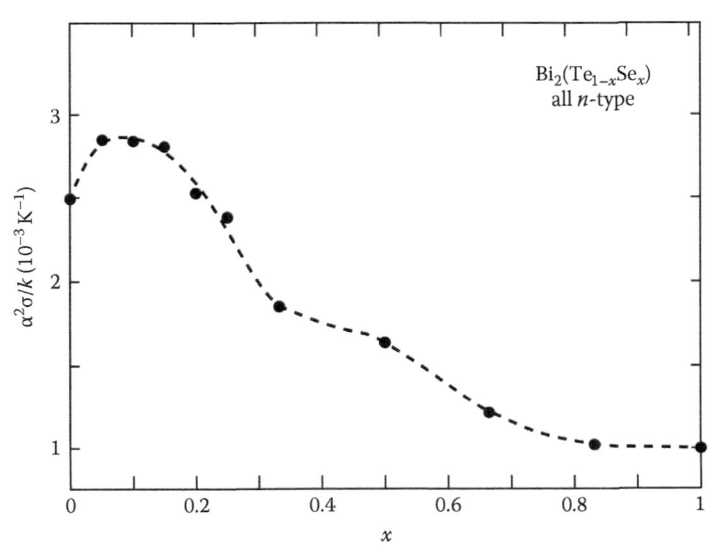

FIGURE 2.28
The composition dependence of the room-temperature figure of merit of pseudobinary *n*-type tetradymite alloys along the (Bi_2Te_3)–(Bi_2Se_3) pseudobinary line at room temperature. (Redrawn from Champness, C.H. et al., *Can. J. Phys.*, 45, 3611, 1967.)

TABLE 2.4

Anisotropy of Single Crystals of the n-Type Alloys (T = 300 K)

	$Bi_2(Te_{0.975}Se_{0.025})$	$Bi_2(Te_{0.975}Se_{0.025})$	$Bi_2(Te_{0.95}Se_{0.05})$	$Bi_2(Te_{0.95}Se_{0.05})$
n (10^{19} cm^{-3})	3.2	12.5	2.5	10
ρ_{33}/ρ_{11}	4.38	6.71	4.29	6.56
α_{33}/α_{11}	1.03	1.07	1.02	1.06
κ_{11}/κ_{33}	2.15	2.41	2.10	2.40
ZT_{11}/ZT_{33}	1.9	2.4	2.0	2.4

Source: After Scherrer, H. and Scherrer, S., Bismuth telluride, antimony telluride and their solid solutions, in: *CRC Handbook of Thermoelectrics*, D.M. Rowe, ed., CRC Press, Boca Raton, FL, 1995, pp. 211–237.

Two alloy compositions are studied and the values given for various doping levels.

reported in Table 2.4. Unlike the p-type material, anisotropy in ZT matters here because, as already mentioned in Section 2.5, it is useful to work with material aligned so that the heat and current flows in the trigonal plane. A loss of anisotropy would result in a decrease of the average ZT that may or may not be compensated for by the nanostructuring. The cause of the anisotropy in ZT is the fact that the anisotropy in the thermal conductivity is less than half that in the electrical conductivity. This added the difficulty that powder processing of all $Bi_2(Te_{1-x}Se_x)_3$ alloys has a strong n-type doping effect that shifts the maximum ZT of such material to above 400 K. Should such operating point be useful, then it is possible to reach ZT = 1.2 at 445 K (172°C) in polycrystalline $Bi_2(Te_{0.77}Se_{0.23})_3$ [126].

Figure 2.29 shows the thermoelectric transport properties of n-type thermoelectric materials from an industrial source, again with a ZT about 10% lower than that of the best materials in use today. The production method resulted in a fine-grained cylindrical ingot, with the grain partially aligned along preferential crystallographic orientations. The results are again shown with the sample aligned so that the electrical and thermal fluxes are either perpendicular or parallel to the cylinder axis. Hall measurements give an electron concentration of (2.8 ± 0.4) × 10^{19} cm^{-3}. As with the p-type sample, the method of the four coefficients [114] was used to determine the DOS effective mass m_D^* and the relaxation time scattering exponent λ. Between 100 and 300 K, λ is around ~+0.5, indicating that the scattering mechanism involves optical phonons and/or defects. The DOS effective mass is $m_D^* \sim 0.55 \pm 0.05 m_e$, where m_e is the free electron mass. Both pieces, the one aligned with the fluxes parallel to cylinder axis and the one with the fluxes perpendicular, gave the same thermopower and m_D^* within the error bars, although the mobilities were different. The band degeneracy is thus much lower than for the p-type material, consistent with the discussion in Section 2.3. The anisotropy is less marked than it is in the single crystals, but the anisotropy in ZT is higher than for the p-type material. The electron mean free paths in the parallel direction is about 25 nm at 80 K and 6 nm at room temperature, a smaller fraction of the Bohr radius than was the case for the p-type material.

Figure 2.30 shows the effect of HDs on the electrical properties of ZM and HP of the n-type $Bi_2(Te_{0.93}\,Se_{0.07})_3$ alloys, whose thermal conductivity was discussed in Figure 2.20 [108]. The introduction of dislocations in the ZM sample has a slight n-type dopant effect: the electron concentration at 300 K increased with HP from n ~ 4.2 × 10^{19} cm^{-3} for the ZM sample to n ~ 5.9 × 10^{19} cm^{-3} after the two hot-deformation process was applied twice (ZM-HD2). The mobility at 300 K also fell slightly, from about ~147 cm^2 V^{-1} s^{-1} before to ~133 cm^2 V^{-1} s^{-1} after, an effect that is much less pronounced than that on the thermal conductivity.

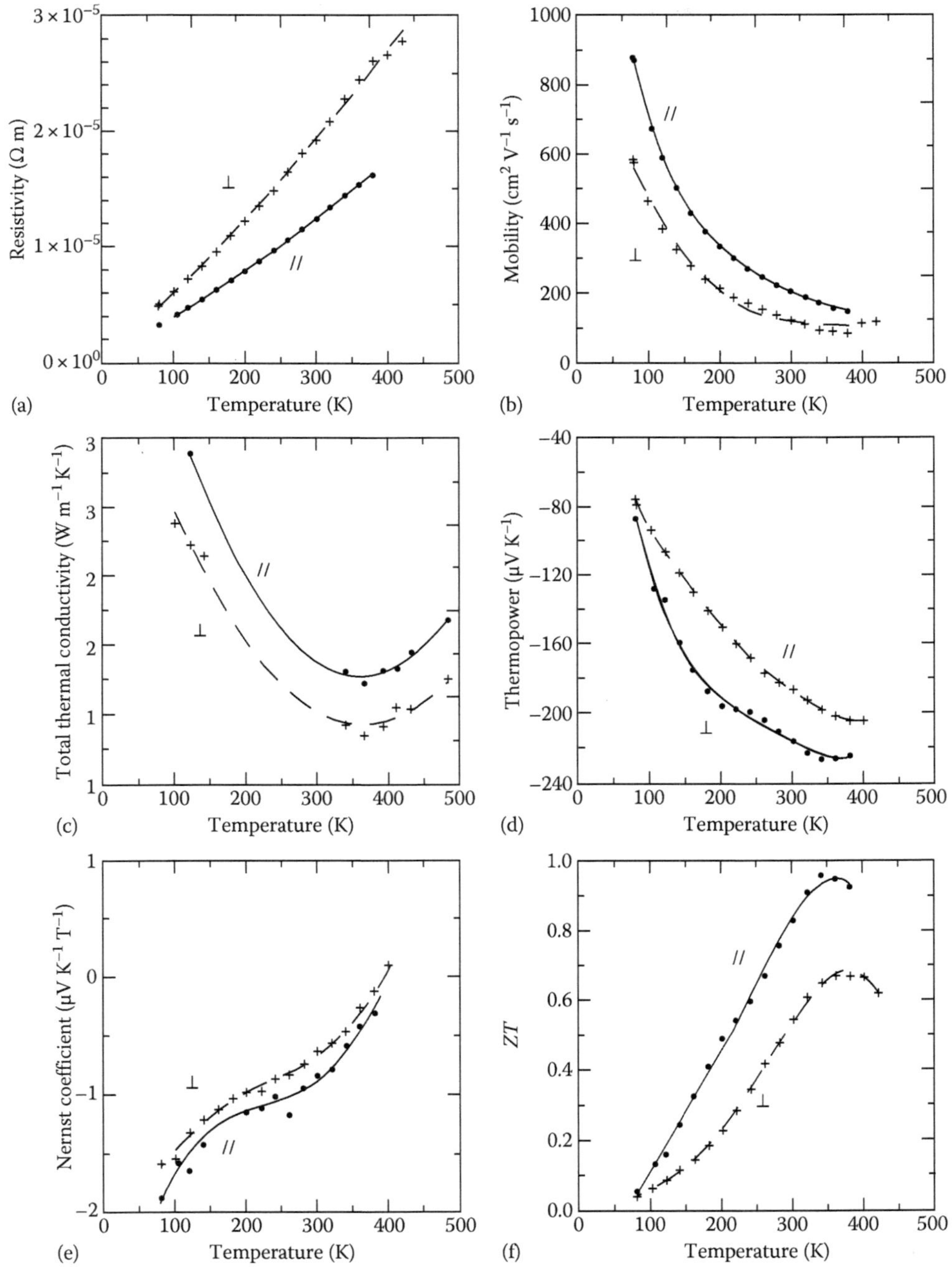

FIGURE 2.29
Thermoelectric transport properties of n-type thermoelectric materials from an industrial source, with a ZT about 10% lower than that of the best materials in use today. Hall measurements give an electron concentration of $(2.8 \pm 0.4) \times 10^{19}$ cm^{-3}. (a) Electrical resistivity, (b) carrier mobility, (c) total thermal conductivity, (d) thermopower, (e) Nernst coefficient, and (f) ZT. The results are shown with the sample aligned so that the electrical and thermal fluxes are either in the direction parallel (closed dots, full lines) to the cylindrical axis of the ingot or perpendicular to it (+ signs, dashed lines).

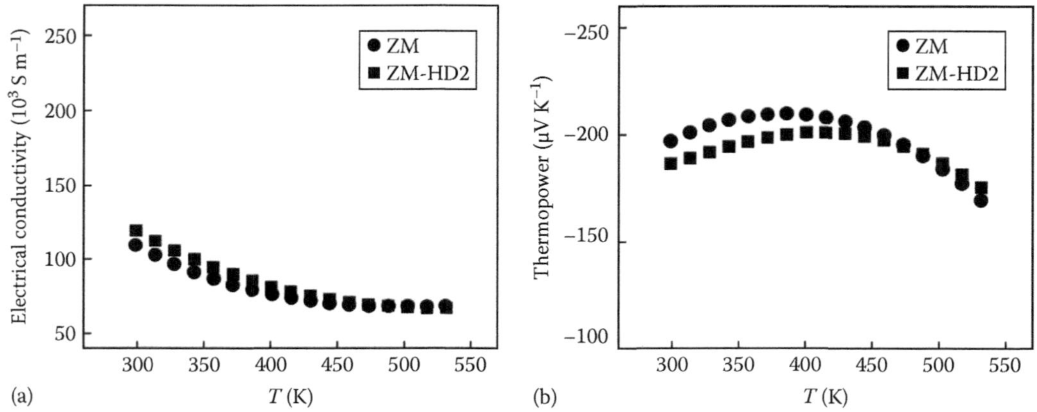

FIGURE 2.30
(a) Electrical conductivity and (b) thermopower of the n-type $Bi_2(Te_{0.93}Se_{0.07})_3$ alloys whose thermal properties are given in Figure 2.20. The sample labels are also in Figure 2.20. The zone-melted sample (ZM) has an electron concentration of $n \sim 4.2 \times 10^{19}$ cm^{-3}, which increased to $n \sim 5.9 \times 10^{19}$ cm^{-3} during the two hot-deformation processes (ZM-HD2). (Redrawn from Hu, L. et al., *Adv. Energy Mater.*, 5, 1500411, 2015.)

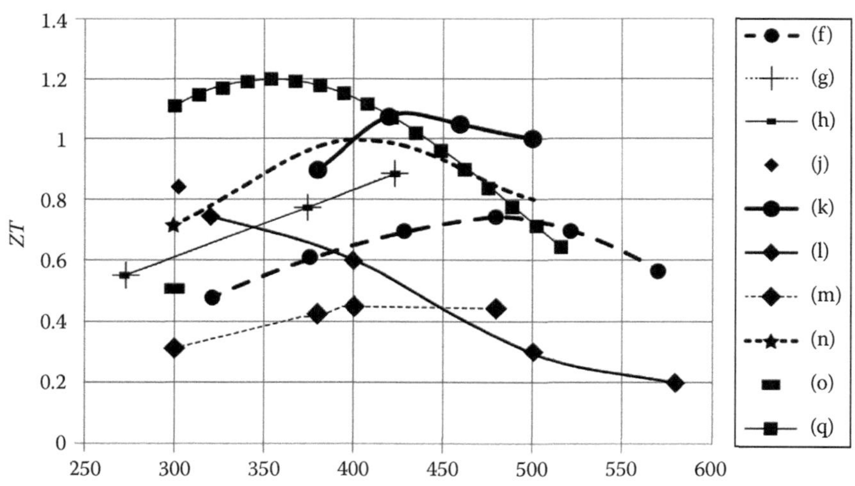

FIGURE 2.31
Compilation of the temperature dependence of the thermoelectric figure of merit of n-type tetradymite alloys. The data are updated from Li et al. [116]. The references for the different curves are (f) Imamuddin and Dupre [121], (g) Matsuura and Rowe [122], (h) Snyder [123], (j) Bogomolov et al. [125], (k) Wu et al. [127], (l) Kim et al. [128], (m) Yu et al. [129], (n) Dashevsky et al. [130], (o) Gerovac et al. [131], and (q) Hu et al. [108].

As a result, a considerable improvement in the ZT was achieved, particularly at higher temperatures (30% at 360 K). The value for ZT for the ZM-HD2 $Bi_2(Te_{0.93}\ Se_{0.07})_3$ alloy is reported in Figure 2.31, along with the values of ZT, which summarize the progress made over the years in n-type alloys. This summary was initially compiled by Goldsmid [115] and is updated here. The references used in Figure 2.31 are (f) [121], (g) [122], (h) [123], (j) [125], (l) [127], (m) [128], (n) [129], (o) [130], (p) [131], (q) [108], and (r) [132].

To conclude this section, three facts need to be pointed out. First, the ZT of n-type tetradymites is systematically lower than that of p-type materials. The coincidences that work in favor of the p-type (Bi_2Te_3)–(Sb_2Te_3) pseudobinary alloys are the fact that at the composition 80% Sb_2Te_3, (a) alloy scattering of phonons is maximal, (b) the valence band density of states has a maximum, and (c) the material has the same anisotropy in electrical and thermal conductivity, leading to an isotropic ZT, which simplifies the powder-metallurgical processes used for nanostructuring. Therefore, it is hardly surprising that such a coincidence is not reproduced in the n-type alloys. Second, the optimum in ZT observed in the n-type (Bi_2Te_3)–(Bi_2Se_3) pseudobinary alloys near 10 at.% Se (Figure 2.28) has a maximum that is only about 20% higher than that shown on the n-type side in Figure 2.24 for the (Bi_2Te_3)–(Sb_2Te_3) pseudobinary alloys. It is conceivable that, with modern nanostructuring and band structure engineering methods, better n-type alloys could be designed around the composition $(Bi_{0.9}Sb_{0.1})_2Te_3$. Third, quaternary alloys have also been studied for n-type alloys: an alloy of $(Bi_{0.95}Sb_{0.05})_2(Te_{0.85}Se_{0.15})_3$ has reached $ZT = 1$ at 460 K [133].

2.7 Conclusions

In summary, the tetradymites form a class of thermoelectric semiconductors that hold the record for the highest thermoelectric figure of merit ZT in the temperature range below 200°C [134]. As a class of materials, they have intrinsic advantages. First is their narrow energy bands, a consequence of the large spin–orbit interactions associated with the heavy p-block elements that are their main constituent atom. This effect leads to large Bohr radii and high mobilities, but also, by itself, would lead to relatively low power factors. This potential disadvantage is circumvented by the high degeneracy of the pockets in the FSs for degenerately doped conduction and valence bands. Second, their phonon properties are also favorable. Intrinsic to this class of materials are a low sound velocity, a small Brillouin zone, and the high degree of anharmonicity of their phonons, leading to an intrinsically low lattice thermal conductivity. To these intrinsic advantages, a number of materials engineering techniques have enabled further thermal conductivity reductions and power factor enhancements. The thermal conductivity is further reduced most effectively by a broad-spectrum approach, where phonons of different frequencies are scattered by different obstacles: high-energy phonons by point defects in alloys, midrange phonons by dislocations, and low-energy phonons by nanoparticles. The power factor is increased by engineering alloys with the largest possible degeneracy of FS pockets. Historically, the doping level was also optimized by using defect doping, which in turn depends on alloy composition. All this has made it possible to increase the maximum ZT for p-type materials from about 1.2 in heritage materials to above 1.8 at 350 K. For n-type materials, the ZT is improved from below 1 in heritage materials to 1.2 at 350 K. As with all thermoelectrics, ZT is higher at higher temperature, making improvements above 400 K easier to achieve. Yet, cooling applications constitute still the vast majority of applications of thermoelectric materials. They require a maximum ZT at temperatures significantly below 400 K, so that the improvements described here are particularly meaningful and relevant to the field of solid-state energy conversion in general.

Acknowledgments

The authors acknowledge the help and support from Drs. Jeff Sharp (II-VI Marlow) and D. A. Broido (Boston College) who constructed Figure 2.16 of this review based on his work [95]. The data in Figures 2.25 and 2.29 were measured by Mr. Michael Adams. JPH is supported by the U.S. National Science Foundation grant 1433467. BW is partly supported by the National Science Center (Poland) (Project No. DEC-2011/02/A/ST3/00124).

References

1. Goldsmid, H. J. and R. W. Douglas. 1954. The use of semiconductors in thermoelectric refrigeration, *Br. J. Appl. Phys.* 5: 386.
2. Heikes, R. R. and R. W. Ure. 1961. *Thermoelectricity: Science and Engineering*, Interscience Publishers, New York, 1961.
3. Hasan, M. Z. and C. L. Kane. 2010. Topological insulators. *Rev. Mod Phys.* 82: 3045.
4. Mellnik, A. R., J. S. Lee, A. Richardella et al. 2014. Spin-transfer torque generated by a topological insulator. *Nature* 511: 449.
5. Scherrer, H. and S. Scherrer. 1995. Bismuth telluride, antimony telluride and their solid solutions. In *CRC Handbook of Thermoelectrics*, ed. D. M. Rowe, CRC Press, Boca Raton, FL, pp. 211–237.
6. Yim, W. M. and F. D. Rosi. 1972. Compound tellurides and their alloys for Peltier cooling. *Solid State Electron.* 15: 1121–1140.
7. Grauer, D. C., Y. S. Hor, A. J. Williams, and R. J. Cava. 2009. Thermoelectric properties of the tetradymite-type Bi_2Te_2S–Sb_2Te_2S solid solution. *Mater. Res. Bull.* 44: 1926–1929.
8. Kuznetsov, V. G. and K. K. Palkina. 1963. Equilibrium diagrams and structures of alloys in the Bi_2Se_3-Sb_2Te_3 and Bi_2Te_3-Sb_2Se_3 systems. *Rus. J. Inorg. Chem.* 8: 624–632.
9. Wiese, J. R. and R. Muldawrer. 1960. Lattice constants of Bi_2Te_3-Bi_2Se_3 solid solution alloys. *J. Phys. Chem. Solids* 15: 13–16.
10. Gao, Y. B., B. He, D. Parker, I. Androulakis, and J. P. Heremans. 2014. Experimental study of the valence band of Bi_2Se_3. *Phys. Rev. B* 90: 125204.
11. Blaha, P., K. Schwarz, G. K. H. Madsen, D. Kvasnicka, and J. Luitz. 2001. WIEN2K, An augmented plane wave local orbitals program for calculating crystal properties, Karlheinz Schwarz, Technical Universitat Wien, Vienna, Austria.
12. Perdew, J. P. and Y. Wang. 1992. Accurate and simple analytic representation of the electron-gas correlation energy. *Phys. Rev. B* 45: 13244.
13. Shi, H., D. Parker, M. H. Du, and D. J. Singh. 2015. Connecting thermoelectric performance and topological-Behavior: Bi_2Te_3 and Bi_2Te_2Se from first principles. *Phys. Rev. Appl.* 3: 014004.
14. Tran, F. and P. Blaha. 2009. Accurate band gaps of semiconductors and insulators with a semilocal exchange-correlation potential. *Phys. Rev. Lett.* 102: 226401.
15. Aguilera, I., C. Friedrich, and S. Blügel. 2013. Spin-orbit coupling in quasiparticle studies of topological insulators. *Phys. Rev. B* 88: 165136.
16. Subvol, F. 1983. *Landolt-Börnstein Numerical Data and Functional Relationships in Science and Technology*, Vol. 17, Subvol F, Springer-Verlag, Berlin, Germany.
17. Kokalj, A. 1999. Graphics generated by XCRYSDEN, XCrySDen—A new program for displaying crystalline structures and electron densities. *J. Mol. Graph. Model* 17: 176. Code available from http://www.xcrysden.org/.
18. Yavorsky, B. Yu., N. F. Hinsche, I. Mertig, and P. Zahn. 2011. Electronic structure and transport anisotropy of Bi_2Te_3 and Sb_2Te_3. *Phys. Rev. B* 84: 165208.

19. Nechaev, I. A. and E. V. Chulkov. 2013. Quasiparticle band gap in the topological insulator Bi_2Te_3. *Phys. Rev. B* 88: 165135.
20. Köhler, H. and A. Fabricius. 1975. Galvanomagnetic properties of Bi_2Se_3 with free carrier densities below 5×10^{17} cm^{-3}. *Phys. Stat. Solidi B* 71: 487.
21. Nechaev, I. A., R. C. Hatch, M. Bianchi et al. 2013. Evidence for a direct band gap in the topological insulator Bi_2Se_3 from theory and experiment. *Phys. Rev. B* 87: 121111(R).
22. Aguilera, I., C. Friedrich, G. Bihlmayer, and S. Blügel. 2013. GW study of topological insulators Bi_2Se_3, Bi_2Te_3, and Sb_2Te_3: Beyond the perturbative one-shot approach. *Phys. Rev. B* 88: 045206.
23. Yazyev, O. V., E. Kioupakis, J. E. Moore, and S. G. Louie. 2012. Quasiparticle effects in the bulk and surface-state bands of Bi_2Se_3 and Bi_2Te_3 topological insulators. *Phys. Rev. B* 85: 161101(R).
24. i.e., this is the same highest (first) valence band, as seen at Γ, but it has second, local maximum in this direction.
25. Parker, D. and D. J. Singh. 2011. Potential thermoelectric performance from optimization of hole-doped Bi_2Se_3. *Phys. Rev. X* 1: 021005.
26. Chen, Y. L., J. G. Analytis, J.-H. Chu et al. 2009. Experimental realization of a three-dimensional topological insulator, Bi_2Te_3. *Science* 325: 178.
27. Chen, Y. L., J.-H. Chu, J. G. Analytis et al. 2010. Massive Dirac fermion on the surface of a magnetically doped topological insulator. *Science* 329: 659.
28. Kong, D., Y. Chen, J. J. Cha et al. 2011. Ambipolar field effect in the ternary topological insulator $(Bi_xSb_{1-x})_2Te_3$ by composition tuning. *Nat. Nanotechnol.* 6: 705.
29. Shoenberg, D. 1984. *Magnetic Oscillations in Metals*, Cambridge University Press, Cambridge, U.K., 1984.
30. Richter, W., H. Köhler, and C. R. Becker. 1977. A Raman and far-infrared investigation of phonons in the rhombohedral V_2-VI_3 compounds Bi_2Te_3, Bi_2Se_3, Sb_2Te_3 and $Bi_2(Te_{1-x}Se_x)_3$ $(0 < x < 1)$, $(Bi_{1-y}Sb_y)_2Te_3$ $(0 < y < 1)$. *Phys. Stat. Solidi (b)* 84: 619.
31. Drope, R. 1975. Zur Gitterdynamik von V_2VI_3 Verbindungen, Dissertation, RWTH, Aachen, Germany.
32. Stordeur, M., K. K. Ketavong, A. Priemuth, H. Sobotta, and V. Ried. 1992. Optical and electrical investigations of *n*-type Bi_2Se_3 single crystals. *Phys. Stat. Solidi (b)* 169: 505.
33. Unkelbach, K. H. 1973. Nachweis optisch aktiver Gitterschwingungen und optische Eigenschaften von Antimontellurid, Dissertation, RWTH, Aachen, Germany.
34. Uemura, O. and T. Satow. 1977. The magnetic susceptibility of antimony and bismuth chalcogenides in the liquid state. *Phys. Stat. Solidi (b)* 84: 353.
35. van Deynse, N., A. van Itterbeek, and R. Dekeyser. 1969. Messungen der magnetischen anisotropie von Bi_2Te_3— und Sb_2Te_3- einkristallen und ihren legierungen. *Z. Angew. Phys.* 26: 174.
36. Sehr, R. and L. R. Testardi. 1962. The optical properties of *p*-type Bi_2Te_3-Sb_2Te_3 alloys between 2–15 microns. *J. Phys. Chem. Solids* 23: 1219–1224.
37. Köhler, H. and J. Hartmann. 1974. Burstein shift of the absorption edge of *n*-Bi_2Se_3. *Phys. Stat. Solidi (b)* 63: 171.
38. Hor, Y. S., A. Richardella, P. Roushan et al. 2009. *P*-type Bi_2Se_3 for topological insulator and low-temperature thermoelectric applications. *Phys. Rev. B* 79: 195208.
39. Köhler, H. 1973. Conduction band parameters of Bi_2Se_3 from Shubnikov-de Haas investigations. *Phys. Stat. Solidi (b)* 58: 91.
40. Köhler, H. 1976. Non-parabolic E(k) relation of the lowest conduction band in Bi_2Te_3. *Phys. Stat. Solidi (b)* 73: 95.
41. Köhler, H. 1976. The application of high magnetic fields in semiconductor physics. In *Proceedings of the International Conference*, Würzburg, Germany.
42. Tichy, L. and J. Horák. 1979. Nonparabolicity of the conduction band and anisotropy of the electron effective mass in *n*-Bi_2Se_3 single crystals. *Phys. Rev. B* 19: 1126.
43. Harman, T. C., B. Paris, S. F. Miller, and H. L. Goering. 1957. Preparation and some physical properties of Bi_2Te_3, Sb_2Te_3, and As_2Te_3. *J. Phys. Chem. Solids* 2: 181.
44. Köhler, H. 1976. Non-parabolicity of the highest valence band of Bi_2Te_3 from Shubnikov-de Haas effect. *Phys. Stat. Solidi (b)* 74: 591.

45. von Middendorff, A., K. Dietrich, and G. Landwehr. 1973. Shubnikov-de Haas effect in *p*-type Sb_2Te_3. *Solid State Commun.* 13: 443.
46. Stordeur, M., M. Stöltzer, H. Sobota, and V. Riede. 1988. Investigation of the valence band structure of thermoelectric $(Bi_{1-x}Sb_x)_2Te_3$ single crystals. *Phys. Stat. Solidi (b)* 150: 165–176.
47. Behnia, K. 2015. On the mobility of electrons in a shallow Fermi sea over a rough seafloor. *J. Phys. Condens. Matter* 27: 375501.
48. Greenaway, D. L. and G. Harbeke. 1965. Band structure of bismuth telluride, bismuth selenide and their respective alloys. *J. Phys. Chem. Solids* 26: 1585–1604.
49. Kutorasinski, K., B. Wiendlocha, J. Tobola, and S. Kaprzyk. 2014. Importance of relativistic effects in electronic structure and thermopower calculations for Mg_2Si, Mg_2Ge, and Mg_2Sn. *Phys. Rev. B* 89: 115205.
50. Kutorasinski, K., B. Wiendlocha, S. Kaprzyk, and J. Tobola. 2015. Electronic structure and thermoelectric properties of *n*- and *p*-type SnSe from first-principles calculations. *Phys. Rev. B* 91: 205201.
51. Mishra, S. K., S. Satpathy, and O. Jepsen. 1997. Electronic structure and thermoelectric properties of bismuth telluride and bismuth selenide. *J. Phys. Condens. Matter* 9: 461.
52. Thonhauser, T., T. J. Scheidemantel, J. O. Sofo, J. V. Badding, and G. D. Mahan. 2003. Thermoelectric properties of Sb_2Te_3 under pressure and uniaxial stress. *Phys. Rev. B* 68: 085201.
53. Hinsche, N. F., B. Yu. Yavorsky, I. Mertig, and P. Zahn. 2011. Influence of strain on anisotropic thermoelectric transport in Bi_2Te_3 and Sb_2 Te_3. *Phys. Rev. B* 84: 165214.
54. Madsen, G. K. H. and D. J. Singh. 2006. BoltzTraP. A code for calculating band-structure dependent quantities. *Comp. Phys. Commun.* 175: 67.
55. Kim, J.-H., J.-Y. Choi, J.-M. Bae, M.-Y. Kim, and T.-S. Oh. 2013. Thermoelectric characteristics of *n*-type Bi_2Te_3 and *p*-type Sb_2Te_3 thin films prepared by co-evaporation and annealing for thermopile sensor applications. *Mater. Trans.* 54: 618.
56. Birkholz, U. and G. Haacke. 1963. Thermoelektrische Eigenschaften von eigenleitendem $Bi_2Te_{3-x}Se_x$. *Z. Naturforsch.* 18a: 638.
57. Analytis, J. G., J.-H. Chu, Y. Chen et al. 2010. Bulk Fermi surface coexistence with Dirac surface state in Bi_2Se_3: A comparison of photoemission and Shubnikov-de Haas measurements. *Phys. Rev. B* 81: 205407.
58. Huang, F.-T., M.-W. Chu, H. H. Kung et al. 2012. Nonstoichiometric doping and Bi antisite defect in single crystal Bi_2Se_3. *Phys. Rev. B* 86: 081104.
59. Navratil, J., Z. Stary, and T. Plechacek. 1996. Thermoelectric properties of *p*-type antimony bismuth telluride alloys prepared by cold pressing. *Mater. Res. Bull.* 31: 1559.
60. Horak, J., M. Matyas, and L. Tichy. 1975. Lattice defects in manganese-doped Sb_2Te_3 crystals. *Phys. Stat. Solidi (a)* 27: 621.
61. Horak, J., P. Lostak, P. Siska, and M. Stordeur. 1982. The nature of silver impurity atoms in antimony telluride. *Phys. Stat. Solidi (b)* 114: 39.
62. Lostak, P., M. Predota, J. Horak, and J. Klikorka. 1986. Point defects in Sb_2Te_3 crystals doped with Cd atoms. *J. Phys. Chem. Solids* 47: 971.
63. Lostak, P., C. Drasar, A. Krejcova et al. 2001. Preparation and some physical properties of tetradymite-type Sb_2Te_3 single crystals doped with CdS. *J. Cryst. Growth* 222: 565–573.
64. Horak, J., L. Tichy, A. Vasko, and M. Frumar. 1972. Reflectivity of iodine-doped Sb_2Te_3 crystals, *Phys. Stat. Solidi (a)* 14: 289.
65. Lostak, P., R. Novotny, J. Horak, and J. Klikorka. 1985. Properties of Sb_2Te_3 single crystals doped with Tl atoms. *Phys. Stat. Solidi (a)* 89: K55.
66. Drasar, C., P. Lostak, J. Navratil, T. Cernohorsky, and V. Mach. 1995. Optical properties of titanium doped Sb_2Te_3 single crystals. *Phys. Stat. Solidi (b)* 191: 523.
67. Stordeur, S. and G. Simon. 1984. Investigation of the weak-field charge transport in semiconducting V_2VI_3 compounds with trigonal symmetry. *Phys. Stat. Solidi (b)* 124: 799.
68. Tichy, L., J. Horak, A. Vasko, and M. Frumar. 1973. Electrical properties of germanium-doped Sb_2Te_3 crystals. *Phys. Stat. Solidi (a)* 20: 717.

69. Rönnlund, B., O. Beckmann, and H. Levy. 1965. Doping properties of Sb_2Te_3 indicating a two valence band model. *J. Phys. Chem. Solids* 26: 1281.
70. Matyas, M. 1974. Electronic properties of $A_2^VB_3^{VI}$ single crystals doped with elements of group IV and VII. In *Proceedings of the 12th International Conference on Physics of Semiconductors* (ed. M. H. Pilkuhn), Stuttgart, Germany.
71. Horak, J., L. Koudelka, J. Klikorka, and L. Sibka. 1982. Transport properties and plasma resonance of bismuth selenide crystals doped with cadmium. *Phys. Stat. Solidi (b)* 111: 575.
72. Weissenstein, J., J. Horak, L. Tichy, and A. Vasko. 1980. Point defects in the crystal lattice of mercury-doped Bi_2Se_3. *Cryst. Lattice Defects* 8: 223.
73. Choi, Y. H., N. H. Jo, K. J. Lee et al. 2012. Simple tuning of carrier type in topological insulator Bi_2Se_3 by Mn doping. *Appl. Phys. Lett.* 101: 152103.
74. Kuroda, K., M. Arita, K. Miyamoto et al. 2010. Hexagonally deformed Fermi surface of the 3D topological insulator Bi_2Se_3. *Phys. Rev. Lett.* 105: 076802.
75. Androulakis, J. and E. Beciragic. 2013. Robust *p*-type behavior in polycrystalline, ball-milled magnesium-doped Bi_2Se_3. *Solid State Commun.* 173: 5.
76. K. Weiss, P. Fielding, and F. A. Kröger. 1960. Untersuchungen an kupferdotiertem wismuttellurid Bi_2Te_3, *Z. Phys. Chem., Neue Folge* 26: 145.
77. Horak, J., L. Tichy, P. Lostak, and A. Vasko. 1976. Lattice defects in iodine-doped Sb_2Te_3 and Bi_2Te_3 crystals. *Cryst. Lattice Defects* 6: 233–238.
78. Kulbachinskii, V. A., N. B. Brandt, P. A. Cheremnykh, S. A. Azou, J. Horak, and P. Lostak. 1988. Magnetoresistance and Hall effect in Bi_2Te_3:Sn in ultrahigh magnetic fields and under pressure. *Phys. Stat. Solidi (b)* 150: 237.
79. Zhitinskaya, M. K., S. A. Nemov, and T. E. Svechnikova. 1998. Specific features of Bi_2Te_3 doping with Sn. *Phys. Solid State* 40: 1297.
80. Jaworski, C. M., V. Kulbachinskii, and J. P. Heremans. 2009. Resonant level formed by tin in Bi_2Te_3 and the enhancement of room-temperature thermoelectric power. *Phys. Rev. B* 80: 233201.
81. Heremans, J. P., B. Wiendlocha, and A. M. Chamoire. 2012. Resonant levels in bulk thermoelectric semiconductors. *Energy Environ. Sci.* 5: 5510.
82. Wiendlocha, B. 2013. Fermi surface and electron dispersion of PbTe doped with resonant Tl impurity from KKR-CPA calculations. *Phys. Rev. B* 88: 205205.
83. Wiendlocha, B. 2014. Localization and magnetism of the resonant impurity states in Ti doped PbTe. *Appl. Phys. Lett.* 105: 133901.
84. Wiendlocha, B., K. Kutorasinski, S. Kaprzyk, and J. Tobola. 2016. Recent progress in calculations of electronic and transport properties of disordered thermoelectric materials. *Scripta Mater.* 111: 33.
85. Heremans, J. P., V. Jovovic, E. S. Toberer et al. 2008. Enhancement of thermoelectric efficiency in PbTe by distortion of the electronic density of states. *Science* 321: 554.
86. Bansil, A., S. Kaprzyk, P. E. Mijnarends, and J. Tobola. 1999. Electronic structure and magnetism of $Fe_{3-x}V_xX$ (X = Si, Ga, and Al) alloys by the KKR-CPA method. *Phys. Rev. B* 60: 13396.
87. Kaprzyk, S. and A. Bansil. 1990. Green's function and a generalized Lloyd formula for the density of states in disordered muffin-tin alloys. *Phys. Rev. B* 42: 7358.
88. Jaworski, C. M., B. Wiendlocha, V. Jovovic, and J. P. Heremans. 2011. Combining alloy scattering of phonons and resonant electronic levels to reach a high thermoelectric figure of merit in PbTeSe and PbTeS alloys. *Energy Environ. Sci.* 4: 4155.
89. Jin, H., B. Wiendlocha, and J. P. Heremans. 2015. *P*-type doping of elemental bismuth with indium, gallium and tin: A novel doping mechanism in solids. *Energy Environ. Sci.* 8: 2027.
90. Ren, Z., A. A. Taskin, S. Sasaki, K. Segawa, and Y. Ando. 2012. Fermi level tuning and a large activation gap achieved in the topological insulator Bi_2Te_2Se by Sn doping. *Phys. Rev. B* 85: 155301.
91. Fuccillo, M. K., S. Jia, M. E. Charles, and R. J. Cava. 2013. Thermoelectric properties of Bi_2Te_2Se compensated by native defects and Sn doping. *J. Electron. Mater.* 42: 1246–1253.

92. Kushwaha, S. K., Q. D. Gibson, J. Xiong et al. 2014. Comparison of Sn-doped and nonstoichiometric vertical-Bridgman-grown crystals of the topological insulator Bi_2Te_2Se. *J. Appl. Phys.* 115: 143708.
93. Wiendlocha, B. 2016. Resonant Levels, Vacancies, and Doping in Bi_2Te_3, Bi_2Te_2Se, and Bi_2Se_3 Tetradymites, *J. Elect. Mater.* 45: 3515.
94. Goldsmid, H. J. 1954. The thermal conductivity of bismuth telluride. *Proc. Phys. Soc. B* 69: 203–209.
95. Hellman, O. and D. A. Broido. 2015. Phonon thermal transport in Bi_2Te_3 from first principles. *Phys. Rev. B* 90: 134309; Hellman, O. and D. A. Broido. 2015. Erratum: Phonon thermal transport in Bi_2Te_3 from first principles. *Phys. Rev. B* 92: 219903(E).
96. Kullmann, W., G. Eichhorn, H. Rauh, R. Geick, G. Eckold, and U. Steigenberger. 1990. Lattice dynamics and phonon dispersion in the narrow gap semiconductor Bi_2Te_3 with sandwich structure. *Phys. Stat. Solidi (b)* 162: 125.
97. Berman, R. 1976. *Thermal Conduction in Solids*, Clarendon Press, Oxford, U.K.
98. Heremans, J. P. 2015. The anharmonicity blacksmith. *Nat. Phys.* 11: 990–991.
99. Lee, S., K. Esfarjani, T. Luo, J. Zhou, Z. Tian, and G. Chen. 2014. Resonant bonding leads to low lattice thermal conductivity. *Nat. Commun.* 5: 3525.
100. Champness, C. H., W. B. Muir, and P. T. Chiang. 1967. Thermoelectric properties of *n*-type Bi_2Te_3-Bi_2Se_3 alloys. *Can. J. Phys.* 45: 3611.
101. Birkholtz, U. 1958. Untersuchung der intermetallischen verbindung Bi_2Te_3 sowie der fersten Losungen $Bi_{2-x}Sb_xTe_3$ und $Bi_2Te_{3-x}Se_x$ hinsichtlich ihrer eignung als material fur halbleiter thermoelemente. *Z. Naturforsch. Teil A* 13: 780.
102. Goldsmid, H. J. 1961. Recent studies of bismuth telluride and its alloys. *J. Appl. Phys.* 32: 2198.
103. Rosi, F. D., B. Abeles, and R. V. Jensen. 1959. Materials for thermoelectric refrigeration. *J. Phys. Chem. Solids* 10: 191.
104. Poudel, B., Q. Hao, Y. Ma et al. 2008. High-thermoelectric performance of nanostructured bismuth antimony telluride bulk alloys. *Science* 320: 634–638.
105. Goldsmid, H. J. 2010. *Introduction to Thermoelectricity*, Springer Verlag, Berlin, Germany.
106. Kim, S. I., K. H. Lee, H. A. Mun et al. 2015. Dense dislocation arrays embedded in grain boundaries for high-performance bulk thermoelectrics. *Science* 348: 109–114.
107. Anderson, A. C. 1983. Phonon Scattering by Dislocations in *Springer Series in Solid-State Sciences*, Vol 51 (Eisenmenger, W., Lassman, K., Doettinger, S., and Cardona, M. Eds.). Springer-Verlag, Heidelberg, Germany, pp. 348–354.
108. Hu, L., H. Wu, T. Zhu et al. 2015. Tuning multiscale microstructures to enhance thermoelectric performance of *n*-type bismuth-telluride-based solid solutions. *Adv. Energy Mater.* 5: 1500411.
109. Champness, C. H. and A. L. Kipling. 1965. The Hall and Seebeck effects in nonstoichiometric bismuth telluride. *Can. J. Phys.* 44: 769–788.
110. Bergman, D. J. and O. Levy. 1991. Thermoelectric properties of a composite medium. *J. Appl. Phys.* 70: 6621.
111. Bergman, D. J. and L. G. Fel. 1999. Enhancement of thermoelectric power factor in composite thermoelectrics. *J. Appl. Phys.* 85: 8205.
112. Rosi, F. D., E. F. Hockings, and N. E. Lindenblad. 1961. Semiconducting materials for thermoelectric power generation. *RCA Rev.* 22: 82.
113. Champness, C. H., P. T. Chiang, and P. Parekh. 1965. Thermoelectric properties of Bi_2Te_3-Sb_2Te_3 alloys. *Can. J. Phys.* 43: 653.
114. Heremans, J. P. 2005. Low-dimensional thermoelectricity. *Acta Phys. Polonica* 108: 609–634.
115. Goldsmid, H. J. 2014. Bismuth telluride and its alloys as materials for thermoelectric generation. *Materials* 7: 2577–2592.
116. Li, H., H. Jing, Y. Han, Y. Xu, G. Q. Lu, and L. Xu. 2013. Microstructure and transport properties of copper-doped *p*-type (BiSbTe) alloy prepared by mechanical alloying and subsequent spark plasma sintering. *J. Alloy. Compd.* 576: 369–374.

117. Bulat, L. P., I. A. Drabkin, V. V. Karatayev et al. 2013. Structure and transport properties of bulk nanothermoelectrics based on $Bi_xSb_{2-x}Te_3$ fabricated by SPS method. *J. Electron. Mater.* 42: 2110–2113.
118. Xie, W., S. Wang, S. Zhu et al. 2013. High performance Bi_2Te_3 nanocomposites prepared by single-element-melt-spinning spark-plasma sintering. *J. Mater. Sci.* 48: 2745–2760.
119. Nguyen, P. K., K. H. Lee, J. Moon et al. 2012. Spark erosion: A high production rate method for producing $Bi_{0.5}Sb_{1.5}Te_3$ nanoparticles with enhanced thermoelectric performance. *Nanotechnology* 23: 415604–415610.
120. Son, J. H., M. W. Oh, B. S. Kim et al. 2013. Effect of ball milling time on the thermoelectric properties of *p*-type $(Bi, Sb)_2Te_3$. *J. Alloy. Compd.* 566: 168–174.
121. Imamuddin, M. and A. Dupre. 1972. Thermoelectric properties of *p*-type Bi_2Te_3-Sb_2Te_3-Sb_2Se_3 alloys and *n*-type Bi_2Te_3-Bi_2Se_3 alloys in the temperature range 300 to 600 K. *Phys. Stat. Solidi (a)* 10: 415–424.
122. Matsuura, K. and D. M. Rowe. 1995. Low-temperature heat conversion. In *CRC Handbook of Thermoelectrics*, ed. D. M. Rowe, CRC Press, Boca Raton, FL.
123. Snyder, G. J. 2006. Thermoelectric power generation: Efficiency and compatibility. In *Thermoelectrics Handbook: Macro to Nano*, ed. D. M. Rowe, CRC Taylor & Francis, Boca Raton, FL, 2006.
124. Goldsmid, H. J. and J. W. Cochrane. 1982. Solar thermoelectric generators. In *Proceedings of the Fourth IEEE International Conference on Thermoelectric Energy Conversion*, Arlington, TX, March 10–12, 1982.
125. Bogomolov, D. I., V. T. Bublik, S. Y. Skipidarov, and N. Y. Tabachkova. 2013. Structure and properties of thermoelectric materials based on $Bi_2(SeTe)_3$ and $(BiSb)_2Te_3$ solid solutions prepared by equal-channel angular pressing. *Inorg. Mater.* 49: 758–775.
126. Hu, L. P., T. J. Zhu, X. H. Liu, and X. B. Zhao. 2014. Point defect engineering of high-performance bismuth-telluride-based thermoelectric materials. *Adv. Funct. Mater.* 24: 5211.
127. Wu, F., H. Song, F. Gao, W. Shi, J. Jia, and X. Hu. 2013. Effects of different morphologies of Bi_2Te_3 nanopowders on thermoelectric properties. *J. Electron. Mater.* 42: 1140.
128. Kim, H., M. K. Han, C. H. Yo, W. Lee, and S. J. Kim. 2012. Effects of Bi_2Se_3 nanoparticle inclusions on the microstructure and thermoelectric properties of Bi_2Te_3-based nanocomposites. *J. Electron. Mater.* 41: 3411–3416.
129. Yu, C., X. Zhang, M. Leng et al. 2013. Preparation and thermoelectric properties of inhomogeneous bismuth telluride alloyed nanorods. *J. Alloy. Compd.* 570: 86–93.
130. Dashevsky, Z., I. Drabkin, V. Korotaev, and D. Rabinovich. 1997. Improved materials for thermoelectric conversion (generation). In *Proceedings of the 16th IEEE International Conference on Thermoelectrics*, Dresden, Germany, August 1997, pp. 26–29.
131. Gerovac, N., G. J. Snyder, and T. Caillat. 2002. Thermoelectric properties of *n*-type polycrystalline $Bi_xSb_{2-x}Te_3$ alloys. In *Proceedings of the 21st IEEE International Conference on Thermoelectrics*, Long Beach, CA, August 2002, pp. 25–29.
132. Prokofieva, L. V., D. A. Pshenay-Severin, P. P. Konstantinov, and A. A. Shabaldin. 2009. Optimum composition of a $Bi_2Te_{3-x}Se_x$ alloy for the *n*-type leg of a thermoelectric generator. *Semiconductors* 43: 973–976.
133. Wang, S. Y., W. J. Xie, H. Li, and X. F. Tang. 2010. High performance *n*-type $(Bi,Sb)_2(Te,Se)_3$ for low temperature thermoelectric generator. *J. Phys. D: Appl. Phys.* 43: 335404.
134. Recently, *p*-type α-MgAgSb has been reported to be able to reach ZT values in excess of unity above 470 K (Ying, P., X. Liu, C. Fu et al. 2015. High performance α-MgAgSb thermoelectric materials for low temperature power generation. *Chem. Mater.* 27: 909) and as low as 370 K (Li, D., H. Zhao, S. Li et al. 2015. Atomic disorders induced by silver and magnesium ion migrations favor high thermoelectric performance in α-MgAgSb-based materials. *Adv. Func. Mater.* 25: 6478). Above 470 K, the ZT reported in the latter reference surpasses the highest ZT reported on *p*-type tetradymites (Rosi, F. D., B. Abeles, and R. V. Jensen. 1959. Materials for thermoelectric refrigeration. *J. Phys. Chem. Solids* 10: 191).

3

Growth and Transport Properties of Tetradymite Thin Films

Hang Chi, Wei Liu, and Ctirad Uher
University of Michigan

CONTENTS

3.1 Introduction

Interest in thermoelectric (TE) properties of tetradymite-type materials, that is, Bi_2Se_3, Bi_2Te_3, and Sb_2Te_3, originated from the work of Goldsmid and coworkers on bulk forms of the structure in the mid-1950s [1]. The excellent dimensionless thermoelectric figure of merit ZT for both n-type and p-type tetradymite solid solutions has enabled this material family to be one of the most successfully commercialized TE systems, especially for low-temperature applications [2,3]. Motivated by the increasing demand for alternative green energy, much research and development have been done in the past two decades toward advancing efficient synthesis processes [4], device fabrication [5], and physical property characterization [6]. These efforts have, in turn, significantly deepened the understanding and improved the performance of TE materials. Indeed, the state-of-the-art p-type ZT has reached ~1.8 at 320 K in the bulk nanocomposite form of this classic TE material system [7–9].

As proposed by Dresselhaus and colleagues in the early 1990s [10–12], materials with lower-dimensional characteristics, such as 2D superlattices (SLs), 1D nanowires, and 0D quantum dots, may possess significantly enhanced TE properties and have been intensively studied to achieve (1) reduction in thermal conductivity (e.g., by all-scale hierarchical phonon scattering [13]) and (2) enhancement in power factor (via formation of band resonant states [14], quantum confinement [15], modulation doping [16], energy filtering [17], and other novel mechanisms [18]). A spectacular $ZT \sim 2.4$ at 300 K was claimed in 2001 for p-type Bi_2Te_3/Sb_2Te_3 SLs [19]. Despite the fact that such impressive ZT values on SLs have never been reproduced by intense efforts worldwide [20], the possibility of enhancing TE performance by decoupling electronic and thermal transport processes with reduced dimensionality has boosted the fabrication of tetradymite-type TE films and SLs.

Recently, topologically nontrivial materials have attracted heightened attention around the globe, owing to their potential of revealing new physical phenomena and prospects for developing next-generation electronics. Ever since their discovery [21–24], tetradymite-based 3D topological insulators (TIs) have stimulated worldwide interest in surface states (SSs) arising from the strong spin-orbital coupling and time-reversal symmetry [25–27]. Immune to perturbations of weak electromagnetic fields and/or nonmagnetic impurities, the relaxation times dominating surface transport can be much enhanced, resulting in a nondissipative surface conduction. The (1) Dirac-like linear energy dispersion and (2) spin-momentum locking are the two key features of these topologically protected SSs, which have been confirmed by angle-resolved photoemission spectroscopy (ARPES) and scanning tunneling microscopy (STM) experiments. Transport-wise, weak antilocalization (WAL) effect and Shubnikov-de Haas oscillations in the TE tensors (resistivity and thermopower under applied magnetic fields) have been effectively used to characterize the SSs of tetradymite-type TIs [28–33].

Despite the success in observing all these transport hallmarks in laboratories, practical applications utilizing SSs still seem rather remote and hindered by the presence of predominant bulk carriers (~10^{19} cm^{-3}, due to intrinsic antisite defects and/or vacancies in near-equilibrium grown tetradymite-type structures). However, remarkable progress is being made in thin films prepared under far-from-equilibrium conditions, whose large surface/volume ratio naturally favors the manifestation of surface conduction. More importantly, it is possible to identify deposition protocols that depress the undesired background carrier density by fine tuning constituent stoichiometry, film thickness, and annealing temperature, that is, the material control parameters that are not easily accessible during the growth of bulk crystals.

In the following sections of this chapter, we will describe in more details the tetradymite-type material structure, which present recent developments regarding the deposition of tetradymite-type thin films and SLs, and summarize their transport properties. Such studies may shed more light on prospects of further enhancing the performance of tetradymite-type thin-film TEs.

3.2 Tetradymite Structure

Combining pnictogen (A^V = Bi, Sb) and chalcogen (B^{VI} = Te, Se, S) atoms in the 2:3 ratio results in the formation of a trigonal $A_2{}^VB_3{}^{VI}$ crystal with the space group $R\bar{3}m$ (D_{3d}^5, No. 166). As shown in Figure 3.1a, $A_2{}^VB_3{}^{VI}$ has a layered structure with a [$B^{VI}(1)$–A^V–$B^{VI}(2)$–A^V–$B^{VI}(1)$] type of quintuple layers (QLs, ~1 nm thick, within which the bonding is of ionic-covalent nature) connected by a weak van der Waals (vdW) interaction between neighboring quintuple layers, often referred to as the tetradymite lattice. The unit cell is the conventional hexagonal cell, which can be adopted for all the 25 trigonal space groups (No. 143–167 [34]). The cations occupy the Wyckoff 6*c* sites A^V (0, 0, z_A ~ 0.4), while the anions are located at 6*c* sites $B^{VI}(1)$ (0, 0, z_B ~ 0.2) and 3*a* sites $B^{VI}(2)$ (0, 0, 0). There are, however, seven space groups in the trigonal crystal system (No. 146, 148, 155, 160, 161, 166, and 167) that possess a rhombohedral lattice, whose unit cells can be reduced from the hexagonal cell (Figure 3.1b, a volume 3 supercell containing three lattice points) to a primitive rhombohedral cell (Figure 3.1c). The lattice parameters ($a, b, c, \alpha, \beta, \gamma$), coordinates ($x, y, z$), directions [$uvw$], and planes ($hkl$) in the hexagonal ($a_H = b_H \neq c_H$, $\alpha_H = \beta_H = 90°$, $\gamma_H = 120°$, using the three-index symbols) and the rhombohedral ($a_R = b_R = c_R$, $\alpha_R = \beta_R = \gamma_R$) representations can be converted back and forth via the following relations:

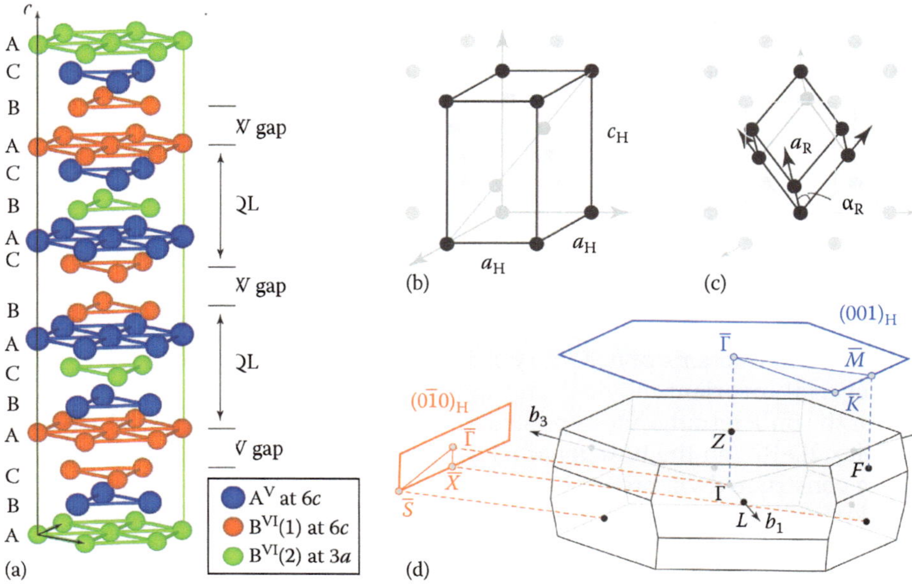

FIGURE 3.1
(a) Tetradymite-type crystal structure with [$\cdots B^{VI}(1)$–A^V–$B^{VI}(2)$–A^V–$B^{VI}(1)\cdots$] quintuple layers that are separated by van der Waals gaps, along with its (b) conventional hexagonal cell, (c) primitive rhombohedral cell, and (d) the first Brillouin zone for 3D bulk and selected 2D surfaces.

hexagonal → rhombohedral

$$a_R = \frac{1}{3}\sqrt{3a_H^2 + c_H^2}, \quad \alpha_R = 2\sin^{-1}\left(\frac{3}{2\sqrt{3+\left(c_H^2/a_H^2\right)}}\right),$$

$$x_R, y_R, z_R = x_H + z_H, -x_H + y_H + z_H, -y_H + z_H,$$

$$[u_R \quad v_R \quad w_R] = [u_H + w_H \quad -u_H + v_H + w_H \quad -v_H + w_H],$$

$$(h_R \quad k_R \quad l_R) = \frac{1}{3}(2h_H + k_H + l_H \quad -h_H + k_H + l_H \quad -h_H - 2k_H + l_H) \tag{3.1}$$

and
rhombohedral → hexagonal

$$a_H = 2a_R \sin\left(\frac{\alpha_R}{2}\right), \quad c_H = a_R\sqrt{3 + 6\cos\alpha_R},$$

$$x_H, y_H, z_H = (0,0,0; \tfrac{2}{3},\tfrac{1}{3},\tfrac{1}{3}; \tfrac{1}{3},\tfrac{2}{3},\tfrac{2}{3}) + (2x_R - y_R - z_R)/3, (x_R + y_R - 2z_R)/3, (x_R + y_R + z_R)/3,$$

$$[u_H \quad v_H \quad w_H] = \frac{1}{3}[2u_R - v_R - w_R \quad u_R + v_R - 2w_R \quad u_R + v_R + w_R],$$

$$(h_H \quad k_H \quad l_H) = (h_R - k_R \quad k_R - l_R \quad h_R + k_R + l_R) \tag{3.2}$$

Specifically, in the Cartesian coordinate system constructed by the binary (x), bisectrix (y), and trigonal (z) axes, the rhombohedral primitive unit vectors $\boldsymbol{a}_i$ (in terms of $\hat{x}, \hat{y}, \hat{z}$) and the reciprocal lattice vectors $\boldsymbol{b}_j$ (in the space of $\hat{k}_x, \hat{k}_y, \hat{k}_z$, where $\boldsymbol{a}_i \cdot \boldsymbol{b}_j = 2\pi\,\delta_{ij}$) are given in the following equation:

$$\boldsymbol{a}_1 = \left(\frac{a}{2}, \frac{a}{2\sqrt{3}}, \frac{c}{3}\right), \quad \boldsymbol{a}_2 = \left(-\frac{a}{2}, \frac{a}{2\sqrt{3}}, \frac{c}{3}\right), \quad \boldsymbol{a}_3 = \left(0, -\frac{a}{\sqrt{3}}, \frac{c}{3}\right);$$
$$\boldsymbol{b}_1 = \frac{2\pi}{a}\left(1, \frac{1}{\sqrt{3}}, \frac{a}{c}\right), \quad \boldsymbol{b}_2 = \frac{2\pi}{a}\left(-1, \frac{1}{\sqrt{3}}, \frac{a}{c}\right), \quad \boldsymbol{b}_3 = \frac{2\pi}{a}\left(0, -\frac{2}{\sqrt{3}}, \frac{a}{c}\right). \tag{3.3}$$

When expressed in fractions of $\boldsymbol{b}_j$, Γ (0, 0, 0), L (½, 0, 0), *F* (½, ½, 0), and Z (½, ½, ½) are the four time-reversal invariant momentum high-symmetry points in the 3D first Brillouin zone (BZ) for the bulk tetradymite-type structure. As illustrated in Figure 3.1d, two important low-index 2D BZ, on the hexagonal $(111)_R$ $[(001)_H]$ and the rectangular $(\bar{1}\bar{1}2)_R$ $[(0\bar{1}0)_H]$ surfaces, are constructed via projection from the 3D BZ.

The best known tetradymite-type structures are Bi_2Se_3 [35], Bi_2Te_3 [36], and Sb_2Te_3 [37], along with ternary Bi_2Te_2Se [38], Bi_2Se_2Te [39], and $Bi_2Te_{1.6}S_{1.4}$ (where Te and S species mix on the anion 6*c* sites) [40] and their solid solutions—for example, $(Bi_{0.58}Sb_{1.42})(Se_{0.12}Te_{2.88})$ [41]. Other binary combinations, that is, Bi_2S_3 [42], Sb_2Se_3 [43], and Sb_2S_3 [44], result in orthorhombic structures with the space group *Pnma* (D_{2h}^{16}, No. 62). Note that elemental Bi [45] and Sb [46] crystals also have the rhombohedral $R\bar{3}m$ (D_{3d}^{5}, No. 166) lattice, which may be used as a buffer layer during the growth of tetradymite-type thin films. Pure Te [47] and Se [48] crystallize in a trigonal $P3_121$ (D_3^4, No. 152) structure (often a convenient choice for capping layers in order to protect as-grown thin films for *ex situ* characterizations), while crystalline S usually takes the S_8 form [49] with orthorhombic *Fddd* (D_{2h}^{24}, No. 70) symmetry. A summary of the crystallographic information for these tetradymite-related structures is tabulated in Table 3.1.

All three binary tetradymite-type compounds are semiconductors with small band gaps of ~0.30 eV (Bi_2Se_3), 0.13 eV (Bi_2Te_3), and 0.28 eV (Sb_2Te_3), respectively. Bulk crystals grown from stoichiometric melts under near-equilibrium conditions (via Bridgman, Czochralski, and/or floating zone type of methods) all possess rather high background charge carrier densities. Due to the presence of selenium vacancies ($V_{Se}^{\cdot\cdot}$, a double-electron donor), the electrical conduction of pristine Bi_2Se_3 is dominated by electrons (*n*-type), with a room temperature electron density ~10^{19} cm^{-3}. Owing to the native antisite defects (Bi'_{Te} or Sb'_{Te}, a single-hole acceptor), pure Bi_2Te_3 and Sb_2Te_3 bulk crystals are usually *p*-type with room temperature hole densities of ~10^{19} and ~10^{20} cm^{-3}, respectively. While Bi_2Se_3 (Bi_2Te_3) may be tuned into a *p*-type (*n*-type) conductor through doping and/or intentional off-stoichiometric growth, there are no reports on preparing an *n*-type Sb_2Te_3. Furthermore, molecular beam epitaxy (MBE)–grown Bi_2Te_3 thin films (as shown in the following) are dominantly *n*-type, while Bi_2Se_3 and Sb_2Te_3 films are still *n*-type and *p*-type, respectively. This is argued as a result of the fact that the optimized growth temperature for Bi_2Te_3 is usually much lower than the melting temperature, which favors interstitial Bi atoms or Te vacancies more than the formation of antisite defects, driving the system *n*-type.

Tetradymite-type compounds have a strong anisotropy in their transport properties inherited from the anisotropy in the crystal structures, see Figure 3.1a. The in-plane

TABLE 3.1

Crystallographic Information[a] for Tetradymite-Related Materials

Materials	SG	ICSD	Reference	a_H (Å)	c_H (Å)	Element	x_H	y_H	z_H
Bi_2Se_3	166	165226	[35]	4.1355	28.6150	Bi	0.0000	0.0000	0.4006
						Se_1	0.0000	0.0000	0.2109
						Se_2	0.0000	0.0000	0.0000
Bi_2Te_3	166	184631	[36]	4.3896	30.5019	Bi	0.0000	0.0000	0.4004
						Te_1	0.0000	0.0000	0.2094
						Te_2	0.0000	0.0000	0.0000
Sb_2Te_3	166	192780	[37]	4.2706	30.4758	Sb	0.0000	0.0000	0.3986
						Te_1	0.0000	0.0000	0.2122
						Te_2	0.0000	0.0000	0.0000
Bi_2Te_2Se	166	43512	[38]	4.2800	29.8600	Bi	0.0000	0.0000	0.3960
						Te	0.0000	0.0000	0.2120
						Se	0.0000	0.0000	0.0000
Bi_2Se_2Te	166	54838	[39]	4.1830	29.1370	Bi	0.0000	0.0000	0.3981
						Se	0.0000	0.0000	0.2116
						Te	0.0000	0.0000	0.0000
Bi	166	64703	[45]	4.5460	11.8620	Bi	0.0000	0.0000	0.2339
Sb	166	64695	[46]	4.3084	11.2740	Sb	0.0000	0.0000	0.2335
Te	152	65692	[47]	4.4560	5.9210	Te	0.2636	0.0000	0.3333
Se	152	40018	[48]	4.3680	4.9580	Se	0.2254	0.0000	0.3333

[a] Space group, inorganic crystal structure database collection code, hexagonal lattice constants and atomic occupation.

electronic properties are superior while the in-plane lattice thermal conductivity is also much higher than that across the plane. For instance, in the Bi_2Te_3–Sb_2Te_3 system, the anisotropic ratio of electrical conductivity and lattice thermal conductivity is in the range of 2.8–6.3 and 2.0–7.1, respectively [2,50]. Typical in-plane transport properties of bulk tetradymite-type single crystals [51–53] have been summarized in Table 3.2. As we will see, when grown in thin-film form, the film orientation and microstructures also remarkably affect the transport properties.

When it comes to TE applications, it is the solid solutions of these end compounds that are of practical interest. The formation of solid solutions introduces short-range disorder into the structure, which strongly scatters heat-conducting phonons and significantly reduces the lattice thermal conductivity. As a result, the best performing *p*-type tetradymite TE is based on a composition near $Bi_{0.5}Sb_{1.5}Te_3$, while the *n*-type counterpart is close to $Bi_2Te_{2.85}Se_{0.15}$. Progress has been made to further reduce the thermal conductivity by fabricating nanostructured composites. Nanostructures introduced by ball milling and melt spinning have demonstrated a ZT ~1.4 (at 373 K) [7] and ~1.5 (at 390 K) [8] in *p*-type BiSbTe nanocomposites. A recent modification of the melt spinning technique (followed by spark plasma sintering) has exploited excessive Te to introduce additional phonon scattering at the grain boundary, which further reduces the thermal conductivity and results in a ZT ~1.8 (at 320 K) in *p*-type $Bi_{0.5}Sb_{1.5}Te_3$ [9]. It is of prime interest to adopt and/or redesign these successful recipes for *p*-type structures and apply them to *n*-type forms of tetradymite-type materials to improve their performance.

TABLE 3.2

Transport Properties of Bulk Bi_2Se_3 [51], Bi_2Te_3 [52], and Sb_2Te_3 [53] in the Basal Plane at 300 K

Properties	Bi_2Se_3	Bi_2Te_3	Sb_2Te_3
E_g	~0.30	0.13	0.28
Type	$n(n)$	$p(n)$	$p(p)$
n_H	2.13	1.12	8.92
μ_H	809	351	330
σ	2.76	0.63	4.72
α	−59	245	79
PF	9.61	37.82	29.46
κ	3.10	2.01	4.95
ZT	0.09	0.56	0.18

The electronic band gap, Eg (eV); carrier type for bulk (film); carrier density, n_H (10^{19} cm^{-3}); Hall mobility, μ_H (cm^2 V^{-1} s^{-1}); electrical conductivity, σ (10^5 S m^{-1}); Seebeck coefficient, α (μV K^{-1}); power factor, $PF \equiv \alpha^2\sigma$ (10^{-4} W m^{-1} K^{-2}); thermal conductivity, κ (W m^{-1} K^{-1}); figure of merit, $ZT \equiv \alpha^2\sigma T/\kappa$.

3.3 Thin-Film Growth

Various methods have been reported to prepare tetradymite-type thin films and SLs with tunable TE properties, such as coevaporation (CE) [54–58], electrochemical deposition [59–62], flash evaporation (FE) [63–66], metal organic chemical vapor deposition (MOCVD) [19,67–77], MBE [78–84], nanoalloying (NA) [85–88], pulsed laser deposition (PLD) [89–95], and sputtering [50,96–106]. Generally, CE, FE, MBE, NA, PLD, and sputtering can be classified as physical vapor deposition (PVD) processes. Although they differ in particulars (e.g., how the atomic/molecular plume/fluxes are generated and delivered onto the substrate), these PVD techniques have much in common; the desired compounds form on a heated substrate by supplying flux of molecules or atoms with an appropriate ratio. Therefore, strategies applicable to tetradymite-type film growth in MBE are also suitable for other PVD techniques, while the appropriate deposition parameters may vary case by case. In MOCVD processes, however, besides the substrate temperature and flux rate, other factors such as the selection of precursors, substrate position, and working pressure are also crucial for the growth and quality control of tetradymite-type films. As for electrochemical deposition, postgrowth annealing at elevated temperatures is often mandatory to improve crystallinity and the orientation of films [60].

The crystallographic building block of the $A_2^VB_3^{VI}$ thin film is the ~1 nm thick B^{VI}-terminated QL (Figure 3.1a). This also implies that *c*-oriented growth of tetradymite-type film is energetically favored due to the weak vdW interaction between adjacent layers. Benefiting from this feature, the fabricated tetradymite-type films are mostly *c*-oriented regardless of the choice of substrates or deposition techniques. A selection of commonly used single crystal substrates are listed in Table 3.3. In fact, the vdW epitaxial nature of tetradymite-type films is so tolerant that tetradymite-type structures can even be grown on amorphous glass substrates with good quality. Such an intermediate

TABLE 3.3

In-Plane Lattice Mismatch between Tetradymite-Type Films (a_{film}) and a Selection of Substrates ($a_{substrate}$), Defined as $\delta = (a_{film}/a_{substrate} - 1) \times 100\%$

Substrate	$a_{substrate}$ (Å)	$\delta_{Bi_2Se_3}$ (%)	$\delta_{Bi_2Te_3}$ (%)	$\delta_{Sb_2Te_3}$ (%)
Al_2O_3 (0001)	4.7602	−13.1	−7.8	−10.3
BaF_2 (111)	4.3815	−5.6	−0.2	−2.5
Bi (0001)	4.5460	−9.0	−3.4	−6.1
CdTe (111)	4.5840	−9.8	−4.2	−6.8
GaAs (111)	3.9966	+3.5	+9.8	+6.9
Si (111)	3.8400	+7.7	+14.3	+11.2
$SrTiO_3$ (111)	5.5202	−25.1	−20.5	−22.6

tetradymite-type thin film has been recently engineered as a buffer layer for producing II–VI films with unprecedented crystallinity [107]. As demonstrated by da Silva et al., *c*-oriented polycrystalline Bi_2Te_3 films were successfully grown on glass substrates using the CE method [55]. The coevaporated Bi_2Te_3 at 130°C was polycrystalline (Figure 3.2b) with an x-ray diffraction pattern similar to that of powder Bi_2Te_3 (Figure 3.2a), while the film grown at 260°C started to show strong *c*-orientation (Figure 3.2c). Evident from the cross-sectional scanning electron microscopy images, the grain boundary of Bi_2Te_3 films grown at 130°C was rather unclear (even with observable pin holes, Figure 3.2d). Increasing the substrate temperature to 260°C significantly promoted the crystal growth and resulted in a large grain size up to ~500 nm (Figure 3.2e). As shown in Figure 3.2f, reduced Te flux led to degraded growth of the Bi_2Te_3 film, advocating for a Te-rich growth condition in order to effectively compensate the reevaporation of Te from the substrate surface at high temperatures. To achieve high crystallinity and thus improve electronic properties in tetradymite-type films, relatively high-substrate temperature and Te flux were preferred, in agreement with observations using other growth techniques [58,76,82,83].

Regardless of the differences of deposition techniques, recent studies on tetradymite-type thin-film TEs have focused on (1) good crystallinity, (2) controllable carrier density, (3) favorable microstructures, (4) stable and functional SLs, and (5) new phenomena for enhanced TE performances. Among these techniques, MOCVD was first selected to grow tetradymite-type SLs with reported extremely high-ZT values in 2001 [19]. However, it was only quite recently that the researchers managed to utilize MBE techniques to reproduce excellent Bi_2Te_3/Sb_2Te_3 SLs with demonstrated good crystallinity [81,108]. Indeed, boosted by the TI research, the MBE technique has produced tetradymite-type films with the best crystallinity, close to that of the best bulk single crystals ever reported. These high-quality films possess strictly *c*-axis orientation, ultrahigh purity in composition, and a noticeably suppressed density of point defects and twin domains [109–114]. The improved film quality is believed to originate from the layer-by-layer growth of films in MBE, which relies on the precise control of growth parameters, such as the ultrahigh vacuum (UHV), flat substrates with extremely clean surfaces, and optimal growth temperature, along with very stable and uniform flux of raw elements.

Due to its unique ability to control an epitaxial growth mode, producing high-quality films using MBE has gained popularity over recent years, despite its expensive nature in terms of construction, operation, and maintenance. As schematically shown in Figure 3.3,

MBE is a thin-film deposition process that takes place in an UHV chamber maintained by a staged continuous pumping system, using a combination of ionic pump, turbomolecular pump, and scroll pump. The chief reason of intentionally choosing only *dry* pumps is to eliminate contaminations from any oil vapors. A cryoshield is often implemented to further improve the vacuum, via the added power of adsorbing atom/molecules on surfaces cooled by flowing liquid nitrogen. Pressure in the growth chamber is read using hot-filament ionization gauges and thermocouple gauges. Potential leaks are monitored by a residue gas analyzer. Resistive baking heaters are installed for regular chamber cleaning. A load/lock entry chamber, separated by a gate valve from the deposition chamber, is designed for fast sample entry and retrieval using a holder/fork mechanism. Chambers with additional functionalities can also be integrated. These include preparation/treatment chambers and characterization chambers. In the most state-of-the-art MBE systems, many advanced characterization probes (e.g., ARPES and STM) are nowadays integrated together with the main MBE chamber, enabling direct *in situ* characterization of films. These recent advances in instrumentation have significantly promoted the study of tetradymite-type thin films.

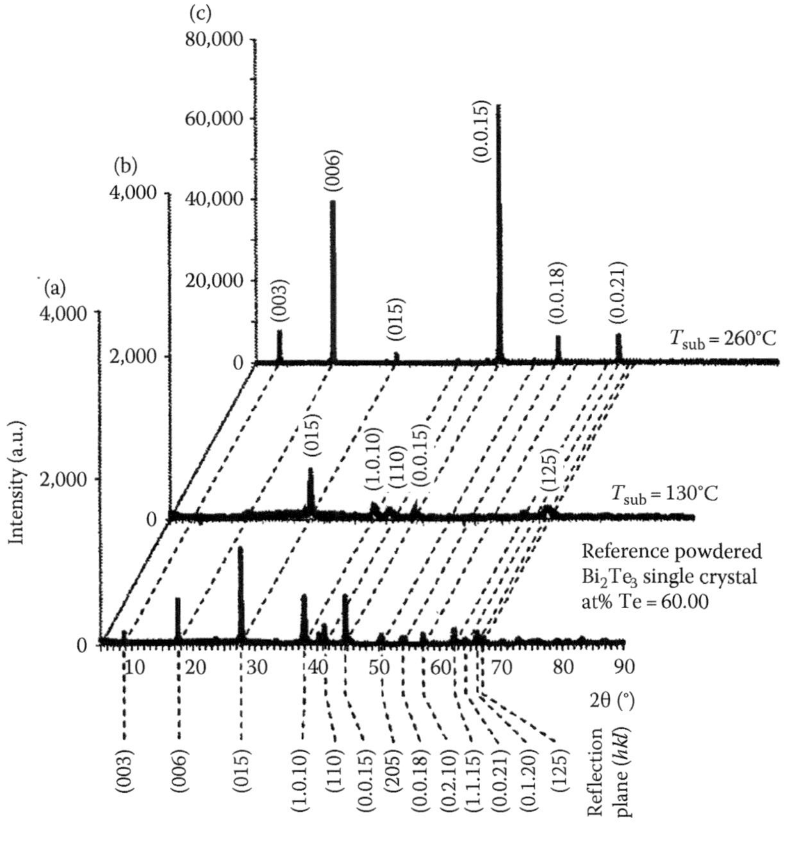

FIGURE 3.2
X-ray diffraction patterns of (a) reference-powdered Bi_2Te_3 and coevaporated Bi_2Te_3 film deposited on a glass substrate at (b) 130°C (at% Te = 60%) and (c) 260°C (at% Te = 60%). *(Continued)*

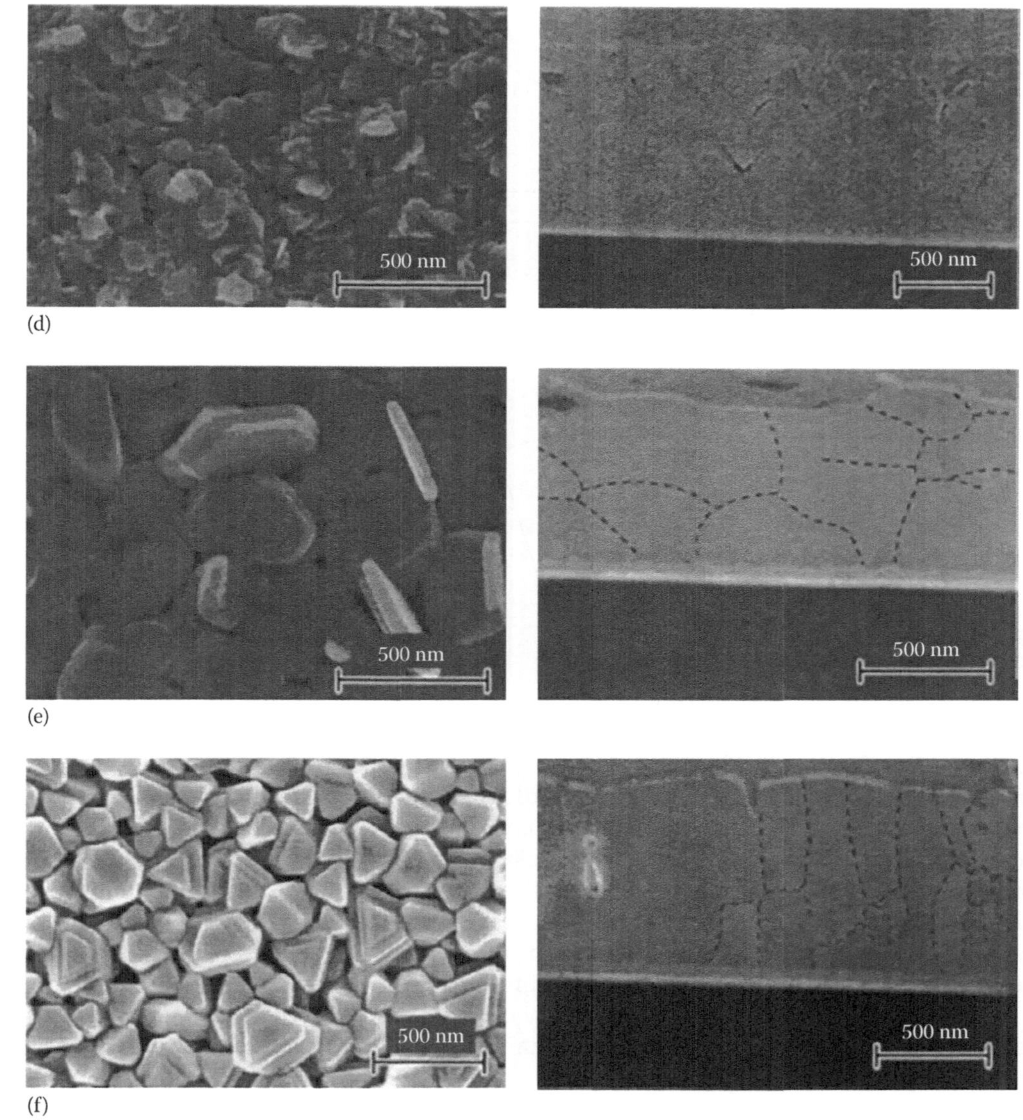

FIGURE 3.2 (*Continued*)
Scanning electron microscopy images showing the top and cross section of the coevaporated Bi_2Te_3 films deposited at (d) 130°C (at% Te = 60%), (e) 260°C (at% Te = 60%), and (f) 260°C (at% Te = 54%), respectively. In (e) and (f), the dashed lines at the cross section views highlight grain boundaries. (Reprinted with permission from da Silva, L.W., Kaviany, M., and Uher, C., Thermoelectric performance of films in the bismuth-tellurium and antimony-tellurium systems, *J. Appl. Phys.*, 97, 114903. Copyright 2005, American Institute of Physics.)

Knudsen effusion cells are used to produce carefully controlled atomic/molecular fluxes. These cells, with crucibles loaded with high-purity source materials, are usually integrated through ports at the bottom side of the chamber. Additional miniature electron beam guns can be incorporated for doping by high-melting-point elements. Gas/plasma-type sources may also be installed if desired. In the case of Bi_2Se_3 (Bi_2Te_3), a Bi-containing

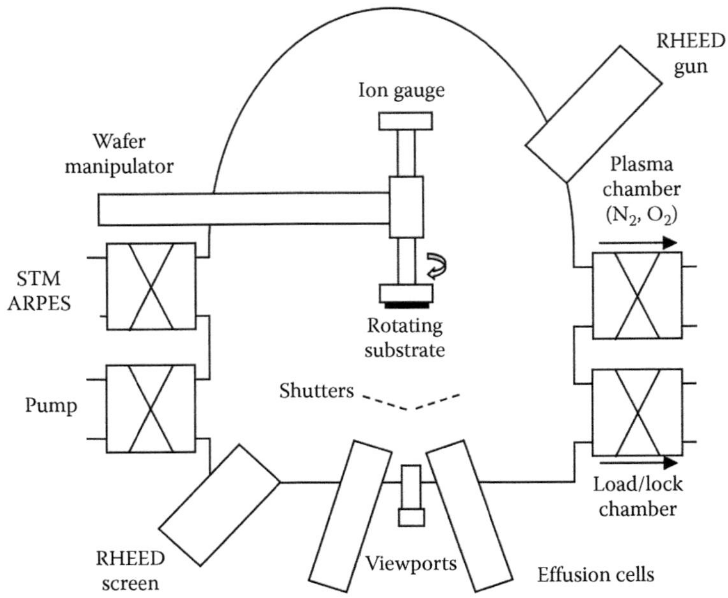

FIGURE 3.3
A schematic of a molecular beam epitaxy system.

cell and a Se- (Te-) containing cell are engaged at the same time. Heated Te often comes out of the effusion cell as a clustered molecule instead of individual atoms [115]. An additional heating filament near the opening may help to *crack* these Te clusters before they escape the effusion cell. Such a specially designed two-stage thermal cracker cell is preferred to achieve a stable and uniform Te flux. Considering that bismuth expands upon solidification, extreme care should be taken when programming the thermal cycling of Bi-cells. Sudden cooling of molten bismuth, which may break the crucible and damage the effusion cell, should be avoided at all times. Instead of using single-elemental fluxes of Sb and Te, Sb_2Te_3 seems to grow better by directly evaporating Sb_2Te_3 compound, with a properly tuned Te compensating flux to maintain the tetradymite phase. Viewports and charge-coupled device (CCD)-based cameras are used for visual examination. Computer-controlled cell shutters are installed near the opening of the cells to realize an on-/off-type of automated manipulation during film deposition. Quartz crystal monitors are helpful in monitoring the flux rate *in situ*.

Since the mean free paths of these atoms/molecules are very long under UHV, they essentially travel as a *beam* toward the substrate. The substrates (atomically smooth single crystal pieces) are mounted upside down, facing toward the incoming fluxes. Upon arrival at the cleaned/treated surface of the substrate, different species of atoms/molecules may react and crystallize. The temperature of the substrate is adjusted accordingly to promote (prohibit) the desired (unwanted) chemical reactions and reevaporation. The substrate may also be placed in a constant rotation mode during the deposition, which improves the uniformity of the film. Postgrowth annealing is sometimes needed in order to further alter the surface morphology and the crystallinity of the as-grown film. A carefully designed combination of sample-shutter/mask movement during the growth can be used to prepare SLs, wedged samples with thickness gradients, etc. All the features mentioned earlier

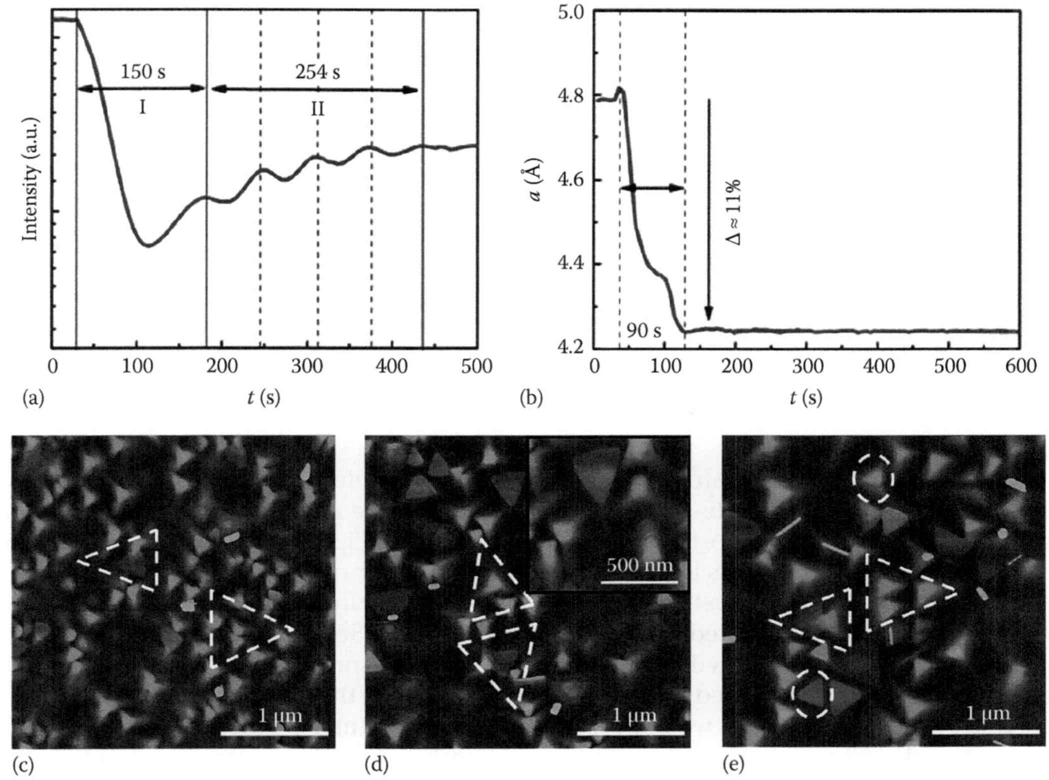

FIGURE 3.4
(a) Reflection high-energy electron diffraction intensity oscillation and (b) in-plane lattice parameter evolution of $BiSbTe_3$ film grown on a sapphire (0001) substrate as a function of deposition time. AFM image ($3 \times 3\ \mu m^2$ in area) of $BiSbTe_3$ films deposited at substrate temperatures of (c) 485 K, (d) 500 K, and (e) 515 K. Inset in (d) shows an enlarged image of Figure 3.4d. The dashed triangles mark the opposite orientation of twin domains in the molecular beam epitaxy grown $BiSbTe_3$ films. The dashed circles in (e) highlight the twin defects in the boundary area among the oppositely oriented domains. (Reprinted from *J. Cryst. Growth,* 410, Liu, W., Endicott, L., Stoica, V.A., Chi, H., Clarke, R., and Uher, C., High-quality ultra-flat $BiSbTe_3$ films grown by MBE, 23–29. Copyright 2015, with permission from Elsevier.)

coupled with an unprecedented control of the flux rate make the MBE synthesis uniquely suitable for the epitaxial thin-film growth.

The epitaxial nature of the growing film can be conveniently monitored by surface sensitive probes such as reflection high-energy electron diffraction (RHEED). An optional differential pumping system may be used to achieve the high vacuum required for energizing the RHEED electron gun. The diffraction pattern can be monitored through the CCD camera situated behind the fluorescent RHEED screen. Depending on the operation voltage of the RHEED gun, users should be aware of and protected from the hazards of potential x-ray generation. Figure 3.4a shows the temporal evolution of the RHEED intensity profile recorded for the growth of a typical $BiSbTe_3$ ternary thin film on a sapphire (0001) substrate [83]. After completing the initial QL, the RHEED intensity starts to develop an oscillating pattern with a well-defined period. Each full cycle indicates the completion of an individual QL. This is a manifestation of the QL-by-QL vdW epitaxial growth mode. Before deposition, the in-plane lattice parameter a deduced from the RHEED pattern is the value of the bare sapphire (0001) substrate. When exposed to the atomic/molecular fluxes, *a* starts to decrease and quickly settles at the bulk value of the tetradymite-type structure (Figure 3.4b). Owing to the weak vdW bonding between the adjacent B^{VI} atomic layers, strains originating from the lattice mismatch can thus be fully relaxed within the first QL. This implies that the requirement of lattice matching, an often difficult to satisfy criterion for high-quality epitaxial growth (e.g., in a cubic crystal system such as GaAs), is somewhat relaxed in the growth of tetradymite-type structures with vdW layers. Note, however, despite the perfectly *c*-oriented growth, one can still have stacks or twin domains in the basal plane that are misoriented by 60°, which are observed even in high-quality MBE films [78,83,110,116]. For example, using atomic force microscopy, Liu et al. have observed 60° rotated domains in as-grown $BiSbTe_3$ MBE thin films at various substrate temperatures (see Figure 3.4c through e). These twin domains may be suppressed by optimizing growth parameters to achieve a stable film growth [110,111]. Up till now, the effect of twin domains on TE transport still remains unclear due to limited reports.

Apart from the general features mentioned earlier of preferred orientation and misoriented domains, a tetradymite-type film also possesses fine microstructures at nanoscale that are important to transport processes, such as nanoprecipitates, structure modulation, dislocations, and twin grains [80,81,85,86,94,99–101,108]. Peranio et al. [85] prepared Bi_2Te_3 and Sb_2Te_3 films on SiO_2 substrates by a NA method utilizing a MBE chamber, where Te and Bi/Sb were first deposited in the sequence of –Te–Bi/Sb–Te–Bi/Sb–Te– at room temperature and the as-deposited films were subsequently annealed at 250°C for 2 h. The obtained Sb_2Te_3 film possessed a strong *c*-orientation, while the Bi_2Te_3 film showed no clear epitaxial growth relation between the substrate and the film. As indicated in Figure 3.5a through e, the cross section high-resolution transmission electron microscopy (HRTEM) analysis on Sb_2Te_3 showed rich nanoscale structures, such as nanograins with size of 350–650 nm and clear fringe structures due to structure modulation with a wavelength of 5–12 nm. A selected area electron diffraction (λ) analysis showed that the two strongly *c*-oriented grains enclosed an angle of 18°, inclining at an angle of up to 37° away from the film growth direction. Figure 3.5f through j shows the HRTEM analysis of Bi_2Te_3 in cross section. The Bi-rich boundary phase, the structure modulation with a wavelength of ~10 nm, and the dislocations and polycrystalline-like SAED ring pattern are evident in the studied area. In addition, the grains in this area were found to be separated by large-angle grain boundaries, and the *c*-axis of the grains deviated by an angle of 39° from the growth direction. The microstructure of tetradymite-type films was found to be strongly related to the growth conditions and compositions. First, structure modulation and twin grains are generally reported [80,94]. Second, a cation-rich boundary phase was found in coevaporated and annealed Bi_2Te_3 film at room temperature but was not observed in the Bi_2Te_3 film prepared by traditional MBE [80] or in coevaporated and annealed Sb_2Te_3 film at room temperature [85]. Other interesting results of basal-plane/nonbasal-plane twin boundary and Te-rich precipitates were observed in PLD-deposited (Bi, Sb)Te_3 films [94] and electric current–stressed Bi_2(Se, Te)$_3$ films [101].

Although the superior *ZT* in Bi_2Te_3/Sb_2Te_3 SLs was reported some 15 years ago, reproducing such Bi_2Te_3/Sb_2Te_3 SL structure was a difficult challenge that has not yet fully succeeded regardless of using NA methods or MOCVD [79,99,100,108]. A cosputtering-assisted NA technique was used to grow *p*-type Sb_2Te_3/(Bi, Sb)$_2Te_3$ and *n*-type Bi_2Te_3/(Bi, Sb)$_2Te_3$ *soft* SLs [99,100]. The film structure closest to Bi_2Te_3/Sb_2Te_3 SLs was recently reported by Hansen et al. [81] utilizing the MBE technique. These authors established that it was possible to deposit Bi_2Te_3/Sb_2Te_3 SLs on BaF_2 (111) substrate at a critical temperature, where

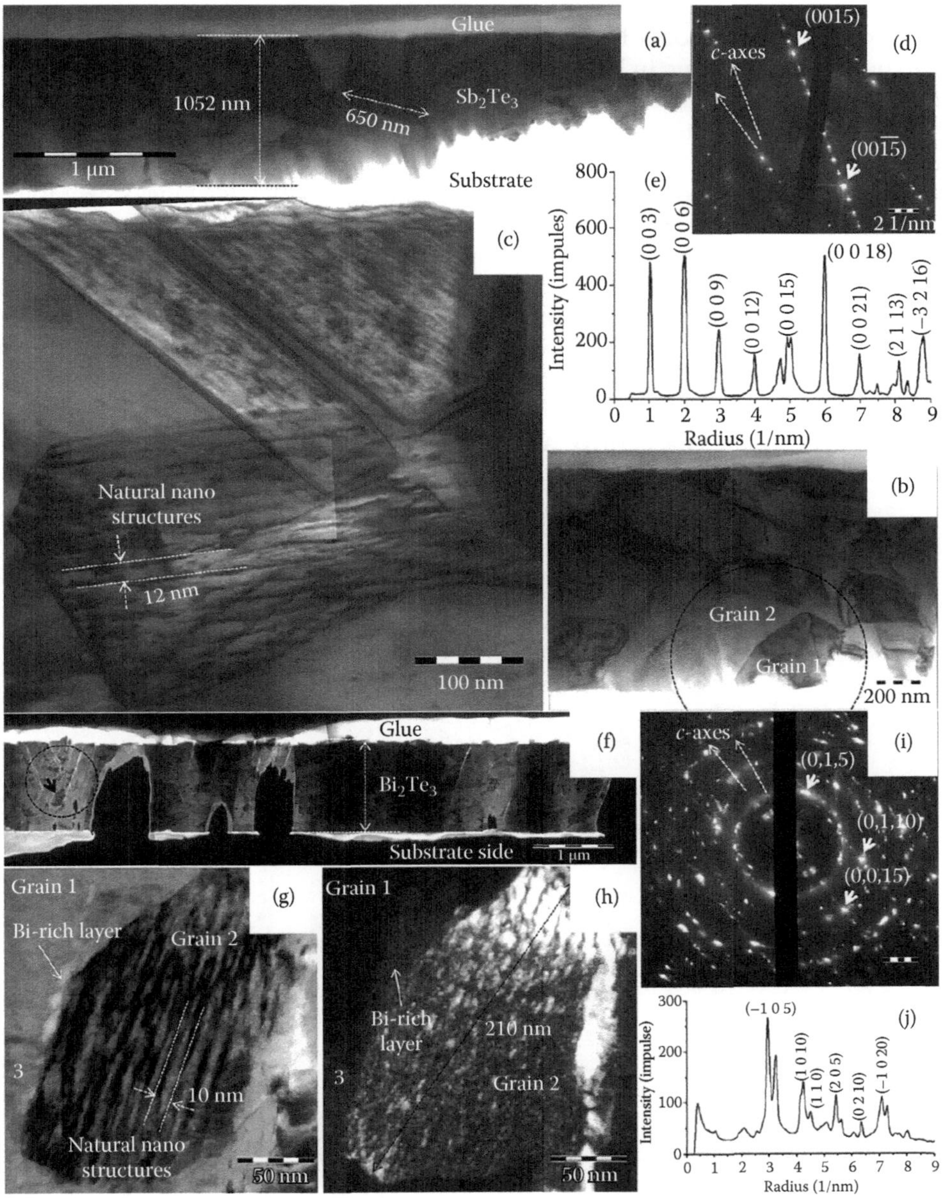

FIGURE 3.5

Transmission electron microscopy cross-sectional results obtained on the annealed (a–e) Sb_2Te_3 and (f–j) Bi_2Te_3 thin film deposited on a SiO_2 substrate. In images (a–d, f–i), the growth direction points toward the top of the images. (a) Overview of bright-field image. (b) Magnified bright-field image. (c) High-magnification bright-field image of grain boundaries. (d) Selected area electron diffraction (SAED) pattern of the encircled area in (b). (e) Normalized radial intensity profile obtained from the diffraction pattern. (f) Overview of bright-field image. (g) High-magnification bright-field image and (h) corresponding dark-field image of a grain. (i) SAED pattern obtained on the encircled area in (f). (j) Normalized radial intensity profile obtained from the diffraction pattern. (Reprinted from *J. Alloy. Compd.*, 521, Peranio, N., Winkler, M., Bessas, D. et al., Room-temperature MBE deposition, thermoelectric properties, and advanced structural characterization of binary Bi_2Te_3 and Sb_2Te_3 thin films, 163–173. Copyright 2012, with permission from Elsevier.)

the thermal energy is sufficient for the formation of SLs from evaporated elements but not high enough to trigger global interdiffusion. The optimized substrate temperature was found to be 350°C. As shown in Figure 3.6, apart from the ordered 1 nm Bi_2Te_3/5 nm Sb_2Te_3, the high-angle annular dark-field (HAADF) scanning transmission electron microscopy (STEM) analysis also identified a strongly bowed SL structure. The subsequent annealing treatments both *ex situ* and *in situ* in the transmission electron microscopy confirmed the strong fading of SL structures above 200°C. Thus, till now, SLs with ideal stacking structure do not seem to be experimentally accessible.

While 10 years ago, the growth of thin films of Bi_2Se_3, Bi_2Te_3, and Sb_2Te_3 was a specialized and rarely pursued research activity, the unprecedented boom in the study of TIs and the fact that the tetradymite-type thin-film structures are perhaps the best realization of the topological medium have resulted in widespread use of MBE growth of such thin films. As a tetradymite-type TI, Bi_2Se_3 stands out due to its large electronic gap. Additionally, its single surface Dirac cone is well defined in the band gap, far away from any interference with bulk states. These competitive advantages have drawn a tremendous attention to

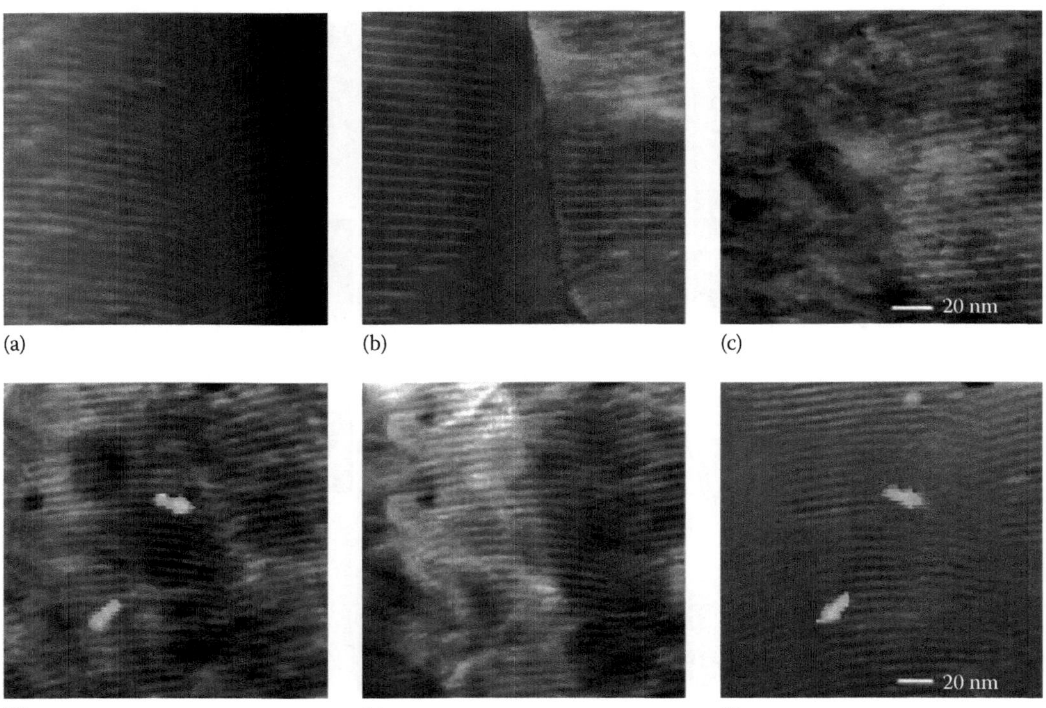

(a) (b) (c) (d) (e) (f)

FIGURE 3.6
High-angle annular dark-field (HAADF) scanning transmission electron microscopy (STEM) micrographs (the same scales) for *ex situ* heating experiments. (a) As-deposited state where strongly bowed superlattice (SL) structures are frequently observed. (b) Sample heated for 1 h at 250°C. The absence of strongly bowed SL structures is evident. (c) Further heat treatment for 1 h at 300°C led to a strong fading of the SL structure. Most of the sample showed no SL structure at all. HAADF STEM micrographs (the same scales) for *in situ* heating experiments at (d) 215°C, (e) 300°C, and (f) 360°C, respectively. Arrows mark positions with strongly bowed SL in as-deposited state. (Reprinted with permission from Hansen, A.-L., Dankwort, T., Winkler, M. et al., Synthesis and thermal instability of high-quality Bi_2Te_3/Sb_2Te_3 superlattice thin film thermoelectric, *Chem. Mater.*, 26, 6518. Copyright 2014 American Chemical Society.)

the growth of Bi_2Se_3 thin films and the topological research has much contributed to the improved quality of such films. To integrate TI/TE with the prevailing Si-based technology, the growth on Si-based substrates was clearly highly desirable, and the readily available and inexpensive Si wafer was one of the first substrates quickly tested for the growth by various groups of researchers [117–119]. An uninitiated reader might find it interesting to learn how one goes about establishing a set of optimized MBE parameters through trial and error.

In order to grow Bi_2Se_3 thin films on Si (111) substrates, earlier attempts required predepositing a monolayer (ML) of Bi (to form a β- $\sqrt{3} \times \sqrt{3}$ Bi surface) and/or an amorphous layer of Bi_2Se_3 at the interface [117,118]. Such intermediate layers were intended to recover the smooth Si (111) 1 × 1 surface from the somewhat rough 7 × 7 surface and improve the film crystallinity. More recently, Bansal et al. demonstrated that these interfacial layers can be dispensed of after all [119]. It turned out that the substrate temperature was the key parameter of producing high-quality Bi_2Se_3 films. To obtain an optimal surface layer that minimizes the formation of defects and secondary phases in the first place, the substrate was treated with an *ex situ* ultraviolet ozone-cleaning procedure and then heated briefly up to ~900°C in the UHV chamber. The undesired Si (111) 7 × 7 surface (Figure 3.7a)

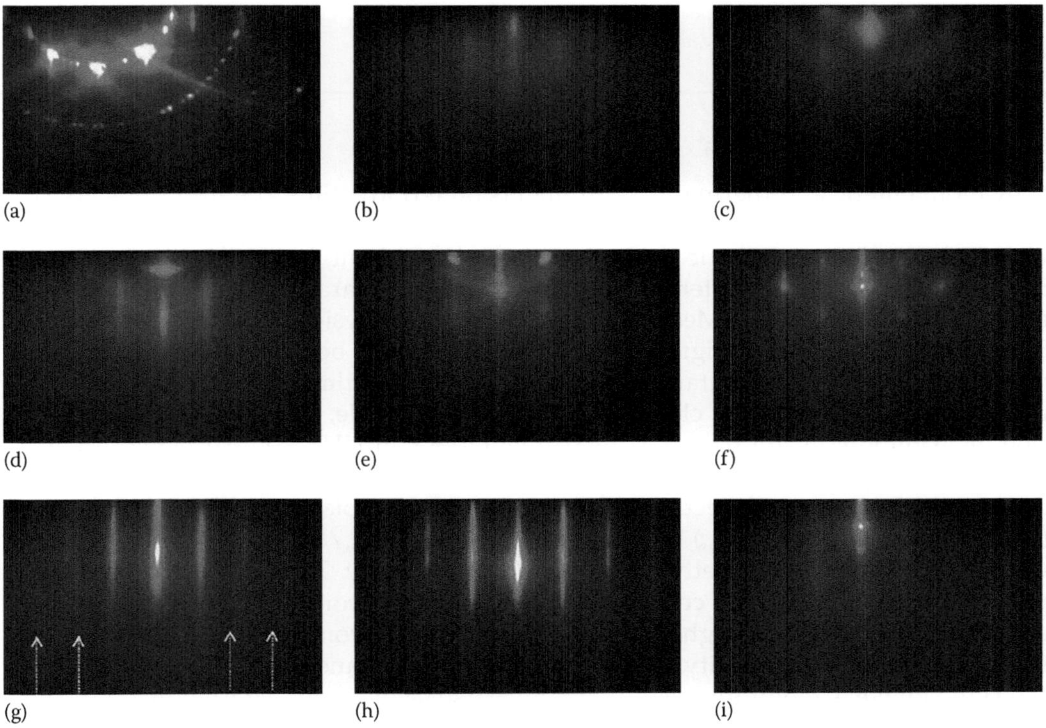

FIGURE 3.7
Reflection high-energy electron diffraction patterns of Bi_2Se_3 on Si (111) under different growth temperatures. (a) Reconstructed Si (111) 7 × 7 surface after cleaning treatment. (b) Short-term (2.5 min) and (c) long-term (10 min) low-temperature growth at 110°C. Short-term (~2.5 min) high-temperature growth at (d) 190°C, (e) 250°C, and (f) 350°C, respectively. Two-temperature growth including an initial 2–3 quintuple layer deposition at 110°C and higher temperature growth at (g) 190°C, (h) 230°C, and (i) 300°C, respectively. (Reprinted from *Thin Solid Films*, 520, Bansal, N., Kim, Y.S., Edrey, E. et al., Epitaxial growth of topological insulator Bi_2Se_3 film on Si(111) with atomically sharp interface, 224–229. Copyright 2011, with permission from Elsevier.)

was unreconstructed upon exposure to Se flux. A ML of Se was formed at ~100°C, while lower (higher) substrate temperature resulted in amorphous Se accumulation ($SiSe_2$ layers). The Se ML passivated the Si surface as well as provided a base for the stacking of the Se-terminated QL of Bi_2Se_3. Attempts of single-step growth at a fixed temperature were not satisfactory. The growth at 110°C resulted in weak streaky patterns (Figure 3.7b) suggesting that the temperature was too low to promote crystallization, while further growth at this temperature led to the formation of polycrystalline ring patterns (Figure 3.7c). Sharper streaks were obtained (Figure 3.7d) when directly grown at a higher temperature (190°C). However, the associated ringlike patterns indicated a rather disordered crystallographic orientation within the film. Further increasing the substrate temperature resulted in the formation of islands/clusters at 250°C (Figure 3.7e) or even secondary phases at 350°C (Figure 3.7f). However, when combining a low-temperature (110°C) growth of the initial 2–3 QLs (to prevent reaction with silicon substrate) and a high-temperature (230°C) growth for subsequent layers of the film (to improve crystallinity), a significantly higher quality was achieved, as demonstrated in the very sharp streaky pattern in Figure 3.7h. As illustrated in Figure 3.7g and i, there was an optimized temperature window for the high-temperature stage of the growth, as too low (high) temperature led to the formation of undesired twin structure (secondary phases).

3.4 Transport Properties

Galvanomagnetic and thermal measurements on tetradymite-type thin films are typically performed using dedicated *ex situ* experimental setups, where the control of temperature, electromagnetic fields, and thermal gradient and laser signals are readily implemented. These systems include the versatile characterization platforms such as the Magnetic Property Measurement System and Physical Property Measurement System from Quantum Design. However, caution should be used in designing experimental protocols so that contamination and oxidation of films are eliminated or at least minimized. Owing to their clear advantages, for example, in overcoming contamination or oxidation, *in situ* transport instrumentations in UHV-compatible cryostats are currently being actively pursued [110,120]. Due to the vdW epitaxial nature, the most commonly characterized electronic properties are the in-plane ($\perp c$) electrical resistivity (ρ_{xx}) and Hall resistivity (ρ_{yx}) or Hall coefficient ($R_H = \rho_{yx}/B$), which can be measured using the van der Pauw method or a Hall bar geometry. The former case requires a perfect square sample with contacts placed on the four corners. In the latter case, the electric current is fed through the outmost longitudinal contacts on a rectangular profile sample, and the resistivity voltage is probed via the inner longitudinal leads, while the Hall voltage is monitored using the transverse leads under external magnetic field. A low frequency ac current is recommended to minimize the Peltier heating effects. With a longitudinal temperature gradient established via heater/sink situated at either end of the Hall bar device, the in-plane Seebeck coefficient (α) can be obtained by probing the longitudinal voltage difference under steady-state conditions. As for measurements across the plane ($||c$), the electronic properties usually require photolithography procedures to provide the necessary length/cross section ratio. While in principle one might be able to measure the in-plane thermal conductivity by the steady-state method, stringent criteria between film and the substrate have to be satisfied first. This makes the steady-state method practically irrelevant for the thermal conductivity measurement of thin films, unless, in the rare cases, the film can be suspended or when the film is very thick and its thermal conductivity is significantly higher than that of the substrate. The across-plane thermal conductivity can be measured by the newly developed time-domain thermoreflectance method [121,122] and the transient 3ω method [123–126]. A modification of the 3ω technique, together with 2D conduction modeling, might be adopted for the estimation of in-plane thermal conductivity. More detailed reviews on the transport measurement techniques were recently provided by Wang et al. [78] and Völklein et al. [127].

The strategies utilized so far to enhance the TE properties of tetradymite-type films have concentrated on (1) enhancing the carrier mobility via fabricating films with preferred orientation, (2) optimizing the carrier density via engineering the defect formation, and (3) reducing the lattice thermal conductivity via strengthening the phonon scattering. We herein focus on the recent progress in tuning the TE transport properties of tetradymite-type films. Specifically, the discussion of in-plane electronic properties and across-plane thermal conductivity will be emphasized.

Using the two-step MBE growth method, Bansal et al. [31] have prepared a series of single crystalline Bi_2Se_3 MBE thin films on sapphire (0001) substrates, with thicknesses ranging from 2 to 256 QLs. The superior quality of these Bi_2Se_3 thin films has allowed the authors to study the thickness-independent surface transport channels. The temperature-dependent sheet resistances $R_\square$ are shown in Figure 3.8a. Overall, these Bi_2Se_3 MBE thin films all exhibit typical metallic temperature dependence—a positive temperature coefficient of resistivity at high temperature due to electron–phonon interaction, which flattens out at the lowest temperature. Upon increasing the thickness, $R_\square$ drops rather rapidly by adding the first several QLs and then essentially saturates at and above 8 QLs. The thickness dependence is more clearly visualized in Figure 3.8b, where the conductance G_{xx} at T = 1.5 K is collected for all the thicknesses. These stoichiometric films possess a nonlinear magnetic field dependence of the Hall resistance, a signature of multiband conduction. Through a two-band model, the analysis identified the existence of two surface channels: a channel composed of topological SSs with carrier concentration ~3.0×10^{13} cm^{-2} (constant down to 2 QL) and another one of the surface accumulation nature with a somewhat lower carrier density ~8.0×10^{12} cm^{-2} (constant only down to ~8 QL).

Furthermore, as shown in Figure 3.8c, the magnetoresistance develops a WAL-type of cusp at low magnetic fields, which can be readily analyzed using the Hikami, Larkin, and Nagaoka (HLN) quantum interference model [128],

$$\Delta G(B) = -A\frac{e^2}{h}\left[\psi\left(\frac{\hbar}{4eL^2B}+\frac{1}{2}\right)-\ln\left(\frac{\hbar}{4eL^2B}\right)\right], \tag{3.4}$$

where ψ is the digamma function, e is the elementary charge, $\hbar$ (= $h/2\pi$) is the Planck constant, and L is the phase coherence length.

Theoretically, the fitting parameter A is predicted to be $1/2\pi$ for each 2D conduction channel (i.e., $2\pi A$ is the effective number of conduction channels). Typical values for A lie between two limiting scenarios in a thin film: (1) $1/\pi$ if the top and bottom surfaces are completely decoupled from each other by an insulating bulk and (2) $1/2\pi$ in the case where the two surfaces are coupled by the bulk or the bulk of the film dominates the conduction.

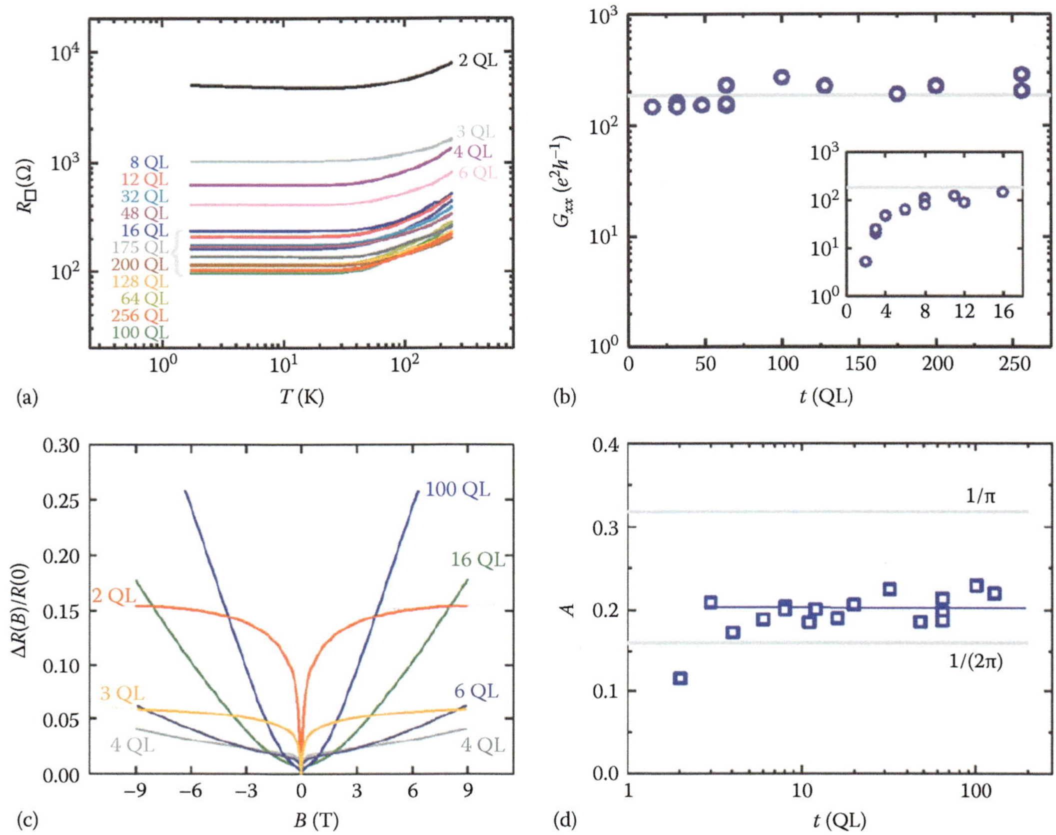

FIGURE 3.8
Transport properties of pure Bi_2Se_3 films with different thickness. (a) Temperature-dependent sheet resistance $R_\square$ at various film thickness. (b) Conductance at 1.5 K as a function of thickness. (c) The weak antilocalization effect at 1.5 K for selected thicknesses. (d) The fitting parameter A in the Hikami, Larkin, and Nagaoka quantum interference model indicating effective number of conduction channels. (Reprinted with permission from Bansal, N., Kim, Y.S., Brahlek, M., Edrey, E., and Oh, S., *Phys. Rev. Lett.*, 109, 116804, 2012. Copyright 2012 by the American Physical Society.)

The observed values of the fitting parameter A are shown in Figure 3.8b and, indeed, reside between the two limits. Nevertheless, the thickness independence of A, above 3QLs, is in agreement with the aforementioned behavior of surface conduction.

By carefully examining the WAL effects using the HLN model in Bi_2Se_3 thin films with different Cu doping and various thicknesses, see Figure 3.9a, Brahlek et al. have recently established a narrow composition window where completely decoupled surface conduction (with $A = 1/\pi$) is realized [129]. It appears that localized bulk states can be achieved in Bi_2Se_3 thin films upon fine tuning the Cu doping level (Figure 3.9b). With such Cu content, the nonlinearity of the field-dependent Hall resistance is greatly suppressed. The optimized Bi_2Se_3 thin films start to show decoupled SSs as the film thickness exceeds an intersurface coupling length (Figure 3.9c). This observation provides an important experimental route of decoupling SSs in tetradymite-type thin films via chemical doping and thickness control.

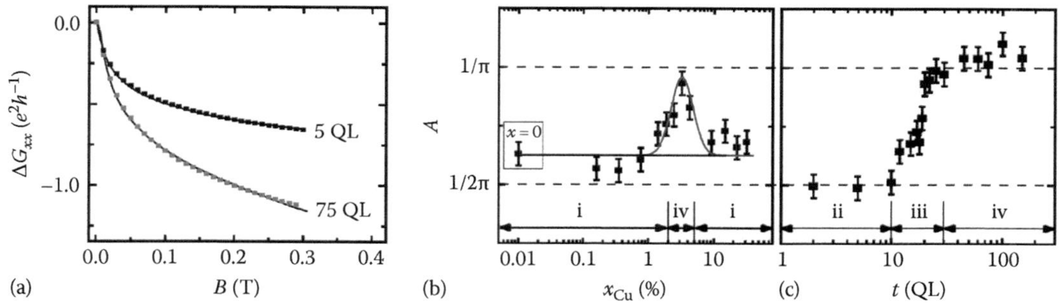

FIGURE 3.9
Weak antilocalization (WAL) effect in Cu-doped Bi_2Se_3 thin films. (a) Typical Hikami, Larkin, and Nagaoka fitting of the WAL-type cusp in conductance at low magnetic fields. Fitting parameter *A*, the effective number of 2D channels, as a function of (b) composition (x_{Cu}) for fixed thickness (20 QLs) and (c) thickness for fixed composition ($x_{Cu} \approx 2.5\%–4.0\%$). (Reprinted with permission from Brahlek, M., Koirala, N., Salehi, M., Bansal, N., and Oh, S., *Phys. Rev. Lett.*, 113, 026801, 2014. Copyright 2014 by the American Physical Society.)

Subjected to a thermal gradient at low temperatures, the charge carrier system in a single crystal may be *dragged* by out-of-equilibrium phonons, leading to significantly enhanced values of the thermopower, the phenomenon known as the phonon-drag effect. Using Bi_2Te_3 as an example, Wang et al. have recently assessed the influence of phonons from the substrates on charge carriers in thin films [130]. As shown in Figure 3.10b, the phonon thermal conductivity of a crystal develops a dielectric peak at low temperatures. The conductivity peak position is related to the Debye temperature of the corresponding substrate/crystal—165, 282, and 980 K for Bi_2Te_3, BaF_2, and Al_2O_3, respectively. It appears that the phonons governing the lattice thermal conductivity are the same as those dominating the phonon-drag thermopower. The peak of the phonon drag (in Figure 3.10a) appears at virtually the same temperature as the peak in the phonon thermal conductivity. As shown in Figure 3.10c, reducing the thickness of the sapphire substrate enhances the grain boundary scattering at the lowest temperatures, which effectively increases the peak temperature of the phonon conductivity (solid lines). This subtle increase also seems to push the phonon-drag peak position to slightly higher temperatures. Figure 3.10d suggests that the influence from substrate phonons diminishes quickly upon increasing the film thickness, and eventually, the bulk value should be restored. These experimental features demonstrate the researchers' capability to tune the position and magnitude of the phonon-drag peak over a large temperature range by simply depositing tetradymite-type thin films on substrates with different Debye temperatures.

Recently, Zhang et al. have grown a series of Bi_2Te_3–Sb_2Te_3 solid solution ultrathin films with 5 QL thickness [131]. As shown in Figure 3.11a, with high Sb content, these $(Bi_{1-x}Sb_x)_2Te_3$ films exhibit somewhat nonmetallic sheet resistance profile. Upon increasing the Sb content x, an n-to-p transition occurs in between $x = 0.90$ and 0.95, according to the Hall coefficient measurement (Figure 3.11b). As shown in Figure 3.11c, due to the bipolar excitation across the bulk band gap, the Seebeck coefficient changes sign near ambient temperature, even for relatively low Sb substitution $x = 0.70–0.80$. In particular, for $x = 0.90$, over the whole temperature range, the Hall coefficient (negative) and the Seebeck coefficient (positive) take different signs, a signature of multiband conduction. Considering that these films are extremely thin, SSs might, indeed, be playing a role in the transport. The possibility and feasibility of developing TE devices utilizing TI concepts have been discussed [132–134].

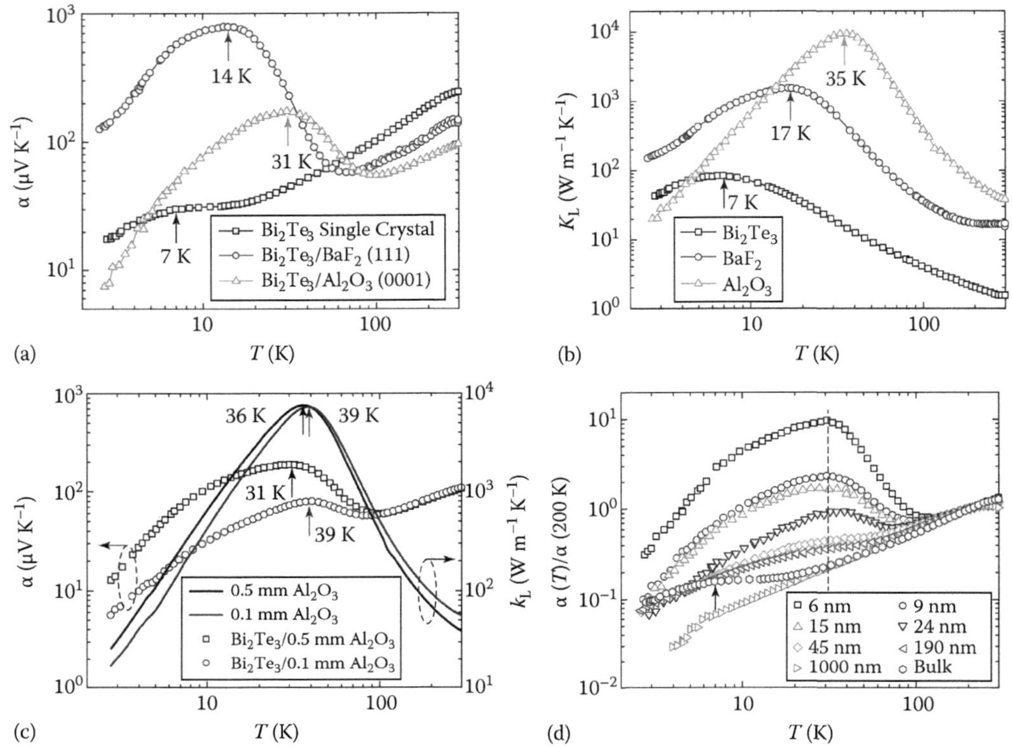

FIGURE 3.10
(a) Temperature-dependent thermopower for Bi_2Te_3 single crystal, 9 nm Bi_2Te_3 films on BaF_2 (111), and sapphire (0001) substrates. (b) The lattice thermal conductivity of Bi_2Te_3 single crystal, bare BaF_2 and sapphire substrates. (c) Thermopower of two 9 nm films grown on sapphire substrates with different thicknesses. The lattice thermal conductivities (right axis) of two substrates are also shown. (d) Normalized thermopower (to the value at 200 K) for Bi_2Te_3 single crystal and films with different thickness on sapphire substrate. (Reprinted with permission from Wang, G., Endicott, L., Chi, H., Lošťák, P., and Uher, C., *Phys. Rev. Lett.*, 111, 046803, 2013. Copyright 2013 by the American Physical Society.)

The anisotropic nature of tetradymite-type structure dictates that *c*-oriented films have superior TE performance [135,136]. The room temperature electronic transport properties of binary Bi_2Te_3 and Sb_2Te_3 films grown using various techniques are summarized in Table 3.4. Bi_2Te_3 and Sb_2Te_3 films with stronger (00*l*) orientation indeed possess a much larger carrier mobility. The carrier mobility of the (015)-oriented Bi_2Te_3 and Sb_2Te_3 films are only about 20 $cm^2\ V^{-1}\ s^{-1}$, which is at least twice smaller than that of (00*l*)-orientated films [103,104]. For the Bi_2Te_3 with high electron density of 6.5–12 × 10^{19} cm^{-3}, it is evident that the (00*l*)-orientated Bi_2Te_3 films possess much higher Seebeck coefficient than the (015)-oriented films, which is likely due to the anisotropy in the Seebeck coefficient [56,103]. Thus, the (00*l*)-oriented Bi_2Te_3 and Sb_2Te_3 films possess an excellent power factor of 4.0 and 3.1 mW $m^{-1}\ K^{-2}$ due to their preferred orientation [56,104]. A similar trend has also been observed in Bi_2Se_3 or ternary tetradymite-type films [66,75,82,85]. As compared to the bulk single crystal counterparts, the carrier mobility obtained in films is still low, especially for the *n*-type Bi_2Te_3 films, and the carrier density is not at an optimal value. Therefore, the improvement in the film orientation and the optimization of carrier density

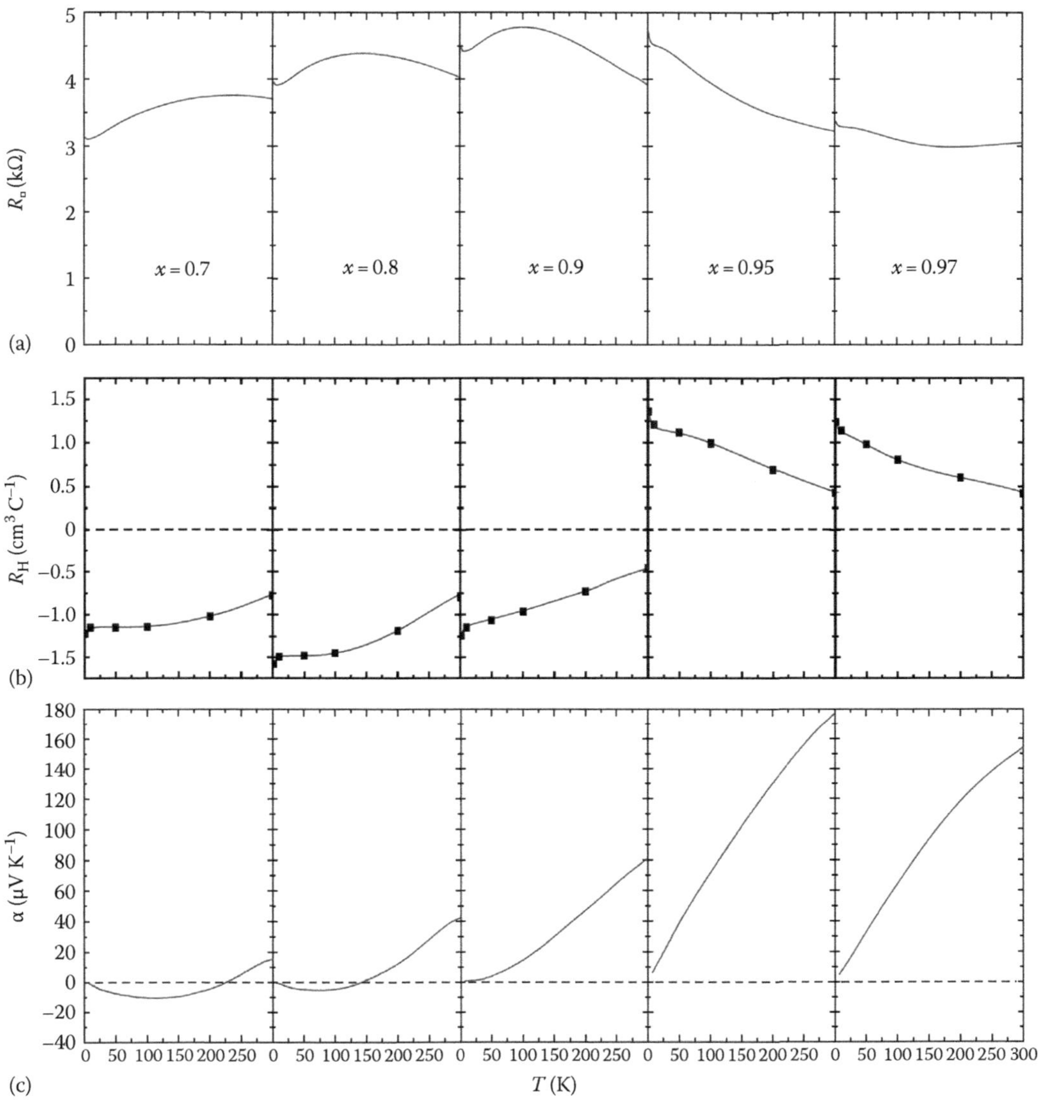

FIGURE 3.11
The (a) sheet resistance, (b) Hall coefficient, and (c) Seebeck coefficient of 5 quintuple layer $(Bi_{1-x}Sb_x)_2Te_3$ films. (Reprinted with permission from Zhang, J., Feng, X., Xu, Y. et al., *Phys. Rev. B*, 91, 075431, 2015. Copyright 2015 by the American Physical Society.)

are still needed to achieve a further improved power factor in tetradymite-type films. In order to strengthen phonon scattering and reduce lattice thermal conductivity, reducing grain size and increasing boundary density are two widely accepted experimental routes [8,13,137]. In tetradymite-type films, decreasing the grain size and forming SL structures are considered to be the most effective approach to reduce the lattice thermal conductivity, while the effect from structure modulation and film thickness is also included in the discussion of grain size effect.

TABLE 3.4

Room Temperature In-Plane Electronic Transport Properties of Binary Bi_2Te_3 and Sb_2Te_3 Films

Methods	Reference	Materials	Orientation	*n*	μ	σ	α	*PF*
Nanoalloying	[85]	Bi_2Te_3	(00*l*) + (1011)	2.7	62	2.7	−182	0.9
		Bi_2Te_3	(00*l*) + (1011)	2.7	79	3.4	−153	0.8
		Sb_2Te_3	(00*l*) + (1011)	2.8	430	19.3	125	3.0
		Sb_2Te_3	(00*l*) + (1011)	2.6	408	17	130	2.9
Sputtering	[103]	Bi_2Te_3	(015)	12	22	4.3	−122	0.6
			(00*l*)	10	52	8.4	−200	3.4
Sputtering	[104]	Sb_2Te_3	(015)	9.1	19	2.8	175	0.9
			(00*l*)	4.2	100	6.7	214	3.1
Thermal evaporation	[54]	Bi_2Te_3	(00*l*) + (015)	3.1	107	5.3	−208	2.3
		Sb_2Te_3	(00*l*)	1.4	348	7.8	160	2.0
Thermal evaporation	[56]	Bi_2Te_3	(00*l*) + (110)	6.5	74	7.7	−228	4.0
		Sb_2Te_3	(00*l*)	3.4	176	9.6	171	2.8

Carrier density, *n* (10^{19} cm^{-3}); mobility, μ (cm^2 V^{-1} s^{-1}); electrical conductivity, σ (10^4 S m^{-1}); Seebeck coefficient, α (μV K^{-1}); power factor, $PF \equiv \alpha^2\sigma$ (10^{-3} W m^{-1} K^{-2}).

It is well accepted that the transport properties of tetradymite-type films are dominated by native point defects such as antisite defects and vacancies, due to their low formation energy. The carrier density of tetradymite-type films grown under out-of-equilibrium conditions may vary in the range of 10^{18}–10^{21} cm^{-3} depending on the material and growth parameters. Varying the ratio of B^{VI}/A^{V} species during the deposition or adding a postannealing process can modify the carrier density of tetradymite-type films. In single crystals, increasing the B^{VI}/A^{V} ratio increases the electron density of Bi_2Te_3 and decreases the hole density of Sb_2Te_3 due to the increasing amount of Te [2]. However, this is not the case in the thin films [73,75,76]. For example, the electron density of Bi_2Te_3 film decreases by one order of magnitude when the Te/Bi ratio increases from 3 to 9 [76], while the hole density of Sb_2Te_3 film increases by 25% when Te/Sb ratio increases from 5 to 7 [73]. These results suggest further efforts are still needed to clarify the different effects of the B^{VI}/A^{V} ratio in bulk crystals and thin films. Postannealing is commonly used to promote the crystallinity and to tune the carrier density of as-deposited films [85]. Postannealing at elevated temperatures, commonly not higher than 300°C, tends to decrease the electron density of *n*-type films of Bi_2Te_3 [82,85] and $Bi_2(Te, Se)_3$ [66]. Annealing in a Te-rich environment follows the same trend [106]. In contrast, annealing in *p*-type films is mostly reported to increase the hole density, as observed in $(Bi, Sb)_2Te_3$ [82,85] films and $Sb_2Te_3/(Bi, Sb)_2Te_3$ SLs [99]. This is usually explained by the Te reevaporation and the creation of $(Bi, Sb)_{Te}$ antisite defects, which generate hole carriers in films.

The thermal conductivity data of selected tetradymite-type films are collected in Table 3.5. Upon decreasing the grain size or film thickness of tetradymite-type films, the across-plane thermal conductivity can be largely reduced. As reported by Chiritescu et al. [138] and Takashiri et al. [139], the across-plane thermal conductivity of the *c*-oriented Bi_2Te_3 film with thickness of 30–150 nm was in the range of 0.55–0.75 W m^{-1} K^{-1}, significantly lower than that of bulk single crystalline Bi_2Te_3. A similar reduction was also observed in 50 nm thick $BiSbTe_3$ films [138]. In contrast, $(Bi_{1-x}Sb_x)_2Te_3$ films, with $x = 0.20$, 0.25, and 0.35 and a grain size of several hundred nanometer, possessed a rather high κ value of ~1.0 W m^{-1} K^{-1}, close to that of their bulk counterparts (0.70–0.88 W m^{-1} K^{-1}) [2,94].

TABLE 3.5

The Film Thickness t (nm), Grain Size or Superlattice Period Δ (nm), and Room Temperature Thermal Conductivity κ (W m^{-1} K^{-1}) or Lattice Thermal Conductivity κ_L (W m^{-1} K^{-1})

Method	Reference	Materials	t	Δ	κ or κ_L	
MOCVD	[19]	Bi_2Te_3/Sb_2Te_3 SL	400–700	5	κ_L, 0.22	Across-plane
			400–700	6	κ_L, 0.25	Across-plane
MBE	[80]	Bi_2Te_3	1000	—	κ, 1.87; κ_L, 1.48	In-plane
		Bi_2Te_3/Sb_2Te_3 SL	1000	10	κ, 1.25; κ_L, 1.01	In-plane
CE	[138]	Bi_2Te_3	30–100	—	κ, 0.55–0.75	Across-plane
		$BiSbTe_3$	50	—	κ, 0.36	Across-plane
FE	[139]	Bi_2Te_3	200	150	κ, 0.6	Across-plane
NA	[100]	$Bi_2Te_3/(Bi_{0.9}Sb_{0.1})_2Te_3$	1500	50	κ_L, 0.53	Across-plane
			1500	50	κ_L, 0.79	Across-plane
NA	[99]	$Sb_2Te_3/(Bi_{0.2}Sb_{0.8})_2Te_3$	1500	50	κ, 0.45	Across-plane
PLD	[94]	$Bi_{0.4}Sb_{1.6}Te_3$	490	—	κ, 1.10	Across-plane
		$Bi_{0.5}Sb_{1.5}Te_3$	690	—	κ, 0.97	Across-plane
		$Bi_{0.7}Sb_{1.3}Te_3$	1250	—	κ, 1.00	Across-plane
MBE	[108]	Bi_2Te_3/Sb_2Te_3 SL	996	6	κ, 0.60; κ_L, 0.28	Across-plane

The thermal conductivity of SLs was obviously lower than that of the nanostructured films, as observed in Bi_2Te_3/Sb_2Te_3 SLs with SL period of 4–6 nm and low lattice thermal conductivity of 0.22–0.25 W m^{-1} K^{-1} [140]. Winkler et al. succeeded in fabricating the 1 nm Bi_2Te_3/5nm Sb_2Te_3 SL film with good crystallinity utilizing a modified MBE process [108], which, probably for the first time, came very close to reproducing the superb ZT values of the Bi_2Te_3/Sb_2Te_3 SL structure [19]. The lattice thermal conductivity of this Bi_2Te_3/Sb_2Te_3 SL film was estimated to be 0.28 W m^{-1} K^{-1} assuming a Lorenz number $L = 2.45 \times 10^{-8}$ V^2 K^{-2} and conductivity anisotropy ratio of 1.12. This was very close to the previously reported value of 0.25 W m^{-1} K^{-1} [19]. p-Type $Sb_2Te_3/(Bi_{0.2}Sb_{0.8})_2Te_3$ and n-type Bi_2Te_3/$(Bi_{0.9}Sb_{0.1})_2Te_3$ *soft* SL structured films prepared by the NA technique were also examined by Winkler et al. [99,100]. Their thermal conductivity was ~0.5 W m^{-1} K^{-1}, which was similar to the values reported in nanostructured films. Therefore, in order to improve TE performance, the tetradymite-type films should combine SL structure (effectively reduced lattice thermal conductivity) with excellent crystallinity (good electronic performance). In addition to the reports in across-plane thermal conductivity, there were also several reports on the in-plane thermal conductivity of tetradymite-type films by Peranio et al. [80] and Tan et al. [57], which considered Bi_2Te_3, Bi_2Te_3/Sb_2Te_3 SL and Sb_2Te_3 films. These results showed that the in-plane thermal conductivity was largely reduced compared to the bulk materials, as a result of the additional phonon scattering from reduced grain size.

In spite of this progress, several issues still remain unclear. First, in contrast to the very high Seebeck coefficient $\alpha \sim 200$–270 μV K^{-1} and mobility $\mu \sim 600$ cm^2 V^{-1} s^{-1} in MOCVD-grown SLs, the MBE-grown SLs possess a much smaller value of $\alpha \sim 120$ μV K^{-1} and $\mu \sim 60$ cm^2 V^{-1} s^{-1} [19,108]. Likely, the carrier density in the SL was not optimized, or there was a significant effect from minority carriers at room temperature. Alternatively, this might be caused by the compensation effect in the electronic transport from n-type Bi_2Te_3 component layer and p-type Sb_2Te_3 layer. Second, the interdiffusion of component layers in SLs during deposition is still controversial [81,141]. Third, the reported anisotropy of

SL structures remains as a surprise in that the across-plane electrical conductivity is very close to, or even larger than, the in-plane values [19,108]. Therefore, systematic investigations of carrier density optimization, controllable growth of SLs and their thermal stability, and TE property characterization in both the in-plane and across-plane orientations, as well as full-scale microstructure characterization, are still needed. These issues are crucial for future Peltier cooling applications of tetradymite-type films.

3.5 Conclusion

Tetradymite-type thin films and SLs are gaining considerable research interest worldwide as promising TEs. We have summarized their material structure and fabrication techniques, along with their transport properties. While lacking behind the bulk TE development, tetradymite-type thin-film TEs hold promise through improving the film quality and developing reliable ways of reducing their thermal conductivity. No doubt, the most recent enthusiasm in tetradymite-based TI thin-film structures will be an added impetus for further study and development of tetradymite-type thin films and SLs.

Acknowledgment

The authors gratefully acknowledge the financial support from the Center for Solar and Thermal Energy Conversion, an energy frontier research center funded by the U.S. Department of Energy, Office of Science, Basic Energy Sciences under Award No. DE-SC0000957.

References

1. Goldsmid, H. J. and R. W. Douglas. 1954. The use of semiconductors in thermoelectric refrigeration. *Br. J. Appl. Phys.* 5: 386.
2. Rowe, D. M., ed. 1995. *CRC Handbook of Thermoelectrics*. Boca Raton, FL: CRC Press.
3. Rowe, D. M., ed. 2006. *Thermoelectrics Handbook: Macro to Nano*. Boca Raton, FL: CRC Press.
4. Yang, J. H., H. L. Yip, and A. K. Y. Jen. 2013. Rational design of advanced thermoelectric materials. *Adv. Energy Mater.* 3: 549–565.
5. Zhang, Q. H., X. Y. Huang, S. Q. Bai, X. Shi, C. Uher, and L. D. Chen. 2016. Thermoelectric devices for power generation: Recent progress and future challenges. *Adv. Eng. Mater.* 18: 194–213.
6. Borup, K. A., J. de Boor, H. Wang et al. 2015. Measuring thermoelectric transport properties of materials. *Energy Environ. Sci.* 8: 423–435.
7. Poudel, B., Q. Hao, Y. Ma et al. 2008. High-thermoelectric performance of nanostructured bismuth antimony telluride bulk alloys. *Science* 320: 634–638.
8. Xie, W., J. He, H. J. Kang et al. 2010. Identifying the specific nanostructures responsible for the high thermoelectric performance of $(Bi, Sb)_2Te_3$ nanocomposites. *Nano Lett.* 10: 3283–3289.

9. Kim, S. I., K. H. Lee, H. A. Mun et al. 2015. Dense dislocation arrays embedded in grain boundaries for high-performance bulk thermoelectrics. *Science* 348: 109–114.
10. Hicks, L. D. and M. S. Dresselhaus. 1993. Effect of quantum-well structures on the thermoelectric figure of merit. *Phys. Rev. B* 47: 12727–12731.
11. Hicks, L. D. and M. S. Dresselhaus. 1993. Thermoelectric figure of merit of a one-dimensional conductor. *Phys. Rev. B* 47: 16631–16634.
12. Dresselhaus, M. S., G. Chen, M. Y. Tang et al. 2007. New directions for low-dimensional thermoelectric materials. *Adv. Mater.* 19: 1043–1053.
13. Biswas, K., J. He, I. D. Blum et al. 2012. High-performance bulk thermoelectrics with all-scale hierarchical architectures. *Nature* 489: 414–418.
14. Heremans, J. P., V. Jovovic, E. S. Toberer et al. 2008. Enhancement of thermoelectric efficiency in PbTe by distortion of the electronic density of states. *Science* 321: 554–557.
15. Heremans, J. P., Ch. M. Thrush, and D. T. Morelli. 2005. Thermopower enhancement in PbTe with Pb precipitates. *J. Appl. Phys.* 98: 063703.
16. Zebarjadi, M., G. Joshi, G. Zhu et al. 2011. Power factor enhancement by modulation doping in bulk nanocomposites. *Nano Lett.* 11: 2225–2230.
17. Faleev, S. V. and F. Léonard. 2008. Theory of enhancement of thermoelectric properties of materials with nanoinclusions. *Phys. Rev. B* 77: 214304.
18. Zhao, L.-D., G. Tan, S. Hao et al. 2016. Ultrahigh power factor and thermoelectric performance in hole-doped single-crystal SnSe. *Science* 351: 141–144.
19. Venkatasubramanian, R., E. Siivola, T. Colpitts, and B. O'Quinn. 2001. Thin-film thermoelectric devices with high room-temperature figures of merit. *Nature* 413: 597–602.
20. Peranio, N., M. Winkler, M. Dürrschnabel, J. König, and O. Eibl. 2013. Assessing antisite defect and impurity concentrations in Bi_2Te_3 based thin films by high-accuracy chemical analysis. *Adv. Funct. Mater.* 23: 4969–4976.
21. Zhang, H., C.-X. Liu, X.-L. Qi, X. Dai, Z. Fang, and S.-C. Zhang. 2009. Topological insulators in Bi_2Se_3, Bi_2Te_3 and Sb_2Te_3 with a single Dirac cone on the surface. *Nat. Phys.* 5: 438–442.
22. Chen, Y. L., J. G. Analytis, J.-H. Chu et al. 2009. Experimental realization of a three-dimensional topological insulator, Bi_2Te_3. *Science* 325: 178–181.
23. Xia, Y., D. Qian, D. Hsieh et al. 2009. Observation of a large-gap topological-insulator class with a single Dirac cone on the surface. *Nat. Phys.* 5: 398–402.
24. Hsieh, D., Y. Xia, D. Qian et al. 2009. A tunable topological insulator in the spin helical Dirac transport regime. *Nature* 460: 1101–1105.
25. Moore, J. E. 2010. The birth of topological insulators. *Nature* 464: 194–198.
26. Hasan, M. Z. and C. L. Kane. 2010. Colloquium: Topological insulators. *Rev. Mod. Phys.* 82: 3045–3067.
27. Qi, X.-L. and S.-C. Zhang. 2011. Topological insulators and superconductors. *Rev. Mod. Phys.* 83: 1057–1110.
28. Qu, D.-X., Y. S. Hor, J. Xiong, R. J. Cava, and N. P. Ong. 2010. Quantum oscillations and Hall anomaly of surface states in the topological insulator Bi_2Te_3. *Science* 329: 821–824.
29. Qu, D.-X., Y. S. Hor, and R. J. Cava. 2012. Quantum oscillations in magnetothermopower measurements of the topological insulator Bi_2Te_3. *Phys. Rev. Lett.* 109: 246602.
30. Shekhar, Ch., C. E. ViolBarbosa, B. Yan et al. 2014. Evidence of surface transport and weak antilocalization in a single crystal of the $BiTe_2Se$ topological insulator. *Phys. Rev. B* 90: 165140.
31. Bansal, N., Y. S. Kim, M. Brahlek, E. Edrey, and S. Oh. 2012. Thickness-independent transport channels in topological insulator Bi_2Se_3 thin films. *Phys. Rev. Lett.* 109: 116804.
32. He, H.-T., G. Wang, T. Zhang et al. 2011. Impurity effect on weak antilocalization in the topological insulator Bi_2Te_3. *Phys. Rev. Lett.* 106: 166805.
33. Analytis, J. G., R. D. McDonald, S. C. Riggs, J.-H. Chu, G. S. Boebinger, and I. R. Fisher. 2010. Two-dimensional surface state in the quantum limit of a topological insulator. *Nat. Phys.* 6: 960–964.
34. Hahn, T., ed. 2005. *International Tables for Crystallography, Volume A: Space-Group Symmetry.* Dordrecht, the Netherlands: Springer.

35. Pérez, V. C., J. L. Tirado, K. Adouby, J. C. Jumas, A. Abba Touré, and G. Kra. 1999. X-ray diffraction and ^{119}Sn Mössbauer spectroscopy study of a new phase in the Bi_2Se_3–SnSe system: $SnBi_4Se_7$. *Inorg. Chem.* 38: 2131–2135.
36. Atuchin, V. V., T. A. Gavrilova, K. A. Kokh, N. V. Kuratieva, N. V. Pervukhina, and N. V. Surovtsev. 2012. Structural and vibrational properties of PVT grown Bi_2Te_3 microcrystals. *Solid State Commun.* 152: 1119–1122.
37. Kokh, K. A., V. V. Atuchin, T. A. Gavrilova, N. V. Kuratieva, N. V. Pervukhina, and N. V. Surovtsev. 2014. Microstructural and vibrational properties of PVT grown Sb_2Te_3 crystals. *Solid State Commun.* 177: 16–19.
38. Bland, J. A. and S. J. Basinski. 1961. The crystal structure of Bi_2Te_2Se. *Can. J. Phys.* 39: 1040–1043.
39. Bindi, L. and C. Cipriani. 2004. The crystal structure of skippenite, Bi_2Se_2Te, from the Kochkar deposit, Southern Urals, Russian Federation. *Can. Mineral.* 42: 835–840.
40. Ji, H., J. M. Allred, M. K. Fuccillo et al. 2012. $Bi_2Te_{1.6}S_{1.4}$: A topological insulator in the tetradymite family. *Phys. Rev. B* 85: 201103.
41. Ivanova, L. D. and Y. V. Granatkina. 1993. Thermoelectric properties of single crystals of the Bi_2Te_3–Sb_2Te_3–Bi_2Se_3 system. *Inorg. Mater.* 29: 182–186.
42. Lundegaard, L. F., E. Makovicky, T. Boffa-Ballaran, and T. Balic-Zunic. 2005. Crystal structure and cation lone electron pair activity of Bi_2S_3 between 0 and 10 GPa. *Phys. Chem. Miner.* 32: 578–584.
43. Tideswell, N. W., F. H. Kruse, and J. D. McCullough. 1957. The crystal structure of antimony selenide, Sb_2Se_3. *Acta Crystallogr.* 10: 99–102.
44. Lundegaard, L. F., R. Miletich, T. Balic-Zunic, and E. Makovicky. 2003. Equation of state and crystal structure of Sb_2S_3 between 0 and 10 GPa. *Phys. Chem. Miner.* 30: 463–468.
45. Cucka, P. and C. S. Barrett. 1962. The crystal structure of Bi and of solid solutions of Pb, Sn, Sb and Te in Bi. *Acta Crystallogr.* 15: 865–872.
46. Barrett, C. S., P. Cucka, and K. Haefner. 1963. The crystal structure of antimony at 4.2, 78 and 298°K. *Acta Crystallogr.* 16: 451–453.
47. Adenis, C., V. Langer, and O. Lindqvist. 1989. Reinvestigation of the structure of tellurium. *Acta Crystallogr. C* 45: 941–942.
48. Keller, R., W. B. Holzapfel, and H. Schulz. 1977. Effect of pressure on the atom positions in Se and Te. *Phys. Rev. B* 16: 4404–4412.
49. Rettig, S. J. and J. Trotter. 1987. Refinement of the structure of orthorhombic sulfur, α-S_8 *Acta Crystallogr. C* 43: 2260–2262.
50. Rothe, K., M. Stordeur, and H. S. Leipner. 2010. Power factor anisotropy of *p*-type and *n*-type conductive thermoelectric Bi–Sb–Te thin films. *J. Electron. Mater.* 39: 1395–1398.
51. Navrátil, J., J. Horák, T. Plecháček et al. 2004. Conduction band splitting and transport properties of Bi_2Se_3. *J. Solid State Chem.* 177: 1704–1712.
52. Chi, H., W. Liu, K. Sun et al. 2013. Low-temperature transport properties of Tl-doped Bi_2Te_3 single crystals. *Phys. Rev. B* 88: 045202–045207.
53. Drasar, C., M. Steinhart, P. Lostak, H. K. Shin, J. S. Dyck, and C. Uher. 2005. Transport coefficients of titanium-doped Sb_2Te_3 single crystals. *J. Solid State Chem.* 178: 1301–1307.
54. Huang, B., Ch. Lawrence, A. Gross et al. 2008. Low-temperature characterization and micropatterning of coevaporated Bi_2Te_3 and Sb_2Te_3 films. *J. Appl. Phys.* 104: 113710.
55. da Silva, L. W., M. Kaviany, and C. Uher. 2005. Thermoelectric performance of films in the bismuth-tellurium and antimony-tellurium systems. *J. Appl. Phys.* 97: 114903.
56. Zou, H., D. M. Rowe, and S. G. K. Williams. 2002. Peltier effect in a co-evaporated Sb_2Te_3(P)–Bi_2Te_3(N) thin film thermocouple. *Thin Solid Films* 408: 270–274.
57. Tan, M., Y. Wang, Y. Deng et al. 2011. Oriented growth of A_2Te_3 (A = Sb, Bi) films and their devices with enhanced thermoelectric performance. *Sens. Actuators A Phys.* 171: 252–259.
58. Sun, Z., S. Liufu, X. Chen, and L. D. Chen. 2011. Tellurization: An alternative strategy to construct thermoelectric Bi_2Te_3 films. *J. Phys. Chem. C* 115: 16167–16171.
59. Lim, S.-K., M.-Y. Kim, and T.-S. Oh. 2009. Thermoelectric properties of the bismuth–antimony–telluride and the antimony–telluride films processed by electrodeposition for micro-device applications. *Thin Solid Films* 517: 4199–4203.

60. Schumacher, Ch., K. G. Reinsberg, R. Rostek et al. 2013. Optimizations of pulsed plated *p* and *n*-type Bi_2Te_3-based ternary compounds by annealing in different ambient atmospheres. *Adv. Energy Mater.* 3: 95–104.
61. Sun, Z., S. Liufu, X. Chen, and L.-D. Chen. 2010. Enhancing thermoelectric performance of bismuth selenide films by constructing a double-layer nanostructure. *Cryst. Eng. Commun.* 12: 2672–2674.
62. Zhu, W., J.-Y. Yang, D.-X. Zhou, C.-J. Xiao, and X.-K. Duan. 2008. Electrochemical aspects and structure characterization of VA-VIA compound semiconductor Bi_2Te_3/Sb_2Te_3 superlattice thin films via electrochemical atomic layer epitaxy. *Langmuir* 24: 5919–5924.
63. Takashiri, M., T. Shirakawa, K. Miyazaki, and H. Tsukamoto. 2007. Fabrication and characterization of bismuth–telluride-based alloy thin film thermoelectric generators by flash evaporation method. *Sens. Actuators A Phys.* 138: 329–334.
64. Takashiri, M., S. Tanaka, and K. Miyazaki. 2010. Improved thermoelectric performance of highly-oriented nanocrystalline bismuth antimony telluride thin films. *Thin Solid Films* 519: 619–624.
65. Das, V. D. and P. G. Ganesan. 1998. Thickness and temperature effects on thermoelectric power and electrical resistivity of $(Bi_{0.25}Sb_{0.75})_2Te_3$ thin films. *Mater. Chem. Phys.* 57: 57–66.
66. Duan, X. and Y. Jiang. 2010. Annealing effects on the structural and electrical transport properties of *n*-type $Bi_2Te_{2.7}Se_{0.3}$ thin films deposited by flash evaporation. *Appl. Surf. Sci.* 256: 7365–7370.
67. Venkatasubramanian, R., T. Colpitts, E. Watko, M. Lamvik, and N. El-Masry. 1997. MOCVD of Bi_2Te_3, Sb_2Te_3 and their superlattice structures for thin-film thermoelectric applications. *J. Cryst. Growth* 170: 817–821.
68. Venkatasubramanian, R., T. Colpitts, B. O'Quinn, S. Liu, N. El-Masry, and M. Lamvik. 1999. Low-temperature organometallic epitaxy and its application to superlattice structures in thermoelectrics. *Appl. Phys. Lett.* 75: 1104–1106.
69. Boulouz, A., S. Chakraborty, A. Giani, F. Pascal-Delannoy, A. Boyer, and J. Schumann. 2001. Transport properties of V–VI semiconducting thermoelectric BiSbTe alloy thin films and their application to micromodule Peltier devices. *J. Appl. Phys.* 89: 5009–5014.
70. Waters, J., D. Crouch, P. O'Brien, and J.-H. Park. 2003. Fabrication of thin films of bismuth selenide using novel single-source precursors by metal organic chemical vapor deposition. *J. Mater. Sci. Mater. Electron.* 14: 599–602.
71. Groshens, T. J., R. W. Gedridge, R. Scheri, and T. Cole. 1996. Low temperature MOCVD growth of V/VI materials via a Me_3SiNMe_2 elimination reaction. *15th International Conference on Thermoelectrics*, Pasadena, CA, March 26–29, 1996.
72. Giani, A., F. Pascal-Delannoy, A. Boyer, A. Foucaran, M. Gschwind, and P. Ancey. 1997. Elaboration of Bi_2Te_3 by metal organic chemical vapor deposition. *Thin Solid Films* 303: 1–3.
73. Giani, A., A. Boulouz, F. Pascal-Delannoy, A. Foucaran, E. Charles, and A. Boyer. 1999. Growth of Bi_2Te_3 and Sb_2Te_3 thin films by MOCVD. *Mater. Sci. Eng. B* 64: 19–24.
74. Giani, A., A. Boulouz, B. Aboulfarah et al. 1999. Effect of antimony concentration on the electrical and thermoelectrical properties of $(Bi_{1-x}Sb_x)_2Te_3$ thin films grown by metal organic chemical vapour deposition (MOCVD) technique. *J. Cryst. Growth* 204: 91–96.
75. Al Bayaz, A., A. Giani, A. Foucaran, F. Pascal-Delannoy, and A. Boyer. 2003. Electrical and thermoelectrical properties of Bi_2Se_3 grown by metal organic chemical vapour deposition technique. *Thin Solid Films* 441: 1–5.
76. You, H. W., S. H. Baek, K.-C. Kim, O. J. Kwon, J.-S. Kim, and Ch. Park. 2012. Growth and thermoelectric properties of Bi_2Te_3 films deposited by modified MOCVD. *J. Cryst. Growth* 346: 17–21.
77. Kang, S.-W., K.-M. Jeon, J.-S. Shin et al. 2013. MOCVD of c-oriented Bi_2Te_3 films on SiO_2 substrates using triethyl bismuth and di-tertiarybutyl tellurium. *Chem. Vapor Depos.* 19: 61–67.
78. Wang, G., L. Endicott, and C. Uher. 2011. Recent advances in the growth of Bi–Sb–Te–Se thin films. *Sci. Adv. Mater.* 3: 539–560.
79. Wang, G., L. Endicott, and C. Uher. 2015. Structure and transport properties of Bi_2Te_3 films. In *Thermoelectric Bi_2Te_3 Nanomaterials*, eds. O. Eibl, K. Nielsch, N. Peranio, and F. Völkline. pp. 73–98. Weinheim, Germany: Wiley-VCH Verlag GmbH & Co. KGaA.

80. Peranio, N., O. Eibl, and J. Nurnus. 2006. Structural and thermoelectric properties of epitaxially grown Bi_2Te_3 thin films and superlattices. *J. Appl. Phys.* 100: 114306.
81. Hansen, A.-L., T. Dankwort, M. Winkler et al. 2014. Synthesis and thermal instability of high-quality Bi_2Te_3/Sb_2Te_3 superlattice thin film thermoelectrics. *Chem. Mater.* 26: 6518–6522.
82. Liu, W., V. A. Stoica, H. Chi, L. Endicott, and C. Uher. 2015. Epitaxial growth and improved electronic properties of $(Bi_{1-x}Sb_x)_2Te_3$ thin films grown on sapphire (0001) substrates: The influence of Sb content and the annealing. *J. Alloys Compd.* 647: 50–56.
83. Liu, W., L. Endicott, V. A. Stoica, H. Chi, R. Clarke, and C. Uher. 2015. High-quality ultra-flat $BiSbTe_3$ films grown by MBE. *J. Cryst. Growth* 410: 23–29.
84. Cho, S., Y. Kim, A. DiVenere, G. K. Wong, J. B. Ketterson, and J. R. Meyer. 1999. Antisite defects of Bi_2Te_3 thin films. *Appl. Phys. Lett.* 75: 1401–1403.
85. Peranio, N., M. Winkler, D. Bessas et al. 2012. Room-temperature MBE deposition, thermoelectric properties, and advanced structural characterization of binary Bi_2Te_3 and Sb_2Te_3 thin films. *J. Alloys Compd.* 521: 163–173.
86. Aabdin, Z., N. Peranio, M. Winkler et al. 2012. Sb_2Te_3 and Bi_2Te_3 thin films grown by room-temperature MBE. *J. Electron. Mater.* 41: 1493–1497.
87. Peranio, N., M. Winkler, Z. Aabdin, J. König, H. Böttner, and O. Eibl. 2012. Room temperature MBE deposition of Bi_2Te_3 and Sb_2Te_3 thin films with low charge carrier densities. *Phys. Status Solidi A* 209: 289–293.
88. Winkler, M., X. Liu, J. D. Konig et al. 2012. Electrical and structural properties of Bi_2Te_3 and Sb_2Te_3 thin films grown by the nanoalloying method with different deposition patterns and compositions. *J. Mater. Chem.* 22: 11323–11334.
89. Haruhiko, O., H. Shun, O. Michihiro, Y. Atsushi, U. Kazuo, and I. Tsutomu. 2009. Thermoelectric properties of Bi_2Te_3-based thin films with fine grains fabricated by pulsed laser deposition. *Jpn. J. Appl. Phys.* 48: 085506.
90. Onose, Y., R. Yoshimi, A. Tsukazaki et al. 2011. Pulsed laser deposition and ionic liquid gate control of epitaxial Bi_2Se_3 thin films. *Appl. Phys. Express* 4: 083001.
91. Zhang, S. X., R. D. McDonald, A. Shekhter et al. 2012. Magneto-resistance up to 60 Tesla in topological insulator Bi_2Te_3 thin films. *Appl. Phys. Lett.* 101: 202403.
92. Gopal, R. K., S. Singh, R. Chandra, and Ch. Mitra. 2015. Weak-antilocalization and surface dominated transport in topological insulator Bi_2Se_2Te. *AIP Adv.* 5: 047111.
93. Symeou, E., M. Pervolaraki, C. N. Mihailescu et al. 2015. Thermoelectric properties of $Bi_{0.5}Sb_{1.5}Te_3$ thin films grown by pulsed laser deposition. *Appl. Surf. Sci.* 336: 138–142.
94. Chang, H.-C., C.-H. Chen, and Y.-K. Kuo. 2013. Great enhancements in the thermoelectric power factor of BiSbTe nanostructured films with well-ordered interfaces. *Nanoscale* 5: 7017–7025.
95. Walachová, J., R. Zeipl, J. Zelinka et al. 2005. High room-temperature figure of merit of thin layers prepared by laser ablation from Bi_2Te_3 target. *Appl. Phys. Lett.* 87: 081902.
96. Fan, P., Z.-H. Zheng, Z.-K. Cai et al. 2013. The high performance of a thin film thermoelectric generator with heat flow running parallel to film surface. *Appl. Phys. Lett.* 102: 033904.
97. Bourgault, D., C. G. Garampon, N. Caillault, L. Carbone, and J. A. Aymami. 2008. Thermoelectric properties of *n*-type $Bi_2Te_{2.7}Se_{0.3}$ and *p*-type $Bi_{0.5}Sb_{1.5}Te_3$ thin films deposited by direct current magnetron sputtering. *Thin Solid Films* 516: 8579–8583.
98. Liao, C.-N., X.-W. Su, K.-M. Liou, and H.-S. Chu. 2011. Electrical and thermal transport properties of electrically stressed Bi–Sb–Te nanocrystalline thin films. *Thin Solid Films* 519: 4394–4399.
99. Winkler, M., X. Liu, J. D. König et al. 2012. Sputtered *p*-type $Sb_2Te_3/(Bi,Sb)_2Te_3$ soft superlattices created by nanoalloying. *J. Electron. Mater.* 41: 1322–1331.
100. Winkler, M., X. Liu, A. L. Hansen et al. 2013. Sputtered *n*-type $Bi_2Te_3/(Bi, Sb)_2Te_3$ superlattice systems. *Nanothermoelectrics* 1: 1.
101. Liou, K. M. and C. N. Liao. 2010. Electric current enhanced defect elimination in thermally annealed Bi–Sb–Te and Bi–Se–Te thermoelectric thin films. *J. Appl. Phys.* 108: 053711.
102. Kim, D.-H., G.-H. Lee, and O.-J. Kim. 2007. The influence of post-deposition annealing on thermoelectric properties of Bi–Sb–Te films prepared by sputtering. *Semicond. Sci. Technol.* 22: 132.

103. Deng, Y., Z. Zhang, Y. Wang, and Y. Xu. 2012. Preferential growth of Bi_2Te_3 films with a nano-layer structure: Enhancement of thermoelectric properties induced by nanocrystal boundaries. *J. Nanopart. Res.* 14: 1–8.
104. Tan, M., Y. Deng, Y. Wang, B. Luo, L. Liang, and L. Cao. 2012. Fabrication of highly (00l)-textured Sb_2Te_3 film and corresponding thermoelectric device with enhanced performance. *J. Electron. Mater.* 41: 3031–3038.
105. Fan, P., T. Chen, Z. Zheng et al. 2013. The influence of Bi doping in the thermoelectric properties of co-sputtering deposited bismuth antimony telluride thin films. *Mater. Res. Bull.* 48: 333–336.
106. Taylor, A., C. Mortensen, R. Rostek, N. Nguyen, and D. C Johnson. 2010. Vapor annealing as a post-processing technique to control carrier concentrations of Bi_2Te_3 thin films. *J. Electron. Mater.* 39: 1981–1986.
107. Stoica, V. A., L. Endicott, H. H. Shen et al. 2014. High-quality II-VI films grown on amorphous substrates using tunable tetradymite templates. *Appl. Phys. Lett.* 105: 221606.
108. Winkler, M. 2015. Nanostructured thermoelectrics: Bi_2Te_3/Sb_2Te_3 based superlattice systems fabricated by MBE and sputtering. PhD dissertation, Eberhard Karls Universität Tübingen, Tübingen, Germany.
109. Chang, C.-Z., J. Zhang, X. Feng et al. 2013. Experimental observation of the quantum anomalous Hall effect in a magnetic topological insulator. *Science* 340: 167–170.
110. Hoefer, K., Ch. Becker, D. Rata, J. Swanson, P. Thalmeier, and L. H. Tjeng. 2014. Intrinsic conduction through topological surface states of insulating Bi_2Te_3 epitaxial thin films. *Proc. Natl. Acad. Sci. USA* 111: 14979–14984.
111. Kampmeier, J., S. Borisova, L. Plucinski, M. Luysberg, G. Mussler, and D. Grützmacher. 2015. Suppressing twin domains in molecular beam epitaxy grown Bi_2Te_3 topological insulator thin films. *Cryst. Growth Des.* 15: 390–394.
112. He, L., X. Kou, M. Lang et al. 2013. Evidence of the two surface states of $(Bi_{0.53}Sb_{0.47})_2Te_3$ films grown by van der Waals epitaxy. *Sci. Rep.* 3: 3406.
113. Zhang, J., C.-Z. Chang, Z. Zhang et al. 2011. Band structure engineering in $(Bi_{1-x}Sb_x)_2Te_3$ ternary topological insulators. *Nat. Commun.* 2: 574.
114. Xu, J.-P., M.-X. Wang, Z. L. Liu et al. 2015. Experimental detection of a Majorana mode in the core of a magnetic vortex inside a topological insulator-superconductor $Bi_2Te_3/NbSe_2$ heterostructure. *Phys. Rev. Lett.* 114: 017001.
115. Lee, J. J., F. T. Schmitt, R. G. Moore, I. M. Vishik, Y. Ma, and Z. X. Shen. 2012. Intrinsic ultrathin topological insulators grown via molecular beam epitaxy characterized by in-situ angle resolved photoemission spectroscopy. *Appl. Phys. Lett.* 101: 013118.
116. He, L., X. Kou, and K. L. Wang. 2013. Review of 3D topological insulator thin-film growth by molecular beam epitaxy and potential applications. *Phys. Status Solidi RRL* 7: 50–63.
117. Li, H. D., Z. Y. Wang, X. Kan et al. 2010. The van der Waals epitaxy of Bi_2Se_3 on the vicinal Si(111) surface: An approach for preparing high-quality thin films of a topological insulator. *New. J. Phys.* 12: 103038.
118. Zhang, G., H. Qin, J. Teng et al. 2009. Quintuple-layer epitaxy of thin films of topological insulator Bi_2Se_3. *Appl. Phys. Lett.* 95: 053114.
119. Bansal, N., Y. S. Kim, E. Edrey et al. 2011. Epitaxial growth of topological insulator Bi_2Se_3 film on Si(111) with atomically sharp interface. *Thin Solid Films* 520: 224–229.
120. Liu, C.-H. and J.-F. Jia. 2015. In situ electrical transport measurement of superconductive ultrathin films. *Chin. Phys. B* 24: 110702.
121. Cahill, D. G., W. K. Ford, K. E. Goodson et al. 2003. Nanoscale thermal transport. *J. Appl. Phys.* 93: 793–818.
122. Cahill, D. G., P. V. Braun, G. Chen et al. 2014. Nanoscale thermal transport. II. 2003–2012. *Appl. Phys. Rev.* 1: 011305.
123. Cahill, D. G. 1990. Thermal conductivity measurement from 30 to 750 K: The 3ω method. *Rev. Sci. Instrum.* 61: 802–808.

124. Cahill, D. G. 2002. Erratum: Thermal conductivity measurement from 30 to 750 K: The 3ω method [*Rev. Sci. Instrum.* 61: 802 (1990)]. *Rev. Sci. Instrum.* 73: 3701–3701.
125. Borca-Tasciuc, T., A. R. Kumar, and G. Chen. 2001. Data reduction in 3ω method for thin-film thermal conductivity determination. *Rev. Sci. Instrum.* 72: 2139–2147.
126. Dames, Ch. 2013. Measuring the thermal conductivity of thin films: 3 omega and related electrothermal methods. *Annu. Rev. Heat Trans.* 16: 7–49.
127. Völklein, F., H. Reith, A. Meier, and M. Schmitt. 2015. Measuring techniques for thermal conductivity and thermoelectric figure of merit of V–VI compound thin films and nanowires. In *Thermoelectric Bi_2Te_3 Nanomaterials*, eds. O. Eibl, K. Nielsch, N. Peranio, and F. Völkline. pp. 225–252. Weinheim, Germany: Wiley-VCH Verlag GmbH & Co. KGaA.
128. Hikami, S., A. I. Larkin, and Y. Nagaoka. 1980. Spin-orbit interaction and magnetoresistance in the two dimensional random system. *Prog. Theor. Phys.* 63: 707–710.
129. Brahlek, M., N. Koirala, M. Salehi, N. Bansal, and S. Oh. 2014. Emergence of decoupled surface transport channels in bulk insulating Bi_2Se_3 thin films. *Phys. Rev. Lett.* 113: 026801.
130. Wang, G., L. Endicott, H. Chi, P. Lošt'ák, and C. Uher. 2013. Tuning the temperature domain of phonon drag in thin films by the choice of substrate. *Phys. Rev. Lett.* 111: 046803.
131. Zhang, J., X. Feng, Y. Xu et al. 2015. Disentangling the magnetoelectric and thermoelectric transport in topological insulator thin films. *Phys. Rev. B* 91: 075431.
132. Xu, Y., Z. Gan, and S.-C. Zhang. 2014. Enhanced thermoelectric performance and anomalous Seebeck effects in topological insulators. *Phys. Rev. Lett.* 112: 226801.
133. Ghaemi, P., R. S. K. Mong, and J. E. Moore. 2010. In-plane transport and enhanced thermoelectric performance in thin films of the topological insulators Bi_2Te_3 and Bi_2Se_3. *Phys. Rev. Lett.* 105: 166603.
134. Shi, H., D. Parker, M.-H. Du, and D. J. Singh. 2015. Connecting thermoelectric performance and topological-insulator behavior: Bi_2Te_3 and Bi_2Te_2Se from first principles. *Phys. Rev. Appl.* 3: 014004.
135. Huang, B.-L. and M. Kaviany. 2008. *Ab initio* and molecular dynamics predictions for electron and phonon transport in bismuth telluride. *Phys. Rev. B* 77: 125209.
136. Xie, W., S. Wang, S. Zhu et al. 2013. High performance Bi_2Te_3 nanocomposites prepared by single-element-melt-spinning spark-plasma sintering. *J. Mater. Sci.* 48: 2745–2760.
137. Zhao, L. D., H. J. Wu, S. Q. Hao et al. 2013. All-scale hierarchical thermoelectrics: MgTe in PbTe facilitates valence band convergence and suppresses bipolar thermal transport for high performance. *Energy Environ. Sci.* 6: 3346–3355.
138. Chiritescu, C., C. Mortensen, D. G. Cahill, D. Johnson, and P. Zschack. 2009. Lower limit to the lattice thermal conductivity of nanostructured Bi_2Te_3-based materials. *J. Appl. Phys.* 106: 073503.
139. Takashiri, M., S. Tanaka, K. Miyazaki, and H. Tsukamoto. 2010. Cross-plane thermal conductivity of highly oriented nanocrystalline bismuth antimony telluride thin films. *J. Alloy. Compd.* 490: L44–L47.
140. Venkatasubramanian, R. 2000. Lattice thermal conductivity reduction and phonon localization-like behavior in superlattice structures. *Phys. Rev. B* 61: 3091–3097.
141. Winkler, M., X. Liu, U. Schürmann et al. 2012. Current status in fabrication, structural and transport property characterization, and theoretical understanding of Bi_2Te_3/Sb_2Te_3 superlattice systems. *Z. Anorg. Allg. Chem.* 638: 2441–2454.

4

All-Scale Hierarchical PbTe: From Nanostructuring to a Panoscopic Material

Gangjian Tan and Mercouri G. Kanatzidis
Northwestern University

CONTENTS

4.1 Introduction

The rock salt cubic semiconductor PbTe material is an iconic thermoelectric system. It has been studied for more than six decades because of its fascinating electronic and thermal transport properties. The material has given the science community many fascinating, intriguing, and even amazing properties to ponder, yet it keeps producing stirring surprises. The pioneering studies on this material began in the 1950s and continued through the 1960s and beyond [1–3]. It was then that many of the fundamental properties were experimentally discovered and theoretically explained. For example, the direct nature of the bandgap, the existence of a second valence band, and the anomalous temperature dependence of the bandgap were documented, and a great deal of the nature of its electronic structure was understood [4–8]. These studies were published in several key papers decades ago [5,6,8].

The appeal of PbTe as a great thermoelectric material to the thermoelectric community has been constant. It has been the premiere thermoelectric material for midrange temperature (600–800 K) applications [9–15]. PbTe melts at around 920°C without phase transitions at any temperature [16]. A narrow bandgap of ~0.3 eV at room temperature is optimal for power generation applications when the hot source is at a temperature of 400°C–600°C [2,4]. An unusual property of the bandgap is that it rises with rising temperature, and this contributes to achieving high performance at high temperatures [4,17]. The PbTe structure has a facile tendency to be doped *n*- and *p*-type with a variety of suitable dopants [9,10,18–21]. The lattice thermal conductivity of PbTe is ~2.0 W m^{-1} K^{-1} at room temperature and falls at higher temperature to about 1.2 W m^{-1} K^{-1} [13,18,19,21]. The maximum thermoelectric figure of merit ($ZT = S^2\sigma T/\kappa_{tot} = S^2\sigma T/(\kappa_{ele} + \kappa_{lat})$, where S is the Seebeck coefficient, σ is the electrical conductivity, and κ_{tot} is the total thermal conductivity, that is, a sum of the electronic contribution κ_{ele} and lattice contribution κ_{lat}) for single crystals or larger polycrystals of PbTe is about 1.0 for *n*-type and 1.2 for *p*-type at ~700–800 K [18–23]. The thermoelectric properties of PbTe are reasonably well understood and have been reviewed previously [24–27]. The most significant recent work has been the major improvement of the thermoelectric properties of PbTe by nanostructuring [10,11,13,20,28–32], and this is a major focus of this chapter. Recent work has also focused on understanding the mechanical properties of PbTe for device fabrication [33–37].

An important feature of PbTe is that it comprises two valence bands [5,6,8], and they are essential in defining its excellent thermoelectric properties. The energy offset in the maxima of these two valence bands is relatively small and decreases with rising temperature. In addition, the energy difference can be further decreased to within a few k_BT (k_B is the Boltzmann constant and T is the absolute temperature) of each other by proper chemical alloying, such as with Mg, Mn, or Cd [13,38,39], thereby leading to effective valence band convergence that strongly enhances the Seebeck coefficient in *p*-type samples. Nanostructured versions of PbTe have very different thermal and charge transport properties and are discussed later in this chapter.

4.2 Electronic Structure

PbTe crystallizes in the rock salt structure under normal pressures, and its electronic structure has been extensively explored experimentally. In addition, a great deal of theoretical studies have been carried out using various different methods, such as the tight-binding method [40], the linear augmented plane wave method [41], the Green's function method [42], the quasiparticle self-consistent GW (QSGW) approach [43], and the local density approximation [44], all of which exhibit very similar results. Figure 4.1a and b show the representative electronic band structure of PbTe theoretically calculated without and with spin–orbit coupling (SOC), respectively [45]. The direct bandgap of PbTe at the L point is ~0.82 eV in the case of non-SOC, while it is reduced to ~0.13 eV when SOC is considered, which is smaller than the experimentally determined bandgap of ~0.19 eV around 0 K for PbTe [2]. Also, one can clearly see that some of the degenerate bands at high symmetry points split due to the SOC. Though these theoretical calculations provide useful insight into the primary electronic band character of PbTe, they are not always accurate enough to describe the thermoelectric transport properties of PbTe. Figure 4.1c represents schematically the well-accepted electronic structure of PbTe at 300 K. Both the conduction band minimum (CBM) and valence band maximum (VBM) of PbTe occur at the L points of the Brillouin zone, which make it a direct bandgap semiconductor with a bandgap E_g of about

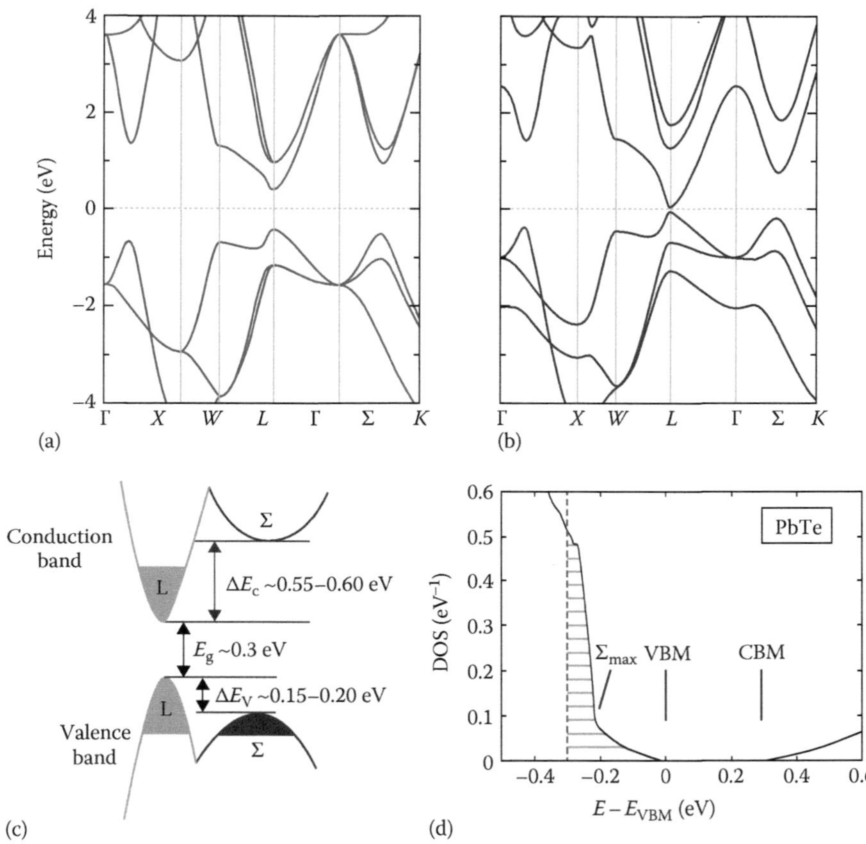

FIGURE 4.1
A representation of theoretically calculated energy band structures of PbTe (a) without and (b) with spin–orbit coupling. The Fermi level is at 0 eV [45]. (c) Schematic electronic structure of PbTe at 300 K. (d) Density of states near the valence band maximum (VBM) of PbTe calculated using quasiparticle self-consistent GW. The energies corresponding to the conduction band minimum, primary valence band maximum (VBM), and the second VBM are marked. (Reprinted with permission from Svane, A., Christensen, N.E., Cardona, M., Chantis, A., Van Schilfgaarde, M., and Kotani, T., *Phys. Rev. B*, 81, 245120, 2010. Copyright 2010 by the American Physical Society.)

0.3 eV (at 300 K). In addition to the CBM and VBM at the L points (the L bands), there is a second conduction and second valence band along the Σ line (the Σ band). This Σ conduction band is separated from the L band by an energy offset of ΔE_C ~0.55–0.60 eV [13] and ΔE_V ~0.15–0.20 eV for the valence band [5,6,8,43,44].

In a rigid band model scenario, the Σ valence band maximum is very close in energy to the L valence band maximum and could be accessed by the Fermi energy when the alloying level is very high, which is feasible experimentally [43]. However, the large ΔE_C makes the Σ conduction band inaccessible in *n*-type samples of PbTe under normal circumstances. Figure 4.1d shows a zoomed-in view of the density of states (DOS) of PbTe around VBM [43]. The DOS shows a kink at the energy of the Σ valence band ($E(\Sigma_{max}) = -0.21$ eV, the energy of VBM is set as zero) representing the onset of a sharp increase of slope, which is a sign of a large DOS effective mass (m^*) of the top of the Σ valence band. Indeed, both theoretical calculations and experimental studies reveal that the Σ bands in PbTe have much larger m^* than the L bands. For example, at 300 K, the carrier effective mass m^* for the L conduction band and the L valence band is ~$0.26m_e$ (m_e is the free electron mass) and $0.31m_e$, respectively, while it is ~$2m_e$ for the Σ valence band [1,5,46].

Another unusual feature of the electronic structure of PbTe is its anomalous bandgap enlargement with rising temperature, which can act to suppress bipolar effects caused by intrinsic carrier activation, thus increasing the optimum ZT in the process [2,4,17]. The positive temperature coefficient of the bandgap in PbTe is not really well understood. It was thought to be due to both lattice expansion and electron–phonon interactions [47]; however, new insights about the origin of this effect have been obtained recently [48] and will be mentioned in the following. Early optical experiments at constant pressure indicated that the direct gap E_g of PbTe at the L point of the Brillouin zone increases almost linearly with increasing temperature in the range of 80–450 K, and, at still higher temperatures, the changes are negligible [2,17] (see Figure 4.2a). A linear fit of the E_g vs. T plot yields slopes of (4.1–4.5) × 10^{-4} eV K^{-1}. The saturation of the bandgap of PbTe takes place at around 450 K. It has been proposed by Tauber et al. [2] that the reason for this is that the shifting of the L point valence band stops as its energy reaches the energy of the Σ valence band and the material becomes an indirect gap semiconductor. Since the energy of the Σ valence band keeps nearly constant with increasing temperature, the increase of E_g involves the lowering of the L band and decrease of ΔE_V, namely, a temperature-induced band convergence. Tauber's work has been cited as evidence of the convergence temperature, T_{cvg}, in PbTe. However, a recent study by Gibbs et al. [47], combining diffuse reflectance infrared Fourier transform spectroscopy and combined with *ab initio* calculations predicted a T_{cvg} of 700 K (assuming the optical

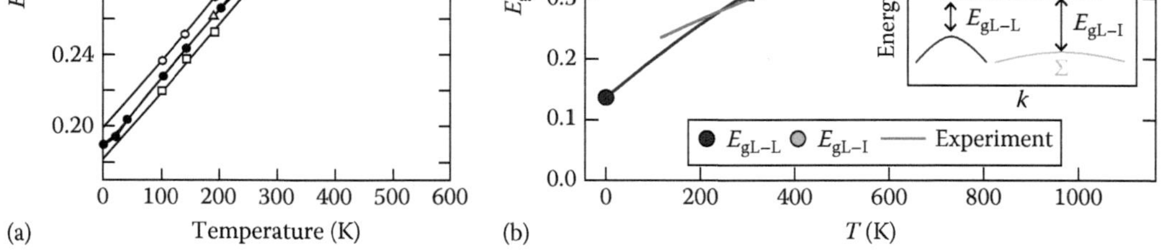

FIGURE 4.2
(a) Calculated (empirical pseudopotential method, solid symbols) and experimental (open symbols) temperature-dependent bandgap of PbTe. (Reprinted with permission from Tsang, Y. and Cohen, M.L., *Phys. Rev. B*, 3, 1254, 1971. Copyright 1971 by the American Physical Society.) (b) *Ab initio* calculated temperature dependence of the bandgap (solid symbols) in PbTe plotted with experimental data (solid gray line). Solid lines are a guide to the eye. (Reprinted with permission from Gibbs, Z.M., Kim, H., Wang, H. et al., Temperature dependent band gap in PbX (X = S, Se, Te), *Appl. Phys. Lett.*, 103, 262109. Copyright 2013, American Institute of Physics.)

bandgap increases up to 673 K). This study provided a direct experimental evidence for a much higher convergence temperature T_{cvg} than assumed previously (Figure 4.2b).

4.2.1 Conduction Band

According to the quality factor B (see Equation 4.1), the theory developed by Chasmar and Stratton [49], which is usually used to evaluate ZT, a large valley band degeneracy (N_V) is important to achieve a high thermoelectric performance:

$$B = \frac{2k_B^2\hbar}{3\pi}\frac{N_V C_l}{m_b^* \Xi^2 \kappa_{lat}} T \tag{4.1}$$

where $\hbar$ is the reduced Planck constant, m_b^* is the inertial effective mass, C_l is the average longitudinal elastic modulus, and Ξ is the deformation potential.

The conduction band at the L point of the Brillouin zone of PbTe is nonparabolic and has a valley degeneracy N_V of 4 [12]. This is the only band involved in n-type charge transport, and a single Kane band model is adequate to describe the electronic transport behavior with reasonable accuracy [50]. The solid line in Figure 4.3a shows the theoretically calculated room temperature Seebeck coefficient vs. electron density relationship, the so-called Pisarenko plot [51] for n-type PbTe, on the basis of the model mentioned earlier [21,52]. It is seen that this Pisarenko plot predicts the Seebeck coefficient of iodine [21] and antimony [53] doped n-type PbTe quite well. As mentioned earlier, the effective mass (m^*) of electrons near the bottom of the L conduction band of PbTe is quite low at ~0.26 m_e. It has been reported that an increase in m^*, giving a higher Seebeck coefficient, is achieved in La-doped PbTe ($Pb_{1-x}La_xTe$, $x = 0$–0.01) [20,52,54]. La doping at Pb sites acts not only as a good n-type dopant, but it also significantly increases the DOS near the lowest conduction band edge, which was demonstrated in the theoretical calculation results for randomly disordered $Pb_{1-x}La_xTe$ performed by the Korringa–Kohn–Rostoker Green function formalism under the so-called coherent potential approximation (KKR-CPA) [54], as shown in Figure 4.3b. The increase in the DOS could seem to arise as a consequence of a resonant impurity, similar to the case observed in Tl-doped p-type PbTe [9]. However, this may not necessarily be true because the DOS peak is mainly from La f-states which do not delocalize sufficiently to result in resonant levels [54]. Despite the increased m^* by La doping in PbTe, the net ZT is not enhanced because of the apparent degradation of the carrier mobility [52,54].

4.2.2 Valence Band

As we mentioned earlier, PbTe features two valence bands, namely, a light hole band with $N_V = 4$ at the L point and a lower lying heavy hole band with $N_V = 12$ at the Σ point, separated by an energy offset (ΔE_V) of ~0.15–0.20 eV at room temperature [6,8,12,27]. In fact, aside from PbTe, other rock salt II–IV analogous compounds, such as PbSe [25], PbS [43,55], and SnTe [56–59], also have two valence bands, but the ΔE_V value is significantly larger. For example, at 300 K, ΔE_V values are ~0.35–0.40 eV and ~0.25–0.30 eV for SnTe [56–59] and PbSe [60], respectively, while it is a much larger value ~0.50 eV for PbS [55]. Similar to PbTe, as the temperature rises, the L valence band in these materials lowers its energy while the Σ valence band stays almost constant in energy [2,61]. As a consequence, the energy difference between the two bands diminishes with the rising temperature, the phenomenon

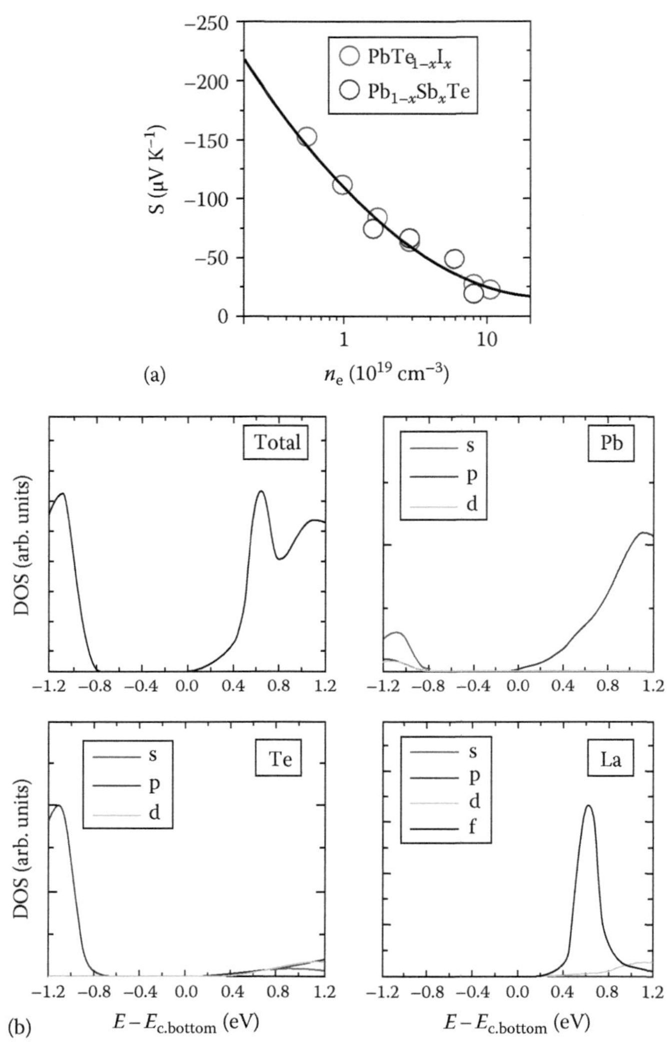

FIGURE 4.3
(a) Room temperature Seebeck coefficient as a function of carrier concentration for *n*-type PbTe. The red and blue open symbols are for iodine-doped [21] and antimony-doped [53] PbTe, respectively. The solid line is the theoretical Pisarenko plot for *n*-type PbTe calculated based on a one band Kane model [21,52]. (b) Total and partial electronic density of states near the bottom of the conduction band of $Pb_{0.97}La_{0.03}Te$. (Reprinted with permission from Takagiwa, Y., Pei, Y., Pomrehn, G., and Snyder, G.J., Dopants effect on the band structure of PbTe thermoelectric material, *Appl. Phys. Lett.*, 101, 092102. Copyright 2012, American Institute of Physics.)

known as temperature-induced band convergence [5,12,26,47] (see Figure 4.4). The temperature at which this band convergence occurs depends on the specific II–VI rock salt chalcogenide.

In these "two-valence-band" materials, a hole carrier redistribution takes place between the two subbands as the temperature increases because of the change in their relative energy. This hole redistribution does not begin at a temperature where the two bands are equal in energy, but at lower temperatures, provided there is enough thermal

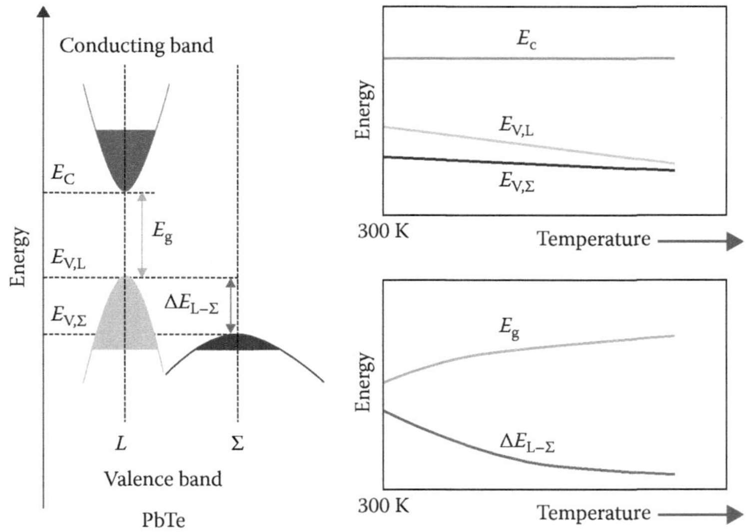

FIGURE 4.4
Schematic representations of the temperature-dependent band energies of the electronic bands of PbTe. As temperature rises, the conduction band (purple) remains almost unchanged in energy, while both the L and Σ valence bands decrease their energy [2,61]. Since the L valence band lowers its energy much faster than the Σ valence band, there is a temperature-induced band convergence.

energy to excite electrons from the top of the Σ band to the top of the L band. This redistribution means that both bands contribute to the transport of holes, and the degree of redistribution depends on the temperature, which scales with the size of ΔE_V. The higher the ΔE_V value, the higher the temperature of the band convergence and, thus, the carrier redistribution. Consequently, the Hall coefficient (R_H) in PbTe and in other II–VI rock salt chalcogenides has very strong temperature dependence (rather than being constant if a single band was involved in the transport) [5,56,60]. Usually, R_H peaks when the contribution to conduction from the two bands is nearly equal. This peak temperature reflects the inherent energy difference (ΔE_V) between the tops of the two valence bands and normally would not be altered by conventional doping unless there is a significant change in the band structure. Commonly, the larger the ΔE_V, the higher the temperature where R_H peaks [56,58]. For instance, as shown in Figure 4.5a, ΔE_V is only ~0.15–0.20 eV at 300 K for PbTe, so R_H peaks at a relatively low temperature around 425 K [5,13]. For *p*-type samples of PbSe [60] and SnTe [56,62], however, with gradually increasing ΔE_V, the peak temperature in the R_H vs. T plots shifts higher, namely, to ~700 K for *p*-PbSe and ~750 K for *p*-SnTe. For *p*-type PbS, ΔE_V is too large, and R_H is basically constant with varying temperature [55], behaving like a single-band material. Likewise, in *n*-type PbTe, where essentially a single band is present, a flat R_H vs. T plot is observed as well [52].

Because of the smaller ΔE_V in PbTe, the heavy Σ valence band can be probed if the samples are heavily alloyed to lower the energy level. Figure 4.5b compares the Pisarenko plots of *n*- and *p*-type PbTe. At the low hole concentration range ($n < 10^{19}$ cm^{-3}), the difference in the absolute Seebeck coefficient between *n*-type PbTe and *p*-type PbTe is fairly small due to single-band conduction and to the similarity of the DOS effective masses of L point conduction (~$0.26m_e$) and valence ($0.31m_e$) bands [1,5,46]. However, a huge

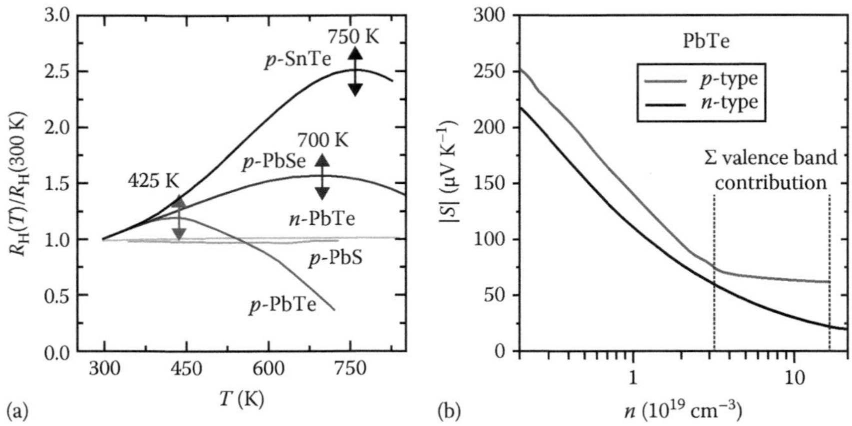

FIGURE 4.5
(a) Temperature dependence of normalized Hall coefficients (R_H/R_{300}) for *p*-type SnTe [56] and lead chalcogenides [13,55,60] and *n*-type PbTe [52]. *n*-type PbTe and *p*-type PbS have temperature-independent R_H, which is typical of single-band conducting materials. *p*-type SnTe, PbTe, and PbSe show strong temperature dependence of R_H, indicating the presence of two bands responsible for conduction. The peak temperature in the R_H–T plot is a measure of the energy separation between the two subbands. (b) Comparison of room temperature Pisarenko plots of *n*- and *p*-type PbTe. The much higher Seebeck coefficients in the *p*-type material at high doping level ($n > 3 \times 10^{19}$ cm^{-3}) are attributed to the onset of the contribution from the heavy Σ valence band.

difference arises when $n > 3 \times 10^{19}$ cm^{-3}, where the Σ valence band with high degeneracy starts to contribute to the Seebeck coefficient [12,27,43]. Therefore, the *p*-type Seebeck coefficient becomes saturated, while the *n*-type Seebeck coefficient continues to decrease with increasing *n*.

4.2.3 Band Engineering through Alloying, Resonance Levels, and Band Convergence

PbTe is the top performing thermoelectric material in the temperature range of 500–900 K, and to a large extent, this is because of its unique electronic structure described earlier. However, the very high performance achieved recently with this material originates from a tremendous flexibility of its electronic band structure that can be significantly modified in a desired direction simply by proper chemical alloying to tune its electronic properties. The band structure modification generally involves three important phenomena: resonant levels [9,63], bandgap opening, and band convergence [12,13,38,39].

4.2.3.1 Resonant Levels

It is well known that some dopant atoms in a semiconductor can introduce electronic states that appear as resonances in the electronic structure [9,63]. Based on a theoretical study [64] of the effect of group 13 dopants on the electronic structure of PbTe, Ga, In, and Tl were substituted in PbTe and $Pb_{1-x}Sn_xTe$. When $Pb_{1-x}Sn_xTe$ was doped with In, the resonance level of In was not located in the conduction or valence bands but acted as a pinning center in the middle of the gap [64]. No enhancement in thermoelectric properties was observed. However, when Tl was substituted in PbTe, the resonant state was located in the valence band, and *p*-type samples exhibited higher Seebeck coefficients than *p*-type samples of Tl-free PbTe [9,63].

According to the Mott expression [65] (see Equation 4.2), which is well accepted to describe the Seebeck coefficient for a system of independent electrons interacting with static scatterers (it is so for most thermoelectric semiconductors), a large Seebeck coefficient S can be realized (1) by increasing the energy dependence of carrier density $n(E)$ near the Fermi level E_F by, for example, introducing resonance levels [9,63] or increasing the band degeneracy [12,13,39,56,58,59] through alloying, and (2) by increasing the energy dependence of mobility $\mu(E)$ using energy filtering [66,67]. In Equation 4.2, k_B is the Boltzmann constant and $\sigma(E)$ is the energy-dependent electrical conductivity. The energy dependence of $\sigma(E)$ is proportional to the DOS.

$$S = \frac{\pi^2 k_B^2}{3e} T \left\{ \frac{d[\ln(\sigma(E))]}{dE} \right\}_{E=E_F} = \frac{\pi^2 k_B^2}{3e} T \left\{ \frac{1}{n}\frac{dn(E)}{dE} + \frac{1}{\mu}\frac{d\mu(E)}{dE} \right\}_{E=E_F} \tag{4.2}$$

If an energy level of an added dopant atom hybridizes with the level near the top of the valence or near the bottom of the conduction band of the host semiconductor, it can form a narrow band called the resonance level and can enhance the DOS. The resonance is meaningful only if its energy is located at a proper place with respect to the host bands. If the impurity level is strongly bound (trap states, such as in Figure 4.6a) or weakly bound (localized states in Figure 4.6b), it is not considered a resonance level. Only when the dopant element's atomic orbitals hybridize strongly with the host band orbitals, to form a type of mini band with high DOS and a small but significant band width, do we get a resonance (see illustrations in Figure 4.6c and d). In Figure 4.6d, however, the "resonant states" are too far from the band edge and usually do not act as resonance levels (i.e., do not impact charge carrier transport at all).

Previous experiments have shown that the group III elements Ga, In, and Tl can create additional energy levels in PbTe, including resonance levels. The position of the additional energy level is still ambiguous in Ga-doped PbTe, while in In-doped PbTe, it favorably resides in the conduction band at very low temperatures but moves into the energy gap around room temperature [63,64,68]. Therefore, indium is not regarded as a good resonant level impurity in PbTe. Instead, it is known for its ability to stabilize the Fermi level within the bandgap of PbTe. Because of this Fermi level pinning effect of indium in PbTe, optimal carrier concentrations in In-doped PbTe samples are not achievable no matter how n-type material is doped. Therefore, the overall thermoelectric performance cannot be easily changed [60,61,65].

There are generally two explanations for this unusual behavior of In in PbTe. One idea is that In exists in mixed oxidation states in the solid, one monovalent and the other trivalent, giving rise to an excited divalent impurity state that is neutral with respect to Pb atoms [67,70]. Mahanti et al. proposed a different picture on the basis of *ab initio* electronic structure calculations [68]. In In-doped PbTe, two electrons of In occupy a localized state whose energy is below the VBM and the other electron occupies a localized state in the gap. These two states are strongly hybridized In s and neighboring Te p states. The spectral weights associated with both of these two localized bands of states are two (including spin degeneracy) per impurity. In this process, the valence band of mostly Te p character loses one state (without spin degeneracy) per impurity. This picture naturally results in the Fermi level pinning in the bandgap.

The addition of Tl, however, does create a resonance level in PbTe, as schematically shown in Figure 4.7a. Optical measurements [71] revealed the presence of several distinct levels associated with Tl in PbTe, and one of them favorably locates at an energy

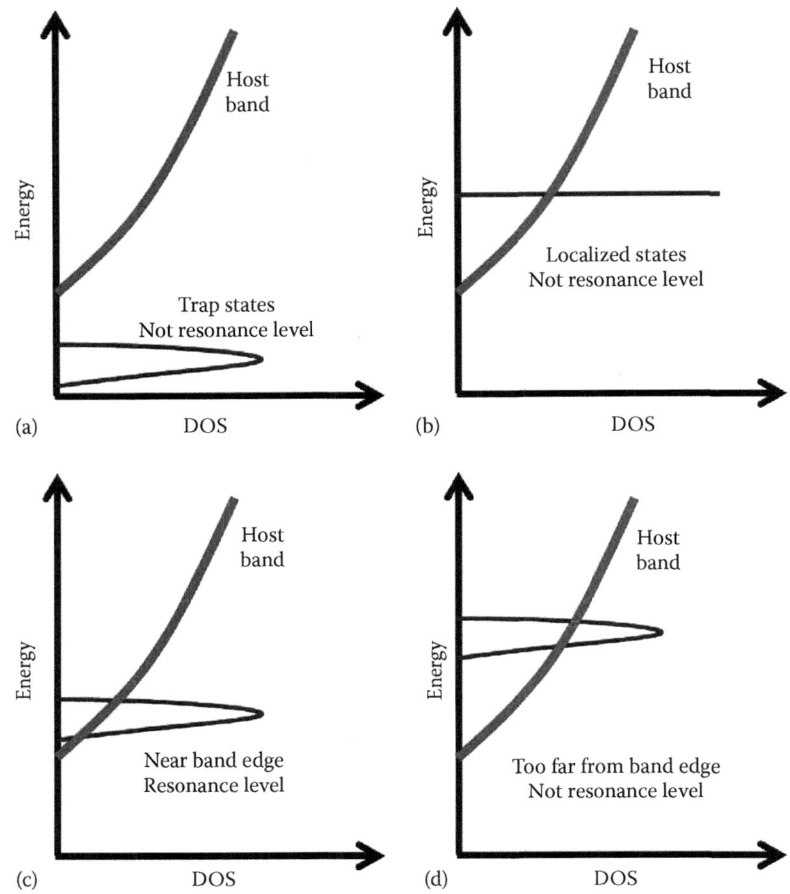

FIGURE 4.6
Schematic presentations of four possible electron energy level states of an impurity in the host material: (a) trap states, (b) localized states with negligible band width, and (c, d) resonant states. In (d), the resonant states are too far from the band edge and may not be thermoelectrically relevant at any realistic doping level.

~0.03–0.06 eV below the valence band edge. It acts as a resonant level and sharply increases the DOS inside the valence band. As a result, the *p*-type Tl-doped PbTe shows much higher Seebeck coefficients than predicted by the theoretical Pisarenko plot, as indicated by the solid curve [9] in Figure 4.7b. Resonance levels are generally effective near the room temperature or below. The beneficial effect of the enhanced Seebeck coefficient tends to disappear at high temperatures as the relative energies of the electronic bands and the resonance levels move around, possibly placing the levels out of resonance [60,63,72].

Theoretical studies have also suggested that Ti [73] and Zn [74] dopants could form resonant levels in the conduction band of PbTe. However, due to the Fermi level pinning effect by Ti, the expected Seebeck coefficient enhancement is not observed. In other words, Ti forms well-localized 3*d* states that do not mix strongly with the valence band orbitals [73]. Zn, on the other hand, was shown to have negligible solubility in PbTe [74].

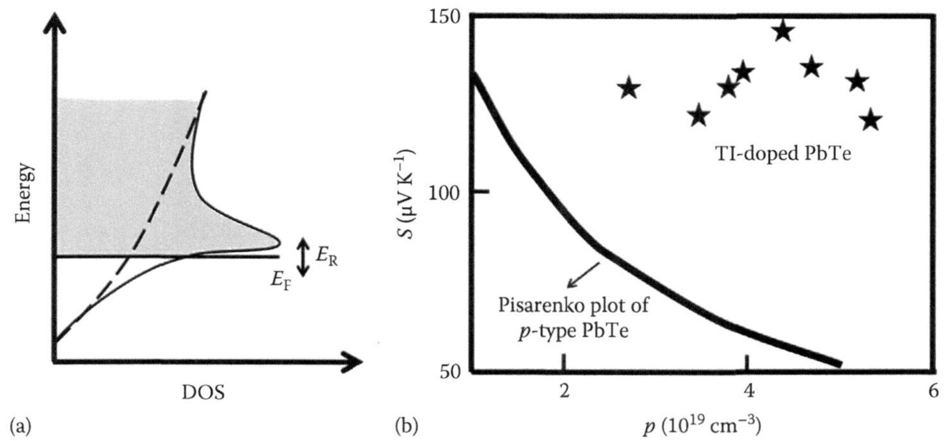

FIGURE 4.7
(a) Schematic representation of the density of states (DOS) of the valence band of pure PbTe (dashed line) with respect to that of Tl-doped PbTe where a Tl-related level increases the DOS. The ZT is enhanced when the Fermi level E_F falls in the energy range E_R of the band distortion. (b) Room temperature Pisarenko plot for PbTe (solid line) in comparison with the experimental results of Tl-doped PbTe samples. (From Heremans, J.P. et al., *Science*, 321, 554, 2008.)

4.2.3.2 Band Convergence

We now consider another important mechanism that may greatly enhance the DOS near the Fermi level. This mechanism is band convergence. It is known that the DOS effective mass m^*_{DOS} is closely related to the band degeneracy N_V through the relation $m^*_{DOS} = (N_V)^{2/3} m^*_b$ with m^*_b being the band effective mass [12]. When two or more degenerate valleys are close to each other in energy (e.g., by compositional alloying), the bands effectively converge, and the total N_V is larger than that of any single band. As a consequence, m^*_{DOS} is dramatically enhanced, which, in turn, leads to an enhancement of the Seebeck coefficient; see the S–m^*_{DOS} relationship [51,75,76] for degenerate semiconductors, as shown in Equation 4.3. For example, both experimental and theoretical studies show that Mg [13], Mn [39], Cd [38,77], and, very recently, Yb [55] substituting for Pb could promote good valence band convergence along with an enlarged bandgap. This can be achieved because, with increasing solid solution fraction, both the L and Σ bands lower their energy, but the maximum of the L band moves faster than that of the Σ band, so that the two bands move closer in energy as the temperature rises [13,26] (Figure 4.8). It is worth noting that to achieve a better band convergence by chemical doping usually requires a high alloying fraction [26,56].

$$S = \frac{8\pi^2 k_B^2}{3eh^2} m^*_{DOS} T \left(\frac{\pi}{3n}\right)^{2/3} \tag{4.3}$$

Figure 4.9a shows the Seebeck coefficient enhancement of p-type PbTe induced by valence band convergence through Mg or Mn alloying [78,79]. The black solid line is the theoretical Pisarenko plot for p-type PbTe [13,39]. Na works as a pure dopant in PbTe, efficiently regulating the hole carrier concentration, and the Seebeck coefficient of Na-doped PbTe

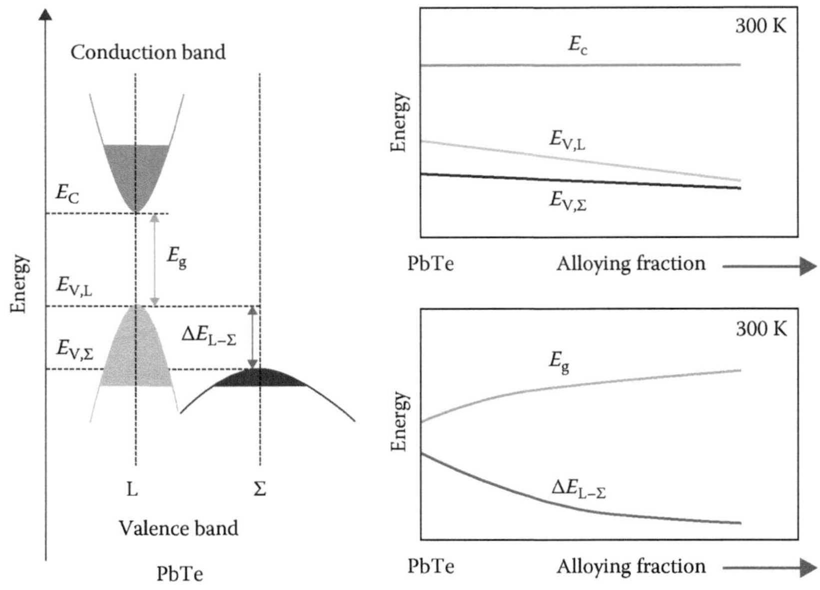

FIGURE 4.8
Schematic representations of the composition-dependent relative energies of the two valence bands (L and Σ) for PbTe alloyed with some specific elements such as Cd [38], Mg [13], and Mn [39]. With increasing alloying fraction, the conduction band remains constant in energy, while both the L and Σ valence bands decrease their energy, leading to an enlarged bandgap. Since the L band lowers its energy much faster than the Σ band, there is an alloying-induced band convergence.

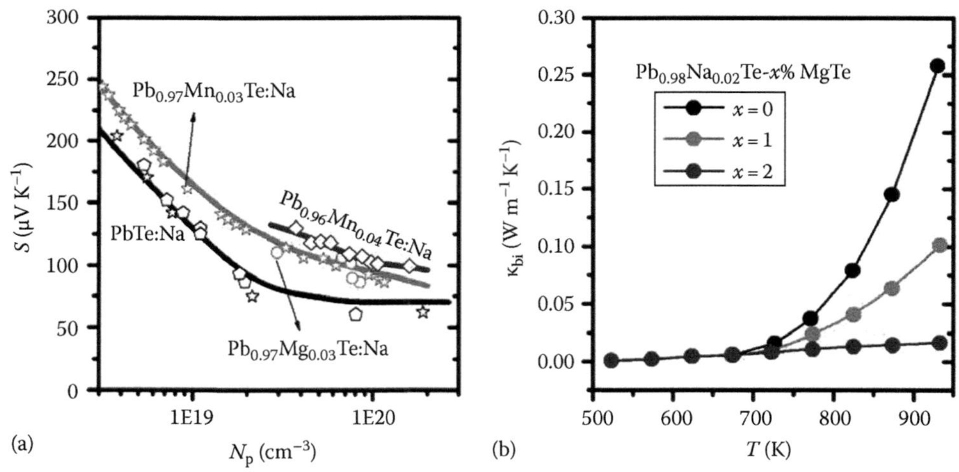

FIGURE 4.9
(a) Room temperature Seebeck coefficients as a function of carrier concentrations for Na- [80,81], Mg- [79], and Mn-alloyed [78] *p*-type PbTe. The black line is a theoretical Pisarenko plot [13,39] for *p*-type PbTe while the gray lines are guides to eyes. (b) Bipolar thermal conductivity (κ_{bi}) as a function of temperature for $Pb_{0.98}Na_{0.02}Te$–*x*%MgTe. (From Zhao, L. et al., *Energy Environ. Sci.*, 6, 3346, 2013.)

can be well described by the theoretical line [80,81]. A 3 mol% Mg [79] or Mn [78] alloying in *p*-type PbTe pushes the Seebeck coefficient well above the black line (see the star symbols and the corresponding line that is a guide to the eyes). Further enhancement of the Seebeck coefficient is achieved when an even higher alloying concentration is used, for example, by 4 mol% Mn alloying [78] (see the diamond symbols and the line for easy reference). The great enhancement of the Seebeck coefficient of *p*-type PbTe upon adding these specific alloying agents has been ascribed to the decreased energy separation (ΔE_V) between the L and Σ valence bands, which is well supported by the high-temperature Hall measurement [13,38] (the peak temperature in the R_H–T plot shifts steadily toward lower temperature as the alloying concentration is increased, which is a signature of a decrease in ΔE_V), as well as the first-principles electronic structure calculations [13,45]. It should be emphasized that, in contrast to resonant levels, which usually seriously deteriorate the carrier mobility of the host material, band convergence usually is not as detrimental to the carrier mobility and, thus, at high temperatures, affords significantly higher power factors than resonant level doping [12,27,39,59].

Another noticeable benefit that is accompanied by the alloying-induced band convergence is bandgap enlargement (see Figure 4.8). When the bandgap widens, the detrimental effect of bipolar conduction on thermoelectric properties at elevated temperatures is suppressed and this allows high power factors and *ZT* values [13,56,58,62,75]. Figure 4.9b shows such an example where the incorporation of 2 mol% MgTe in PbTe significantly suppresses the bipolar thermal conductivity κ_{bi} from ~0.25 W m^{-1} K^{-1} for PbTe to almost zero at ~923 K [13]. This bandgap enlargement has been well demonstrated by optical bandgap measurements [13].

Materials can achieve high thermoelectric performance if there are multiple bands within a narrow energy range holding the same carrier type contributing to transport. This is often referred to as "band convergence". As we mentioned earlier, the valence band convergence temperature T_{cvg} in PbTe was initially reported to be around 450 K [5,6,8], but new evidence shows that the actual T_{cvg} may be significantly higher than previously thought, perhaps as much as ~700 K [47]. To push T_{cvg} of PbTe to this higher (ideal) working temperature range, a good way could be through forming solid solutions between PbTe and PbSe. Se alloying in PbTe increases ΔE_V at room temperature, so that the expected T_{cvg} is shifted to higher temperatures. Coupled with the lower thermal conductivity, a very high *ZT* of ~1.8 at about 850 K was achieved in the sample of $PbTe_{0.85}Se_{0.15}$ doped with 2% Na [12].

Band convergence has now become the most practical concept of designing high-performance bulk thermoelectrics. It has been demonstrated that there are two conduction bands in Mg_2Si–Mg_2Sn solid solutions and the relative energy of these two bands can be easily controlled by adjusting the Si/Sn atomic ratio [82]. An effective conduction band convergence appears around the chemical composition of $Mg_2Si_{0.3}Sn_{0.7}$ [82,83], which leads to the great enhancement of the Seebeck coefficient over a broad temperature range. Likewise, in the rock salt–structured analogs of PbTe, namely, SnTe and PbSe, two similar valence bands are recognized. It has been shown that alloying with Cd [59,84], Hg [62], Mg [85], and Mn [45,56] can effectively decrease the energy difference of the two valence bands in these II–VI compounds, leading to valence band convergence for improved thermoelectric performance. Notably, SnSe, which is a newly recognized high-performance thermoelectric material with intrinsically low thermal conductivity [86], features at least three valence bands that lie very close in energy [87]. In other words, the valence bands of SnSe are already nearly converged, even as the material is a pure unalloyed phase. It is shown that Na doping in SnSe pushes the Fermi level deep into the band structure, activating multiple valence band maxima involved in electron transport, leading to enhanced Seebeck coefficients and power factors. Coupled with its low thermal conductivity, ZT values ranging from 0.7 to 2.0 at 300–773 K were achieved in 1.5 mol% Na-doped SnSe single crystals, which represents the highest average ZT ~1.35 in this working temperature range [87]. It makes SnSe a very promising and robust candidate for low- to mid-temperature thermoelectric power generation.

4.3 Thermal Conductivity

Though band structure modification provides an efficient way of enhancing the power factor of thermoelectric materials, it is equally important to decrease the thermal conductivity so as to ensure a high ZT [10,11,30,32,58,88]. Prior to the analysis of the thermal conductivity of PbTe, it is necessary to say a few words about the basic and common phonon scattering mechanisms in solids. According to the modified Callaway's model [89], the lattice thermal conductivity (κ_{lat}) of a solid can be expressed as

$$\kappa_{lat} = \frac{k_B}{2\pi^2 \upsilon}\left(\frac{k_B T}{\hbar}\right)^3 \int_0^{\Theta_D/T} \tau_c \frac{x^4 e^4}{(e^x - 1)^2} dx \tag{4.4}$$

where υ is the average phonon group velocity, and τ_c is the total relaxation time of phonons.

There are multiple phonon scattering mechanisms in solids, each of which contributes to the total lattice thermal conductivity [89]. These scattering sources include point defects, dislocations, grain boundaries, displacement layers, nanoscale precipitates, electron–phonon interaction, and phonon–phonon scattering processes. Therefore, τ_c is a combination of the contributions from all of these scattering mechanisms and can be written as [89]

$$\tau_c^{-1} = \tau_U^{-1} + \tau_N^{-1} + \tau_{GB}^{-1} + \tau_{NP}^{-1} + \tau_D^{-1} + \tau_S^{-1} + \tau_{PD}^{-1} + \tau_{DL}^{-1} \tag{4.5}$$

where τ_U, τ_N, τ_{GB}, τ_{NP}, τ_D, τ_S, τ_{PD}, and τ_{DL} represent the relaxation times corresponding to scattering from Umklapp processes, normal processes, grain boundaries, nanoprecipitates, dislocations, strains, point defects, and displacement layers, respectively. It would be ideal to incorporate all of these phonon scattering mechanisms in one single material (the so-called all-scale hierarchical architectures [11,13]) to achieve the lowest possible thermal conductivity, which should work for PbTe as well [11,13,26].

Most of the early studies of PbTe focused mainly on minute doping and its influence on the electronic structures [1–3]. In the early period, the two-valence-band character of PbTe was already well documented, but its thermal transport properties were not fully understood [4–8]. In fact, due to the problem with the steady-state method of measurement for high-temperature thermal property evaluation, the thermal conductivity of PbTe has been historically overestimated [90]. Reinvestigations of both the n-type [21] and the p-type [18] doped PbTe using the laser flash method (a technology commonly utilized today) show surprisingly lower thermal conductivities, only about 60% of the thermal conductivity measured using steady-state methods, for example, ~2 W m^{-1} K^{-1} at room temperature [14,20,21]. The corresponding thermal conductivities of PbSe and PbS are similarly low at 1.8 and 3 W m^{-1} K^{-1}, respectively [56,84]. These values are only about one-fifth of polycrystalline $Si_{0.7}Ge_{0.3}$ (10 W m^{-1} K^{-1} at 300 K, a thermoelectric alloy with strong random mass disorder [48]) and 1/4–1/10 of other binary chalcogenides, such as CdTe (~7.5 W m^{-1} K^{-1} at 300 K [91]), CdSe (~9 W m^{-1} K^{-1} at 300 K [92]), and CdS (~20 W m^{-1} K^{-1} at 300 K [93]). Generally, such low thermal conductivities are expected in highly disordered alloys [75,94,95] or crystals with rather complex unit cells having loosely bound atoms "rattling" in open cages, for example, clathrates [96,97] and filled skutterudites [76,98–102]. The anomalously low thermal conductivity seen in such simple high-symmetry rock salt–structured binary PbTe, PbSe, and PbS compounds remained a mystery over a long period.

Recently, neutron diffraction studies on lead chalcogenides uncovered the physical origin of their abnormally high thermal resistance and positive temperature dependence of the bandgap, that is, the giant anharmonic chemical bonding [48,103].

Recently, Božin et al. [48] discovered a surprise in these materials using the atomic pair distribution function (PDF) analysis technique based on synchrotron x-ray and neutron diffraction data. The PDF analysis is a powerful local structure-based approach that is capable of seeing aperiodic defects and imperfections in crystalline solids. The most striking finding in this work, which went undetected before, is that they observed an abnormal thermal broadening of the nearest neighbor Pb–Te peak in PbTe, which could only be interpreted as a differentiation on the uniformity of the six Pb–Te bonds of the octahedron in the PbTe structure. In other words, instead of six equal bonds around each Pb atom, the PDF data revealed that the Pb atoms were off-center in the $PbTe_6$ octahedron, randomly along the three symmetry equivalent ⟨100⟩ directions. This off-center Pb displacement is absent at $T = 0$ K but appears around 100 K and increases with temperature up to 500 K, where it saturates (Figure 4.10). This result was also observed in PbSe, PbS, and SnTe, and it has changed the way we look at PbTe and its analogs. This Pb off-centering effect, which is favored on warming, has been called emphanisis. It is caused by the well-known tendency of the lone pair of electrons in group VI divalent cations (e.g., Pb^{2+} $6s^2$, Sn^{2+} $5s^2$, Ge^{2+} $4s^2$) to stereochemically express themselves and occupy a defined region of space. In the perfect octahedral coordination of these divalent ions, the lone pair of electrons is not stereochemically expressed, occupies no defined space, and is pushed up in energy very close to the top of the valence band. By moving off-center, stereochemical expression occurs and the energy level moves down, thereby opening the energy bandgap. This local disorder can account for some of the peculiar properties of the lead chalcogenides, for example, the very low lattice thermal conductivity at elevated temperatures and the positive temperature dependence of the bandgap [2,7,9–13,27,55,84].

Motivated by the PDF analysis results, Delaire et al. [103] further examined the strong anharmonicity in PbTe using a combination of inelastic neutron scattering (INS) measurements and first-principles computations of the phonons. Figure 4.11a is the INS data for the phonon wave vector $\mathbf{q} = (0, 0, 1/3)$ along (00*L*) in the (*HKL*) = (113) Brillouin zone at 300 K, where (HKL) denote reciprocal lattice units (rlu). The white lines plotted on top of the INS data are the harmonic phonon dispersions from *ab initio* phonon calculations using density functional theory (DFT), and the corresponding schematic representations of the labeling branches are shown in Figure 4.11b. The huge difference between the INS data and DFT calculations in the harmonic approximation comes from the TO branch. The TO branch has a minimum $E_{TO} = 4$ meV at Γ and a steep dispersion to ~10 meV at $q = 0.2$ rlu and does

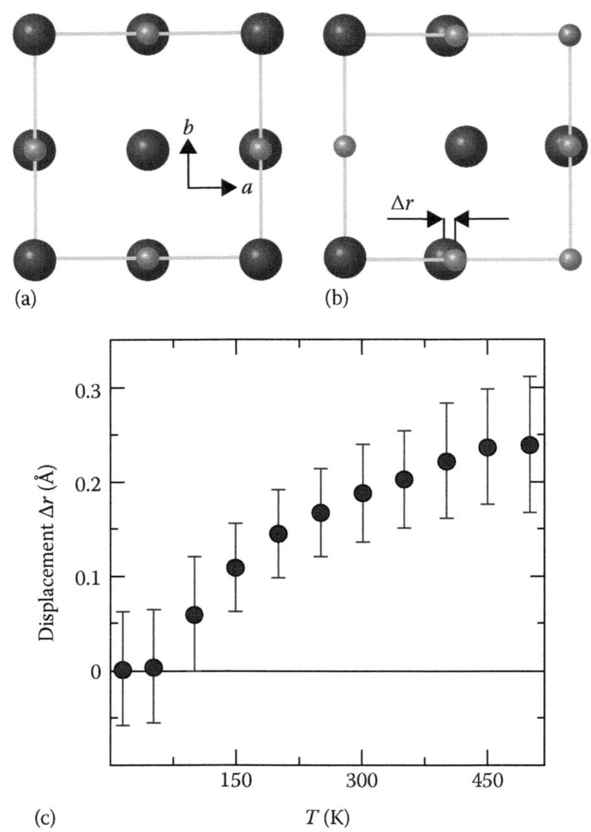

FIGURE 4.10
(a) Schematic representation of the rock salt structure shown in projection down the *c* axis, where Pb atoms are in large spheres and Te in small spheres. The Pb is in a perfect octahedron of six Te atoms. (b) Same view of the distorted rock salt structure above room temperature. (c) Amplitude of Pb local off-centering refined from the ⟨100⟩ displaced model. (From Božin, E.S., Malliakas, C.D., Souvatzis, P. et al., Entropically stabilized local dipole formation in lead chalcogenides, *Science*, 330, 1660–1663, 2010. Reprinted with permission of AAAS.)

not cross the LA branch. Instead, a strong repulsion between the two branches is observed. The TO branch quickly rises above the top of the LA branch, whereas the intensity of the LA branch is strongly suppressed around q = 1/3. The same effect was measured in all other zones where the TO mode was observable. It is noted that the LA–TO crossing would normally occur between noninteracting modes of different symmetries along the symmetry lines, and thus the observed avoided crossing must stem from an anharmonic interaction. This strong anharmonicity in bonding originates from the Pb emphanisis effect and gives rise to a large Grüneisen parameter of about 2 for PbTe, which should account for its low thermal conductivity.

In parallel with the understanding of the intrinsic low thermal conductivity in PbTe, approaches to further decrease the thermal conductivity have been an important topic in the past several decades. These approaches include synergistic alloying [10,15,28], second phase nanostructuring [88,104], and mesostructuring (grain boundary scattering) [11,13,26], the integration of which constructs all-scale hierarchical architectures

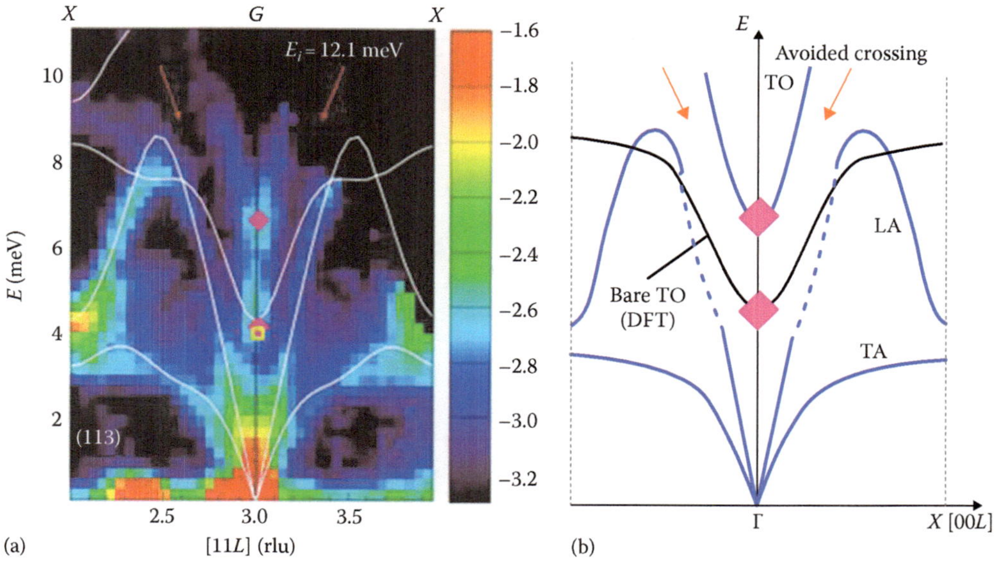

FIGURE 4.11
(a) Inelastic neutron scattering data of PbTe at 300 K, showing the avoided crossing behavior of LA and TO phonon branches. Solid white lines are harmonic dispersions calculated with density functional theory. (b) Schematic representations of the dispersions (blue lines), with the bare TO branch as a thin black line. Pink diamonds indicate the positions of the peaks in the TO scans at Γ (113). (Reprinted by permission from Macmillan Publishers Ltd., *Nat. Mater.*, Delaire, O., Ma, J., Marty, K. et al., Giant anharmonic phonon scattering in PbTe, 10, 614–619, copyright 2011.)

that are able to scatter a wide spectrum of heat-carrying phonons with different mean free paths (MFPs) [11,13].

4.3.1 Atomic-Scale Structuring: Solid Solutions and Carrier Doping

Creating lattice imperfections in a perfect crystal by forming a solid solution is a well-established approach to lower the thermal conductivity [89,105,106]. These are atomic-scale point defects and are particularly effective in scattering short-wavelength phonons. In PbTe, the frequently used dopants are Na [12,13], K [107], Tl [9], In [108], Ag [10], Cd [38], Hg [32], Mg [13], Mn [39], Sb [109], Bi [110], Sn [111], Ge [112], La [20] at the cationic sites, and Se [12], S [115], as well as halogens [21,113] at the anionic sites. To realize the strongest point defect scattering of phonons, the mass and size difference between the guest and host atoms should be as large as possible; also, the doping fraction should be as high as possible [89,105,106]. In a thermoelectric optimization strategy, however, an excessive amount of point defects may not be desirable despite the strong reduction in the thermal conductivity. This is because they also cause a strong reduction in the carrier mobility, thereby diminishing any benefits and preventing an improvement in *ZT*. In practice, most of these dopants have limited solubility in PbTe due to charge imbalance or atomic size mismatch. It is known that to exert stronger influence on the lattice thermal conductivity by alloying, the alloying fraction (or point defects concentration) should be as high as possible. Therefore, a cross-substitution approach was proposed to maximize the alloying limit. Cross substitution means replacing one or more of the host elements by pairs from other groups of the periodic table, while keeping the overall

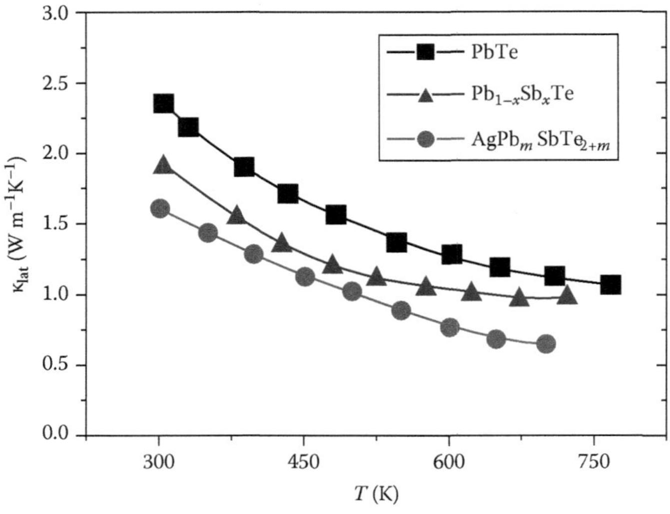

FIGURE 4.12
Comparison of temperature-dependent lattice thermal conductivity (κ_{lat}) for pure PbTe [104], Sb singly doped PbTe [109], and Ag/Sb cross-substituted PbTe ($AgPb_mSbTe_{2+m}$) [10]. For the sake of comparison under equal conditions in the calculation of the electronic thermal conductivity contribution, $\kappa_{ele} = L\sigma T$, the Lorenz number L was unified by fitting the Seebeck coefficient to the reduced chemical potential. (From Biswas, K. et al., *Nature*, 489, 414, 2012; Tan, G. et al., *J. Mater. Chem. C*, 3, 8372, 2015; Tan, G. et al., *J. Mater. Chem. A*, 2, 20849, 2014.)

valence electron count constant. For instance, two Pb^{2+} atoms were concurrently replaced by Ag^{1+} and Sb^{3+}, yielding $Pb_{1-2x}Ag_xSb_xTe$ (more frequently written as $AgPb_mSbTe_{2+m}$ and known as the LAST alloy [10]). The charge-balanced cross-substituted compound has significantly higher alloying fraction than what can be achieved with single doping, thereby displaying a much lower thermal conductivity (note that nanostructuring may also contribute to the very low thermal conductivity in the LAST system [10], as we discuss later) (Figure 4.12).

It should be emphasized that using Vegard's law in predicting the lattice parameters to gauge the alloying fraction [38,75,76] may be misleading, as has been pointed out in many previous studies [56,114–116]. The most well-known example is the LAST system [10], which has been viewed as a solid solution for a long time because the Vegard's law between the two end members was apparently obeyed [117,118]. However, more detailed and refined structural characterization (e.g., transmission electron microscopy [TEM] [13,58,119,120]) beyond the basic x-ray powder diffraction experiments revealed extensive nanostructuring caused by compositional fluctuations [114]. Later, many other systems that were previously believed to be solid solutions were demonstrated to be actually nanostructured, such as PbTe–PbSe [121], PbTe–PbS [14,122], PbTe–Na [123], and so on. Notably, in the p-type K-doped $PbTe_{1-x}S_x$ system [122], it is not a complete solid solution but a spinodally decomposed system with large PbTe-rich and PbS-rich regions, where each region exhibits dissimilar types of nanostructures. The rich microstructures in this system allow a very low lattice thermal conductivity ~0.4 W m^{-1} K^{-1} at 923 K, while the same S incorporation on the Te sites modifies the band structure of PbTe by pushing the two valence bands closer in energy (thus valence band convergence) yielding a significantly enhanced Seebeck coefficient. Concurrently, a maximum ZT of ~2.2 at 923 K, together with the broadest plateau of $ZT > 2$ from 673 to 923 K was achieved in this system [122].

4.3.2 Nanostructuring

When materials are nanostructured, the high density of interfaces between the matrix and the second phase are expected to strongly scatter the midwavelength phonons with MFPs similar in size to the nanostructures [11,13,14,20,25,29–31,59]. It has now been shown that thermal conductivity values well below the "alloy limit" can be achieved by the nanostructuring approach [10,11]. The early studies of nanostructured thermoelectrics mainly focused on thin-film superlattices [124], quantum dots [125], and nanowires [126]. However, many of these low-dimensional materials are unstable and tend to diffuse and homogenize at high temperatures and difficult to measure accurately. Also, they are not practical for large-scale commercial use because they are prepared by atomic layer deposition processes, such as molecular beam epitaxy [127], making the process slow and expensive to fabricate and scale up.

Bulk nanostructured materials can be fabricated in large quantities and in a form that is compatible with existing thermoelectric measurement techniques and device configurations. There are generally two ways of introducing nanostructures into bulk materials. The first kind consists of a single phase material assembled from nanosized particles or grains that can be produced by ball milling or wet chemical synthesis [128,129]. The other kind is systems with a major bulk phase matrix embedded by a minor second phase, typically *via* precipitation from a solid solution. In the latter, the second phase is nanoscaled but the matrix does not necessarily need to be. This intuitive approach has been mostly employed in recent studies with great success [10,11,32]. Such bulk nanostructured materials are, by their nature, thermodynamically stable because of the insolubility of the second phases.

A general approach of introducing second phase nanostructures in a parent matrix is through nucleation and growth or spinodal decomposition. To do so, the second phase candidates should be almost entirely soluble with the parent phase when molten but have limited or no solubility in their solid state [26]. Figure 4.13a schematically shows the formation process of a nanostructured second phase B in the matrix A by a metallurgy method. When the temperature is raised above the melting points of A and B, they are homogeneous melts, meaning A and B are fully dissolved in one another (point 1). When such melts are cooled to become solids, the B phase will precipitate out since its solid state solubility limit is reached or exceeded; see the typical morphology of these nanostructures in Figure 4.13b. To ensure that the B phase is nanostructured and well distributed in the matrix of A, one usually chooses to first fast quench the melts to room temperature and then anneal the resultant solids at different temperatures (e.g., points 2 and 3) or for different times [20,29]. In this fashion, the size and volume concentration of the second phase can be better controlled.

The PbTe–$AgSbTe_2$ system ($AgPb_mSbTe_{2+m}$, LAST [10]) was the first example in which this type of nanostructuring was discovered in a bulk thermoelectric material and was shown to have a significant impact on lowering the lattice thermal conductivity. It was the powerful tool of TEM that clearly demonstrated that LAST is nanostructured and that the nanoscale inclusions of minor phases in LAST exhibit coherent or semicoherent interfaces with the matrix [10] (see Figure 4.14). This coherency involves the good lattice matching (or very small lattice mismatch) around the precipitated nanoparticles, which we have called endotaxy. This is an important and favorable feature for facile carrier transport throughout the sample and for creating lattice strain to scatter phonons [10,11,30,31]. The high density of spontaneous and ubiquitous nanoscale Ag–Sb-rich precipitates in the PbTe matrix, together with the structural complexity of the LAST system, gives rise to a very low thermal conductivity of ~2.3 W m^{-1} K^{-1} at room temperature when $m = 18$. As a result, a high ZT

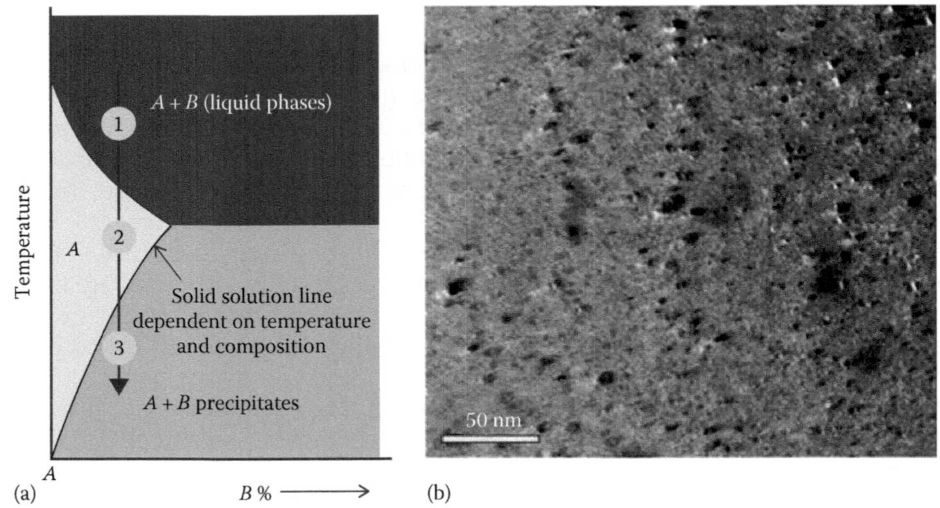

FIGURE 4.13
(a) Schematic representation of pseudobinary phase diagram, indicating a strongly temperature-dependent solubility of the second phase of *B* in the matrix *A*. (b) Typical nanoprecipitates observed by transmission electron microscopy. (From Zhao, L.-D., Dravid, V.P., and Kanatzidis, M.G., The panoscopic approach to high performance thermoelectric, *Energy Environ. Sci.*, 7, 251–268, 2014. Reproduced by permission of The Royal Society of Chemistry.)

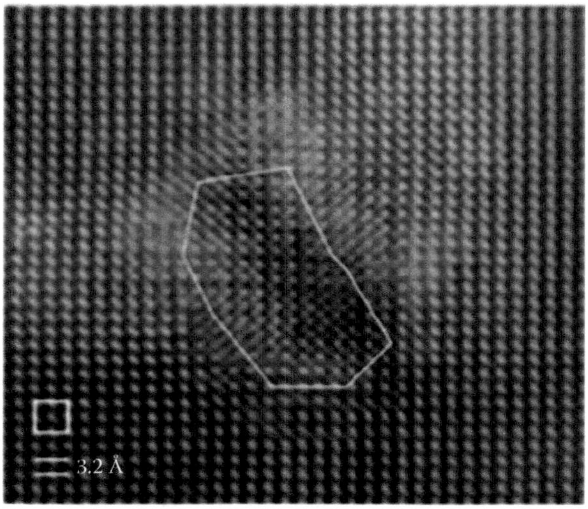

FIGURE 4.14
Transmission electron microscopy image [10] of a LAST system with $m = 18$ showing a Ag–Sb-rich nanoparticle highlighted by the closed white curve. (From Hsu, K.F., Loo, S., Guo, F. et al., Cubic AgPb(m)SbTe(2+m): Bulk thermoelectric materials with high figure of merit, *Science*, 303, 818–821, 2004. Reprinted with permission of AAAS.)

~1.5–1.7 can be achieved at ~700 K by optimizing the composition (*m* value) [10]. After this first observation, nanostructures were also discovered in a number of other PbTe-based bulk materials, including $Ag(Pb_{1-x}Sn_x)_mSbTe_{2+m}$(LASTT) [130], $NaPb_mSbTe_{2+m}$ (SALT) [29], KPb_mSbTe_{2+m} (PLAT) [131], PbTe–PbS [132], and, most recently, PbTe–MTe (M = Ca, Sr, Ba) [11,30,88]. Extensive studies have shown that the precipitation phenomena in these quaternary systems are very complex and more than one second phase can precipitate in the matrix. This causes complications in understanding the nanostructuring and controlling it. By far simpler and more instructive are the pseudobinary systems PbTe–MTe, which will be discussed in the following.

The PbTe-MTe systems proved to be the most interesting from the point of view of nanostructuring and achieving exceptional thermoelectric performance [11,30,88]. In this series, the minor phases have a strong tendency to precipitate out from the PbTe matrix because of their limited solid state solubility (<1 mol%). Even a very small amount of these second phases can yield a high concentration of nanostructures with quite a narrow size distribution (1–10 nm), which exert strong disturbance on the phonon propagation. Meanwhile, these MTe second phases crystallize in the same rock salt structure as PbTe does and have quite small lattice mismatches (<10%) with respect to the matrix. Therefore, these nanostructures usually exhibit good endotaxy, which is beneficial to retain high carrier mobility. As a consequence, at room temperature, the lattice thermal conductivity of PbTe can be reduced from ~2.0–2.5 to ~1.0 W m^{-1} K^{-1}, to ~1.2 W m^{-1} K^{-1}, and to ~1.2 W m^{-1} K^{-1} when 5% CaTe, 3% BaTe, and 2% SrTe are introduced as nanostructures, respectively. The lattice thermal conductivity is, in fact, reduced at all temperatures and, correspondingly, the highest *ZT* of ~1.5 at 765 K, ~1.3 at 760 K, and ~1.7 at 800 K has been obtained [11,30,88].

In some systems, a strong reduction of the lattice thermal conductivity of PbTe has also been observed with relatively large nanostructures. Platelike Ag_2Te precipitates [20] with long dimensions of 100–200 nm and short dimensions of 50–100 nm were finely distributed in the PbTe matrix by a facile control of material synthesis according to the equilibrium phase diagram. These large nanostructures could reduce the thermal conductivity of PbTe below that of a macroscale composite of saturated alloys with the Kapitza-type interfacial thermal resistance at the same overall composition. Another such example was demonstrated in the PbTe–$PbSnS_2$ composite [104]. TEM study showed that $PbSnS_2$ segregates into coherent lamellar structures of 50–100 nm in thickness and 100 nm to 15 μm in length (Figure 4.15a and b). Despite the relatively large size of the $PbSnS_2$ precipitates, the incorporation of $PbSnS_2$ in PbTe led to a strong reduction in the lattice thermal conductivity down to 0.4–0.65 W m^{-1} K^{-1} over the temperature range of 300–700 K, a 50%–70% smaller thermal conductivity than in the bulk PbTe (see Figure 4.15c). As a result, a maximum *ZT* of 1.1 was obtained for ingot samples of PbTe–6%$PbSnS_2$ [104] (Figure 4.15d).

Despite the important progress made in lowering the thermal conductivity with the nanostructuring approach, it is useful to emphasize that not all nanostructures are effective in scattering phonons. For example, Bi and Pb nanostructures [119] and Na- or K-containing platelet-like nanostructures [123] are very inefficient in decreasing the lattice thermal conductivity of PbTe. Therefore, the mere presence of nanostructures in the materials is not enough to increase phonon scattering. It is believed that the use of appropriate nanoprecipitates that create large mass contrast, significant strain fields (elastic strain), and dislocations (plastic and elastic strain) may be more important for blocking the phonon transport [119]. In practice, the interactions of phonons and interfaces in different structures can be quite complex and the effects of interfacial structure, composition, and

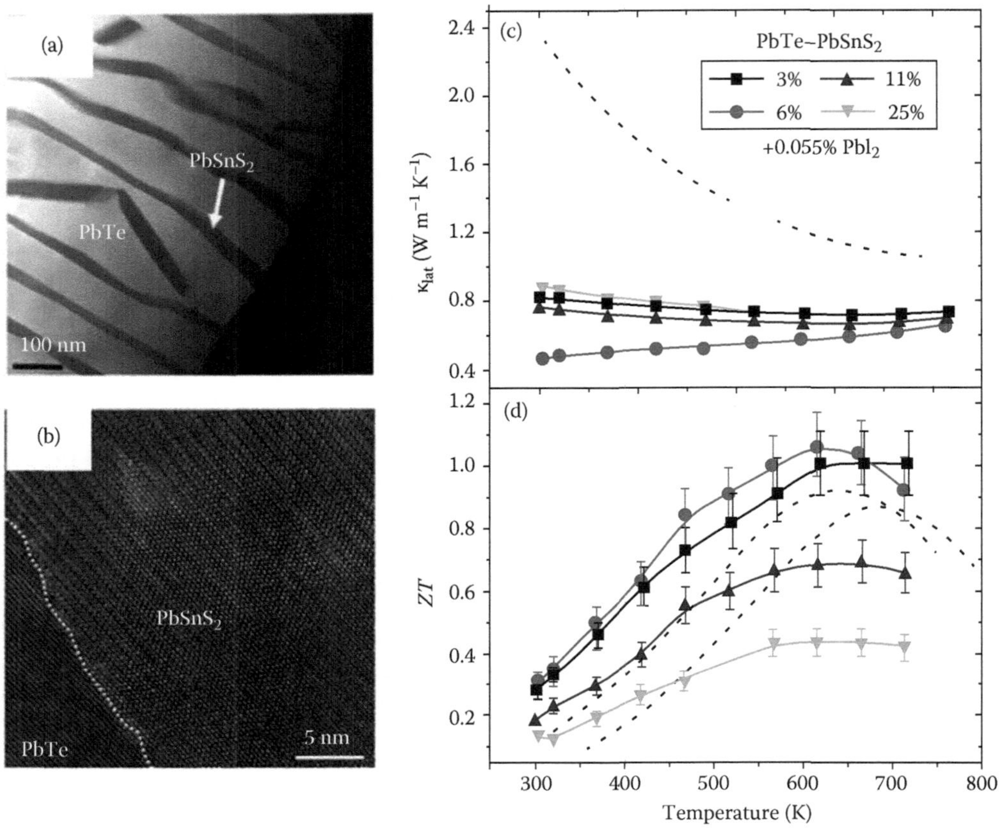

FIGURE 4.15
(a) Low-magnification transmission electron microscopy (TEM) image of PbTe–PbSnS$_2$ showing large PbSnS$_2$ lamellae throughout the PbTe matrix. (b) A high-resolution TEM image of PbTe–PbSnS$_2$ indicating the coherent interface between the matrix PbTe and the second phase PbSnS$_2$. (c) Lattice thermal conductivity and (d) ZT value as a function of temperature for the 0.055% PbI$_2$-doped n-type PbTe-PbSnS$_2$ nanocomposite. In (c) and (d), values for pure PbTe are indicated by the dashed line. (From Girard, S.N., Chasapis, T.C., He, J. et al., PbTe–PbSnS$_2$ thermoelectric composites: Low lattice thermal conductivity from large microstructures, *Energy Environ. Sci.*, 5, 8716–8725, 2012. Reproduced by permission of The Royal Society of Chemistry.)

morphology on the phonon and electron transports are not yet fully understood. More in-depth analysis and understanding of interfacial physics would facilitate a better control and design of the most desirable microstructures for advanced thermoelectrics.

4.3.3 Mesostructuring and Spark Plasma Sintering

Grain boundaries can be viewed as defects that modify the stiffness (or acoustic impedance), thereby providing additional mechanisms for phonon scattering [25]. To investigate the influence of grain sizes (L) on the thermal conductivity, Rowe et al. [133] prepared a series of fully dense SiGe polycrystalline samples with L in the range of $10 < L < 25$ μm, $5 < L < 10$ μm, and $L < 5$ μm. The thermal conductivities of these polycrystals were compared with that of their single crystal counterpart (where L can be viewed as infinity). It was found that the lattice thermal conductivity decreased steadily with decreasing L, and, in the sample with $L < 5$ μm, this reduction was ~28% in comparison with the single crystal material. This work highlighted the importance the grains have in impeding the phonon transport in thermoelectric materials. Later, this conclusion was verified in several systems, including PbTe [11,13].

Similarly, in the as-cast PbTe ingots, the grains are quite coarse, and their sizes are typically larger than 10 μm. When they were pulverized to reduce the grain sizes to less than 5 μm and then compacted by powder processing techniques, such as spark plasma sintering (SPS), it was found that the lattice thermal conductivity was significantly smaller with respect to the cast ingot. For example, the lattice thermal conductivity of $Pb_{0.98}Na_{0.02}Te$-4%SrTe decreased from ~1 W m^{-1} K^{-1} for the cast ingot to well below 0.6 W m^{-1} K^{-1} at ~900 K by SPS treatment and, correspondingly, the maximum ZT increased from ~1.7 to ~2.2 at ~913 K [11].

SPS is an advanced sintering technology producing mesostructured specimens. In comparison with the conventional hot pressing method, where the heat is provided by externally heating elements over long periods of time (e.g., hours), the SPS technique is rapid (tens of min) and makes use of pulsed direct current that produces internal Joule heating. SPS allows the production of specimens with near theoretical density at relatively low sintering temperatures. Moreover, the high heating and cooling rate (up to 1000 K min^{-1}) of the SPS technique has a great advantage of densifying powders while avoiding the coarsening of grains, a very important feature in the field of thermoelectric materials research [13,26].

4.3.4 All-Scale Architecture

Heat in solids is carried by phonons with different MFPs. The different contributions to phonon scattering by defects at all length scales (i.e., atomic-, nano-, and mesoscale) have been recently calculated for PbTe [134] and Si [135] (Figure 4.16a). For these systems,

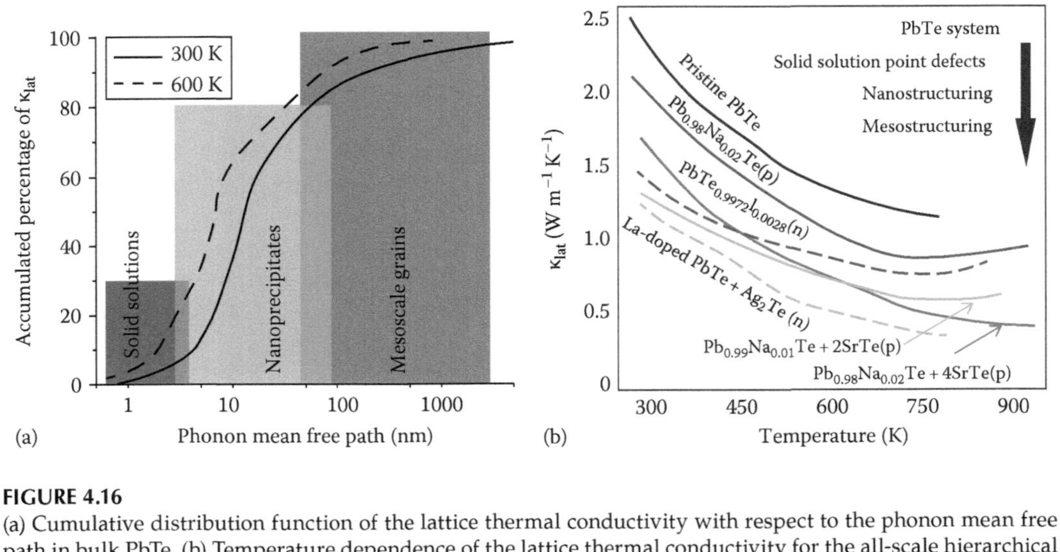

FIGURE 4.16
(a) Cumulative distribution function of the lattice thermal conductivity with respect to the phonon mean free path in bulk PbTe. (b) Temperature dependence of the lattice thermal conductivity for the all-scale hierarchical architecture PbTe system. (From Zhao, L.-D., Dravid, V.P., and Kanatzidis, M.G., The panoscopic approach to high performance thermoelectric, *Energy Environ. Sci.*, 7, 251–268, 2014. Reproduced by permission of The Royal Society of Chemistry.)

it is shown that ~25% of the lattice thermal conductivity of PbTe is contributed by phonons with MFP <5 nm, which can be attributed to scattering by a combination of atomic-scale solid solution alloying and/or dislocations, and ~55% is contributed by phonons with MFPs of 5–100 nm, which can be scattered by nanoparticles embedded in PbTe. The remaining ~20% is contributed by phonons with MFPs of 0.1–1 μm. Therefore, structures of all length scales, if present in a single bulk material, can strongly scatter a wider spectrum of heat-carrying phonons and therefore lead to extremely low lattice thermal conductivity [11,13].

Figure 4.16b compares the lattice thermal conductivities of both *n*-type and *p*-type PbTe systems with undoped PbTe [11,20,21]. The point defects by Na and I doping lower the room temperature lattice thermal conductivity of PbTe ingots by ~25% and ~55%, respectively. When they are nanostructured with SrTe or Ag_2Te, additional phonon scattering mechanisms are introduced, and the lattice thermal conductivities of *p*-type and *n*-type ingots are further decreased by ~50% and ~70%, respectively. The mesostructuring in SrTe-nanostructured PbTe is created by the SPS treatment and leads to an even lower lattice thermal conductivity. With all-scale hierarchical architectures in one material, the lowest lattice thermal conductivity approaches ~0.6 W m^{-1} K^{-1} at elevated temperature, which is getting close to the amorphous limit of PbTe of ~0.4 W m^{-1} K^{-1} [136]. This all-scale architecture concept of reducing thermal conductivity has been now widely applied in many thermoelectric systems [55,56].

4.4 Second Phase Band Alignment

Second phase nanostructuring is very effective in reducing the thermal conductivity. However, in most cases, the power factors are also decreased by this nanostructuring process because the carriers are also scattered at the matrix/precipitate interfaces [56,88]. Therefore, the net enhancement in *ZT* by nanostructuring in bulk materials is only moderate and, in some cases, can even become negative. It is possible to minimize or eliminate the carrier mobility deterioration caused by the interface scattering by paying attention to the electronic band alignment of the two phases.

Previous studies on GaAs/InAs (001) heterojunctions indicated that charge carriers can transmit freely when the band offsets are minimal [137]. The matrix and the second phase may show a huge difference in their bandgaps. However, as long as their valence or conduction band energy offsets are small enough and comparable to the magnitude of the thermal energy at elevated temperatures of a few k_BT (k_B is the Boltzmann constant and T is the absolute temperature), the carriers can easily flow across the interfaces without encountering too much scattering [11,30].

The first successful example of band alignment in a bulk system was *p*-type PbTe embedded with SrTe nanocrystals [11,30]. As shown in Figure 4.17a, the introduction of the SrTe nanoscale second phase steadily decreases the lattice thermal conductivity (dotted lines) of PbTe, while the hole mobility (solid lines) remains largely unaffected. First-principles electronic band structure calculations of PbTe–SrTe composite structures indicated that, at 0 K, the VBM of PbTe lies slightly lower than that of SrTe, separated by a small energy offset of only 0.17 eV [11]. With increasing temperature, due to the peculiar decrease in the energy of the VBM of PbTe (as discussed earlier), this small band offset between the two phases becomes even smaller, resulting in almost aligned valence bands of PbTe and SrTe

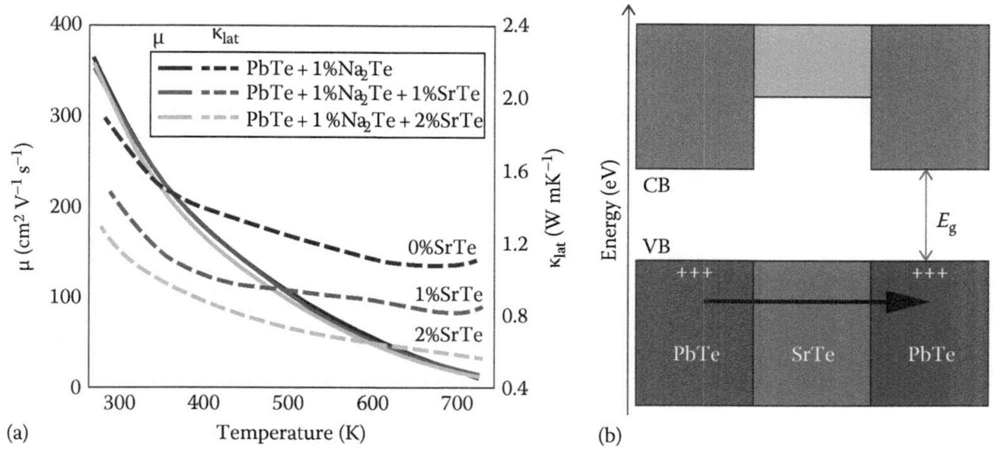

FIGURE 4.17
(a) Temperature-dependent carrier mobility and lattice thermal conductivity for the hole doped PbTe–SrTe nanostructured system. (b) Schematic representation of the alignment of the valence band and the conduction band energies of SrTe precipitates in the PbTe matrix. (From Zhao, L.-D., Dravid, V.P., and Kanatzidis, M.G., The panoscopic approach to high performance thermoelectric, *Energy Environ. Sci.*, 7, 251–268, 2014. Reproduced by permission of The Royal Society of Chemistry.)

(at $T > 300$ K) (Figure 4.17b). The matrix/precipitate valence band alignment preserves the high carrier mobility of *p*-type PbTe [11].

This stratagem of enhancing the thermoelectric performance of bulk nanostructured materials has been extended to other lead chalcogenide systems [84,120]. Four candidate nanostructures (CdS, ZnS, CaS, and SrS) were chosen to examine their effects on the electronic and thermal transport properties of *p*-type PbS [120]. These four nanostructures are equally effective in reducing the lattice thermal conductivity of PbS to a level of ~0.5 W m^{-1} K^{-1} at high temperature ($T > 800$ K). However, the CdS- and ZnS-nanostructured PbS samples exhibit much higher mobilities than those achievable with SrS and CaS nanostructuring and therefore achieve significantly larger ZT values. The theoretical electronic structure calculations indicate that, with respect to PbS, the valence band offsets are ~0.13, ~0.16, ~0.53, and ~0.63 eV for CdS, ZnS, CaS, and SrS, respectively. The small valence band energy offsets between PbS and CdS or ZnS are comparable to the magnitude of the thermal energy even at elevated temperatures, thereby resulting in intact hole mobilities. In the cases of CaS and SrS nanostructuring, the valence band offsets are too large and hole transport is deteriorated. Therefore, theoretical electronic structure calculations can be used to judiciously select materials as second phases for nanostructuring PbTe and other materials for best band alignment.

It is not always possible to find an ideal second phase candidate that has well-aligned electronic bands with the matrix. In this regard, band alignment using compositionally alloyed nanostructures seems to be a promising route [84]. For example, as shown in Figure 4.18b, none of CdSe, CdS, ZnSe, and ZnS has a perfect valence band alignment with PbSe. However, by forming a solid solution with the host matrix PbSe through interface diffusion, the CdS/ZnS changes its composition to $CdS_{1-x}Se_x/ZnS_{1-x}Se_x$. These new compositions have effective valence band alignment with that of PbS (Figure 4.18b). As a result, the CdS/ZnS-nanostructured PbSe has much higher carrier mobilities than CdSe/ZnSe-nanostructured PbSe (Figure 4.18a).

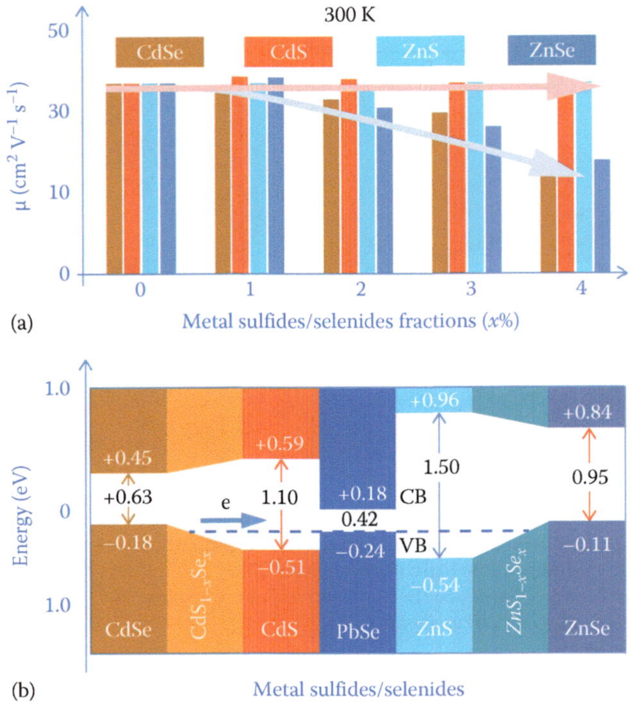

FIGURE 4.18
(a) Room temperature carrier mobility as a function of second phase fractions for hole-doped PbSe nanostructured with metal selenides/sulfides [84]. All samples have a similar carrier concentration of ~3×10^{20} cm^{-3}. (b) Conduction band and valence band energy levels for PbSe, CdS, CdSe, ZnS, and ZnSe. $CdS_{1-x}Se_x$ and $ZnS_{1-x}Se_x$ solid solutions were determined by density functional theory calculations. (From Zhao, L.-D., Dravid, V.P., and Kanatzidis, M.G., The panoscopic approach to high performance thermoelectric, *Energy Environ. Sci.*, 7, 251–268, 2014. Reproduced by permission of The Royal Society of Chemistry.)

4.5 Conclusions

PbTe is a paradigmatic thermoelectric material that has been studied for a long time. It combines many interesting features for its outstanding thermoelectric properties. Its anharmonic chemical bonding and the ease with which its microstructure can be engineered at all length scales enable a very low lattice thermal conductivity that almost approaches its amorphous limit. In the meanwhile, it has two valence bands, the energy offset of which can be feasibly controlled by alloying for effective band convergence, leading to extraordinary electronic properties. More importantly, the electron and phonon transports in PbTe can be partly decoupled by using the matrix/precipitate band alignment effect, allowing the highest possible ZT to be achieved.

To date, most of the advances occurring in the PbTe system mainly focus on the p-type structure, with the maximum ZT reaching 2.2 at 913 K. In comparison, the progress in n-type seems slow. Of course, thermoelectric modules require both n- and p-type materials with equally high thermoelectric performance to maximize the efficiency. Therefore, the future research should be aimed at increasing the ZT values of n-type PbTe.

From the earlier discussion, it is evident that to achieve the highest possible thermoelectric performance, all property enhancing mechanisms aiming at high *ZT*s need to be integrated in one single material. These mechanisms should work together and not interfere with one another. We call this a "panoscopic" approach. Its validity has been well demonstrated in the PbTe system and is suitable for use in many other thermoelectric systems. In the 1960s, the Na-doped PbTe was shown to have a relatively low maximum *ZT* of ~0.7 because the material was too lightly doped [138]. Later, the two-valence-band character of PbTe was better understood, and heavy doping was brought into play in making use of the heavy hole band that greatly benefited the Seebeck coefficient. As a result, a significantly higher maximum *ZT* of ~1.4 was realized at 750 K as long as the proper level of doping (~2×10^{20} cm^{-3}) was maintained [18]. By combining with nanostructuring (PbTe–MTe, M = Ca, Sr, Ba) [11,30,88] or intra-matrix band engineering (PbTe–MTe, M = Mg, Mn, and Cd) [13,38,39], the *ZT* of PbTe was further improved to ~1.7–1.8 at ~ 800 K. More recently, studies using the all-scale hierarchical architecture concept (the so-called panoscopic approach) yielded a record high *ZT* of ~2.2 at 913 K [11]. Similarly, a high *ZT* of ~2 at 875 K was reported for the all-scale hierarchically architectured PbTe–MgTe system, where Mg alloying in PbTe facilitated valence band convergence and created point defects, while excess MgTe emerged as endotaxial nanostructures [13].

In addition to concentrating on increasing the thermoelectric performance of PbTe, it is also highly desirable to assess and understand the mechanical properties of this material as they are of great importance for device fabrication. Although some articles focusing on this point already exist [33–37], much more work is still needed to pave the path for device assembly, testing, and, ultimately, commercialization.

Acknowledgment

This work was supported by the Department of Energy, Office of Science Basic Energy Sciences under grant DE-SC0014520.

References

1. Lyden, H. A. 1964. Temperature dependence of the effective masses in PbTe. *Phys. Rev.* 135: 514–521.
2. Tauber, R., A. Machonis, and I. Cadoff. 1966. Thermal and optical energy gap in PbTe. *J. Appl. Phys.* 37: 4855–4860.
3. Greig, D. 1960. Thermoelectricity and thermal conductivity in the lead sulfide group of semiconductors. *Phys. Rev.* 120: 358–365.
4. Keffer, C., T. M. Hayes, and A. Bienenstock. 1970. Debye-Waller factor and the PbTe band gap temperature dependence. *Phys. Rev. B* 2: 1966–1976.
5. Allgaier, R. and B. Houston, Jr. 1966. Hall coefficient behavior and the second valence band in lead telluride. *J. Appl. Phys.* 37: 302–309.
6. Sitter, H., K. Lischka, and H. Heinrich. 1977. Structure of the second valence band in PbTe. *Phys. Rev. B* 16: 680–687.
7. Kohn, S., P. Yu, Y. Petroff, Y. Shen, Y. Tsang, and M. Cohen. 1973. Electronic band structure and optical properties of PbTe, PbSe, and PbS. *Phys. Rev. B* 8: 1477–1488.

8. Allgaier, R. 1961. Valence bands in lead telluride. *J. Appl. Phys.* 32: 2185–2189.
9. Heremans, J. P., V. Jovovic, E. S. Toberer et al. 2008. Enhancement of thermoelectric efficiency in PbTe. *Science* 321: 554–557.
10. Hsu, K. F., S. Loo, F. Guo et al. 2004. Cubic AgPb(m)SbTe(2+m): Bulk thermoelectric materials with high figure of merit. *Science* 303: 818–821.
11. Biswas, K., J. He, I. D. Blum et al. 2012. High-performance bulk thermoelectrics with all-scale hierarchical architectures. *Nature* 489: 414–418.
12. Pei, Y., X. Shi, A. LaLonde et al. 2011. Convergence of electronic bands for high performance bulk thermoelectrics. *Nature* 473: 66–69.
13. Zhao, L., H. Wu, S. Hao et al. 2013. All-scale hierarchical thermoelectrics: MgTe in PbTe facilitates valence band convergence and suppresses bipolar thermal transport for high performance. *Energy Environ. Sci.* 6: 3346–3355.
14. Girard, S. N., J. He, X. Zhou et al. 2011. High performance Na-doped PbTe–PbS thermoelectric materials: Electronic density of states modification and shape-controlled nanostructures. *J. Am. Chem. Soc.* 133: 16588–16597.
15. Korkosz, R. J., T. C. Chasapis, S.-H. Lo et al. 2014. High *ZT* in p-type $(PbTe)_{1-2x}(PbSe)_x(PbS)_x$ thermoelectric materials. *J. Am. Chem. Soc.* 136: 3225–3237.
16. Brebrick, R. and R. Allgaier. 1960. Composition limits of stability of PbTe. *J. Chem. Phys.* 32: 1826–1831.
17. Tsang, Y. and M. L. Cohen. 1971. Calculation of the temperature dependence of the energy gaps in PbTe and SnTe. *Phys. Rev. B* 3: 1254–1261.
18. Pei, Y., A. LaLonde, S. Iwanaga, and G. J. Snyder. 2011. High thermoelectric figure of merit in heavy hole dominated PbTe. *Energy Environ. Sci.* 4: 2085–2089.
19. Gelbstein, Y., Z. Dashevsky, and M. Dariel. 2005. High performance n-type PbTe-based materials for thermoelectric applications. *Phys. B* 363: 196–205.
20. Pei, Y., J. Lensch-Falk, E. S. Toberer, D. L. Medlin, and G. J. Snyder. 2011. High thermoelectric performance in PbTe due to large nanoscale Ag_2Te precipitates and La doping. *Adv. Funct. Mater.* 21: 241–249.
21. LaLonde, A. D., Y. Pei, and G. J. Snyder. 2011. Reevaluation of $PbTe_{1-x}I_x$ as high performance n-type thermoelectric material. *Energy Environ. Sci.* 4: 2090–2096.
22. Sootsman, J. R., H. Kong, C. Uher et al. 2008. Large enhancements in the thermoelectric power factor of bulk PbTe at high temperature by synergistic nanostructuring. *Angew. Chem.* 120: 8746–8750.
23. Ahn, K., C. Li, C. Uher, and M. G. Kanatzidis. 2009. Improvement in the thermoelectric figure of merit by La/Ag cosubstitution in PbTe. *Chem. Mater.* 21: 1361–1367.
24. LaLonde, A. D., Y. Pei, H. Wang, and G. J. Snyder. 2011. Lead telluride alloy thermoelectrics. *Mater. Today* 14: 526–532.
25. He, J., M. G. Kanatzidis, and V. P. Dravid. 2013. High performance bulk thermoelectrics via a panoscopic approach. *Mater. Today* 16: 166–176.
26. Zhao, L.-D.,V. P. Dravid, and M. G. Kanatzidis. 2014. The panoscopic approach to high performance thermoelectrics. *Energy Environ. Sci.* 7: 251–268.
27. Pei, Y., H. Wang, and G. Snyder. 2012. Thermoelectric materials: Band engineering of thermoelectric materials. *Adv. Mater.* 24: 6125–6135.
28. Kanatzidis, M. G. 2009. Nanostructured thermoelectrics: The new paradigm? *Chem. Mater.* 22: 648–659.
29. Poudeu, P. F., J. D'Angelo, A. D. Downey, J. L. Short, T. P. Hogan, and M. G. Kanatzidis. 2006. High thermoelectric figure of merit and nanostructuring in bulk p-type $Na_{1-x}Pb_mSb_yTe_{m+2}$. *Angew. Chem.* 118: 3919–3923.
30. Biswas, K., J. He, Q. Zhang et al. 2011. Strained endotaxial nanostructures with high thermoelectric figure of merit. *Nat. Chem.* 3: 160–166.
31. Vineis, C. J., A. Shakouri, A. Majumdar, and M. G. Kanatzidis. 2010. Nanostructured thermoelectrics: Big efficiency gains from small features. *Adv. Mater.* 22: 3970–3980.

32. Ahn, K., K. Biswas, J. He, V. Dravid, and M. G. Kanatzidis. 2013. Enhanced thermoelectric properties of p-type nanostructured PbTe–MTe (M = Cd, Hg) materials. *Energy Environ. Sci.* 6: 1529–1537.
33. Gelbstein, Y., G. Gotesman, Y. Lishzinker, Z. Dashevsky, and M. Dariel. 2008. Mechanical properties of PbTe-based thermoelectric semiconductors. *Scripta Mater.* 58: 251–254.
34. Sootsman, J. R., J. He, V. P. Dravid, C.-P. Li, C. Uher, and M. G. Kanatzidis. 2009. High thermoelectric figure of merit and improved mechanical properties in melt quenched PbTe–Ge and PbTe–$Ge_{1-x}Si_x$PbTe–$Ge_{1-x}Si_x$ eutectic and hypereutectic composites. *J. Appl. Phys.* 105: 083718.
35. Gelbstein, Y., Z. Dashevsky, and M. P. Dariel. 2008. The search for mechanically stable PbTe based thermoelectric materials *J. Appl. Phys.* 104: 033702.
36. Sootsman, J. R., J. He, V. P. Dravid, S. Ballikaya, D. Vermeulen, C. Uher, and M. G. Kanatzidis. 2009. Microstructure and thermoelectric properties of mechanically robust PbTe-Si eutectic composites. *Chem. Mater.* 22: 869–875.
37. Ni, J. E., E. D. Case, K. N. Khabir et al. 2010. Room temperature Young's modulus, shear modulus, Poisson's ratio and hardness of PbTe–PbS thermoelectric materials. *Mater. Sci. Eng. B* 170: 58–66.
38. Pei, Y., A. D. LaLonde, N. A. Heinz, and G. J. Snyder. 2012. High thermoelectric figure of merit in PbTe alloys demonstrated in PbTe–CdTe. *Adv. Energy Mater.* 2: 670–675.
39. Pei, Y., H. Wang, Z. M. Gibbs, A. D. LaLonde, and G. J. Snyder. 2012. Thermopower enhancement in $Pb_{1-x}Mn_xTe$ alloys and its effect on thermoelectric efficiency. *NPG Asia Mater.* 4: e28.
40. Mitchell, D. and R. Wallis. 1966. Theoretical energy-band parameters for the lead salts. *Phys. Rev.* 151: 581–595.
41. Andersen, O. K. 1975. Linear methods in band theory. *Phys. Rev. B* 12: 3060–3083.
42. Overhof, H. and U. Rössler. 1979. Electronic structure of PbS, PbSe, and PbTe. *Phys. Stat. Solidi (b)* 37: 691–698.
43. Svane, A., N. E. Christensen, M. Cardona, A. Chantis, M. Van Schilfgaarde, and T. Kotani. 2010. Quasiparticle self-consistent *GW* calculations for PbS, PbSe, and PbTe: Band structure and pressure coefficients. *Phys. Rev. B* 81: 245120.
44. Lach-Hab, M., M. Keegan, D. Papaconstantopoulos, and M. Mehl. 2000. Electronic structure calculations of PbTe. *J. Phys. Chem. Solids* 61: 1639–1645.
45. Tan, X., H. Shao, T. Hu, G.-Q. Liu, and S.-F. Ren. 2015. Theoretical understanding on band engineering of Mn-doped lead chalcogenides PbX (X = Te, Se, S). *J. Phys. Condens. Mater.* 27: 095501.
46. Martinez, G., M. Schlüter, and M. L. Cohen. 1975. Electronic structure of PbSe and PbTe. I. Band structures, densities of states, and effective masses. *Phys. Rev. B* 11: 651–659.
47. Gibbs, Z. M., H. Kim, H. Wang et al. 2013. Temperature dependent band gap in PbX (X=S, Se, Te). *Appl. Phys. Lett.* 103: 262109.
48. Božin, E. S., C. D. Malliakas, P. Souvatzis et al. 2010. Entropically stabilized local dipole formation in lead chalcogenides. *Science* 330: 1660–1663.
49. Chasmar, R. and R. Stratton. 1959. The thermoelectric figure of merit and its relation to thermoelectric generators. *J. Electron. Control* 7: 52–72.
50. Foley, G. and D. Langenberg. 1977. Electronic structure of PbTe near the band gap. *Phys. Rev. B* 15: 4850–4864.
51. Ioffe, A. F. 1957. *Semiconductor Thermoelements and Thermoelectric Cooling*, Infosearch, Ltd., London, U.K.
52. Pei, Y., Z. M. Gibbs, A. Gloskovskii, B. Balke, W. G. Zeier, and G. J. Snyder. 2014. Optimum carrier concentration in n-type PbTe thermoelectrics. *Adv. Energy Mater.* 4: 1400486.
53. Dow, H., M. Oh, B. Kim et al. 2010. Effect of Ag or Sb addition on the thermoelectric properties of PbTe. *J. Appl. Phys.* 108: 113709.
54. Takagiwa, Y., Y. Pei, G. Pomrehn, and G. J. Snyder. 2012. Dopants effect on the band structure of PbTe thermoelectric material. *Appl. Phys. Lett.* 101: 092102.
55. Zhao, L.-D., J. He, C.-I. Wu et al. 2012. Thermoelectrics with earth abundant elements: High performance p-type PbS nanostructured with SrS and CaS. *J. Am. Chem. Soc.* 134: 7902–7912.

56. Tan, G., F. Shi, S. Hao et al. 2015. Valence band modification and high thermoelectric performance in SnTe heavily alloyed with MnTe. *J. Am. Chem. Soc.* 137: 11507–11516.
57. Brebrick, R. and A. Strauss. 1963. Anomalous thermoelectric power as evidence for two-valence bands in SnTe. *Phys. Rev.* 131: 104–110.
58. Tan, G., L.-D. Zhao, F. Shi et al. 2014. High thermoelectric performance of p-type SnTe via a synergistic band engineering and nanostructuring approach. *J. Am. Chem. Soc.* 136: 7006–7017.
59. Tan, G., F. Shi, S. Hao et al. 2015. Codoping in SnTe: Enhancement of thermoelectric performance through synergy of resonance levels and band convergence. *J. Am. Chem. Soc.* 137: 5100–5112.
60. Wang, H., Z. M. Gibbs, Y. Takagiwa, and G. J. Snyder. 2014. Tuning bands of PbSe for better thermoelectric efficiency. *Energy Environ. Sci.* 7: 804–811.
61. Andreev, A. and V. Radionov. 1967. Extraordinary role of Hg in enhancing the thermoelectric performance of p-type SnTe. *Sov. Phys. Semicond.* 1: 145–148.
62. Tan, G., F. Shi, J. W. Doak et al. 2015. Extraordinary role of Hg in enhancing the thermoelectric performance of p-type SnTe. *Energy Environ. Sci.* 8: 267–277.
63. Heremans, J. P., B. Wiendlocha, and A. M. Chamoire. 2012. Resonant levels in bulk thermoelectric semiconductors. *Energy Environ. Sci.* 5: 5510–5530.
64. Kaidanov, V. and Y. I. Ravich. 1985. Deep and resonance states in $A^{IV}B^{VI}$ semiconductors. *Sov. Phys. Usp.* 28: 31–53.
65. Mott, N. F. and E. A. Davis. 1979. *Electronic Processes in Non Crystalline Materials*, Oxford University Press, Oxford, U.K.
66. Paul, B., A. Kumar, and P. Banerji. 2010. Embedded Ag-rich nanodots in PbTe: Enhancement of thermoelectric properties through energy filtering of the carriers. *J. Appl. Phys.* 108: 064322.
67. Ko, D.-K., Y. Kang, and C. B. Murray. 2011. Enhanced thermopower via carrier energy filtering in solution-processable Pt–Sb_2Te_3 nanocomposites. *Nano Lett.* 11: 2841–2844.
68. Ahmad, S., K. Hoang, and S. Mahanti. 2006. *Ab initio* study of deep defect states in narrow band-gap semiconductors: Group III impurities in PbTe. *Phys. Rev. Lett.* 96: 056403.
69. Drabkin, I. and B. Y. Moizhes. 1983. *Sov. Phys. Semicond.* 17: 611–613.
70. Khokhlov, D. and B. Volkov. 1996. Mixed valence, electrical activity and metastable states in doped IV–VI compounds: Experiment and theory. *Proceedings of the 23rd International Conference on the Physics of Semiconductors*, Berlin, Germany, July 21–26, 1996, eds. M. Scheffer and R. Zimmermann, *World Sci.*
71. Nemov, S. and Y. I. Ravich. 1998. Thallium dopant in lead chalcogenides. *Phys. Usp.* 41: 735–759.
72. Kaidanov, V., S. Nemov, and Y. I. Ravich. 1992. Resonance scattering current carriers in IV-VI semiconductors. *Sov. Phys. Semicond.* 26: 113–125.
73. König, J. D., M. D. Nielsen, Y.-B. Gao et al. 2011. Titanium forms a resonant level in the conduction band of PbTe. *Phys. Rev. B* 84: 205126.
74. Rawat, P., B. Paul, and P. Banerji. 2014. Exploration of Zn resonance levels and thermoelectric properties in I-doped PbTe with ZnTe nanostructures. *ACS Appl. Mat. Interfaces* 6: 3995–4004.
75. Tan, G., W. Liu, H. Chi et al. 2013. Realization of high thermoelectric performance in p-type unfilled ternary skutterudites $FeSb_{2+x}Te_{1-x}$ via band structure modification and significant point defect scattering. *Acta Mater.* 61: 7693–7704.
76. Tan, G., S. Wang, H. Li, Y. Yan, and X. Tang. 2012. Enhanced thermoelectric performance in zinc substituted p-type filled skutterudites $CeFe_{4-x}Zn_xSb_{12}$. *J. Solid State Chem.* 187: 316–322.
77. Ahn, K., M.-K. Han, J. He et al. 2010. Exploring resonance levels and nanostructuring in the PbTe–CdTe system and enhancement of the thermoelectric figure of merit. *J. Am. Chem. Soc.* 132: 5227–5235.
78. Vinogradova, M., N. Kolomoets, and L. Sysoeva. 1971. Influence of manganese on energy spectrum of p-PbTe. *Sov. Phys. Semicond.* 5: 186–189.
79. Pei, Y., A. D. LaLonde, N. A. Heinz et al. 2011. Stabilizing the optimal carrier concentration for high thermoelectric efficiency. *Adv. Mater.* 23: 5674–5678.
80. Airapetyants, S., M. Vinogradova, I. Dubrovskaya, N. Kolomoets, and I. Rudnik. 1966. Structure of the valence band of heavily doped lead telluride. *Sov. Phys. Solid State* 8: 1069–1072.

81. Chernik, I., V. Kaidanov, M. Vinogradova, and N. Kolomoets. 1968. Investigation of the valence band of lead telluride using transport phenomena. *Sov. Phys. Semicond.* 2: 645–651.
82. Liu, W., X. Tan, K. Yin et al. 2012. Convergence of conduction bands as a means of enhancing thermoelectric performance of n-type $Mg_2Si_{1-x}Sn_x$ solid solutions. *Phys. Rev. Lett.* 108: 166601.
83. Liu, X., T. Zhu, H. Wang et al. 2013. Low electron scattering potentials in high performance $Mg_2Si_{0.45}Sn_{0.55}$ based thermoelectric solid solutions with band convergence. *Adv. Energy Mater.* 3: 1238–1244.
84. Zhao, L.-D., S. Hao, S.-H. Lo et al. 2013. High thermoelectric performance via hierarchical compositionally alloyed nanostructures. *J. Am. Chem. Soc.* 135: 7364–7370.
85. Banik, A., U. S. Shenoy, S. Anand, U. V. Waghmare, and K. Biswas. 2015. Mg Alloying in SnTe facilitates valence band convergence and optimizes thermoelectric properties. *Chem. Mater.* 27: 581–587.
86. Zhao, L.-D., S.-H. Lo, Y. Zhang et al. 2014. Ultralow thermal conductivity and high thermoelectric figure of merit in SnSe crystals. *Nature* 508: 373–377.
87. Zhao, L.-D., G. Tan, S. Hao et al. 2016. Ultra-high power factor and thermoelectric performance in hole doped single crystal SnSe. *Science* 351: 141–144.
88. Biswas, K., J. He, G. Wang et al. 2011. High thermoelectric figure of merit in nanostructured p-type PbTe-MTe (M = Ca, Ba). *Energy Environ. Sci.* 4: 4675–4684.
89. Callaway, J. and H. C. von Baeyer. 1960. Effect of point imperfections on lattice thermal conductivity. *Phys. Rev.* 120: 1149–1154.
90. Parker, W., R. Jenkins, C. Butler, and G. Abbott. 1961. Flash method of determining thermal diffusivity, heat capacity, and thermal conductivity. *J. Appl. Phys.* 32: 1679–1684.
91. Slack, G. A. and S. Galginaitis. 1964. Thermal conductivity and phonon scattering by magnetic impurities in CdTe. *Phys. Rev.* 133: 253–268.
92. Slack, G. A. 1972. Thermal conductivity of II-VI compounds and phonon scattering by Fe^{2+} impurities. *Phys. Rev. B* 6: 3791–3800.
93. Holland, M. 1964. Phonon scattering in semiconductors from thermal conductivity studies. *Phys. Rev.* 134: 471–480.
94. Tan, G., S. Wang, and X. Tang. 2013. High thermoelectric figure of merit of p-type ternary unfilled skutterudite $FeSb_2$ Te via Ge doping. *Sci. Adv. Mater.* 5: 1974–1982.
95. Tan, G., H. Chi, W. Liu et al. 2015. Toward high thermoelectric performance p-type $FeSb_{2.2}Te_{0.8}$ via *in situ* formation of InSb nanoinclusions. *J. Mater. Chem. C* 3: 8372–8380.
96. Nolas, G., J. Cohn, G. Slack, and S. Schujman. 1998. Semiconducting Ge clathrates: Promising candidates for thermoelectric applications. *Appl. Phys. Lett.* 73: 178–180.
97. Deng, S., X. Tang, and Q. Zhang. 2007. Synthesis and thermoelectric properties of p-type $Ba_8Ga_{16}Zn_xGe_{30-x}Ba_8Ga_{16}Zn_xGe_{30-x}$ type-I clathrates. *J. Appl. Phys.* 102: 043702.
98. Sales, B. C., D. Mandrus, and R. K. Williams. 1996. Filled skutterudite antimonides: A new class of thermoelectric materials. *Science* 272: 1325–1328.
99. Tan, G., Y. Zheng, Y. Yan, and X. Tang. 2014. Preparation and thermoelectric properties of p-type filled skutterudites $Ce_yFe_{4-x}Ni_xSb_{12}$. *J. Alloys Compd.* 584: 216–221.
100. Tan, G., W. Liu, S. Wang et al. 2013. Rapid preparation of $CeFe_4Sb_{12}$ skutterudite by melt spinning: Rich nanostructures and high thermoelectric performance. *J. Mater. Chem. A* 1: 12657–12668.
101. Tan, G., Y. Zheng, and X. Tang. 2013. High thermoelectric performance of nonequilibrium synthesized $CeFe_4Sb_{12}$ composite with multi-scaled nanostructures. *Appl. Phys. Lett.* 103: 183904.
102. Tan, G., S. Wang, X. Tang, H. Li, and C. Uher. 2012. Preparation and thermoelectric properties of Ga-substituted p-type fully filled skutterudites $CeFe_{4-x}Ga_xSb_{12}$. *J. Solid State Chem.* 196: 203–208.
103. Delaire, O., J. Ma, K. Marty et al. 2011. Giant anharmonic phonon scattering in PbTe. *Nat. Mater.* 10: 614–619.
104. Girard, S. N., T. C. Chasapis, J. He et al. 2012. PbTe–$PbSnS_2$ thermoelectric composites: Low lattice thermal conductivity from large microstructures. *Energy Environ. Sci.* 5: 8716–8725.

105. Klemens, P. 1960. Thermal resistance due to point defects at high temperatures. *Phys. Rev.* 119: 507–509.
106. Abeles, B. 1963. Lattice thermal conductivity of disordered semiconductor alloys at high temperatures. *Phys. Rev.* 131: 1906–1911.
107. Androulakis, J., I. Todorov, D.-Y. Chung et al. 2010. Thermoelectric enhancement in PbTe with K or Na codoping from tuning the interaction of the light- and heavy-hole valence bands. *Phys. Rev. B* 82: 115209.
108. Dashevsky, Z., S. Shusterman, M. Dariel, and I. Drabkin. 2002. Thermoelectric efficiency in graded indium-doped PbTe crystals. *J. Appl. Phys.* 92: 1425–1430.
109. Zhu, P., Y. Imai, Y. Isoda, Y. Shinohara, X. Jia, and G. Zou. 2005. Enhanced thermoelectric properties of PbTe alloyed with Sb_2Te_3. *J. Phys. Condens. Matter.* 17: 7319–7326.
110. Zhang, Q., T. Sun, F. Cao et al. 2010. Tuning the shape and thermoelectric property of PbTe nanocrystals by bismuth doping. *Nanoscale* 2: 1256–1259.
111. Hiscocks, S. and P. West. 1968. Crystal pulling and constitution in $Pb_{1-x}Sn_xTe$. *J. Mater. Sci.* 3: 76–79.
112. Li, S., J. Li, Q. Wang, L. Wang, F. Liu, and W. Ao. 2011. Synthesis and thermoelectric properties of the $(GeTe)_{1-x}(PbTe)_x$ alloys. *Solid State Sci.* 13: 399–403.
113. Dong, Y., M. A. McGuire, A.-S. Malik, and F. J. DiSalvo. 2009. Transport properties of undoped and Br-doped PbTe sintered at high-temperature and pressure ≥4.0 GPa. *J. Solid State Chem.* 182: 2602–2607.
114. Quarez, E., K.-F. Hsu, R. Pcionek, N. Frangis, E. Polychroniadis, and M. G. Kanatzidis. 2005. Nanostructuring, compositional fluctuations, and atomic ordering in the thermoelectric materials $AgPb_mSbTe_{2+m}$. The myth of solid solutions. *J. Am. Chem. Soc.* 127: 9177–9190.
115. Tan, G., F. Shi, H. Sun et al. 2014. SnTe–$AgBiTe_2$ as an efficient thermoelectric material with low thermal conductivity. *J. Mater. Chem. A* 2: 20849–20854.
116. Tan, G., W. G. Zeier, F. Shi et al. 2015. High thermoelectric performance SnTe–In_2Te_3 solid solutions enabled by resonant levels and strong vacancy phonon scattering. *Chem. Mater.* 27: 7801–7811.
117. Irie, T., T. Takahama, and T. Ono. 1963. The thermoelectric properties of $AgSbTe_2$-$AgBiTe_2$, $AgSbTe_2$-PbTe and –SnTe systems. *Jpn. J. Appl. Phys.* 2: 72–82.
118. Ono, T., T. Takahama, and T. Irie. 1962. The thermoelectric properties of $AgSbTe_2$–$AgBiTe_2$, –PbTe and –SnTe systems. *J. Phys. Soc. Jpn.* 17: 1070–1071.
119. He, J., J. R. Sootsman, S. N. Girard et al. 2010. On the origin of increased phonon scattering in nanostructured PbTe based thermoelectric materials. *J. Am. Chem. Soc.* 132: 8669–8675.
120. Zhao, L.-D., J. He, S. Hao et al. 2012. Raising the thermoelectric performance of p-type PbS with endotaxial nanostructuring and valence-band offset engineering using CdS and ZnS. *J. Am. Chem. Soc.* 134: 16327–16336.
121. Poudeu, P. F., J. D'Angelo, H. Kong et al. 2006. Nanostructures versus solid solutions: low lattice thermal conductivity and enhanced thermoelectric figure of merit in $Pb_{9.6}Sb_{0.2}Te_{10-x}Se_x$ bulk materials. *J. Am. Chem. Soc.* 128: 14347–14355.
122. Wu, H., L. Zhao, F. Zheng et al. 2013. Broad temperature plateau for thermoelectric figure of merit ZT > 2 in phase-separated $PbTe_{0.7}S_{0.3}$. *Nat. Commun.* 5: 4515.
123. He, J., J. Androulakis, M. G. Kanatzidis, and V. P. Dravid. 2012. Seeing is believing: Weak phonon scattering from nanostructures in alkali metal-doped lead telluride. *Nano Lett.* 12: 343–347.
124. Beyer, H., J. Nurnus, H. Böttner, A. Lambrecht, T. Roch, and G. Bauer. 2002. PbTe based superlattice structures with high thermoelectric efficiency. *Appl. Phys. Lett.* 80: 1216–1218.
125. Harman, T., P. Taylor, M. Walsh, and B. LaForge. 2002. Quantum dot superlattice thermoelectric materials and devices. *Science* 297: 2229–2232.
126. Yan, Q., H. Chen, W. Zhou, H. H. Hng, F. Y. C. Boey, and J. Ma. 2008. A simple chemical approach for PbTe nanowires with enhanced thermoelectric properties. *Chem. Mater.* 20: 6298–6300.
127. Springholz, G., G. Bauer, and G. Ihninger. 1993. MBE of high mobility PbTe films and PbTe/$Pb_{1-x}Eu_xTe$ heterostructures. *J. Cryst. Growth* 127: 302–307.

128. Yu, B., Q. Zhang, H. Wang et al. 2010. Thermoelectric property studies on thallium-doped lead telluride prepared by ball milling and hot pressing. *J. Appl. Phys.* 108: 016104.
129. Qiu, X., Y. Lou, A. Samia et al. 2005. PbTe nanorods by sonoelectrochemistry. *Angew. Chem. Int. Ed.* 44: 5855–5857.
130. Androulakis, J., K. Hsu, R. Pcionek et al. 2006. Nanostructuring and high thermoelectric efficiency in p-type $Ag(Pb_{1-y}Sn_y)_mSbTe_{2+m}$. *Adv. Mater.* 18: 1170.
131. Poudeu, P. F., A. Guéguen, C.-I. Wu, T. Hogan, and M. G. Kanatzidis. 2009. High figure of merit in nanostructured n-type KPb_mSbTe_{m+2} thermoelectric materials. *Chem. Mater.* 22: 1046–1053.
132. Androulakis, J., C.-H. Lin, H.-J. Kong et al. 2007. Spinodal decomposition and nucleation and growth as a means to bulk nanostructured thermoelectrics: enhanced performance in $Pb_{1-x}Sn_xTe$–PbS. *J. Am. Chem. Soc.* 129: 9780–9788.
133. Rowe, D. M., V. S. Shukla, and N. Savvides. 1981. Phonon scattering at grain boundaries in heavily doped fine-grained silicon–germanium alloys. *Nature* 290: 765–766.
134. Qiu, B., H. Bao, G. Zhang, Y. Wu, and X. Ruan. 2012. Molecular dynamics simulations of lattice thermal conductivity and spectral phonon mean free path of PbTe: Bulk and nanostructures. *Comp. Mater. Sci.* 53: 278–285.
135. Tian, Z., K. Esfarjani, J. Shiomi, A. S. Henry, and G. Chen. 2011. On the importance of optical phonons to thermal conductivity in nanostructures. *Appl. Phys. Lett.* 99: 053122.
136. Pei, Y., A. F. May, and G. J. Snyder. 2011. Self-tuning the carrier concentration of PbTe/Ag_2Te composites with excess Ag for high thermoelectric performance. *Adv. Energy Mater.* 1: 291–296.
137. Tit, N., M. Peressi, and S. Baroni. 1993. *Ab initio* calculation of the band offset at strained GaAs/InAs (001) heterojunctions. *Phys. Rev. B* 48: 17607–17610.
138. Rogers, L. 1967. The Hall mobility and thermoelectric power of *p*-type lead telluride. *Br. J. Appl. Phys.* 18: 1227–1235.

5

Thermoelectric Properties of Magnesium Silicide–Based Solid Solutions and Higher Manganese Silicides

Johannes de Boor*, Titas Dasgupta, and Eckhard Müller***

**German Aerospace Center*

***Indian Institute of Technology*

CONTENTS

5.1 Mg_2Si and its Solid Solutions

Mg_2Si-based solid solutions are an ideal material class with respect to thermoelectric material optimization. Several *n*-type dopants have been identified that can efficiently tune the charge carrier density to the ideal level. Isovalent substitution of Si by Sn provides means to alter the band structure of the material, resulting in a convergence of the two conduction bands for compositions close to $Mg_2Si_{0.4}Sn_{0.6}$. Band degeneracy is an option to break the usual coupling between the Seebeck coefficient and electrical conductivity, allowing for increasing the Seebeck coefficient without decreasing the electrical conductivity. Isovalent substitution can also be used to suppress the lattice thermal conductivity in the material as it increases phonon scattering.

The chapter is organized as follows: following the introductory Section 5.1, we will review and compare different synthesis and fabrication approaches for Mg_2X (with X = Si, Ge, Sn) in Section 5.1.2. Section 5.1.3 discusses (intrinsic) defects in Mg_2Si and their influence on the thermoelectric properties. The thermoelectric transport properties of (doped) binary Mg_2Si are analyzed in more detail in Section 5.1.4.1. We focus on *n*-type material, comparing several dopants and compiling available literature data. As secondary phases are often observed in Mg_2Si, we also evaluate their influence on the thermoelectric properties. A discussion of *p*-type Mg_2Si concludes this section. The best thermoelectric properties are obtained for solid solutions of the form $Mg_2(Si, Sn)$ due to band convergence and decreased thermal conductivity. The effect of the solid solution formation on electronic and thermal transport is analyzed in Section 5.1.4.2.

5.1.1 Introduction

As it is a very promising material, Mg_2Si and its solid solutions have been addressed in previous publications, including the *CRC Handbook of Thermoelectrics* [1,2]; we will therefore summarize the main features only briefly. The constituting elements of Mg_2Si are among the most abundant in the earth's crust: Si is the second most and Mg the eighth most abundant element [3]. Mg_2Si as well as Mg_2Ge and Mg_2Sn crystallize in the (cubic) antifluorite structure, with Si in face-centered cubic positions and Mg in tetrahedral sites; see Figure 5.1. One of the advantages of Mg_2Si-based solid solutions is their low mass density, which varies between 2 g/cm^3 for Mg_2Si and 3.6 g/cm^3 for Mg_2Sn. This is roughly a factor of 3 lower than other thermoelectric materials like PbTe and $CoSb_3$-based skutterudites that work in the same temperature region. Mg_2X (with X = Si, Ge, Sn) compounds are therefore especially promising where weight is crucial, such as space and airborne applications. Due to the good availability of Mg and Si, Mg_2Si-based solid solutions are among the most promising materials also from an economical point of view [4].

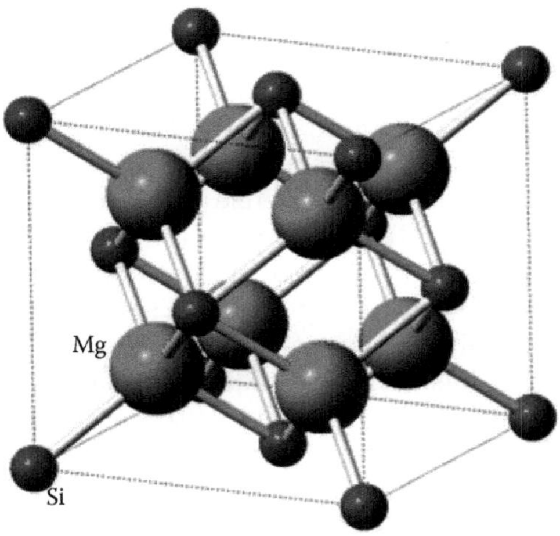

FIGURE 5.1
Cubic structure of Mg_2Si (space group $Fm\bar{3}m$). The larger spheres correspond to Mg, while the smaller ones indicate Si atoms. (From Yi, T.H., Chen, S.P., Li, S. et al., Synthesis and characterization of Mg_2Si/Si nanocomposites prepared from MgH_2 and silicon, and their thermoelectric properties, *J. Mater. Chem.*, 22, 24805–24813, 2012. Reproduced by permission of The Royal Society of Chemistry.)

5.1.2 Fabrication Routes

Various synthesis techniques to prepare Mg_2Si have been reported and evaluated, each with its own advantages and characteristics. The aim of this section is to provide an overview of the most relevant existing synthesis approaches as well as to evaluate their potential benefits and shortcomings. For simplicity, this section will focus on (doped) Mg_2Si; however, almost all techniques can be applied to Mg_2X, with slightly adjusted parameters.

Synthesis of Mg_2Si faces two general challenges. The first is the large difference in the melting point between Mg (923 K) and Si (1687 K) and the proximity of the boiling point of Mg (1383 K) to the melting point of Mg_2Si (~1355 K), which results in a high Mg vapor pressure at the synthesis temperature; see, for example, [5]. This can lead to Mg loss and thus nonstoichiometric compounds. This can have dramatic consequences on the thermoelectric properties. The second issue is the reactivity of elemental Mg. This requires suitable crucibles during the synthesis procedure, and, moreover, this often leads to the formation of MgO impurities in the material. The role of MgO and other secondary phases is discussed in detail in Section 5.1.4.1.3.

The fabrication of Mg_2Si compacts usually requires two steps: the synthesis of the compound, often obtained as powder or polycrystalline ingot material, and the compaction step.

We first discuss synthesis routes where the educts go into a liquid state and then routes where the educts (partially) remain in a solid state.

Melting of the elements is an often employed and traditional approach for the synthesis of Mg_2Si [6–11]. In this approach, the elements are brought together in a sealed vessel and heated above the melting point of Mg_2Si (~1355 K). For the Mg-Si phase diagram, see, for example, [12,13]. Heating is often provided by inductive heating of the vessel (graphite, boron nitride). The liquid is held above 1355 K for some minutes to ensure a complete reaction and is then cooled, resulting in a polycrystalline ingot material.

This approach is relatively simple and has been used to yield materials with excellent thermoelectric properties [6,14]. However, it also has some drawbacks. First, due to the high synthesis temperature, very good vacuum is required to prevent Mg from oxidation. Second, again due to the high temperature, the evaporation of Mg is inevitable. Even if the reaction is performed in a closed vessel, some Mg will evaporate due to its high vapor pressure and possibly condense at colder parts of the vessel, yielding a Mg-deficient Mg_2Si compound. To compensate for that, excess Mg (5%–10%) is typically added [10,14]. However, identifying the right amount of excess Mg requires some amount of trial and error and is system specific. This is clearly disadvantageous from a reproducibility and comparability point of view.

An interesting modification has been presented by Bernard-Granger et al. [15], who used gas-phase atomization to synthesize Mg_2(Si, Sn). In this approach, the elements are inductively heated to temperatures between 1573 and 1723 K. The melt is then guided through a nozzle where it is atomized into micron-sized droplets by an argon gas jet. This jet additionally quenches the material. The material is not a single phase after the gas phase atomization, but homogenizes during the compaction step in a spark plasma sintering (SPS) setup. For a nominal composition of $Mg_2Si_{0.3875}Sn_{0.6}Sb_{0.0125}$, Bernard-Granger et al. obtained excellent thermoelectric properties and also demonstrated good sample homogeneity. While requiring some expensive infrastructure, the technique has some clear advantages like scalability, compatibility with industrial-scale processes, and some control over the samples' micro- and nanostructure due to rapid cooling rates and the process parameters. The authors also claimed to use stoichiometric amounts of the elements, indicating that Mg loss is not crucial. More work and reproduction is required for a thorough analysis of this method.

Solid-state syntheses generally have the advantage that they require lower temperatures than liquid synthesis routes. One approach is to start with elemental powders and use a thermal treatment to facilitate the reaction to Mg_2Si.

Liu et al. [16–18] used a two-step solid-state reaction method to synthesize Mg_2(Si, Sn). Starting from elemental powders, the constituents were ground and exposed to a thermal treatment at 600°C–700°C twice, followed by consolidation using SPS. They obtained thermoelectric material with very good properties for different Si/Sn ratios and dopant types and amounts. Tani and Kido [19] employed a single-step solid-state reaction under an Ar–H_2 stream to test different dopants in Mg_2Si. In both approaches, excess Mg was added to account for Mg loss during the synthesis. Other research groups also employed a single-step solid-state reaction process, usually with reasonable thermoelectric properties [20].

One drawback of the approach suggested by Liu et al. is the relatively long annealing time required for the reaction to be completed. Tani and Kido combined synthesis and compaction in one step, directly sintering the elemental powders together in the SPS to obtain Al-doped Mg_2Si and Mg_2(Si, Sn) [21,22]. A similar approach has been employed for synthesis of binary Mg_2Si [23]. However, no results on sample homogeneity or microstructure were given.

Two other fast synthesis routes are combustion synthesis and microwave synthesis. Godlewska et al. [5] mixed elemental powders in stoichiometric proportions, compacted the mixture at room temperature, and started a fast, self-propagating reaction by global heating to 823–852 K. This is possible as the standard formation enthalpy of Mg_2Si is −77.8 kJ/mol. The reaction product was Mg_2Si powder that was compacted using hot-pressing. The compacted material was almost single phase, with small Si and MgO impurities, but no thermoelectric properties were reported in the manuscript. It is possible that the approach by Tani and Kido [21,22] is related to the self-propagating reaction studied by Godlewska et al. as the SPS temperature of Tani and Kido was above the starting temperature for the self-propagating reaction reported by Su et al. [24] for Mg_2Si; however, this has not been discussed in the paper.

Savary et al. [25] employed microwave heating to synthesize nanoscale Mg_2Si powder. They found that Si is susceptible to microwave radiation and can thus be used to heat up the mixed elemental powders and start the solid-state formation reaction of Mg_2Si. With a heating time of 2 min, the reaction is, indeed, very fast; however, the method offers little control of the local temperature. This might lead to Mg loss during the synthesis and could explain the relatively poor thermoelectric properties of doped Mg_2Si and $Mg_2(Si, Sn)$ that were synthesized by the same approach [26].

Gao et al. [27] presented a modified solid-state synthesis with better control over Mg loss. In the so-called flux method, they mixed the elemental powders in an alumina crucible and covered this with B_2O_3 powder before cold precompaction and the heating step at 973 K that starts the solid-state synthesis. The B_2O_3 has a low melting point (723 K) and therefore forms a liquid seal above the Mg, Si, and Sn mixture during the synthesis, in principle inhibiting Mg loss. The method is scalable [28], does not require vacuum conditions during the synthesis, and has been used to fabricate $Mg_2(Si, Sn)$ with excellent thermoelectric properties [28,29]. On the other hand, all authors use excess Mg to compensate for Mg loss to various extents, and precise sample stoichiometry seems to be hard to control [27,28]. Interestingly, results have been reported only for Sn-rich $Mg_2(Si, Sn)$, but not for binary Mg_2Si. We found large amounts of MgO impurities when attempting to synthesize Mg_2Si with the flux method.

Another approach to solve the oxidation problem of Mg is to use MgH_2 rather than elemental Mg as a reaction educt. Yi et al. [30] showed that the endothermic reaction $2MgH_2 + Si \rightarrow Mg_2Si + 2H_2(g)$ can be used to synthesize Mg_2Si. Annealing the MgH_2/Si mixture for 12 h at only 623 K yielded an almost single-phase material. The gaseous H_2 kept the reaction environment oxygen-free, and the very low synthesis temperature furthermore hindered MgO formation. In this and a comparable study, no MgO was found in the powder after synthesis. Nevertheless, MgO might also form during compaction [31]. This relatively simple approach yields MgO-free powder of (doped) Mg_2Si; however, the electrical mobility and the thermoelectric figure of merit at low temperatures is drastically reduced [30,31] compared to other synthesis methods; see, for example, [32]. At high temperatures, the values become comparable, possibly indicating additional scattering mechanisms in the material. This might be related to the nanopore formation that is observed in the compacts.

Instead of relying on thermal energy, Mg_2Si can also be synthesized employing mechanical energy from high-energy ball milling (BM) ("mechanical alloying"). While Si is hard and brittle, Mg is relatively soft and tends to agglomerate during the BM process. Furthermore, due to the long BM times, oxidation is a severe problem, making mechanical alloying a nontrivial process [33]. Attempts to overcome this problem include the use of organic media like hexane as milling aids. However, while this might prevent agglomeration, incomplete removal of the organic substance often lowers the electrical conductivity of the resulting samples [33]. Bux et al. [32] used dry BM in an inert atmosphere to synthesize Mg_2Si. They started the milling process with Si chunks and a small amount of Mg, adding Mg during the milling process in discrete steps until stoichiometry was reached. In this way they prevented Mg agglomeration and achieved completely reacted Mg_2Si after only 8 h with no detectable amount of MgO. As this is a "room temperature" synthesis, Mg loss is not an issue, and they reported very good thermoelectric properties, making the relatively laborious Mg adding process the only drawback of the method. Vracar et al. [34] employed a longer (60 h) wet milling process with an unspecified processing agent to synthesize $Mg_2(Si, Sn)$. They did not observe a massive MgO formation and obtained reasonable thermoelectric properties of the material. Some reduction of the electrical conductivity was discernable, especially at lower temperatures, which might have been related to residues from the milling agent.

As discussed earlier, high-energy BM can be employed for Mg_2Si synthesis. Besides that, BM can also be used to incorporate secondary phases or dopants into the material, as well as reduce grain sizes and thus influence material microstructure and thermoelectric properties. That is particularly relevant because presynthesized Mg_2Si is available from several commercial companies, and the tedious synthesis step can thus be skipped [35–37].

Fiameni et al. [37] used wet BM to incorporate Bi into undoped commercial Mg_2Si. While the resulting material was two-phased, with a presumably undoped phase and a Bi-doped Mg_2Si phase, the resulting thermoelectric properties of the compound were good, with a $ZT_{max} = 0.8$ at 873 K. Similarly, Battiston et al. [38] used BM to incorporate Al into Mg_2Si; here, the material appeared single-phased. Others used BM to obtain nanocrystalline Mg_2Si powder [36,39] with crystallite sizes decreasing down to 10 nm with increasing BM time. Nanocrystalline material might enhance phonon scattering and thus suppress the lattice thermal conductivity. However, it is debatable whether this might actually enhance the overall thermoelectric properties [40]; see also the discussion in Section 5.1.4.1.3. Moreover, grain growth is usually observed during compaction and thermal treatment/measurement, which (partially) cancels the effect of BM [41]. Finally, an increase in the MgO fraction is usually reported with increasing BM time, which is expected to deteriorate the thermoelectric properties of the material [36,39,41].

Compaction of Mg_2Si samples is usually achieved by sintering under elevated temperature and pressure. Traditionally, a hot-press has been applied for compaction [6,7,10,32,42,43], while in recent papers current-assisted sintering was employed in the majority of cases. Current-assisted sintering heats the sample material and/or the surrounding die by Joule heating, which can be much faster than the external heating employed in a hot-press. Current-assisted sintering therefore has the advantage of much shorter sintering times (minutes vs. hours), which helps to maintain the microstructure of the samples and reduces Mg loss during sintering. The most popular system is the SPS that utilizes pulsed current, while systems with constant current have also been reported. Effects of the current flow beyond Joule heating, for example, electromigration of atoms in the lattice or inhomogeneity of the compacts due to asymmetry of Peltier heat, are discussed in the literature but have not been clarified yet.

Besides chemical composition and material structure, compaction parameters play a major role in determining the material properties. It is therefore worthwhile to discuss the influence of experimentally accessible compaction parameters on the microstructure and the resulting thermoelectric properties. The numbers discussed in the following are related to Mg_2Si, but the general trends are valid for the whole Mg_2X family. Usually, lower sintering temperatures are employed with high Sn content. The analysis is also restricted to current-assisted sintering. The easily accessible compaction parameters are sintering temperature, sintering time, sintering pressure, and powder particle size. All of these parameters show some interdependency, making a simple comparison between different reports challenging. Most reports give compaction parameters that are presumably optimized; nevertheless, they vary considerably between the authors. A somewhat systematic study has been performed by de Boor et al. [44], who used Sb-doped Mg_2Si from a single batch and studied the influence of sintering temperature, pressure, and powder

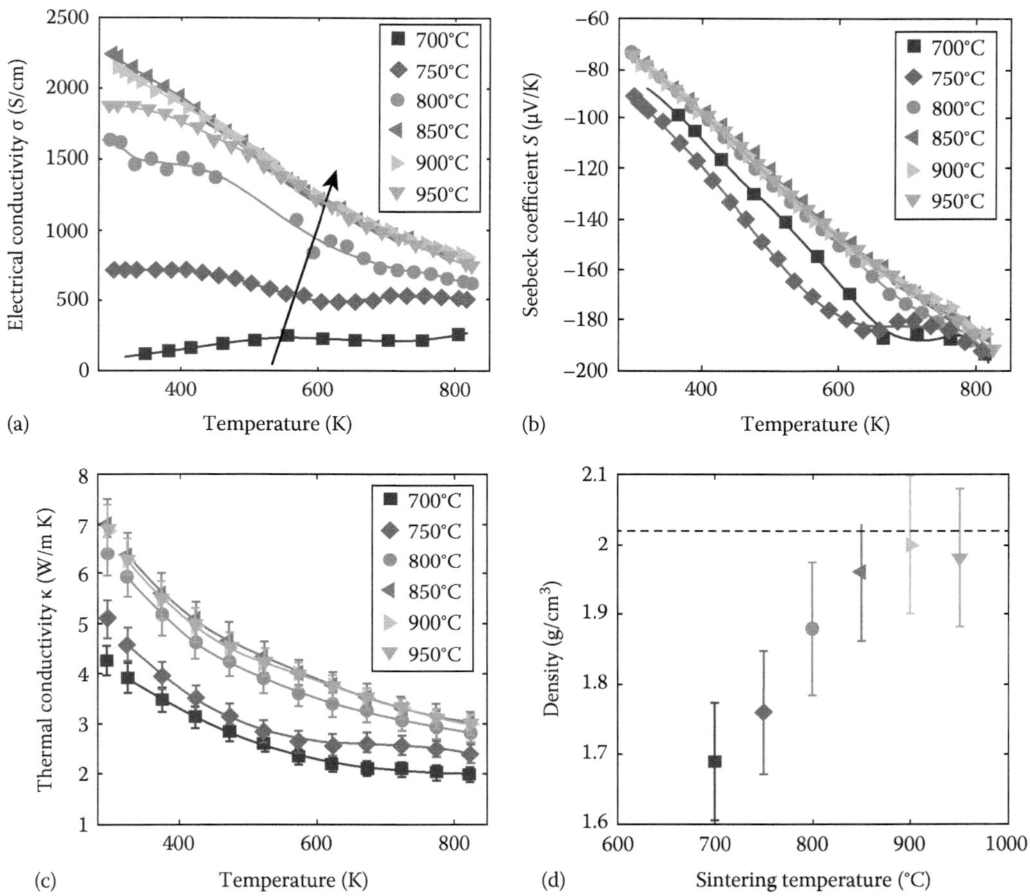

FIGURE 5.2
Thermoelectric properties ((a)-(c), heating curves) and density (d) of $Mg_2Si_{0.9875}Sb_{0.0125}$ as a function of sintering temperature. Samples with $T_{sinter} \geq 850°C$ were approximately stable during measurements, while the low temperature samples showed a hysteresis. (With kind permission from Springer Science+Business Media: *J. Mater. Sci.*, Fabrication parameters for optimized thermoelectric Mg_2Si, 49, 2014, 3196–3204, de Boor, J., Compere, C., Dasgupta, T. et al.)

particle size on the thermoelectric properties. With respect to temperature, they found an optimum of 900°C ± 50°C, with lower processing temperatures resulting in reduced electrical conductivity; see Figure 5.2. This is mainly due to incomplete sintering and, hence, higher porosity, but they also found a reduction in carrier concentration, presumably due to carrier pinning at free interfaces within the material. Most other articles report optimal sintering temperatures between 750°C and 850°C [19–21,26,35,37,39], with higher sintering temperatures often resulting in higher MgO content or increased Mg loss, both deteriorating the thermoelectric properties.

With respect to sintering pressure, the authors found an increase in thermoelectric performance with increasing sintering pressure. Part of the explanation is again the density, which increases with sintering pressure and reduces porosity. However, increasing ZT was still observed beyond 50 MPa, where the density was basically constant. The same phenomenon was observed by Chen et al. [31], who went to higher pressures and observed improvement above 100 MPa, at constant density. One explanation might be a pressure-induced change of the boundaries between the grains that enhances carrier transport. Powder particle size can be influenced by (ball) milling duration and sieving [36] and might influence the thermoelectric properties (and sintering conditions), although it is debated whether this might be beneficial in Mg_2Si; see Section 5.1.4.1.3. de Boor et al. observed improving thermoelectric properties with increasing average particle size [44], but found later that this is due to increasing MgO content with decreasing particle size and not an intrinsic material effect [45]. For this reason, no clear indication of optimal particle size can be given at this stage.

5.1.3 Defect Chemistry and Off-Stoichiometry

5.1.3.1 Defects in Intrinsic Mg_2Si

While different methods have been tried to synthesize Mg_2Si, the inherent problem of high Mg vapor pressure at the reaction temperature often results in the starting nominal composition to be overstoichiometric in Mg. Additionally, Mg loss during synthesis is difficult to control and can result in the product containing either a slight excess or deficiency of Mg. Several papers report phase pure material with nonstoichiometric compositions, indicating that binary Mg_2Si and its solid solutions are not line phases with respect to the Mg/X ratio; see, for example, [46,47]. Limiting values are not clear, but appear to depend on synthesis conditions and exact starting composition. The possible effects of this off-stoichiometry have been studied by different authors and it has been shown that this off-stoichiometry results in the formation of various stable defects in intrinsic as well as doped Mg_2Si. The presence of these defects, in turn, affects the thermoelectric transport, and, thus, their understanding and control plays an important role in optimizing the thermoelectric properties.

Vacancy formation in intrinsic Mg_2Si has been studied by Staab [48]. The energetics of vacancy formation at the respective sites (V_{Mg}, V_{Si}) and also neighboring di-/trivacancies (V_{MgSi}, $V_{Mg-Si-Mg}$) were theoretically studied. Results show that it is easier to form di-/trivacancies compared to monovacancies. Also, it is easier to form a monovacancy at the Si site compared to the Mg site. The possibility of other types of defects, namely, antisites (Si_{Mg}, Mg_{Si}) and interstitials (Mg_i, Si_i), was studied by Kato et al. [49]. The interstitial position was taken to be the dodecahedral 4b site that is surrounded by 8 Mg atoms; see Figure 5.1. The effects of processing conditions (i.e., Si or Mg rich) have also been studied. Their results indicate the Mg_i defect to be the most stable under all processing conditions. These results have been further verified by theoretical studies [50,51]. In Table 5.1 the reported formation energy of different defects in intrinsic Mg_2Si is given. The results consistently indicate interstitial Mg to be the favored defect in intrinsic Mg_2Si under all processing conditions. Among the other defects, V_{Mg} and Si_{Mg} have low formation energies under Si-rich conditions. It is to be noted that di-/trivacancies also have low formation energies but significant variation exists among the reported formation energy values. The presence of defects in intrinsic Mg_2Si has also been experimentally verified by Kubouchi et al. [23] based on single-crystal x-ray diffraction (XRD) measurements. Refinement of the single-crystal XRD data was carried out using standard programs, and multiple structural models were analyzed to obtain the best fit. Their results show the presence of Mg_i in specimens both rich and depleted in Mg. Thus, it is clear from both experimental and theoretical observations that some amount of interstitial Mg (depending on the synthesis conditions) is inherent in intrinsic Mg_2Si.

TABLE 5.1

Defect Types and Their Formation Energies in Intrinsic Mg_2Si

	Formation Energy (eV)			
Defect Type	**Stoichiometric**	**Mg Rich**	**Si Rich**	**Doping Type**
V_{Mg}	1.71 (Han)			*p*
	1.54 (Jund)	1.72	1.44	
	1.74 (Staab)			
V_{Si}	2.35 (H)			*n*
	2.26 (J)	1.89	2.44	
	1.04 (S)			
Mg_i	1.12 (H)			*n*
	1.40 (J)	1.21	1.49	
Si_i	2.26 (H)			*n*
	2.57 (J)	2.94	2.38	
Si_{Mg}	1.6 (H)			*n*
	1.66 (J)	2.22	1.39	
Mg_{Si}	2.76 (H)			*p*
	2.83 (J)	2.27	3.10	
V_{MgSi}	1.16 (J)	1.07	1.21	—
	0.68 (S)			
V_{MgSiMg}	1.42 (J)	1.42	1.42	*p*
	0.78 (S)			

5.1.3.2 Effect of Doping on Defects

The data in Table 5.1 indicate that the formation energies of some of the defects are not too different. Thus changes in the chemical composition are expected to affect the formation energy landscape. In this regard, the effect of Sb and Ag doping on defect formation in Mg_2Si has been reported. Experimental studies by Nolas et al. [42] and Dasgupta et al. [7] on Sb-doped Mg_2Si pointed out the presence of vacancies at the Mg site (V_{Mg}). In both reports, the synthesis was carried out in the presence of excess Mg, thus indicating a change in the defect formation energies compared to intrinsic Mg_2Si. Vacancy formation in Sb-doped Mg_2Si has also been theoretically studied [52], and the increase in the formation energy of Mg_2Si with the increasing content of Sb has been interpreted as the reason for vacancy formation.

Theoretical studies of the effect of Ag doping on defects in Mg_2Si suggested [53] that Ag doping at the Mg site makes it energetically easier to incorporate more defects (Mg_i, Si_{Mg}). Moreover, the doped Ag formed linear clusters Ag–Mg_i–Ag (under Si-rich conditions) and Ag–Mg_i (under Mg-rich condition) with neighboring interstitial Mg atoms. From calculations of the migration barriers, the authors concluded that the Ag–Mg_i–Ag linear clusters, once formed, are kinetically stable.

5.1.3.3 Effect of Defects on Conduction Type

In Table 5.1, the doping type for different defects in intrinsic Mg_2Si is given. Interstitial Mg (Mg_i), which is the most stable defect under all processing conditions, is seen

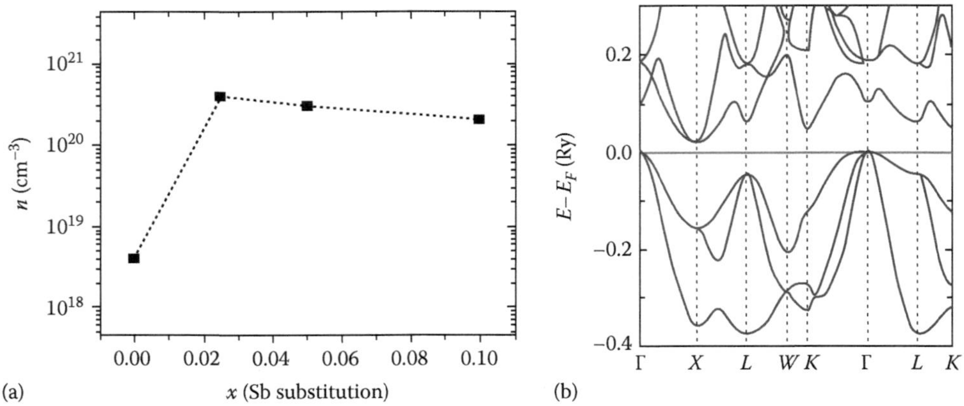

FIGURE 5.3
(a) The charge compensation effect in $Mg_2Si_{1-x}Sb_x$ due to V_{Mg} formation. (Data taken from Dasgupta, T. et al., *Phys. Rev. B*, 83, 235207, 2011.) (b) Degeneracy of the two bands at the conduction band minimum (*X* point) in $Mg_2Si_{0.6}Sb_{0.4}$. (With kind permission from Springer Science+Business Media:. *J. Electron. Mater.*, Mg-vacancy-induced semiconducting properties in Mg_2Si_{1-x} Sb_x from electronic structure calculations, 39, 2010, 2064–2069, Tobola, J., Kaprzyk, S., and Scherrer, H.)

to act as an electron donor. Thus, intrinsic Mg_2Si is always *n*-type as observed in numerous experimental studies. V_{Mg} is seen to act as an acceptor, and thus in Sb-doped specimens (where V_{Mg} formation is observed), a charge compensation takes place [7,42], as shown in Figure 5.3a. The effect of Ag doping is similar with the formation of Ag–Mg_i–Ag linear clusters [53]. The acceptor nature of Ag is compensated by the donor Mg_i atoms making it difficult to achieve significant hole concentrations in the doped specimens.

5.1.3.4 Effect on Thermoelectric Transport

Point defects act as scattering centers for both electrons and phonons; thus the beneficial effect on thermoelectric properties is determined by their relative scattering strengths. As a consequence, in Sb-doped Mg_2Si where V_{Mg} formation is observed, significant reduction in the lattice thermal conductivity is measured [7,42]. However, the negative effect of the vacancies on the electron mobility results in an overall decrease in the *ZT* in the studied composition range. The vacancies also result in a shift in the two conduction band sub-edges that tend to degenerate (Figure 5.3b). This is expected to result in an enhancement in the power factor due to the increased density-of-state effective mass. However, this has not been experimentally verified.

The linear Ag–Mg_i–Ag clusters that form in Ag-doped Mg_2Si result in reduced hole concentrations due to their compensating effect [53]. Thus, hole concentrations that are two orders of magnitude less (compared to expected) are observed in heavily doped specimens and prevent achieving the optimum charge carrier concentration necessary for optimizing *ZT*. The linear clusters also enhance electron scattering resulting in reduced mobility. Consequently, the thermoelectric performance of Ag-doped *p*-type Mg_2Si is poor compared to the *n*-doped specimens.

5.1.4 Thermoelectric Transport

5.1.4.1 Mg_2Si

To achieve optimum carrier concentration, Mg_2Si and its solid solutions with Sn and Ge have to be doped. *n*-type doping has been documented repeatedly with different dopants, and we will analyze the results in Sections 5.1.4.1.1 and 5.1.4.1.2. On the other hand, *p*-type Mg_2Si with good thermoelectric properties remains a challenge; see Section 5.1.4.1.4. Desired or undesired secondary phases in Mg_2X can have a considerable impact on the thermoelectric properties; their role is discussed in Section 5.1.4.1.3. Due to reduced complexity, we will again focus on Mg_2Si, while specifics for $Mg_2(Si, Sn)$ are discussed where appropriate.

5.1.4.1.1 n-Type Doping

The aim of this section is to provide an overview of experimental and theoretical results for *n*-type Mg_2Si. Questions to be addressed include available dopants, their solubility limits, and the doping efficiencies. Detailed influence of the dopants on the thermoelectric transport is given in Section 5.1.4.1.2. All (relevant) dopants discussed here have also been employed in doping $Mg_2(Si, Sn)$.

As a group 15 element, Sb can replace Si and act as an electron donor. Using density functional theory (DFT), Tani and Kido [22] showed that it is energetically more favorable for Sb to replace Si rather than Mg. Bourgeois et al. [54] showed that the incorporation of Sb shifts the Fermi level toward the conduction band, while basically maintaining the band structure. Several groups varied the Sb concentration in $Mg_2Si_{1-y}Sb_y$ and obtained very good thermoelectric properties [8,22]. Studies with a very high Sb content ($y \leq 0.37$) found that, while Sb is completely soluble, the deficiency of Mg increases [42]. On the other hand, Farahi et al. [43] found Mg_3Sb_2 at the grain boundaries in $Mg_2Si_{0.98}Sb_{0.02}$.

Similar to Sb, Bi too can potentially replace Si as a substitutional dopant, donating one electron into the conduction band for each Bi atom entering the $Mg_2Si_{1-y}Bi_y$ lattice. Furthermore, Bi is significantly larger than Mg making it unlikely to substitute for Mg or enter interstitial sites. Due to its large size, Bi has a finite solubility limit, and several authors reported that Bi segregates in the samples [32,41]. The effect of such segregates on the thermoelectric properties remains a contentious issue. Tani and Kido [55] found the best thermoelectric properties for Bi concentrations beyond the solubility limit they identified. Similarly, Farahi et al. [43] obtained better thermoelectric properties (at a given T and nominal stoichiometry) for samples with Mg_3Sb_2 aggregates than Bux et al. [32] for samples without such segregates. On the other hand, Fiameni et al. [37] reported that samples with Bi concentrations beyond 1 at% became fragile and often broke during synthesis.

The third popular dopant is Al. Being a group 3 element, it might, in principle, replace Mg while acting as an *n*-dopant or replace Si acting as a *p*-dopant. Tani and Kido [21] showed that Mg substitution is generally energetically more favorable, in agreement with their experimental results and the work of other groups [38]. Green's function calculations indicated [56] that, while Al might substitute Si at low Al concentrations, it preferentially substitutes Mg at higher concentrations.

Figure 5.4 compares literature data for the dopant content versus measured Hall carrier concentration. The dashed line shows the expected behavior for simple electron counting, that is, where each dopant atom donates one electron to the carrier density. We have listed literature data for the three dopants discussed earlier and employed a theoretical atomic density of 4.7×10^{22} cm^{-3}, that is, disregarding the lattice parameter increase due to dopant incorporation. The relatively small difference between the measured Hall carrier

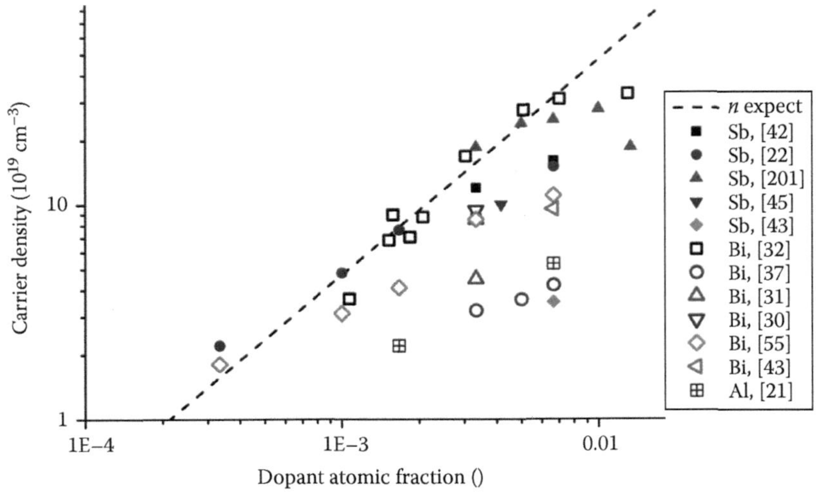

FIGURE 5.4
Atomic fraction of the dopant atoms versus measured (Hall) carrier concentration for the mainly used dopants Sb (solid symbols), Bi (open symbols), and Al (crossed symbols). The line corresponds to the expected behavior when each dopant atom generates one free electron. Note that a fraction of 1 at% corresponds to $Mg_2Si_{0.97}X_{0.03}$; this is not consistent throughout the literature.

concentration n_H and the "true" carrier concentration n is also ignored here. It can be seen that for small dopant concentrations, the expected and the measured n can match quite well for Sb and Bi, indicating that the simple electron counting picture is appropriate. On the other hand, the few existing data for Al do not fit this picture. This might be due to synthesis differences, but it also might have fundamental reasons such as Al possibly showing an amphoteric nature, where substitutions on Mg and Si sites would lead to electron hole compensation [54,56]. One can also clearly deduce that dopant activation and incorporation depend strongly on synthesis conditions. While Bux et al. [32] showed a full incorporation and activation up to 0.7 at% for Bi, using mechanical alloying from the elements, Fiameni et al. [37] found very incomplete incorporation/activation after wet BM from Mg_2Si and elemental Bi. The effect of synthesis and fabrication conditions is furthermore demonstrated by the data of Chen et al. [31], who observed an increasing carrier concentration for higher sintering temperatures in the SPS. From the existing data, an experimental solubility limit of 0.7 at% is extracted for Bi and Sb. Intuitively, a better solubility for Sb might be expected since Sb has a smaller atomic radius and is thus expected to "fit" better in the crystal lattice than Bi. However, this observed limit is purely empirical and not fundamental. Moreover, Nolas et al. [42] showed that increasing the Sb content increases the number of compensating Mg vacancies so that, while up to 12 at% Sb could be dissolved in Mg_2Si, the simple electron counting picture breaks down much earlier.

Several other dopants have been tested experimentally. A measureable increase in n and σ upon incorporation of La and In was detected [19]. Futhermore improved electrical conductivity upon addition of TiO_2 nanoparticles was observed, indicating that Ti works as an electron [57]. This result was reproduced by Fiameni et al. [58], who also showed some doping effect of Cu and Zr. Nevertheless, good thermoelectric properties for n-type Mg_2Si were found for Al, Sb, and Bi dopants only.

There are a few reports on double doping. Better results were obtained for (Bi, Al) double-doped samples compared to just Bi-doped samples, but it is not clear whether this is a fundamental effect or merely a consequence of an optimized charge carrier concentration [59]. Double doping Mg_2Si was also tried with (Pb, Sb), (Pb, Bi), and (Bi, Sb), and good thermoelectric properties independent of dopant combination were obtained [60]. Combined doping of Al or Sb with Zn incorporation was also attempted [11]. Although Zn is considered an isovalent substitution for Mg [54,56], an increased carrier concentration upon Zn substitution was observed. Furthermore, a significant reduction of the lattice thermal conductivity in Zn-doped samples was obtained, resulting in a high $ZT = 0.96$ [11].

5.1.4.1.2 Analysis of Thermoelectric Transport Data for n-Mg_2Si

Many authors have studied the thermoelectric properties of Mg_2Si. The reported data sets differ significantly, which can be due to subtle differences in sample composition, processing parameters, or other, potentially not fully understood, experimental conditions. One aim of the section is therefore to give an overview of existing literature results and, where possible, indicate reasons for the observed variations. The second aim is to analyze the reported transport data in the framework of relatively simple models, compare different studies, and, finally, try to extract fundamental material parameters. As it is impossible to analyze all data, we will restrict the discussion to selected, relevant, complete, and most recent data sets.

A sketch of the band structure of Mg_2Si is shown in Figure 5.5. Mg_2X is an indirect semiconductor with the valence band maximum at the Γ-point. The bandgap E_g is approximately 0.77 eV for Mg_2Si, 0.74 eV for Mg_2Ge, and 0.35 eV for Mg_2Sn [6]. $Mg_2Si_{1-x}Sn_x$ has two threefold degenerate conduction bands CB_L and CB_H with an energy difference E_0 that depends on x. For small x, the light conduction band is at a lower energy, while at large x, the heavy band is at a lower energy. The crossover occurs at around $x = 0.6$ [17].

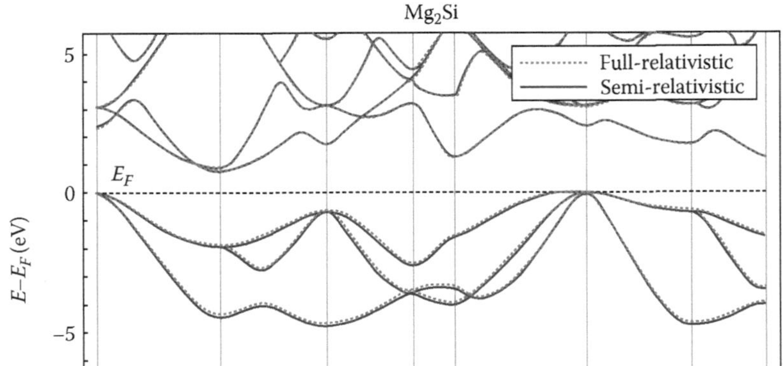

FIGURE 5.5
Band structure of Mg_2Si. (Reprinted with permission from Kutorasinski, K., Wiendlocha, B., Tobola, J., and Kaprzyk, S., Importance of relativistic effects in electronic structure and thermopower calculations for Mg_2Si, Mg_2Ge, and Mg_2Sn, *Phys. Rev. B*, 89, 115205, 2014. Copyright 2014 by the American Physical Society.)

The calculated value for E_0 falls between 0.2 and 0.5 eV [6,17,54,61]. Assuming a single parabolic band (SPB) for the moment, a chemical potential of 0.2 eV corresponds roughly to $n \sim 4 \times 10^{20}$ cm^{-3}, which is beyond the experimentally observed optimum doping. This estimation shows that the second, heavy band should not contribute significantly to the electronic transport in n-Mg_2Si (the situation is clearly different for $Mg_2Si_{1-x}Sn_x$; see Section 5.1.4.2) in the experimentally relevant charge carrier regime.

In the following, we will therefore analyze the transport data in the framework of a SPB model. The model has been employed by several authors and has been presented in detail, for example, by May and Snyder [62] and Liu et al. [18]. Using the experimental data for the Seebeck coefficient, the reduced chemical potential η can be extracted from

$$S = -\frac{k_B}{e}\left(\frac{2F_1(\eta)}{F_0(\eta)} - \eta\right), \tag{5.1}$$

where $F_j(\eta)$ is the Fermi integral of order i, k_B is the Boltzmann's constant, and e is the elementary charge.

The reduced chemical potential, in turn, can be employed to calculate the density-of-state effective mass m_D^* using

$$n = 4\pi\left(\frac{2m_D^* k_B T}{h^2}\right)^{3/2} F_{1/2}(\eta). \tag{5.2}$$

The true carrier concentration n differs from the experimentally accessible Hall carrier density $n_H = n/r_H$ by the Hall factor $r_H = (1.5F_{0.5}F_{-0.5})/(2F_0^2)$.

From n, the Lorenz number $L = \left(k_B^2/e^2\right)\left(3F_0F_2 - 4F_1^2/F_0^2\right)$ can be calculated, which allows for the determination of the lattice conductivity $\kappa_{lat} = \kappa - L\sigma T$, provided that the bipolar contribution is negligible.

The thermoelectric figure of merit can be written in terms of more fundamental quantities

$$ZT = \frac{S^2}{L + (\psi\beta)^{-1}} \tag{5.3}$$

where

$$\psi = \frac{8\pi e}{3}\left(\frac{2m_0k_b}{h^2}\right)^{1.5} F_0$$

$$\beta = \frac{\mu_0\left(m_D^*/m_0\right)^{1.5} T^{2.5}}{\kappa_{lat}}.$$

β and μ_0 are material parameters independent of carrier concentration, and μ_0 is related to the measured Hall mobility by $\mu_0 = \mu_H * 2F_0/F_{-0.5}$.

The equations given earlier are valid in the limit that the scattering of the charge carriers by acoustic phonons (APs) is the dominant scattering mechanism. This is a reasonable assumption for n-Mg_2Si at room temperature and above. The incorporation of dopants (with different atomic masses) will initiate some ionized impurity scattering, but several calculations show that this is very small compared to AP scattering for Mg_2Si at the usual doping levels [63,64].

For comparison of data sets from several sources, we have extracted the experimental results (σ, S, κ, n_H, μ) graphically and employed the equations given earlier to calculate $ZT, \kappa_{lat}, m_D^*, \mu_0$. Sometimes it was necessary to back-calculate the experimental results, for example, if κ_{lat} was given or S and the power factor. For most data sets, only room temperature Hall data are available. In that case, it was assumed that n_H is constant with temperature. This assumption is generally valid for highly doped Mg_2Si before the onset of intrinsic conduction; see, for example, the experimental results in [31,32,44]. The calculated values can differ from the original reports for two reasons. First, small deviations due to the graphical extraction and interpolation between the individual data points are possible but are not expected to be larger than a few percent. Second, the authors might have assumed different models for the analysis. For example, Nolas et al. [42] assumed mixed ionized impurity and AP scattering mechanisms, which is probably valid for the very high doping levels they employed for some of their samples but not for the moderately doped ones.

While its simplicity makes the data analysis relatively easy, the model also has clear limitations that should be kept in mind. First, it does not account for the contribution of the valence band, which becomes important at high temperatures and low doping levels due to the thermal excitation of intrinsic charge carriers. For the following analysis, the use of such data has been omitted. Second, as discussed previously, the model also ignores the contribution from the second conduction band that might become relevant at high temperatures and high doping levels. Third, the model is based on a rigid band picture, that is, the assumptions that the band structure does not change significantly with temperature and that the incorporation of dopants shifts the chemical potential but does not modify the band structure itself. Neither assumption is entirely true; however, several calculations show that the effect of adding the most popular dopants Sb and Bi does not affect the band structure [43,54]. Finally, as the SPB is a model for the electronic transport, it uses the lattice thermal conductivity as an input parameter and cannot itself account for differences in κ_{lat}, for example, due to nanostructuring.

Results: The thermoelectric figure of merit is plotted as a function of n_H at 770 K, compiling data from different sources in Figure 5.6. Some reports measure to higher temperatures; however, at 770 K, most of the data are available.

The optimum carrier concentration is $n_H \approx 1 \times 10^{20}$ cm^{-3}, in agreement with SPB modeling results from Bux et al. [32]. This corresponds approximately to a composition of $Mg_2Si_{0.994}X_{0.006}$ and is well within the experimentally accessible composition range. However, as pointed out in Figure 5.4, dopant activation is experimentally often below unity and depends on the fabrication route. The optimum carrier concentration depends only weakly on the temperature.

A fundamental material parameter, the density of states (DOS) effective mass is presented in Figure 5.7. When authors used several samples, the average m_D^* is calculated and presented. The data show considerable spread around $m_D^* \approx 1.1 m_0$ at room temperature, where m_0 is the free electron mass. The scatter is presumably not representative of variations within the material but is rather the footprint of measurement uncertainties of the Seebeck coefficient and the Hall data.

Most of the data in Figure 5.7a show a slight increase of $m_D^*(T)$, indicating a very moderate change of the band structure with temperature. One of the assumptions of the SPB is that $m_D^* \neq m_D^*(n)$. This is basically confirmed in Figure 5.7b where extracted effective

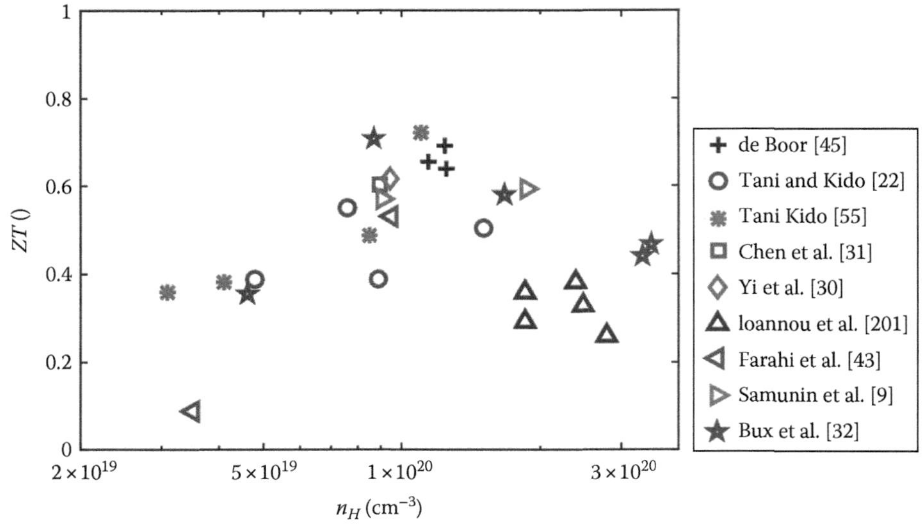

FIGURE 5.6
Thermoelectric figure of merit at 770 K from several sources. The maximum $ZT \approx 0.75$ at 770 K is found at the concentration $n_H \sim 1 \times 10^{20}$ cm^{-3}.

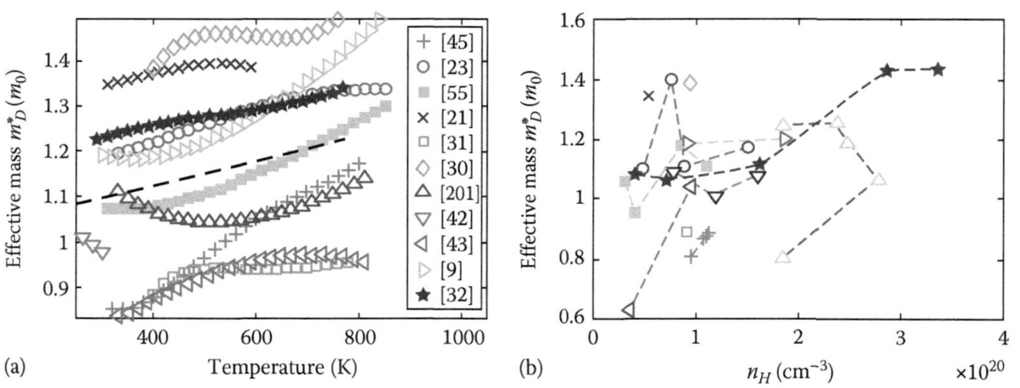

FIGURE 5.7
(a) Effective mass versus temperature extracted from several recent publications using a single parabolic band model. If the publications contain data sets for several samples, the average over those samples is plotted. The data show significant scatter with $m_D^* \approx 1.1m_0$ at room temperature and a small increase with temperature. (b) Effective mass versus Hall carrier concentration n_H at room temperature (295–330 K). The data do not show a significant change with n_H, indicating a good parabolicity for the (lowest lying) Mg_2Si conduction band. The legend holds for the following figures and markers do not represent measurement points but are chosen for clarity.

masses for individual samples are given as a function of the measured Hall carrier concentration near room temperature. While the data again show some scatter, no trend is discernable either from the complete data or from individual data sets.

This indicates relatively good parabolicity of the lower conduction band and excludes a significant contribution of the higher, heavier conduction band for the relevant carrier densities, in agreement with the initial assumptions.

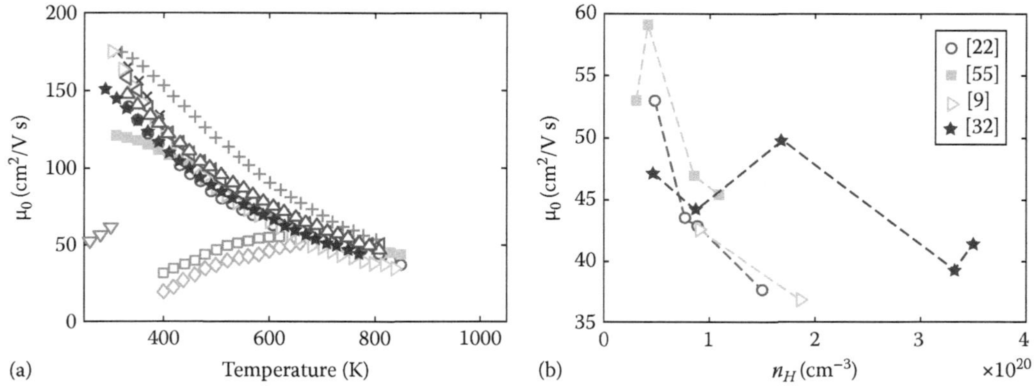

FIGURE 5.8
(a) Carrier density corrected mobility μ_0 versus temperature and (b) versus n_H at 770 K. At low temperatures, different synthesis techniques can lead to different mobilities that tend to converge at higher temperatures. A small reduction with increasing carrier concentration can be deduced from the data. Legend as in Figure 5.7.

The mobility parameter μ_0 is presented in Figure 5.8. Most of the data sets show the expected decay with temperature ($\mu_0 \propto T^{-p}$, with $p \approx 1.5$), and the data show less scatter than the results for m_D^*, especially at higher temperature, where all data sets converge. While, at high temperatures, AP scattering is the dominant scattering mechanism, at lower temperatures, additional scattering due to grain boundaries or porosity can be visible; see the discussion in Section 5.1.4.1.3.

The mobility μ_0 is corrected for the usual reduction due to carrier–carrier scattering with increasing carrier density; it is thus in principle supposed to be independent of n. As can be seen from Figure 5.8b, this is fulfilled to a large degree. On the other hand, the data by several authors indicate a small reduction of $\mu_0(n_H)$ presumably due to additional ionized impurity scattering by the Bi/Sb ions. This mechanism is not included in the SPB model and shows its limitations; however, the effect is relatively small. This effect is negative with respect to ZT optimization, albeit relatively minor. The lattice thermal conductivity is plotted versus temperature and n_H in Figure 5.9. Again, the average over all samples is shown. It was calculated using $L(\eta)$ and might therefore differ from the results of the authors, who sometimes use a constant L [21,22,55]. If the degenerate limit is employed ($L = L_0 = 2.45 \times 10^{-8}$ V^2/K^2), the electronic contribution to the thermal conductivity is overestimated and, hence, κ_{lat} is underestimated. This can lead to an apparent decrease of κ_{lat} with increasing n_H.

It roughly holds that $\kappa_{lat} \propto T^{-1}$ for all data sets; however, the absolute values differ considerably. The κ_{lat} data of [21,22,55,65] exhibit relatively high values; the samples were produced by BM and direct SPS. These synthesis routes are known to result in MgO formation, and MgO impurities can increase κ_{lat} [66]. However, this can only be part of the explanation, given that all samples employed here showed average to excellent σ, indicating that MgO content will be of the order of a few percent at max. Chen et al. [31] recorded very low κ_{lat} and attributed this to the nanograins and nanopores resulting from the employed synthesis approach. In addition to low κ_{lat}, the samples also exhibited a very low μ, supporting the argument qualitatively. Figure 5.9b presents $\kappa_{lat}(n_H)$ at ~600 K, which was chosen because at higher temperatures the influence of bipolar thermal conduction affects the lower doped samples. Several authors observed a reduction of κ_{lat} with increasing dopant concentration, compensating the similarly observed effect on μ_0; see Figure 5.8b.

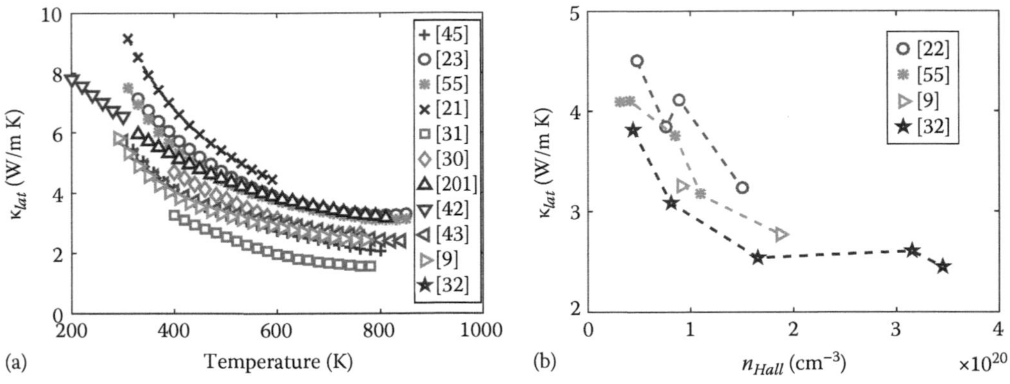

FIGURE 5.9
(a) Lattice thermal conductivity $\kappa_{lat} = \kappa - \sigma L(\eta)T$ as function of temperature and (b) Hall carrier concentration. Legend as in Figure 5.7.

Overall, the influence of synthesis routes on sample microstructure and, hence, on κ_{lat} is clearly observed. It would be expected that the footprint of this is most clearly observed at lower temperatures, and it should somewhat weaken at higher temperatures as Umklapp scattering becomes progressively more important with increasing temperature than other possible, microstructure-induced scattering mechanisms. The fact that this is not observed indicates either that different/additional scattering mechanisms are also active at high temperatures in some of the samples or that the data are affected by significant measurement uncertainties.

It was shown that the main features of the electronic transport in Mg_2Si can be well understood in the framework of a SPB model with AP scattering as the dominant mechanism. This makes this relatively simple model a powerful tool to analyze and evaluate experimental data. Data sets from different authors show considerable differences. These differences may, in part, be related to physical differences between the samples but are to some extent presumably the footprint of measurement inaccuracies. We have therefore compiled data from several authors, aiming at a more reliable extraction of parameters like the effective mass and the optimal carrier concentration.

5.1.4.1.3 Influence of Secondary Phases and the Microstructure

The introduction of secondary phases into a bulk material is a popular approach to enhance the thermoelectric properties and has been tested with varying success for several material classes [67]. There are several reasons why (nanostructured) secondary phases might be beneficial. First, they create additional interfaces that might scatter phonons more efficiently than charge carriers. Second, interfaces with relatively low barrier heights can enhance the power factor of a material by energy filtering [64,68]. Third, secondary phases can also enhance the mechanical properties, which is crucial from an application point of view. Apart from the intentional addition of secondary phases, they can also be created during the synthesis or fabrication of the material. This is particularly true for Mg_2Si, where synthesis often involves high temperatures. Due to the high vapor pressure of Mg and its tendency to oxidize, the occurrence of MgO and Si as secondary phases is usually reported; see Section 5.1.2. For $Mg_2(Si, Ge, Sn)$, partial immiscibility between the binary compounds can furthermore lead to the formation of phases with different Si/Ge/Sn ratios.

We will discuss the influence of secondary phases that form during the synthesis first. Of these, MgO is probably the most relevant. It is observed using XRD or SEM in the majority of publications on Mg_2Si; see, for example, [36,39,45]. MgO formation is most significant when exposing large surface areas and when working at high temperatures. For example, Riffel and Schilz [33] reported an increase in MgO content with BM time, while Saleemi et al. [39] pointed out the MgO increase with increasing sintering temperature, and Fiameni et al. [41] observed considerable MgO increase after each thermal treatment. MgO formation can be partially suppressed by using a solid-state synthesis approach rather than a melting approach, at best working under reducing atmosphere; see Section 5.1.2. However, even if the material is synthesized at lower temperatures, compaction usually still has to be performed at $T > 1000$ K, so that MgO is often found in the compacted samples, even if the powder is apparently MgO-free [16,20]. MgO is often found in combination with elemental Si, indicating decomposition of Mg_2Si *via* $Mg_2Si + O_2 \rightarrow 2MgO + Si$. Oxidation of Mg_2Si might be puzzling at first glance, given that most people work under "inert" conditions. However, traces of O_2 might still be present in the material or get into it if not the whole fabrication route is inert, leading to MgO formation in the compaction step.

As the contamination of Mg_2Si with MgO is hard to avoid, understanding of the influence of MgO is important. Studying *n*-type Mg_2Si samples from a large batch process, that is, with identical composition, de Boor et al. [45] found that the electrical conductivity is drastically affected by the presence of MgO in the samples. Using high-temperature Hall measurements, it was shown that the carrier concentration is unaffected by MgO, but the carrier mobility is reduced, especially at lower temperatures; see Figure 5.10.

The behavior can be understood in the framework of a SPB model with the combination of scattering of the charge carriers by APs and grain boundaries. Employing a simple grain boundary (GB) scattering picture [69], reasonable estimates for physical parameters such as the grain size and the energy barrier height were obtained. As GB scattering is

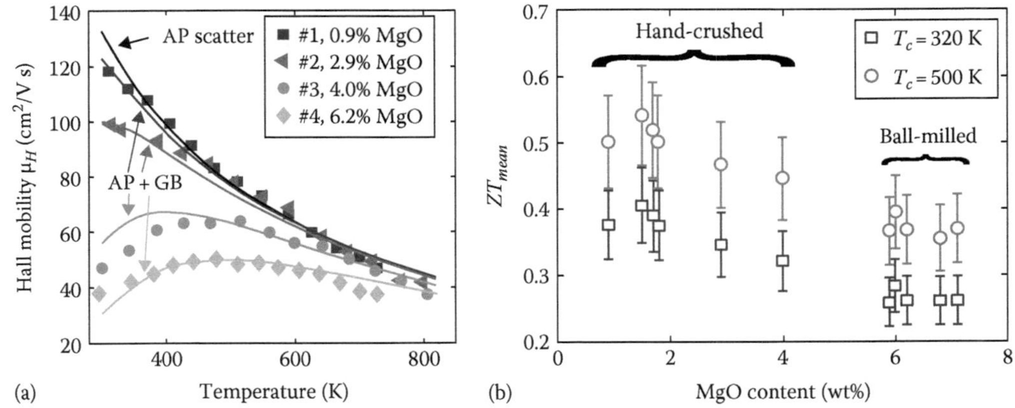

FIGURE 5.10
(a) Hall mobility of several samples of nominal stoichiometry of $Mg_2Si_{0.9875}Sb_{0.0125}$, but different MgO content due to different processing parameters. Increasing MgO content causes a reduction of μ_H *via* grain boundary scattering at low temperatures, while at higher temperatures scattering by acoustic phonons dominates. (b) A clear correlation between average ZT and MgO content can be observed. (Reprinted from *Acta Mater.*, 77, de Boor, J., Dasgupta, T., Kolb, H., Compere, C., Kelm, K., and Müller, E., Microstructural effects on thermoelectric efficiency: A case study on magnesium silicide, 68–75, Copyright 2014, with permission from Elsevier.)

most effective at lower temperatures, σ and μ data of samples with the varying MgO content tend to converge at higher temperatures; thus, the effect on the maximum ZT is often small. However, for applications, it is the average ZT_{mean} across a finite temperature range that is relevant, and thus MgO clearly has a detrimental influence on the thermoelectric performance of Mg_2Si. This is summarized in Figure 5.10b, where ZT_{mean} is plotted versus MgO content. Several authors reported a nonmonotonous behavior of $\sigma(T)$ in reasonably doped samples where, in principle, a monotonically decreasing $\sigma(T)$ would be expected [37,41,70] and MgO-induced GB scattering can explain the observed data. Nevertheless, open questions remain. While there seems to be a clear correlation between MgO content and mobility reduction, the microscopic picture is unclear. MgO is often observed as localized particles between Mg_2Si grains [36], which does not really fit the picture of MgO acting as energy barriers for the carriers between the grains. Furthermore, it should also be noted that a temperature-activated behavior has also been observed for apparently (XRD-wise) MgO-free samples [30,31]. In these reports, the authors compacted nanopowder with relatively small crystallite sizes. Moreover, nanopore formation was also observed in the samples [31]. This indicates that a high interface density can lead to a similar temperature dependence of μ and σ as caused by MgO at the grain boundaries.

So far, we have restricted the discussion to binary Mg_2Si. In $Mg_2(Si, Sn)$ solid solutions, MgO is also often observed [10,27,46,54], and its effect is, in principle, expected to be the same. However, the impact on the thermoelectric performance is less severe for two reasons. First, the synthesis and compaction temperatures of $Mg_2(Si, Sn)$ tend to decrease with increasing Sn content, reducing MgO formation. Second, the carrier mobility in $Mg_2(Si, Sn)$ is reduced due to alloy scattering. As scattering lengths add inversely, additional scattering due to MgO will cause a smaller degradation in the mobility compared to binary Mg_2Si.

If one assumes that MgO is formed during the synthesis or by decomposition of Mg_2Si, a correlation between the contents of MgO and Si would be expected. This is, indeed, observed [20,54], but often there is no linear relation between MgO and Si contents [37,45,58]. Furthermore, some papers report only MgO but not elemental Si [10]. Several explanations are plausible for that behavior. First, MgO and Si are usually detected using XRD that is sensitive to crystalline materials, and it is possible that MgO and Si in the samples have a different degree of crystallinity. Second, Mg_2Si has a finite tolerance for a nonstoichiometric composition [7,42,65]; see also Section 5.1.3.

Aiming for improved thermoelectric or mechanical properties, several authors have intentionally added secondary phases to Mg_2Si. Low melting glass frit added to Mg_2Si was observed to increase $ZT_{\max}$ from 0.55 to 0.65 as well as enhance the tensile strength of the samples [71]. However, the authors also pointed out that the reference samples had a relatively low density (~90%), while the samples with the glass frit achieved almost theoretical density. An increase in the density usually results in improved thermoelectric and mechanical properties; see, for example, Figure 5.2. It is thus not clear, whether the glass frit really improves the properties or only enhances an otherwise imperfect compaction process. Cederkrantz et al. [57] added small amounts of TiO_2 nanoparticles to undoped Mg_2Si and reported improved thermoelectric properties. Their Hall measurements indicated that a fraction of TiO_2 was reduced and Ti acted as a poor electron donor. They did not observe a change in κ upon TiO_2 addition. As a consequence, the effect of the TiO_2 nanoparticles is essentially an improved electronic transport in the same fashion as achieved by typical dopants like Bi and Sb, and not an improvement due to the secondary phases itself. Mg_2Si/Si nanocomposites were synthesized with nanosized Mg_2Si grains and Si inclusions [30]. The authors found a small reduction in the thermal conductivity and a massive reduction in the room temperature carrier mobility, partly due to the synthesis process and partly due to the high interface density. Comparing samples with and without Si nanoparticles, no significant improvement in the thermoelectric properties was obtained. Adding carbon nanostructures, some improvement in the thermoelectric properties was obtained, however, on a relatively low level [41]. By incorporating a metallic binder into Sb-doped Mg_2Si, mechanical properties were improved and sample cracking was significantly decreased [35]. Furthermore, the authors found an improved power factor for 5 wt% of Ni microparticles, while κ_{lat} remained almost unaffected, leading to a significant improvement in ZT. The mechanism for this improvement was not discussed in detail, and the authors' attempts to reproduce the results were not successful. As the approach is simple and the first results appeared promising, future research and reproduction of the results are highly desired.

Except for the data of Hayatsu et al. [35], no experimental studies are known where the addition of a secondary phase would lead to improved thermoelectric properties. "Improved" here means either that the power factor was enhanced due to energy filtering or that κ_{lat} was decreased more than μ due to additional scattering at the interfaces. This is in agreement with theoretical calculations. Pshenai-Severin et al. [61] calculated the band structure of Mg_2Si using DFT and fitted the solutions of the Boltzmann transport equations for a two-band model to experimental data of n-Mg_2Si. Using the obtained scattering rates and interaction potentials, they calculated the ratio of σ/κ versus grain size and found only moderate improvement ~10%, indicating similar mean free paths for electrons and phonons in Mg_2Si. This was supported by the calculations of Satyala and Vashaee [40,63] who used a multiband model to show that the dominant phonons have a mean free path comparable to that of the charge carriers. The authors also pointed out [72] that a possible power factor enhancement from energy filtering is expected to be small, concluding that nanostructuring is not a promising approach for Mg_2Si. This is in agreement with the majority of the experimental findings.

5.1.4.1.4 *p-Type Doping*

n-Type Mg_2Si-based solid solutions have excellent thermoelectric properties. Thermoelectric generators require n- and p-type materials, ideally with similar compositions and lattice structures in order to facilitate contacting and assuring comparable mechanical properties and thermal expansion under thermal stress. The development of p-type Mg_2Si-based solid solutions is thus highly desired and has been addressed in many studies.

From a band structure point of view, a performance comparable to n-Mg_2Si can be expected for p-Mg_2Si. Calculations indicate [73] that for p-$Mg_2(Si, Ge, Sn)$, the valence band degeneracy at the Γ-point is removed, which is likely to be detrimental. Nevertheless, p-type material with a reasonable performance can be expected with the predicted ZT_{max} ~0.5 at 700 K, assuming k_{lat} = 2 W/m K [64] and values of ZT_{max} ~0.8 at 850 K [63]. In contrast to this, experimental results for (binary) p-Mg_2Si usually show $ZT_{max} < 0.15$, with some partially debated exceptions [26,74,75].

Verified p-dopants for Mg_2Si are Li, Ag, and Ga [19,59,76–79]. Common observations in all studies are that the carrier concentrations are orders of magnitude lower than expected from the nominal dopant concentration and that the hole mobility is very low ≤10 cm^2/V s. This results in a poor electrical conductivity as well as a significant impact of intrinsic conduction at relatively low temperatures. The onset of intrinsic conduction causes a drastic decrease of the Seebeck coefficient, which might even change its sign [59]. Moreover, it has been shown that Ag can precipitate out of the Mg_2Si lattice at intermediate temperatures [80]. These effects lead to the observed small values of *ZT*. The reason for the observed low carrier concentration can partially be insufficient dopant activation. A further reason is given by a theoretical study [53] that showed that the *p*-type conductivity of Ag-doped Mg_2Si is achieved through linear clustering of Ag with Mg interstitials (Ag–Mg_{int}–Ag). The positive charge of the Ag substitutions is compensated to a large extent by the negative charge of Mg_{int}. If a similar argument is valid for other *p*-dopants, it would explain why the hole concentration necessary for good *p*-type material (~(2–4) × 10^{20} cm^{-3}) [63,64] has not been obtained experimentally. It has also been argued [53] that the formation of the Ag–Mg_{int}–Ag clusters reduces the hole mobility, which is in agreement with the experimental observations.

Better thermoelectric properties can be achieved if Si is partially replaced by Ge or Sn. In Ga-doped Mg_2Si and $Mg_2Si_{0.6}Ge_{0.4}$, for the same Ga content, the concentration of holes doubled and the mobility was significantly improved in Ge-containing samples [77]. This might indicate that defect formation is less significant when Si is partially replaced. This finding is in agreement with a report by Liu et al. [46] who systematically varied the Ga content in $Mg_{2.1}(Si_{0.3}Sn_{0.7})_{1-y}Ga_y$. They were also able to achieve high hole concentrations n_{max} ~3 × 10^{20} cm^{-3} and carrier mobilities of ~20 cm^2/V s. They also pointed out that a high hole concentration can be achieved more easily in Sn-rich solid solutions than in Si-rich Mg_2(Si, Sn), in agreement with a study by Fedorov et al. [81]. Although they were able to obtain high carrier concentration, the onset of intrinsic conduction limited the achievable *ZT* to 0.35. Further progress has been shown by using Li as a dopant to replace Mg [82]. This relatively effective doping resulted in a high carrier concentration up to n ~5 × 10^{20} cm^{-3} and an increased figure of merit *ZT* = 0.5. The carrier concentration–dependent effective mass as well as DFT calculations moreover indicated that, for $Mg_{2-x}Li_xSi_{0.3}Sn_{0.7}$, two valence bands are close to convergence, which is an additional reason for the improved performance. These relatively new results suggest promising directions for future research, with further improvements to be expected.

5.1.4.2 Solid Solutions of Mg_2Si and Mg_2Sn

The thermoelectric performance of solid solutions of the isostructural Mg_2X (X = Si, Ge, Sn) compounds show significant improvement over binary Mg_2Si. The Mg_2Si–Mg_2Sn system in particular has been studied in detail and is the focus of this section.

5.1.4.2.1 Electronic Transport

In Figure 5.11, *ZT* values of optimized binary Mg_2Si are compared with that of Mg_2(Si, Sn). A near doubling of the *ZT* is observed in the solid solutions due to a combination of the reduced lattice thermal conductivity (alloying effect) and the increased power factor (band convergence). Originally suggested by Zaitsev et al. [6], the occurrence of band convergence was predicted theoretically by Tan et al. [83] and subsequently verified experimentally [17]; see Figure 5.11b. The conduction band minima in the Mg_2X compounds consist of two near degenerate bands whose orbital contributions and band mass are different for Mg_2Si (also Mg_2Ge) and Mg_2Sn. Thus, in Mg_2Si the light band (with Si character) is the lowest lying conduction band while in Mg_2Sn the lowest lying conduction band is the heavy band having Mg character.

In the case of a solid solution, at some intermediate composition (Si/Sn ratio), the two bands converge, resulting in an increase in the absolute Seebeck coefficient value (due to the increased density-of-state effective mass) while retaining high electrical conductivity. Theoretical estimates of the Si/Sn ratio for band convergence vary from around

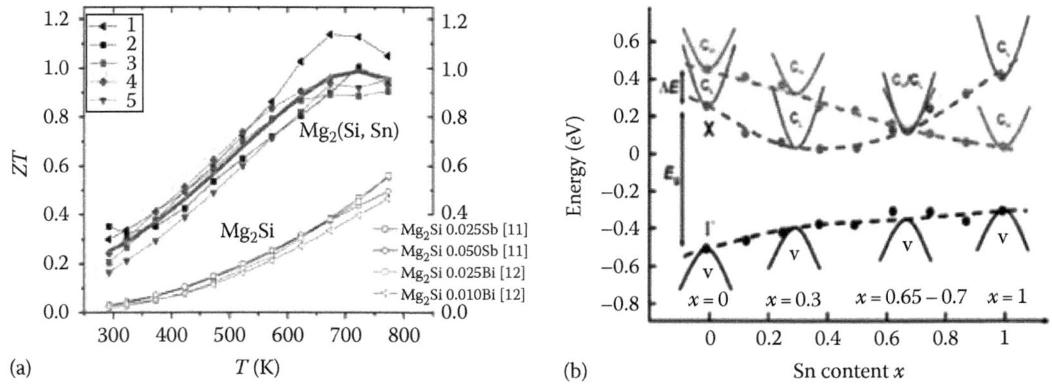

FIGURE 5.11
(a) Comparison of experimental *ZT* values in optimized Mg_2Si and $Mg_2Si_{0.4}Sn_{0.6}$ based materials. (From Dasgupta, T., Stiewe, C., de Boor, J., and Müller, E.: Influence of power factor enhancement on the thermoelectric figure of merit in $Mg_2Si_{0.4}Sn_{0.6}$ based materials. *Phys. Stat. Solidi (a)*. 21. 2014. 1250–1254. Copyright Wiley-VCH Verlag GmbH & Co. KGaA. Reproduced with permission.) (b) Convergence of the two conduction bands as a function of Si/Sn ratio in $Mg_2Si_{1-x}Sn_x$. (Reprinted with permission from Liu, W., Tan, X.J., Yin, K. et al., Convergence of conduction bands as a means of enhancing thermoelectric performance of n-type $Mg_2Si_{1-x}Sn_x$ solid solutions, *Phys. Rev. Lett.*, 108, 166601, 2012. Copyright 2012 by the American Physical Society.)

40:60 [6,84] to 35:65 [17,83]. Thus, compositions close to $Mg_2Si_{0.4}Sn_{0.6}$ are expected to have high *ZT* values.

5.1.4.2.1.1 Doping Studies Traditional dopants for optimization of $Mg_2Si_{0.4}Sn_{0.6}$ are antimony and bismuth. In addition, it has been reported that Mg acts as a self-dopant. A systematic study of the influence of Mg content on the electronic transport properties in the Sb-doped Mg_2(Si, Sn) system has been carried out by Liu et al. [47]. Their result showed that an excess of Mg at interstitial positions acts as a donor while a substoichiometric amount of Mg results in Mg site vacancies, which compensate the donor effect of Sb, resulting in low charge carrier concentration. Moreover, they also showed that a modest excess of Mg produces the best *ZT* value, which was interpreted as due to the increased weighted mobility $\left(\mu m_D^{*\,3/2}\right)$. Thus, the actual Mg content in the specimen (apart from the dopant concentration) seems to play an important role in determining the resulting TE properties. However, as discussed in Section 5.1.2, the exact content of Mg is difficult to control due to the volatility of Mg at the synthesis temperature, and this is a contributing factor to a wide range of reported *ZT* values; see in the following text.

In Figure 5.12, the TE properties of optimized $Mg_{2+z}Si_{0.4}Sn_{0.6}$ compositions reported by different research groups are compared. The maximum obtained *ZT* value varies from 0.9 to 1.55, a difference of over 70%. To get a better perspective, the synthesis parameters and room temperature transport properties reported by different groups are given in Table 5.2.

From the table, it is noted that the maximum *ZT* values were obtained over a wide range of carrier concentrations that can be sorted into two groups: (1.5–1.7) × 10^{20} cm^{-3} and (2.35–3.2) × 10^{20} cm^{-3}. This is also reflected in the electrical conductivity data (Figure 5.12b) that form two distinct clusters. Theoretical calculations [84] of the optimal electron concentrations to achieve the maximum *ZT* indicate a range of (1.1–1.6) × 10^{20} cm^{-3}, which coincides with the observed experimental data of the low *n* group [84]. Also, the experimental

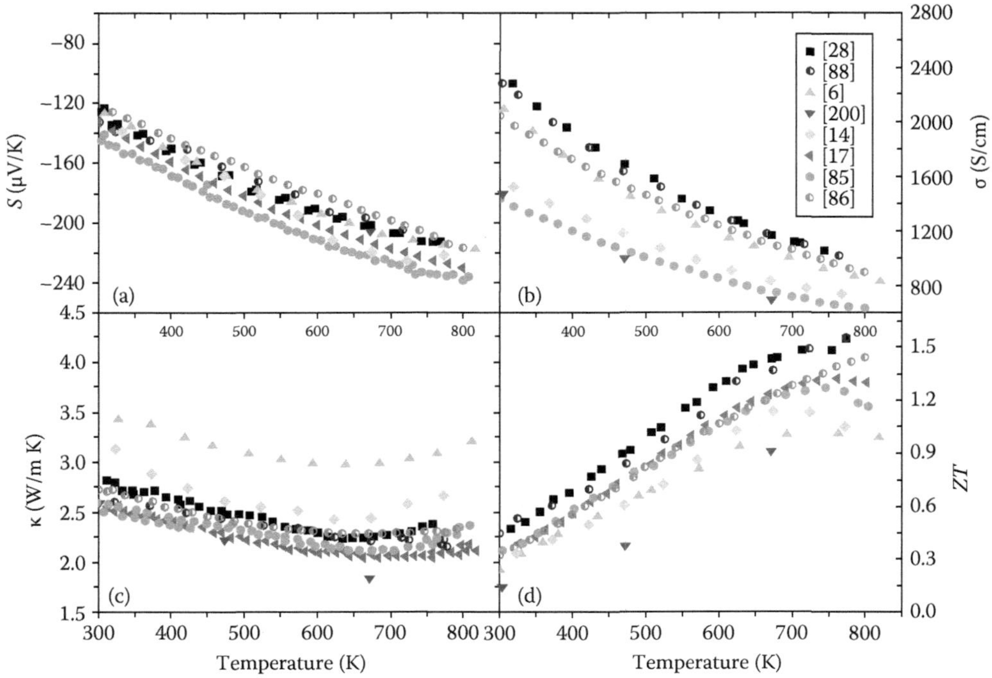

FIGURE 5.12
Comparison of the reported temperature dependence of (a) Seebeck coefficient, (b) electrical conductivity, (c) thermal conductivity, and (d) *ZT* in doped $Mg_2Si_{0.4}Sn_{0.6}$. Dopants used in the reports were either Sb or Bi (the data with Bi doping are shown with half-filled markers).

TABLE 5.2
Synthesis Conditions and Room Temperature Transport Parameters of $Mg_zSi_{0.4-x}Sn_{0.6}(Sb/Bi)_x$

Reference	Mg (z)	Doping (x)	n (cm^{-3}) × 10^{20}	μ (cm^2/V s)	m_D^*	ZT_{max} (T)	Synthesis Technique
[28]	2.08	0.036 (Sb)	2.35	42	2.55	1.51 (716 K)	Flux method
[88]	2.08	0.03 (Bi)	3.2	42	—	1.55 (773 K)	Flux method
[6]	—	— (Sb)	2.99	—	—	1.1 (763 K)	Induction melting
[200]	2.2	0.0075 (Sb)	1.65	51	—	0.9 (673 K)	Induction melting
[14]	2.06	0.0075 (Sb)	1.50	80	1.90	1.15 (700 K)	Induction melting
[17]	2.14	0.009 (Sb)	1.70	56	1.51	1.3 (700 K)	Two-step solid-state
[85]	2.14	0.0075 (Sb)[a]	1.67	56	1.51	1.3 (740 K)	Two-step solid-state
[86]	2.12	0.025 (Bi)[a]	2.36	51	—	1.4 (800 K)	Two-step solid-state

[a] Composition considered as $Mg_z(Si_{0.4}Sn_{0.6})_{1-x}(Sb/Bi)_x$.

data [14,85] for solid solutions with high charge carrier concentration (~2.5 × 10^{20} cm^{-3}) indicate a lower *ZT*. These results hint at the high *ZT*s obtained at higher n values to be due to changes in the lattice thermal conductivity, the topic discussed in the next section.

Doping with Sb and Bi seems to produce similar results, indicating a rigid bandlike behavior in the presence of these dopants. This is also observed from the Pisarenko plot in [14,18,29] and in agreement with experimental and theoretical findings on doped Mg_2Si; see Section 5.1.4.1.1. An increased density-of-state effective mass $\left(m_D^*\right)$ of the solid solution compared to binary Mg_2Si in these reports is an experimental evidence of band convergence in the solid solutions. It is to be noted that an increase in m_D^* with increasing n has also been reported by some authors [85,86], and further work is required to understand the effect of the dopants on the interband separation E_0 and shape.

The temperature dependence of mobility from room temperature and above indicates two regions with temperature exponent of −0.5 (near room temperature) and −1 (at higher temperature), representing alloy scattering and AP scattering, respectively, as the dominant electron scattering mechanism. Calculations of the alloy scattering potential, U (0.7, 0.32–0.39 eV), and the deformation potential, Ξ (13 eV, 8.7–9.4 eV), from the temperature dependence of mobility data indicate values that are on the lower side compared to other TE solid solutions [18,29]. This results in an increased weighted mobility and has been interpreted as an additional reason for the high ZT value of the solid solutions. The cause of the low U values remains unclear and provides scope for further work.

The bandgap (E_G) in the solid solutions is a function of the Si/Sn ratio and decreases with the increasing content of Sn. Estimated E_G values at 0 K lie between 0.54 and 0.61 eV [6,28,64] for $Mg_2Si_{0.4}Sn_{0.6}$. The bandgap is also reported to decrease linearly with temperature. Thus, bipolar heat conduction is observed at high temperatures, with the onset at around 500 K for undoped $Mg_2Si_{0.4}Sn_{0.6}$. The presence of bipolar heat flow tends to decrease the ZT value and narrow the operational range of the material. From Table 5.2, it is observed that, in general, the heavily doped specimens have a higher ZT_{max} value and peak temperature. One of the reasons for this is the suppression of bipolar transport with increasing charge carrier concentrations. Finding effective ways how to mitigate bipolar thermal transport is important and the topic provides a relevant scope for future research [18,86].

From Table 5.2, it is also observed that high-temperature synthesis using induction melting results in lower ZT_{max} values. This is possibly associated with difficulties to control the Mg stoichiometry at higher temperatures. Formation of stable defects in binary Mg_2Si is well studied (see Section 5.1.3), and the presence of interstitial Mg in $Mg_2Si_{0.4}Sn_{0.6}$ is reported to improve the thermoelectric performance [47]. A point defect chemistry approach for improving the thermoelectric performance in Mg_2(Si, Sn) solid solutions has been studied [87], and tuning the doping concentration of antimony and Mg vacancies and interstitials was shown to improve ZT. Likewise, the presence of interstitial Bi atoms in $Mg_2Si_{0.4}Sn_{0.6}$ was reported to increase the ZT value [88]. Thus, it would seem that point defects are helpful in improving the TE performance in the solid solutions, though further studies are required to understand their effect on the electronic transport parameters. It also seems beneficial to take the synthesis conditions into account because studies such as the kinetics of defect formation and defect stability are dependent on the local chemical composition and process parameters. The presence of secondary phases has also been reported by different groups and has been shown to affect the lattice thermal conductivity (covered in the next section). A theoretical study [64] on the effect of secondary phases in enhancing ZT showed filtering of low-energy electrons to provide an opportunity for significant enhancements, with ZT greater than 3 being possible in $Mg_2Si_{0.4}Sn_{0.6}$ by filtering electrons with energies lower than 0.4 eV.

5.1.4.2.2 Thermal Transport

The formation of solid solutions results in a decrease of the lattice thermal conductivity (κ_{lat}) due to the additional scattering of phonons by mass and strain contrast.

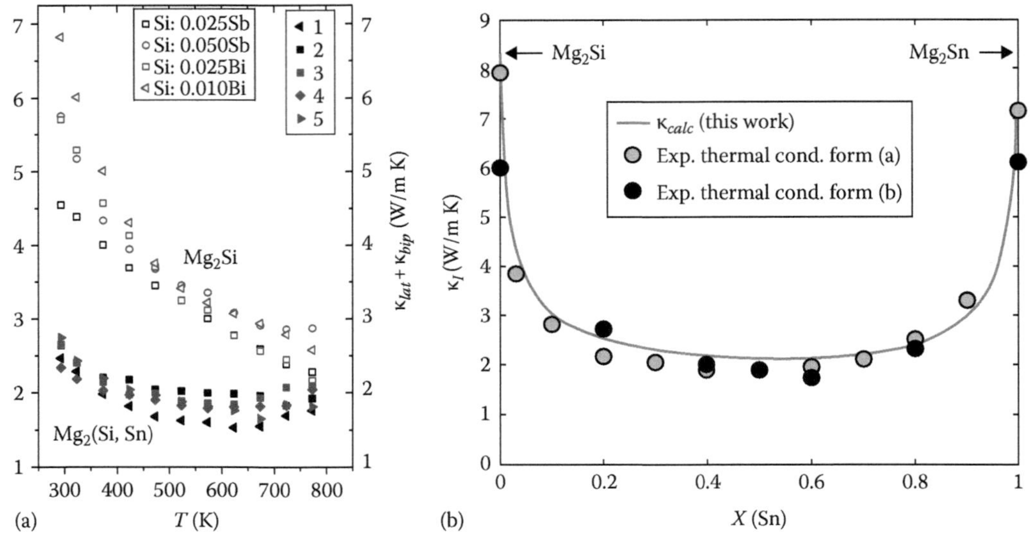

FIGURE 5.13
(a) Comparison of experimental lattice thermal conductivity (κ_L) in optimized binary Mg_2Si and $Mg_2Si_{0.4}Sn_{0.6}$. (b) Variation of κ_L in Mg_2(Si, Sn) solid solutions at room temperature as a function of Sn content. (Reprinted from *Acta Mater.*, 95, Bellanger, P., Gorsse, S., Bernard-Granger, G., Navone, C., Redjaimia, A., and Vives, S., Effect of microstructure on the thermal conductivity of nanostructured Mg_2(Si,Sn) thermoelectric alloys: An experimental and modeling approach, 102–110. Copyright 2015, with permission from Elsevier.)

Thus, alloying is a traditional technique utilized to improve the performance of thermoelectric materials. The effect of alloying on the lattice thermal conductivity in the Mg_2(Si, Sn) solid solutions is shown in Figure 5.13a. The extent of decrease in κ_{lat} depends on the mass difference between the substituting and substituted atoms and the amount of substitution [6,89,90]. This is shown in Figure 5.13b where about 30% of Sn is needed to reach a plateau in the thermal conductivity of the Mg_2(Si, Sn) solid solutions. Since alloying causes defects at the atomic scale, it results in efficient scattering of short-wavelength phonons, and the effect on κ_{lat} is most notable at lower temperatures where the Umklapp processes are frozen out; see Figure 5.13a. To reduce κ_{lat} further, additional defects at different length scales are required. The beneficial effect of such multiscale scattering on the overall thermoelectric performance has been demonstrated in other thermoelectric material systems [67].

The presence of secondary phases in Mg_2(Si,Sn) solid solutions has been reported by various authors, and the studies encompass nanometer-scale inclusions, micron-sized inclusions, and mesoscale phases. Spherical nanoinclusions with a composition around $Mg_2Si_{0.3}Sn_{0.7}$ (based on transmission electron microscopy [TEM] imaging) within the matrix ($Mg_2Si_{0.4}Sn_{0.6}$) were reported by Liu et al. [86] and Bellanger et al. [90]. The latter authors applied an aging treatment at 500°C to check the stability of the nanophase. Results showed these nanoparticles to be unchanged up to a time period of 4 weeks, indicating their stability once formed. Micron-sized Si-rich precipitates have also been reported [14,86]. A possible reason for the formation of secondary phases is the presence of a miscibility gap in the Mg_2Si–Mg_2Sn phase diagram; see, for example, Jung et al. [13] and references therein. A quantitative analysis of the effect of secondary nanoparticles on κ_{lat} in $Mg_2Si_{0.4}Sn_{0.6}$ has been carried out [90] and is shown in Figure 5.14a. As finely

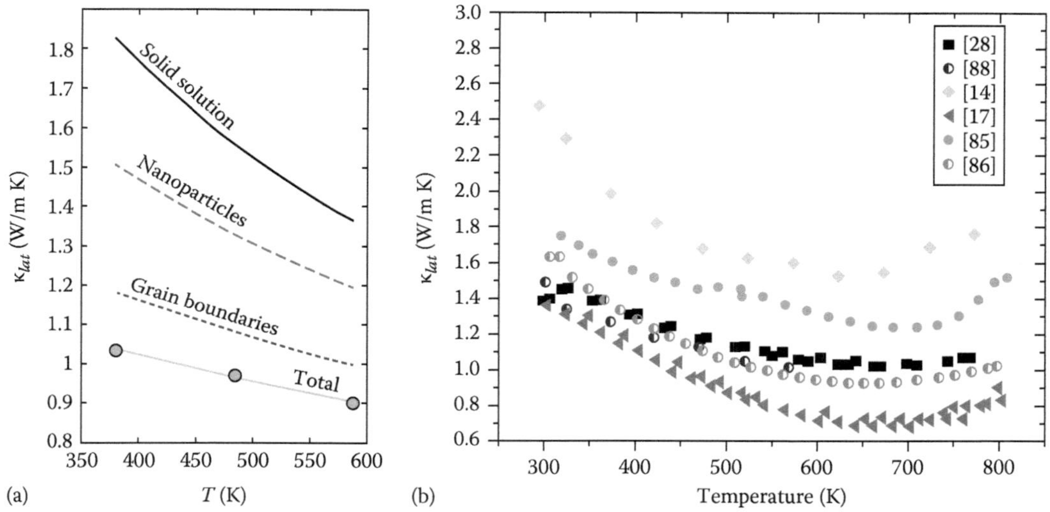

FIGURE 5.14
(a) Theoretical estimate of the contribution to the lattice thermal conductivity by different phonon scattering mechanisms in $Mg_2Si_{0.4}Sn_{0.6}$ (experimental data shown by circles). (b) Comparison of the reported experimental κ_L by various groups. ([a]: Reprinted from *Acta Mater.*, 95, Bellanger, P., Gorsse, S., Bernard-Granger, G., Navone, C., Redjaimia, A., and Vives, S., Effect of microstructure on the thermal conductivity of nanostructured Mg_2(Si,Sn) thermoelectric alloys: An experimental and modeling approach, 102–110. Copyright 2015, with permission from Elsevier.)

dispersed nano-inclusions scatter longer wavelength phonons, a decrease, compared to the solid solution effect, is observed even at high temperatures. In this work, the contribution of grain boundaries (mesoscale defects) was also studied in bulk nanostructured materials and is shown in Figure 5.14a. The overall result is a significant reduction of the total κ_{lat}. One should note that defects also scatter electrons and whether their overall effect is beneficial depends on the relative scattering strength. A theoretical estimate of the effect of GB scattering on the electronic and thermal transport in Sn-rich Mg_2(Si, Sn) has been made [61] and the results indicate the suitability of Mg_2(Si, Sn) solid solutions for nanostructuring with an expected increase in *ZT* of around 15%. The authors also pointed out that this approach might be more beneficial in Mg_2(Si, Sn) solid solutions than in binary Mg_2Si.

In Figure 5.14b, we plot the reported κ_{lat} by various research groups. A wide variation in the κ_{lat} values is observed, indicating it to be the major reason for the reported variations in the *ZT* values (Table 5.2). We caution that a highly underestimated value of κ_{lat} can be obtained if the degenerate limit of the Lorenz number $L = 2.44 \times 10^{-8}$ V^2/K^2 is used. Fortunately, in most of the reports mentioned earlier, the Lorenz number was obtained from the SPB model (see Section 5.1.4.1.2). Theoretical estimates of the minimum thermal conductivity κ_{lat}(min) in $Mg_2Si_{0.4}Sn_{0.6}$ based on Cahill's formula indicate a value of 0.5 W/m K [85]. From Figure 5.14b, it follows that the data of [17] are close to this limit, while the data of [14], which indicate the highest values of κ_{lat}, are close to the alloy limit. We should point out that, from the reported transport data in [17], it seems the calculation of κ_{lat} was done with a fixed *L* (degenerate limit value), resulting in an underestimated κ_{lat}. The compositions in all the specimens are similar, and no special heat treatments were applied in any of these reports to precipitate secondary phases. This indicates the importance of the process parameters in the study of these solid solutions. From Table 5.2, it also follows that the specimens with lower κ_{lat} values do not necessarily achieve the maximum reported ZT values. Thus, a proper combination of the best power factor and the lowest κ_{lat} is expected to raise the ZT even further. Theoretical estimates in this regard indicate a ZT_{max} value close to 2 in $Mg_2Si_{0.4}Sn_{0.6}$ [84].

5.2 HMS

Environmentally friendly, low-cost materials made of abundant elements are a precondition for large-scale application of thermoelectric devices for waste heat recovery and autarkic electricity supplies. Accordingly, focused interest has been dedicated to the class of silicides. Among other semiconducting silicides, such as Mg_2Si, $CrSi_2$, β-$FeSi_2$, $ReSi_{1.75}$, and Ru_2Si_3, higher manganese silicide (HMS) has been found to show promising thermoelectric performance. Whereas n-type $Mg_2Si_xSn_{1-x}$ has reached high figure of merit ZT, the values of p-type silicides have, so far, remained moderate with the highest values of about 0.7 reported for HMS, resulting in a persisting need to further improve thermoelectric p-type silicides. Apart from the application-driven focus, HMS is one of the most interesting material groups regarding its peculiar lattice structure and unique functional features arising from it. Good thermal and chemical stability of crystalline material makes HMS an ideal candidate for thermoelectric mass technology.

From a first glance on, this intriguing material attracts scientific attention due to its typical peculiarities, setting it apart from many other known TE materials. It starts with its odd stoichiometry and the experience of the crystal growers that it is not feasible to obtain a phase pure homogeneous macrosample with intrinsic semiconducting characteristics. The composition of a pure HMS phase, which has a very narrow homogeneity range if any, is linked to an inherently high hole concentration and transport behavior characteristic of a degenerate semiconductor. Often described as crystallizing in a Nowotny chimney–ladder (NCL) structure [91], HMS is characterized by incommensurability of the c axes of the Mn and Si sublattices in small integers, obscuring description of the lattice structure. An inherent layered microstructure occurring in all samples crystallized from the melt complicates determination of functional properties. The anisotropy of the lattice structure is reflected by anisotropic mechanical and transport properties. The thermal conductivity is rather low and exhibits unusual temperature dependence. The complex diffraction patterns have inspired the assumption that several of the suggested commensurate phases might coexist in the bulk state. However, the functional properties of undoped crystallized and annealed samples with low content of side phases show little variation in their TE properties, and the structure and properties remain stable up to above 800°C and, with that, over the whole temperature range of the maximum thermoelectric performance. In the range of thermal excitation of carriers, the system behaves unusually, as the electron's mobility is far below that of the holes, and electronic and bipolar conductivity contributions remain negligible, whereas the Seebeck coefficient reacts strongly to the change in the energy distribution of the carriers. Also, efficient n-type conduction cannot be achieved. Mn substitution by high concentration of aliovalent elements like Cr and Fe results in an inadequately weak change of the hole density, while Ge as an isoelectronic substitution of Si acts as an efficient electronic acceptor. It has been stated recently that the diffraction patterns can be explained much more consistently by a strong lattice disorder than by any of the numerous proposed ordered models of the NCL lattice structure. Nevertheless, HMS crystallizes in a very narrow Si/Mn concentration range irrespective of the starting composition. Despite this line phase character, the thermal conductivity, in particular, may vary considerably, showing values between 1.9 and 4.5 W/m K.

Although comprehensive reviews of HMS have been presented during recent years [92–94], substantial progress has been achieved recently in the understanding of some of the listed peculiarities, and a lot of new fundamental insight has been gained that makes it compelling to review the subject matter.

5.2.1 Crystal Structure

The crystal structure of HMS has been described for some time in terms of ordered structures, that is, assuming fully occupied lattice positions. Only recently, it has been shown that a consistent explanation of XRD diffraction patterns can be given in terms of a disordered lattice structure that is characterized by a high concentration of vacancies at the lattice sites [95]. The concept of commensurate phases in the NCL structure has guided the structural interpretation over decades. On the other hand, if the ratio γ between the c lattice parameters of the two intergrown Mn and Si subsystems cannot be expressed by a rational number, the phases are referred to as incommensurate.

5.2.1.1 HMS Nowotny Chimney–Ladder Phases

HMS in a stoichiometry of $MnSi_r$ with the Si/Mn atomic ratio r slightly below 1.75, mostly indicated in the range of 1.72–1.75, was often described to crystallize in a NCL structure, the unit cell of which consists of a tetragonal sublattice occupied by Mn atoms based on the β-Sn structure (resembling a chimney) and a "ladder" sublattice of pairs of Si atoms forming two coupled helices inside the chimney; see Figure 5.15. Based on this building principle, a number of commensurate structures of increasing complexity, according to the formula Mn_nSi_{2n-m}, have been reported from XRD and electron diffraction patterns. In these structures, each crystallographic unit cell contains an integral number n of Mn subcells containing four atoms each and another integral number m of Si subcells containing 24–28 atoms each. Accordingly, the lattice constant along the c axis of the HMS primitive cell, $c = n\,c_{Mn} = mc_{Si}$, with c_{Mn},c_{Si} being the lattice constants of the Mn and Si sublattices, respectively, gets elongated with increasing n values. A number of different structural formulae of such commensurate phases have been reported, including (in the order of ascending Si/Mn ratio) Mn_7Si_{12}, $Mn_{11}Si_{19}$, $Mn_{26}Si_{45}$, $Mn_{15}Si_{26}$, $Mn_{19}Si_{33}$, $Mn_{27}Si_{47}$, $Mn_{39}Si_{68}$, and Mn_4Si_7. Structural solutions including atomic positions have been reported for $Mn_{11}Si_{19}$ (m = 3) [96], for $Mn_{15}Si_{26}$ (m = 4) [97], for Mn_4Si_7 (m = 1) [98], and for $Mn_{27}Si_{47}$ (m = 7) [99]. All reported HMS phases exhibit nearly identical a lattice parameters (a = 5.53 Å), slightly varying Si/Mn atomic ratio with a maximum of r = 1.75 (for m = 1) but greatly varying c lattice parameters ranging from about 17 Å for Mn_4Si_7 to as much as 118 Å for $Mn_{27}Si_{47}$, and possibly larger ones. Roughly, the c parameter can be read from high-resolution TEM images (see Figure 5.16), which show typical regular lattice fringes. Some structures (Mn_7Si_{12}, $Mn_{19}Si_{33}$, $Mn_{39}Si_{68}$) have only been detected in nanowires or nanoribbons [100,101] or as microscopic domains in the bulk. It was suggested that several among these commensurate phases may coexist in bulk materials, but clear evidence is difficult to provide since the structures are hardly distinguishable by conventional XRD or energy-dispersive x-ray analysis (EDX). From a general point of view, it is questionable whether two or more

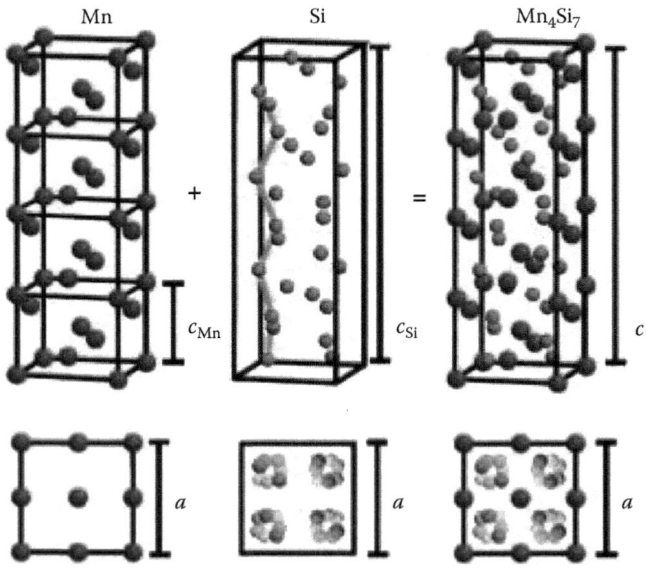

FIGURE 5.15
Schematic illustrations of the crystal structure of Mn_4Si_7. c, c_{Mn}, and c_{Si} are the lattice constants along the c axis for the higher manganese silicide primitive cell, Mn sublattice, and Si sublattice, respectively. (Reprinted by permission from Macmillan Publishers Ltd., *Nat. Commun.*, Chen, X., Weathers, A., Carrete, J. et al., Twisting phonons in complex crystals with quasi-one-dimensional substructures, 6, 6723, copyright 2015.)

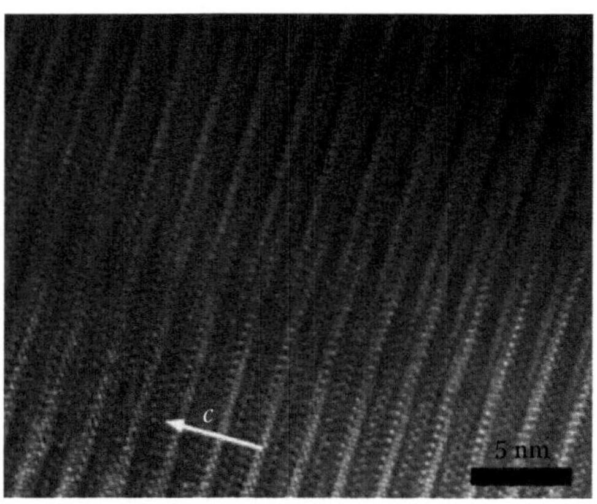

FIGURE 5.16
High-resolution transmission electron microscopy lattice fringe image of the higher manganese silicide matrix. The direction of the crystallographic c axis is indicated. (Reprinted by permission from Macmillan Publishers Ltd., *Nat. Commun.*, Chen, X., Weathers, A., Carrete, J. et al., Twisting phonons in complex crystals with quasi-one-dimensional substructures, 6, 6723, copyright 2015.)

distinct HMS phases could coexist in thermodynamic equilibrium in the presence of a further side phase (MnSi or Si, as mostly observed).

The X/M atomic ratio of a chimney–ladder compound M_nX_{2n-m} (here, the Si/Mn ratio of a commensurate phase Mn_nSi_{2n-m}) is related to the c_X/c_M length ratio of the unit cells of the X and M substructures [91], respectively.

$$\frac{X}{M} = \frac{2n-m}{n} = 2 - \left(\frac{c_X}{c_M}\right)^{-1} \quad (5.4)$$

Here, the Si subcell is about four times as long as the Mn subcell.

Governed by the parity of the *m* and *n*, the commensurate HMS phases have been assigned to the space groups $P\bar{4}n2, P\bar{4}c2, I\bar{4}2d$.

5.2.1.2 Incommensurate Phases

Focused effort was spent to identify the correct phase or phases actually occurring in prepared samples. A comparison has been made [97] between the structures of $Mn_{15}Si_{26}$, $Mn_{11}Si_{19}$, and Mn_4Si_7 by structural refinement of single-crystal XRD data comprising more than a thousand reflections. The fit between calculated and experimental diffraction patterns improved with increasing *c* parameter among these structures, favoring $Mn_{15}Si_{26}$ as the best fitting structure. However, the authors concluded that a final decision regarding a pure one-phase nature of the material could not be made. The $Mn_{27}Si_{47}$ phase, which was later identified as the only stable room temperature phase, was not included in their analysis.

TEM studies indicated that the Mn and Si sublattices can appear as incommensurate, whereas commensurate phases have to be assumed for structural solutions in diffraction [102–104]. A further kind of fringes observed in TEM images, much more widely spaced than the ones mentioned previously, was shown to be related to selected electron diffraction peaks, giving evidence of spacing and orientation anomalies, with the latter also being concluded from an occasionally observed inclination of the series of satellite reflections in the electron diffraction patterns versus the *z* direction. Usually, the satellites appear on lines parallel to the (001) direction: see Figure 5.17.

The clear evidence of phase changes [105] between distinct phases closely neighbored in composition (see in the following text) is a prominent step in the lasting controversy whether HMS represents a variety of coexisting (commensurate or incommensurate)

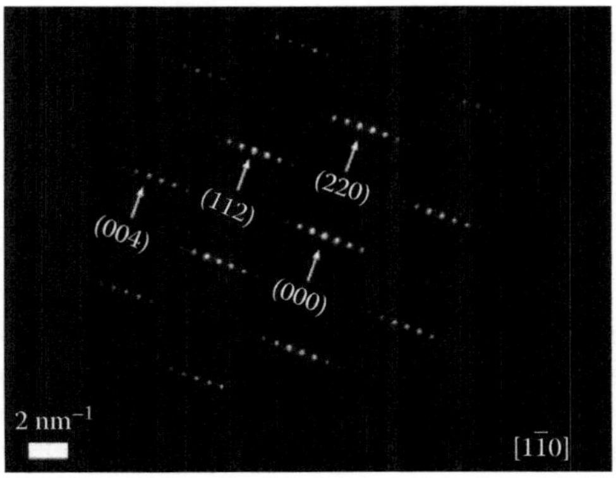

FIGURE 5.17
Selected area electron diffraction of a higher manganese silicide crystal along the zone axis indexed with respect to the Mn sublattice. (Reprinted by permission from Macmillan Publishers Ltd., *Nat. Commun.*, Chen, X., Weathers, A., Carrete, J. et al., Twisting phonons in complex crystals with quasi-one-dimensional substructures, 6, 6723, copyright 2015.)

phases. With the observed sharp transitions spanning over relatively wide concentration differences in the range of the HMS NCL phases (from $r = 1.741$ to 1.733 and finally to 1.727), according to the identified commensurate structures, it is obvious that the bulk system in thermodynamic equilibrium prefers just one among a small number of distinct phases.

The spontaneous transition between distinct phases of different, although closely related, composition provokes kinetic considerations. It seems that the structure provides channels along the chimneys that enable balancing the Si concentration over tens of micrometers along the *c* axis during phase changes, whereas the chimney walls seem to pose a barrier for Si motion, as can be concluded from the omnipresent wide-spaced layered MnSi microstructure (see in Section 5.2.1.3-5.2.1.5 and Figure 5.19). An easier motion of Si atoms along the *c* than along the *a* axis was also concluded from the higher thermal expansion of the *c* lattice parameter compared to the *a* lattice parameter [106]. If Si is trapped in the chimneys, this could preserve out-of-equilibrium phases or compositions under certain circumstances, in particular if the material is cooled below a temperature that activates the interdiffusion of Si along the chimneys. Thus, the Si/Mn ratio *r* would be frozen in the structure, whereas the equilibrium Si/Mn ratio might depend on temperature, leading to metastable nonequilibrium compositions.

It is worthwhile to discuss which structure might be set up in a chimney where a Si concentration, apart from an equilibrium phase, was frozen. The high vacancy concentration on the Si sites in the recently revealed disordered structure [95] (see in the following text) suggests that a gradient of the Si concentration along *c* could balance facing low-energy barriers or could spread over a certain length overlaying the γ ratio and modulation of the equilibrium phases. This would cause incommensurate domains to appear locally, which have been detected by electron microscopy and electron diffraction. Lattice defects, structural reordering near interfaces, thermal history, and irradiation under electron microscopy or neutron studies could initiate partial relaxation of such metastable states, causing a length change of the Si sublattice and leading to the coexistence of different Si/Mn ratios and observations of incommensurate structures.

Doping studies revealed that the preferably formed phase is sensitive to minor defect changes. Substitutions such as Cr or Fe at Mn sites changed the identified value of *n*. Because of the minor energetic differences between the commensurate HMS phases, Nowotny predicted that the formation of different commensurate phases is strongly influenced by thermal and mechanical treatments. However, annealing at elevated temperature should heal this by homogenizing the material.

5.2.1.3 Modulated Lattice Structure

Based on a concept and a suggested superspace group [107], the crystal structure of HMS was described [108] by means of a (3 + 1)-dimensional superspace group approach that allows the description of the whole series of the proposed NCL HMS phases in one integral group where the reported commensurate phases are included as particular cases of a generally incommensurate ratio γ of the *c* parameters of the Mn and Si subunits. These subunits, containing four atoms each, form the base blocks of the structure; see Figure 5.18. Note that the Si subunit is different from and much shorter than the introduced unit cell of the Si subsystem mentioned earlier. The modulated approach takes into account a displacement of any atom in the lattice from its base position in the subunit, which is calculated as a steady periodic function of the respective shift between the Mn and Si subunits using a Fourier series of eighth order. This shift, in principle, is introduced as a fourth "superspace"

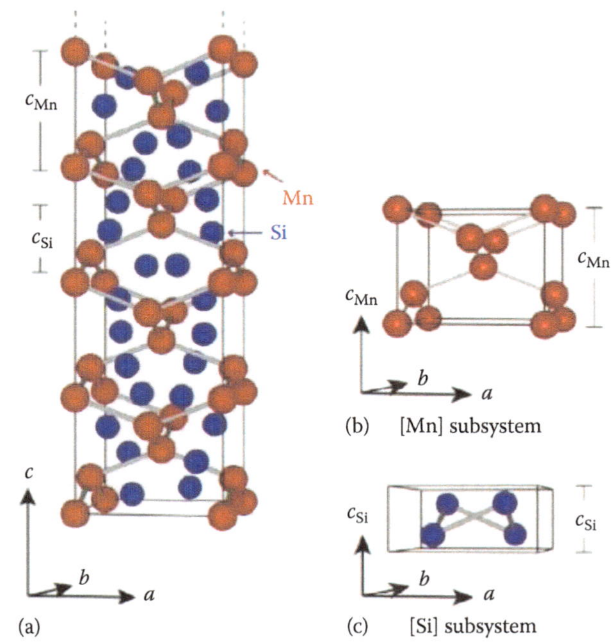

FIGURE 5.18
(a) Incommensurate crystal structure of $MnSi_\gamma$ showing Mn and Si atoms in the range of $a \times a \times 4\ c_{Mn}$, (b) unit cells of the [Mn] subsystem, and (c) the [Si] subsystem. (Reprinted from *J. Alloys Compd.*, 616, Kikuchi, Y., Nakajo, T., Hayashi, K., and Miyazaki, Y., High temperature X-ray diffraction study on incommensurate composite crystal $MnSi_\gamma$ (3+1)-dimensional superspace approach, 263–267. Copyright 2014, with permission from Elsevier.)

coordinate. It was revealed that the displacement of the Mn atoms remains small and is restricted mainly to the *z* direction, whereas the position of the Si atoms is modulated in a helical shape with a strong rotational modulation in the *xy* plane. Neutron diffraction data of an arc-melted sample annealed at 1273 K have been refined by the superspace group and gave an incommensurate axes ratio $\gamma = c_{Mn}/c_{Si}$ of 1.736, which is close to the commensurate $Mn_{19}Si_{33}$ structure and just halfway between the structures $Mn_{27}Si_{47}$ [109,110] and $Mn_{15}Si_{26}$ [97,111] that were identified as the stable-room-temperature phases before. From a related full potential linearized augmented planewave (FLAPW) band structure calculation [108] based on a commensurate approximation to $\gamma = 1.736$, it was concluded that not only slight changes in γ but also stacking faults lead to strong changes in the Si position modulation and thus to different gap width, density of states near the Fermi level, and, hence, to changed TE properties. This could explain the considerable variation of the reported experimental gap values. Accordingly, controlled substitution on the Mn and Si sites is a suitable handle to modify the band structure and TE properties of HMS.

5.2.1.4 Disordered Lattice Structure of HMS

A recent XRD analysis of the lattice structure based on a disordered modulated composite model [95] came to a new structural interpretation that is essentially different from all previously suggested structures. Striking evidence of a substantially improved refinement of the experimental diffraction patterns could be presented when the superspace group proposed by Yamamoto [107] was replaced by the group $I4_1/amd(00ss)$. By a direct phase determination technique, a yet unknown Si position at $x \approx 0.18$, $y \approx 0.38$, and $z \approx 0.83$ was identified in the Mn sublattice in addition to the basic position of Mn at (0, ¼, ⅜). The high quality of the refinement was confirmed by repeating the calculations with another refinement program, providing identical results. Modulation waves of first order (containing only one sine and one cosine function) to describe the displacement of the Si and Mn atoms from their base positions turned out to be sufficient for good agreement between the calculated and experimental diffraction patterns.

Samples of various starting compositions exposed to a variety of different thermal treatments were compared [95], and the width of selected diffraction peaks was evaluated as an indicator of the degree of lattice disorder. All samples had the same lattice structure with no significant variation with the starting composition, whereas the observed peak width was related to the thermal history. The lowest peak width (related to the least lattice disorder) was found for annealing temperatures at or below 800°C, whereas even very long annealing at 900°C and above produced wider peaks in the Si sublattice. With regard to the thermodynamic study of Allam et al. [105], this could be understood from the observed phase transitions between 800°C and 900°C. Ball-milled samples reached a similar degree of disorder after compaction by SPS as crystallized nonannealed material. The Mn sublattice turned out to be less sensitive to the annealing conditions than the Si subsystem.

The identified structural solution is characterized by a much higher density of Si lattice positions per volume than in the previously favored structures (which would be in disagreement with the experimental mass density) in conjunction with a high degree of disorder including both antisite defects of Mn on the Si position in the Mn subsystem and a high number of vacancies at the Si sites.

In view of this new structural understanding, many of the previous discussions carried out in terms of the ordered commensurate NCL structures and, in particular, conclusions drawn on the related chemical composition values have to be reviewed. Specifically, observations regarding the diffraction peak rearrangement during high-temperature phase transitions call for reinterpretation in line with the new concept. Probably, techniques evaluating selected individual diffraction peak positions [105,110] detect length changes of the Si sublattice that are linked to specific γ values. Naturally, the equivalence between the Si/Mn atomic ratio r and the c parameter ratio γ of the Mn and Si base cells of the modulated structure model (which has been used to assess the HMS composition in the ordered models) cannot be adopted anymore in the disordered structure. However, the atomic ratio r can be obtained if the occupancy of the lattice positions is refined reliably. From the refinement within the disordered model [95], the value of $\gamma = 1.737$ was obtained. This value coincided with the γ value found by Miyazaki et al. [108], but the ratio r turned out to be equal to $r = 1.748$. The γ ratio is found, however, to be very close to chemical analysis values reported in [111,112], which yielded $r = 1.733$ for the HMS phase after correcting for the gross composition shift by the MnSi layers. The disordered nature of HMS will also make band structure calculations more complicated where suitable ordered representations of occupation patterns need to be selected.

5.2.1.5 Role of the Atomic Packing Ratio

HMS is representing an end member of a series of NCL phases extending over a wide composition range of $Ru_{1-x}Mn_xSi_y$ solid solutions [113]. It was observed that the X/M values (Si/metal atomic ratio) for these chimney–ladder phases deviate from the ideal values expected from the valence electron count (VEC) = 14 rule even more strongly than

HMS does, and these include a change of sign, as was likewise observed before in the $Ru_{1-x}Re_xSi_y$ series of NCL phases [114]. The explanation is based on the main statement that chimney–ladder phases can exist when the atomic packing factor values are in a certain range, and these phases are destabilized once the atomic packing falls out of this range [113]. No hint was given of the physical nature of the governing mechanism for the stability criterion behind the atomic packing factor, but it can be assumed that this is a quantity related to bonding lengths or interatomic distances in the lattice.

Following the orbital schematic of HMS according to Fredrickson et al. [115], the Si atomic orbitals are split into occupied and empty ones when constructing the VEC = 14 scheme, whereas the Mn d orbitals are completely filled. Thus, it is supposed that the Si-Si interatomic distances may sensitively affect the band structure and, consequently, the bonding energy in HMS [116]. Accordingly, this would represent a second driving force governing the lattice stability, in competition with the VEC = 14 rule. The insensitivity of the a axis to variations in the Si/Mn ratio may also be understood from the orbital scheme because the Mn–Mn interactions remain unaffected by small VEC changes. With the chimney forming a stiff frame, a change in the Si–Si interatomic distances will be equivalent to a change of the length of the Si subcell along the c axis and thus to the Si/Mn ratio. As a consequence, the Si/Mn ratio r is tuned by a balance between the VEC rule driving toward $r = 1.75$ and optimum Si–Si distance.

Detailed future analysis of changes in the lattice structure of HMS alloyed with substitutes at the Mn site, such as Fe and Cr, may reveal whether the observed electrical compensation mechanism for aliovalent substitution at the Mn site can be understood from this balance. Widening of the Mn frame caused by a substitute would thus increase the Si/metal ratio and reduce hole concentration, and *vice versa*. An XRD peak shift to higher angles according to a shrinking of the a parameter was observed for increasing Fe addition [117]. Additionally, a decrease of the Si/Mn ratio was detected. It was concluded that the lattice tends to reject silicon as the Fe concentration is increased. According to the VEC rule, this will cause an increase of the hole concentration, counteracting the donor action when substituting Mn, as was experimentally observed.

5.2.2 Electronic Structure

5.2.2.1 Valence Electron Count Concept

According to the VEC concept, HMS obeys the 14-electron rule, predicting a stable bonding if an average of 14 valence electrons (VEs) is provided per each transition metal atom. An explanation given in [115] can be applied similarly to Mn_4Si_7; see, for example, [118]. The VE count is based on the consideration of atomic and molecular orbitals (MOs), where each Mn atom is contributing 5 d orbitals and each Si 4 orbitals of s and p electrons, amounting to 20Mn d and 28Si sp orbitals per Mn_4Si_7 formula unit. A number of xMOs formed from xMn d and xSi sp orbitals and split into x bonding and x antibonding states will act as a strong Mn–Si bonding. With all of the 20 Mn d and $8 + x$ out of the 28 Si sp orbitals involved in the bonding, $20–x$ nonbonding Mn d orbitals and 8 orbitals of Si–Si bonds plus x bonding Mn–Si MO will be involved in the crystalline bonding, summing up to 28 MO in total. Assuming that each Si atom will contribute 4 VE and each Mn 7 VE leads to a number of 56 VE to be distributed over the available states just completely filling the listed 28 MO. With the energy gap occurring between the bonding and antibonding states, all states below the gap are occupied, whereas the ones mentioned earlier are empty. Thus, Mn_4Si_7 represents an intrinsic semiconductor with the Fermi level located in the gap. Assuming the energy spectrum as fixed, a reduction of the Si/Mn ratio *r* below the value of 1.75 will reduce the number of VE, leading to partial filling of the states below the gap, which represents the situation of a *p*-type extrinsic semiconductor. This situation applies to all of the considered commensurate NCL HMS structures except Mn_4Si_7. The applicability of the VEC = 14 rule has been qualitatively confirmed by the results of DFT calculations [119–121]. However, the origin of the observed slight deviation from the rule toward VEC < 14 is not obvious from this consideration.

If it is assumed that the Si atoms are confined in the chimneys, then a change in the equilibrium Si concentration can only be balanced by a shift of the Si ladder along the chimney. The high vacancy concentration facilitates position change of Si along the *c* axis. In such a situation, obviously, the lattice structure itself does not provide a cause for a narrow line phase nature. Rather, its origin is assumed in the optimization of the VE concentration. With Si atoms easily moving along *c*, the system can optimize its bonding energy by a quasicontinuous adjustment of the Si concentration. Based on the VEC rule, the concentration of free carriers is proportional to the deviation of the Si concentration from 1.75.

5.2.2.2 Band Structure

Migas et al. [121] calculated from the first principles the electronic band structure of several commensurate HMS NCL structures: Mn_4Si_7, $Mn_{11}Si_{19}$, $Mn_{15}Si_{26}$, and $Mn_{27}Si_{47}$. Mn_4Si_7 was found to be an intrinsic semiconductor with the Fermi level in the bandgap. It has an indirect bandgap of 0.77 eV and a direct bandgap of 0.79 eV, whereas $Mn_{11}Si_{19}$, $Mn_{15}Si_{26}$, and $Mn_{27}Si_{47}$ behave like degenerate *p*-type semiconductors with the Fermi level located slightly below the top of the valence band. The degenerate characteristics are in accord with the VEC rule that predicts an intrinsic semiconductor for VEC = 14 according to Mn/Si = 1.75 (applying to Mn_4Si_7) but degenerate semiconducting behavior for VEC < 14 according to Mn/Si = 1.72···1.74 (applying to all other suggested commensurate HMS phases). Likewise, it is in accordance with the facts that crystallized HMS was always found to be a degenerate semiconductor experimentally and, in addition, that no evidence of the pure Mn_4Si_7 structure was reported in any of the studies that could distinguish between the commensurate phases. Furthermore, the calculations revealed that the band structure changes and the gap is reduced significantly when stacking faults are introduced into the structure. However, *ab initio* methods tend to overestimate the gap width. Another FLAPW calculation [108] obtained a gap of about 0.6 eV in the band structure of $MnSi_\gamma$ with $\gamma = 1.736$ based on a commensurate approximation of an incommensurate structure refined to diffraction data.

5.2.2.3 Energy Gap and Carrier Effective Mass

A direct bandgap of about 0.7 eV and an indirect gap near 0.4 eV at room temperature were determined for HMS by optical measurements [122,123]. Absorption edge measurements gave the direct bandgap of 0.66 eV [124] and 0.68 eV [125]. A comparable bandgap of 0.70 eV was obtained from the temperature slope of the electrical conductivity [111]. However, other measurements of the electrical resistivity and Hall effect yielded much smaller values of 0.42 and 0.45 eV, respectively [126], similar to early values of Nishida [127].

The hole effective mass has been determined from Seebeck and Hall measurements in extrinsic conditions and from optical reflection on Ge- and Cr-doped HMS [128], giving high values of about $10m_0$. It has been concluded that the valence band maximum is located at the center of the Brillouin zone, which was confirmed by *ab initio* calculations

[121]. Comparably high values were obtained along the principle axes of the energy ellipsoid, $m_{xx} = 11m_0$ and $m_{zz} = 15m_0$ [111] and also a mean value of $12m_0$ [127]. However, much lower values were obtained from DFT: $m_{xx} = 1.6m_0$ and $m_{zz} = 5.7m_0$ [121]. The discrepancy between the theoretical and experimental data was explained by crystal imperfections. Also, the reduced experimental gap of 0.40 eV, compared to 0.77–0.79 eV as calculated, was explained by the presence of defects. It was found [129] that the electrons have a mobility far below that of the holes but are much heavier, which was in principle confirmed in [127]. Qualitatively, the very different mobility of holes and electrons can be explained based on the disordered modulated lattice structure. The Mn sublattice, with little deviation from periodicity due to small modulation amplitude, may support a sufficient bandlike mobility of holes, whereas strong disorder at the Si sites is the reason why regular band conduction cannot be established for electrons.

5.2.3 Phases and Layered Microstructure

5.2.3.1 *Layered Microstructure of Manganese Monosilicide*

Crystallized HMS has always been found to contain traces of manganese monosilicide. MnSi is an intermetallic phase with interesting magnetic properties, crystallizing in the B20 structure with the cubic $P2_13$ space group [130], and resembles a distorted rock salt structure with the atoms displaced along the [110] direction. MnSi shows metallic behavior related to a high hole concentration on the order of 10^{22} cm^{-3} and is a good electrical and thermal conductor [131]. With a typical resistivity of 0.2 mΩ cm and a Seebeck coefficient near 40 μV/K, its *ZT* reaches the maximum of about 0.04 at 450 K [132]. High thermodynamic stability compared to HMS has been stated for MnSi [133].

All HMS samples crystallized from the melt exhibit a regular but not strictly periodic microstructure of thin MnSi layers [134,135]; see Figure 5.19. The layers are oriented perpendicularly to the *c* axis of the HMS matrix. Notably, these layers are embedded in the HMS matrix phase with identical crystallographic orientation of the matrix at both

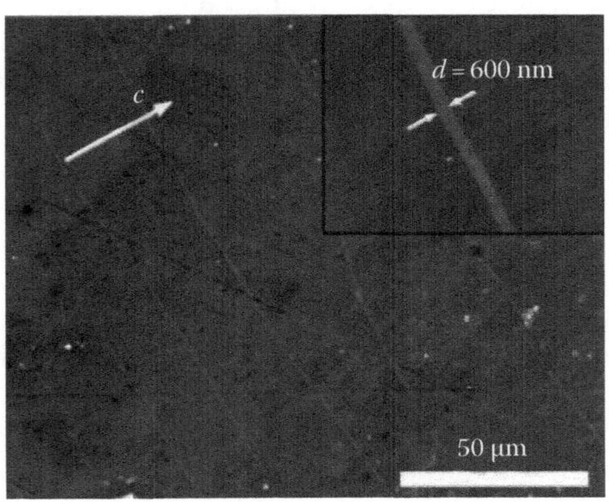

FIGURE 5.19
Scanning electron microscopy image of a Bridgman grown higher manganese silicide (HMS) crystal showing a regular pattern of MnSi lines, here with a spacing of 20–50 μm. The inset is a magnified region showing the presence of MnSi layers with a thickness of about 600 nm in the HMS matrix. (Reprinted by permission from Macmillan Publishers Ltd., *Nat. Commun.*, Chen, X., Weathers, A., Carrete, J. et al., Twisting phonons in complex crystals with quasi-one-dimensional substructures, 6, 6723, copyright 2015.)

sides of the layers pointing to their precipitate nature [111,135,136]. Although the layers are depleted in Si, compared to the HMS matrix, the formation of the layers cannot be suppressed by excess silicon. Rather, the layers do coexist with Si secondary phase inclusions [134]. It was observed that the MnSi layers cannot be resolved by long annealing at 1000°C. A survey regarding the evolution of this peculiar microstructure in undoped and alloyed HMS [2] reveals that while the form of the microstructure can be modified by various annealing conditions, the regular pattern is restored by annealing above 1080°C. A constant volume ratio between the MnSi layers and the matrix was found to be independent of the starting composition of undoped HMS and also independent of the gradually variable spacing of the MnSi layers [111]. This points to a formation mechanism by decomposition of a high-temperature phase of adjacent composition with slightly lower Si/Mn ratio and small width of the homogeneity region under restricted local concentration balance. When HMS was crystallized by chemical vapor transport at 800°C, zero or negligible MnSi precipitation was found [112,137]. Good agreement between the transport properties of crystallized and CVT samples was found [112,138], showing that the thermoelectric properties of HMS are not significantly altered by the layered microstructure. However, magnetic properties [110,139] are dominated by MnSi secondary phase inclusions.

The microstructure will be affected by the growth conditions of the crystals and reacts strongly to alloying additives, for example, the addition of Cr will weaken the phenomenon [140]. Adding Ge may cause grid patterns of MnSi stripes. An initial increase and then decrease of the MnSi volume ratio with increasing Ge addition was found and was correlated to a change of TE properties [141].

From these observations, the peculiar MnSi layers in any melt crystallized HMS can be most probably understood as a result of a phase transition occurring below the peritectic temperature but above 800°C, being related to a small change in the Mn/Si ratio. Assuming melt solidifies into a homogeneous phase, which is depleted in Si compared to the room temperature stable HMS, would explain the constant volume fraction of MnSi layers if, in addition, facilitated Si diffusion along the chimneys is also assumed.

Functionally, the microstructure can be treated as a regular structure of energy barriers that may influence the transport properties. A description of the transport in such a structure was given based on the formation of a soliton system related to the incommensurability of the Mn and Si sublattices, leading to coexistence of regular nano- and micro-configurations of fractal characteristics [2,93].

5.2.3.2 Phase Diagram

The phase diagram in the composition range of HMS was assumed for a long time to be quite complex [2] and has hardly been studied in detail in the elevated temperature range below the peritectic temperature. The early drawn phase diagram near the HMS concentration [142] was in good accord with first measurements on Mn–Si melts made by Dörinckel [143]. According to a more recent phase diagram [136], HMS is formed in a peritectic reaction at a temperature of 1428 K with a composition of $MnSi_{1.72}$. This peritectic temperature has been confirmed quite recently [105]; see Figure 5.20. A slightly higher value, still assuming HMS as Mn_3Si_5, has also been reported [144]. The Mn–Si binary phase diagram was calculated in Calphad [145], in agreement with the experimental diagrams. The homogeneity region of HMS has been found to be very narrow (not more than 0.3 at%) [136]. A similar width of a homogeneity range, but near $r = 1.74$, was adopted

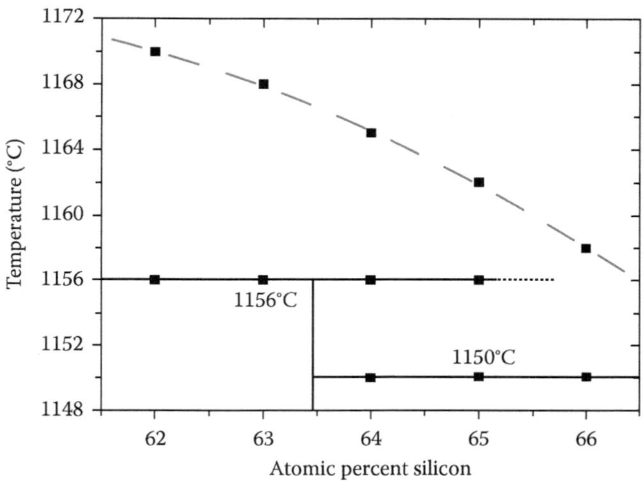

FIGURE 5.20
Section from the higher manganese silicide phase diagram near the peritectic temperature. (Reprinted from *J. Alloys Compd.*, 551, Allam, A., Boulet, P., Nunes, C.A., Sopousek, J., Broz, P., and Record, M.C., Phase transformations in higher manganese silicides, 30–36. Copyright 2013, with permission from Elsevier.)

recently [95] from the scatter in the subcell axes ratio. Also, the observed dependence of phase change temperatures on annealing composition [105] seems to prove a finite homogeneity range.

Different composition values of HMS ranging from $MnSi_{1.70}$ to $MnSi_{1.77}$ were reported in several early works. However, in a series of Bridgman growth experiments with a varying starting Mn/Si ratio [111], it was found that, independent of the starting composition and growth conditions, the HMS matrix was always close to $MnSi_{1.73}$, in good accord with the composition $Mn_{15}Si_{26}$. Accordingly, the phase constitution (i.e., the fraction of secondary phases) varied, reflecting the starting concentration. As constant matrix composition of HMS was repeatedly found in coexistence with both neighboring phases MnSi and Si (see, e.g., [95,109] and [97,105,111,112,148,162,194]), it is well accepted today that there is only one single phase in crystalline HMS at room temperature and it is rather sharp and stable up to above 800°C.

When a melt of gross composition close to $MnSi_{1.75}$ is cooled, MnSi will start to solidify about 10° above the peritectic temperature. At 1428 K, solid MnSi and the melt enriched in Si begin to react to form HMS, controlled by the dissolution rate of MnSi. If this reaction is slow, upon further cooling, the residual melt would eutectically decompose into Si and HMS about 5 K below the peritectic line. Then, the growth of HMS proceeds by the peritectic reaction (MnSi + Si → HMS) controlled by solid-state diffusion from the precipitated Si [141]. Although solidification from the melt is necessarily related to the precipitation of MnSi, which has occasionally been given as the reason why MnSi is always found in crystallized HMS, the peculiar layered MnSi microstructure cannot be explained in this manner. Rather, it is plausible that the MnSi layers are precipitating from a homogeneous Si-depleted HMS phase in a further transformation below the peritectic temperature.

The constant difference between gross concentration and matrix concentration of Si stated in [112] amounts to about 1.5 at% Si, resulting in a composition of the high-temperature phase close to $r \approx 1.718$.

5.2.3.3 Phase Transformations at High Temperature

High-temperature phase transformations in HMS have been recently studied with XRD and differential scanning calorimetry [105,106,116]. Analysis of selected peak position differences and also recording the MnSi phase content over temperature by energy-dispersive XRD turned out to be suitable for tracking phase transformations of HMS. A detailed study at elevated temperatures was made by *in situ* XRD on arc-melted, annealed (at 773 and 1073 K for 2 weeks and 1 week, respectively), and finally quenched samples containing controlled Si deficit or excess [109]. A subtle angular difference between diffraction peaks near 2θ between 42° and 43° (using Cu K_α radiation) was found to be a specific feature to distinguish between commensurate HMS NCL phases. It was concluded that the only stable phase at room temperature and up to above 800°C was $Mn_{27}Si_{47}$ ($\gamma = 1.741$), irrespective of the starting composition and thermal history. When heating up this phase, $Mn_{15}Si_{26}$ ($\gamma = 1.733$) and $Mn_{11}Si_{19}$ ($\gamma = 1.727$) were identified to form subsequently [109], whereas the transformation temperature was dependent on the starting composition. For the Si-deficient material, 820°C and 880°C were found as transition temperatures, whereas for the Si excess these were 860°C and 960°C. This significant difference might be due to a finite homogeneity range existing at least near 800°C. In the picture of a disordered structure, this could be understood as a variation in the Si vacancy concentration that is established during long annealing under Si excess or deficit. From the concept of the HMS structure being stabilized by an optimum VE concentration slightly deviating from the VEC = 14 stability criterion, higher thermal stability of the Si-enriched structure is plausible. In this concept, the high-temperature phase transitions are seen in a context with thermal excitation of carriers over the gap that reduces the occupation of the bonding valence band states, thus weakening the bonding stability of the crystal [116]. Higher Si concentration, that is, lower extrinsic hole concentration, would shift the point of loss of stability to higher temperature. On the other hand, the shift of transition temperatures can be understood in this picture also as a shift of the gap width related to the variation of the Si concentration. Also, microstructural aspects may play a role since long-term annealing below the transition temperature and under Si excess could resolve the MnSi striations.

With ongoing thermal excitation of free electrons at higher temperatures, the bonding situation may shift further and cause more phase transitions. A phase transition at 1123°C was reported [146].

Although the Si/Mn ratio, in terms of the NCL ordered phases $Mn_{15}Si_{26}$ ($\gamma = 1.733$) and $Mn_{11}Si_{19}$ ($\gamma = 1.727$) as assigned to the phases involved in the transformations above 800°C [109], shows that both phase transitions should be accompanied by the consumption of MnSi or release of Si during heating, the opposite was found experimentally for the lower step; see Figure 5.21. Probably, the origin of this discrepancy lies in the assumption of Si/Mn concentration values according to the stoichiometry of the identified NCL phases. Given that the involved phases are disordered structures, this does not hold anymore. Rather, both indicators—peak position to identify the structure and measure the cell lengths on the one hand and peak intensity to quantify silicide phase concentrations on the other—provide independent information linked to each other by defect (vacancy) concentrations in the structure.

Another unique method, comparable to the one mentioned earlier for the identification of individual commensurate HMS NCL phases, is based on an evaluation of precision XRD $\theta - 2\theta$ scans of the first two reflections $(0\ 0\ 2(2n - m))$ and $(0\ 0\ 4n)$ using Cu K_α and K_β radiation [110]. From the angular difference of the two peaks, $\gamma = 1.746$ according to $Mn_{27}Si_{47}$

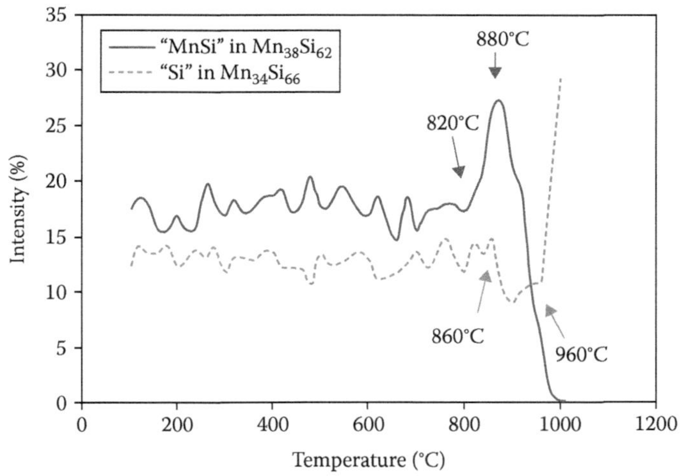

FIGURE 5.21
Evolution of the intensity of the main x-ray diffraction peaks corresponding to MnSi and Si versus temperature. The MnSi and Si intensities were recorded in a sample with Si deficit and Si excess, respectively, according to the indicated nominal starting compositions. (Reprinted from *J. Alloys Compd.*, 512, Allam, A., Nunes, C.A., Zalesak, J., and Record, M.C., On the stability of the higher manganese silicides, 278–281. Copyright 2012, with permission from Elsevier.)

was determined for Czochralski-grown single crystals. Thus, both methods gave a strong hint to believe that high-quality single-crystalline bulk material adopts the $Mn_{27}Si_{47}$ NCL phase at room temperature. However, the recent findings [95] are shedding new light here; in view of its superior refinement quality, the disordered modulated structure model is most probably representing the only stable room temperature HMS.

The kinetics of the upper phase transition at 860°C or 960°C reported in [109], which shifts toward lower Si content when heated and is considered as the origin of the layered morphology, could be qualitatively explained as follows. When reaching the transition temperature, the MnSi layers will be flooded by the excess Si from the matrix slipping along the chimneys, enabling them to transform into HMS. *Vice versa*, during cooling, the homogeneous high-temperature phase will have to overcome a considerable nucleation threshold. Regions of reduced Si density may occur in the chimneys when the Si-rich equilibrium phase is forming locally by consecutive site change of Si atoms along the chimney. With an increasing length of the reordered structure of the forming phase, the Si-depleted regions will grow likewise, and, when they reach a critical size, the chimney structure will break down due to local Si deficit. This mechanism may explain the large and relatively regular spacing of the MnSi layer patterns. Initiated by a fluctuation or defect, happening first in a single chimney, this will act as a nucleus where Si-depleted regions of neighboring chimneys can condensate. The anisotropy of the lattice, together with the facilitated mobility of the Si atoms along the chimney, controls the lateral spread of the discontinuity perpendicular to the *c* axis, resembling crack growth in brittle solids. Due to the nucleation energy, the formation of the MnSi layers may be delayed, leading to undercooling effects and hysteresis in thermal cycling during thermal analysis.

A second-order phase transition was detected [104] in the first thermal cycle of a thermal analysis experiment at 750°C, but was not observed in the second and following cycles.

From thermoelectric measurements, no correlated change of the properties at any of the phase change temperatures was reported. With the structural changes being small, functional effects may remain small likewise.

5.2.3.4 Phase Stability

From the studies previously mentioned, it was concluded that crystalline undoped HMS is stable between room temperature and about 800°C. HMS shows high sublimation stability up to above 800°C in vacuum, and oxidation of HMS crystals in air could only be observed above 760°C, decomposing into MnSi, γ-Mn_2O_3, and SiO_2 [147]. Phase stability was studied also on micro- and nanopowders using *in situ* energy-dispersive XRD at high temperatures and/or high pressure [106]. Phase transformations linked to formation or consumption of MnSi were observed in different temperature regions and were interpreted as transitions between different HMS phases. In nanopowdered HMS, the effects were found stronger, accelerated, and occurring at lower temperatures (430°C and 725°C) than for micropowders. High pressure in the GPa range affected the observed phase transformations. HMS polycrystals made by arc melting showed a change of the γ ratio with temperature starting at 500°C, which was interpreted as an inherent origin of mechanical instability related to the onset of thermal excitation of the carriers [116].

From the partly different observations for micro- and nanopowders, a general trend can be read. Due to the long lattice cell, the phase stability of HMS is closely related to the amount of lattice defects, in particular to the concentration of interfaces and grain boundaries. Reordering of the lattice structure near interfaces is a well-known phenomenon, with the equilibrium phase constitution near the surface being different from the bulk. With the elongated cells and low-energy barriers for site changes of Si atoms, reordering may range deep into the grains and comprise a considerable portion of the grain volume, possibly strongly shifting the equilibrium composition in nanopowders. Similarly, external strain induced by plastic or milling deformation could favor different modified phases. This is emphasized by finding HMS structures such as Mn_7Si_{12} and $Mn_{39}Si_{68}$ in nanowires that have never been detected in the bulk [101]. Phase transformations at interfaces would be facilitated by low energy differences between slightly different occupation patterns in a disordered structure. Thus, the concentration equilibrium, according to a phase change, can be easily reached by Si position shifts along the chimneys, and the interfaces can act as Si dumps or sources. Decomposition of HMS from the surface can be understood as a corrosion mechanism driven by the lower stability of HMS compared to MnSi and the local disturbance of the HMS lattice stability at the interface by the change of the Si concentration, in contrast to the situation in a perfect lattice where Si is trapped in the structure. Accelerated decomposition of nanopowdered HMS at moderate temperatures as well as the observed mechanochemical decomposition [148] indicate that boundaries and mechanically induced strong defects in the HMS structure act as nucleation centers for decomposition, whereas the ideal extended crystalline structure remains stable over a wide temperature range.

5.2.4 Synthesis Methods

Several techniques of single-crystal growth have been used to prepare HMS samples, including the Bridgman method [112,149,150], Czochralski growth [150,151], chemical transport [112,138,152], and molten metal flux methods [147,153]. Single-crystalline samples up to several mm^3 in volume have been obtained from Ga or Sn flux. Adding Mn to a Na–Si melt, polycrystalline bulk HMS samples were prepared at 800°C [154].

A variety of studies preferred a less tedious route *via* arc melting, crushing, and hot-pressing or SPS to produce polycrystalline samples. Certain conditions, such as vibration-assisted hot-pressing, have been tested [155]. Furthermore, polycrystalline material has been prepared by solid-state reaction or mechanical alloying; see, for example, [156,157]. Furthermore, rapid solidification by melt spinning [158] and gas atomization [159] have been shown to be suitable to prepare HMS of good quality, potentially also in amounts beyond laboratory scale.

Wherever studied in detail, no significant differences in the composition of the HMS matrix could be shown to be correlated to the preparation method of undoped HMS. Neither could characteristic differences in their TE properties be correlated. Chemical vapor transport [138], representing a method of synthesis near 800°C and thus below the reported high-temperature phase transitions, ends up with an almost negligible concentration of the MnSi layers. However, it is difficult to measure all of the TE properties on such samples because of their small size.

Routes employing hard BM or mechanical alloying face the problem of decomposition of the HMS into MnSi accompanied by an often unexplained loss of Si [148,160,161]. There were also claims of a tremendous improvement of ZT up to 0.83 of undoped HMS fabricated by solid-state reaction of premilled elemental powders in a pulse discharge sintering facility [162].

Thin films of HMS made by epitaxial growth on Si substrates [122,123,125] have been studied and potential optoelectronic applications have been discussed.

5.2.5 Thermoelectric Properties of Undoped and Doped HMS

5.2.5.1 Anisotropy of Transport Properties

Linked to their structural anisotropy, HMS single crystals possess anisotropic electronic and thermal transport properties [2,141,151]. With regard to the peculiar microstructure of layered metallic MnSi precipitates that spread perpendicularly to the crystallographic c axis, it was assumed that the measured properties and anisotropy could be significantly shifted. Remarkably, the lattice thermal conductivity along the c axis is merely half as large as that along the a axis at room temperature. Similarly strong anisotropy is found for all of the transport properties over a wide temperature range [150]. However, an estimate [135] of the anisotropy shift of the electrical conductivity and the Seebeck coefficient expected from the microstructure turned out to be much smaller than the experimental anisotropy ratios. Hence, anisotropy of the TE properties is inherent to the HMS phase itself. This was confirmed time and again [112], most recently from the comparison to measurements on layer-free crystals grown by chemical vapor transport [138]. Early reports on the anisotropy of the Hall coefficient [150] were, however, not confirmed [111].

A survey of the temperature dependence of the Seebeck coefficient, electrical, and thermal conductivity from numerous studies was given by Fedorov and Zaitsev [2]. Although the microstructure of the samples was often not reported, the obvious similarity of the observed temperature dependencies of the thermoelectric properties leads to the conclusion that the dependence of the thermoelectric properties of undoped crystalline HMS on the fabrication method and small variations in the starting composition is generally weak; for the thermal conductivity, see Figure 5.22. This is plausible from the single phase nature and very narrow homogeneity range. Variations in gross composition will accordingly

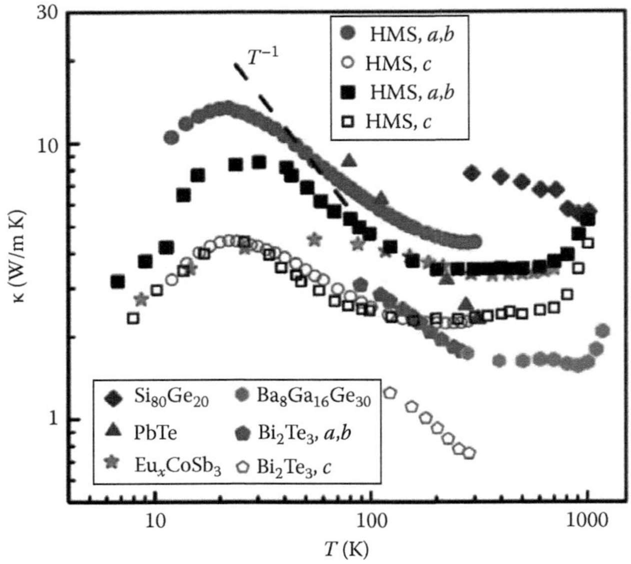

FIGURE 5.22
Temperature dependence and anisotropy of the thermal conductivity of crystallized higher manganese silicide (HMS). Measured HMS data along the *c*- and *a*-axes are shown in a comparison between two references and are compared to reported thermal conductivity data of further relevant thermoelectric materials. (Reprinted by permission from Macmillan Publishers Ltd., *Nat. Commun.*, Chen, X., Weathers, A., Carrete, J. et al., Twisting phonons in complex crystals with quasi-one-dimensional substructures, 6, 6723, copyright 2015.)

result in variations in the microstructure that, of course, will affect the effective properties, gradually depending on the amount and shape of secondary phase inclusions.

Figure 5.22 shows the anisotropy and temperature dependence of the thermal conductivity of Bridgman grown HMS crystals. The thermal conductivity along the *c* axis measured from 4 to 1300 K displays a plateau region from 130 K to above 600 K with an onset of bipolar conduction at about 650 K [163]. A hole mass around 3 m_0 and a gap of about 0.5 eV were deduced from the transport data.

The anisotropy of the Seebeck coefficient in a Si/Mn ratio range of the starting composition of 1.72–1.75 was found as a_c/a_a up to 1.7 independent of the composition [164]. Remarkably, a separate local maximum was observed in the temperature curve of the Seebeck coefficient along the *a* axis (but not along the *c* axis) at $T < 100$ K, which was attributed to a phonon drag [93]. The temperature dependence of the thermal conductivity shows several unusual features. Whereas its rise from the lowest temperatures follows a regular T^3 law, the subsequent descent is more shallow than T^{-1} [2].

Several studies [93,163,165] found that from about 200 K up to the onset of bipolar effects near 800 K, the thermal conductivity remains constant in both crystallographic directions (see Figures 5.22 and 5.25d), whereas a more uniform curvature was measured by others.

The phonon dispersion of HMS, which is essential for understanding thermal transport, has remained unexplored until studied experimentally and theoretically in detail recently [166]. Inelastic neutron scattering measurements and density functional theory calculations of the phonon spectrum found numerous low-energy optical vibrational modes shown in Figure 5.23. These optical modes, including unusual low-frequency twisting motions of the Si ladders inside the Mn chimneys, provide a large phase space

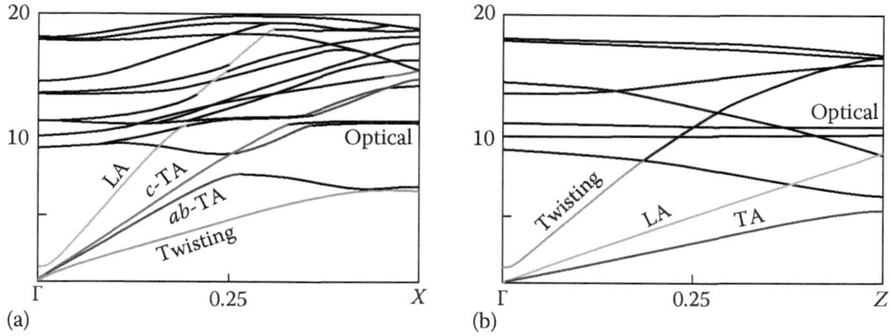

FIGURE 5.23
Calculated phonon dispersion of Mn_4Si_7 along (a) the Γ–*X* direction and along (b) the Γ–*Z* direction. Low-energy harmonic dispersions indexed by the polarization as LA (longitudinal acoustic), TA (transversal acoustic), twisting or other optical modes. Along the Γ–*X* direction the modes show avoided crossing near 1 meV. (Reprinted by permission from Macmillan Publishers Ltd., *Nat. Commun.*, Chen, X., Weathers, A., Carrete, J. et al., Twisting phonons in complex crystals with quasi-one-dimensional substructures, 6, 6723, copyright 2015.)

for scattering APs. A hybrid phonon and diffusion model was proposed to explain the inherently very low thermal conductivity of HMS and its anisotropy. Based on this hybrid model, nanostructuring was estimated as a promising approach to further suppress the thermal conductivity in HMS.

5.2.5.2 Thermal Expansion

From first measurements of the coefficient of thermal expansion (CTE) of HMS, irregularities were reported at low heating rates and the occurrence of phase transitions was supposed [2]. *In situ* recording of the relative strain in HMS micropowders by XRD [105,106] showed a mainly continuous expansion of the *a* and *c* lattice parameters with temperature with an apparent discontinuity in the *c* parameter near 950°C. Based on several works (see a review by Fedorov and Zaitsev [2]), the temperature dependence of the CTE $a_a = (5 + 7 \times 10^{-3}\ T/K) \times 10^{-6}\ K^{-1}$ and $a_c = (9 + 5 \times 10^{-3}\ T/K) \times 10^{-6}\ K^{-1}$ is in good agreement with the powder XRD values [167] of $a_a = (7 + 8 \times 10^{-3}\ T/°C) \times 10^{-6}\ K^{-1}$ and $a_c = (10 + 5 \times 10^{-3}\ T/°C) \times 10^{-6}\ K^{-1}$. From [106], $a_a = 11 \times 10^{-6}\ K^{-1}$ and $a_c = 9 \times 10^{-6}\ K^{-1}$ were found from room temperature up to about 700°C. At temperatures above about 700°C, the thermal expansion is slightly higher for α_a, but much higher for α_c, as read from curves of the relative strain obtained by XRD up to 1000°C. Allam et al. [105] reported $a_a = (15 - 7 \times 10^{-3}\ T/°C) \times 10^{-6}\ K^{-1}$ and $a_c = (7 + 13 \times 10^{-3}\ T/°C) \times 10^{-6}\ K^{-1}$. Distinguishing between the CTE of the Mn and Si subcells, based on the (3 + 1)-dimensional superspace approach and refined lattice constants, $\alpha_a = 11.9 \times 10^{-6}\ K^{-1}$ and $\alpha_{cMn} = 12.5 \times 10^{-6}\ K^{-1}$ have been found, both almost independent of temperature, whereas the Si sublattice is changing its CTE at 500°C, that is, near the onset temperature for bipolar excitation. From $\alpha_{cSi} = 10.7 \times 10^{-6}\ K^{-1}$ at room temperature, it increases to $\alpha_{cSi} = 32.8 \times 10^{-6}\ K^{-1}$ at 900°C [116]. Accordingly, $\gamma = 1.739$ was found to be stable up to 500°C with a subsequent linear decrease to about 1.724 at 900°C; see Figure 5.24.

Thermal expansion was also studied for HMS with various amounts of Ge or Fe additives [168], and prominent irregularities were observed above 800 K, showing strong dependence on the sample composition, the sample orientation, and the cycling mode (heating or cooling). The observations were assumed to be linked to the kinetics of phase transitions.

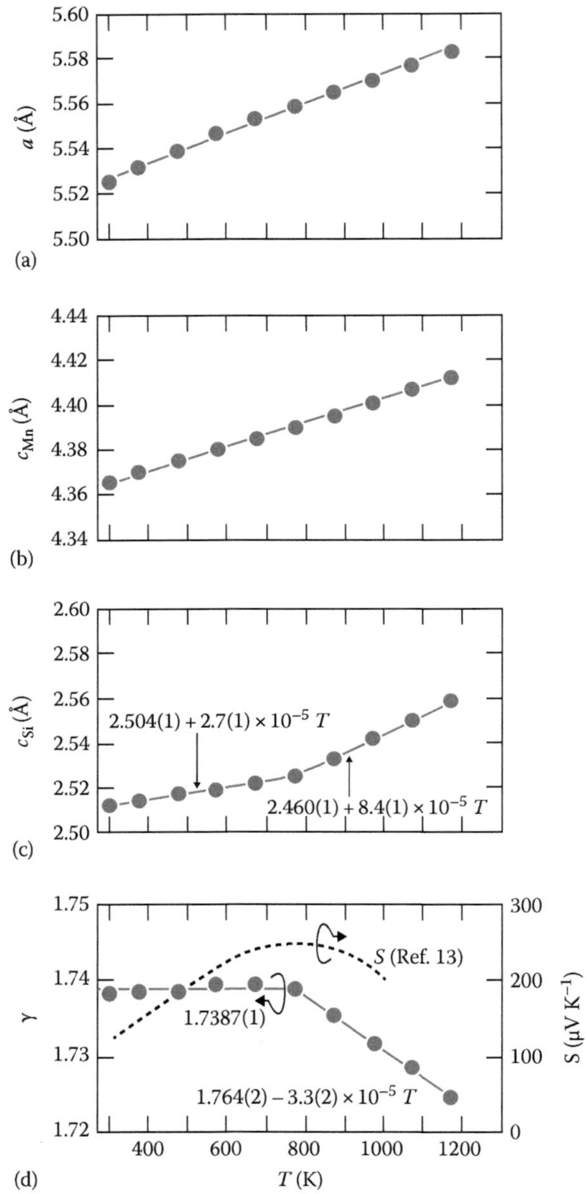

FIGURE 5.24
Temperature dependence of the lattice parameters (a) a, (b) c_{Mn}, (c) c_{Si}, and (d) the subcell length ratio γ complemented by a typical Seebeck curve showing its maximum near the onset of bipolar thermal carrier excitation. (Reprinted from *J. Alloys Compd.*, 616, Kikuchi, Y., Nakajo, T., Hayashi, K., and Miyazaki, Y., High temperature X-ray diffraction study on incommensurate composite crystal $MnSi_\gamma$ (3+1)-dimensional superspace approach, 263–267. Copyright 2014, with permission from Elsevier.)

5.2.5.3 Solid Solutions

The transport properties of HMS can be modified by forming solid solutions. Numerous studies were dedicated to the addition of Ge (e.g., [140,141,165,169,170]), Fe or $FeSi_2$ (e.g., [129,171,172]), Cr (e.g., [117,173,174]), Al (e.g., [156,175]), Ru (e.g., [113]), and many more. Ru doping was found to be detrimental as it reduces the hole mobility [176].

Addition of Ge in Bridgman-grown HMS single crystals [141] led to a significant increase of the figure of merit *ZT* along the *c* axis with best results at 1% Ge addition. The electrical conductivity in Ge-doped polycrystalline HMS prepared by induction melting followed by hot-pressing was continuously enhanced by Ge addition, while the Seebeck coefficient was slightly decreased [170]. The thermal conductivity initially decreased and then, above 1% of Ge addition, increased again. Maximum *ZT* close to 0.6 was achieved with 0.8% Ge at 550°C.

The effects of Al shown in Figure 5.25 and (Al, Ge) doping on the microstructure and TE properties of polycrystalline HMS were revisited recently [165]. Samples prepared by solid-state reaction, BM, and SPS showed an increase in the hole concentration *p* by Al doping that further rose upon additional Ge doping. The maximum power factor was found

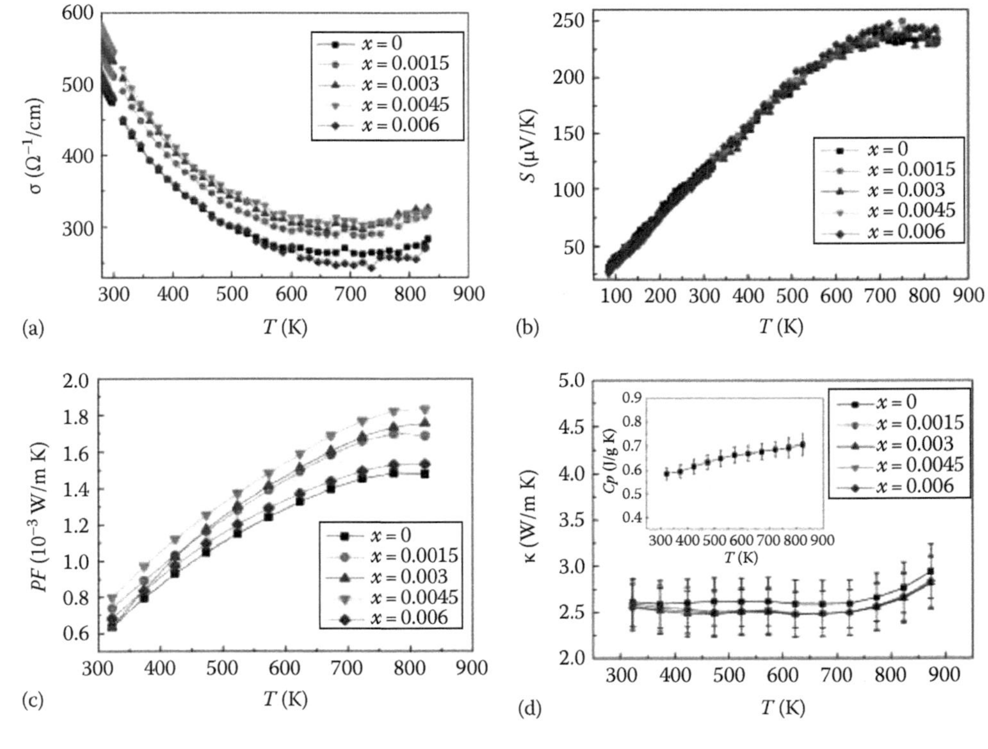

FIGURE 5.25
Temperature dependence of transport properties for $Mn(Al_xSi_{1-x})_{1.8}$. (a) Electrical conductivity, (b) Seebeck coefficient, (c) power factor (obtained by polynomial fitting to the measured Seebeck coefficient and electrical conductivity), and (d) thermal conductivity. The inset of (d) shows the average specific heat of six different samples. (Reprinted with permission from Chen, X., Weathers, A., Salta, D. et al., Effects of (Al, Ge) double doping on the thermoelectric properties of higher manganese silicides, *J. Appl. Phys.*, 114, 173705. Copyright 2013, American Institute of Physics.)

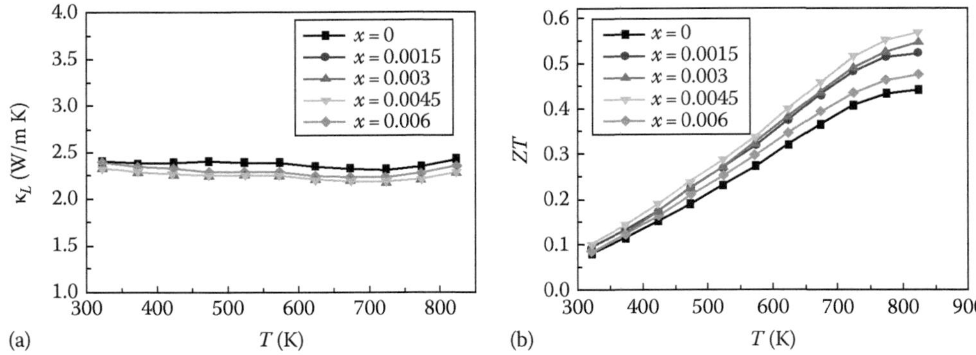

FIGURE 5.26
(a) Lattice thermal conductivity and (b) dimensionless figure of merit of $Mn(Al_xSi_{1-x})_{1.8}$. The relative uncertainty in the ZT values was indicated as 15%. (Reprinted with permission from Chen, X., Weathers, A., Salta, D. et al., Effects of (Al, Ge) double doping on the thermoelectric properties of higher manganese silicides, *J. Appl. Phys.*, 114, 173705. Copyright 2013, American Institute of Physics.)

at a hole concentration of $p = 1.8–2.2 \times 10^{21}$ cm^{-3}, whereas (Al, Ge)-doped HMS showed lower figure of merit than just Al-doped material. The thermal conductivity was found not to change significantly by Al and (Al, Ge) doping as compared to undoped HMS. The lattice thermal conductivity was separated based on a two-band analysis and showed an almost constant value of 2.3 W/m K, independent of temperature and doping concentration between 300 and 823 K; see Figure 5.26a.

Double and triple substitution including Mo, W, and Al as dopants in softly ball-milled and spark plasma sintered Ge-doped HMS ($MnSi_{1.75}Ge_{0.02}$) was also studied [177]. For all substitutions, $(Mn, Mo, W)_{11}(Si, Ge, Al)_{19}$ was identified as the matrix phase from the length ratio of the Mn and Si subcells. Due to its wider solubility, Mo substitution enlarged the unit cell more than W did, whereas an addition of Al hardly changed the cell parameters. Substantial reduction of the lattice thermal conductivity (up to 25% for Mo) was observed.

Addition of Cr or Fe results in a monotonic variation of the a lattice parameter and a change of the NCL structure according to a noncontinuous change of c [178]. An interrelation between the Si/Mn ratio and VE concentration was concluded. With Cr addition, the Si/Mn ratio decreases while it increases with Fe addition. Alloying HMS with $FeSi_2$ resulted in the compensation of free holes down to a sign change in the conduction type [179] at about 25 mol% of $FeSi_2$ addition.

Density functional theory calculations have been used to study the effects of substitution on the electronic transport properties of HMS [180] predicting a significant increase of the Seebeck coefficient by In and Tl doping due to the high contribution of d states to the DOS near the Fermi level.

The qualitative picture of bonding strength maintained by obeying the VEC = 14 rule in competition to the adjustment of an optimum Si–Si interatomic distance as the guiding principle tuning Si concentration and hole density also helps to understand the effect of Ge doping. Whereas the effect of Cr and Fe as an acceptor and donor, respectively, can be understood in principle as an aliovalent substitution on the Mn lattice site (although it turned out that the doping effect is much weaker than expected [117]), Ge represents an isoelectronic substitution of Si. Its electrical action as an acceptor, however, can be understood in terms of changes in the lattice structure that are induced by Ge incorporation. With its larger size (atomic radius of 0.125 nm compared to 0.111 nm for Si), the lattice expansion can be expected. As found in [170], the *c* value increased steadily with the Ge content, whereas *a* remained constant. Looking closer reveals that the Si subcell expands more than the Mn subcell, which results in a reduced Si/Mn ratio. This is equivalent to an increasing deviation from the VEC = 14 criterion and, with that, an increase of the hole concentration, which was evidenced by Hall measurements. This mechanism, in turn, may also explain the vanishing of the inherent MnSi layer pattern at about 1% of the Si replaced by Ge [181]. The corresponding shift of the subcells' length ratio is about 0.2%. With a reduced Si/Mn ratio, the concentration difference *vis-à-vis* the high-temperature Si-depleted phase gets smaller, and, hence, a decreasing volume ratio of MnSi layers is expected. Accordingly, the intensity and position of the peak near $2\theta = 43°$ (Cu K_α), which was found to be sensitive to the subcell length ratio, was reported to depend on the doping concentration of Ge and of Re [170,182].

In a study of the doping influence of Cr and Fe [117], the carrier concentration measured was much lower than expected from the high degree of Mn substitution (up to 30%). It was supposed that $MnSi_\gamma$ phases tend to preserve the hole concentration by rejecting Si, which supports the idea that HMS are carrier concentration stabilized phases.

5.2.5.4 Composites

Composite materials, in particular if nanostructured, represent an effective option to improve the energy conversion performance of HMS because the lattice contribution has a dominant share in its thermal conductivity [183]. Additional grain boundaries will increase phonon scattering and can be introduced by milling down to nanosize. However, nanograins tend to coarsen during thermal treatment, leading to functional instability and loss of nanostructure [106] unless grain boundaries are stabilized, for example, by inert nanoscaled additives. Additives of the silicide nature, such as SiGe [181], MnSi [184], dispersed Si [185], and of other TE phases, such as Ag_2Te and PbTe [186], as well as inert ceramic nanopowders like $CeSi_2$ [187] and alumina [188], Yb [189], and carbon nanotubes (CNT) [190] have been intermixed to HMS, often in a milling step. However, some of these composites did not result in improved TE performance since the combined change of electrical and thermal transport did not lead to an increase in the figure of merit.

A reduction in the thermal conductivity down to about 1.7 W/m K, far below the value of pure HMS, was observed for a HMS/Si micro-/nanocomposite prepared by BM and SPS, as an effect of enhanced phonon scattering [185].

The electrical conductivity of polycrystalline HMS was increased by adding SiGe while the Seebeck coefficient changed insignificantly [181]. The lattice thermal conductivity was effectively reduced, mainly due to enhanced phonon scattering by substituting Si atoms by Ge, and, as a consequence, a maximum ZT of 0.5 was achieved at 550°C upon adding 2 at% of SiGe. Although SiGe micro-inclusions precipitated, the effect of substituting Ge in the HMS matrix was the dominant reason behind the improvement.

During the hard BM, HMS was found to decompose into MnSi [148] as a result of a mechanochemical effect induced by the high impact energy of mechanical collisions. Some loss of Si during milling to nanosized powders was also observed [191] but remained unexplained. After optimization of the milling parameters, HMS powders down to a few tens of nanometers without noticeable MnSi byproduct could be achieved [148].

5.2.5.5 Figure of Merit

The thermoelectric performance of undoped HMS is quite moderate but can be increased by doping or alloying. Despite its limited peak ZT values, the average ZT over the full operational range, the parameter that really matters for conversion efficiency, is, nevertheless, not that far from values of the established thermoelectric materials.

A comprehensive compilation of reported ZT values can be found in [118] on pages 28–31. For undoped HMS, the highest ZT values are reported between 0.3 and 0.65 (in one case even up to 0.83 [162]) with a majority of values below 0.45. The temperatures where the peak ZT is observed fall between 673 and 890 K. Along with a wide scatter of the peak ZT values, the spread of the peak temperatures over an interval of more than 200 K is remarkable. For doped or alloyed p-type HMS and their composites, the maximum ZT values between 0.3 and 0.76, with many of them above 0.5 (see, e.g., Figure 5.26b for Al-doped HMS), were found in the same range of temperatures as for the undoped HMS. A clear link to any specific preparation method cannot be deduced, nor is there a dependence on the reported Si/Mn ratio. However, many of the best values were obtained on samples compacted by SPS. It should be kept in mind that a comparison of values from different references is subject to measurement uncertainty that is usually not better than 20% for ZT.

Anisotropy of ZT is rarely found in the literature. Aoyama et al. [192] reported values of ZT_c/ZT_a near 0.9 for $(Mn_{0.98}Mo_{0.02})(Si_{0.9865}Al_{0.0035}Ge_{0.01})_{1.74}$ with peak temperatures along the a axis at 880 K and along the c axis at 780 K. Particularly high ZT values were found with Re and Ru doping, furthermore with Cr, Ge, and Al. Multiple doping turned out efficient in certain cases. Sophisticated combination doping, such as $Mn_{0.98}Mo_{0.02}[(Si_{0.98}Ge_{0.02})_{1.75}]_{0.99}Al_{0.01}$, $Mn_{0.97}Re_{0.03}(Si_{1.745}Ge_{0.005})_{0.99}Al_{0.01}$, $Mn(Si_{0.98}Ge_{0.02})_{1.75}$, and $Mn_{0.99}W_{0.01}[(Si_{0.99}Ge_{0.01})_{1.75}]_{0.99}Sc_{0.01}$, was elaborated on in [193], with the last mentioned alloy reaching ZT up to 0.9 at 850 K.

For n-type HMS, low ZT values were expected because of the very low electron mobility. A peak ZT of 0.071 at 644 K was measured for $(Mn_{0.7}Fe_{0.3})Si_{1.68}$ [172].

Boltzmann transport calculations were implemented [183] assuming a semiclassical two-band model and considering carrier and phonon transport subjected to several scattering mechanisms, in particular the effect of grain boundaries. The model was fed with the data from band structure calculations [121] and scattering potentials chosen to fit experimental thermoelectric properties of HMS. The model was used to predict achievable ZT of bulk and nanostructured HMS. The authors assumed a sufficiently small grain size, and the model returned the maximum theoretical ZT values up to 0.7–0.8.

Further improvement in the thermoelectric performance of HMS is expected from the implementation of multiple doping to decrease the lattice thermal conductivity (particularly across the c axis), as well as optimization of the carrier concentration and microstructure. Also, there is no clear statement about the optimum carrier concentration level. In the particular situation of a carrier concentration stabilized structure, substitution doping is not a straightforward strategy toward optimized carrier density.

5.2.5.6 Nanostructuring

Nanostructuring has been increasingly explored over the last decade with the intent to reduce the lattice thermal conductivity of HMS, for example, [100,101,157,184,186,194]. A theoretical study on the effect of grain size on the TE properties of HMS [183] revealed a remarkable loss of the electrical conductivity when the grain size was reduced. Hence, ZT values cannot be improved significantly in this way. Accordingly, HMS/PbTe and

HMS/Ag_2Te nanocomposites [186] yielded not only a reduced lattice thermal conductivity but also a much lower power factor. Nevertheless, a *ZT* improvement could be achieved for HMS/MnSi [184], HMS/carbon nanotubes [190], and HMS/Yb composites [189].

Nanocomposites of $MnSi_{1.75}Ge_{0.02}$ and multiwalled carbon nanotubes were prepared by mechanical milling under mild conditions followed by reactive SPS [190]. Homogeneous dispersion of the CNTs in the HMS matrix led to a reduction of the lattice thermal conductivity by 30%, while the power factor remained substantially unchanged.

HMS nanowires have been successfully synthesized *via* a vapor phase conversion approach by reacting arrays of Si nanowires at 950°C [101]. *In situ* synchrotron PXRD heating experiments showed that Mn silicides start to form around 750°C. The resistivity of the HMS nanowires was found similar to bulk.

In spite of a large volume of research, many open questions remain on the interaction of the peculiarities of the HMS lattice structure with various doping approaches and the related stability issues. These are of substantial relevance for the practical use of the best-found doped HMS material to date and, thus, require further research and evaluation.

5.2.6 Technological Application

Technical usage of HMS in thermoelectric generators was proposed decades ago. Thermogenerator elements made of HMS have been patented by Bienert and Gillen in 1968 [132]. Unicouple-type TEGs made of *n*-type $FeSi_2$ and *p*-type HMS were demonstrated in the laboratory [195]. Important technological issues, such as diffusion at HMS/Cr and HMS/Cr/Ni contacts, have been studied [196], and a contacting technique to $TiSi_2$ bridges was developed [197]. Kaibe et al. [198] demonstrated 6.5% efficiency by a silicide-based TEG module combining HMS and Mg_2Si materials operating between 30°C and 550°C. The module was stacked with a bismuth telluride–based lower stage aiming at an overall 15% efficiency. Recently, TEG modules with 2–31 pairs of TE legs based on *n*-type Mg_2(Si, Sn) and *p*-type HMS have been assembled and tested [199]. Development of mass applicable material fabrication is being pursued by institutes and companies worldwide.

Acknowledgements

J. de Boor thanks the Helmholtz Association for the financial support and Mrs. Bölt from the DLR library for getting all the papers.

References

1. Zaitsev, V. K., M. I. Federov, I. S. Eremin, and E. A. Gurieva. 2005. Thermoelectrics on the base of solid solutions of Mg_2B IV compounds. In *Thermoelectrics Handbook: Macro to Nano*, ed. D. M. Rowe. CRC Press, Boca Raton, FL, Chapter 29.
2. Fedorov, M. I. and V. K. Zaitsev. 2005. Thermoelectrics of transition metal silicides. In *Thermoelectrics Handbook: Macro to Nano*, ed. D. M. Rowe. CRC Press, Boca Raton, FL.
3. Abundance in Earth's crust. 2015. webelements.com. Accessed November 18, 2015.

4. LeBlanc, S., S. K. Yee, M. L. Scullin, C. Dames, and K. E. Goodson. 2014. Material and manufacturing cost considerations for thermoelectrics. *Renew. Sust. Energ. Rev.* 32: 313–327.
5. Godlewska, E., K. Mars, R. Mania, and S. Zimowski. 2011. Combustion synthesis of Mg_2Si. *Intermetallics* 19: 1983–1988.
6. Zaitsev, V. K., M. I. Fedorov, E. A. Gurieva, I. S. Eremin, P. P. Konstantinov, A. Y. Samunin, and M. V. Vedernikov. 2006. Highly effective $Mg_2Si_{1-x}Sn_x$ thermoelectrics. *Phys. Rev. B* 74: 045207.
7. Dasgupta, T., C. Stiewe, R. Hassdorf, A. J. Zhou, L. Boettcher, and E. Müller. 2011. Effect of vacancies on the thermoelectric properties of $Mg_2Si_{1-x}Sb_x$ ($0 \le x \le 0.1$). *Phys. Rev. B* 83: 235207.
8. de Boor, J., S. Gupta, H. Kolb, T. Dasgupta, and E. Müller. 2015. Thermoelectric transport and microstructure of optimized $Mg_2Si_{0.8}Sn_{0.2}$. *J. Mater. Chem. C.* 3: 10467–10475.
9. Samunin, A. Y., V. K. Zaitsev, P. P. Konstantinov et al. 2013. Thermoelectric properties of hot-pressed materials based on $Mg_2Si_nSn_{1-n}$. *J. Electron. Mater.* 42: 1676–1679.
10. Zhang, Q., J. He, T. J. Zhu, S. N. Zhang, X. B. Zhao, and T. M. Tritt. 2008. High figures of merit and natural nanostructures in $Mg_2Si_{0.4}Sn_{0.6}$ based thermoelectric materials. *Appl. Phys. Lett.* 93: 102109.
11. Zhao, J. B., Z. X. Liu, J. Reid et al. 2015. Thermoelectric and electrical transport properties of Mg_2Si multi-doped with Sb, Al and Zn. *J. Mater. Chem. A* 3: 19774–19782.
12. Nayeb-Hashemi, A. A. and J. B. Clark. 1984. The Mg–Si (Magnesium–Silicon) system. *Bull. Alloy Phase Diag.* 5: 584–592.
13. Jung, I. H., D. H. Kang, W. J. Park, N. J. Kim, and S. Ahn. 2007. Thermodynamic modeling of the Mg-Si-Sn system. *Calphad* 31: 192–200.
14. Dasgupta, T., C. Stiewe, J. de Boor, and E. Müller. 2014. Influence of power factor enhancement on the thermoelectric figure of merit in $Mg_2Si_{0.4}Sn_{0.6}$ based materials. *Phys. Stat. Solidi (a)* 21: 1250–1254.
15. Bernard-Granger, G., C. Navone, J. Leforestier et al. 2015. Microstructure investigations and thermoelectrical properties of an N-type magnesium-silicon-tin alloy sintered from a gas-phase atomized powder. *Acta Mater.* 96: 437–451.
16. Liu, W., X. Tang, and J. Sharp. 2010. Low-temperature solid state reaction synthesis and thermoelectric properties of high-performance and low-cost Sb-doped $Mg_2Si_{0.6}Sn_{0.4}$. *J. Phys. D: Appl. Phys.* 43: 085406.
17. Liu, W., X. J. Tan, K. Yin et al. 2012. Convergence of conduction bands as a means of enhancing thermoelectric performance of n-type $Mg_2Si_{1-x}Sn_x$ solid solutions. *Phys. Rev. Lett.* 108: 166601.
18. Liu, W., H. Chi, H. Sun et al. 2014. Advanced thermoelectrics governed by a single parabolic band: $Mg_2Si_{0.3}Sn_{0.7}$, a canonical example. *Phys. Chem. Chem. Phys.* 16: 6893–6897.
19. Tani, J.-I. and H. Kido. 2013. Fabrication and thermoelectric properties of Mg_2Si-based composites using reduction reaction with additives. *Intermetallics* 32: 72–80.
20. Hu, X. K., D. Mayson, and M. R. Barnett. 2014. Synthesis of Mg_2Si for thermoelectric applications using magnesium alloy and spark plasma sintering. *J. Alloys Compd.* 589: 485–490.
21. Tani, J.-I. and H. Kido. 2008. Thermoelectric properties of Al-doped $Mg_2Si_{1-x}Sn_x$ ($x \le 0.1$). *J. Alloys Compd.* 466: 335–340.
22. Tani, J.-I. and H. Kido. 2007. Thermoelectric properties of Sb-doped Mg_2Si semiconductors. *Intermetallics* 15: 1202–1207.
23. Kubouchi, M., K. Hayashi, and Y. Miyazaki. 2014. Quantitative analysis of interstitial Mg in Mg_2Si studied by single crystal X-ray diffraction. *J. Alloys Compd.* 617: 389–392.
24. Su, X., F. Fu, Y. Yan et al. 2014. Self-propagating high-temperature synthesis for compound thermoelectrics and new criterion for combustion processing. *Nat. Commun.* 5: 4908.
25. Savary, E., F. Gascoin, and S. Marinel. 2010. Fast synthesis of nanocrystalline Mg_2Si by microwave heating: A new route to nano-structured thermoelectric materials. *Dalton Trans.* 39: 11074–11080.
26. Berthebaud, D. and F. Gascoin. 2013. Microwaved assisted fast synthesis of n and p-doped Mg_2Si. *J. Solid State Chem.* 202: 61–64.

27. Gao, H. L., T. J. Zhu, X. X. Liu, L. X. Chen, and X. B. Zhao. 2011. Flux synthesis and thermoelectric properties of eco-friendly Sb doped $Mg_2Si_{0.5}Sn_{0.5}$ solid solutions for energy harvesting. *J. Mater. Chem.* 21: 5933–5937.
28. Gao, P., I. Berkun, R. Schmidt et al. 2013. Transport and mechanical properties of high-ZT $Mg_{2.08}Si_{0.4-x}Sn_{0.6}Sb_x$ thermoelectric materials. *J. Electron. Mater.* 43: 1790–1803.
29. Liu, X. H., T. J. Zhu, H. Wang et al. 2013. Low electron scattering potentials in high performance $Mg_2Si_{0.45}Sn_{0.55}$ based thermoelectric solid solutions with band convergence. *Adv. Energy Mater.* 3: 1238–1244.
30. Yi, T. H., S. P. Chen, S. Li et al. 2012. Synthesis and characterization of Mg_2Si/Si nanocomposites prepared from MgH_2 and silicon, and their thermoelectric properties. *J. Mater. Chem.* 22: 24805–24813.
31. Chen, S. P., X. Zhang, W. H. Fan et al. 2015. One-step low temperature reactive consolidation of high purity nanocrystalline Mg_2Si. *J. Alloys Compd.* 625: 251–257.
32. Bux, S. K., M. T. Yeung, E. S. Toberer, G. J. Snyder, R. B. Kaner, and J. P. Fleurial. 2011. Mechanochemical synthesis and thermoelectric properties of high quality magnesium silicide. *J. Mater. Chem.* 21: 12259–12266.
33. Riffel, M. and J. Schilz. 1997. Influence of production parameters on the thermoelectric properties of Mg_2Si. In Paper presented at the *Proceedings ICT'97—XVI International Conference on Thermoelectrics*, New York.
34. Vracar, R., G. Bernard-Granger, C. Navone et al. 2014. Microstructure and thermoelectrical investigations of an N-type magnesium–silicon–tin alloy. *J. Alloys Compd.* 598: 272–277.
35. Hayatsu, Y., T. Iida, T. Sakamoto et al. 2012. Fabrication of large sintered pellets of Sb-doped n-type Mg_2Si using a plasma activated sintering method. *J. Solid State Chem.* 193: 161–165.
36. Ioannou, M., E. Hatzikraniotis, C. Lioutas et al. 2012. Fabrication of nanocrystalline Mg_2Si via ball milling process: Structural studies. *Powder Technol.* 217: 523–532.
37. Fiameni, S., A. Famengo, F. Agresti et al. 2014. Effect of synthesis and sintering conditions on the thermoelectric properties of n-doped Mg_2Si. *J. Electron. Mater.* 43: 2301–2306.
38. Battiston, S., S. Fiameni, M. Saleemi et al. 2013. Synthesis and characterization of Al-doped Mg_2Si thermoelectric materials. *J. Electron. Mater.* 42: 1956–1959.
39. Saleemi, M., M. S. Toprak, S. Fiameni et al. 2013. Spark plasma sintering and thermoelectric evaluation of nanocrystalline magnesium silicide (Mg_2Si). *J. Mater. Sci.* 48: 1940–1946.
40. Satyala, N. and D. Vashaee. 2012. The effect of crystallite size on thermoelectric properties of bulk nanostructured magnesium silicide (Mg_2Si) compounds. *Appl. Phys. Lett.* 100: 073107.
41. Fiameni, S., S. Battiston, S. Boldrini et al. 2012. Synthesis and characterization of Bi-doped Mg_2Si thermoelectric materials. *J. Solid State Chem.* 193: 142–146.
42. Nolas, G. S., D. Wang, and M. Beekman. 2007. Transport properties of polycrystalline $Mg_2Si_{1-y}Sb_y$ ($0 \leq y < 0.4$). *Phys. Rev. B* 76: 235204.
43. Farahi, N., M. VanZant, J. B. Zhao et al. 2014. Sb- and Bi-doped Mg_2Si: Location of the dopants, micro- and nanostructures, electronic structures and thermoelectric properties. *Dalton Trans.* 43: 14983–14991.
44. de Boor, J., C. Compere, T. Dasgupta et al. 2014. Fabrication parameters for optimized thermoelectric Mg_2Si. *J. Mater. Sci.* 49: 3196–3204.
45. de Boor, J., T. Dasgupta, H. Kolb, C. Compere, K. Kelm, and E. Müller. 2014. Microstructural effects on thermoelectric efficiency: A case study on magnesium silicide. *Acta Mater.* 77: 68–75.
46. Liu, W., K. Yin, X. Su et al. 2013. Enhanced hole concentration through Ga doping and excess of Mg and thermoelectric properties of p-type $Mg_{2(1+z)}(Si_{0.3}Sn_{0.7})_{(1-y)}Ga_y$. *Intermetallics* 32: 352–361.
47. Liu, W., X. F. Tang, H. Li, J. Sharp, X. Y. Zhou, and C. Uher. 2011. Optimized thermoelectric properties of Sb-doped $Mg_{2(1+z)}Si_{0.5-y}Sn_{0.5}Sb_y$ through adjustment of the Mg content. *Chem. Mater.* 23: 5256–5263.
48. Staab, T. E. M. 2009. Vacancies in magnesium silicide—Stoichiometric vacancies preferred? *Phys. Stat. Solidi B-Basic Solid State Phys.* 246: 1587–1589.
49. Kato, A., T. Yagi, and N. Fukusako. 2009. First-principles studies of intrinsic point defects in magnesium silicide. *J. Phys. Condens. Matter* 21: 205801.

50. Viennois, R., P. Jund, C. Colinet, and J. C. Tedenac. 2012. Defect and phase stability of solid solutions of Mg_2X with an antifluorite structure: An ab initio study. *J. Solid State Chem.* 193: 133–136.
51. Jund, P., R. Viennois, C. Colinet, G. Hug, M. Fevre, and J. C. Tedenac. 2013. Lattice stability and formation energies of intrinsic defects in Mg_2Si and Mg_2Ge via first principles simulations. *J. Phys. Condens. Matter* 25: 035403.
52. Tobola, J., S. Kaprzyk, and H. Scherrer. 2010. Mg-vacancy-induced semiconducting properties in Mg_2Si_{1-x} Sb_x from electronic structure calculations. *J. Electron. Mater.* 39: 2064–2069.
53. Han, X. P. and G. S. Shao. 2015. Interplay between Ag and interstitial Mg on the p-type characteristics of Ag-doped Mg_2Si: Challenges for high hole conductivity. *J. Mater. Chem. C* 3: 530–537.
54. Bourgeois, J., J. Tobola, B. Wiendlocha et al. 2013. Study of electron, phonon and crystal stability versus thermoelectric properties in Mg_2X (X = Si, Sn) compounds and their alloys. *Funct. Mater. Lett.* 06: 1340005.
55. Tani, J.-I. and H. Kido. 2005. Thermoelectric properties of Bi-doped Mg_2Si semiconductors. *Phys. B: Condens. Matter* 364: 218–224.
56. Zwolenski, P., J. Tobola, and S. Kaprzyk. 2011. A theoretical search for efficient dopants in Mg_2X (X = Si, Ge, Sn) thermoelectric materials. *J. Electron. Mater.* 40: 889–897.
57. Cederkrantz, D., N. Farahi, K. A. Borup, B. B. Iversen, M. Nygren, and A. E. C. Palmqvist. 2012. Enhanced thermoelectric properties of Mg_2Si by addition of TiO_2 nanoparticles. *J. Appl. Phys.* 111: 023701.
58. Fiameni, S., A. Famengo, S. Boldrini et al. 2013. Introduction of metal oxides into Mg_2Si thermoelectric materials by spark plasma sintering. *J. Electron. Mater.* 42: 2062–2066.
59. Sakamoto, T., T. Iida, A. Matsumoto et al. 2010. Thermoelectric characteristics of a commercialized Mg_2Si source doped with Al, Bi, Ag, and Cu. *J. Electron. Mater.* 39: 1708–1713.
60. Muthiah, S., B. Sivaiah, B. Gahtori et al. 2014. Double-doping approach to enhancing the thermoelectric figure-of-merit of n-type Mg_2Si synthesized by use of spark plasma sintering. *J. Electron. Mater.* 43: 2035–2039.
61. Pshenai-Severin, D. A., M. I. Fedorov, and A. Y. Samunin. 2013. The influence of grain boundary scattering on thermoelectric properties of Mg_2Si and $Mg_2Si_{0.8}Sn_{0.2}$. *J. Electron. Mater.* 42: 1707–1710.
62. May, A. F. and G. J. Snyder. 2012. Introduction to modeling thermoelectric transport at high temperatures. In *Thermoelectrics and Its Energy Harvesting*, ed. D. M. Rowe. CRC Press, Taylor & Francis Group, Boca Raton, FL, Chapter 11.
63. Satyala, N. and D. Vashaee. 2012. Modeling of thermoelectric properties of magnesium silicide (Mg_2Si). *J. Electron. Mater.* 41: 1785–1791.
64. Bahk, J. H., Z. X. Bian, and A. Shakouri. 2014. Electron transport modeling and energy filtering for efficient thermoelectric $Mg_2Si_{1-x}Sn_x$ solid solutions. *Phys. Rev. B* 89: 075204.
65. Ioannou, M., G. S. Polymeris, E. Hatzikraniotis, K. M. Paraskevopoulos, and T. Kyratsi. 2014. Effect of Bi-doping and Mg-excess on the thermoelectric properties of Mg_2Si materials. *J. Phys. Chem. Solids* 75: 984–991.
66. Riffel, M. and J. Schilz. 1995. Mechanical alloying of Mg_2Si. *Scr. Metall. Mater.* 32: 1951–1956.
67. Biswas, K., J. Q. He, I. D. Blum et al. 2012. High-performance bulk thermoelectrics with all-scale hierarchical architectures. *Nature* 489: 414–418.
68. Heremans, J. P., C. M. Thrush, and D. T. Morelli. 2004. Thermopower enhancement in lead telluride nanostructures. *Phys. Rev. B* 70: 115334.
69. Seto, J. Y. W. 1975. The electrical properties of polycrystalline silicon films. *J. Appl. Phys.* 46: 5247–5254.
70. Kajikawa, T., K. Shida, K. Shiraishi, T. Ito, M. Omori, and T. Hirai. 1998. Thermoelectric figure of merit of impurity doped and hot-pressed magnesium silicide elements. In *Proceeding of the XVII International Conference on Thermoelectrics*, Piscataway, NJ, IEEE Catalaog Number 98TH8365, pp. 362–369.
71. Satyala, N., J. S. Krasinski, and D. Vashaee. 2014. Simultaneous enhancement of mechanical and thermoelectric properties of polycrystalline magnesium silicide with conductive glass inclusion. *Acta Mater.* 74: 141–150.

72. Satyala, N. and D. Vashaee. 2012. Detrimental influence of nanostructuring on the thermoelectric properties of magnesium silicide. *J. Appl. Phys.* 112: 093716.
73. Kutorasinski, K., B. Wiendlocha, J. Tobola, and S. Kaprzyk. 2014. Importance of relativistic effects in electronic structure and thermopower calculations for Mg_2Si, Mg_2Ge, and Mg_2Sn. *Phys. Rev. B* 89: 115205.
74. Noda, Y., H. Kon, Y. Furukawa, I. A. Nishida, and K. Masumoto. 1992. Temperature-dependence of thermoelectric properties of $Mg_2Si_{0.6}Ge_{0.4}$. *Mater. Trans. JIM* 33: 851–855.
75. Noda, Y., H. Kon, Y. Furukawa, N. Otsuka, I. A. Nishida, and K. Masumoto. 1992. Preparation and thermoelectric properties of $Mg_2Si_{1-x}Ge_x$ (x = 0.0–0.4) solid-solution semiconductors. *Mater. Trans. JIM* 33: 845–850.
76. Tani, J. I. and H. Kido. 2008. First-principles and experimental studies of impurity doping into Mg_2Si. *Intermetallics* 16: 418–423.
77. Ihou-Mouko, H., C. Mercier, J. Tobola, G. Pont, and H. Scherrer. 2011. Thermoelectric properties and electronic structure of p-type Mg_2Si and $Mg_2Si_{0.6}Ge_{0.4}$ compounds doped with Ga. *J. Alloys Compd.* 509: 6503–6508.
78. Akasaka, M., T. Iida, A. Matsumoto et al. 2008. The thermoelectric properties of bulk crystalline n- and p-type Mg_2Si prepared by the vertical Bridgman method. *J. Appl. Phys.* 104: 013703.
79. Isoda, Y., S. Tada, T. Nagai, H. Fujiu, and Y. Shinohara. 2010. Thermoelectric properties of p-type $Mg_{2.00}Si_{0.25}Sn_{0.75}$ with Li and Ag double doping. *J. Electron. Mater.* 39: 1531–1535.
80. Prytuliak, A., E. Godlewska, K. Mars, and D. Berthebaud. 2014. Synchrotron study of Ag-doped Mg_2Si: Correlation between properties and structure. *J. Electron. Mater.* 43: 3746–3752.
81. Fedorov, M. I., V. K. Zaĭtsev, I. S. Eremin et al. 2006. Transport properties of $Mg_2X_{0.4}Sn_{0.6}$ solid solutions (X = Si, Ge) with p-type conductivity. *Phys. Solid State* 48: 1486–1490.
82. Zhang, Q., L. Cheng, W. Liu et al. 2014. Low effective mass and carrier concentration optimization for high performance p-type $Mg_{2(1-x)}Li_{2x}Si_{0.3}Sn_{0.7}$ solid solutions. *Phys. Chem. Chem. Phys.* 16: 23576–23583.
83. Tan, X. J., W. Liu, H. J. Liu, J. Shi, X. F. Tang, and C. Uher. 2012. Multiscale calculations of thermoelectric properties of n-type $Mg_2Si_{1-x}Sn_x$ solid solutions. *Phys. Rev. B* 85: 205212.
84. Kutorasinski, K., J. Tobola, and S. Kaprzyk. 2013. Calculating electron transport coefficients of disordered alloys using the KKR-CPA method and Boltzmann approach: Application to $Mg_2Si_{1-x}Sn_x$ thermoelectrics. *Phys. Rev. B* 87: 195205.
85. Liu, W., X. F. Tang, H. Li et al. 2012. Enhanced thermoelectric properties of n-type $Mg_{2.16}(Si_{0.4}Sn_{0.6})_{1-y}Sb_y$ due to nano-sized Sn-rich precipitates and an optimized electron concentration. *J. Mater. Chem.* 22: 13653–13661.
86. Liu, W., Q. Zhang, K. Yin et al. 2013. High figure of merit and thermoelectric properties of Bi-doped $Mg_2Si_{0.4}Sn_{0.6}$ solid solutions. *J. Solid State Chem.* 203: 333–339.
87. Jiang, G. Y., J. He, T. J. Zhu et al. 2014. High performance Mg_2(Si, Sn) solid solutions: A point defect chemistry approach to enhancing thermoelectric properties. *Adv. Funct. Mater.* 24: 3776–3781.
88. Gao, P., X. Lu, I. Berkun, R. D. Schmidt, E. D. Case, and T. P. Hogan. 2014. Reduced lattice thermal conductivity in Bi-doped $Mg_2Si_{0.4}Sn_{0.6}$. *Appl. Phys. Lett.* 105: 202104.
89. Bhandari, C. M. 1995. Minimizing the thermal conductivity. In *CRC Handbook of Thermoelectrics*, ed. D. M. Rowe. CRC Press, Boca Raton, FL, Chapter 6, pp. 55–66.
90. Bellanger, P., S. Gorsse, G. Bernard-Granger, C. Navone, A. Redjaimia, and S. Vives. 2015. Effect of microstructure on the thermal conductivity of nanostructured Mg_2(Si,Sn) thermoelectric alloys: An experimental and modeling approach. *Acta Mater.* 95: 102–110.
91. Nowotny, H. 1970. Crystal chemistry of transition element defect silicides and related compounds. In *The Chemistry of Extended Defects in Non-Metallic Solids*, eds. M. OKeeffe and L. Eyring. North-Holland, Amsterdam, the Netherlands, p. 223.
92. Fedorov, M. I. and V. K. Zaitsev. 2012. Silicide thermoelectrics: State of the art and prospects. In *Thermoelectrics and Its Energy Harvesting*, 2-Vol. Set: Modules, Systems, and Applications in Thermoelectrics, ed. D. M. Rowe. CRC Press, Boca Raton, FL, Chapter 11.

93. Zaitsev, V. K. 1995. Thermoelectric properties of anisotropic $MnSi_{1.75}$. In *CRC Handbook of Thermoelectrics*, ed. D. M. Rowe. CRC Press, New York, p. 299.
94. Miyazaki, Y. and Y. Kikuchi. 2013. Higher manganese silicide $MnSi_\gamma$. In *Thermoelectric Nanomaterials: Materials Design and Applications*, eds. K. Koumoto and T. Mori. Springer, Berlin, Germany, pp. 141–156.
95. Akselrud, L., R. C. Gil, M. Wagner-Reetz, and Y. Grin. 2015. Disorder in the composite crystal structure of the manganese "disilicide" $MnSi_{1.73}$ from powder X-ray diffraction data. *Acta Crystallogr. B* 71: 707–712.
96. Schwomma, O., A. Preisinger, H. Nowotny, and A. Wittmann. 1964. Die Kristallstruktur von $Mn_{11}Si_{19}$ und deren Zusammenhang mit Disilicid-Typen. *Monatshefte für Chemie und verwandte Teile anderer Wissenschaften* 95: 1527–1537.
97. Knott, H. W., M. H. Mueller, and L. Heaton. 1967. The crystal structure of $Mn_{15}Si_{26}$. *Acta Crystallogr.* 23: 549–555.
98. Karpinskij, O. G. and V. A. Evseev. 1969. Crystal structure of Mn_4Si_7. *Izvestiya AN SSSR (Neorganicheskie Materialy)* 5: 525.
99. Zwilling, G. and H. Nowotny. 1971. Crystal structure of manganese silicides with $MnSi_{\sim 1.7}$ stoichiometry. *Monatshefte für Chemie/Chem. Monthly* 102: 672–677.
100. Higgins, J. M., A. L. Schmitt, I. A. Guzei, and S. Jin. 2008. Higher manganese silicide nanowires of Nowotny chimney ladder phase. *J. Am. Chem. Soc.* 130: 16086–16094.
101. Pokhrel, A., Z. P. Degregorio, J. M. Higgins, S. N. Girard, and S. Jin. 2013. Vapor phase conversion synthesis of higher manganese silicide ($MnSi_{1.75}$) nanowire arrays for thermoelectric applications. *Chem. Mater.* 25: 632–638.
102. De Ridder, R., G. Van Tendeloo, and S. Amelinckx. 1976. Electron microscopic study of the chimney ladder structures $MnSi_{2-x}$ and $MoGe_{2-x}$. *Phys. Stat. Solidi (a)* 33: 383–393.
103. Ye, H. and S. Amelinckx. 1986. High-resolution electron microscopic study of manganese silicides $MnSi_{2-x}$. *J. Solid State Chem.* 61: 8–39.
104. De Ridder, R., G. Van Tendeloo, and S. Amelinckx. 1975. Incommensurate superstructures in $MnSi_{2-x}$. *Phys. Stat. Solidi (a)* 30: K99–K101.
105. Allam, A., P. Boulet, C. A. Nunes, J. Sopousek, P. Broz, and M. C. Record. 2013. Phase transformations in higher manganese silicides. *J. Alloy. Compd.* 551: 30–36.
106. Zhou, A. J., T. J. Zhu, X. B. Zhao, and E. Mueller. 2011. Grain size effect on the phase transformations of higher manganese silicide thermoelectric materials: An in situ energy dispersive X-ray diffraction study. *J. Mater. Res.* 26: 1900–1906.
107. Yamamoto, A. 1993. Determination of composite crystal structures and superspace groups. *Acta Crystallogr. A* 49: 831–846.
108. Miyazaki, Y., D. Igarashi, K. Hayashi, T. Kajitani, and K. Yubuta. 2008. Modulated crystal structure of chimney-ladder higher manganese silicides MnSiγ (γ similar to 1.74). *Phys. Rev. B* 78: 214104.
109. Allam, A., C. A. Nunes, J. Zalesak, and M. C. Record. 2012. On the stability of the higher manganese silicides. *J. Alloy. Compd.* 512: 278–281.
110. Nakajima, T. and J. Schelten. 1980. Peculiarities of the band magnetism in the higher manganese silicide. *J. Magn. Magn. Mater.* 21: 157–166.
111. Kawasumi, I., M. Sakata, I. Nishida, and K. Masumoto. 1981. Crystal-growth of manganese silicide, MnSi approximately 1.73 and semiconducting properties of $Mn_{15}Si_{26}$. *J. Mater. Sci.* 16: 355–366.
112. Kojima, T., I. Nishida, and T. Sakata. 1979. Crystal-growth of $Mn_{15}Si_{26}$. *J. Cryst. Growth* 47: 589–592.
113. Okamoto, N. L., T. Koyama, K. Kishida, K. Tanaka, and H. Inui. 2009. Crystal structure and thermoelectric properties of chimney–ladder compounds in the Ru_2Si_3_Mn_4Si_7 pseudobinary system. *Acta Mater.* 57: 5036–5045.
114. Simkin, B. A., A. Ishida, N. L. Okamoto, K. Kishida, K. Tanaka, and H. Inui. 2006. Formation of a series of chimney–ladder compounds in the Ru–Re–Si system. *Acta Mater.* 54: 2857–2865.

115. Fredrickson, D. C., S. Lee, and R. Hoffmann. 2004. The Nowotny chimney ladder phases: Whence the 14 electron rule? *Inorg. Chem.* 43: 6159–6167.
116. Kikuchi, Y., T. Nakajo, K. Hayashi, and Y. Miyazaki. 2014. High temperature X-ray diffraction study on incommensurate composite crystal $MnSi_\gamma$ (3+1)-dimensional superspace approach. *J. Alloy. Compd.* 616: 263–267.
117. Ponnambalam, V. and D. T. Morelli. 2011. Effect of Cr and Fe substitution on the transport properties of the Nowotny chimney–ladder $MnSi_\delta$ ($1.73 < \delta < 1.75$) compounds. *J. Electron. Mater.* 41: 1389–1394.
118. Truong, D. Y. N. 2015. Thermoelectric properties of higher manganese silicides. Thesis. University of Waterloo, Waterloo, Ontario, Canada.
119. Imai, Y. and A. Watanabe. 2005. Consideration of the validity of the 14 valence electron rule for semiconducting chimney-ladder phase compounds. *Intermetallics* 13: 233–241.
120. Allam, A., P. Boulet, and M. C. Record. 2014. DFT calculations of electronic and transport properties of substituted Mn_4Si_7. *J. Alloy. Compd.* 584: 279–288.
121. Migas, D. B., V. L. Shaposhnikov, A. B. Filonov, V. E. Borisenko, and N. N. Dorozhkin. 2008. Ab initio study of the band structures of different phases of higher manganese silicides. *Phys. Rev. B* 77: 075205.
122. Rebien, M., W. Henrion, H. Angermann, and S. Teichert. 2002. Interband optical properties of higher manganese silicide thin films. *Appl. Phys. Lett.* 81: 649–651.
123. Gao, Y. S., G. S. Shao, Q. Li et al. 2007. Microstructural and optical properties of semiconducting $MnSi_{1.7}$ synthesized by ion implantation. *Jpn. J. Appl. Phys.* 46: 5777–5779.
124. Zaitsev, V. K., S. V. Ordin, V. I. Tarasov, and M. I. Fedorov. 1979. Optical properties of higher manganese silicide. *Sov. Phys. Solid State* 21: 1454–1455.
125. Bost, M. and J. E. Mahan. 1987. An optical determination of the bandgap of the most silicon-rich manganese silicide phase. *J. Electron. Mater.* 16: 389–395.
126. Krontiras, C., K. Promoni, and M. Roilos. 1988. Resistivity and the Hall effect for thin $MnSi_{1.73}$ films. *J. Phys. D* 21: 509–512.
127. Nishida, I. 1972. Semiconducting properties of nonstoichiometric manganese silicides. *J. Mater. Sci.* 7: 435–440.
128. Zaitsev, V. K., M. I. Fedorov, K. A. Rakhimov, A. E. Engalychev, and V. V. Popov. 1984. Lattice thermal conductivity and reflection spectra of the higher manganese silicide and some solutions based on it. *Sov. Phys. Solid State* 26: 494.
129. Nikitin, E. N. and V. I. Tarasov. 1971. Electrical properties of the solid solutions based on higher silicides of manganese. *Sov. Phys. Solid State* 13: 2938.
130. Jørgensen, J.-E. and S. E. Rasmussen. 1991. Refinement of the structure of MnSi by powder diffraction. *Powder Diffract.* 6: 194–195.
131. Hortamani, M. 2006. Theory of adsorption, diffusion and spin polarization of Mn on Si (001) and Si(111) substrates. Thesis, Freie Universität Berlin, Berlin, Germany.
132. Bienert, W. B. and F. M. Gillen. 1968. Process of making MnSi thermoelectric element and product of said process. U.S. Patent No. 3407037.
133. Kanibolotskii, D. S. and V. V. Lesnyak. 2006. Thermodynamic properties of Mn-Si alloys. *Russ. Metall.* 2006: 199–205.
134. Nishida, I., K. Masumoto, I. Kawasumi, and M. Sakata. 1980. Striations and crystal structure of the matrix in the MnSi-Si alloy system. *J. Less Common Metals* 71: 293–301.
135. Levinson, L. M. 1973. Investigation of the defect manganese silicide Mn_nSi_{2n-m}. *J. Solid State Chem.* 6: 126–135.
136. Morokhovets, M. A., E. I. Elagina, and N. K. Abrikosov. 1966. Phase diagram of the system Mn–Si in the region of higher manganese silicide. *Izvestiya AN SSSR (Neorganicheskie Materialy)* 2: 650.
137. Kojima, T. and I. Nishida. 1975. Crystal-growth of $MnSi_{1.73}$ by chemical-transport method. *Jpn. J. Appl. Phys.* 14: 141–142.
138. Girard, S. N., X. Chen, F. Meng et al. 2014. Thermoelectric properties of undoped high purity higher manganese silicides grown by chemical vapor transport. *Chem. Mater.* 26: 5097–5104.

139. Vinokurova, L., V. Ivanov, E. Kulatov, and A. Vlasov. 1990. Magnetic phase transitions and electronic structure of the manganese silicides. *J. Magn. Magn. Mater.* 90: 121–125.
140. Abrikosov, N. K., L. D. Ivanova, and V. G. Murav'ev. 1972. Preparation and study of single crystals and solid solutions of higher manganese silicide with Ge and $CrSi_2$. *Izvestiya AN SSSR (Neorganicheskie Materialy)* 8: 1194.
141. Aoyama, I., M. I. Fedorov, V. K. Zaitsev et al. 2005. Effects of Ge doping on micromorphology of MnSi in $MnSi_{\sim 1.7}$ and on their thermoelectric transport properties. *Jpn. J. Appl. Phys.* 44: 8562.
142. Korshunov, V. A., F. A. Sidorenko, P. V. Gel'd, and K. N. Davydov. 1961. Phase components of the system MnSi–Si. *Fiz. Metal. Metalloved.* 12: 277–284.
143. Dörinckel, F. 1906. Über die Verbindungen des Mangans mit Silicium. *Z. Anorg. Chem.* 50: 117–132.
144. Dudkin, L. D. and E. S. Kuznetsova. 1962. Investigation of the electrophysical properties of alloys on the basis of semiconductor disilicides of chromium and manganese. *Poroshk. Metall.* 6: 20.
145. Shukla, A., Y.-B. Kang, and A. D. Pelton. 2008. Thermodynamic assessment of the Si–Zn, Mn–Si, Mg–Si–Zn and Mg–Mn–Si systems. *Calphad* 32: 470–477.
146. Ugai, Y. A., V. Z. Anokhin, and E. M. Averbakh. 1972. Chemical bonds and properties of the higher silicide of manganese. In *Chemical Bonds in Solids*, ed. N. N. Sirota. Consultants Bureau, New York, p. 16.
147. Okada, S., T. Shishido, Y. Ishizawa et al. 2001. Crystal growth by molten metal flux method and properties of manganese silicides. *J. Alloy. Compd.* 317: 315–319.
148. Zhou, A. J., X. B. Zhao, T. J. Zhu et al. 2010. Mechanochemical decomposition of higher manganese silicides in the ball milling process. *Intermetallics* 18: 2051–2056.
149. Nikitin, E. N., V. I. Tarasov, A. A. Andreev, and L. N. Shumilova. 1969. Electrical properties of single crystal highest silicide of manganese. *Sov. Phys. Solid State* 11: 2757.
150. Ivanova, L. D., N. K. Abrikosov, E. I. Elagina, and V. D. Khvostikova. 1969. Production and the study of properties of higher manganese silicide single crystals. *Izvestiya AN SSSR (Neorganicheskie Materialy)* 5: 1933.
151. Voronov, B. K., L. D. Dudkin, and N. N. Trusova. 1967. Anisotropy of thermoelectric properties in single crystals of chromium disilicide and higher manganese silicide. *Kristallografia* 12: 1137.
152. Nikitin, E. N. and V. K. Zaitsev. 1965. Obtaining the higher manganese silicide by chemical transport. *Izvestiya AN SSSR (Neorganicheskie Materialy)* 1: 1526.
153. Okada, S., T. Shishido, M. Ogawa et al. 2001. MnSi and $MnSi_{2-x}$ single crystals growth by Ga flux method and properties. *J. Cryst. Growth* 229: 532–536.
154. Yamada, T., Y. Miyazaki, and H. Yamane. 2011. Preparation of Higher Manganese Silicide (HMS) bulk and Fe-containing HMS bulk using a Na–Si melt and their thermoelectrical properties. *Thin Solid Films* 519: 8524–8527.
155. Avetisyan, A. O., Y. M. Goryachev, and T. M. Yarmola. 1978. Application of vibration during the sintering the products of higher manganese silicide. *Poroshk. Metall.* 10: 38.
156. Luo, W., H. Li, F. Fu, W. Hao, and X. Tang. 2011. Improved thermoelectric properties of Al-doped higher manganese silicide prepared by a rapid solidification method. *J. Electron. Mater.* 40: 1233–1237.
157. Chen, X., A. Weathers, A. Moore, J. Zhou, and L. Shi. 2012. Thermoelectric properties of cold-pressed higher manganese silicides for waste heat recovery. *J. Electron. Mater.* 41: 1564–1572.
158. Shi, X., X. Shi, Y. Li, Y. He, L. Chen, and Q. Li. 2014. Enhanced power factor of higher manganese silicide via melt spin synthesis method. *J. Appl. Phys.* 116: 245104.
159. Bernard-Granger, G., M. Soulier, H. Ihou-Mouko et al. 2015. Microstructure investigations and thermoelectrical properties of a p-type polycrystalline higher manganese silicide material sintered from a gas-phase atomized powder. *J. Alloy. Compd.* 618: 403–412.
160. Shin, D.-K., K.-W. Jang, S.-C. Ur, and I.-H. Kim. 2013. Thermoelectric properties of higher manganese silicides prepared by mechanical alloying and hot pressing. *J. Electron. Mater.* 42: 1756–1761.

161. Sadia, Y., L. Dinnerman, and Y. Gelbstein. 2013. Mechanical alloying and spark plasma sintering of higher manganese silicides for thermoelectric applications. *J. Electron. Mater.* 42: 1926–1931.
162. Yoshikura, M. and T. Itoh. 2010. Thermoelectric properties of higher manganese silicide compounds synthesized by MG-PDS method. *J. Jpn. Soc. Powder Powder Metall.* 57: 242–246.
163. Nikitin, E. N., V. I. Tarasov, and T. P. V. 1969. Thermal and electrical properties of the higher manganese silicide from 4.2 to 1300 K and its structure. *Sov. Phys.—Solid State* 11: 187–189.
164. Zaitsev, V. K., S. V. Ordin, K. A. Rakhimov, and A. E. Engalychev. 1981. Characteristics of crystal structure and thermoelectric power of higher manganese silicide. *Sov. Phys. Solid State* 23: 353.
165. Chen, X., A. Weathers, D. Salta et al. 2013. Effects of (Al,Ge) double doping on the thermoelectric properties of higher manganese silicides. *J. Appl. Phys.* 114: 173705.
166. Chen, X., A. Weathers, J. Carrete et al. 2015. Twisting phonons in complex crystals with quasi-one-dimensional substructures. *Nat. Commun.* 6: 6723.
167. Engström, I. and B. Lönnberg. 1988. Thermal expansion studies of the group IV-VII transition-metal disilicides. *J. Appl. Phys.* 63: 4476–4484.
168. Zaitsev, V. K., S. A. Ktitorov, A. E. Kalyazin, N. D. Marchuk, and S. V. Ordin. 1992. Thermal-expansion of the higher manganese silicide. *Fizika tverdogo tela* 34: 2589–2594.
169. Nikitin, E. N., A. F. Sidorov, V. I. Tarasov, and A. I. Zaslavskij. 1970. Higher manganese silicide doping according to the results of microprobe analysis. *Izvestiya AN SSSR* (*Neorganicheskie Materialy*) 6: 604.
170. Zhou, A. J., T. J. Zhu, X. B. Zhao et al. 2010. Improved thermoelectric performance of higher manganese silicides with Ge additions. *J. Electron. Mater.* 39: 2002–2007.
171. Sadia, Y., N. Madar, I. Kaler, and Y. Gelbstein. 2014. Thermoelectric properties of the quasi-binary $MnSi_{1.73}$–$FeSi_2$ system. *J. Electron. Mater.* 44: 1637–1643.
172. Miyazaki, Y., Y. Saito, K. Hayashi, K. Yubuta, and T. Kajitani. 2011. Preparation and thermoelectric properties of a chimney-ladder $(Mn_{1-x}Fe_x)Si_\gamma$ (γ~1.7) solid solution. *Jpn. J. Appl. Phys.* 50: 035804.
173. Abrikosov, N. K., L. D. Ivanova, and L. V. Gromova. 1973. Determination of the distribution coefficients of Cr, Fe and Ge in higher manganese silicide. *Izvestiya AN SSSR* (*Neorganicheskie Materialy*) 9: 489.
174. Kikuchi, Y., Y. Miyazaki, Y. Saito, K. Hayashi, K. Yubuta, and T. Kajitani. 2012. Enhanced thermoelectric performance of a chimney-ladder $(Mn_{1-x}Cr_x)Si_\gamma$ (γ ~1.7) solid solution. *Jpn. J. Appl. Phys.* 51: 085801.
175. Barczak, S. A., R. A. Downie, S. R. Popuri, R. Decourt, M. Pollet, and J. W. G. Bos. 2015. Thermoelectric properties of Fe and Al double substituted $MnSi_\gamma$ (γ ~1.73). *J. Solid State Chem.* 227: 55–59.
176. Ponnambalam, V., D. T. Morelli, S. Bhattacharya, and T. M. Tritt. 2013. The role of simultaneous substitution of Cr and Ru on the thermoelectric properties of defect manganese silicides $MnSi_\delta$ ($1.73 < \delta < 1.75$). *J. Alloy. Compd.* 580: 598–603.
177. Truong, D. Y., D. Berthebaud, F. Gascoin, and H. Kleinke. 2015. Molybdenum, tungsten, and aluminium substitution for enhancement of the thermoelectric performance of higher manganese silicides. *J. Electron. Mater.* 44: 3603–3611.
178. Flieher, G., H. Völlenkle, and H. Nowotny. 1968. Neue Abkömmlinge der $TiSi_2$-Struktur. *Monatshefte für Chemie/Chem. Monthly* 99: 2408–2415.
179. Zaitsev, V. K., V. K. Tarasov, and A. Adilbekov. 1975. Metal—Nonmetal transition in compensated higher manganese silicide. *Sov. Phys. Solid State* 17: 370.
180. Allam, A., P. Boulet, and M. C. Record. 2014. Substitutional atom influence on the electronic and transport properties of Mn_4Si_7. *J. Electron. Mater.* 43: 761–773.
181. Zhou, A. J., X. B. Zhao, T. J. Zhu et al. 2010. Microstructure and thermoelectric properties of SiGe-added higher manganese silicides. *Mater. Chem. Phys.* 124: 1001–1005.
182. Chen, X., S. N. Girard, F. Meng et al. 2014. Approaching the minimum thermal conductivity in rhenium-substituted higher manganese silicides. *Adv. Energy Mater.* 4: 1400452.

183. Norouzzadeh, P., Z. Zamanipour, J. S. Krasinski, and D. Vashaee. 2012. The effect of nanostructuring on thermoelectric transport properties of p-type higher manganese silicide $MnSi_{1.73}$. *J. Appl. Phys.* 112: 124308.
184. Luo, W., H. Li, Y. Yan et al. 2011. Rapid synthesis of high thermoelectric performance higher manganese silicide with in-situ formed nano-phase of MnSi. *Intermetallics* 19: 404–408.
185. Perumal, S., S. Gorsse, U. Ail et al. 2015. Low thermal conductivity of endogenous manganese silicide/Si composites for thermoelectricity. *Mater. Lett.* 155: 41–43.
186. Zhou, A. J., X. B. Zhao, T. J. Zhu et al. 2009. Composites of higher manganese silicides and nanostructured secondary phases and their thermoelectric properties. *J. Electron. Mater.* 38: 1072–1077.
187. Zhou, A. J., T. J. Zhu, H. L. Ni, Q. Zhang, and X. B. Zhao. 2008. Preparation and transport properties of $CeSi_2$/HMS thermoelectric composites. *J. Alloy. Compd.* 455: 255–258.
188. Itoh, T. and N. Ono. 2013. Influence of addition of alumina nanoparticles on thermoelectric properties of higher manganese silicide. *MRS Online Proc. Library Archive* 1490: 127–132.
189. Saleemi, M., A. Famengo, S. Fiameni et al. 2015. Thermoelectric performance of higher manganese silicide nanocomposites. *J. Alloy. Compd.* 619: 31–37.
190. Truong, D. Y. N., H. Kleinke, and F. Gascoin. 2014. Thermoelectric properties of higher manganese silicide/multi-walled carbon nanotube composites. *Dalton Trans.* 43: 15092–15097.
191. Umemoto, M., Z. G. Liu, R. Omatsuzawa, and K. Tsuchiya. 2000. Production and characterization of Mn-Si thermoelectric material. *Mater. Sci. Forum* 343(3): 918–923.
192. Aoyama, I., H. Kaibe, L. Rauscher et al. 2005. Doping effects on thermoelectric properties of Higher Manganese Silicides (HMSs, $MnSi_{1.74}$) and characterization of thermoelectric generating module using p-type (Al, Ge and Mo)-doped HMSs and n-type $Mg_2Si_{0.4}Sn_{0.6}$ legs. *Jpn. J. Appl. Phys.* 44: 4275.
193. Fedorov, M. I. 2007. Physical principles of development of thermoelectric materials based on silicon compounds—Author referate (in Russian). Russian State Pedagogical University A. I. Herzen, St. Petersburg, Russia.
194. Sadia, Y. and Y. Gelbstein. 2012. Silicon-rich higher manganese silicides for thermoelectric applications. *J. Electron. Mater.* 41: 1504–1508.
195. Gross, E., M. Riffel, and U. Stöhrer. 1995. Thermoelectric generators made of $FeSi_2$ and HMS: Fabrication and measurement. *J. Mater. Res.* 10: 34–40.
196. Petrova, L. I., L. D. Dudkin, M. I. Fedorov, F. Y. Solomkin, V. K. Zaitsev, and I. S. Eremin. 2002. Physicochemical interaction at the contact of higher manganese silicide with chromium. *Tech. Phys.* 47: 550–554.
197. Riffel, M., E. Gross, and U. Stöhrer. 1995. Electrical contacts for $FeSi_2$ and higher manganese silicide thermoelectric elements. *J. Mater. Sci.: Mater. Electron.* 6: 182–185.
198. Kaibe, H., L. Rauscher, I. Aoyama et al. 2004. Development of the thermoelectric generating modules using silicide. In *Proceedings of the 23rd International Conference on Thermoelectrics, ICT 2004*, Adelaide, South Australia, Australia.
199. Tarantik, K. R., J. D. König, M. Jägle et al. 2015. Thermoelectric modules based on silicides—Development and characterization. *Mater. Today Proc.* 2: 588–595.
200. Sondergaard, M., M. Christensen, K. A. Borup, H. Yin, and B. B. Iversen. 2013. Thermoelectric properties of the entire composition range in $Mg_2Si_{0.9925-x}Sn_xSb_{0.0075}$. *J. Electron. Mater.* 42: 1417–1421.
201. Ioannou, M., G. Polymeris, E. Hatzikraniotis, A. U. Khan, K. M. Paraskevopoulos, and T. Kyratsi. 2013. "Solid-State Synthesis and Thermoelectric Properties of Sb-Doped Mg2Si Materials." J. Electron. Mater. 42 (7): 1827–1834.

6

Clathrate-Based Thermoelectrics

Toshiro Takabatake and Koichiro Suekuni
Hiroshima University

CONTENTS

6.1 Introduction

The term "clathrate" originates from the Latin word *clathratus*, which means "protected by cross bars of a grating." Archetypal compounds are the so-called clathrate hydrates enclosing gaseous or liquid guest species. Intermetallic clathrates are inorganic inclusion compounds structurally related to the hydrates. Over 300 clathrates have been synthesized so far. Among them, clathrates composed of the group 14 elements are regarded as prospective thermoelectric materials that realize the *phonon–glass electron–crystal* (PGEC) concept proposed by G. A. Slack in 1995 [1]. The PGEC materials are expected to have low thermal conductivity κ like a glass and simultaneously low electrical resistivity ρ like a metal. Such a material may achieve a high value of the dimensionless thermoelectric figure of merit, defined as $ZT = \alpha^2 T/\rho\kappa$, if the Seebeck coefficient α is large enough.

Slack mentioned compounds encapsulating guest atoms whose translational positions or angular orientations are not exactly defined [1]. In such a system, the guest atom can "rattle" around the center in the cage. This rattling motion is expected to scatter the low-energy acoustic phonons and, thus, decrease the thermal conductivity to values typically seen in glass. The electrons, on the other hand, can flow through the network of the cage as in a crystalline metal. According to this guiding principle, intermetallic clathrates have been synthesized, and their physical and structural properties have been investigated extensively. Their thermoelectric properties have been reviewed previously in *Thermoelectrics Handbook* and related texts [2–7]. To avoid overlapping the contents of this chapter with the previous works, this review presents a survey of recent works published

TABLE 6.1
Structural Types and Maximum *ZT* Values for Intermetallic Clathrate Compounds

Structural Type	Compounds	Carrier Type	ZT_{max}	Temperature (°C)	Reference
I	$Ba_8Ga_{16-x}Ge_{30+x}$	*p*	0.50	500	[12]
	$Ba_8Ga_{16-x}Ge_{30+x}$	*n*	0.86	677	[13]
	$Ba_8Au_{5.3}Ge_{40.7}$	*p*	0.90	407	[14]
	$Ba_8Ni_{0.31}Zn_{0.52}Ga_{13.06}Ge_{32.2}$	*n*	1.2	727	[15]
	$Ba_8Ga_{15}Si_{31}$	*n*	0.55	627	[16]
	$Ba_8Al_{15}Si_{31}$	*n*	0.42	627	[17]
	$Ba_8Ga_{16-x}Sn_{30+x}$	*p*	0.58	177	[18]
	$Ba_8Ga_{16-x}Sn_{30+x}$	*n*	0.5	177	[18]
VIII	$Ba_8Ga_{16.7}Sn_{28.6}Sb_{0.7}$	*p*	0.98	207	[19]
	$Ba_8Ga_{15.7}Cu_{0.125}Sn_{30.3}$	*n*	1.45	247	[20]
II	$K_8Ba_{16}Ga_{40}Sn_{96}$	*n*	0.93	364	[21]
IX	$Ba_{24}Ga_{15}Ge_{85}$	*n*	1.25	670	[22]

after 2006. The close relation between crystal structures and thermoelectric properties will be discussed on example systems.

In common with other thermoelectric Zintl compounds [8,9], some intermetallic clathrates have an advantage that the carrier type and carrier density can be tuned by adjusting the composition of cage atoms and vacancies. For example, the carrier type and its density in type I clathrate $Ba_8Ga_{16+x}Ge_{30-x}$ were tuned to either *p*- or *n*-type in a wide range (10^{19}–10^{21} cm^{-3}) by adjusting x [10–13]. This tunability may enable us to construct an efficient thermoelectric module using *p*- and *n*-type legs made of one compound. Typical clathrate compounds with high *ZT* values are listed in Table 6.1 [12–22]. Their structural and thermoelectric properties will be described in the following.

6.2 Crystal Structure of Intermetallic Clathrates

Intermetallic clathrates are classified into five types: I, II, III, VIII, and IX [2,4,6,23]; the structures of which are displayed in Figure 6.1 [24]. The general formulae are described, respectively, as A_8E_{46}, $A_{24}E_{136}$, $A_{30}E_{172}$, A_8E_{46}, and $A_{24}E_{100}$. One important subgroup is those where A represents the cationic guests and E represents the cage atoms of group 14 elements Si, Ge, and Sn. Another subgroup with anionic guests, such as Cl, Br, and I, are called inverse clathrates [5]. The cage atoms are partially substituted by E′ atoms of the group 13 elements Al, Ga, and In in order to balance the cation charge transfer to the cage. For example, ternary derivatives of these type I clathrates adopt the chemical formula $A_8E'_{16}E_{30}$ and $A'_8E'_8E_{38}$, in which A is an atom of group 2 (Ba, Sr) and A′ is an atom of group 1 (Na, K, Rb, and Cs).

The guest A atoms in $A_8E'_{16}E_{30}$ are caged in tetrakaidecahedra and dodecahedra, as shown in Figure 6.2. It should be noted that cage atoms E′ and E are distributed on the three crystallographic sites (6*c*, 24*k*, and 16*i*) by a certain rule. X-ray and neutron diffraction experiments proved the preferential occupation of the E′ atom at the 6*c* sites [4,23]. Extended x-ray absorption fine structure studies of $Ba_8Ga_{16}E_{30}$ have revealed that the

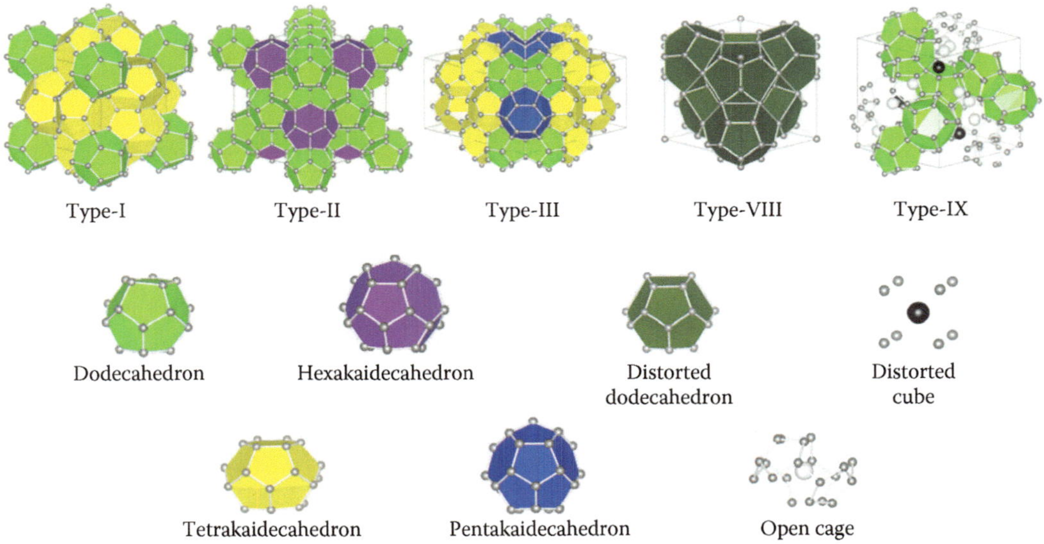

FIGURE 6.1
Crystal structures of intermetallic clathrates: type I A_8E_{46}, type II $A_{24}E_{136}$, type III $A_{30}E_{172}$, type VIII A_8E_{46}, and type IX $A_{24}E_{100}$. (Reprinted with permission from Takabatake, T., Suekuni, K., Nakayama, T., and Kaneshita, E., *Rev. Mod. Phys.*, 86, 669, 2014. Copyright 2014 by the American Physical Society.)

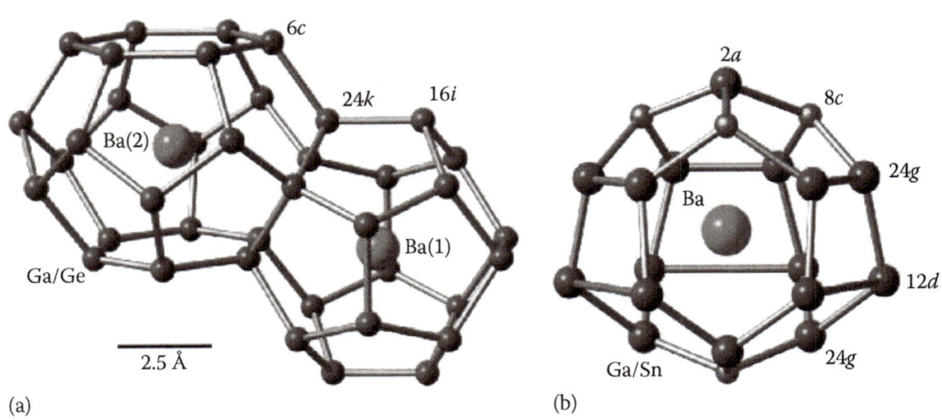

FIGURE 6.2
Atomic configurations of cages and guests in (a) type I and (b) type VIII clathrates. (Modified from Avila, M.A. et al., *Phys. Rev. B*, 74, 125109, 2006.)

Ga–Ga bond is avoided due to electrostatic repulsion so that only 20% of the Ga neighbors are Ga [25]. Because the average pair distances within the cages depend on the bond lengths, the environment of the Ba guest becomes increasingly disordered from E = Ge to Si and to Sn. The local distortions of the cages scatter phonons as well as charge carriers will be shown in Section 6.4.

Among the type I clathrates, only four are known to also have type VIII structure. They are $Ba_8Ga_{16}Sn_{30}$ [26,27], $Eu_8Ga_{16}Ge_{30}$ [28], $Sr_8(Al,Ga)_{16}Si_{30}$ [29], and $Sr_8(Al,Ga)_{16}Ge_{30}$ [30].

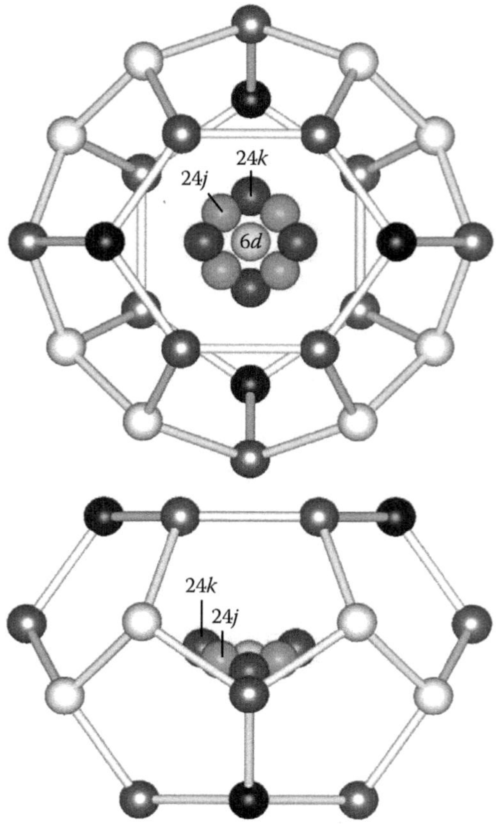

FIGURE 6.3
Split 24*j* and 24*k* sites for the guest atom around 6*d* site in the tetrakaidecahedron of type I $Eu_8Ga_{16}Ge_{30}$ and $Ba_8Ga_{16}Sn_{30}$. The split sites 24*k* are 0.4 Å away from the cage center 6*d*. (Reprinted with permission from Takabatake, T., Suekuni, K., Nakayama, T., and Kaneshita, E., *Rev. Mod. Phys.*, 86, 669, 2014. Copyright 2014 by the American Physical Society.)

The dodecahedron in type I has high symmetry, whereas that in type VIII is distorted, as shown in Figure 6.2. Therefore, the guest atom in the former occupies the center of a dodecahedron, while that in the latter is vibrating nearly on the center. Moreover, the guest(2) in the tetrakaidecahedron of type I occupies sites close to the 6*d* site at the cage center. The occupation at off-centered 24*k* and 24*j* positions occurs due to the disorder of cage atom occupations [23], as shown in Figure 6.3. Definite splitting of the guest site was first revealed in type I $Eu_8Ga_{16}Ge_{30}$ by neutron diffraction experiments [31]. The Eu atom moves off the 6*d* site about 0.4 Å to one of the four equivalent 24*k* positions. This splitting gives the Eu guest atom the freedom to move among the four sites, which brings about a strong interaction with the acoustic phonons [31].

Similar splitting was found in type I $Ba_8Ga_{16}Sn_{30}$ by single-crystal x-ray and neutron diffraction analysis [32,33]. Figure 6.4 shows the probability density function (PDF) for the Ba atoms of type I (a) and type VIII (b) $Ba_8Ga_{16}Sn_{30}$ with both carrier types. The upper one is the PDF of Ba(2) in the cross section at $z = 0.005$, which is the relative distance from the cage center along the hexagonal face of the tetrakaidecahedron. The PDF for both *n*- and

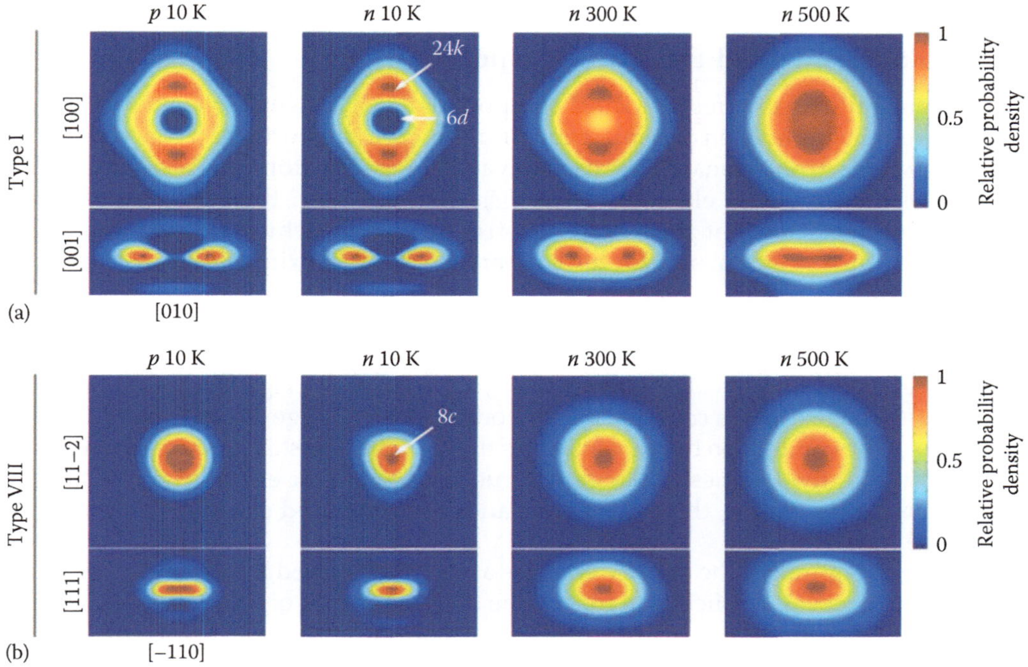

FIGURE 6.4
Probability density function for the Ba atom of type I (a) and type VIII (b) $Ba_8Ga_{16}Sn_{30}$ at different temperature. (From Christensen, S., Avila, M.A., Suekuni, K., Piltz, R.O., Takabatake, T., and Christensen, M., Combined X-ray and neutron diffraction study of vacancies and disorder in the dimorphic clathrate $Ba_8Ga_{16}Sn_{30}$ of type I and VIII, *Dalton Trans.*, 42, 14766–14775, 2013. Reproduced by permission of The Royal Society of Chemistry.)

p-types has the probability density accumulated in the 24*k* site displaced about 0.4 Å from the cage center at the 6*d* site. For type VIII with *n*-type carriers, the anharmonic nature of the Ba guest atom is visible at 10 K as the atomic displacement increases toward the distorted pentagonal faces of the cage. The PDF of the *p*-type at 10 K does not have the directional features but has a shallow local minimum in the center of the PDF, which indicates that Ba is slightly displaced from the 8*c* site. On heating, the Ba guest atoms move away from the 2*a* framework position to the nearest framework site 8*c*. The anharmonic nature of the Ba vibration in type VIII also manifests itself in the increase of the vibration energy by 12% from 4.9 to 5.6 meV on heating from 4 to 300 K [34].

Guest site splitting also occurs in type II and type IX compounds based on Ba–Ga–Sn and Ba–Ga–Ge. In type II compound $(K,Ba)_{24}(Ga,Sn)_{136}$, the guest K^+ and Ba^{2+} ions are preferentially incorporated into the hexakaidecahedra and dodecahedra, respectively [35]. The guest site in the former cage splits into four sites, which is consistent with the presence of four minima in the electrostatic potential in the hexakaidecahedron. In type IX compound $Ba_{24}Ga_yGe_{100-y}$, the Ba guest in the open dodecahedron occupies two split sites [36]. The guest site splitting in type II and type IX compounds allows rattling of the guest among the split sites, which may effectively reduce the lattice thermal conductivity as in type I compounds with off-centered guest rattling.

6.3 Band Structure and Electronic Properties

Most intermetallic clathrates are Zintl compounds [8,9], in which the charge balance is well maintained between the guest ion and the cage network. In ternary clathrates $A_8E'_{16}E_{30}$, two electrons donated by each divalent A guest atom are accepted by two E′ atoms of the group 13 element to form sp^3-like bonding. First-principles calculations have pointed out that the electronic structures of stoichiometric compounds of $A_8Ga_{16}Ge_{30}$ with A = Sr, Ba, and Eu have a bandgap of approximately 0.2–0.9 eV at the Fermi level [37,38].

The band calculations for type I and type VIII $Ba_8Ga_{16}Sn_{30}$ were performed by the full-potential linearized augmented plane wave method [39]. The results showed the presence of indirect gaps of 0.51 and 0.32 eV, respectively, as displayed in Figure 6.5. For both types, the top of the valence bands consists of a sp^3 bonding state of cage atoms Ga and Sn, while the bottom of the conduction bands consists of the antibonding state of the cage atoms and atomic orbitals of the Ba guest. Because the guests donate their electrons to antibonding states of the cage networks, the conduction bands are composed of orbitals of both guest and cage atoms.

As shown in Figure 6.5, the density of states at the band-gap edges has steep slopes that lead to a high Seebeck coefficient α. In fact, the absolute values of α at 300 K are typically around 300 μV K^{-1} for both *p*- and *n*-type samples of type I $Ba_8Ga_{16}Sn_{30}$ [32], as shown in Figure 6.6. It should be noted that the *p*- and *n*-type crystals were grown from Ga and Sn flux, respectively. The values of $\rho(T = 300\ K)$ are in the range 20–40 mΩ cm, which is too large to be used as thermoelectric materials. This large resistivity is caused by the rather low carrier density of 10^{19} cm^{-3} and rather low carrier mobility of 20–40 $cm^2\ V^{-1}\ s^{-1}$. The low mobility is a result of random distortions of the cage atoms [25]. Carrier doping by

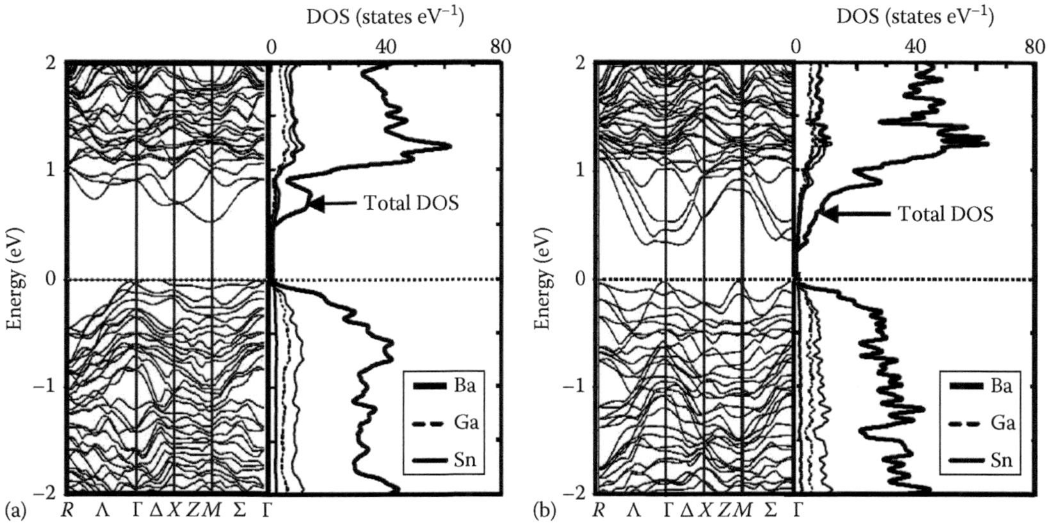

FIGURE 6.5
Electronic band structure for (a) type I and (b) type VIII $Ba_8Ga_{16}Sn_{30}$. (Modified from Kono, Y. et al., *J. Appl. Phys.*, 107, 123720, 2010.)

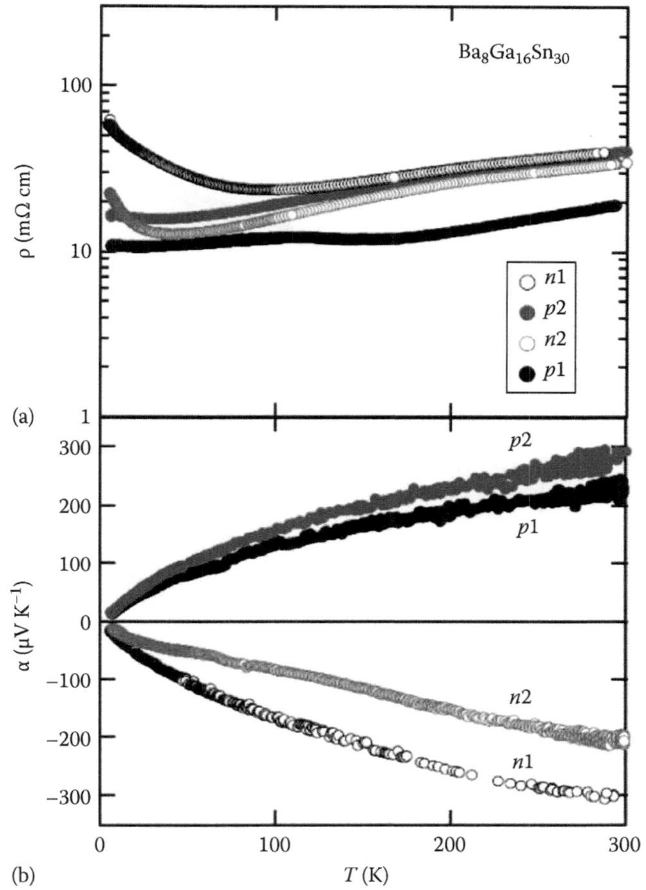

FIGURE 6.6
Temperature dependence of (a) electrical resistivity ρ and (b) Seebeck coefficient α for type I $Ba_8Ga_{16}Sn_{30}$ single crystals. The samples *p*1 and *p*2 are grown from Ga–Sn flux, while *n*1 and *n*2 are grown from Sn flux. (Modified from Suekuni, K. et al., *Phys. Rev. B*, 77, 235119, 2008.)

substitution of cage atoms would be effective to reduce the resistivity of type I material. As for type VIII samples, the thermoelectric properties above 300 K will be presented and discussed in Section 6.5.

6.4 Rattling and Thermal Conductivity

Before going into the "phonon–glass" behavior of the thermal conductivity κ(*T*) in intermetallic clathrates, it is instructive to recall the characteristics of κ(*T*) in a glass. The data of a typical glass, vitreous SiO_2 [40], are displayed in Figure 6.7b. It is characterized by T^2 dependence at $T < 1$ K, a plateau at around 5 K, and roughly *T*-linear dependence above 10 K. The three characteristics are observed in the lattice thermal conductivity $\kappa_L(T)$ for

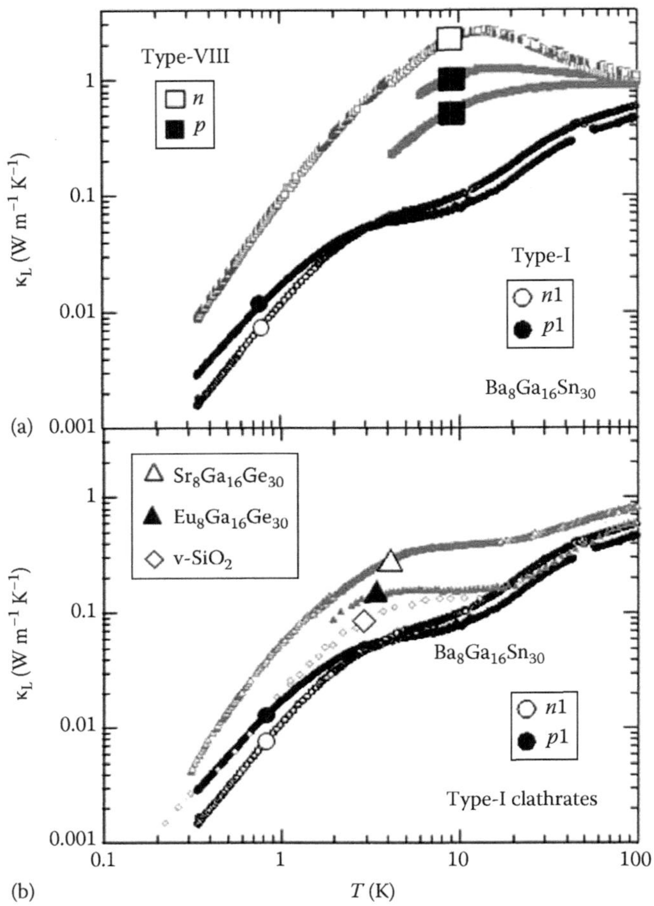

FIGURE 6.7
Temperature dependence of lattice thermal conductivity κ_L for (a) type I [29] and type VIII [39] $Ba_8Ga_{16}Sn_{30}$ and (b) type I $Ba_8Ga_{16}Sn_{30}$, $X_8Ga_{16}Ge_{30}$ (X = Sr [38], Eu [25]), and vitreous SiO_2 [37]. (Modified from Suekuni, K. et al., *Phys. Rev. B*, 77, 235119, 2008.)

type I $Ba_8Ga_{16}Sn_{30}$ [32,41], $Eu_8Ga_{16}Ge_{30}$ [28], and $Sr_8Ga_{16}Ge_{30}$ [42]. Here, κ_L was obtained by subtracting the electronic contribution κ_e from the measured total thermal conductivity κ. It is worth noting that the absolute value of κ_L of type I $Ba_8Ga_{16}Sn_{30}$ at the plateau is actually lower than κ of vitreous SiO_2. By contrast, the $\kappa_L(T)$ for type VIII $Ba_8Ga_{16}Sn_{30}$ with *n*-type carriers shows a peak at around 15 K, which is the behavior of a crystalline material [32,41].

Let us consider the relation between the phonon–glass behavior in $\kappa_L(T)$ and structural properties. For type I $Ba_8Ga_{16}Sn_{30}$, $Eu_8Ga_{16}Ge_{30}$, and $Sr_8Ga_{16}Ge_{30}$, the guest atom in the tetrakaidecahedron occupies four split sites that are 0.3–0.4 Å away from the center, as shown in Figure 6.3. This splitting offers the freedom to move among the four sites. This "off-center rattling" violates the translational symmetry and strongly interacts with the acoustic phonons, which in turn reduces the κ_L. This scenario was first proposed based on the fact that the smaller guest ions lead to smaller $\kappa_L(T)$ among $A_8Ga_{16}Ge_{30}$ with A = Ba^{2+} (1.61 Å), Sr^{2+}(1.44 Å), and Eu^{2+} (1.35 Å) [31]. To examine this model, the $\kappa_L(T)$ data at

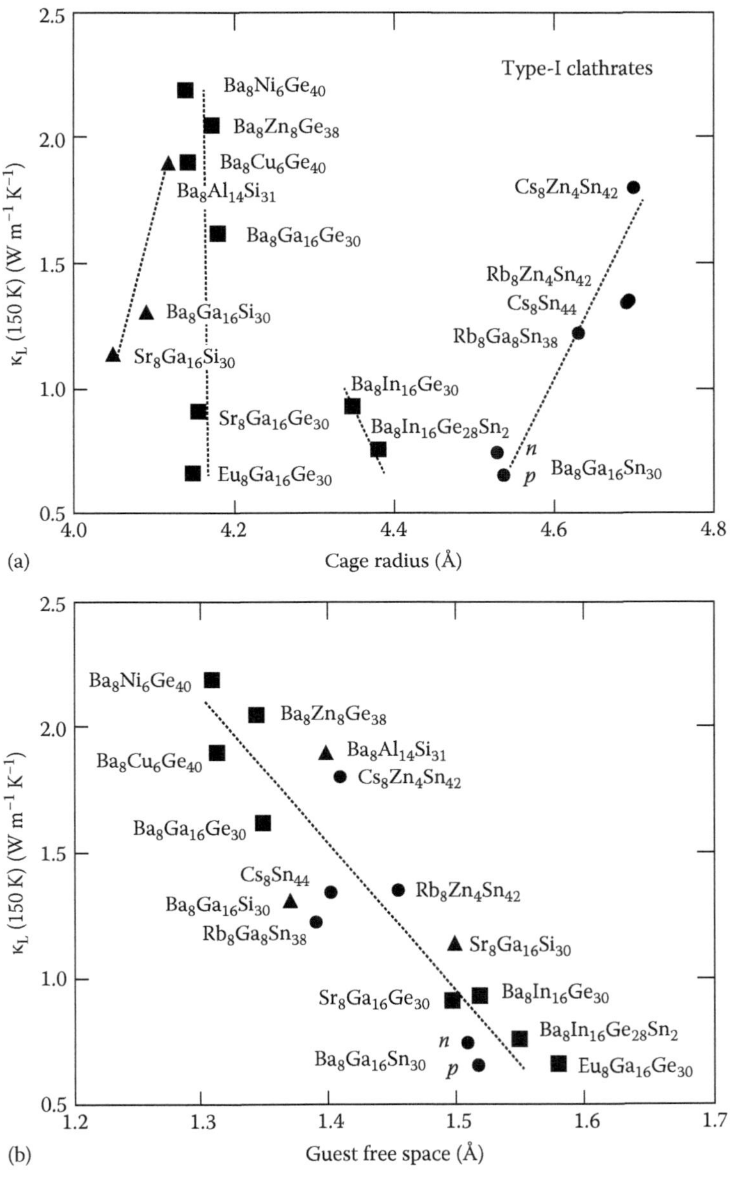

FIGURE 6.8
Lattice thermal conductivity κ_L of various type I clathrates at 150 K as a function of (a) cage radius and (b) guest free space. (Modified from Suekuni, K. et al., *Phys. Rev. B*, 77, 235119, 2008.)

150 K are plotted in Figure 6.8a for various type I clathrates reported against the cage radius [32]. The cage radius R_{cage} of the tetrakaidecahedron is defined as the distance from the 6*d* site at the center to the 24*k* site of the cage. The cage radius increases from 4.1 to 4.6 Å as the majority cage atom changes from Si, Ge to Sn, in the sequence of the increasing covalent radius r_{cage} from 1.11 to 1.22 and to 1.41 Å. Among each series of compounds with Si, Ge, and Sn cages, κ_L decreases with the decrease in the guest ionic

radius, as illustrated by the guidelines in Figure 6.8a. This relation in turn indicates that the relevant structural parameter for κ_L is not the cage radius R_{cage} but the "free space," which is defined as $R_{free} = R_{cage} - r_{cage} - r_{guest}$, where r_{guest} is the ionic radius of the guest. In Figure 6.8b, the data of $\kappa_L(T = 150\ K)$ are replotted against R_{free} [32]. All data fall on a narrow band, indicating that the guest free space is the key parameter that determines the magnitude of κ_L for type I clathrates. Among them, the lowest κ_L is realized in $Ba_8Ga_{16}Sn_{30}$ and $Eu_8Ga_{16}Ge_{30}$ with definite off-center rattling guests. This fact confirms that the off-center rattling effectively scatters the acoustic phonons and reduces the thermal conductivity at low temperatures.

Recently, the glass-like thermal conductivity of off-center rattling systems has been theoretically investigated [24,43]. The dependence of $\kappa_L(T) \propto T^{2-\delta}$ at $T < 1$ K was explained by the two-level tunneling states generated by interacting electric dipoles. The plateau at around 4 K is a direct consequence of the flattening of acoustic modes due to the hybridization with local vibrations of off-center guest atoms. The subsequent *T*-linear rise above the plateau originates from the hopping of local modes assisted by acoustic modes of the network cage. Even in the on-center rattling system, it has been pointed out that the phonon lifetime is strongly reduced in a wide frequency range by the interaction between the guest and cage network [44,45].

6.5 Thermoelectric Properties of Intermetallic Clathrates

In this section, the thermoelectric properties above room temperature are compared among type I clathrates $Ba_8Ga_{16}E_{30}$ (E = Si, Ge, Sn), their substituted compounds, type VIII clathrate $Ba_8Ga_{16}Sn_{30}$, and type II and type IX clathrates based on Ba–Ga–Ge and Ba–Ga–Sn. Compared with Ge- and Sn-based clathrates, the Si-based counterparts have higher stability in air at high temperatures. However, as mentioned in Section 6.4, the smaller cage compared with Ge and Sn cages is unfavorable for reducing κ_L. The carrier-tuned *n*-type samples of type I $Ba_8Ga_{15}Si_{31}$ and $Ba_8Al_{15}Si_{31}$ exhibited the maximum *ZT* values of 0.55 and 0.42 at 627°C [16,17]. As for *p*-type samples, $Rb_8Ga_8Si_{38}$ and $Cs_8Al_8Si_{38}$ were synthesized, but the low power factors α^2/ρ make them less promising [46].

As mentioned in Section 6.1, carrier tuning in $Ba_8Ga_{16+x}Ge_{30-x}$ was performed by adjusting the composition x [10–13]. As shown in Figure 6.9, the sign of α actually changed from negative to positive by increasing the amount of Ga. The *p*-type sample had the minimum value of $\kappa = 0.9\ W\ m^{-1}\ K^{-1}$, which was lower than *n*-type samples. The upswing in $\kappa(T)$ at $T > 500$°C was attributed to the carrier excitations from the top of the valence band to the conduction bands, which is called the bipolar effect [47]. The $ZT = \alpha^2 T/\rho\kappa$ reaches the maximum of 0.50 at 500°C for *p*-type and 0.86 at 677°C for *n*-type [12,13].

Thermoelectric properties of $Ba_8Ga_{16}Ge_{30}$ were improved by substitutions for the framework elements. The Au atom can fully substitute for Ga to form *p*-type samples of $Ba_8Au_{5.3}Ge_{40.7}$, where the Ba atom in the tetrakaidecahedron occupies the 24*k* off-center site [14]. The low κ of $1.0\ W\ m^{-1}\ K^{-1}$, together with metallic conduction, results in the high *ZT* of 0.90 at 407°C. A still higher *ZT* of 1.2 was achieved at 727°C for an *n*-type sample of $Ba_8Ni_{0.31}Zn_{0.52}Ga_{13.06}Ge_{32.2}$ [15]. It was suggested that the cross substitution of Ni and Zn for the cage element Ga induces ionized impurities and lattice defects, which scatter acoustic phonons and thus reduce κ_L.

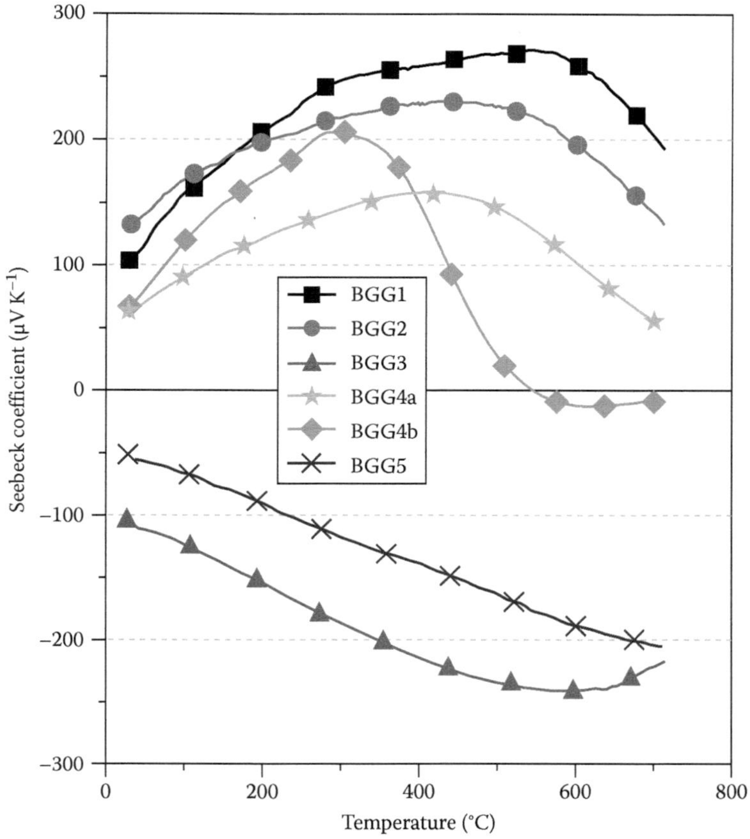

FIGURE 6.9
Seebeck coefficient α as a function of temperature for samples of $Ba_8Ga_{16}Ge_{30}$ prepared from different compositions and heat treatments. (Reprinted with permission from Cederkrantz, D., Saramat, A., Snyder, G.J., and Palmqvist, A.E.C., Thermal stability and thermoelectric properties of *p*-type $Ba_8Ga_{16}Ge_{30}$ clathrates, *J. Appl. Phys.*, 106, 074509. Copyright 2009, American Institute of Physics.)

A systematic carrier tuning was done for type VIII $Ba_8Ga_{16}Sn_{30}$. The growth of single crystals from Ga flux led to *p*-type carriers and that from Sn flux led to *n*-type carriers. Furthermore, the carrier density was tuned by controlling the excess flux amount. In fact, the value of $\alpha(T = 300\ K)$ changes systematically as a function of the excess flux amount, as shown in Figure 6.10 [19]. Here, the sample labeled p-30, for example, means the starting composition of Ga was 30. On the right-hand side of Figure 6.10, the magnitude of α decreases linearly with increasing the excess amount of Ga from 14 (p-30) to 29 (p-45). Meanwhile, the *p*-type carrier density gradually increases, while no significant change in the Ga composition of the crystal was observed. Conversely, for the crystals grown from Sn flux, it was found that both the Sn composition and the *n*-type carrier density gradually increased as the excess amount was increased from 10 (n-40) to 30 (n-60) [19].

For both type I and type VIII $Ba_8Ga_{16}Sn_{30}$, the thermoelectric properties have been assessed above room temperature. The values of κ were 0.4 and 0.7 W m^{-1} K^{-1} at 300 K for type I and type VIII, respectively [18]. Type I, with lower value of κ, seemed to be more attractive for thermoelectric applications. The values of *ZT* increased from 0.37 at 300 K to

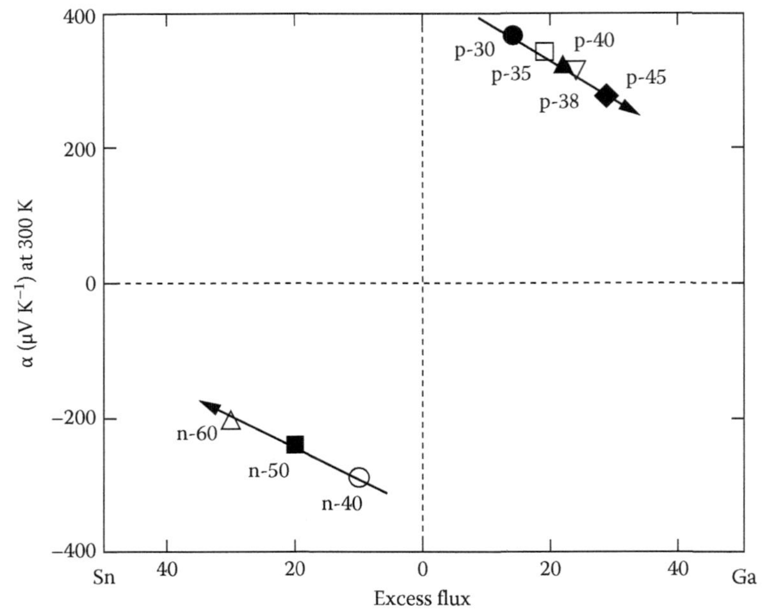

FIGURE 6.10
Seebeck coefficient α at 300 K versus excess flux amounts for single crystals of type VIII $Ba_8Ga_{16}Sn_{30}$. The data for crystals grown from Sn and Ga fluxes are shown in the left and the right, respectively. (Redrawn from Saiga Y. et al., *J. Alloys Compd.*, 507, 1, 2010.)

the broad maxima of 0.58 and 0.50 at around 450 K, respectively, for *p*- and *n*-types [18]. At higher temperatures, however, type I phase became unstable and transformed into type VIII above 739 K [48]. This structural instability makes type I $Ba_8Ga_{16}Sn_{30}$ unfavorable for application at high temperatures. For type VIII, on the other hand, the κ value of 0.7 W m^{-1} K^{-1} at 300 K is still comparable with that of the state-of-the-art materials based on Bi–Te and Pb–Te [8]. A fraction of the reduction in κ originates from the complex unit cell containing a large number of atoms [7]. Another reason is that the probability of umklapp scattering is enhanced by the combination of the rattling modes of the Ba atom at low-energy and acoustic phonon modes [49]. The anharmonic nature of the rattling modes was confirmed by the optical conductivity measurements [34].

The effect of carrier doping is manifested in the gradual decrease in $\rho(T = 300\text{K})$ from above 10 mΩ cm to a few mΩ cm for both *p*- and *n*-types. In this sense, type VIII $Ba_8Ga_{16}Sn_{30}$ is transformed into an electron crystal. The values of *ZT* are shown in Figure 6.11. Among the *p*-type and *n*-type samples, p-40 and n-40 have the maximum *ZT* of 1.0 and 0.90 at around 450 K, respectively [19].

Let us briefly describe the thermoelectric properties of type II and type IX compounds based on Ba–Ga–Sn and Ba–Ga–Ge with guest split sites. In type II compound $(K, Ba)_{24}(Ga, Sn, \square)_{136}$, where □ stands for the vacancy, the off-center rattling of K ions in the hexakaidecahedra was expected to reduce the κ. In fact, κ stays at a low level of 1.0–1.3 W m^{-1} K^{-1} in a temperature range 300–700 K, leading to a rather high *ZT* of 0.93 at 637 K [21]. In type IX compound $Ba_{24}Ga_yGe_{100-y}$, the Ba guest in the open dodecahedron occupies two split sites [36]. Similar to type I clathrates mentioned earlier, the $\kappa(T = 300\text{ K})$ was found to decrease with increasing the off-center displacement. The low κ in $Ba_{24}Ga_{15}Ge_{85}$ resulted in a high value of $ZT = 1.25$ at 943 K [22].

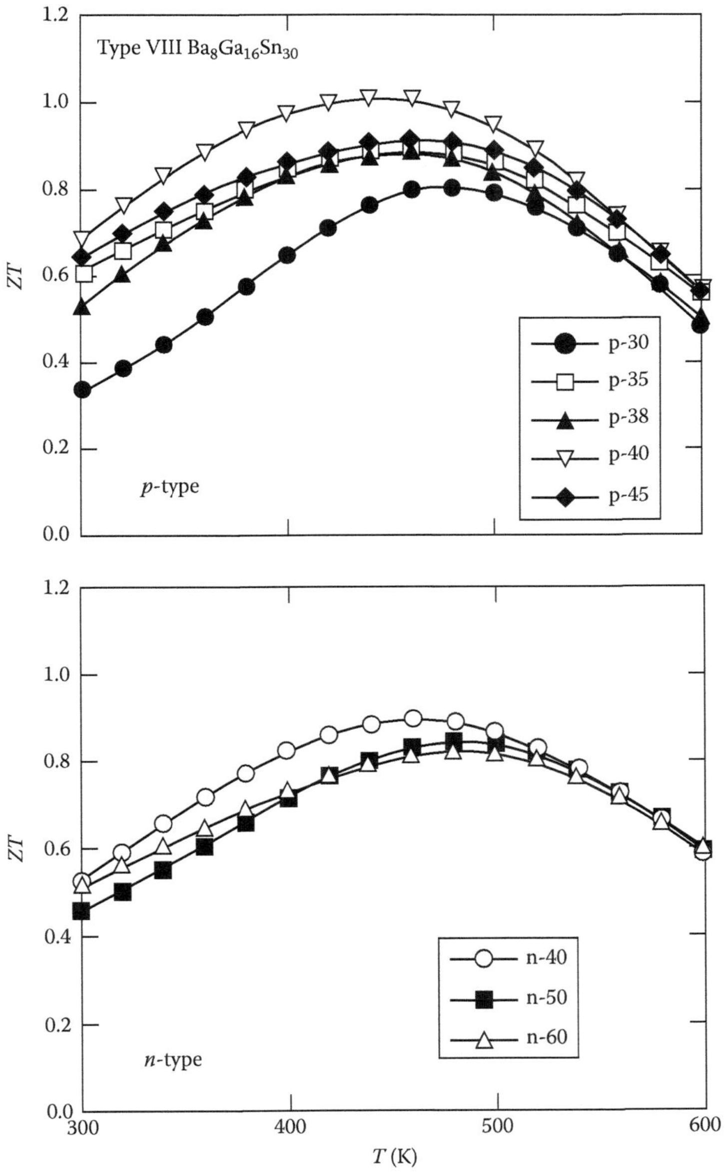

FIGURE 6.11
Temperature dependence of the dimensionless figure of merit *ZT* values for *p*-type (upper) and *n*-type (lower) samples of type VIII $Ba_8Ga_{16}Sn_{30}$. (Redrawn from Saiga Y. et al., *J. Alloys Compd.*, 507, 1, 2010.)

6.6 Substitution Effect on the Figure of Merit of Type VIII $Ba_8Ga_{16}Sn_{30}$

A number of works have reported the improvement of thermoelectric properties of clathrates by substitutions of cage atoms [50]. We concentrate here on the substitutions of type VIII $Ba_8Ga_{16}Sn_{30}$. The substitutes were chosen as isovalent elements Al and In for Ga [51], electron-deficient elements Cu and Zn for Ga [20,52,53], isovalent Ge for Sn [54,55], and electron excess element Sb for Sn [56]. The significant effects of Al and Cu substitutions are described here. Because Ga is classified as a rare element, the substitution by Al has the advantage of reducing the cost, as well as decreasing the weight of the material. The samples of $Ba_8Ga_{16-x}Al_xSn_{30}$ were prepared by the Sn flux method; thus, the dominant carriers were *n*-type [51]. The lattice parameter increases linearly with increasing the Al content x up to the solubility limit of 10.5. The ρ first decreases with increasing x up to 2 and, then, gradually increases for $x > 2$. The absolute value of α changes likewise. The initial decreases in ρ and $|\alpha|$ with x up to 2 are attributed to the increase in the carrier mobility. The κ of all samples is almost the same up to 500 K. For $x = 6$, the upturn in $\kappa(T)$ due to the bipolar effect becomes weaker. As a result, the ZT is largely enhanced to 1.2 at 500 K [51].

Hole doping in type VIII $Ba_8Ga_{16}Sn_{30}$ was tried by substituting Cu, which has two less electrons than Ga [20,52]. However, the solubility of Cu was found to be much lower than that of Al. In the crystals of $Ba_8Ga_{16-x}Cu_xSn_{30}$ grown from Ga flux, x is smaller than 0.014. For crystals grown from Sn flux, y is smaller than 0.125 in $Ba_8Ga_{15.9-y}Cu_ySn_{30.1+y}$, where the Sn composition is self-compensated. Both ρ and κ for *p*-type samples increased, but the carrier density was almost the same, meaning that no holes were doped by Cu substitution. On the other hand, for the *n*-type samples, a more systematic change occurred in both ρ and α. The decrease in ρ was attributed to the doubling of the mobility because the carrier density did not change. The value of κ for the *n*-type sample was lower than for *p*-type samples. Furthermore, the bipolar effect on $\kappa(T)$ at $T > 500$ K was suppressed in the *n*-type samples. By the Cu substitution, all three quantities α, ρ, and κ were decreased by different rates. As a result of the largest reduction in ρ by 50%, the ZT of *n*-type samples was increased to 1.45 at 520 K. This value is the highest among all substituted systems of type VIII $Ba_8Ga_{16}Sn_{30}$ [20].

6.7 Summary

In this chapter, we have described the relation between the structural and thermoelectric properties of intermetallic clathrates. Emphasis was placed on how the concept of PGEC has guided the development of thermoelectric clathrates. Among type I clathrates, $Ba_8Ga_{16}Sn_{30}$ has been proven to be the best example of phonon glass due to the strong interaction between the acoustic phonons and off-center rattling in the tetrakaidecahedron. Type VIII $Ba_8Ga_{16}Sn_{30}$, on the other hand, shows better thermoelectric and structural properties above room temperature. The ZT values of type VIII $Ba_8Ga_{16}Sn_{30}$ and the substituted samples are plotted together with other thermoelectric materials presented in Reference 8. As shown in Figure 6.12, there was a valley in ZT at around 200°C for both *p*-type and *n*-type materials, which had been the bottleneck for applications. Now, this valley is covered by the data for type VIII $Ba_8Ga_{16}Sn_{30}$ and the substituted systems. This material has the advantage that a module can be made from the same compound with both *p*- and *n*-type forms of transport, preventing the module from degrading by thermal hysteresis. It is worthy to note that the weak point associated with the low melting point of 495°C has been overcome by the segmentation with type I $Ba_8Ga_{16}Ge_{30}$, which has a higher melting point of 960°C. In fact, this segmented structure has achieved a rather high conversion efficiency of 7.5% for the temperature difference of 570°C [57].

We have shown how the concept of PGEC has proven effective for developing thermoelectric clathrates with rattling guests. Recently, this route has been extended to

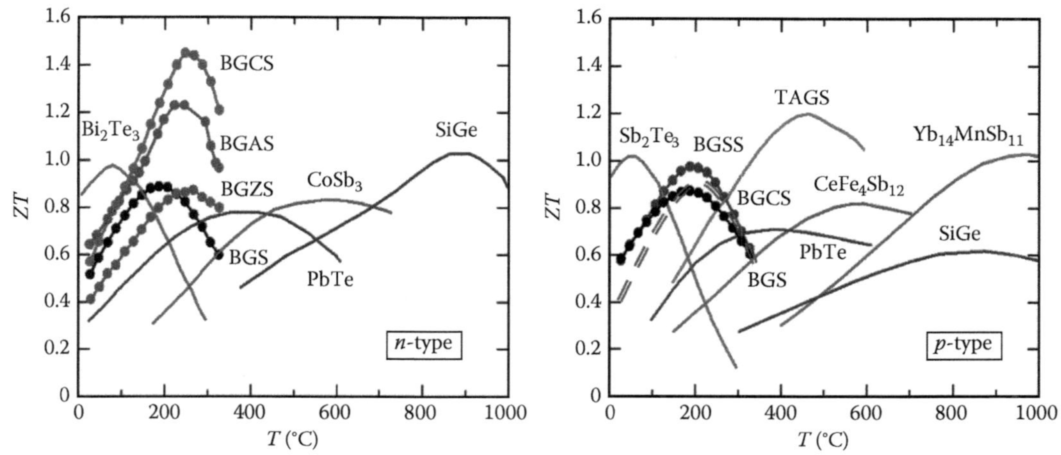

FIGURE 6.12
Temperature dependence of the thermoelectric dimensionless figure merit ZT for type VIII $Ba_8Ga_{16}Sn_{30}$ (BGS), $Ba_8Ga_{15.8-2x}Zn_xSn_{30.2+x}$ (BGZS), $Ba_8Ga_{16-x}Al_xSn_{30y}$ (BGAS), $Ba_8Ga_{16-x}Cu_xSn_{30}$ (BGCS), and $Ba_8Ga_{16+x}Sn_{30-x-y}Sb_y$ (BGSS) plotted with the data of conventional materials. (From Snyder, G.J. and Toberer, E.S., *Nat. Mater.*, 7, 105, 2008.)

complex structure compounds with disordered atomic positions, such as tetrahedrites $Cu_{12-x}M_xSb_4S_{13}$ (M is the transition metal) with Cu rattlers in trigonal bipyramids [58,59] and the $Na_{2+x}Ga_{2+x}Sn_{4-x}$ system with Na rattlers in helical tunnels [60]. This fact indicates that materials with low-energy rattling among multiple positions, even in noncaged structures, have inherently low thermal conductivity and thus can be good candidates for thermoelectric materials.

Acknowledgments

The authors gratefully acknowledge D. Huo, M. A. Avila, S. K. Deng, B. Du, Y. Saiga, and Y. X. Chen for their fruitful collaboration in the study of thermoelectric clathrates. They thank E. Kaneshita and T. Nakayama for scientific discussions and M. A. Avila for a careful review of this chapter. This work was supported by a NEDO grant no. 09002139-0, Grants-in-Aid for Scientific Research from MEXT of Japan with grant no. 20102004, and the International Joint Research Program for Innovative Energy Technology funded by METI, Japan.

References

1. Slack, G. A. 1995. New materials and performance limits for thermoelectric cooling. In *CRC Handbook of Thermoelectrics*, ed. D.M. Rowe, Chapter 34, CRC Press, Boca Raton, FL.
2. Rogl, P. 2006. Formation and crystal chemistry of clathrates. In *Thermoelectrics Handbook: Macro to Nano*, ed. D.M. Rowe, Chapter 32, CRC Press, Boca Raton, FL.

3. Nolas, G. S. 2006. Structure, thermal conductivity, and thermoelectric properties of clathrate compounds. In *Thermoelectrics Handbook: Macro to Nano*, ed. D.M. Rowe, Chapter 33, CRC Press, Boca Raton, FL.
4. Iversen, B. B. 2006. Structural studies of thermoelectric materials. In *Thermoelectrics Handbook: Macro to Nano*, ed. D.M. Rowe, Chapter 21, CRC Press, Boca Raton, FL.
5. Falmbigl, M. and P. Rogl. 2012. Thermoelectric inverse clathrates. In *Thermoelectrics and Its Energy Harvesting: Modules, Systems, and Applications in Thermoelectrics*, ed. D. M. Rowe, Chapter 9, CRC Press, Boca Raton, FL.
6. Shevelkov, A. V. and K. Kovnir. 2011. Zintl clathrates. *Struct. Bond.* 139: 97–142.
7. Nolas G. S., ed. 2014. *The Physics and Chemistry of Inorganic Clathrates*, Springer, New York.
8. Snyder, G. J. and E. S. Toberer. 2008. Complex thermoelectric materials. *Nat. Mater.* 7: 105–114.
9. Toberer, E. S., A. F. May, and G. J. Snyder. 2010. Zintl chemistry for designing high efficiency thermoelectric materials. *Chem. Mater.* 22: 624–634.
10. Bentien, A., M. Christensen, J. D. Bryan et al. 2004. Thermal conductivity of thermoelectric clathrates. *Phys. Rev. B* 69: 045107.
11. Avila, M. A., K. Suekuni, K. Umeo, H. Fukuoka, S. Yamanaka, and T. Takabatake. 2006. Glasslike versus crystalline thermal conductivity in carrier-tuned $Ba_8Ga_{16}X_{30}$ clathrates (X = Ge, Sn). *Phys. Rev. B* 74: 125109.
12. Cederkrantz, D., A. Saramat, G. J. Snyder, and A. E. C. Palmqvist. 2009. Thermal stability and thermoelectric properties of *p*-type $Ba_8Ga_{16}Ge_{30}$ clathrates. *J. Appl. Phys.* 106: 074509.
13. May, A. F., E. S. Toberer, A. Saramat, and G. J. Snyder. 2009. Characterization and analysis of thermoelectric transport in *n*-type $Ba_8Ga_{16-x}Ge_{30+x}$. *Phys. Rev. B* 80: 125205.
14. Zhang, H., H. Borrmann, N. Oeschler et al. 2011. Atomic interactions in the *p*-type clathrate I $BaAu_{5.3}Ge_{40.7}$. *Inorg. Chem.* 50: 1250–1257.
15. Shi, X., J. Yang, S. Bai et al. 2010. On the design of high-efficiency thermoelectric clathrates through a systematic cross-substitution of framework elements. *Adv. Funct. Mater.* 20: 755–763.
16. Anno, H., H. Yamada, T. Nakabayashi, M. Hokazono, and R. Shirataki. 2012. Compositional dependence of thermoelectric properties in polycrystalline type-I $Ba_8Ga_xSi_{46-x}$ (nominal x = 14–18). *AIP Conf. Proc.* 1449: 259–262.
17. Anno, H., M. Hokazono, R. Shirataki, and Y. Nagami. 2013. Crystallographic, thermoelectric, and mechanical properties of polycrystalline type-I $Ba_8Al_{16}Si_{30}$-based clathrates. *J. Mater. Sci.* 48: 2846–2854.
18. Saiga, Y., K. Suekuni, B. Du, and T. Takabatake. 2012. Thermoelectric properties and structural instability of type-I clathrate $Ba_8Ga_{16}Sn_{30}$ at high temperatures. *Solid State Commun.* 152: 1902–1905.
19. Saiga Y., K. Suekuni, S. K. Deng et al. 2010. Optimization of thermoelectric properties of type-VIII clathrate $Ba_8Ga_{16}Sn_{30}$ by carrier tuning. *J. Alloys Compd.* 507: 1–5.
20. Saiga, Y., B. Du, S. K. Deng, K. Kajisa, and T. Takabatake. 2012. Thermoelectric properties of type-VIII clathrate $Ba_8Ga_{16}Sn_{30}$ doped with Cu. *J. Alloys Compd.* 537: 303–307.
21. Koda, S., K. Kishimoto, K. Akai, H. Asada, and T. Koyanagi. 2014. Thermoelectric and transport properties of sintered *n*-type $K_8Ba_{16}Ga_{40}Sn_{96}$ with type-II clathrate structure. *J. Appl. Phys.* 116: 023710.
22. Kim, J. H., N. L. Okamoto, K. Kishida, K. Takata, and H. Inui. 2006. High thermoelectric performance of type-III clathrate compounds of the Ba-Ge-Ga system. *Acta Mater.* 54: 2057–2062.
23. Christensen, M., S. Johnsen, and B. B. Iversen. 2010. Thermoelectric clathrates of type I. *Dalton Trans.* 39: 978–992.
24. Takabatake, T., K. Suekuni, T. Nakayama, and E. Kaneshita. 2014. Phonon-glass electron-crystal thermoelectric clathrates: Experiments and theory. *Rev. Mod. Phys.* 86: 669–716.
25. Keiber, T., P. Nast, S. Medling et al. 2015. Comparison of local distortion in $Ba_8Ga_{16}X_{30}$ (X= Si, Ge, Sn): An EXAFS study. *J. Mater. Chem. C* 3: 10574–10582.
26. Eisenmann, B., H. Schäfer, and R. Zagler. 1986. Die Verbindungen $A^{II}{}_8B^{III}{}_{16}B^{IV}{}_{30}$ ($A^{II} \equiv$ Sr, Ba; $B^{III} \equiv$ Al, Ga; $B^{IV} \equiv$ Si, Ge, Sn) und ihre Käfigstrukturen. *J. Less-Common Met.* 118: 43–55.

27. von Schnering, H. G., W. Carrillo-Cabrera, R. Kröner et al. 1998. Crystal structure of the clathrate β-$Ba_8Ga_{16}Sn_{30}$. *Z. Kristallogr. New Cryst. Struct.* 213: 679.
28. Paschen, S., W. Carrillo-Cabrera, A. Bentien et al. 2001. Structural, transport, magnetic, and thermal properties of $Eu_8Ga_{16}Ge_{30}$. *Phys. Rev. B* 64: 214404.
29. Kishimoto, K., N. Ikeda, K. Akai, and T. Koyanagi. 2008. Synthesis and thermoelectric properties of silicon clathrates $Sr_8Al_xGa_{16-x}Si_{30}$ with the type-I and type-VIII Structures. *Appl. Phys. Express* 1: 031201.
30. Sasaki, Y., K. Kishimoto, T. Koyanagi, H. Asada, and K. Akai, 2009. Synthesis and thermoelectric properties of type-VIII germanium clathrates $Sr_8Al_xGa_yGe_{46-x-y}$. *J. Appl. Phys.* 105: 073702.
31. Sales, B. C., B. C. Chakoumakos, R. Jin, J. R. Thompson, and D. Mandrus. 2001. Structural, magnetic, thermal, and transport properties of $X_8Ga_{16}Ge_{30}$ (X = Eu, Sr, Ba) single crystals. *Phys. Rev. B* 63: 245113.
32. Suekuni, K., M. A. Avila, K. Umeo et al. 2008. Simultaneous structure and carrier tuning of dimorphic clathrate $Ba_8Ga_{16}Sn_{30}$. *Phys. Rev. B* 77: 235119.
33. Christensen, S., M. A. Avila, K. Suekuni, R. O. Piltz, T. Takabatake, and M. Christensen. 2013. Combined X-ray and neutron diffraction study of vacancies and disorder in the dimorphic clathrate $Ba_8Ga_{16}Sn_{30}$ of type I and VIII. *Dalton Trans.* 42: 14766–14775.
34. Iwamoto, K., T. Mori, S. Kajitani et al. 2013. Optical conductivity spectra of rattling phonons and charge carriers in the type-VIII clathrate $Ba_8Ga_{16}Sn_{30}$. *Phys. Rev. B* 88: 104308.
35. Mano, S., T. Onimaru, S. Yamanaka, and T. Takabatake. 2011. Off-center rattling and thermoelectric properties of type-II clathrate $(K, Ba)_{24}(Ga, Sn, \square)_{136}$ single crystals. *Phys. Rev. B* 84: 214101.
36. Okamoto, N. L., J. H. Kim, K. Tanaka, and H. Inui. 2006. Splitting of guest atom sites and lattice thermal conductivity of type-I and type-III clathrate compounds in the Ba-Ga-Ge system. *Acta Mater.* 54: 5519–5528.
37. Madsen, G. K. H., K. Schwarz, P. Blaha, and D. J. Singh. 2003. Electronic structure and transport in type-I and type-VIII clathrates containing strontium, barium, and europium. *Phys. Rev. B* 68: 125212.
38. Nenghabi, E. N. and C. W. Myles. 2008. First principles calculations of the structural and electronic properties of the type-I semiconductor clathrate alloys $Ba_8Ga_{16}Si_xGe_{30-x}$ and $Sr_8Ga_{16}Si_xGe_{30-x}$. *Phys. Rev. B* 77: 205203.
39. Kono, Y., N. Ohya, T. Taguchi et al. 2010. First-principles study of type-I and type-VIII $Ba_8Ga_{16}Sn_{30}$ clathrates. *J. Appl. Phys.* 107: 123720.
40. Zeller, R. C. and R. O. Pohl. 1971. Thermal conductivity and specific heat of noncrystalline solids. *Phys. Rev. B* 4: 2029–2041.
41. Avila, M. A., K. Suekuni, K. Umeo, H. Fukuoka, S. Yamanaka, and T. Takabatake. 2008. $Ba_8Ga_{16}Sn_{30}$ with type-I clathrate structure: Drastic suppression of heat conduction. *Appl. Phys. Lett.* 92: 041901.
42. Umeo, K., M. A. Avila, T. Sakata, K. Suekuni, and T. Takabatake. 2005. Probing glasslike excitations in single-crystalline $Sr_8Ga_{16}Ge_{30}$ by specific heat and thermal conductivity. *J. Phys. Soc. Jpn.* 74: 2145–2148.
43. Nakayama, T. and E. Kaneshita. 2011. Significance of off-center rattling for emerging low-lying THz modes in type-I clathrates. *J. Phys. Soc. Jpn.* 80: 104604.
44. Pailhès, S., H. Euchner, V. M. Giordano et al. 2014. Localization of propagative phonons in a perfectly crystalline solid. *Phys. Rev. Lett.* 113: 025506.
45. Tadano, T., Y. Gohda, and S. Tsuneyuki. 2015. Impact of ratters on thermal conductivity of a thermoelectric clathrate: A first-principles study. *Phys. Rev. Lett.* 114: 095501.
46. Sui, F., H. He, S. Bobev et al. 2015. Synthesis, structure, thermoelectric properties, and band gaps of alkali metal containing type-I clathrates: $A_8Ga_8Si_{38}$ (A = K, Rb, Cs) and $K_8Al_8Si_{38}$. *Chem. Mater.* 27: 2812–2820.
47. Goldsmid, H. J. 2010. *Introduction to Thermoelectricity*, Springer, Heidelberg, Germany.
48. Du, B., Y. Saiga, K. Kajisa, T. Takabatake, E. Nishibori, and H. Sawa. 2012. Study of $\alpha \leftrightarrow \beta$ transformation in the dimorphic clathrate $Ba_8Ga_{16}Sn_{30}$. *Philos. Mag.* 92: 2541–2552.

49. Lee, C. H., I. Hase, H. Sugawara, H. Yoshizawa, and H. Sato. 2006. Low-lying optical phonon modes in the filled skutterudite $CeRu_4Sb_{12}$. *J. Phys. Soc. Jpn.* 75: 123602.
50. Stefanoski, S., M. Beekman, and G. S. Nolas. 2014. Inorganic clathrates for thermoelectric applications. In *The Physics and Chemistry of Inorganic Clathrates*, ed. G.S. Nolas, Chapter 6, Springer, New York.
51. Deng, S., Y. Saiga, K. Suekuni, and T. Takabatake. 2010. Enhancement of thermoelectric efficiency in type-VIII clathrate $Ba_8Ga_{16}Sn_{30}$ by Al substitution for Ga. *J. Appl. Phys.* 108: 073705.
52. Deng, S., Y. Saiga, K. Kajisa, and T. Takabatake. 2011. High thermoelectric performance of Cu substituted type-VIII clathrate $Ba_8Ga_{16-x}Cu_xSn_{30}$ single crystal. *J. Appl. Phys.* 109: 103704.
53. Du, B., Y. Saiga, K. Kajisa, and T. Takabatake. 2012. Thermoelectric performance of Zn-substituted type-VIII clathrate $Ba_8Ga_{16}Sn_{30}$ single crystals. *J. Appl. Phys.* 111: 013707.
54. Kishimoto K., H. Yamamoto, K. Akai, and T. Koyanagi. 2012. Effect of Ge substitution on carrier mobilities and thermoelectric properties of sintered *p*-type $Ba_8Ga_{16+x}Sn_{30-x-y}Ge_y$ with the type-VIII clathrate structure. *J. Phys. D: Appl. Phys.* 45: 445306.
55. Chen, Y. X., B. Du, Y. Saiga, K. Kajisa, and T. Takabatake. 2013. Crystal growth and thermoelectric properties of type-VIII clathrate $Ba_8Ga_{15.9}Sn_{30.1-x}Ge_x$ with *p*-type charge carriers. *J. Phys. D: Appl. Phys.* 46: 205302.
56. Kono, Y., N. Ohya, Y. Saiga et al. 2011. Carrier doping in the type VIII clathrate $Ba_8Ga_{16}Sn_{30}$ through Sb substitution. *J. Electron. Mater.* 40: 845–850.
57. Kishimoto, K., Y. Tsuchiya, T. Koyanagi et al. 2015. Segmented thermoelectric modules using type-VIII Ba-Ga-Sn and type-I Ba-Ga-Ge clathrate compounds. *Abstract of 34th Annual International Conference on Thermoelectrics* (*ICT 2015*), Vol. 7A.3.
58. Suekuni, K., K. Tsuruta, M. Kunii et al. 2013. High-performance thermoelectric mineral $Cu_{12-x}Ni_xSb_4S_{13}$. *J. Appl. Phys.* 113: 043712.
59. Lu. X., D. M. Morelli, Y. Xia et al. 2013. High performance thermoelectricity in earth-abundant compounds based on natural mineral tetrahedrites. *Adv. Energy. Mater.* 3: 342–348.
60. Yamada, T., H. Yamane, and H. Nagai. 2015. A thermoelectric Zintl phase $Na_{2+x}Ga_{2+x}Sn_{4-x}$ with disordered Na atoms in helical tunnels. *Adv. Mater.* 27: 4708–4713.

7

Advances in Nanostructured Half-Heusler Alloys for Thermoelectric Applications

Pierre F.P. Poudeu, Ruiming Lu, Yuanfeng Liu, Pranati Sahoo, and Alexander Page
University of Michigan

CONTENTS

7.1 Introduction

Energy production from clean, renewable, and sustainable sources with efficient energy conversion is necessary to tackle the fast continuing growth of the global demand for energy. For instance, there is an increasing demand for low-power consumption, high-performance miniaturized devices such as cellular phones, digital cameras, and MP3 players, which is driving research activities to develop nonvolatile memory devices and ultrahigh-density information storage spin-based devices for computing [1,2] and spintronic [3] applications. In addition, substantial increases in research activities in the areas of photovoltaic and thermoelectric energy production technologies over the past six decades have led to significant improvements in materials performance. However, the development of thermoelectric materials with the figure of merit $ZT \geq 3$ ($ZT = T\sigma S^2/\kappa$, where σ is the electrical conductivity; S is the Seebeck coefficient, often called the thermopower; and κ is the thermal conductivity) necessary for the realization of cost-effective solid-state energy conversion devices capable of competing with traditional mechanical energy conversion systems has proven extremely difficult, mainly because of the interdependence and coupling between electronic and thermal parameters. In conventional semiconductors, the electrical conductivity (σ) and thermopower (S) are fundamentally adversely coupled through the concentration of charge carriers, n. For instance, in most good thermoelectric materials,

S is inversely proportional to $n^{2/3}$, while σ is proportional to n. Therefore, the maximization of one parameter by tuning the carrier concentration (n) through doping and/or substitutional chemistry inevitably results in the minimization of the other.

Among the various strategies employed to enhance the ZT of traditional and emerging thermoelectric materials the concept of nanostructuring as an effective means to achieve drastic reductions in the lattice thermal conductivity (κ_L) while minimizing reductions in the power factor (PF) has dominated the thermoelectric research over the past 15 years [4–11]. The mechanism by which phonons are scattered at matrix/inclusion interfaces in various nanocomposite systems is now better understood and has been intensively discussed in numerous excellent reviews [5–6,8,12,13]. However, the effect of the embedded inclusions on the electronic properties and the mechanism of the carriers' transport at the matrix/inclusion interfaces in thermoelectric nanocomposite materials remains poorly understood and is therefore generally ignored in most published work. Unfortunately, ZT enhancement through large reductions in κ using nanostructuring so far has resulted in an overall maximum $ZT < 2$ in most material systems (except possibly for the Bi_2Te_3/Sb_2Te_3 superlattices with their ZT ~2.3 at 300 K) [7,14–27]. Further enhancements to reach the targeted goal of $ZT \geq 3$ must now come from drastic increases in the PF simultaneously with significant decreases in κ. It is therefore essential to gain a better understanding of the mechanism of the charge carriers' transport in various nanostructured thermoelectric materials. This will be the main topic of discussion in this chapter. We will limit our discussion to electronic and thermal transport in bulk nanostructured half-Heusler (HH) composites consisting of an HH matrix containing various types of *in situ* grown second phases.

7.2 Half-Heusler Alloys as Promising Thermoelectric Materials

The HH alloys represent a fascinating large class of intermetallic phases with general composition TMX (T and M are transition metal, or rare-earth metals, and X is a main-group element) that crystallize in the MgAgAs structure type (space group [SG]: $F\bar{4}3m$). Their structure can be described as consisting of four interpenetrating face-centered cubic sublattices of T, M, and X atoms occupying, respectively, the crystallographic positions (0,0,0), (1/4, 1/4, 1/4), and (1/2, 1/2,1/2), whereas the fourth position (3/4, 3/4, 3/4) is vacant [28,29]. Materials parameters, such as the crystal structure type, the stoichiometry, the nature of the constituent elements, and the valence electron concentration (VEC), dictate the physical and chemical properties of HH compounds. With careful selection of T, M, and X atoms to satisfy a VEC = 18, HH phases with semiconducting behavior suitable for thermoelectric applications can be fabricated. Upon varying the chemical composition, one can tune the electronic bandgaps (e.g., ~0–0.4 eV), as well as other electronic and thermal properties of HH compounds [30–38]. This unique feature has led to a surge of research interest for the discovery of new physics and the development of multifunctional materials and devices for various important technological applications. For example, HH compounds have attracted tremendous attention as potential candidates for thermoelectrics [24,30–33,35–37,39–53], as half-metallic dilute magnetic semiconductors (spintronics) [54–57], and as topological insulators [58–60].

Among HH phases, (Ti, Hf, Zr) NiSn and (Ti, Hf, Zr) CoSb are the most popular compositions investigated so far for high-temperature power generation. These compositions attracted tremendous attention not only because the elements involved are cheap, abundant, lightweight, and environmentally friendly but also due to their combination of high Seebeck coefficients with moderately low electrical resistivities [32,34,37,43,52,53,61–63], which give rise to intrinsically high PFs. However, the very large thermal conductivities associated with these HH compositions limit our ability to synthesize materials with decent figures of merit ($ZT > 1$). Significant research efforts have been invested in the optimization of the thermoelectric figure of merit of HH alloys.

Earlier strategies consisted of optimizing the PF ($PF = \sigma S^2$) through substitutional doping at the Sn (Sb) sites, as observed in (Ti, Hf, Zr) $NiSn_{1-x}M_x$ (M = Sb, Bi) [46,48,61] and (Ti, Hf, Zr)$CoSb_{1-x}M_x$ (M = Sn, Ge) [43,64], and reducing the thermal conductivity using mass-fluctuation phonon scattering *via* solid-solution alloying of atoms with different masses within various atomic positions in the HH structure [34,61,62] or phonon scattering at grain boundaries in nanometer-sized polycrystalline samples fabricated by mechanical alloying [34,44,65,66]. The former approach generally induced large enhancements in the electrical conductivity due to large increases in the carrier density. This often leads to a reduction in the thermopower and an increase in the electronic contribution to the thermal conductivity. The latter approach also generally results in a large drop in the PF due to a large decrease in the carrier mobility arising from increased scattering of charge carriers at the grain boundaries. Therefore, marginal enhancements in the figure of merit are generally observed. To achieve meaningful improvements in the thermoelectric figure of merit of HH-based materials, electronic and thermal transport must be decoupled in order to achieve on one hand, improvement of the PF without increasing the total thermal conductivity, or to drastically reduce the total thermal conductivity without sacrificing the electronic properties on the other hand. To realize such independent manipulation of the electronic and thermal properties in HH materials, novel strategies consisting of embedding various types of the nanostructures in the bulk HH matrix have been implemented [39–42,49,67–73]. The resulting nanocomposites generally exhibit thermal, electronic, and mechanical properties that are markedly different from those of the pristine HH matrix. These properties are strongly influenced by the internal structure, at various length scales (micrometer, nanometer, and atomic scale), of the resulting composites which, in turn, is controlled by the materials processing parameters. For example, the matrix/inclusion interfaces in the composites can be coherent, semicoherent, or incoherent depending on the structural relations between the HH matrix and the embedded inclusion, as well as the synthesis and processing approach used. The nature and complexity of matrix/inclusion interfaces also depend on the solid-state phase transformation mechanism, such as spinodal decomposition [19,74], nucleation and growth [7,16,18,75], or matrix encapsulation [76,77], governing the interaction between the nanoinclusions in the bulk matrix. Although acoustic phonons are effectively scattered by all three types of interfaces [4,7,16,18,19,75,76,78,79], the carrier mobility across the interfaces strongly depends on the lattice coherency. The electronic, thermal, and mechanical properties of these composites are also affected by the volume fraction of the inclusion phases; the size, shape, and distribution of the nanoinclusions within the matrix; and the charge transfer at the semiconductor/semimetal/metal interfaces, which is strongly energy dependent [77,80].

Recent seminal works on nanocomposite HH-based materials can be grouped into three main categories: (1) bulk nanostructured HH composites containing *in situ* grown inclusions, (2) bulk HH nanocomposites formed by physical mixing of presynthesized HH matrix and various nanoinclusions, and (3) nanobulk HH materials obtained by grain size refinement of the single-phase HH matrix using mechanical alloying. These efforts led to significant improvement in the thermoelectric properties when compared to the pristine

HH matrix. However, the observed enhancement in the figure of merit is generally attributed to a large reduction in the lattice thermal conductivity arising from enhanced phonon scattering at interfaces [7,14,16,18], ignoring or neglecting the impact of the nanostructure on the electronic properties. In this chapter, we will attempt to discuss correlations between the microstructure and observed electronic properties in selected examples of HH-based composites containing *in situ* grown full-Heusler (FH) inclusions.

7.3 Bulk Nanostructured Half-Heusler Composites

7.3.1 Synthesis, Processing, and Microstructure

As described earlier, half of the tetrahedral sites in the structure of HH are vacant. Upon filling these empty sites (3/4, 3/4, 3/4), within the structure of an HH (TMX) with additional M atoms, one directly generates the related ternary FH phase with the general composition TM_2X (Figure 7.1) [39]. For a given combination of T, M, and X elements, the conversion of an HH structure to the related FH generally induces only marginal increases (2%–5%) in the unit cell parameter. However, there occurs the drastic change in the crystal symmetry, switching from the noncentrosymmetric cubic MgAgAs structure type (SG: $F\bar{4}3m$) to the centrosymmetric cubic $MnCu_2Al$ structure type (SG: $Fm\bar{3}m$), which indicates that both HH and FH structures cannot form solid solutions and, therefore, should coexist as separate phases with a high degree of lattice coherency. This prediction, based on chemical intuition, was ultimately confirmed by first-principles calculations [81,82]. For example, Kirievsky et al. [81] demonstrated, using *ab initio* calculations and statistical thermodynamics, that the decomposition at elevated temperatures of the off-stoichiometric Ni-rich HH alloy into the stoichiometric TiNiSn phase and Ni-rich Heusler phase $TiNi_2Sn$ is energetically favorable. Page et al. [82] investigated the thermodynamic

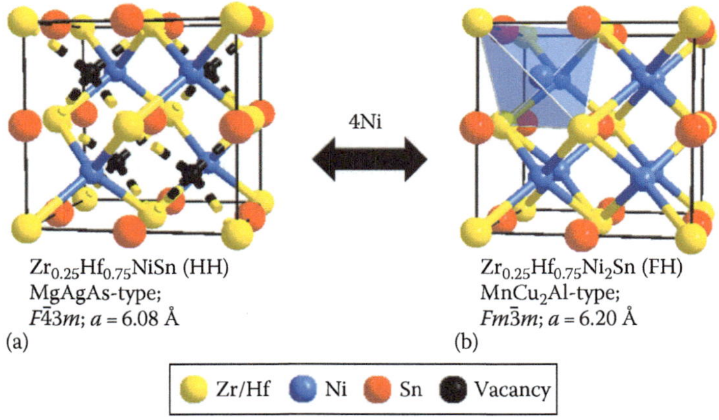

FIGURE 7.1
Relationship between the crystal structures of half-Heusler (HH) and full-Heusler (FH) alloys. (a) $Zr_{0.25}Hf_{0.75}NiSn$ (HH): MgAgAs type; $F\bar{4}3m$; $a = 6.10$ Å. (b) $Zr_{0.25}Hf_{0.75}Ni_2Sn$ (FH): $MnCu_2Al$ type; $Fm\bar{3}m$; $a = 6.24$ Å. (Reprinted from *J. Solid State Chem.*, 184, Makongo, J.P.A., Misra, D.K., Salvador, J.R. et al., Thermal and electronic charge transport in bulk nanostructured $Zr_{0.25}Hf_{0.75}NiSn$ composites with full-Heusler inclusions, 2948–2960, Copyright 2011, with permission from Elsevier.)

phase equilibria of the TNiSn–TNi_2Sn (T = Ti, Zr, Hf) pseudobinary systems using density functional theory combined with a cluster expansion and Monte Carlo simulations and found that there is a very small solubility of Ni in HH alloys even at high temperatures (Figure 7.2). This indicates that any Ni excess in the HH structure will decompose into a two-phase mixture of HH and FH compounds [82]. In addition, Do et al. [83] suggested that excess Ni atoms in HH tend to stay close to each other and form nanoclusters [18]. The resulting Ni atoms and vacancies compete to occupy different sites in FH and thus are distributed randomly.

Interestingly, recently published works showed that the internal structure of the resulting HH/FH composites (i.e., shape, size, and dispersion of the FH inclusion in the HH matrix) strongly depends on the synthesis and processing methods [39,40,67–73,84–90]. For example, taking advantage of the structural relationship between HH and FH phases, Poudeu et al. introduced the concept of atomic-scale structural engineering of the HH matrix using coherently embedded FH inclusions as a mean to tailor electronic and phonon transport in the HH matrix [38–40,71,72]. Starting from a presynthesized HH matrix, it was demonstrated that HH composites containing various mole fractions of coherent FH inclusions can be created by solid-state reaction of the HH matrix, TMX, with an incremental amount x of excess metal N to generate the starting mixture with the nominal composition TMN_xX. It was found that the reaction of this mixture at temperatures below the melting points of both starting materials (900°C < T < 1100°C) unambiguously leads to the composites HH(1 − x)/FH(x) within which both HH (TMX) and FH (TMNX) phases coexist [39,40,67–73,84–90]. The microstructure of the HH(1 − x)/FH(x) composites was found to vary strongly with the fraction x of the FH phase. For instance, Figure 7.3 gives the microstructure for composites HH(1 − x)/FH(x) with $0.06 \leq x \leq 0.6$ that have been synthesized by reacting an HH matrix with composition $Zr_{0.25}Hf_{0.75}NiSn$ with an incremental amount x of elemental Ni [39]. For the HH(0.91)/FH(0.09) composite, the FH phase exists as nanoprecipitates with the size ranging from 5 to 10 nm. Some of the precipitates occasionally agglomerate to form larger particles with sizes up to 20 nm. When the mole fraction of FH increases to 30%, FH precipitates tend to agglomerate severely and the resulting large particles form interconnected networks. Regardless of the composition, the precipitates form coherent interfaces with the matrix due to the similarity in their lattice parameters. A full reversal of the composite with the matrix as FH and the HH phase as the inclusion is obtained when the mole fraction of FH reaches 60%. In this composite, a *snakelike* pattern is observed in local areas of the HH(0.4)/FH(0.6) composite, which is considered as spatial modulation of the local composition reminiscent of a spinodal decomposition. Similar types of microstructures were obtained through solid-state reaction of the $Zr_{0.25}Hf_{0.75}NiSn$ matrix with an incremental amount x of elemental Co [91].

It is interesting to note that the observed nanometer-scale FH inclusions have been achieved through solid-state diffusion of Ni or Co into the HH structure at 1073 K, according to the reaction scheme in the following equation:

$$Zr_{0.25}Hf_{0.75}NiSn + xN \rightarrow (1-x)\ Zr_{0.25}Hf_{0.75}NiSn\ (\text{HH matrix}) + xZr_{0.25}Hf_{0.75}NiNSn\ (\text{FH}) \quad (7.1)$$

where N = Ni or Co.

Considering the high melting temperatures of both starting reagents (HH matrix and elemental Ni or Co) and the observed distribution of the FH phase in the TEM images of the resulting composites, one can speculate that the transformation of a fraction of the

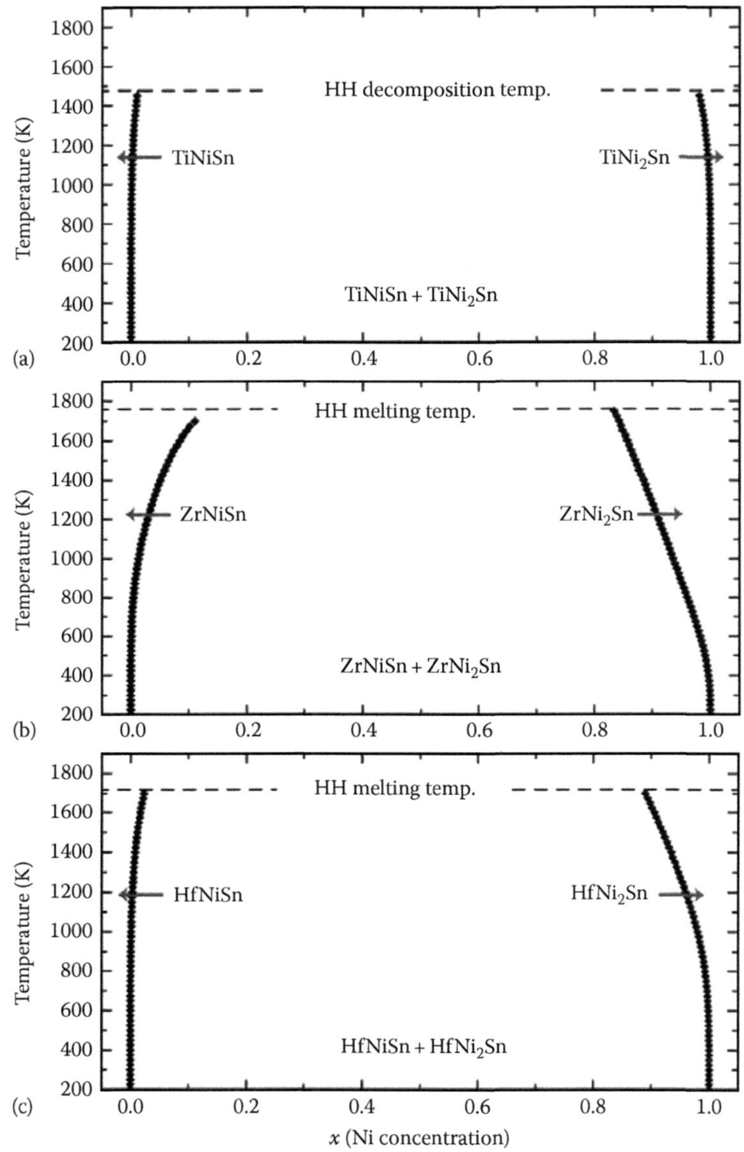

FIGURE 7.2
Temperature-concentration pseudobinary phase diagrams of $TiNi_{1+x}Sn$ (a), $ZrNi_{1+x}Sn$ (b), and $HfNi_{1+x}Sn$ (c). The small black diamonds represent the calculated points along the phase boundary and the horizontal dashed line indicates the experimental melting/decomposition points for the half-Heusler compounds. (Reprinted with permission from Page, A., Uher, C., Poudeu, P.F., and Van der Ven, A., Phase separation of full-Heusler nanostructures in half-Heusler thermoelectrics and vibrational properties from first-principles calculations, *Phys. Rev. B*, 92, 174102. Copyright 2015 by the American Physical Society.)

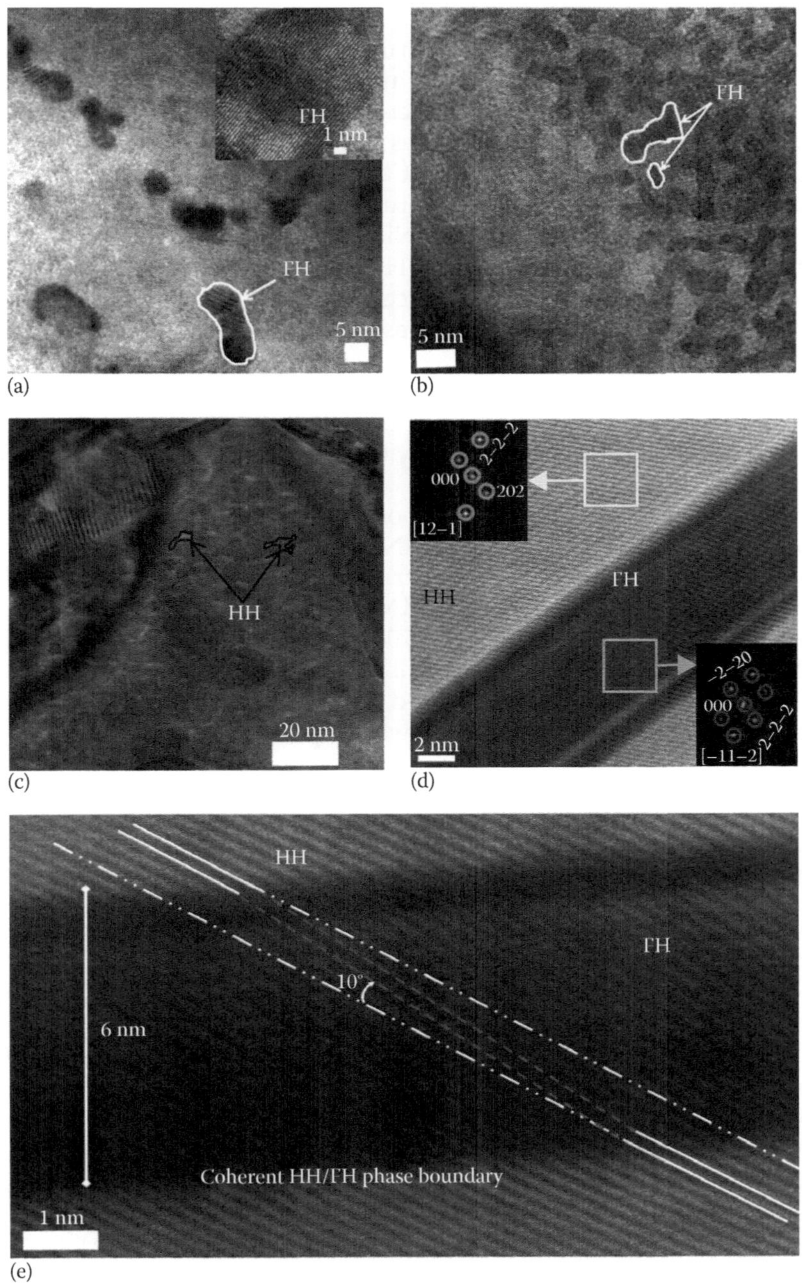

FIGURE 7.3
TEM images of selected HH(1 − x)/FH(x) composites from solid-state reaction between the presynthesized $Zr_{0.25}Hf_{0.75}NiSn$ matrix and x at % excess Ni. (a) HH(0.91)/FH(0.09), (b) HH(0.7)/FH(0.3), (c) HH(0.4)/FH(0.6), and (d) high-magnification TEM images of the HH(0.7)/FH(0.3) composite showing alternating layers of the HH matrix and the FH precipitate. (e) Details of the coherent phase boundary at the HH/FH interfaces. (Reprinted from *J. Solid State Chem.*, 184, Makongo, J.P.A., Misra, D.K., Salvador, J.R. et al., Thermal and electronic charge transport in bulk nanostructured $Zr_{0.25}Hf_{0.75}NiSn$ composites with full-Heusler inclusions, 2948–2960, Copyright 2011, with permission from Elsevier.)

HH matrix to the FH inclusion proceeds *via* solid-state diffusion of elemental Ni or Co into adjacent vacant sites within the HH structure. The random distribution of the FH phase and the wide range of particle sizes in the resulting composite suggests that the solid-state transformation starts with the nucleation of FH seed crystals at the Ni(Co)/HH interfaces between micrometer-size particles of Ni(Co) and the HH matrix (Figure 7.4). Here, the HH phase serves as both reactant and substrate on which endotaxial growth of FH inclusions occurs because of the structural similarity between HH and FH phases. The Ni/HH reaction front, driven by the Ni concentration gradient, subsequently diffuses, forming the first few layers of the FH phase between Ni(Co) and HH particles. This results in the formation of new interfaces, such as Ni(Co)/FH and FH/HH. At this point, further growth of the FH layer requires the diffusion of Ni(Co) from the FH phase to the HH matrix, leaving behind vacancies that are filled through Ni(Co) diffusion at the Ni(Co)/FH interfaces. Such Ni(Co) diffusion through the FH phase presumably continues until complete depletion of Ni(Co) atoms at the Ni(Co)/FH interfaces. Further diffusion of the FH/HH interfaces should be driven by the relative stability of the FH seed crystals within the HH matrix.

As discussed earlier, a solid-state reaction of an HH matrix, TMX, with an incremental amount of elemental Ni or Co generates TMN_xX composites in which large inclusions of the FH phases are embedded in the HH matrix. However, Poudeu et al. also showed that a drastically different microstructure of the TMN_xX composites is observed when elemental metals are used as starting reagents [40,70–73]. For example, as shown in Figure 7.5a, widely dispersed nanometer-scale (less than 1–30 nm) FH particles were observed in various (HH(1 − x)/FH(x)) composites synthesized through solid-state reaction of the elements according to the reaction scheme in the following equation:

$$0.25Zr + 0.75Hf + (1 + x)Ni + Sn \rightarrow Zr_{0.25}Hf_{0.75}Ni_{1+x}Sn \tag{7.2}$$

where $0 \le x \le 0.1$ and $Zr_{0.25}Hf_{0.75}Ni_{1+x}Sn$ indicates the HH(1 − x)/FH(x) composites with composition (1 − x) $Zr_{0.25}Hf_{0.75}NiSn$ (HH) + x $Zr_{0.25}Hf_{0.75}Ni_2Sn$ (FH).

The spherical shape of the precipitates suggests their nucleation and isotropic growth, as well as their endotaxial insertion within the HH matrix. In addition to spherical precipitates, 2–8 nm thick and up to 30 nm long lamellar structures are also observed in Figure 7.5b. The high-magnification image of one of the lamellar fine features revealed details of the HH/FH phase boundary structure (Figure 7.5c). The FH and HH phases form a tilted coherent interface with tilt angle of approximately $\theta = 30°$. Fast Fourier transform (FFT) diffractograms of both HH and FH regions showed that the FH inclusions with unit cell parameter $a = 6.29(3)$ Å are oriented along the $[1\bar{1}\bar{2}]$ zone axis (Figure 7.5d), while the HH matrix with the unit cell parameter $a = 6.06(3)$ Å is oriented along $[\bar{1}21]$ (Figure 7.5e).

Similar types of microstructures were also observed for the HH(1 − x)/FH(x) nanocomposites $Ti_{0.1}Zr_{0.9}Ni_{1+x}Sn$ [72], $Ti_{0.1}Zr_{0.9}Ni_{1+x}Sn_{0.975}Sb_{0.025}$ [71], $Ti_{0.5}Hf_{0.5}Co_{1+x}Sb_{0.9}Sn_{0.1}$ [84], and $Zr_{0.25}Hf_{0.75}Ni_{1+x}Sn_{0.975}Bi_{0.025}$ [40] synthesized through solid-state reaction of the elements at ~1173 K for 2 weeks.

This drastic change in the microstructure upon alteration of the starting reagents suggests a completely different reaction mechanism. The high degree of similarity of the structures of the HH and FH phases with similar chemical compositions implies a very small difference in their formation energy. This suggests that during solid-state reaction from the elements, both phases conucleate at various regions of the reaction medium.

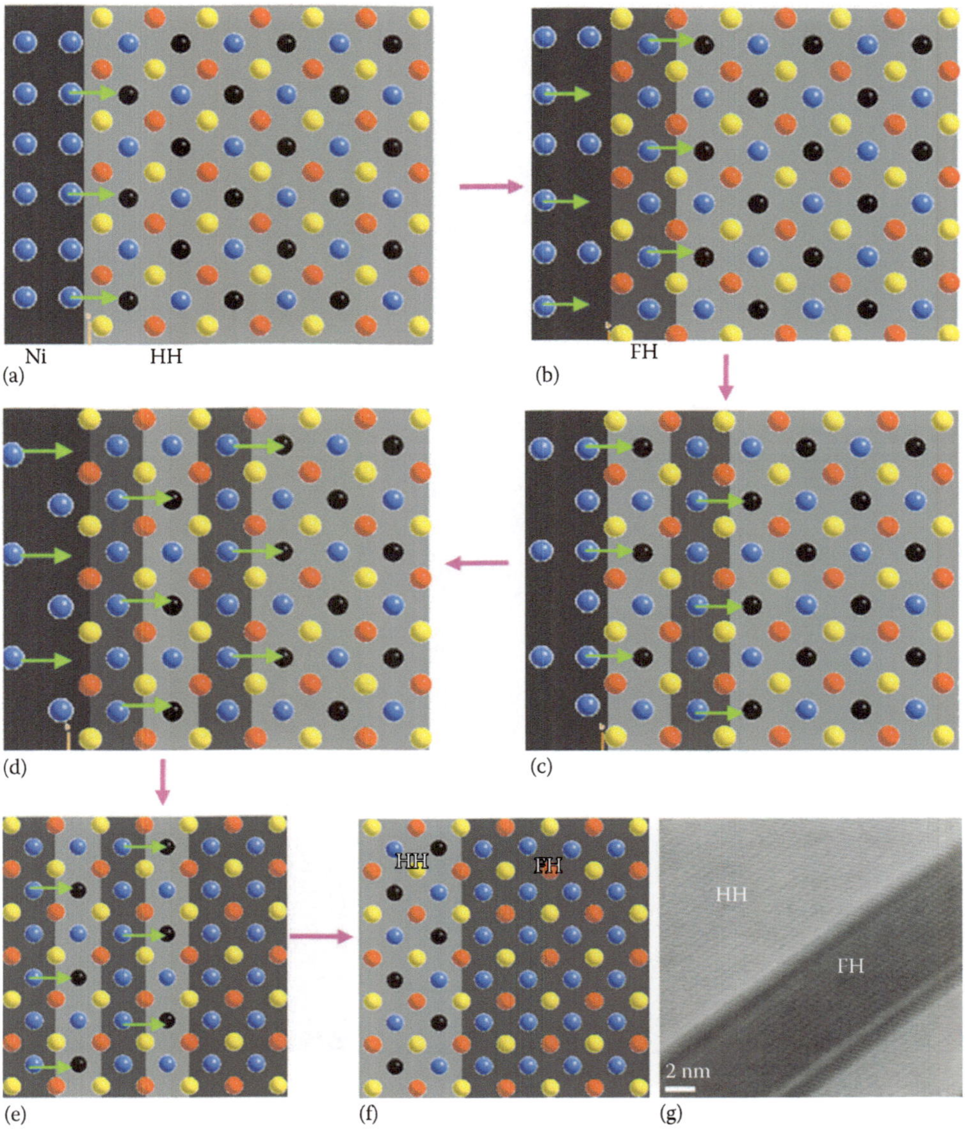

FIGURE 7.4
Proposed Ni(Co) diffusion mechanism at the Ni(Co)/HH matrix interface and throughout the HH matrix (at the FH/HH interface), leading to the formation of an HH/FH composite *via* solid-state phase transformation. (a) Condition of the Ni(Co)/HH interface at the beginning of the diffusion process; (b) formation of the first seed layer of the FH phase, which triggers the diffusion of more Ni(Co) toward the Ni(Co)/FH interface; (c) simultaneous diffusion at both Ni(Co)/HH and FH/HH interfaces; (d) formation of the second FH layer and continuous Ni(Co) diffusion at all interfaces (Ni(Co)/FH; FH/HH); (e) Ni(Co) diffusion at the FH/HH interfaces within the composite leading to the migration and agglomeration of unstable FH seed layers; and (f) formation of a stable FH/HH interface away from the initial Ni(Co)/HH interface. Blue spheres, Ni(Co); red spheres, Zr/Hf; yellow spheres, Sn; black spheres, vacant sites within the HH structure. (g) High-magnification TEM image of the Ni containing HH(0.7)/FH(0.3) composite illustrating alternating layers of the HH matrix and the FH precipitate. (Reprinted from *J. Solid State Chem.*, 184, Makongo, J.P.A., Misra, D.K., Salvador, J.R. et al., Thermal and electronic charge transport in bulk nanostructured $Zr_{0.25}Hf_{0.75}NiSn$ composites with full-Heusler inclusions, 2948–2960, Copyright 2011, with permission from Elsevier.)

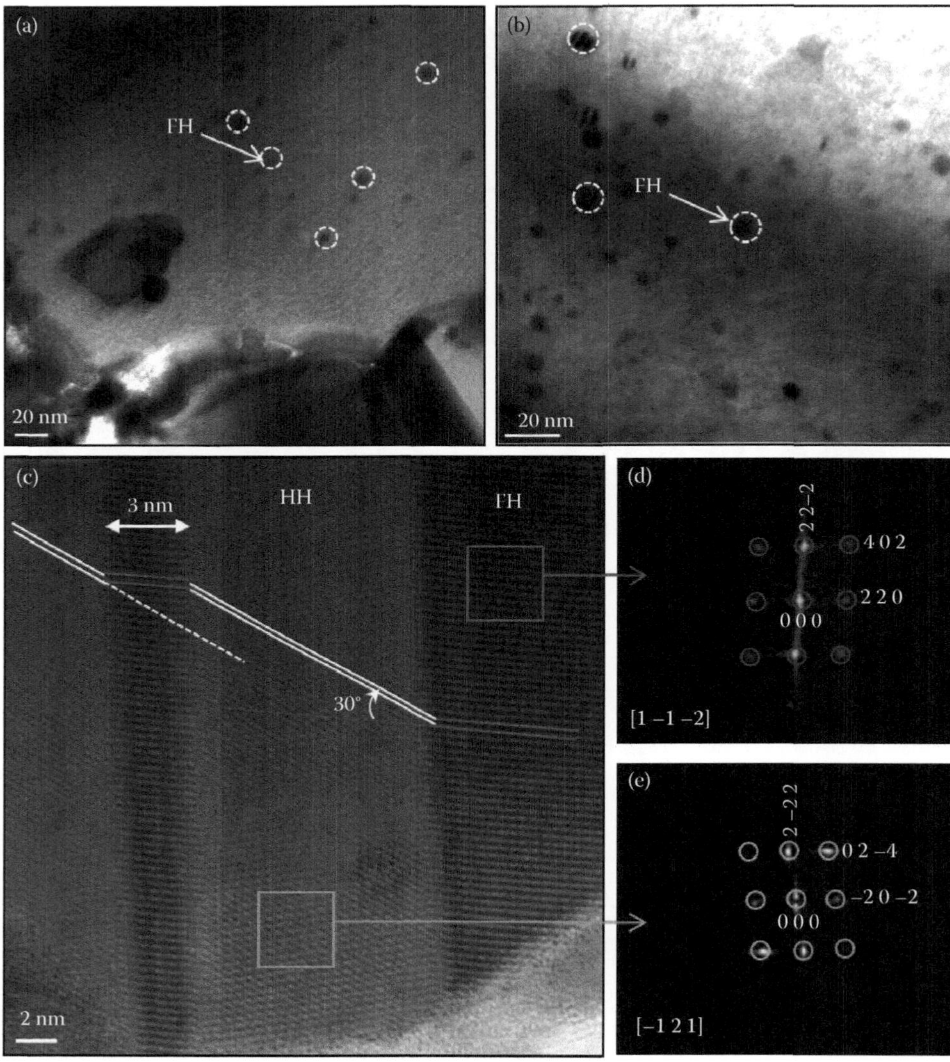

FIGURE 7.5
TEM images of spark plasma sintered pellets of HH(1 − x)/FH(x) bulk nanocomposites containing 2 mole% (a) and 5 mole% (b) FH inclusions. (c) High-magnification image revealing details of the HH/FH phase boundary structure. The FH (d) and HH (e) structures are confirmed by fast Fourier transform diffractograms. (Reprinted with permission from Makongo, J.P.A., Misra, D.K., Zhou, X. et al., Simultaneous large enhancements in thermopower and electrical conductivity of bulk nanostructured half-Heusler alloys, *J. Am. Chem. Soc.*, 133, 18843–18852. Copyright 2011 American Chemical Society.)

The formation of either HH or FH seed crystals depends on the local compositions of the reaction medium. This conucleation and growth reaction mechanism is supported by the extremely small size of the FH inclusions as well as their endotaxial insertion with atomic-scale coherency in the grains with HH structure and composition (Figure 7.5). While the volume fraction of the FH phase can be directly controlled through adjustment of the excess of elemental Ni in the starting mixture, control over the dispersion and homogeneity of FH particle size are quite challenging.

Chai et al. [86–88] also showed that HH/FH composites synthesized through arc melting of the elements followed by directional solidification exhibit microstructures dominated by a high density from 2 to 7 nm wide FH nanobands embedded in the HH matrix. Figure 7.6 shows the bright-field TEM images of the microstructure of HH matrices with compositions ZrNiSn, ($Zr_{0.5}$, $Hf_{0.5}$)NiSn and Zr(Ni, $Co_{0.2}$)Sn and their corresponding inverse FFT (IFFT) images. One can observe from the IFFT images that the FH nanobands embedded in the HH matrix are interconnected into networks. The nanobands form coherent interfaces with the HH matrix and show no preferential orientation. The nanobands bear rather irregular morphology and interband spacing of <10 nm, while the nanoprecipitates obtained through solid-state reaction of the elements at temperatures below the melting points of both HH and FH phases tend to maintain some orientation relationship with the HH matrix and their size ranges from 5 to 30 nm [40,72]. The absence of nanoprecipitates in HH/FH composites prepared by arc melting followed by directional solidification indicates that the excess Ni and (Ni + Co) in the ZrNiSn, ($Zr_{0.5}Hf_{0.5}$)NiSn, and Zr(Ni, $Co_{0.2}$)Sn alloys must be below the supersaturation limit for FH-phase precipitation. Douglas et al. [89] have drawn a similar conclusion that excess Ni atoms tend to stay close to each other, forming nanoclusters rather than dispersing randomly. The structural excess of Ni in the TiNiSn alloy has been studied through a combination of synchrotron x-ray and neutron scattering.

7.3.2 Electronic Transport in Half-Heusler Nanocomposites with Full-Heusler and InSb Inclusions

The physical and chemical properties of HH and FH compounds are dictated by the crystal structure, the stoichiometry, the nature of the constituting elements, and the valence electron count. On one hand, the HH compounds (1:1:1) with careful selection of T, M, and X atoms to satisfy a total valence electron count of 18 are semiconductors with bandgaps that are tunable (e.g., ~0–0.4 eV) upon varying the chemical composition. Therefore, HH compounds have attracted tremendous attention as potential candidates for solar cells, thermoelectrics [24,30–33,35–37,39–53], half-metallic dilute magnetic semiconductors (spintronics) [54–57], and topological insulators [58–60]. The FH compounds (1:2:1), on the other hand, are magnetic and semiconductors for compositions with 18 or 24 valence electrons or nonmagnetic superconducting with approximately 27 valence electrons [92]. Therefore, a large number of FH (TM_2X) compounds with multifunctional magnetic properties, such as magnetooptical [93], magnetoelectrical [94], and magnetocaloric properties [95,96], have been reported. Although HH and FH phases are two distinct classes of structurally related materials exhibiting dissimilar physical and chemical properties specific to each class, the ability to interchange between the HH and FH structure upon addition or removal of M atoms (at [3/4, 3/4, 3/4] sites) makes possible the atomic-scale manipulation of the properties (chemical, physical, mechanical, electrical, optical, etc.) of one material by the other.

7.3.2.1 HH Matrix with Large FH Inclusions

Recent published works showed that the incorporation of FH nanoinclusions in HH matrices with various compositions drastically alters the electronic properties of the resulting HH/FH composites [39,40,67–73,84–90]. Interestingly, the net effect of the embedded FH inclusions on the electronic properties seems to be related largely to the average size of the embedded FH inclusions. For example, it was observed that the incorporation of FH

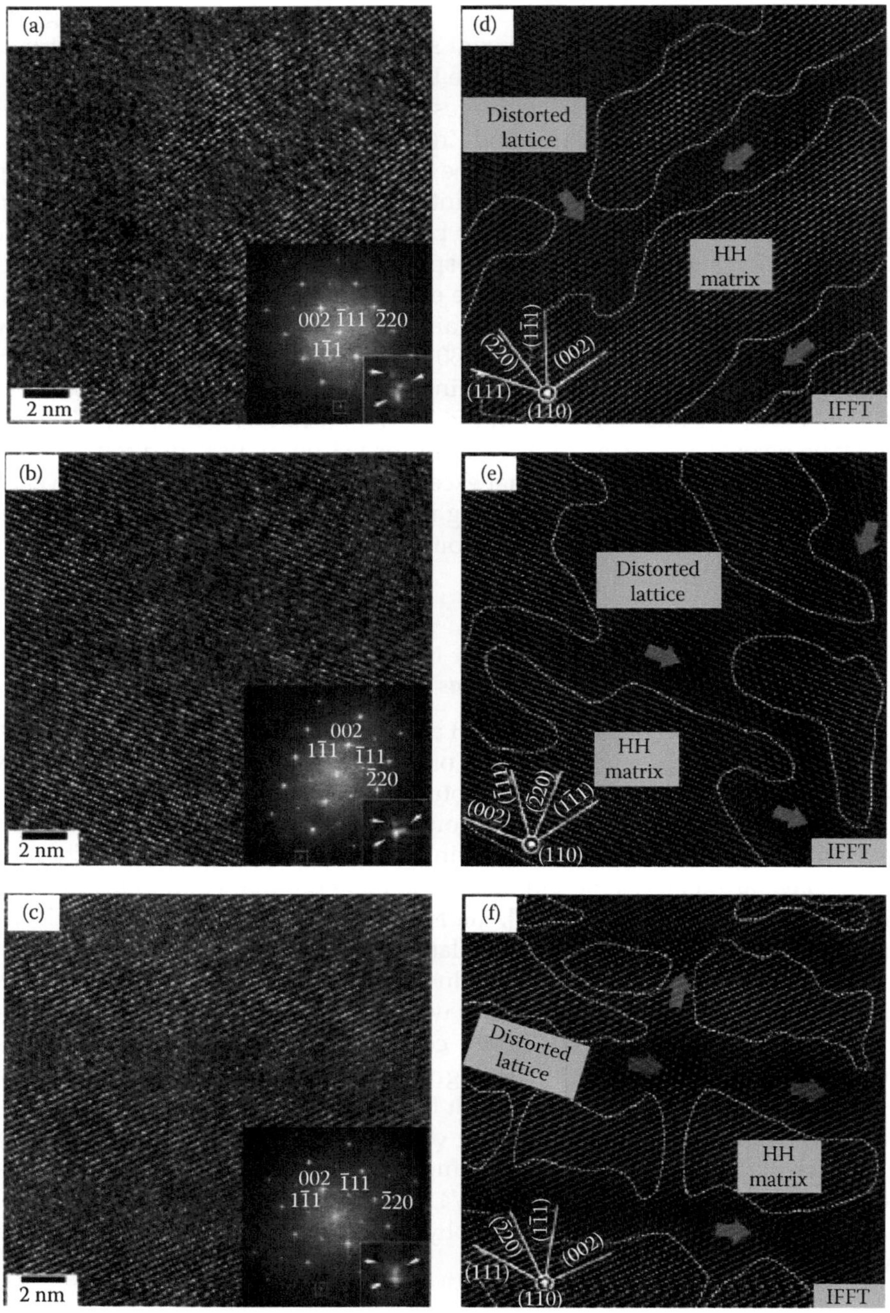

FIGURE 7.6
HRTEM images revealing the half-Heusler (HH) lattice structure of (a) ZrNiSn, (b) $(Zr_{0.5}Hf_{0.5})NiSn$, and (c) $Zr(NiCo_{0.2})Sn$. The insets show the corresponding [110] HH fast Fourier transform (FFT) patterns. The arrows indicate splitting of the diffraction spots. The inverse FFT images corresponding to (a–c) are shown in (d–f), respectively. The distorted lattice bands in the HH matrix are enclosed by dashed lines and highlighted by arrows. (Reprinted from *Scr. Mater.*, 83, Chai, Y.W., Yoshioka, K., and Kimura, K., Intrinsic point defects in thermoelectric half-Heusler alloys, 13–16, Copyright 2014, with permission from Elsevier.)

inclusions with large particle size (>50 nm) into an *n*-type HH matrix with composition $Zr_{0.25}Hf_{0.75}NiSn$ and the agglomeration of such inclusions into networks result in a large increase in the carrier density and electrical conductivity. This leads to a drastic drop in the thermopower and carrier mobility (see Figure 7.7). The observed changes in the electronic properties suggest that the embedded large particles of FH inclusions act as dopants to the HH matrix, dumping additional electrons into the existing ensemble of conduction electrons. The electrical conductivity increases with the fraction of FH inclusions. Although this trend is not so dramatic at low fraction of the FH inclusion, larger increases in the electrical conductivity are observed for HH(0.7)/FH(0.3) and HH(0.4)/FH(0.6) composites. Composites with low fraction of FH inclusion display semiconducting behavior, whereas HH(0.7)/FH(0.3) composite displays semimetal behavior and HH(0.4)/FH(0.6) exhibits a typical metallic behavior.

Accordingly, the thermopower decreases with increasing fraction of FH, with the HH(0.7)/FH(0.3) and HH(0.4)/FH(0.6) composites showing dramatic degradation of the thermopower. The observed trend in the change of electronic transport properties suggests that at low fraction, FH inclusions exist as isolated nanoprecipitates and their contribution to the electronic transport is marginal due to the lack of electronic percolation between FH precipitates. When the fraction of FH inclusions reaches above 0.2, these FH precipitates form an interconnected network that substantially enhances the electrical transport. This explanation has been confirmed by the Hall result, as indicated in Figure 7.7. The observed increase in the carrier density is consistent with the alteration in the electronic transport that could be anticipated from the intermixing of metallic elements in a semiconductor. This result is also in line with the theoretical investigation of the influence of clustering and nanostructure in Zr–Ni–Sn alloys on electronic structure [83]. It was found that Ni vacancies have a modest effect on the electronic structure in FH alloys, while excess Ni dramatically changes the electronic structure of HH alloys through the formation of impurity states within the bandgap of HH, resulting in a finite density of states below the conduction bands (CBs).

Douglas et al. [89] investigated the effect of an FH second phase on the electronic transport of $TiNi_{1+x}Sn$ composites, with $x = 0.00$, 0.05, 0.10, 0.15, 0.20, and 0.25, synthesized by levitation melting followed by annealing. This synthesis approach also resulted in a high density of FH precipitates, with a size ranging from a few nanometers to several micrometers, embedded within the HH matrix (Figure 7.8). Isolated islands of the FH phase form in the continuous HH phase and bear similar grain sizes to that of the HH phase, as indicated in Figure 7.8. The FH phase generally has dendritic morphology, suggesting that upon cooling from the melt, the FH phase forms before the HH phase. This intermixing between large and small FH precipitates led to some interesting electronic properties.

As shown in Figure 7.9, all $TiNi_{1+x}Sn$ samples display semiconducting-like behavior. The electrical conductivity monotonically increases with temperature even for compositions containing a large mole fraction of the FH phase. This is in sharp contrast with the result discussed earlier for the $Zr_{0.25}Hf_{0.75}Ni_{1+x}Sn$ nanocomposites where compositions with a large fraction of FH inclusions exhibit semimetal (HH(0.7)/FH(0.3)) or metallic (HH(0.4)/FH(0.6)) behavior [39]. The electrical conductivity of $TiNi_{1+x}Sn$ samples increases with increasing fraction of Ni, which is consistent with the results obtained from the $Zr_{0.25}Hf_{0.75}Ni_{1+x}Sn$ nanocomposites, although the magnitude of the increase in the electrical conductivity is smaller for the $TiNi_{1+x}Sn$ samples. The Seebeck coefficient decreases with increasing fraction of Ni in $TiNi_{1+x}Sn$ samples, with a substantial drop observed for samples of $x = 0.20$ and $x = 0.25$. However, the reduction in the thermopower with increasing FH content is not

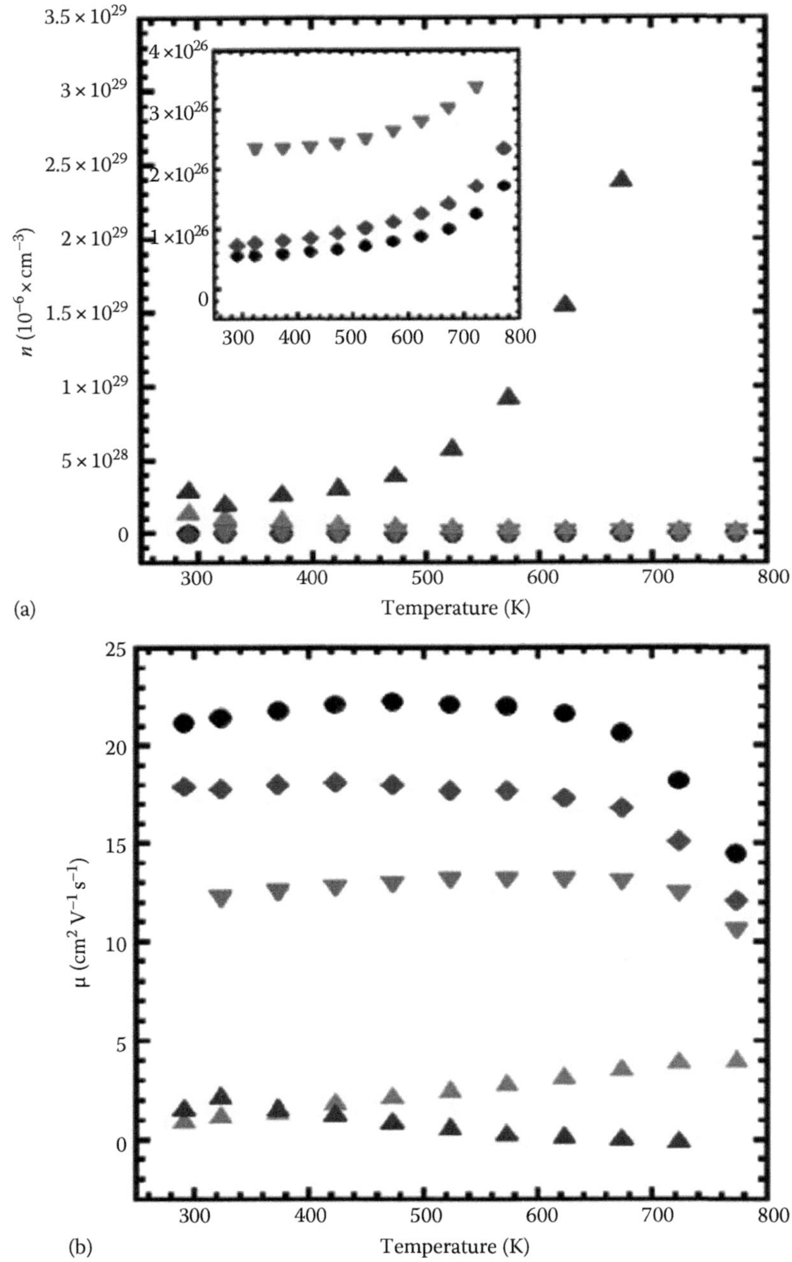

FIGURE 7.7
Temperature-dependent electronic transport properties of HH(1 − x)/FH(x) composites. (a) Carrier density, (b) carrier mobility. *(Continued)*

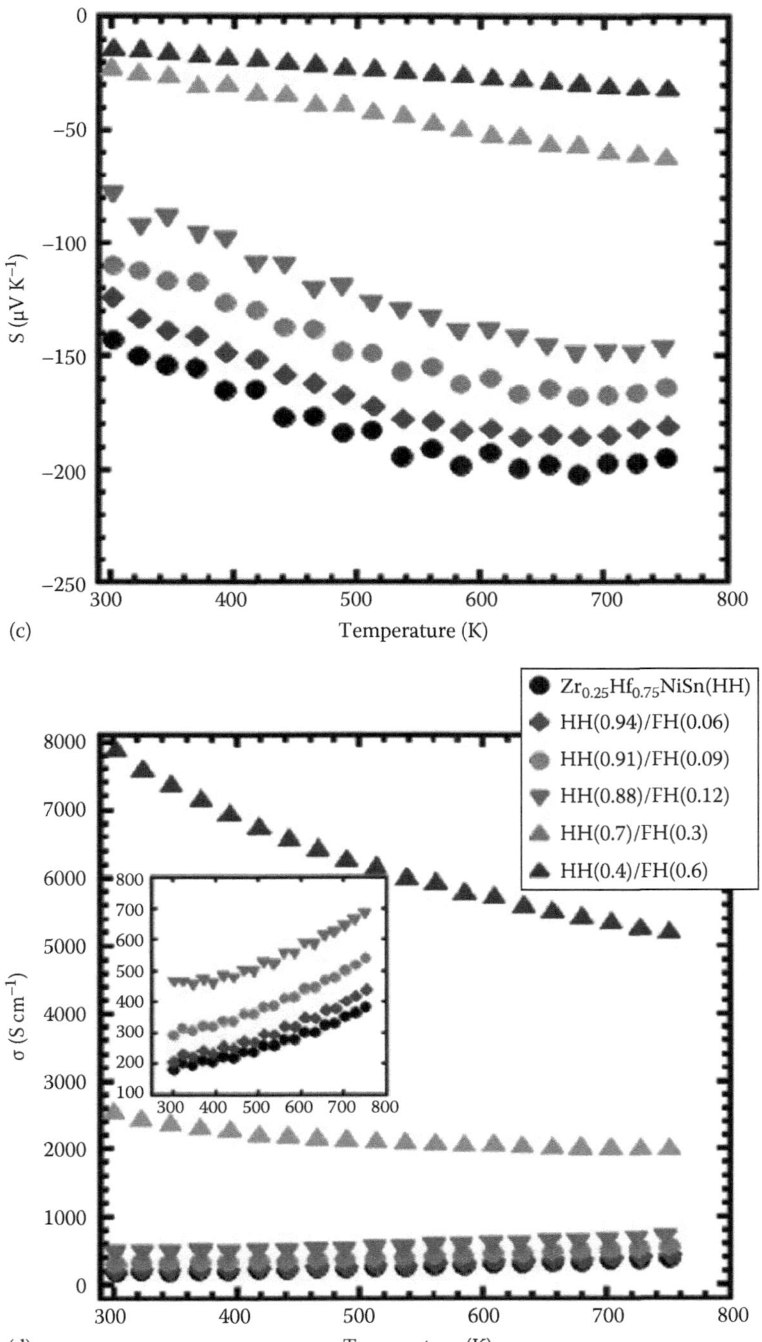

FIGURE 7.7 ***(Continued)***
Temperature-dependent electronic transport properties of HH(1 − x)/FH(x) composites. (c) Seebeck coefficient, and (d) electrical conductivity. (Reprinted from *J. Solid State Chem.*, 184, Makongo, J.P.A., Misra, D.K., Salvador, J.R. et al., Thermal and electronic charge transport in bulk nanostructured $Zr_{0.25}Hf_{0.75}NiSn$ composites with full-Heusler inclusions, 2948–2960, Copyright 2011, with permission from Elsevier.)

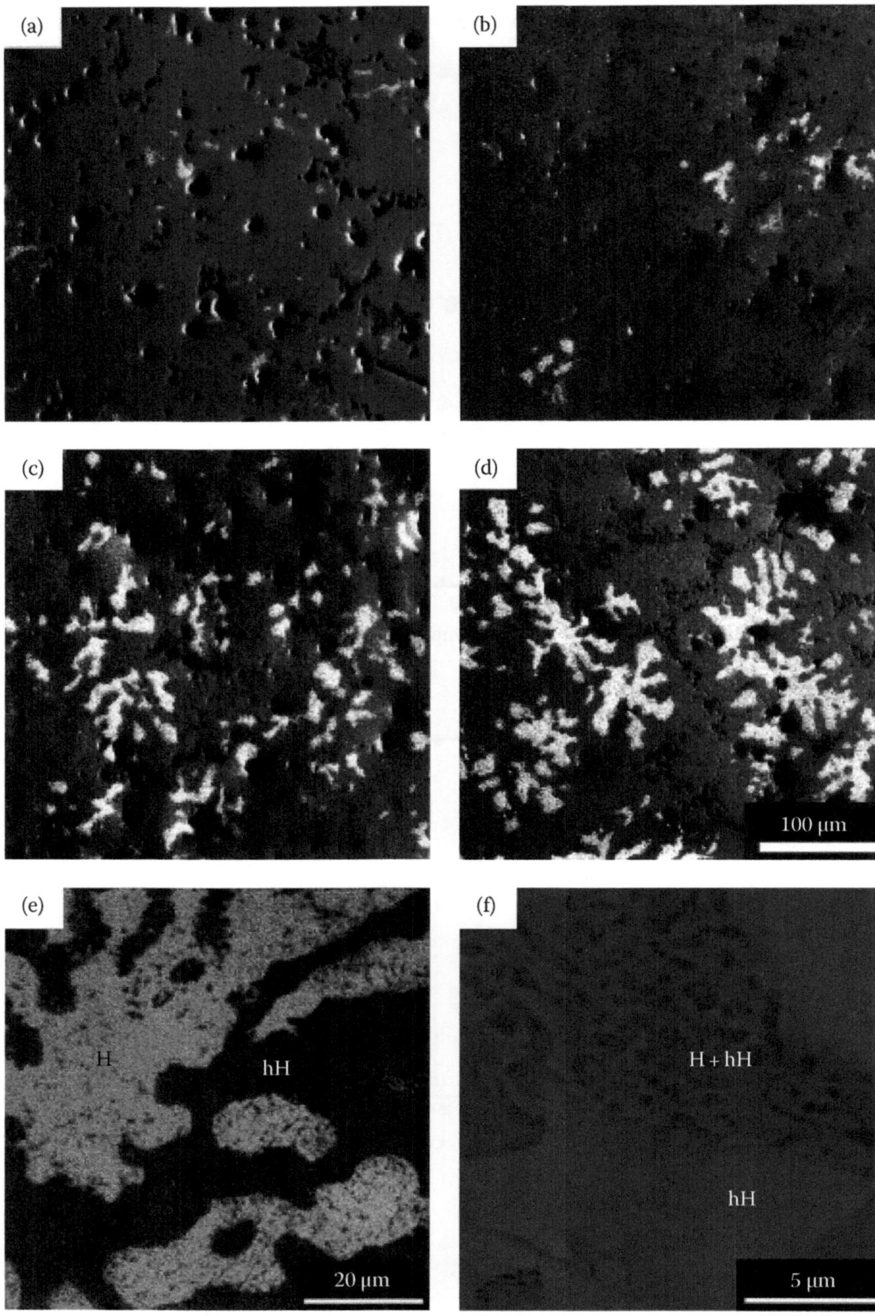

FIGURE 7.8
Optical microscope images of polished specimens (a) $x = 0.00$, (b) $x = 0.05$, (c) $x = 0.15$, and (d) and (e) $x = 0.20$. Light regions are the full-Heusler (FH) phase and charcoal is half-Heusler (HH). Black circles are pores. A Ti–Sn minor phase is also observable in (a) dark-gray impressions. (f) Backscatter detection scanning electron microscopy (SEM), x¼0.15. HH precipitates (dark) within the $TiNi_2Sn$, on the order of 100 nm. (Reprinted with permission from Douglas, J.E., Birkel, C.S., Verma, N. et al., Phase stability and property evolution of biphasic Ti-Ni-Sn alloys for use in thermoelectric applications, *J. Appl. Phys.*, 115, 043720. Copyright 2014, American Institute of Physics.)

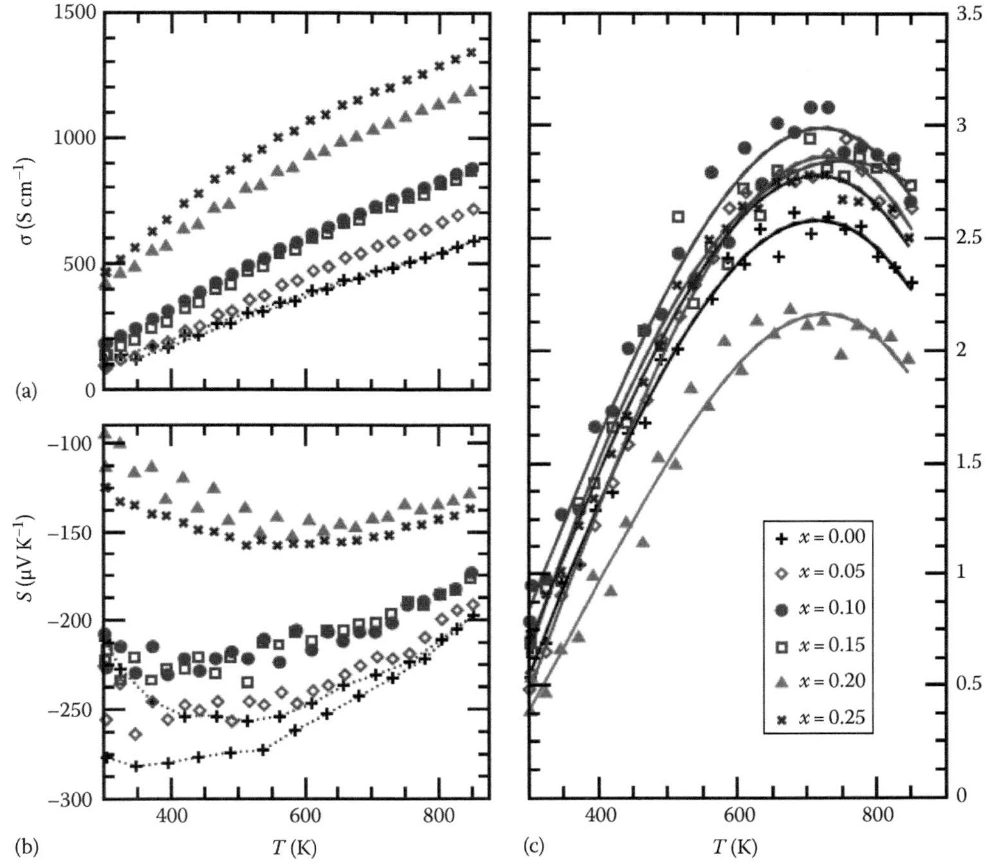

FIGURE 7.9
Temperature-dependent electronic properties of $TiNi_{1+x}Sn$ samples. (a) The electrical conductivity, (b) the Seebeck coefficient, and (c) the power factor. (Reprinted with permission from Douglas, J.E., Birkel, C.S., Verma, N. et al., Phase stability and property evolution of biphasic Ti-Ni-Sn alloys for use in thermoelectric applications, *J. Appl. Phys.*, 115, 043720. Copyright 2014, American Institute of Physics.)

as drastic as the trend observed for the $Zr_{0.25}Hf_{0.75}Ni_{1+x}Sn$ nanocomposites. This suggests that nanometer-scale and micrometer-scale FH inclusions in $TiNi_{1+x}Sn$ samples play different roles in the alteration of the electronic transport properties. Micron-size FH inclusions are metallic and therefore are expected to dope the matrix leading a large increase in the electrical conductivity and drastic decrease in the thermopower, as observed for the $Zr_{0.25}Hf_{0.75}Ni_{1+x}Sn$ nanocomposites. However, in the nanometer-scale regime, FH inclusions embedded into an HH matrix induced a completely unexpected electronic transport behavior.

7.3.2.2 HH Matrix with Small FH Inclusions

Makongo et al. [39] demonstrated that the thermoelectric properties of $Zr_{0.25}Hf_{0.75}Ni_{1+x}Sn$ nanocomposites can be drastically improved if the large FH inclusions are broken down to smaller (<1–3 nm) particles that are widely dispersed in the HH matrix. This type of microstructure is obtained through a synthesis approach that enables competing and simultaneous crystallization of both HH and FH phases during a solid-state reaction of the elements, while also avoiding melting of the product. As shown in Figure 7.10, a drastic enhancement in the thermopower is observed for the $Zr_{0.25}Hf_{0.75}Ni_{1.02}Sn$ (HH(0.98)/FH(0.02)) composite. The thermopower value at 300 K, −252 μV K^{-1}, corresponds to a 67% increase compared to the value of −148 μV K^{-1} observed for the $Zr_{0.25}Hf_{0.75}NiSn$ matrix.

Interestingly, only a marginal reduction in the electrical conductivity is observed at 300 K. Furthermore, the electrical conductivity rapidly increases with temperature, leading to a simultaneous enhancement of both thermopower and electrical conductivity at 775 K. This surprising behavior of electronic transport in $Zr_{0.25}Hf_{0.75}Ni_{1.02}Sn$ was attributed to the presence of extremely small isolated FH nanoinclusions coherently embedded in the HH matrix (Figure 7.5a). Such nanoinclusions are believed to act as an energy filter to low-energy (LE) carriers at the HH/FH interfaces, while promoting the transport of high-energy (HE) carriers across the interface. This results in a drastic drop in the carrier density and an increase in the carrier mobility (Figure 7.10).

Similar enhancement of the electronic transport was also observed in Bi-doped HH/FH nanocomposites $Zr_{0.25}Hf_{0.75}Ni_{1+x}Sn_{0.975}Bi_{0.025}$ (see Figure 7.11). The thermopower increases with increasing mole fraction of FH inclusions, reaching a maximum value of −235 μV K^{-1} at 300 K for the $Zr_{0.25}Hf_{0.75}Ni_{1.06}Sn_{0.975}Bi_{0.025}$ HH(0.94)/FH(0.06) composite. This corresponds to a ~42% increase compared to that of the Bi-doped matrix and a ~64% increase compared to that of the undoped $Zr_{0.25}Hf_{0.75}NiSn$ matrix. Further increasing the fraction of FH led to degradation of the thermopower. When the mole fraction of FH increases up to 10%, the FH tends to form large metallic inclusions and its doping effect becomes dramatic. Interestingly, the carrier density of $Zr_{0.25}Hf_{0.75}Ni_{1+x}Sn$ and $Zr_{0.25}Hf_{0.75}Ni_{1+x}Sn_{0.975}Bi_{0.025}$ samples increases rapidly with temperature and matches those of the respective HH matrices at high temperature. The substantial enhancement in thermopower, especially around room temperature, has been attributed to a large reduction in the carrier density due to the energy-filtering effect, which screens out the LE carriers at the HH/FH interfaces. Those filtered out carriers gradually gain enough energy to cross the potential barrier with increasing temperature and thus increase the overall carrier density and electrical conductivity, as illustrated in Figure 7.12. Surprisingly, these composites display higher electrical conductivity than do the corresponding matrices at high temperature though they have similar carrier densities. Even at 300 K, only a marginal decrease in the electrical conductivity is observed compared to the HH matrices though the carrier density has been substantially reduced. These results imply that these composites should have substantially increased carrier mobility, which is confirmed by the Hall effect results. Increased carrier mobility is expected when carrier density is reduced at room temperature. However, the carrier mobilities slowly decrease (significantly deviating from the typical $T^{-\lambda}$ law) and remain ~80% higher than those of their corresponding matrices at 775 K, despite the exponentially increasing carrier density with rising temperature. This unexpected temperature dependence of the carrier mobility in various $Zr_{0.25}Hf_{0.75}Ni_{1+x}Sn$ and $Zr_{0.25}Hf_{0.75}Ni_{1+x}Sn_{0.975}Bi_{0.025}$ samples is rationalized by taking into account the change in the fraction of LE carriers to HE carriers, LE/HE, arising from the varying of the CB offset, ΔE, at the HH/FH interfaces through the alteration of the size of the FH inclusion (Figure 7.12).

Generally, HE electrons have lower mobility because of their high electron–electron and acoustic-electron scattering. The overall carrier mobility at a given temperature strongly depends on the fraction of LE and HE electrons as well as the average mobility for each

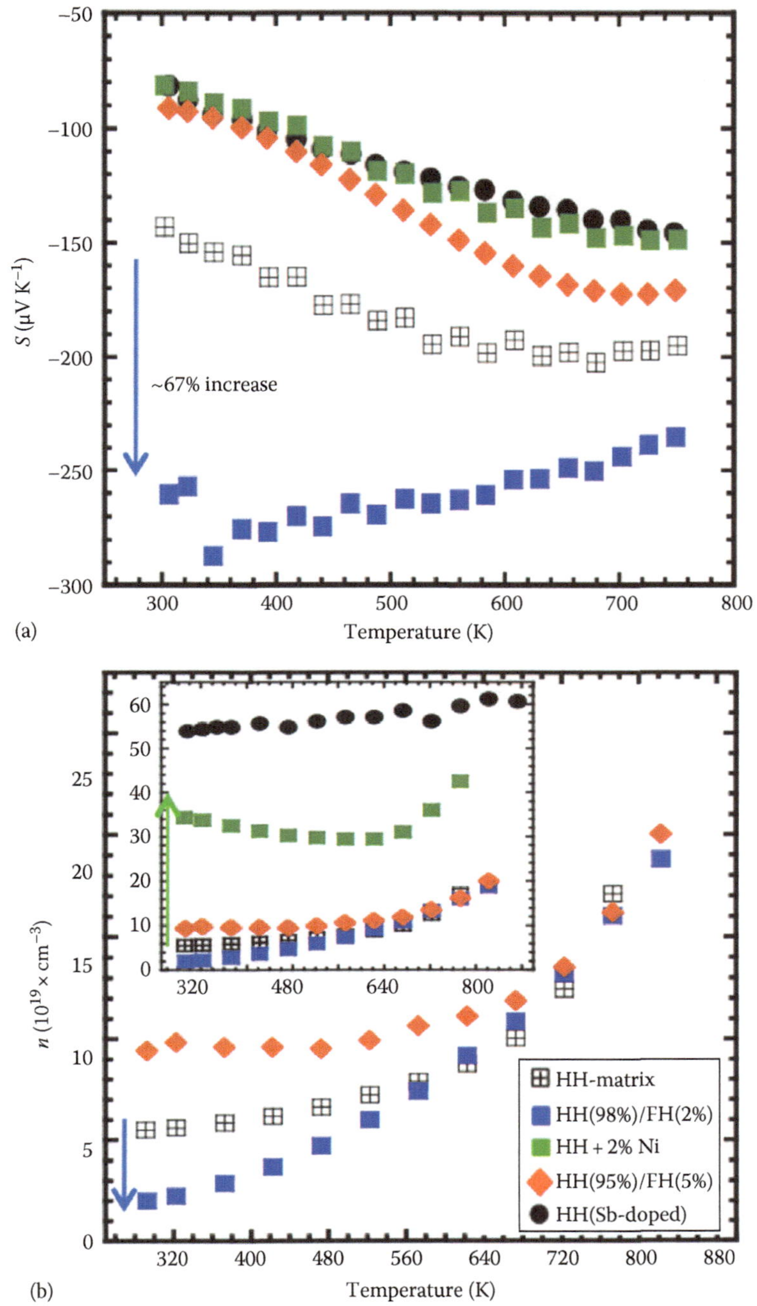

FIGURE 7.10
Temperature-dependent electronic properties of HH(1 – x)/FH(x) composites (x = 0.02, 0.05). (a) The Seebeck coefficient, (b) the carrier concentration.

(Continued)

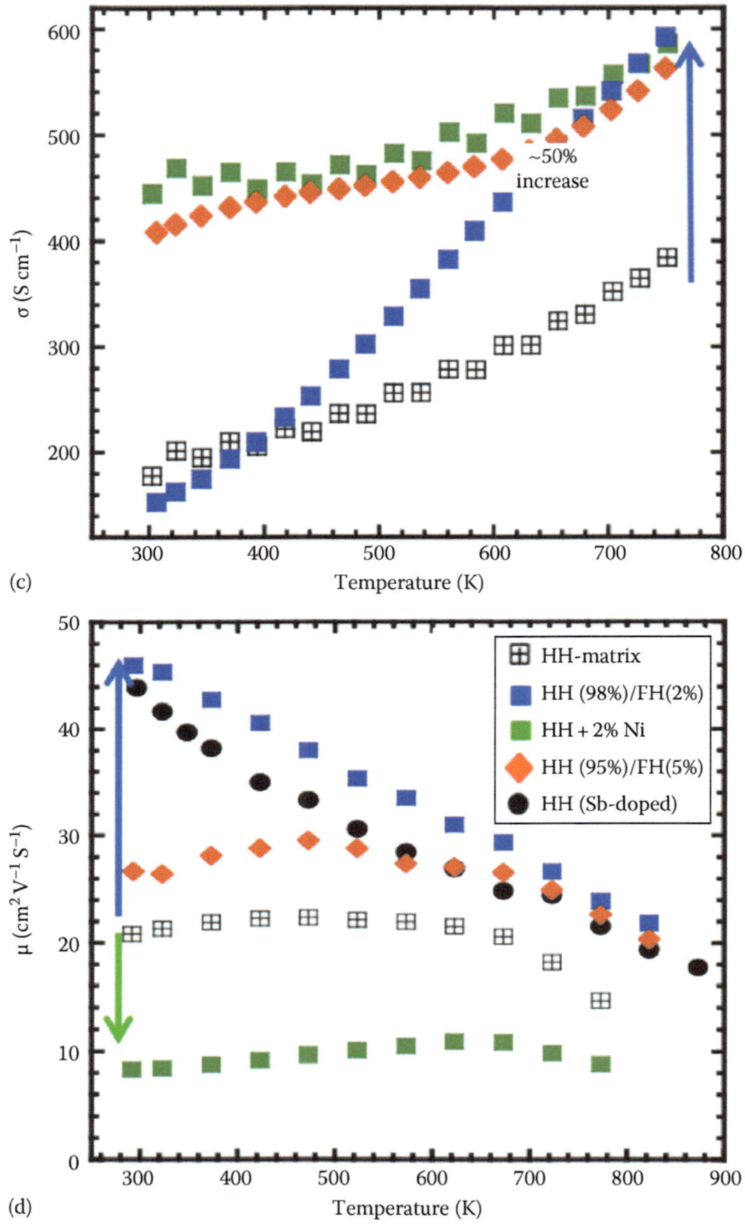

FIGURE 7.10 (*Continued*)
Temperature-dependent electronic properties of HH$(1 - x)$/FH(x) composites ($x = 0.02, 0.05$). (c) the electrical conductivity, and (d) the carrier mobility. (Reprinted with permission from Makongo, J.P.A., Misra, D.K., Zhou, X. et al., Simultaneous large enhancements in thermopower and electrical conductivity of bulk nanostructured half-Heusler alloys, *J. Am. Chem. Soc.*, 133, 18843–18852. Copyright 2011 American Chemical Society.)

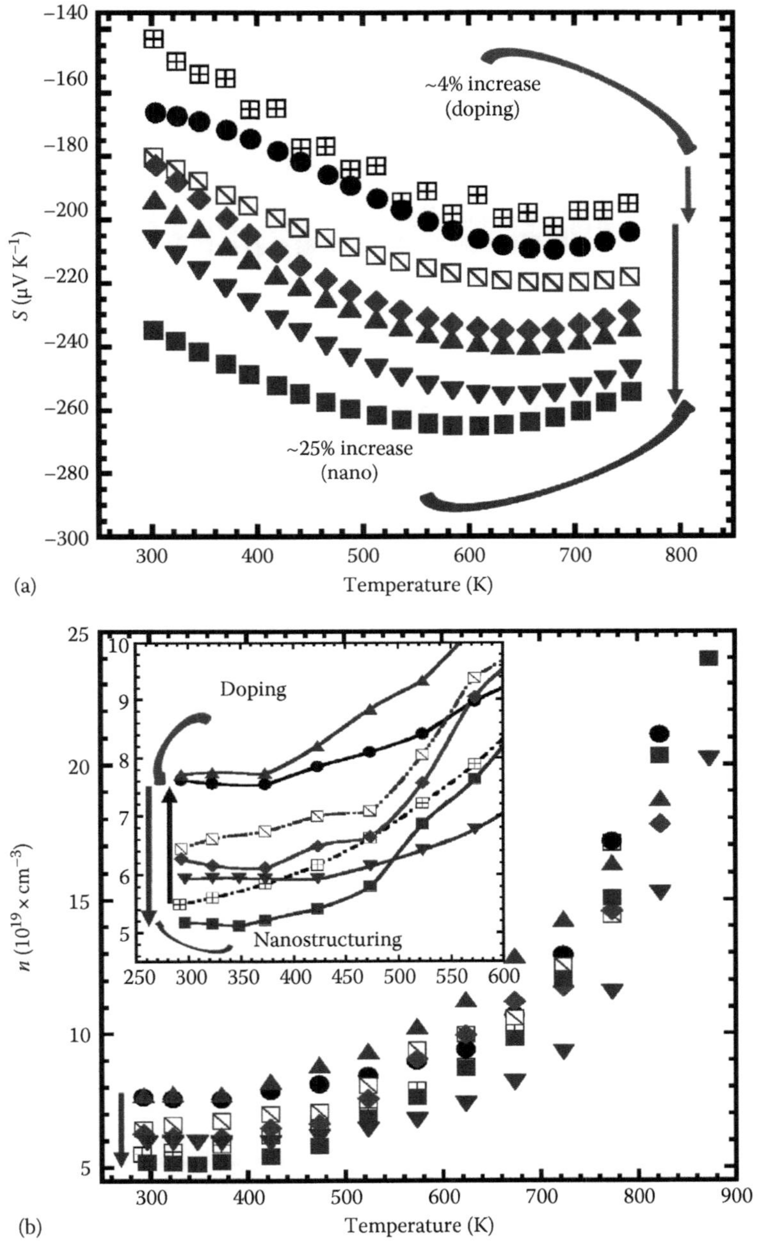

FIGURE 7.11
Temperature-dependent electronic properties of Bi-doped HH(1 − x)/FH(x) composites ($0 \le x \le 0.10$). (a) The Seebeck coefficient, (b) the carrier concentration. *(Continued)*

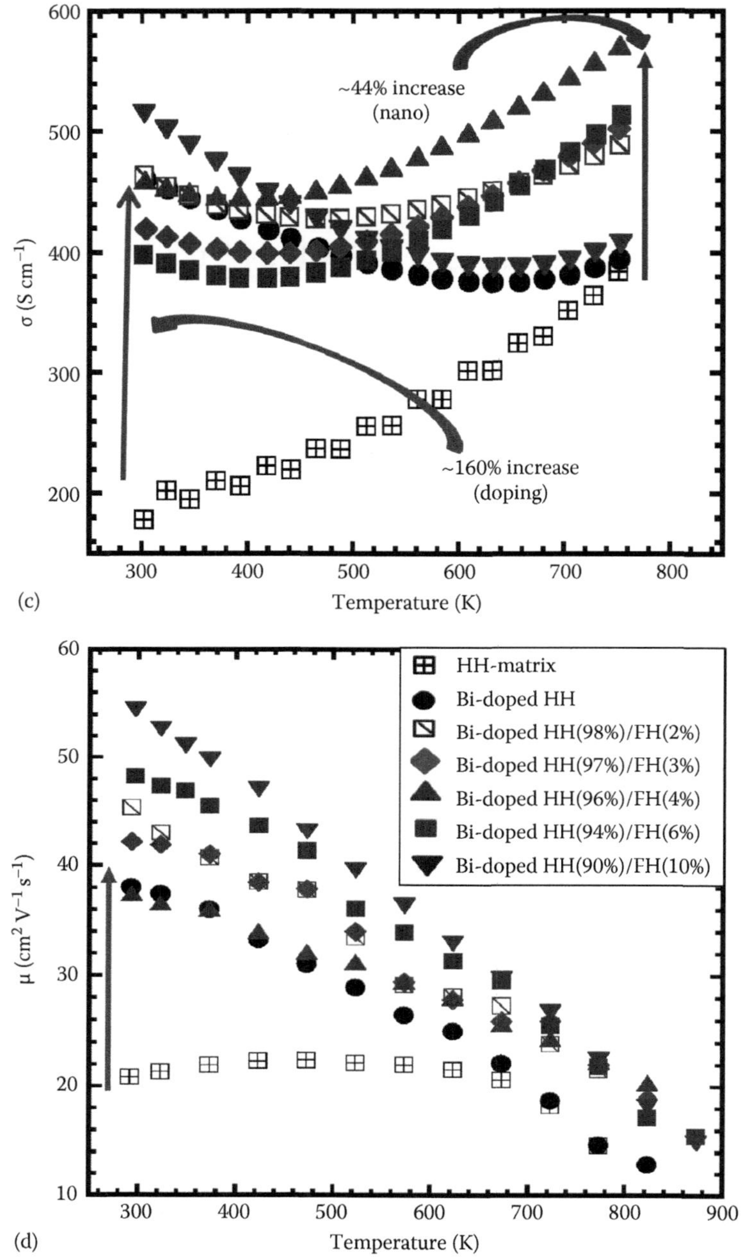

FIGURE 7.11 ***(Continued)***
Temperature-dependent electronic properties of Bi-doped HH(1 − x)/FH(x) composites ($0 \le x \le 0.10$). (c) the electrical conductivity, and (d) the carrier mobility. (Reprinted with permission from Makongo, J.P.A., Misra, D.K., Zhou, X. et al., Simultaneous large enhancements in thermopower and electrical conductivity of bulk nanostructured half-Heusler alloys, *J. Am. Chem. Soc.*, 133, 18843–18852. Copyright 2011 American Chemical Society.)

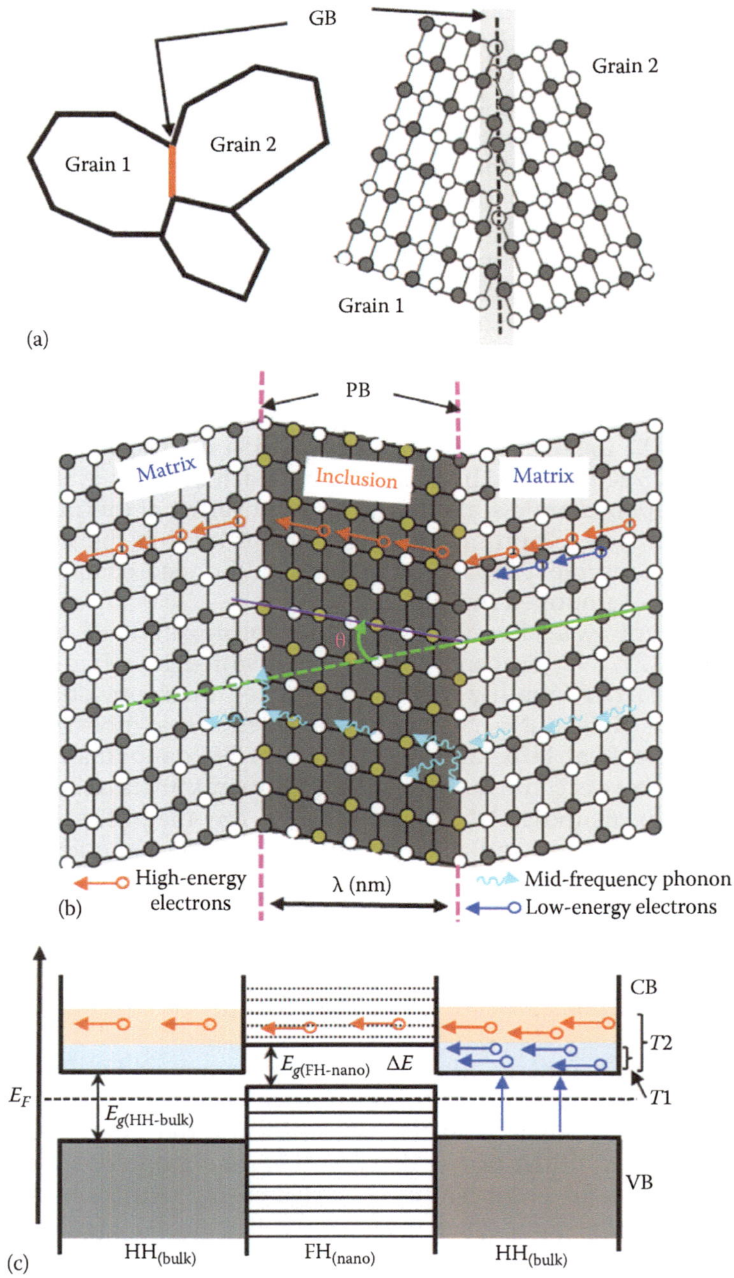

FIGURE 7.12
Schematic diagram of the energy-filtering effect at half-Heusler/full-Heusler (HH/FH) interfaces. (a) Incoherent interface at grain boundaries, (b) coherent phase boundaries between matrix and inclusion phases with similar structure, and (c) nanometer-scale heterojunction between bulk HH matrix and nanometer-scale FH inclusion highlighting the proposed low-energy electron filtering. (Reprinted with permission from Makongo, J.P.A., Misra, D.K., Zhou, X. et al., Simultaneous large enhancements in thermopower and electrical conductivity of bulk nanostructured half-Heusler alloys, *J. Am. Chem. Soc.*, 133, 18843–18852. Copyright 2011 American Chemical Society.)

subgroup of charge carriers. For the $Zr_{0.25}Hf_{0.75}Ni_{1+x}Sn$ and $Zr_{0.25}Hf_{0.75}Ni_{1+x}Sn_{0.975}Bi_{0.025}$ nanocomposites reported here, the carrier density increases exponentially with temperature since the LE electrons gain enough energy to cross the energy offset, which gradually increases the portion of LE electron in the overall electron population and thus minimizes the drop of the overall carrier mobility with rising temperature. Decoupling the temperature dependence of the mobility of LE and HE electrons has been achieved using sub-10 nm scale coherent interfaces between the HH matrix and FH inclusions. Because of the simultaneous enhancement in both thermopower and electrical conductivity, a dramatic improvement in the PF was observed for selected $Zr_{0.25}Hf_{0.75}Ni_{1+x}Sn$ and $Zr_{0.25}Hf_{0.75}Ni_{1+x}Sn_{0.975}Bi_{0.025}$ nanocomposites [39]. Since the enhancement in the PF mainly results from the large increase in thermopower rather than from the increase in the electrical conductivity, minimal increase in the thermal conductivity is observed for these nanocomposites. Therefore, a figure of merit as high as ZT ~0.7 at 775 K has been achieved for the $Zr_{0.25}Hf_{0.75}Ni_{1.02}Sn$ composite. This represents a 250% enhancement compared to ZT of 0.2 for the undoped $Zr_{0.25}Hf_{0.75}NiSn$ matrix. This slightly surpasses the ZT of ~0.6 at 775 K for $Zr_{0.25}Hf_{0.75}NiSn_{0.975}Sb_{0.025}$ achieved by a Sb-doping strategy, indicating the concept of atomic-scale structural engineering of thermoelectrics (ASSET) as a promising approach to optimize performance of leading thermoelectric material systems. Similar enhancement in ZT is achieved for Bi-doped composites compared to the Bi-doped matrix or undoped matrix.

The ASSET concept was also applied to the optimization of the thermoelectric performance of $Ti_{0.1}Zr_{0.9}Ni_{1+x}Sn$ composites. This resulted in a large enhancement of thermopower and carrier mobility in the synthesized quantum dot engineered bulk semiconductors [72]. $Ti_{0.1}Zr_{0.9}Ni_{1+x}Sn$ nanocomposites are synthesized by solid-state reaction of the elements at 1173 K. The formation of FH nanodots coherently embedded in the HH matrix was confirmed by high resolution transmission electron microscopy (HRTEM) study, as indicated in Figure 7.13. A low-magnification TEM image of the $Ti_{0.1}Zr_{0.9}Ni_{1.04}Sn$ composite indicates the presence of FH nanodots with broad size distribution ranging from 10 to 60 nm. The high-magnification TEM image of the $Ti_{0.1}Zr_{0.9}Ni_{1.10}Sn$ composite shows excellent lattice coherency with no noticeable strain at the HH/FH interfaces, suggesting an endotaxial growth mechanism, which is facilitated by the 3D structural similarity between FH and HH. This high degree of coherency between FH inclusions and the HH matrix is beneficial to carrier transfer across the interface.

As anticipated from the energy-filtering effect of the embedded FH nanoinclusions, lower carrier density was measured at 300 K for various $Ti_{0.1}Zr_{0.9}Ni_{1+x}Sn$ nanocomposites (Figure 7.14) when compared to the nanoinclusion-free matrix. In addition, the temperature dependence of the carrier density in various samples showed a thermally activated behavior with the onset temperature decreasing with increasing FH content. The carrier density of the composites decreases with increasing fraction of FH inclusions. The observed drop in carrier density with increasing fraction of FH inclusions cannot be explained by an increase in the bandgap but is attributed to the energy-filtering effect by the energy offset formed at the interfaces between FH nanodots and the HH matrix. The fraction of LE electrons trapped by the energy offset increases with the density of HH/FH interfaces in the $Ti_{0.1}Zr_{0.9}Ni_{1+x}Sn$ nanocomposites but also decreases with increasing size of FH nanostructures, which determines the energy offset at the interfaces. Thus, as the fraction of FH inclusions increases, a larger density of small FH nanodots or a mixture of small and large FH inclusions is present, leading to a high density of the interfaces and therefore an increase in the strength of the

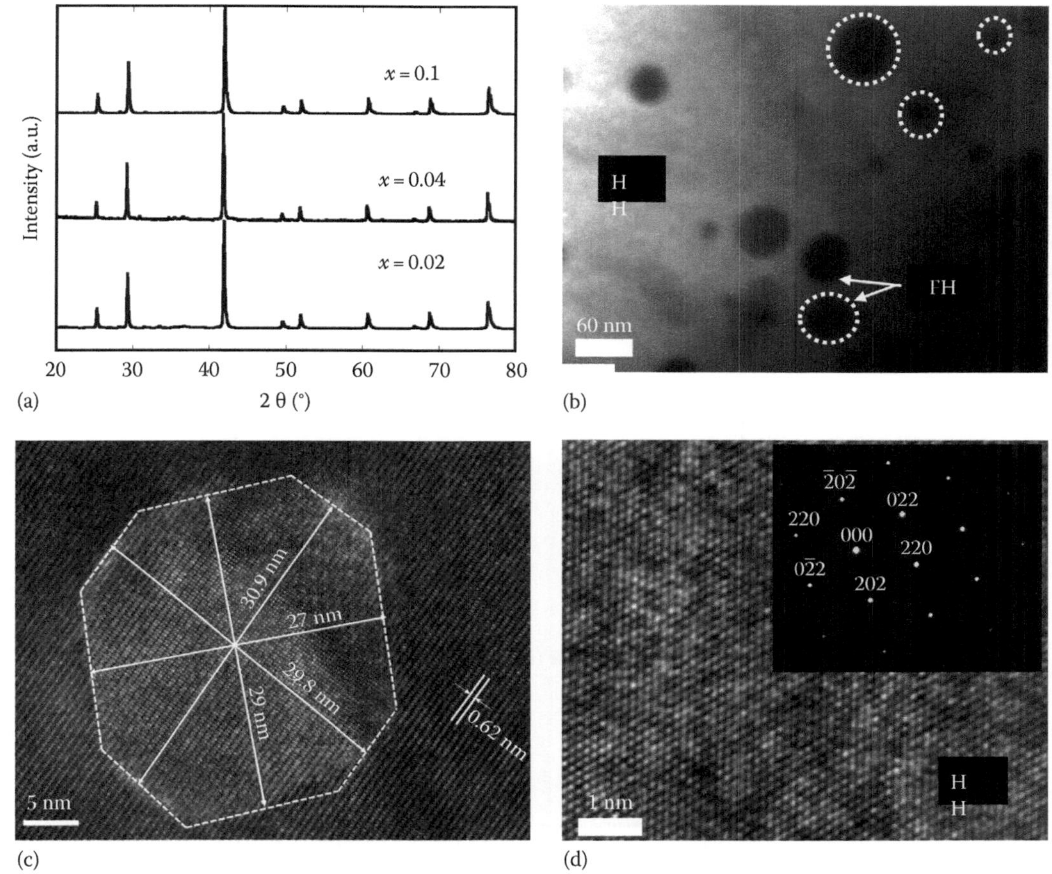

FIGURE 7.13
Structural characteristics of $Ti_{0.1}Zr_{0.9}Ni_{1+x}Sn$ nanocomposites. (a) X-ray diffraction (XRD) patterns of the as-synthesized $Ti_{0.1}Zr_{0.9}Ni_{1+x}Sn$ material showing only diffraction peaks of pure half-Heusler (HH) phase. (b) Low-magnification image of the nanocomposite with $x = 0.04$ showing full-Heusler (FH) nanoparticles with broad size distribution (10–60 nm) coherently embedded inside the HH ($Ti_{0.1}Zr_{0.9}NiSn$) matrix. (c) High-magnification image of a typical nearly spherical FH nanoparticle (QD) from the nanocomposite with $x = 0.1$, showing detailed lattice structure. (d) High-magnification image of the HH matrix showing excellent crystallinity and the inset showing selected-area electron diffraction pattern indicating the face-centered cubic structure of the HH matrix. (Reprinted with permission from Liu, Y.F., Sahoo, P., Makongo, J.P.A. et al., Large enhancements of thermopower and carrier mobility in quantum dot engineered bulk semiconductors, *J. Am. Chem. Soc.*, 135, 7486–7495. Copyright 2013 American Chemical Society.)

energy-filtering effect. As the temperature increases, the trapped electrons gradually gain enough energy to cross the energy barrier, which accounts for the exponential increase of the carrier density with temperature.

As a result of the drop in the carrier density with increasing fraction of FH inclusions, the thermopower also increases, especially for the HH(0.90)/FH(0.10) composites. However, the reduction of the carrier density is not the only factor contributing to thermopower enhancement because the marginal increase of carrier concentration for the HH(0.90)/FH(0.10) composite should not incur such dramatic enhancement in thermopower compared to the HH(0.96)/FH(0.04) composite. Therefore, it is also anticipated that the carrier effective mass, m^*, in various $Ti_{0.1}Zr_{0.9}Ni_{1+x}Sn$ nanocomposites also increases [72]. Regardless of the composition, the thermopower slowly increases with increasing temperature and reaches a maximum at temperatures between 600 and 750 K. The temperature for the maximum thermopower decreases with an increasing amount of FH inclusion. The existence of maximum thermopower indicates the thermal excitation at sufficiently high temperatures of LE electrons initially trapped by the energy barrier. Furthermore, increasing the excess of elemental Ni resulted in larger-size FH inclusions. This implies a reduction in the heights of the energy barrier and thus a drop in the temperature of thermal excitation for trapped electrons (Figure 7.15). It was also found that the embedded FH nanoinclusions significantly increase the carrier mobility in various $Ti_{0.1}Zr_{0.9}Ni_{1+x}Sn$ nanocomposites. This large increase in carrier mobility was attributed to

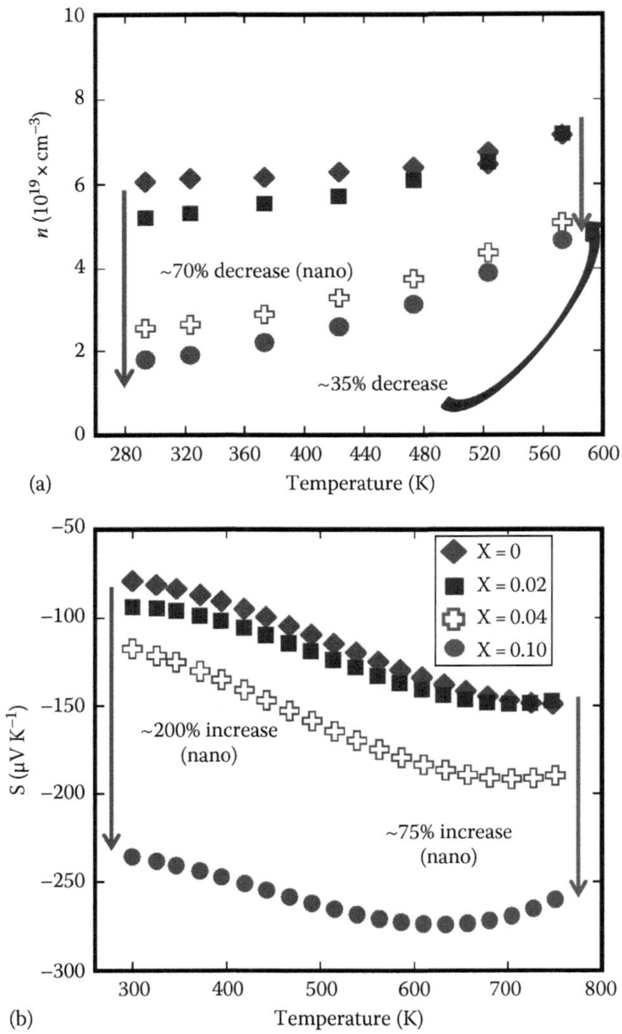

FIGURE 7.14
Temperature-dependent electronic properties of $Ti_{0.1}Zr_{0.9}Ni_{1+x}Sn$ composites. (a) The carrier concentration, (b) the Seebeck coefficient.

(Continued)

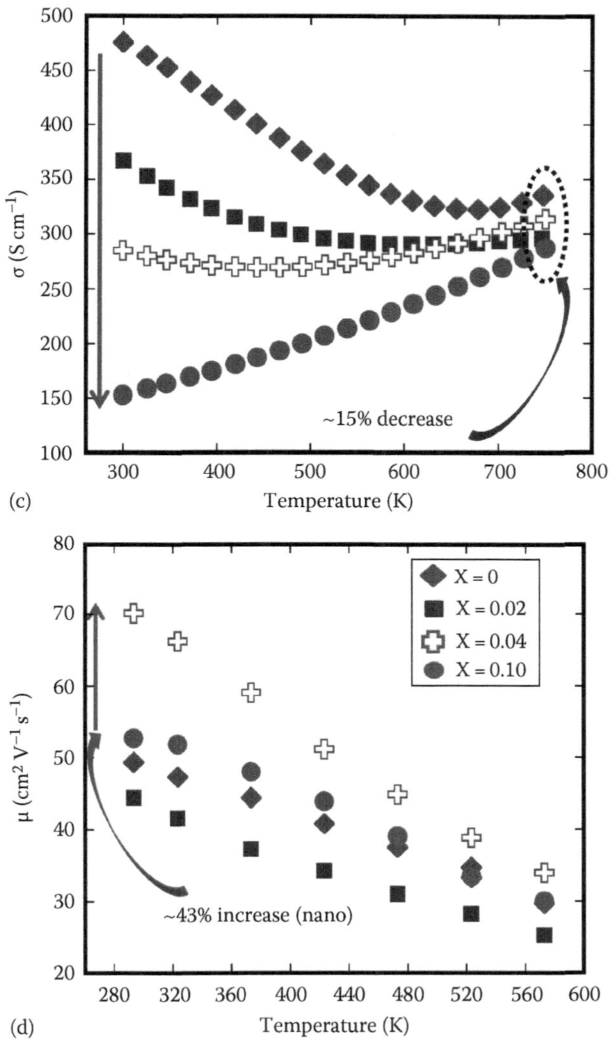

FIGURE 7.14 *(Continued)*
Temperature-dependent electronic properties of $Ti_{0.1}Zr_{0.9}Ni_{1+x}Sn$ composites. (c) the electrical conductivity, and (d) the carrier mobility. (Reprinted with permission from Liu, Y.F., Sahoo, P., Makongo, J.P.A. et al., Large enhancements of thermopower and carrier mobility in quantum dot engineered bulk semiconductors, *J. Am. Chem. Soc.*, 135, 7486–7495. Copyright 2013 American Chemical Society.)

both the increase in relaxation time τ and decrease in carrier effective mass ($\mu = e\tau/m^*$) [72]. The reduction in the carrier density by the energy-filtering effect partly accounts for the increase in relaxation time since a lower frequency of electron–electron scattering is expected in this situation. However, the drastic difference in the carrier mobility of HH(0.96)/FH(0.04) and HH(0.90)/FH(0.10) composites, despite the similarity in their carrier density, implies that there are additional contributing factors, such as carrier scattering by ionized impurities and structural defects (dislocations, vacancies), affecting the carrier mobility in these composites (Figure 7.15).

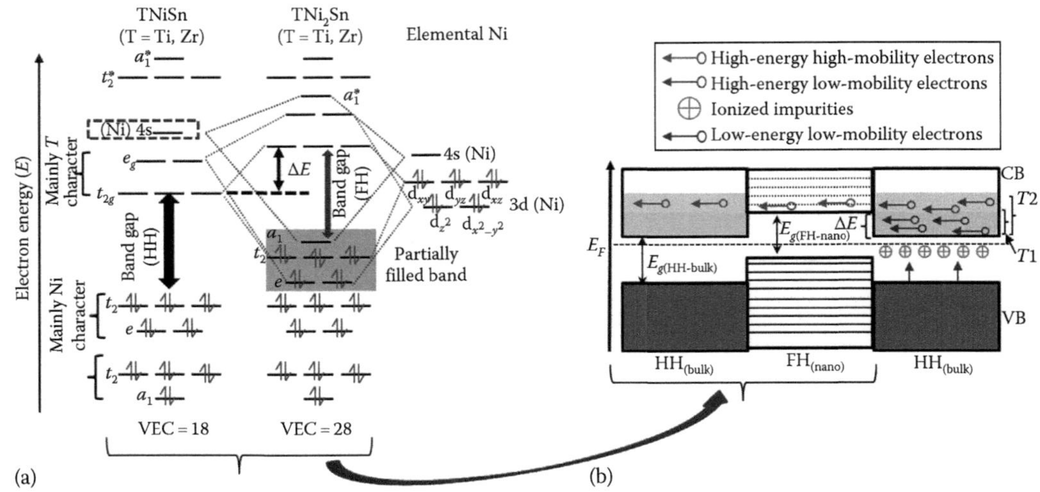

FIGURE 7.15
Illustrations of the half-Heusler (HH) and full-Heusler (FH) band structures alignment at the HH/FH interfaces. (a) Molecular orbital diagram of FH (TNi_2Sn) generated from that of the HH phase (TNiSn) upon insertion of elemental Ni into the vacant tetrahedral site. Comparison of the HH and FH molecular orbital diagrams reveals an energy offset of both the valence-band maxima and the CBMs leading to nanometer-scale type II (staggered gap) heterojunctions at the HH/FH interface. The CB energy offset, ΔE, acts as an energy filter for conduction electrons within the HH matrix. (b) Proposed mechanism of the filtering of low-energy electrons (from the CB of HH) at the potential barrier, ΔE (reducing effective carrier density), and spatial separation of high-energy electrons (within the CB of FH) from ionized impurities at nanometer-scale HH/FH heterojunctions. Cyan and orange colors represent distributions of low- and high-energy electrons at temperatures $T1$ and $T2$ ($T2 > T1$) within the CB of the HH matrix. (Reprinted with permission from Liu, Y.F., Sahoo, P., Makongo, J.P.A. et al., Large enhancements of thermopower and carrier mobility in quantum dot engineered bulk semiconductors, *J. Am. Chem. Soc.*, 135, 7486–7495. Copyright 2013 American Chemical Society.)

The large drop in the carrier density and marginal increase in the carrier mobility at 300 K for $Ti_{0.1}Zr_{0.9}Ni_{1+x}Sn$ nanocomposites resulted in a reduction in the electrical conductivity with increasing fraction of FH inclusions. Interestingly, upon increasing the mole fraction of FH inclusions, the electronic transport gradually switched from heavily doped semiconductivity ($x = 0$) to the typical intrinsic semiconducting behavior. The observed increase in the electrical conductivity for HH/FH composites at high temperatures stems from the exponential increase in the carrier density and the linear decrease in the carrier mobility with rising temperature, which ensures a marginal decrease in the electrical conductivity for HH/FH composites compared to the HH matrix at high temperatures. This combination of large enhancement (up to 200% at 300 K) in the thermopower and marginal decrease (~15%) in the electrical conductivity at high temperatures led to a dramatic increase in the PF. The PF value of 1150 $\mu W\ m^{-1}\ K^{-2}$ at 780 K for the HH(0.96)/FH(0.04) composite corresponds to a ~53% enhancement compared to the value of 750 $\mu W\ m^{-1}\ K^{-2}$ for the $Ti_{0.1}Zr_{0.9}NiSn$ HH matrix. Even larger enhancement of the PF (1960 $\mu W\ m^{-1}\ K^{-2}$ at 780 K) was observed for the HH(0.90)/FH(0.10) composite. In addition to PF enhancement, the embedded FH nanoinclusions also contribute significantly to the reduction (~30% for HH(0.90)/FH(0.10)) in the lattice thermal conductivity. This is attributed to the enhancement of phonon scattering at the high density of HH/FH interfaces.

The earlier discussion of electronic transport in $Ti_{0.1}Zr_{0.9}Ni_{1+x}Sn$ nanocomposites showed that the incorporation of FH nanodots results in a change of electronic characteristics, such as the band structure, bandgap, carrier effective mass, and carrier density of the composite. The magnitude of the alteration of these parameters strongly depends on the density, size, and distribution of FH nanoinclusions in the HH matrix. These results once again demonstrate the effectiveness of the ASSET approach for the optimization of the thermoelectric performance of HH alloys.

To explore the suitability of the ASSET approach for doped semiconductors, several Sb-doped $Ti_{0.1}Zr_{0.9}Ni_{1+x}Sn_{0.975}Sb_{0.025}$ nanocomposites were synthesized, and the effect of the HH/FH interfacial energy potential on the transport of extrinsic carriers was investigated [71]. The presence of FH nanoprecipitates in Sb-doped $Ti_{0.1}Zr_{0.9}Ni_{1+x}Sn_{0.975}Sb_{0.025}$ nanocomposites was confirmed by the HRTEM images (see Figure 7.16). Well-dispersed

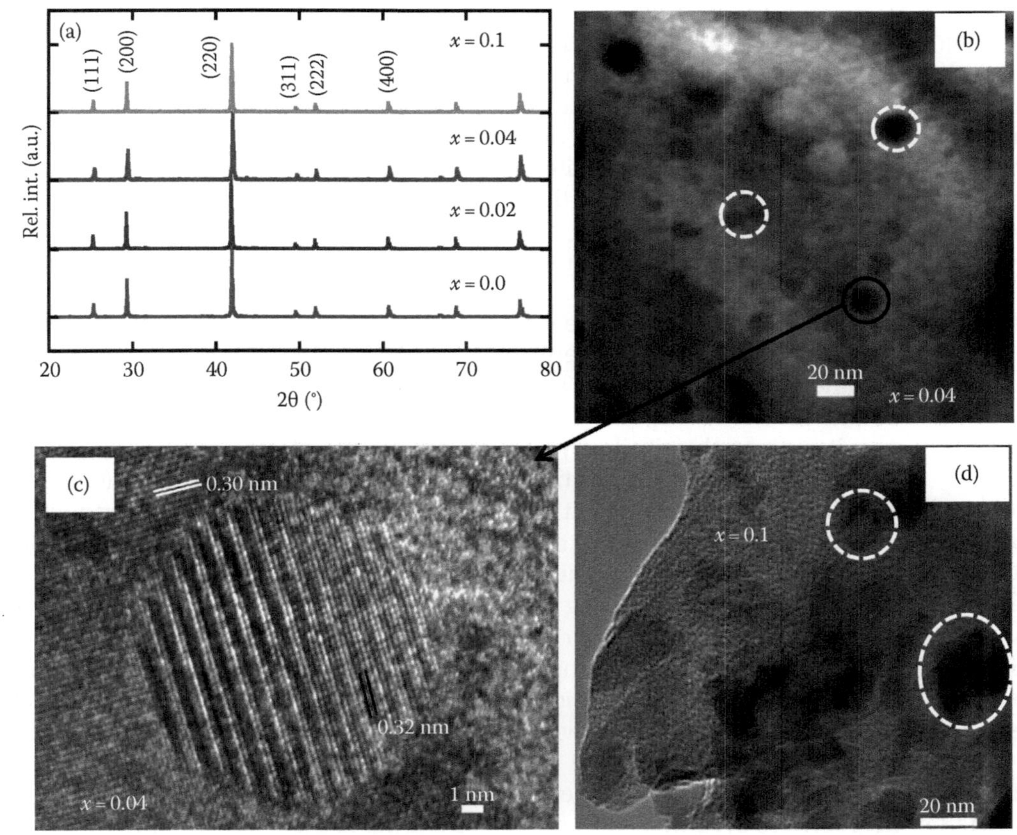

FIGURE 7.16
Structural characteristics of $Ti_{0.1}Zr_{0.9}Ni_{1+x}Sn_{0.975}Sb_{0.025}$ ($x=0$, 0.02, 0.04, 0.1) nanocomposites. (a) Powder x-ray diffraction patterns showing only peaks corresponding to the half-Heusler (HH) matrix. (b) Low-magnification TEM image of the sample with $x=0.04$, revealing the presence of spherical FH precipitates embedded in the HH matrix. (c) High-magnification image of a typical full-Heusler (FH) nanostructure from the sample with $x=0.04$, showing a highly coherent phase boundary with the surrounding HH matrix. (d) Low-magnification TEM image of the sample with $x=0.1$, showing high agglomeration of FH nanostructures into large particles. (From Liu, Y.F., Page, A., Sahoo, P. et al., Electronic and phonon transport in Sb-doped $Ti_{0.1}Zr0.9Ni_{1+x}Sn_{0.975}Sb_{0.025}$ nanocomposites, *Dalton Trans.*, 43, 8094–8010, 2014. Reproduced by permission of The Royal Chemical Society.)

spherical precipitates of FH with size ranging from <5 to 20 nm are observed within the HH(0.96)/FH(0.04) composite. The precipitates form coherent interfaces with the HH matrix. As the fraction of FH increased to 10% (HH(0.90)/FH(0.10) composite), FH precipitates agglomerated into larger particles with size ranging from 20 to 100 nm. Figure 7.17 gives the temperature dependence of the electronic transport properties of $Ti_{0.1}Zr_{0.9}Ni_{1+x}Sn_{0.975}Sb_{0.025}$ nanocomposites. A large increase (870%) in the carrier density is observed upon substitution of 2.5% Sn by Sb in the undoped matrix, $Ti_{0.1}Zr_{0.9}NiSn$. Starting from this heavily doped matrix, the carrier density remains unchanged upon incorporating up to 4% FH nanoinclusions. This suggests that the FH inclusions in these samples are transparent to charge carriers and also do not contribute additional carriers to the matrix. However, increasing the volume fraction of the FH inclusions to 10% [HH(0.90)/FH(0.10)] induced a 38% increase in carrier concentration. This indicates that the FH inclusions in this samples act as dopants. The observed difference in the behavior of FH inclusions in the HH(0.90)/FH(0.10) and HH(0.96)/FH(0.04)] samples is associated with the difference in the average particle size of FH inclusions in these composites. In the HH(0.96)/FH(0.04)] sample, FH precipitates with an average size smaller than 20 nm are well dispersed in the HH matrix and are not electronically inert with respect to the matrix. However, in the HH(0.90)/FH(0.10) samples, the FH precipitates agglomerate into larger particles and therefore recover the metallic behavior associated with bulk FH phase. It is interesting to note that despite the 38% higher carrier concentration observed for the HH(0.90)/FH(0.10) composite, the thermopower is comparable to the values observed for the remaining $Ti_{0.1}Zr_{0.9}Ni_{1+x}Sn_{0.975}Sb_{0.025}$ nanocomposite samples. This trend in the thermopower is attributed to an increase in the carrier effective mass ($m^* = 3.5\ m_e$) for the HH(0.90)/FH(0.10) composite.

Sb doping substantially enhanced the electrical conductivity due to the large increase in the carrier density. However, the difference in the electrical conductivity among all the composites and the Sb-doped matrix is marginal, especially at high temperature. This is easy to understand for the Sb-doped HH matrix and the HH(0.98)/FH(0.02) and HH(0.96)/FH(0.04) composites given the similarity in their carrier densities. However, one should normally expect an increase in the electrical conductivity of the HH(0.90)/FH(0.10) composite due to the increased carrier density. Therefore, the observed electrical conductivity for the HH(0.90)/FH(0.10) composite is attributed to a drop in the effective carrier mobility, which offset the increase in the carrier density.

It can be derived from the preceding analysis that the incorporation of coherent FH nanoinclusions in a heavily doped matrix has a marginal effect on the electronic transport of $Ti_{0.1}Zr_{0.9}Ni_{1+x}Sn_{0.975}Sb_{0.025}$ nanocomposites. This suggests that extrinsic carriers from Sb dopants dominate electrical conduction in such nanocomposites. Such extrinsic carriers presumably occupy shallow states with energy higher than the CB minimum (CBM) of the FH inclusions and are therefore not trapped by the potential barrier, ΔE, at the HH/FH inclusion. To probe this hypothesis, the composites $Zr_{0.25}Hf_{0.75}Ni_{1+x}Sn_{1-y}Sb_y$ doped to different degree were synthesized in order to investigate the behavior of extrinsic carriers at the HH/FH interfaces [73]. Similar to previous results on HH/FH composites, the formation of FH nanoparticles embedded into the Sb-doped HH matrix was confirmed by HRTEM study, as indicated in Figure 7.18. Regardless of doping level, increasing the amount of excess elemental Ni leads to an increase in the density and average size (ranging from 5 to 100 nm) of FH nanoinclusions. As discussed earlier, smaller and larger FH nanoinclusions have very different electronic behavior in the HH matrix. Large FH inclusions display doping behavior as a metal phase, while the small precipitates behave like semiconductors.

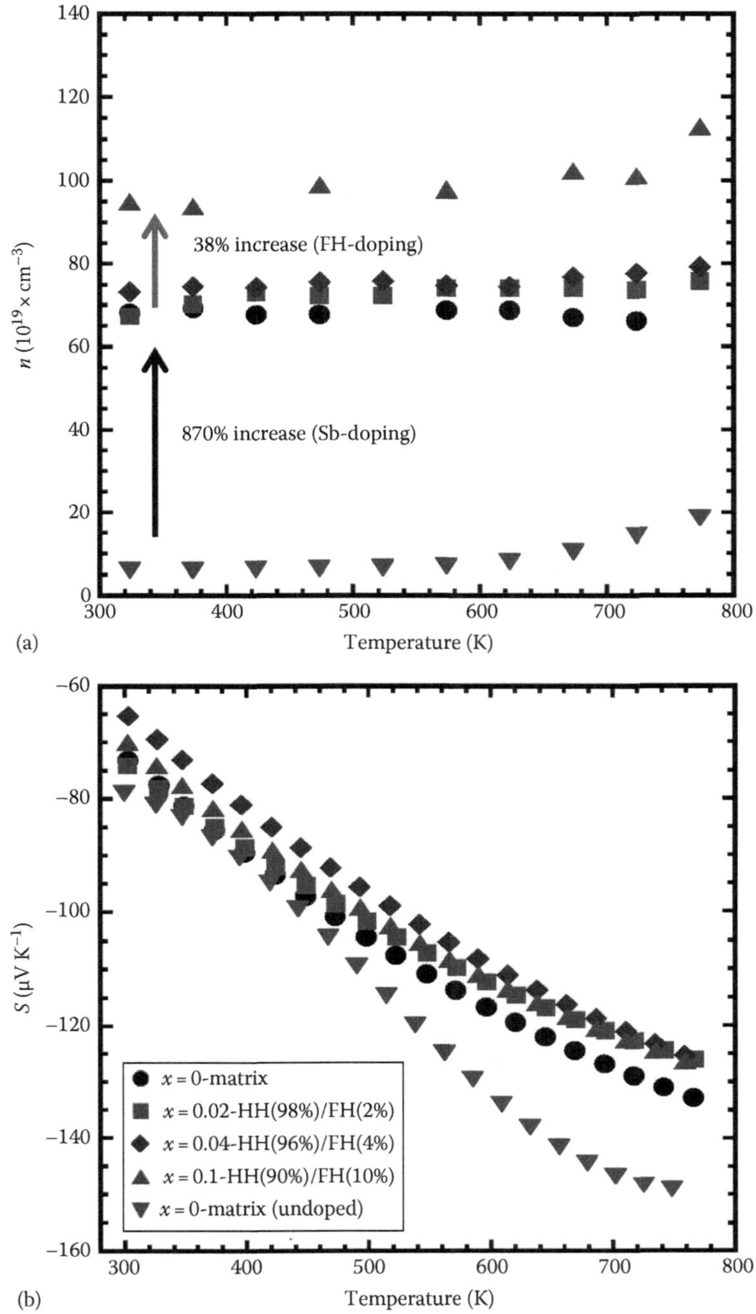

FIGURE 7.17
Temperature-dependent electronic properties of $Ti_{0.1}Zr_{0.9}Ni_{1+x}Sn_{0.975}Sb_{0.025}$ composites. (a) The carrier concentration, (b) the Seebeck coefficient. *(Continued)*

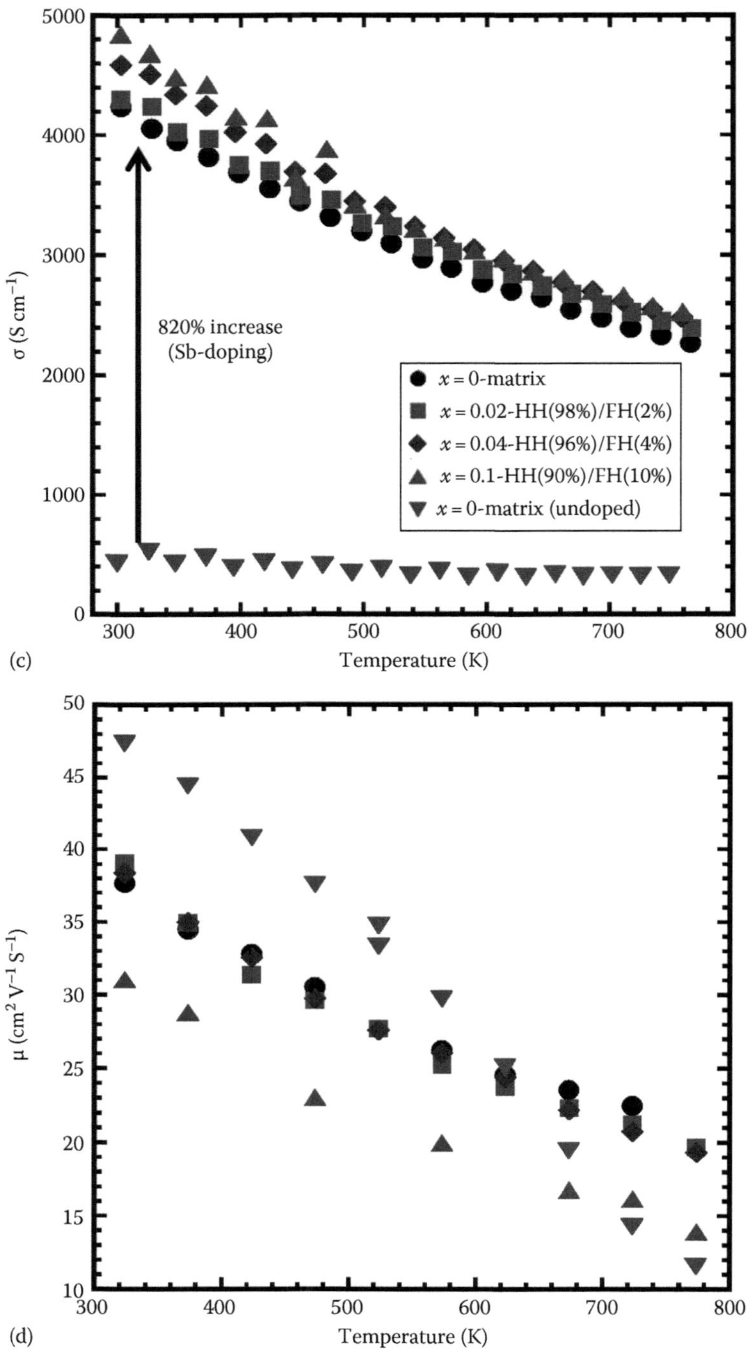

FIGURE 7.17 *(Continued)*
Temperature-dependent electronic properties of $Ti_{0.1}Zr_{0.9}Ni_{1+x}Sn_{0.975}Sb_{0.025}$ composites. (c) the electrical conductivity, and (d) the carrier density. (From Liu, Y.F., Page, A., Sahoo, P. et al., Electronic and phonon transport in Sb-doped $Ti_{0.1}Zr0.9Ni_{1+x}Sn_{0.975}Sb_{0.025}$ nanocomposites, *Dalton Trans.*, 43, 8094–8010, 2014. Reproduced by permission of The Royal Chemical Society.)

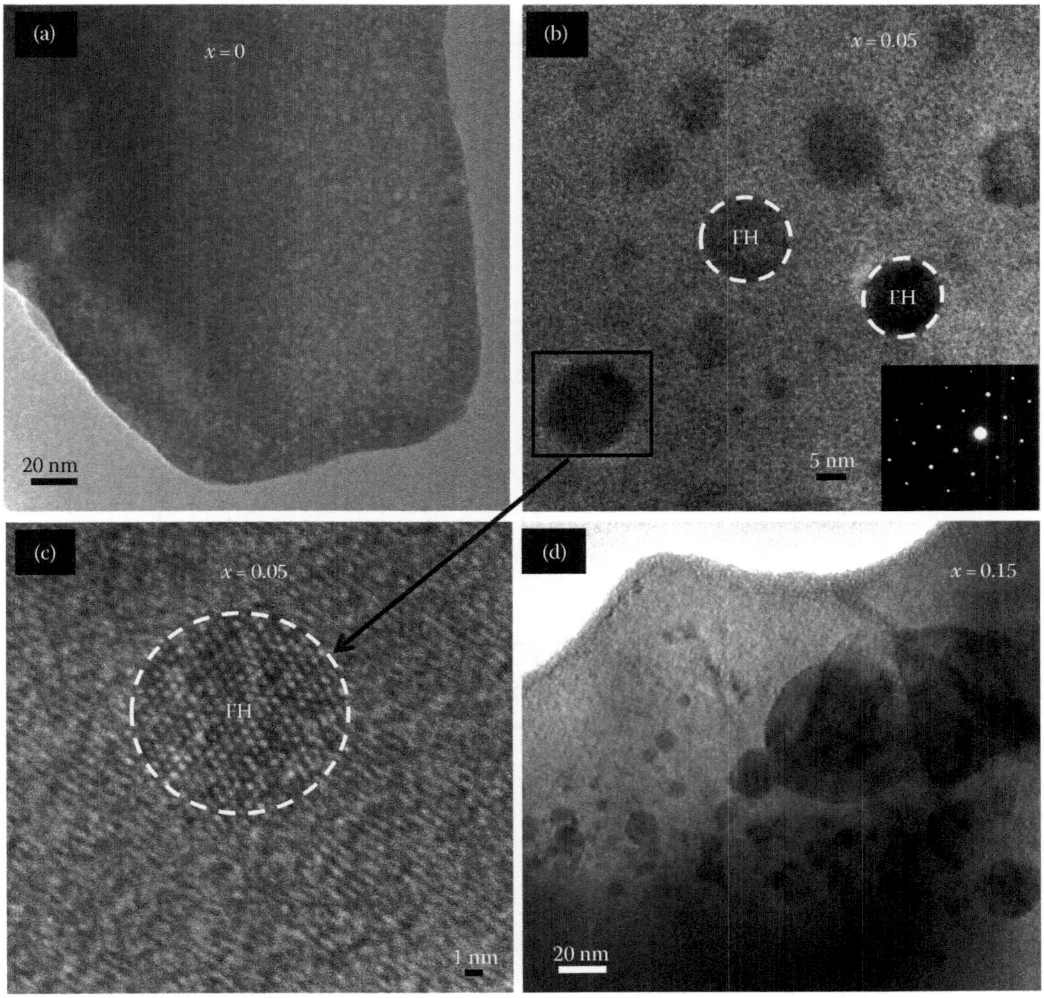

FIGURE 7.18
Structural characteristics of $Zr_{0.25}Hf_{0.75}Ni_{1+x}Sn_{0.995}Sb_{0.005}$ nanocomposites. (a) Low-magnification TEM image of the half-Heusler (HH) matrix, revealing the matrix is clear with no inclusion or impurities. (b) Medium-magnification TEM image of the sample with $x = 0.05$ showing full-Heusler (FH) particles with size range from 5 to 20 nm coherently embedded inside the HH matrix and selected-area electron diffraction pattern (inset of b) indicating the face-centered cubic structure of the HH matrix. (c) High-magnification TEM image of a selected FH particle from the sample with $x = 0.05$, showing nearly spherical FH nanoinclusion lattice details. (d) Low-magnification TEM image of the sample with $x = 0.15$, showing high agglomeration of the FH nanophase into large particles. (Reprinted from *J. Solid State Chem.*, 234, Liu, Y.F., Makongo, J.P.A., Page, A. et al., Distribution of impurity states and charge transport in $Zr_{0.25}Hf_{0.75}Ni_{1+x}Sn_{1-y}Sb_y$ nanocomposites, 72–86, Copyright 2016, with permission from Elsevier.)

Figure 7.19 shows the temperature dependence of the electrical properties of $Zr_{0.25}Hf_{0.75}Ni_{1+x}Sn_{0.975}Sb_{0.025}$ composites. Regardless of the composition, the electrical conductivities decrease with rising temperature, displaying the typical behavior of heavily doped semiconductors. These doped composites showed slightly higher electrical conductivity and lower thermopower when compared to the doped matrix. These changes in the electrical conductivity and thermopower suggest a marginal doping effect

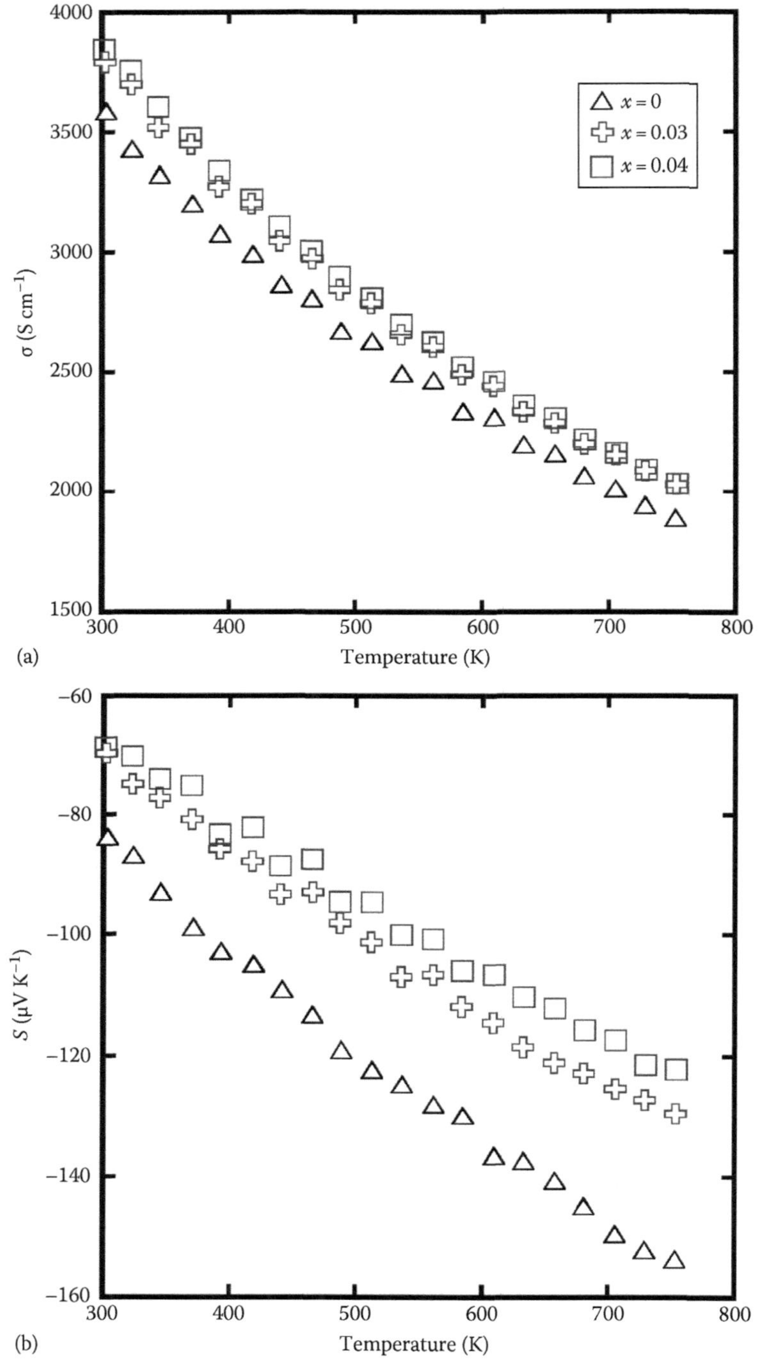

FIGURE 7.19
Temperature-dependent electronic properties of $Zr_{0.25}Hf_{0.75}Ni_{1+x}Sn_{0.975}Sb_{0.025}$ samples. (a) The electrical conductivity, (b) the Seebeck coefficient.

(Continued)

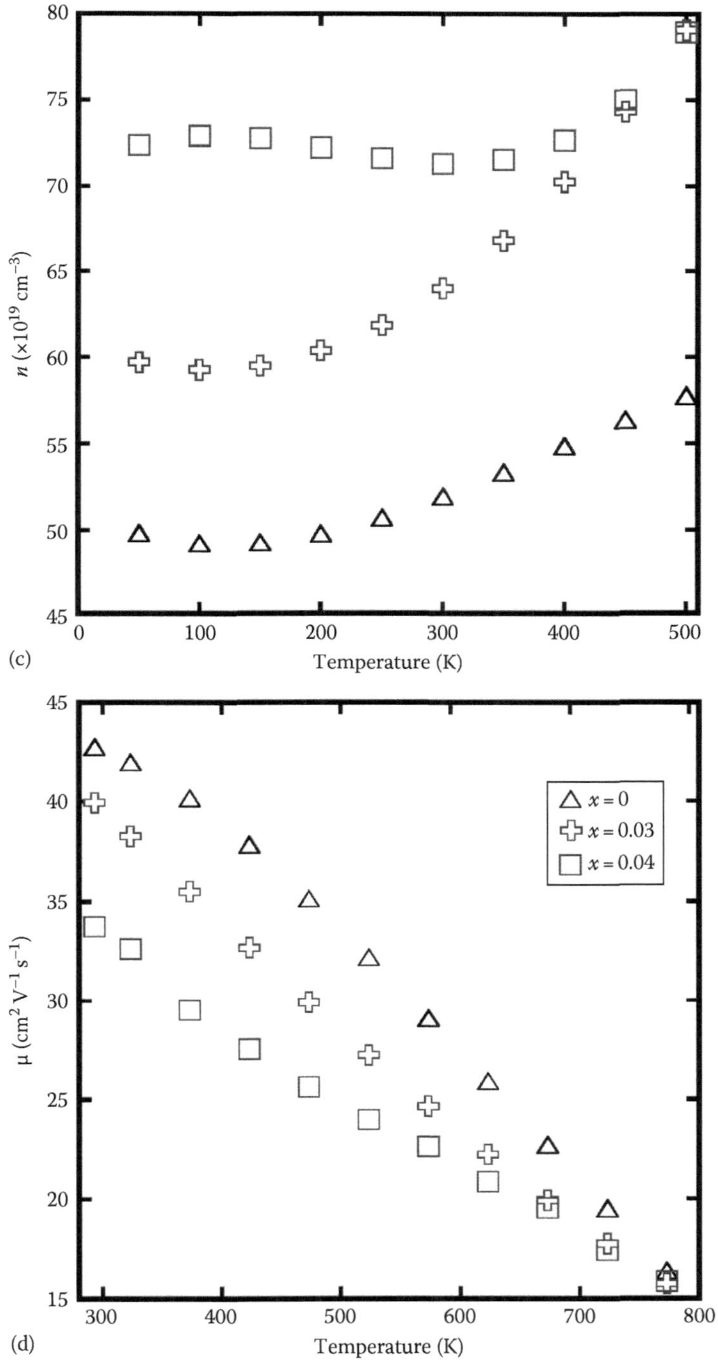

FIGURE 7.19 (*Continued*)
Temperature-dependent electronic properties of $Zr_{0.25}Hf_{0.75}Ni_{1+x}Sn_{0.975}Sb_{0.025}$ samples. (c) the carrier concentration, and (d) the carrier mobility. (Reprinted from *J. Solid State Chem.*, 234, Liu, Y.F., Makongo, J.P.A., Page, A. et al., Distribution of impurity states and charge transport in $Zr_{0.25}Hf_{0.75}Ni_{1+x}Sn_{1-y}Sb_y$ nanocomposites, 72–86, Copyright 2016, with permission from Elsevier.)

from FH inclusions. This analysis is supported by the Hall measurement, as shown in Figure 7.19c and d. The carrier density increases with increasing FH content, and correspondingly the carrier mobility decreases. Upon increasing the temperature, the carrier density initially remains almost constant up to a critical temperature and thereafter begins to increase with rising temperature. The onset temperature of the increase in carrier density increases with increasing FH content from 373 K ($x = 0$) to 423 K ($x = 0.03$) and to 623 K ($x = 0.04$). Similar thermopower values were observed for the HH(0.97)/FH(0.03) composite and HH(0.96)/FH(0.04) composite despite their dramatic difference in carrier density. This suggests a larger carrier effective mass for the HH(0.96)/FH(0.04) composite.

It is interesting to note the simultaneous increase in the carrier density and mobility for the Sb-doped HH matrix compared to the undoped HH matrix (Figure 7.20). This indicates that Sb doping contributes a large fraction of high-mobility extrinsic conducting electrons to the matrix and thus increases the overall carrier mobility. For this particular heavily doped system, the energy-filtering effect is not obvious since no dramatic decrease in carrier density (LE electrons) is observed, suggesting that the electrical transport in $Zr_{0.25}Hf_{0.75}Ni_{1+x}Sn_{0.975}Sb_{0.025}$ is determined by the large fraction of HE extrinsic carriers, which are contributed by Sb doping and additional doping from metallic FH inclusions. These dopants occupy shallow states in the bandgap, which facilitates their thermal excitation across the potential barrier at the interfaces between FH inclusions and the HH matrix. Since the energy offset only filters the LE electrons, the energy-filtering effect does not affect the carrier density, which is dominated by HE electrons. From this analysis, it appears that the combination of doping and energy-filtering strategies requires a good balance between the fraction of high-mobility extrinsic carriers gained by doping and the fraction of LE intrinsic carriers filtered at the inclusion/matrix interfaces.

To verify this approach, we have investigated the electrical properties of more lightly doped $Zr_{0.25}Hf_{0.75}Ni_{1+x}Sn_{0.99}Sb_{0.01}$ composites (Figure 7.21). As anticipated from the low doping level, the electrical conductivity is lower compared to 2.5% Sb-doped composites. However, the 1% Sb-doped samples still display heavily doped semiconducting behavior. Interestingly, the electrical conductivity decreases with the addition of FH inclusions, which is in stark contrast to the behavior observed for the 2.5% Sb-doped composites. This behavior is attributed to both the decrease in the fraction of extrinsic carriers (doping level) and the energy filtering of LE electrons at FH/HH interfaces. This latter factor also accounts for the observed drop in the electrical conductivity with increasing FH content. This is supported by the Hall carrier concentration and mobility data. The room temperature carrier concentration is decreased to 35×10^{19} cm^{-3} for the 1% Sb-doped HH matrix from 50×10^{19} cm^{-3} for the 2.5% Sb-doped matrix, as indicated in Figure 7.21c. Among the 1% Sb-doped composites, the carrier concentration is further reduced to ~23×10^{19} cm^{-3} for the HH(0.97)/FH(0.03) composite. Increasing the fraction of FH inclusions to 4% and 5% resulted in a marginal increase in carrier concentration to ~26×10^{19} and 30×10^{19} cm^{-3} at room temperature, respectively. This marginal increase in carrier density indicates a doping effect from large FH inclusions formed when the FH content increases. Despite the doping effect from FH inclusions, the carrier density in composites with 4% and 5% FH inclusions is still lower than that of the corresponding matrix (Figure 7.21). The carrier density remains almost constant with rising temperature until a critical temperature T_c, above which it displays a nearly exponential increase with temperature. Interestingly, this critical temperature decreases with rising

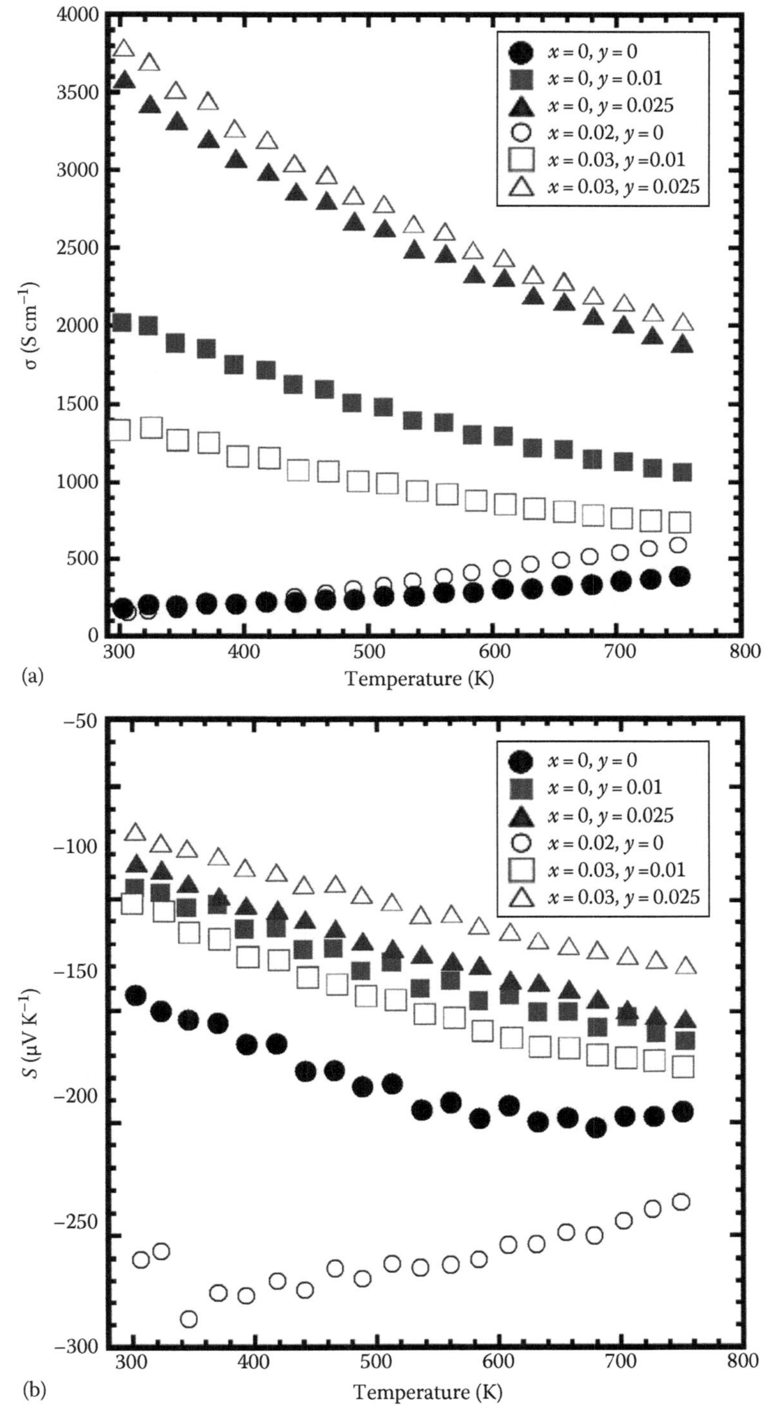

FIGURE 7.20
Comparison of electronic transport in $Zr_{0.25}Hf_{0.75}Ni_{1+x}Sn_{1-y}Sb_y$ with various doping levels ($y = 0$, 0.01 and 0.025). (a) The electrical conductivity, (b) the Seebeck coefficient. *(Continued)*

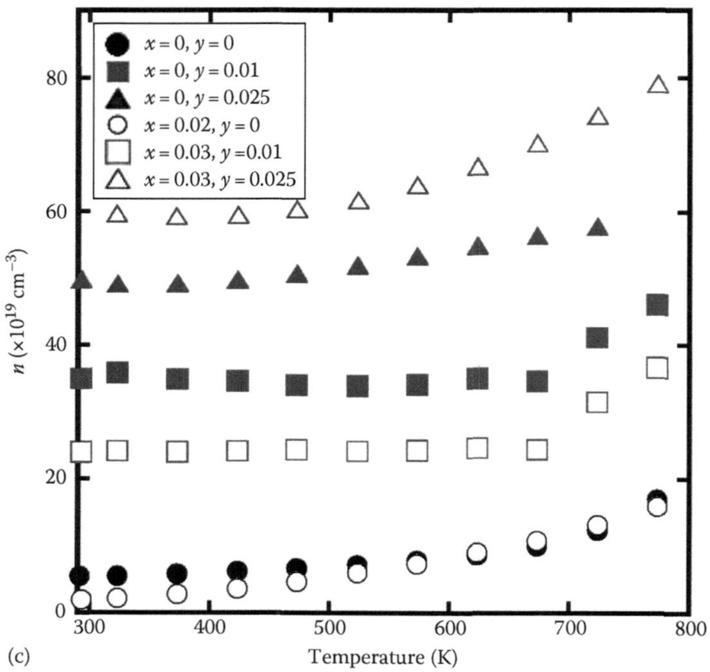

(c)

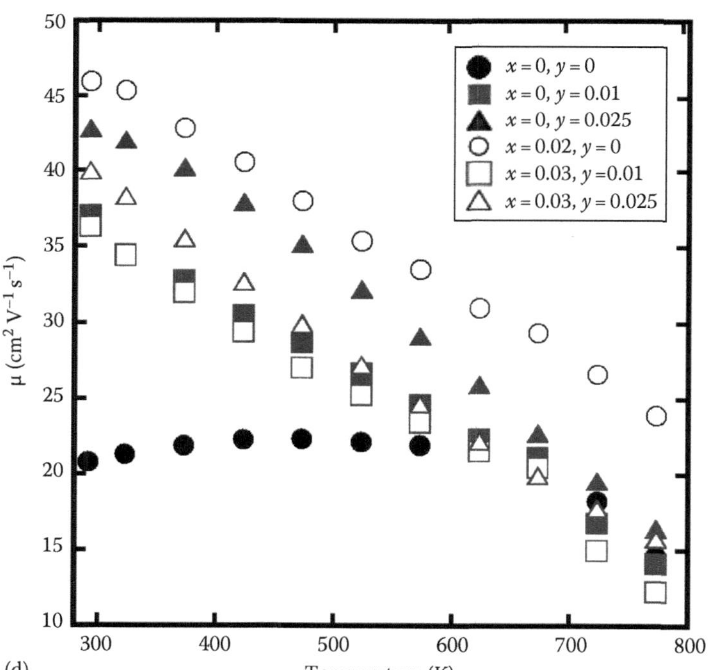

(d)

FIGURE 7.20 (*Continued*)
Comparison of electronic transport in $Zr_{0.25}Hf_{0.75}Ni_{1+x}Sn_{1-y}Sb_y$ with various doping levels (y = 0, 0.01 and 0.025). (c) the carrier concentration, and (d) the carrier mobility. (Reprinted from *J. Solid State Chem.*, 234, Liu, Y.F., Makongo, J.P.A., Page, A. et al., Distribution of impurity states and charge transport in $Zr_{0.25}Hf_{0.75}Ni_{1+x}Sn_{1-y}Sb_y$ nanocomposites, 72–86, Copyright 2016, with permission from Elsevier.)

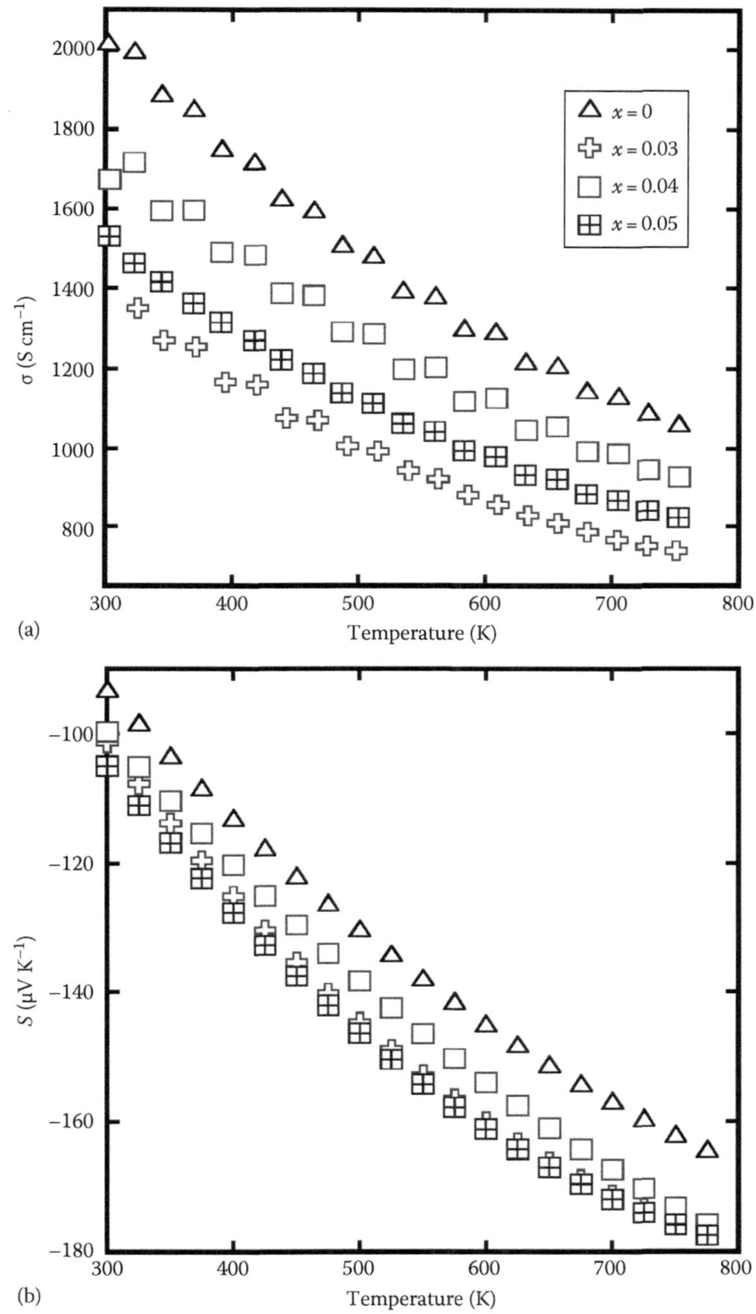

FIGURE 7.21
Temperature-dependent electronic properties of $Zr_{0.25}Hf_{0.75}Ni_{1+x}Sn_{0.99}Sb_{0.01}$ samples. (a) The electrical conductivity, (b) the Seebeck coefficient.

(Continued)

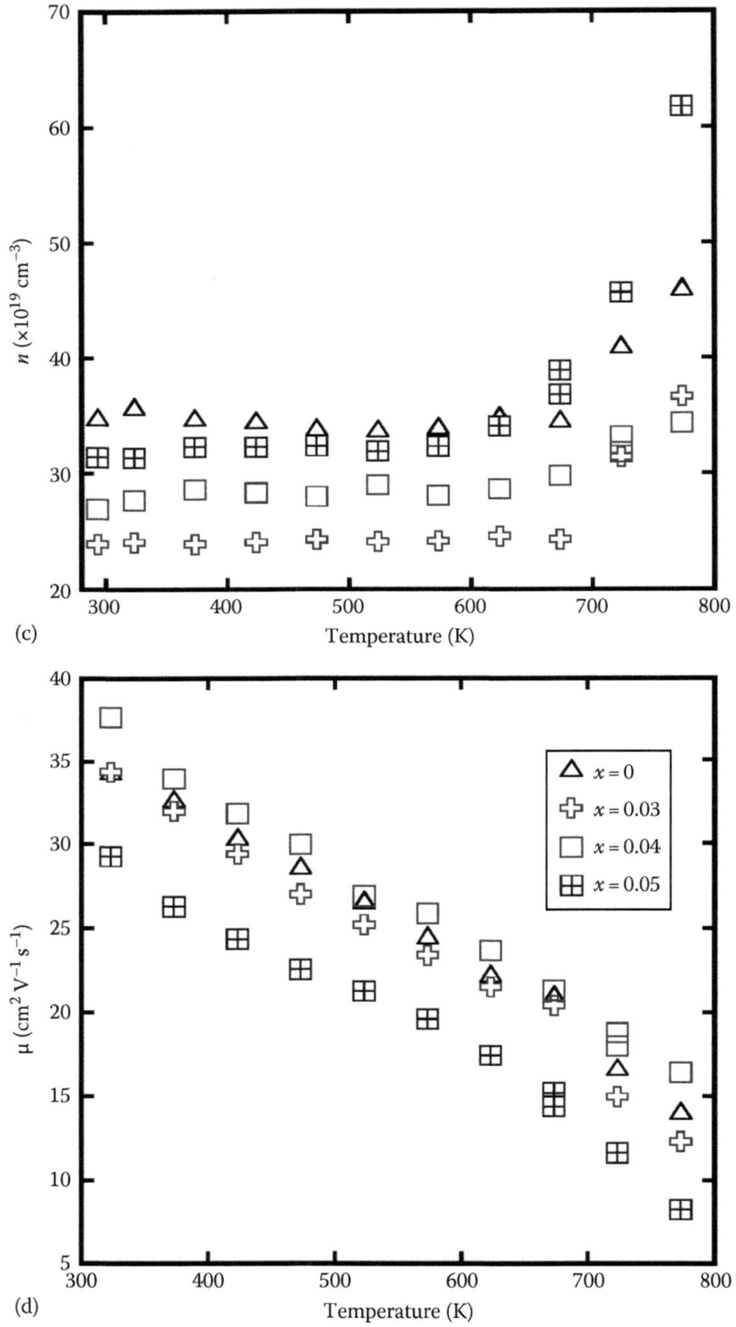

FIGURE 7.21 *(Continued)*
Temperature-dependent electronic properties of $Zr_{0.25}Hf_{0.75}Ni_{1+x}Sn_{0.99}Sb_{0.01}$ samples. (c) the carrier concentration, and (d) the carrier mobility. (Reprinted from *J. Solid State Chem.*, 234, Liu, Y.F., Makongo, J.P.A., Page, A. et al., Distribution of impurity states and charge transport in $Zr_{0.25}Hf_{0.75}Ni_{1+x}Sn_{1-y}Sb_y$ nanocomposites, 72–86, Copyright 2016, with permission from Elsevier.)

FH content ($T_c = 675$ K for HH(0.97)FH(0.03), $T_c = 625$ K for HH(0.96)/FH(0.04), and $T_c = 575$ K for HH(0.95)/FH(0.05)). The nearly exponential increase in the carrier density for the composites arises from thermal excitation of existing extrinsic carriers, from shallow states, across the energy barrier rather than the thermal excitation of intrinsic carriers. With increasing FH content, a decreasing energy barrier height is expected and thus a decreasing critical temperature for thermal excitation is observed. An enhancement in thermopower is observed with increasing FH content due to the increase in the energy filtering of carriers. The carrier mobility of the 1% Sb samples decreases linearly with rising temperature. Fitting the temperature-dependence curves indicates that acoustic phonon scattering at high temperature dominates among all the scattering mechanisms. The energy-filtering effect at the HH/FH interfaces observed in the 1% Sb-doped composites suggests that a significant fraction of intrinsic carriers (low energy) contributes to the electrical conductivity.

However, the only marginal change in the carrier density with increasing FH content observed for the $Zr_{0.25}Hf_{0.75}Ni_{1+x}Sn_{0.975}Sb_{0.025}$ and $Zr_{0.25}Hf_{0.75}Ni_{1+x}Sn_{0.99}Sb_{0.01}$ composites, when compared to the pristine matrix, suggests that donor states formed by Sb doping and metallic FH inclusions dominate the electronic conduction in these composites. These donor states occupy high energy levels or shallow states within the bandgap of the HH matrix, which results in easy thermal excitation across the energy barrier formed at the HH/FH interfaces. Thus, the energy-filtering effect does not impact carriers contributed by these shallow donor states. Therefore, composites with deep donor states should be studied to clarify the relationship between the energy-filtering effect and donor states.

To explore this concept, $Zr_{0.25}Hf_{0.75}Ni_{1+x}Sn_{0.995}Sb_{0.005}$ samples unintentionally doped by impurities such as Fe, introducing a large fraction of LE electrons, were synthesized by mechanical alloying. The electrical conductivities for all $Zr_{0.25}Hf_{0.75}Ni_{1+x}Sn_{0.995}Sb_{0.005}$ samples, shown in Figure 7.22, decrease with increasing temperature, displaying typical behavior of heavily doped semiconductors. The electrical conductivity decreases with increasing FH inclusions with room temperature values ranging from 2700 S cm^{-1} for the matrix to ~1500 S cm^{-1} for the HH(0.95)/FH(0.05) composite. A marginal increase in the electrical conductivity is observed when the fraction of FH inclusions increases to 8% and 15%. Correspondingly, the thermopower is dramatically enhanced for the HH(0.95)/FH(0.05) composite compared to the matrix and a marginal decrease occurs when FH content increases to 8% and 15%. Surprisingly, the carrier densities for $Zr_{0.25}Hf_{0.75}Ni_{1+x}Sn_{0.995}Sb_{0.005}$ samples are one order of magnitude higher than that of $Zr_{0.25}Hf_{0.75}Ni_{1+x}Sn_{0.975}Sb_{0.025}$ samples despite their lower Sb-doping level. The carrier density decreases with increasing FH content and reaches a minimum for HH(0.95)/HF(0.05) composite. Further increasing the FH content leads to a marginal increase in carrier density, which is still much lower than that of the corresponding HH matrix. It is suggested that impurities, such as metallic Fe, coming from mechanically alloying act as donors and contribute a large fraction of carriers. Additionally, donor states originating from Fe impurities are deep states within the bandgap, and thus, the resulting carriers are influenced by the energy-filtering effect. These LE carriers are screened out by the energy barrier formed at the HH/FH interfaces. The marginal increases in the carrier concentration and electrical conductivity for samples with 8% and 10% FH inclusions are ascribed to the doping effect from large FH inclusions, which presumably form shallow donor states within the bandgap and thus are not influenced by the energy-filtering effect. The drastic increase in the carrier density resulted in

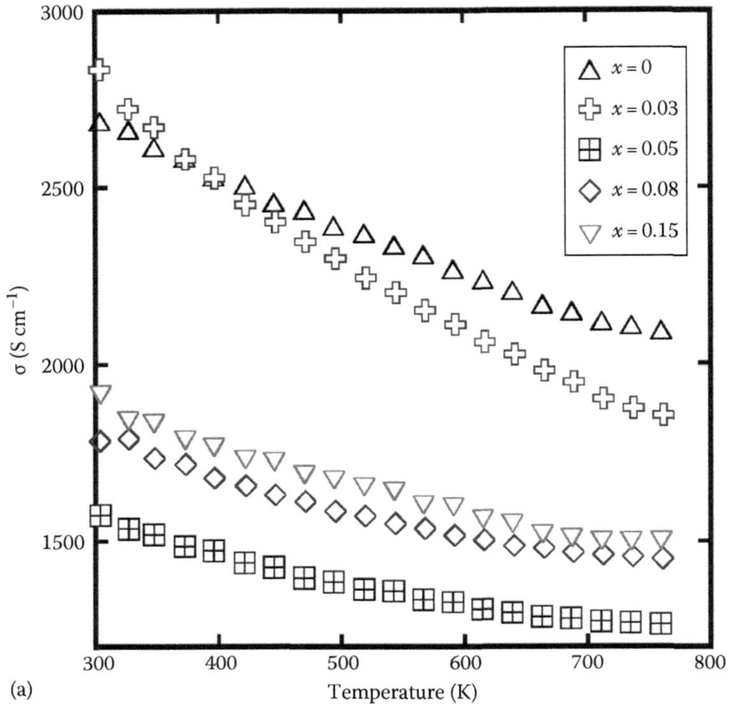

(a)

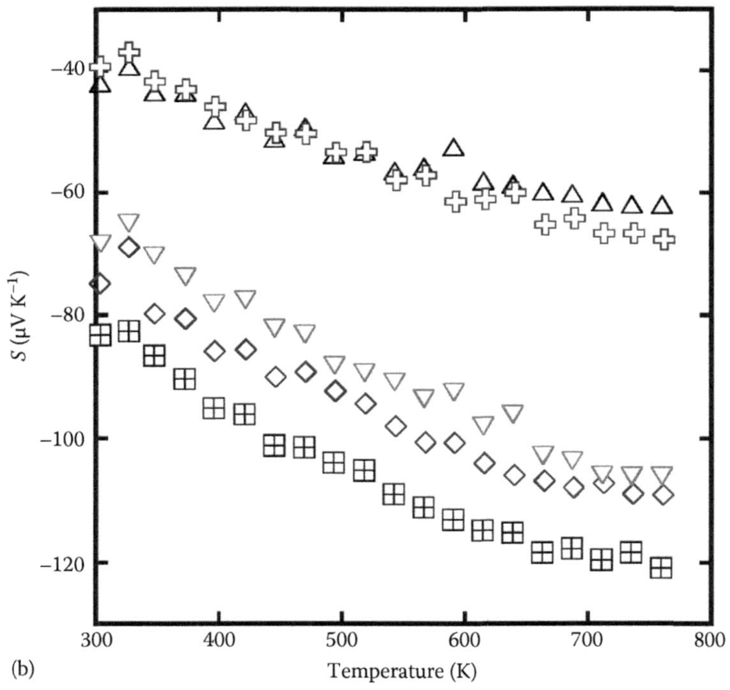

(b)

FIGURE 7.22
Temperature-dependent electronic properties of $Zr_{0.25}Hf_{0.75}Ni_{1+x}Sn_{0.995}Sb_{0.005}$. (a) The electrical conductivity, (b) the Seebeck coefficient. *(Continued)*

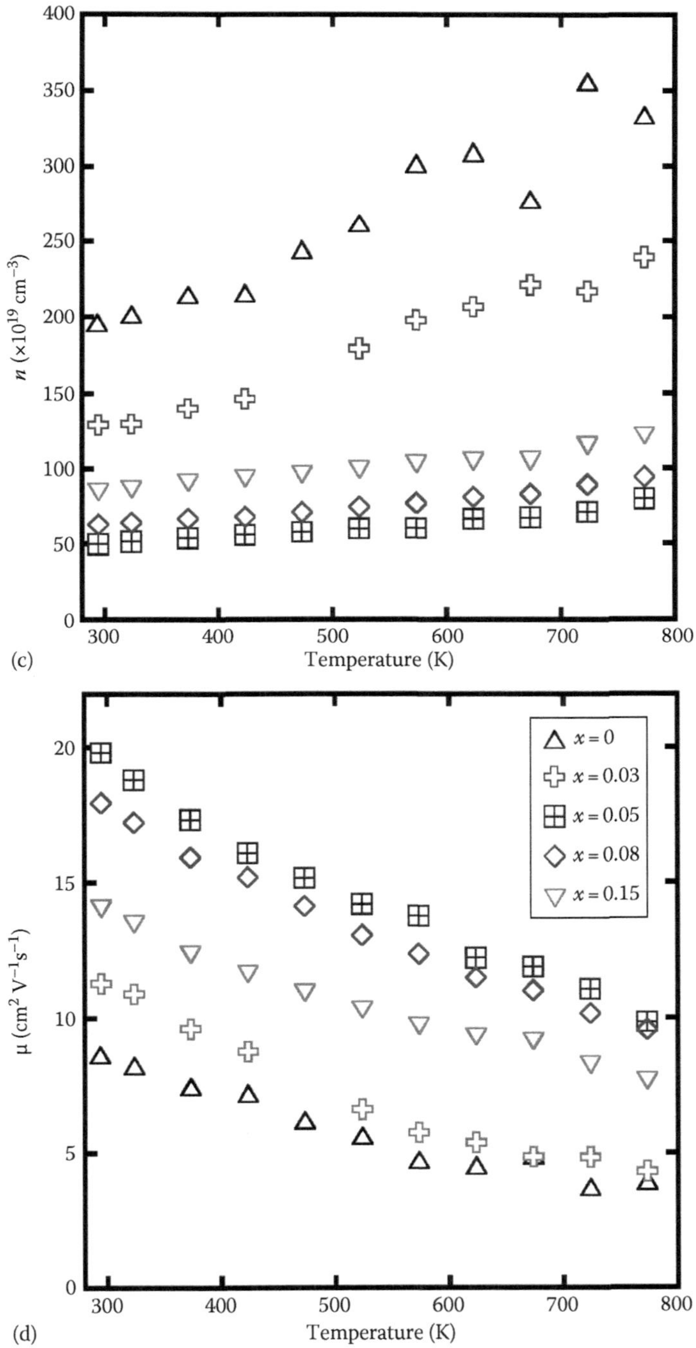

FIGURE 7.22 (*Continued*)
Temperature-dependent electronic properties of $Zr_{0.25}Hf_{0.75}Ni_{1+x}Sn_{0.995}Sb_{0.005}$. (c) the carrier concentration, and (d) the carrier mobility. (Reprinted from *J. Solid State Chem.*, 234, Liu, Y.F., Makongo, J.P.A., Page, A. et al., Distribution of impurity states and charge transport in $Zr_{0.25}Hf_{0.75}Ni_{1+x}Sn_{1-y}Sb_y$ nanocomposites, 72–86, Copyright 2016, with permission from Elsevier.)

a severe decrease in carrier mobility for $Zr_{0.25}Hf_{0.75}Ni_{1+x}Sn_{0.995}Sb_{0.005}$ samples compared to $Zr_{0.25}Hf_{0.75}Ni_{1+x}Sn_{0.975}Sb_{0.025}$ samples. The $Zr_{0.25}Hf_{0.75}NiSn_{0.995}Sb_{0.005}$ matrix displays the lowest carrier mobility of ~8 $cm^2 V^{-1} s^{-1}$ at 300 K. Interestingly, the carrier mobility is significantly enhanced upon incorporation of FH inclusions. The HH(0.95)/FH(0.05) composite shows the highest carrier mobility of ~20 $cm^2 V^{-1} s^{-1}$ at 300 K due to its lowest carrier density.

In summary, the effectiveness of the energy-filtering effect at HH/FH interfaces strongly depends on both the height of the energy barrier and the energy level of donor states within the bandgap of the HH matrix. The height of the energy barrier is influenced by the size of FH inclusions, whereas the energy level of donor states is related to the chemical nature of the dopant. On one hand, the magnitude of the energy-filtering effect is influenced by the distribution and size of FH inclusions. High density of small FH inclusions leads to more efficient energy filtering. On the other hand, the effectiveness of the energy-filtering effect under constant height of the energy barrier is determined by the energy level of donor states in the bandgap of the HH matrix. Shallow donor states lead to a large fraction of thermally excited HE carriers, which are marginally influenced by the energy barriers and therefore dominate the electrical transport behavior of the materials. However, deep states lead to a large fraction of LE carriers, which are severely influenced by the energy barriers.

7.3.2.3 HH Matrix with InSb Nanoinclusions

In addition to nanostructured HH alloys with *in situ* grown FH inclusions, the effect of *in situ* grown second phases, such as InSb, on the electronic transport of the HH matrix was also investigated. Xie et al. [49] demonstrated a simultaneous optimization of the electrical conductivity, thermopower, and thermal conductivity of (TiZrHf)(CoNi)Sb alloys using *in situ* formed InSb nanoinclusions. The InSb nanoinclusions are mainly distributed at the grain boundaries, as indicated in Figure 7.23, and the nanoinclusions grow rapidly with an increasing amount of InSb. The average size of InSb nanoinclusions increases from 10 to 30 nm for the sample with 1% InSb to 200–300 nm for the sample with 7% InSb. Regardless of the temperature, the nanocomposites show enhanced electrical conductivity compared to that of the matrix (Figure 7.24a). In addition, the electrical conductivity increases with increasing InSb content. All composites show a transformation from semiconducting behavior to semimetal behavior at around 460 K. The electrical conductivity of the matrix initially decreases with rising temperature and then increases after 700 K, indicating the onset of thermal activation of intrinsic carriers. Meanwhile, the Seebeck coefficient of the nanocomposites is substantially enhanced compared to the matrix. In addition, the Seebeck coefficient initially increases with increasing amount of InSb up to 1% and decreases thereafter for higher InSb content (Figure 7.24b). The resulting PF is significantly improved due to the enhancement of both electrical conductivity and Seebeck coefficient. The incorporation of InSb inclusions substantially increases the carrier density due to the *electron injection* effect and thus enhances the carrier mobility and the electrical conductivity. However, an increase in the carrier density is expected to have a negative effect on the Seebeck coefficient. Therefore, the surprising enhancement of the Seebeck coefficient observed for these samples can be attributed to the energy filtering of LE intrinsic carriers at the interfaces between the nanoinclusions and the matrix, which offsets the reduction in the Seebeck coefficient arising from the increase of the carrier density.

FIGURE 7.23
Microstructures of as-prepared HH-x nanocomposites, where x is the atomic % InSb in the HH matrix with composition $Ti_{0.5}Zr_{0.25}Hf_{0.25}Co_{0.95}Ni_{0.05}Sb$. (a, b) HH-1 and (c–e) HH-7. (Reprinted from *Acta Mater.*, 58, Xie, W.J., He, J., Zhu, S. et al., Simultaneously optimizing the independent thermoelectric properties in (Ti, Zr, Hf)(Co, Ni)Sb alloy by in situ forming InSb nanoinclusions, 4705–4713, Copyright 2010, with permission from Elsevier.)

7.4 Conclusion

Recent advances in HH alloys revealed that enhancements in the thermoelectric performance through large improvement of the PF using nanostructuring, as opposed to the traditional doping strategy, could also be achieved in bulk nanostructured HH [39,40,70–72,84,97–100]. Improvements in the PF in such nanocomposites mainly occur when enhancements in the thermopower and/or carrier mobility can be realized, while avoiding additional doping of the matrix. Careful analysis of HH matrix/nanoinclusions investigated so far suggests that simultaneous enhancement of both thermopower and carrier mobility is observed when the embedded nanostructures (1) have a high degree of lattice coherency with the matrix so as to minimize reduction in the carrier mobility by eliminating scattering of charge carriers during transport across the matrix/inclusion interfaces; (2) have a suitable CB or valence-band energy offset, ΔE, with the matrix, so as to enable enhancement of the thermopower and overall carrier mobility by trapping or filtering undesirable conducting carriers at the matrix/inclusion interfaces; and (3) are electronically inert with respect to the matrix so as to avoid dumping additional

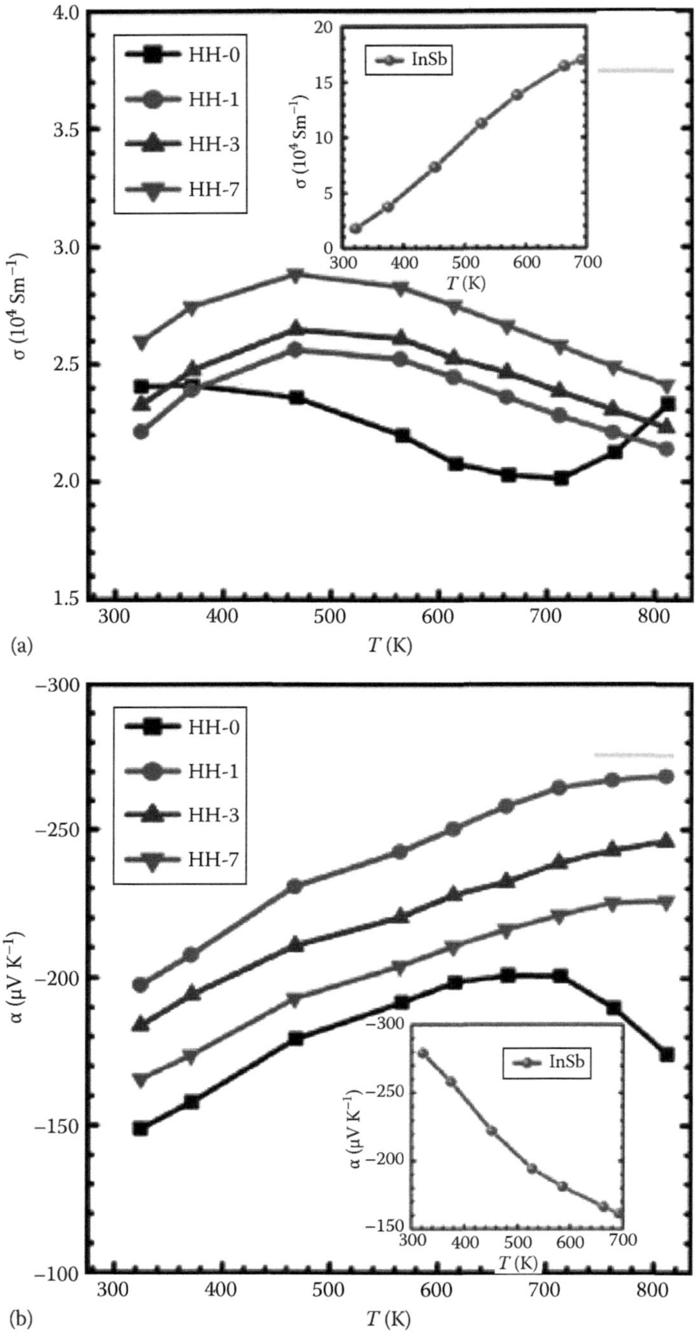

FIGURE 7.24
Temperature-dependent (a) electrical conductivity and (b) the Seebeck coefficient of as-prepared HH-x nanocomposites, where x is the atomic % of InSb in the HH matrix with composition $Ti_{0.5}Zr_{0.25}Hf_{0.25}Co_{0.95}Ni_{0.05}Sb$. (Reprinted from *Acta Mater.*, 58, Xie, W.J., He, J., Zhu, S. et al., Simultaneously optimizing the independent thermoelectric properties in (Ti, Zr, Hf)(Co, Ni)Sb alloy by in situ forming InSb nanoinclusions, 4705–4713, Copyright 2010, with permission from Elsevier.)

charge carriers into the existing ensemble of free carriers within the matrix. For a given HH/FH system, it appears that the given conditions are realized when the FH inclusions are sufficiently small in size, are widely dispersed, and are coherently embedded in the HH matrix. This desirable microstructure strongly depends on the materials synthesis and processing methods. For example, it was found that large FH inclusions are generally formed in the HH matrix when the composite is obtained through slow or fast cooling of the molten phase. Such large and interconnected FH inclusions are metallic in nature, resulting in partial doping of the HH matrix. However, very small semiconducting-like FH inclusions are formed when the HH/FH composites are formed through solid-state reaction of the elements at temperatures below the melting temperature of either phase. A synergistic integration of the concepts of doping and energy filtering through nanostructuring can lead to further enhancement of the overall thermoelectric performance of HH alloys and other promising thermoelectric materials systems. This may, however, require a good match between (1) the type of doping elements (energy levels occupied by impurity states) and (2) the average size and distribution (the heights of the potential energy barrier) of the nanostructures coherently embedded inside the semiconducting matrix.

Acknowledgment

We gratefully acknowledge the financial support from the Department of Energy, Office of Basic Energy Science under Award # DE-SC-0008574.

References

1. Leuenberger, M. N. and D. Loss. 2001. Quantum computing in molecular magnets. *Nature* 410: 789–793.
2. Mannini, M., F. Pineider, P. Sainctavit et al. 2009. Magnetic memory of a single-molecule quantum magnet wired to a gold surface. *Nat. Mater.* 8: 194–197.
3. Bogani, L. and W. Wernsdorfer. 2008. Molecular spintronics using single-molecule magnets. *Nat. Mater.* 7: 179–186.
4. Kanatzidis, M. G. 2010. Nanostructured thermoelectrics: The new paradigm? *Chem. Mater.* 22: 648–659.
5. Sootsman, J. R., D. Y. Chung, and M. G. Kanatzidis. 2009. New and old concepts in thermoelectric materials. *Angew. Chem. Int. Ed.* 48: 8616–8639.
6. Vineis, C. J., A. Shakouri, A. Majumdar, and M. G. Kanatzidis. 2010. Nanostructured thermoelectrics: Big efficiency gains from small features. *Adv. Mater.* 22: 3970–3980.
7. Poudeu, P. F. P., A. Gueguen, C. I. Wu, T. Hogan, and M. G. Kanatzidis. 2010. High figure of merit in nanostructured n-type KPb_mSbTe_{m+2} thermoelectric materials. *Chem. Mater.* 22: 1046–1053.
8. Dresselhaus, M. S., G. Chen, M. Y. Tang et al. 2007. New directions for low-dimensional thermoelectric materials. *Adv. Mater.* 19: 1043–1053.
9. Minnich, A. J., M. S. Dresselhaus, Z. F. Ren, and G. Chen. 2009. Bulk nanostructured thermoelectric materials: Current research and future prospects. *Energy Environ. Sci.* 2: 466–479.

10. He, J. Q., J. R. Sootsman, L. Q. Xu et al. 2011. Anomalous electronic transport in dual-nanostructured lead telluride. *J. Am. Chem. Soc.* 133: 8786–8789.
11. Johnsen, S., J. Q. He, J. Androulakis et al. 2011. Nanostructures boost the thermoelectric performance of PbS. *J. Am. Chem. Soc.* 133: 3460–3470.
12. Radousky, H. B. and H. Liang. 2012. Energy harvesting: An integrated view of materials, devices and applications. *Nanotechnology* 23: 502001.
13. Zhu, T. J., C. G. Fu, H. H. Xie, Y. T. Liu, and X. B. Zhao. 2015. High efficiency half-Heusler thermoelectric materials for energy harvesting. *Adv. Energy Mater.* 5: 1500588.
14. Hsu, K. F., S. Loo, F. Guo et al. 2004. Cubic $AgPb_mSbTe_{2+m}$: Bulk thermoelectric materials with high figure of merit. *Science* 303: 818–821.
15. Wang, H., J. F. Li, C. W. Nan et al. 2006. High-performance $Ag_{0.8}Pb_{18+x}SbTe_{20}$ thermoelectric bulk materials fabricated by mechanical alloying and spark plasma sintering. *Appl. Phys. Lett.* 88: 092104.
16. Poudeu, P. F. P., J. D'Angelo, A. D. Downey et al. 2006. High thermoelectric figure of merit and nanostructuring in bulk p-type $Na_{1-x}Pb_mSb_yTe_{m+2}$. *Angew. Chem. Int. Ed.* 45: 3835–3839.
17. Gueguen, A., P., F. P. Poudeu, C. P. Li et al. 2009. Thermoelectric properties and nanostructuring in the p-type materials $NaPb_{18-x}Sn_xMTe_{20}$ (M = Sb, Bi). *Chem. Mater.* 21: 1683–1694.
18. Androulakis, J., K. F. Hsu, R. Pcionek et al. 2006. Nanostructuring and high thermoelectric efficiency in p-type $Ag(Pb_{1-y}Sn_y)_{(m)}SbTe_{2+m}$. *Adv. Mater.* 18: 1170.
19. Androulakis, J., C. H. Lin, H. J. Kong et al. 2007. Spinodal decomposition and nucleation and growth as a means to bulk nanostructured thermoelectrics: Enhanced performance in $Pb_{1-x}Sn_xTe$-PbS. *J. Am. Chem. Soc.* 129: 9780–9788.
20. Harman, T. C., P. J. Taylor, M. P. Walsh, and B. E. LaForge. 2002. Quantum dot superlattice thermoelectric materials and devices. *Science* 297: 2229–2232.
21. Lan, Y. C., B. Poudel, Y. Ma et al. 2009. Structure study of bulk nanograined thermoelectric bismuth antimony telluride. *Nano Lett.* 9: 1419–1422.
22. Poudel, B., Q. Hao, Y. Ma et al. 2008. High-thermoelectric performance of nanostructured bismuth antimony telluride bulk alloys. *Science* 320: 634–638.
23. Venkatasubramanian, R., E. Siivola, T. Colpitts, and B. O'Quinn. 2001. Thin-film thermoelectric devices with high room-temperature figures of merit. *Nature* 413: 597–602.
24. Yan, X., G. Joshi, W. Liu et al. 2011. Enhanced thermoelectric figure of merit of p-type half-Heuslers. *Nano Lett.* 11: 556–560.
25. Mi, J. L., X. B. Zhao, T. J. Zhu, and J. P. Tu. 2007. Improved thermoelectric figure of merit in n-type $CoSb_3$ based nanocomposites. *Appl. Phys. Lett.* 91: 172116.
26. Zide, J. M. O., D. Vashaee, Z. X. Bian et al. 2006. Demonstration of electron filtering to increase the Seebeck coefficient in $In_{0.53}Ga_{0.47}As/In_{0.53}Ga_{0.28}Al_{0.19}As$ superlattices. *Phys. Rev. B* 74: 205335.
27. Sootsman, J. R., H. Kong, C. Uher et al. 2008. Large enhancements in the thermoelectric power factor of bulk PbTe at high temperature by synergistic nanostructuring. *Angew. Chem. Int. Ed.* 47: 8618–8622.
28. Jeitschko, W. 1970. Transition metal stannides with MgAgAs and $MnCu_2Al$ type structure. *Metall. Trans.* 1: 3159.
29. Webster, P. J. and K. R. A. Ziebeck. 1988. *Landolt-Börnstein-Group III Condensed Matter*, Vol. 19C, pp. 75–184 (Berlin, Germany: Springer).
30. Aliev, F. G., N. B. Brandt, V. V. Moshchalkov et al. 1989. Gap at the Fermi level in the intermetallic vacancy system TiNiSn, ZrNiSn, HfNiSn. *Z. Phys. B Condens. Matter* 75: 167–171.
31. Aliev, F. G., V. V. Kozyrkov, V. V. Moshchalkov, R. V. Scolozdra, and K. Durczewski. 1990. Narrow-band in the intermetallic compounds TiNiSn, ZrNiSn, HfNiSn. *Z. Phys. B Condens. Matter* 80: 353–357.
32. Poon, S. J. 2000. Electronic and thermoelectric properties of half-Heusler alloys, Chapter 2. In: Tritt T. M. (ed.), *Semiconductors and Semimetals*, Recent Trends in Thermoelectric Materials Research, Vol. 70, Academic Press, San Diego, CA, pp. 37–76.
33. Simonson, J. W. and S. J. Poon. 2008. Electronic structure of transition metal-doped XNiSn and XCoSb (X = Hf, Zr) phases in the vicinity of the band gap *J. Phys. Condens. Matter* 20: 255220.

34. Yu, C., T. J. Zhu, R. Z. Shi et al. 2009. High-performance half-Heusler thermoelectric materials $Hf_{1-x}Zr_xNiSn_{1-y}Sb_y$ prepared by levitation melting and spark plasma sintering. *Acta Mater.* 57: 2757–2764.
35. Takas, N. J., P. Sahoo, D. Misra et al. 2011. Effects of Ir substitution and processing conditions on thermoelectric performance of *p*-type $Zr_{0.5}Hf_{0.5}Co_{1-x}Ir_xSb_{0.99}Sn_{0.01}$ half-Heusler alloys. *J. Electron. Mater.* 40: 662–669.
36. Takas, N. J., M. R. Shabetai, and P. F. P. Poudeu. 2011. Effect of Sn doping on the thermoelectric performance of the complex *p*-type $Zr_{0.5}Hf_{0.5}Co_{0.3}Ir_{0.7}Sb_{1-y}Sn_y$ half-Heusler system. *Sci. Adv. Mater.* 3: 571–576.
37. Maji, P., N. J. Takas, D. K. Misra et al. 2010. Effects of Rh on the thermoelectric performance of the *p*-type $Zr_{0.5}Hf_{0.5}Co_{1-x}Rh_xSb_{0.99}Sn_{0.01}$ half-Heusler alloys. *J. Solid State Chem.* 183: 1120–1126.
38. Maji, P., J. P. A. Makongo, X. Y. Zhou et al. 2013. Thermoelectric performance of nanostructured *p*-type $Zr_{0.5}Hf_{0.5}Co_{0.4}Rh_{0.6}Sb_{1-x}Sn_x$ half-Heusler alloys. *J. Solid State Chem.* 202: 70–76.
39. Makongo, J. P. A., D. K. Misra, J. R. Salvador et al. 2011. Thermal and electronic charge transport in bulk nanostructured $Zr_{0.25}Hf_{0.75}NiSn$ composites with full-Heusler inclusions. *J. Solid State Chem.* 184: 2948–2960.
40. Makongo, J. P. A., D. K. Misra, X. Zhou et al. 2011. Simultaneous large enhancements in thermopower and electrical conductivity of bulk nanostructured half-Heusler alloys. *J. Am. Chem. Soc.* 133: 18843–18852.
41. Misra, D. K., J. P. A. Makongo, P. Sahoo et al. 2011. Microstructure and thermoelectric properties of mechanically alloyed $Zr_{0.5}Hf_{0.5}Ni_{0.8}Pd_{0.2}Sn_{0.99}Sb_{0.01}/WO_3$ half-Heusler composites. *Sci. Adv. Mater.* 3: 607–614.
42. Yaqub, R., P. Sahoo, J. P. A. Makongo et al. 2011. Investigation of the effect of NiO nanoparticles on the transport properties of $Zr_{0.5}Hf_{0.5}Ni_{1-x}Pd_xSn_{0.99}Sb_{0.01}$ ($x=0$ and 0.2). *Sci. Adv. Mater.* 3: 633–638.
43. Culp, S. R., J. W. Simonson, S. J. Poon et al. 2008. (Zr, Hf)Co(Sb, Sn) half-Heusler phases as high-temperature (>700°C) p-type thermoelectric materials. *Appl. Phys. Lett.* 93: 022105.
44. Joshi, G., X. Yan, H. Wang et al. 2011. Enhancement in thermoelectric figure-of-merit of an *n*-type half-Heusler compound by the nanocomposite approach. *Adv. Energy Mater.* 1: 643–647.
45. Lan, Y. C., A. J. Minnich, G. Chen, and Z. F. Ren. 2010. Enhancement of thermoelectric figure-of-merit by a bulk nanostructuring approach. *Adv. Funct. Mater.* 20: 357–376.
46. Tritt, T. M. and H. Böttner. 2011. Focus issue: Advances in thermoelectric materials introduction. *J. Mater. Res.* 26: 1743–1744.
47. Simonson, J. W., D. Wu, W. J. Xie, T. M. Tritt, and S. J. Poon. 2011. Introduction of resonant states and enhancement of thermoelectric properties in half-Heusler alloys. *Phys. Rev. B* 83: 235211.
48. Poudeu, P. F., J. Salvador, and J. Sakamoto. 2011. A special issue on advanced thermoelectric materials and devices. *Sci. Adv. Mater.* 3: 515–516.
49. Xie, W. J., J. He, S. Zhu et al. 2010. Simultaneously optimizing the independent thermoelectric properties in (Ti, Zr, Hf)(Co, Ni)Sb alloy by *in situ* forming InSb nanoinclusions. *Acta Mater.* 58: 4705–4713.
50. Barth, J., B. Balke, G. H. Fecher et al. 2009. Thermoelectric properties of CoTiSb based compounds. *J. Phys. D: Appl. Phys.* 42: 308–311.
51. Chaput, L., J. Tobola, P. Pecheur, and H. Scherrer. 2006. Electronic structure and thermopower of $Ni(Ti_{0.5}Hf_{0.5})Sn$ and related half-Heusler phases. *Phys. Rev. B* 73: 045121.
52. Sakurada, S.and N. Shutoh. 2005. Effect of Ti substitution on the thermoelectric properties of (Zr, Hf)NiSn half-Heusler compounds. *Appl. Phys. Lett.* 86: 082105.
53. Uher, C., J. Yang, S. Hu, D. T. Morelli, and G. P. Meisner. 1999. Transport properties of pure and doped MNiSn (M = Zr, Hf). *Phys. Rev. B* 59: 8615–8621.
54. Nanda, B. R. K. and I. Dasgupta. 2005. Electronic structure and magnetism in doped semiconducting half-Heusler compounds. *J. Phys. Condens. Matter* 17: 5037–5048.
55. Kroth, K., B. Balke, G. H. Fecher et al. 2006. Diluted magnetic semiconductors with high Curie temperature based on C1(b) compounds: $CoTi_{1-x}Fe_xSb$. *Appl. Phys. Lett.* 89: 202509.

56. Ouardi, S., G. H. Fecher, B. Balke et al. 2010. Electronic transport properties of electron- and hole-doped semiconducting C1b Heusler compounds: $NiTi_{1-x}M_xSn$ (M = Sc, V). *Phys. Rev. B* 82: 085108.
57. Sanyal, B., O. Eriksson, K. G. Suresh et al. 2006. Ferromagnetism in Mn doped half-heusler NiTiSn: Theory and experiment. *Appl. Phys. Lett.* 89: 212502.
58. Lin, H., L. A. Wray, Y. Q. Xia et al. 2010. Half-Heusler ternary compounds as new multifunctional experimental platforms for topological quantum phenomena. *Nat. Mater.* 9: 546–549.
59. Chadov, S., X. L. Qi, J. Kubler et al. 2010. Tunable multifunctional topological insulators in ternary Heusler compounds. *Nat. Mater.* 9: 541–545.
60. Xiao, D., Y. G. Yao, W. X. Feng et al. 2010. Half-Heusler compounds as a new class of three-dimensional topological insulators. *Phys. Rev. Lett.* 105: 096404.
61. Culp, S. R., S. J. Poon, N. Hickman, T. M. Tritt, and J. Blumm. 2006. Effect of substitutions on the thermoelectric figure of merit of half-Heusler phases at 800°C. *Appl. Phys. Lett.* 88: 042106.
62. Shen, Q., L. Chen, T. Goto et al. 2001. Effects of partial substitution of Ni by Pd on the thermoelectric properties of ZrNiSn-based half-Heusler compounds. *Appl. Phys. Lett.* 79: 4165–4167.
63. Yang, J., G. P. Meisner, and L. Chen. 2004. Strain field fluctuation effects on lattice thermal conductivity of ZrNiSn-based thermoelectric compounds. *Appl. Phys. Lett.* 85: 1140–1142.
64. Liu, Y. F. and P. F. P. Poudeu. 2015. Thermoelectric properties of Ge doped n-type $Ti_xZr_{1-x}NiSn_{0.975}Ge_{0.025}$ half-Heusler alloys. *J. Mater. Chem. A* 3: 12507–12514.
65. Chen, L. D., X. Y. Huang, M. Zhou, X. Shi, and W. B. Zhang. 2006. The high temperature thermoelectric performances of $Zr_{0.5}Hf_{0.5}Ni_{0.8}Pd_{0.2}Sn_{0.99}Sb_{0.01}$ alloy with nanophase inclusions. *J. Appl. Phys.* 99: 064305.
66. Yu, C., T.-J. Zhu, K. Xiao et al. 2010. Reduced grain size and improved thermoelectric properties of melt spun (Hf, Zr)NiSn half-Heusler alloys. *J. Electron. Mater.* 39: 2008–2012.
67. Birkel, C. S., J. E. Douglas, B. R. Lettiere et al. 2013. Influence of Ni nanoparticle addition and spark plasma sintering on the TiNiSn-Ni system: Structure, microstructure, and thermoelectric properties. *Solid State Sci.* 26: 16–22.
68. Birkel, C. S., J. E. Douglas, B. R. Lettiere et al. 2013. Improving the thermoelectric properties of half-Heusler TiNiSn through inclusion of a second full-Heusler phase: Microwave preparation and spark plasma sintering of $TiNi_{1+x}Sn$. *Phys. Chem. Chem. Phys.* 15: 6990–6997.
69. Douglas, J. E., P. A. Chater, C. M. Brown, T. M. Pollock, and R. Seshadri. 2014. Nanoscale structural heterogeneity in Ni-rich half-Heusler TiNiSn. *J. Appl. Phys.* 116: 163514.
70. Sahoo, P., Y. F. Liu, and P. F. P. Poudeu. 2014. Nanometer-scale interface engineering boosts the thermoelectric performance of n-type $Ti_{0.4}Hf_{0.6}Ni_{1+z}Sn_{0.975}Sb_{0.025}$ alloys. *J. Mater. Chem. A* 2: 9298–9305.
71. Liu, Y. F., A. Page, P. Sahoo et al. 2014. Electronic and phonon transport in Sb-doped $Ti_{0.1}Zr_{0.9}Ni_{1+x}Sn_{0.975}Sb_{0.025}$ nanocomposites. *Dalton Trans.* 43: 8094–8101.
72. Liu, Y. F., P. Sahoo, J. P. A. Makongo et al. 2013. Large enhancements of thermopower and carrier mobility in quantum dot engineered bulk semiconductors. *J. Am. Chem. Soc.* 135: 7486–7495.
73. Liu, Y. F., J. P. A. Makongo, A. Page et al. 2016. Distribution of impurity states and charge transport in $Zr_{0.25}Hf_{0.75}Ni_{1+x}Sn_{1-y}Sb_y$ nanocomposites. *J. Solid State Chem.* 234: 72–86.
74. Gelbstein, Y., B. Dado, O. Ben-Yehuda et al. 2010. High thermoelectric figure of merit and nanostructuring in bulk *p*-type $Ge_x(Sn_yPb_{1-y})_{(1-x)}Te$ alloys following a spinodal decomposition reaction. *Chem. Mater.* 22: 1054–1058.
75. Poudeu, P. F. P., J. D'Angelo, H. J. Kong et al. 2006. Nanostructures versus solid solutions: Low lattice thermal conductivity and enhanced thermoelectric figure of merit in $Pb_{9.6}Sb_{0.2}Te_{10-x}Se_x$ bulk materials. *J. Am. Chem. Soc.* 128: 14347–14355.
76. Sootsman, J. R., R. J. Pcionek, H. J. Kong, C. Uher, and M. G. Kanatzidis. 2006. Strong reduction of thermal conductivity in nanostructured PbTe prepared by matrix encapsulation. *Chem. Mater.* 18: 4993–4995.
77. Heremans, J. P., C. M. Thrush, and D. T. Morelli. 2005. Thermopower enhancement in PbTe with Pb precipitates. *J. Appl. Phys.* 98: 063703.

78. Quarez, E., K. F. Hsu, R. Pcionek et al. 2005. Nanostructuring, compositional fluctuations, and atomic ordering in the thermoelectric materials $AgPb_mSbTe_{2+m}$. The myth of solid solutions. *J. Am. Chem. Soc.* 127: 9177–9190.
79. Sootsman, J. R., J. Q. He, V. P. Dravid et al. 2010. Microstructure and thermoelectric properties of mechanically robust PbTe-Si eutectic composites. *Chem. Mater.* 22: 869–875.
80. Faleev, S. V. and F. Leonard. 2008. Theory of enhancement of thermoelectric properties of materials with nanoinclusions. *Phys. Rev. B* 77: 214304.
81. Kirievsky, K., Y. Gelbstein, and D. Fuks. 2013. Phase separation and antisite defects in the thermoelectric TiNiSn half-Heusler alloys. *J. Solid State Chem.* 203: 247–254.
82. Page, A., C. Uher, P. F. Poudeu, and A. Van der Ven. 2015. Phase separation of full-Heusler nanostructures in half-Heusler thermoelectrics and vibrational properties from first-principles calculations. *Phys. Rev. B* 92: 174102.
83. Do, D. T., S. D. Mahanti, and J. J. Pulikkoti. 2014. Electronic structure of Zr-Ni-Sn systems: Role of clustering and nanostructures in half-Heusler and Heusler limits. *J. Phys. Condens. Matter* 26: 275501.
84. Sahoo, P., Y. F. Liu, J. P. A. Makongo et al. 2013. Enhancing thermopower and hole mobility in bulk p-type half-Heuslers using full-Heusler nanostructures. *Nanoscale* 5: 9419–9427.
85. Hazama, H., M. Matsubara, R. Asahi, and T. Takeuchi. 2001. Improvement of thermoelectric properties for half-Heusler TiNiSn by interstitial Ni defects. *J. Appl. Phys.* 110: 063710.
86. Chai, Y. W., K. Yoshioka, and Y. Kimura. 2014. Intrinsic point defects in thermoelectric half-Heusler alloys. *Scr. Mater.* 83: 13–16.
87. Chai, Y. W. and Y. Kimura. 2013. Microstructure evolution of nanoprecipitates in half-Heusler TiNiSn alloys. *Acta Mater.* 61: 6684–6697.
88. Chai, Y. W. and Y. Kimura. 2012. Nanosized precipitates in half-Heusler TiNiSn alloy. *Appl. Phys. Lett.* 100: 033114.
89. Douglas, J. E., C. S. Birkel, N. Verma et al. 2014. Phase stability and property evolution of biphasic Ti-Ni-Sn alloys for use in thermoelectric applications. *J. Appl. Phys.* 115: 043720.
90. Douglas, J. E., M. P. Echlin, W. C. Lenthe, R. Seshadri, and T. M. Pollock. 2015. Three-dimensional multimodal imaging and analysis of biphasic microstructure in a Ti-Ni-Sn thermoelectric material. *Appl. Mater.* 3: 096107.
91. Poudeu, P. F., J. P. A. Makongo, Y. F. Liu et al. 2015. Thermoelectric behavior of nanostructured $Zr_{0.25}Hf_{0.75}NiCo_xSn$ half-Heusler alloys. *Sci. Adv. Today* 1: 25209.
92. Graf, T., C. Felser, and S. S. P. Parkin. 2011. Simple rules for the understanding of Heusler compounds. *Prog. Solid State Chem.* 39: 1–50.
93. Vanengen, P. G., K. H. J. Buschow, R. Jongebreur, and M. Erman. 1983. PtMnSb, a material with very high magneto-optical Kerr effect. *Appl. Phys. Lett.* 42: 202–204.
94. Wurmehl, S., G. H. Fecher, V. Ksenofontov et al. 2006. Half-metallic ferromagnetism with high magnetic moment and high Curie temperature in Co_2FeSi. *J. Appl. Phys.* 99: 032503.
95. Krenke, T., M. Acet, E. F. Wassermann et al. 2005. Martensitic transitions and the nature of ferromagnetism in the austenitic and martensitic states of Ni-Mn-Sn alloys. *Phys. Rev. B* 72: 014412.
96. Krenke, T., E. Duman, M. Acet et al. 2005. Inverse magnetocaloric effect in ferromagnetic Ni-Mn-Sn alloys. *Nat. Mater.* 4: 450–454.
97. Zhao, L. D., V. P. Dravid, and M. G. Kanatzidis. 2014. The panoscopic approach to high performance thermoelectrics. *Energy Environ. Sci.* 7: 251–268.
98. Zhao, L. D., J. Q. He, S. Q. Hao et al. 2012. Raising the thermoelectric performance of *p*-type PbS with endotaxial nanostructuring and valence-band offset engineering using CdS and ZnS. *J. Am. Chem. Soc.* 134: 16327–16336.
99. Zhao, L. D., J. Q. He, C. I. Wu et al. 2012. Thermoelectrics with earth abundant elements: High performance *p*-type PbS nanostructured with SrS and CaS. *J. Am. Chem. Soc.* 134: 7902–7912.
100. Yu, B., M. Zebarjadi, H. Wang et al. 2012. Enhancement of thermoelectric properties by modulation-doping in silicon germanium alloy nanocomposites. *Nano Lett.* 12: 2077–2082.

8

Thermoelectric Properties of $Cu_{2-\delta}X$ (X = S, Se, and Te)

Pengfei Qiu, Xun Shi, and Lidong Chen
Shanghai Institute of Ceramics

CONTENTS

8.1 Overview of $Cu_{2-\delta}X$ (X = S, Se, and Te)

Binary Cu-based chalcogenides $Cu_{2-\delta}X$ (X = S, Se, and Te) have gained a great attention recently because of their excellent thermoelectric (TE) performance and interesting physical properties. In 1827, A. C. Becquerel used candles to heat a ring consisting of platinum and copper wires, and the electricity was generated in the circuit [1]. Burning some sulfur powders atop the Cu wire significantly enhanced the generation of electricity, which is the first time that Cu-based chalcogenides showed the possibility of possessing good TE properties. In 1866, the first *TE battery* made by Cu-based TE materials (Cu_2S) was fabricated by A. E. Becquerel, the son of A. C. Becquerel [2]. In 1947, Hirahara carried out a study on the electrical performance of Cu_2S and was the first to report the three polymorphic phases of Cu_2S possessing different electrical properties [3]. This result was confirmed in the middle of the 1960s by Ogorelec et al. and Sorokin et al. [4,5]. In 1959, the first patent for copper selenide (Cu_2Se) was filed by the Monsanto Company, claiming that Cu_2Se was

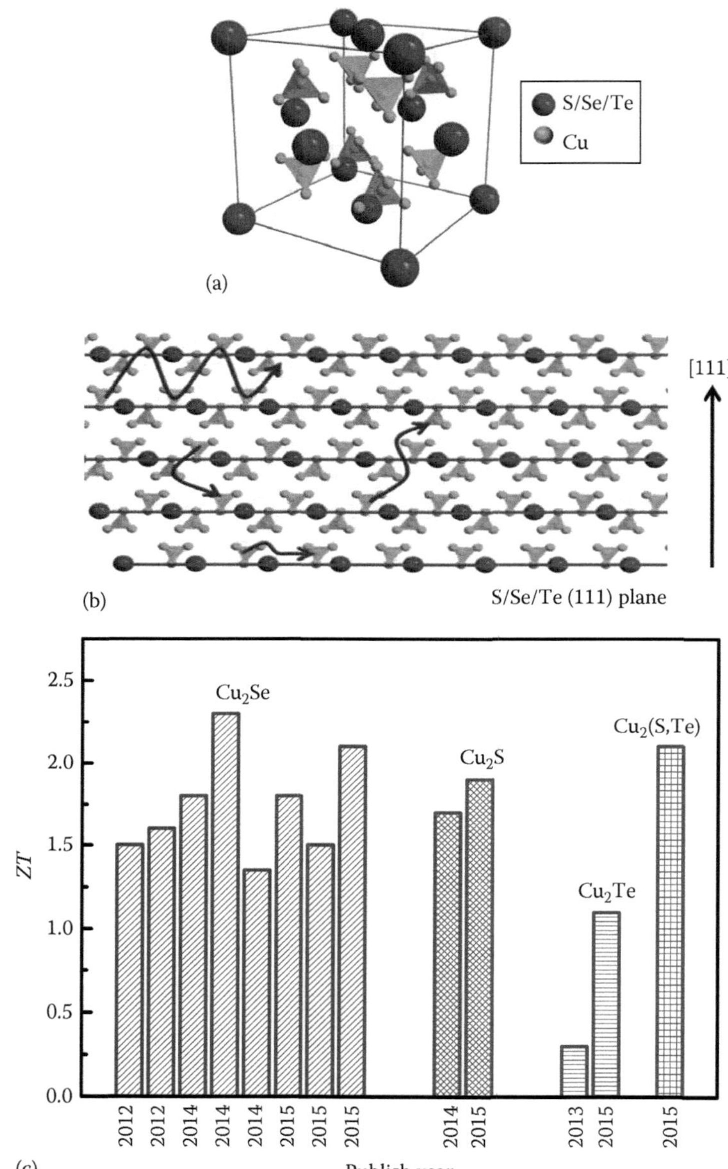

FIGURE 8.1
(a) Unit cell of high-temperature cubic $Cu_{2-\delta}X$ (X = S, Se, and Te) phase where only the 8c and 32f interstitial positions are shown with Cu atoms. (b) Projected plane representation of the crystal structure along the cubic $[1\bar{1}0]$ direction. The arrows indicate that the Cu ions can freely travel among the interstitial sites. (c) The representative TE figure of merit for $Cu_{2-\delta}X$ (X = S, Se, and Te) compounds reported since 2012. ([a,b]: Reprinted by permission from Macmillan Publishers Ltd., *Nat. Mater.*, Liu, H., Shi, X., Xu, F. et al., Copper ion liquid-like thermoelectric, 11, 422, copyright 2012.)

a promising TE material with high TE performance [6]. Generally, the TE performance of a material can be evaluated by its dimensionless figure of merit (*ZT*), which is defined as $ZT = S^2\sigma T/\kappa$, where *S* is the Seebeck coefficient, σ is the electrical conductivity, κ is the thermal conductivity, and *T* is the absolute temperature [7]. In the 1960s–1970s, the development of Cu_2Se-based radioisotope thermal generators (RTGs) was carried out by 3M Corporation, Teledyne Energy Systems, General Atomics Corporation, and the NASA Jet Propulsion Laboratory [8]. A peak *ZT* of 1.2 at 1000 K had been achieved in $Cu_{1.97}Ag_{0.03}Se_{1+y}$ ($y < 0.01$) by 3M Corporation [9]. Notably, such good TE performance in $Cu_{1.97}Ag_{0.03}Se_{1+y}$ can be maintained over 4000 h for stable materials [10]. These fascinating results made Cu_2Se a potential candidate to compete with SiGe and PbTe in NASA's space program. However, research efforts to develop Cu_2Se were finally stopped in 1979 due to the inferior TE performance caused by selenium evaporation and Cu-ion migrations [11]. Three decades later, a resurgence of interest in Cu-based chalcogenides $Cu_{2-\delta}X$ (X = S, Se, and Te) was inspired by the *liquid-like* behavior of Cu ions resulting in the ultralow thermal conductivity and reduced specific heat at high temperatures, as well as other interesting and abnormal physical properties. A high *ZT* of about 1.5 at 1000 K for Cu_2Se was obtained by Liu et al. in 2012 and a new concept named "phonon-liquid electron-crystal" (PLEC) was proposed to explain such high TE performance [12]. This study successfully extended the TE research to superionic conductors and enlarged the scope of prospective high performance TE materials. After that, high *ZT*s, such as 1.3–2.1 for $Cu_{2-\delta}Se$ [13–23], 1.7–1.9 for $Cu_{2-\delta}S$ [24–28], 0.3–1.0 for $Cu_{2-\delta}Te$ [15–30], and 2.1 for $Cu_2(S, Te)$ [31], have been reported. A more detailed history of $Cu_{2-\delta}X$ (X = S, Se, and Te), especially that before the twenty-first century, can be found in the papers published by Brown et al. and Dennler et al. [8,32]. Figure 8.1 shows the *ZT*s for $Cu_{2-\delta}X$ (X = S, Se, and Te)–based compounds reported since 2012. More importantly, a few new physical phenomena and mechanisms, such as the *liquid-like* behavior and critical scattering [12,16], have been discovered in $Cu_{2-\delta}X$ (X = S, Se, and Te), which shed light also on the research of other TE materials.

8.2 Phase Diagrams and Crystal Structures of $Cu_{2-\delta}X$ (X = S, Se, and Te)

In spite of a simple chemical formula, the phase diagrams and crystal structures of $Cu_{2-\delta}X$ (X = S, Se, and Te) are quite complex, and they depend sensitively on the amount of Cu deficiency δ. For example, besides the stoichiometric Cu_2S (chalcocite), there exist several other kinds of $Cu_{2-\delta}S$ compounds, termed $Cu_{1.93}S$–$Cu_{1.96}S$ (djurleite), $Cu_{1.8}S$ (digenite), $Cu_{1.75}S$ (anilite), and CuS (covellite), which are experimentally proven as stable structures at room temperature [33]. After at least one phase transition, the room temperature $Cu_{2-\delta}X$ phase gradually converts to the final high-temperature cubic phase, with the actual temperature and number of phase transitions determined by δ. The high-temperature $Cu_{2-\delta}X$ cubic phases are stable over a wide composition range, for example, $\delta = 0$–0.25 for $Cu_{2-\delta}Se$[34] and $\delta = 0$–0.27 for $Cu_{2-\delta}S$ [33]. Now, the crystal structures of the high-temperature $Cu_{2-\delta}X$ cubic phase have been already clearly determined, while the crystal structures of their low-temperature phases have not been completely clarified yet. Specifically, the Cu deficiency would destroy the initial lattice periodicity in stoichiometric Cu_2X phases and induce more complex atomic arrangements, enhancing the difficulty and uncertainty when determining the crystal structure of low-temperature $Cu_{2-\delta}X$ phase. Thus, here we only discuss the phase diagrams and crystal structures of the stoichiometric Cu_2X (X = S, Se, and Te) compounds.

8.2.1 Cu_2Se

Cu_2Se has two temperature-dependent phases, termed as low-temperature α-phase and high-temperature β-phase [34]. As shown in Figure 8.2a, the high-temperature Cu_2Se β-phase exhibits a face-centered cubic cell for Se atoms, while Cu atoms are randomly distributed at either the body-centered (8c) or the face-centered (32f) sites of the Se tetrahedral [16]. In 1974, Murray and Heyding proposed a monoclinic structure for the Cu_2Se α-phase *via* x-ray powder diffraction analyses [35]. Another monoclinic cell was proposed by Vucic et al. who analyzed electron diffraction patterns [36]. Later, Kashida and Akai identified α-Cu_2Se as pseudomonoclinic and characterized the structure by a periodic array of vacant tetrahedral and interstitial copper atoms, forming a ladderlike structure [37]. The theoretical calculation carried out by Nguyen et al. further indicated that the structures of the Cu_2Se α-phase preferred a unique but unspecified layered structure [38]. In 2012, Liu et al. confirmed a lamellar fashion for the structure of the Cu_2Se α-phase, as shown in Figure 8.2b [12]. Their theoretical and experimental investigations concluded that different forms of room temperature structures of Cu_2Se may exist and the uncertainty may arise from their possible multiformity, which is hard to clarify by diffractometry methods like x-ray diffraction. In 2015, Lu et al. further investigated the structure of the Cu_2Se α-phase *via* transmission electron microscopy (TEM) [39]. It revealed a specific self-independent binary sublattice in Cu_2Se for Cu and Se. At room temperature, Cu atoms congregate alternately into Se interlayers and bear an ordered configuration for energy consideration. Ordering of Cu employs a variety of fashions with little energy difference, leading to multiformity of Cu ordering embedded in the extensive and unique Se fcc sublattice. Therefore, on the microscale, the structure consists of a large number of domains of several nanometers but arranged in a highly coherent fashion [39].

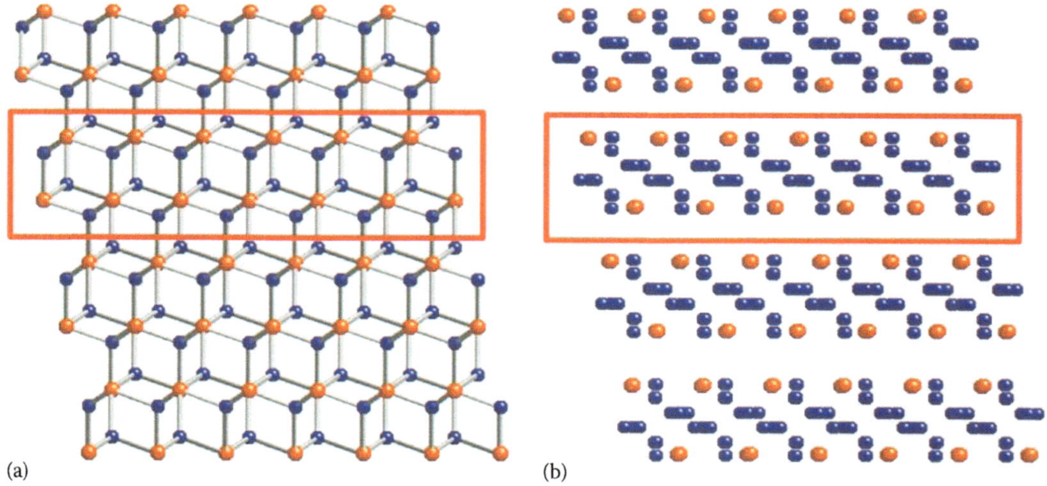

FIGURE 8.2
Crystal structures of Cu_2Se: (a) high-temperature cubic β-phase and (b) room-temperature α-phase. (From Liu, H.L., Yuan, X., Lu, P. et al.: Ultrahigh thermoelectric performance by electron and phonon critical scattering in $Cu_2Se_{1-x}I_x$. *Adv. Mater.* 2013. 25. 6607–6612. Copyright Wiley-VCH Verlag GmbH & Co. KGaA. Reproduced with permission.)

8.2.2 Cu_2S

The stoichiometric compound Cu_2S possesses two phase transitions [33]. One is at about 370 K and the other is around 700 K. Below 370 K, it is called the low-chalcocite γ-phase (L-chalcocite). The second phase is between 370 and 700 K, called the high-chalcocite β-phase (H-chalcocite). Since 1963, the crystal structures of L- and H-chalcocite have been studied extensively [40]. As shown in Figure 8.3, it is now widely accepted that the H-chalcocite has a $P6_3/mmc$ hexagonal symmetry crystal structure with two molecule units $(Cu_2S)_2$ per unit cell [40]. While S atoms occupy a hexagonal lattice, Cu atoms randomly occupy a few possible high-symmetry Wyckoff sites with certain probabilities, and different authors report somewhat different occupied positions. For example, Wuensch and Buerger included Cu at 2b, 4f, and 6g positions [41], while Sadanaga et al. included Cu at 2b, 4f, and 12k positions [42]. In 2002, Will et al. used the Debye–Scherrer film technique to characterize the crystal structure of Cu_2S and concluded that Cu atoms are located at 2b, 4f, 6g, 6h, and 2a sites, while S atoms are located at 2c sites [43]. In contrast, the L-chalcocite has a large orthorhombic unit cell with 96 molecule units $(Cu_2S)_{96}$, but each Cu atom has a unique site [44]. The lattice parameters at room temperature for L-chalcocite are a = 15.27 Å, b = 11.93 Å, c = 13.45 Å, and β = 115.6° [45]. In fact, the crystal structure of L-chalcocite is closely related to that of H-chalcocite. It is essentially a superstructure of the hexagonal lattice with slight distortions [46]. Above 700 K, Cu_2S transforms into the α-phase with an fcc cubic structure. Similar to the Cu_2Se β-phase, S atoms in the Cu_2S α-phase maintain a rigid sublattice, while

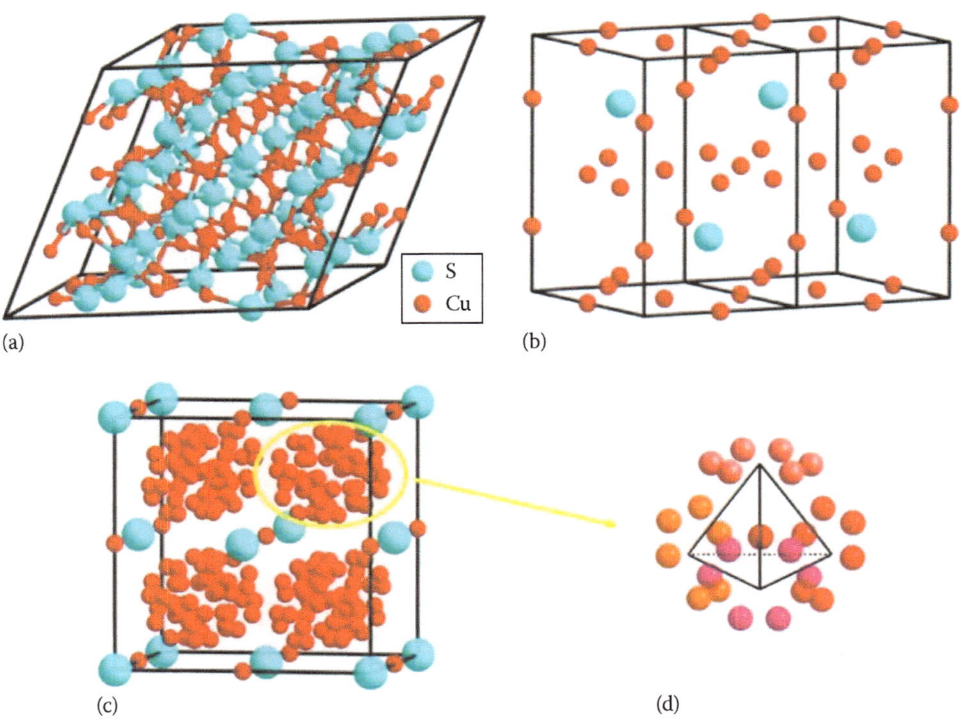

FIGURE 8.3
Crystal structures of Cu_2S: (a) low-chalcocite γ-phase; (b) high-chalcocite β-phase with all possible Cu ions at 2b, 4f, and 6g interstitial sites; (c) cubic-chalcocite α-phase with all possible Cu ions at 8c, 4b, and 192l interstitial positions; and (d) the enlarged view around the position of (3/4, 1/4, 3/4).

Cu atoms are distributed throughout many possible positions, which is indicative of the high degree of disorder and low threshold for ion motion, that is, characteristic of *liquid-like* behavior [43]. The temperature dependence of the lattice parameter (a) for the Cu_2S α-phase is given by $a = 0.5514 + 2.857 \times 10^{-5} T$ [47].

8.2.3 Cu_2Te

The phase sequence in Cu_2Te is more complicated than that in Cu_2Se and Cu_2S, as it has five successive phase transitions between 300 and 900 K [48]. These phases are described as the α-phase (below 548 K), β-phase (548–593 K), γ-phase (593–638 K), δ-phase (638–848 K), and ε-phase (above 848 K). The Cu_2Te α-phase, β-phase, and γ-phase crystallize in orthorhombic structure, while the σ-phase crystallizes in hexagonal structure. There are two kinds of α-phases coexisting at room temperature, named the α_{II}-phase and the α_I-phase [48]. The α_{II}-phase is a metastable orthorhombic superstructure of the Nowotny's hexagonal basic structure. The α_I-phase is stable at room temperature, and it is sluggishly formed even at room temperature by an interpolytypic transformation of the α_{II}-phase. The crystal structure of the Cu_2Te ε-phase is the same as that of the cubic Cu_2Se β-phase or the Cu_2S γ-phase. Te atoms occupy positions on the face-centered cubic sublattice, while Cu atoms are distributed over tetrahedral interstitial sites. The detailed information about the crystal structure for each Cu_2Te phase is shown in Table 8.1.

8.3 First Principles Calculations on $Cu_{2-\delta}X$ (X = S, Se, and Te)

Due to the lack of accurate information on the Cu occupations, it is difficult to obtain the precise band structure of $Cu_{2-\delta}X$ (X = S, Se, and Te) from the first principles calculations. This is the reason why some calculation results show that stoichiometric Cu_2X (X = S, Se, and Te) are zero-gap materials or even semimetals [23,28,49–52], while the optical experimental studies usually suggest that they are semiconductors with a bandgap around 1.0–1.2 eV [45,53–54]. Chi et al. calculated the band structure of the Cu_2Se α-phase by first principles calculations and found that the bandgap of Cu_2Se is around 1.03 eV [55], which agrees well with the experimental value of 1.23 eV reported by Sorokin et al. [53]. However, Zhao et al. calculated the band structure of Cu_2Se β-phase and found Cu_2Se β-phase is a

TABLE 8.1

Phase Transformations of Stoichiometric Cu_2Te

Temperature (K)	Phase	Structure	Lattice Parameters
RT–548	α_I-Phase (*RT*–448 K)	Monoclinic	$a_m = c_m = 1.28$ nm, $b_m = 4.47$ nm, $\beta = 120°$
	α_{II}-Phase	Orthorhombic	$a_o = 0.75$ nm, $b_o = 0.43$ nm, $c_o = n \times 0.74$ nm ($n = 2, 4, 5, 7$)
548–593	β-Phase	Orthorhombic	$a_\beta = 0.45$ nm, $b_\beta = 0.78$ nm, $c_\beta = 1.48$ nm
593–638	γ-Phase	Orthorhombic	$a_\gamma = 0.45$ nm, $b_\gamma = 0.78$ nm, $c_\gamma = 11.10$ nm
638–848	δ-Phase	Hexagonal *P6/mmm*	$a_\delta = 0.43$ nm, $b_\delta = 0.74$ nm
>848	ε-Phase	Cubic	$a_\varepsilon = a_c \cong 0.61$ nm

Source: Data are adapted from Vouroutzis, N. et al., *Phys. State Solidi*, 202, 271, 2005.

zero-gap material [23]. Cu 3d states make the dominant contribution to the total density of states (DOS) from −4 eV to the Fermi level (E_F), while the total DOS in the conduction bands is mainly built up by Cu 3p states and Se 4s states. Kim et al. used *ab initio* molecular dynamics (AIMD) to analyze the thermal transport of the high-temperature Cu_2Se β-phase [56]. It shows that the atomic trajectories of Cu atoms become random with increasing temperature and some migration among different sites occurs at elevated temperature. Lukashev et al. calculated the electronic structure of Cu_2S by using both the local density approximation method and the quasiparticle self-consistent GW (QSGW) method [49]. However, as shown in Figure 8.4, both of the methods suggest that Cu_2S is a semimetal and the Cu d valence bands and Cu s-like bands are degenerate at the Γ point. Interestingly, small deviations from the ideal antifluorite positions for the Cu atoms open a small gap because the symmetry breaking allows the Cu s and Cu d bands to hybridize. The total energies of the structures are found to increase from monoclinic to hexagonal to cubic [49]. Their calculation also suggested that the creation of Cu vacancies in Cu_2S is actually favorable even when the system is assumed to be in thermodynamic equilibrium with bulk Cu [49]. This shows that Cu_2S is intrinsically unstable toward the formation of $Cu_{2-\delta}S$.

Although optical studies show that Cu_2Te is a semiconductor with a bandgap of 1.04 eV [53], the calculation carried out by Kashida et al. suggests that Cu_2Te is a semimetal with the valence band and conduction band overlapped severely near the Γ point [52]. The calculated electron densities show a charge transfer from Te to Cu atoms. The estimated valences of Cu and Te are −0.20 and +0.40, respectively, suggesting that there is strong covalent bonding in Cu_2Te. This is in contrast to the case of Cu_2S, where the charge transfer occurs from Cu to S atoms [53]. The higher degree of covalent bonding in Cu_2Te can well

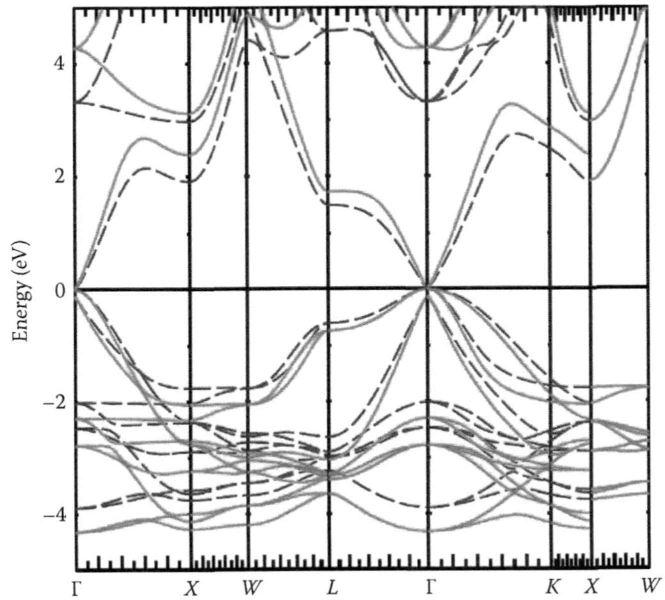

FIGURE 8.4
Band structure of antifluorite Cu_2S. Solid lines represent the result calculated by the quasiparticle self-consistent GW method and the dashed lines represent the result calculated by the local density approximation +Cu s shift method. (Reprinted with permission from Lukashev, P., Lambrecht, W.L., Kotani, T., and Schilfgaarde, M., *Phys. Rev. B*, 76, 195202, 2007. Copyright 2007 by the American Physical Society.)

explain the larger mobility in Cu_2Te than that in Cu_2S [30]. In 2014, Zhang et al. calculated the electronic structures of $Cu_{2-\delta}X$ (X = S, Se, and Te) using the modified Becke-Johnson potential plus an on-site Coulomb U approach (the mBJ+U approach).[50] Different with the previous results, semiconducting band structures and normal band orderings are successfully obtained, but the calculated band gaps of Cu_2S and Cu_2Se still seem to be underestimated compared with experimental results. The differences between calculations and experiments in $Cu_{2-\delta}X$ (X = S, Se, and Te) system suggest that further improvements on calculation approaches are required.

8.4 Abnormal Physical Phenomena and Mechanisms in $Cu_{2-\delta}X$ (X = S, Se, and Te)

8.4.1 *Liquid-Like* Behavior

The Cu-ion diffusivities in the stoichiometric Cu_2Se β-phase, with the order of 10^{-5} cm^2 s^{-1}, are similar to or larger than that of liquid water, making it one of the best superionic conductors [57]. Such high Cu-ion diffusivities enable the Cu ions to consecutively jump from one equilibrium position to another within a very short time, exhibiting a *liquid-like* behavior. Thus, the crystal structure of the Cu_2Se β-phase can be regarded as a rigid and crystalline sublattice made by Se atoms around which a *liquid-like* charged fluid of Cu ions diffuses. This makes the Cu_2Se β-phase fit well with the PLEC concept [12], which may be considered an extension of the PGEC concept proposed by Slack [58]. Structurally, a liquid is similar to glass, as both lack long-range periodicity, but a glass has lost the ability to flow. Thermal transport in disordered systems can be considered as the random diffusion of energy located at fixed atomic positions, with consecutive atomic jumps from one equilibrium position to another [59]. Local atomic jumps and rearrangement of the *liquid* inhibit the propagation of transverse waves and disrupt heat propagation by phonons, thereby leading to a significant reduction in specific heat (C_v) [12]. C_v is directly related to the excitation of lattice vibrational modes (phonons). In a typical solid, C_v is $3Nk_B$ (the Dulong–Petit limit) at high temperatures, where N is the number of particles and k_B is the Boltzmann constant. However, in a liquid, C_v is reduced to 2–2.5Nk_B because most transverse vibrational waves do not propagate [60,61]. As presented in Figure 8.5a and b, when the frequency is below the cutoff frequency, the transverse phonons are softened and disappear, leading to the Cu_2Se β-phase exhibiting a C_v value between that of a liquid and a solid [12]. In addition, the *liquid-like* mobile Cu ions also introduce extra scattering to longitudinal phonons to further disrupt heat propagation or alter the phonon transfer and thereby shorten the phonon mean free path, which approaches the minimum wavelength around 1.5 Å [39]. As shown in Figure 8.5c, this wavelength is almost the lowest value among the reported TE materials so far. Such *liquid-like* behavior results in ultralow lattice thermal conductivity (see Figure 8.5d) in the Cu_2Se β-phase of only around 0.4–0.6 W m^{-1} K^{-1} and, consequently, yields the high *ZT*s around 1.3–2.1 at 1000 K [13–23].

Besides the Cu_2Se β-phase, both the Cu_2S α/β-phase and the Cu_2Te ε-phase are also superionic conductors with high ionic conductivities (around 0.1–2 S cm^{-1}) [62,63]. The superionicity comes from the fact that Cu atoms have the ability to move among the different possible Cu positions and occupy these positions with equal probability. Wang et al. carried out long time *ab initio* Born–Oppenheimer molecular dynamics simulations on Cu_2S and obtained an averaged Cu diffusion constant of 3.2×10^{-6} cm^2 s^{-1} for the Cu_2S β-phase [64]. Their calculations also found that the potential barrier for the Cu movement from one minimum to a nearby minimum is only about 50 meV, in the same order as the k_BT. Importantly, the Cu movement can involve similar magnitude displacements from several Cu atoms rather than just one Cu atom, which suggests that the Cu movement fits well with the picture of a typical liquid. Similar to that for Cu_2Se mentioned earlier, the *liquid-like* behavior of Cu ions also contributes to the ultralow thermal conductivity and high *ZT* for Cu_2S and Cu_2Te [24–30].

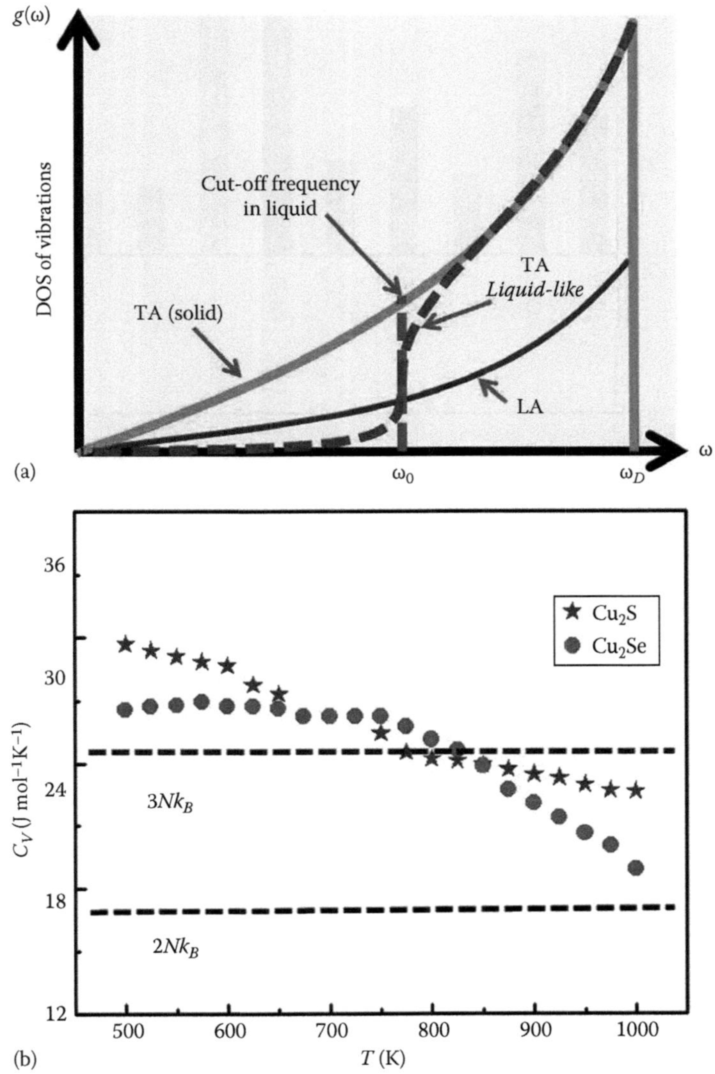

FIGURE 8.5
Diagrams of (a) phonon density of states and (b) specific heat in stoichiometric Cu_2S and Cu_2Se. The transverse phonons are softened and disappear when the frequency is below the cutoff frequency in *liquid-like* materials, leading to the reduced C_v values. $3Nk_B$ is the limit value in a solid while $2Nk_B$ is the limit value in a liquid.
(Continued)

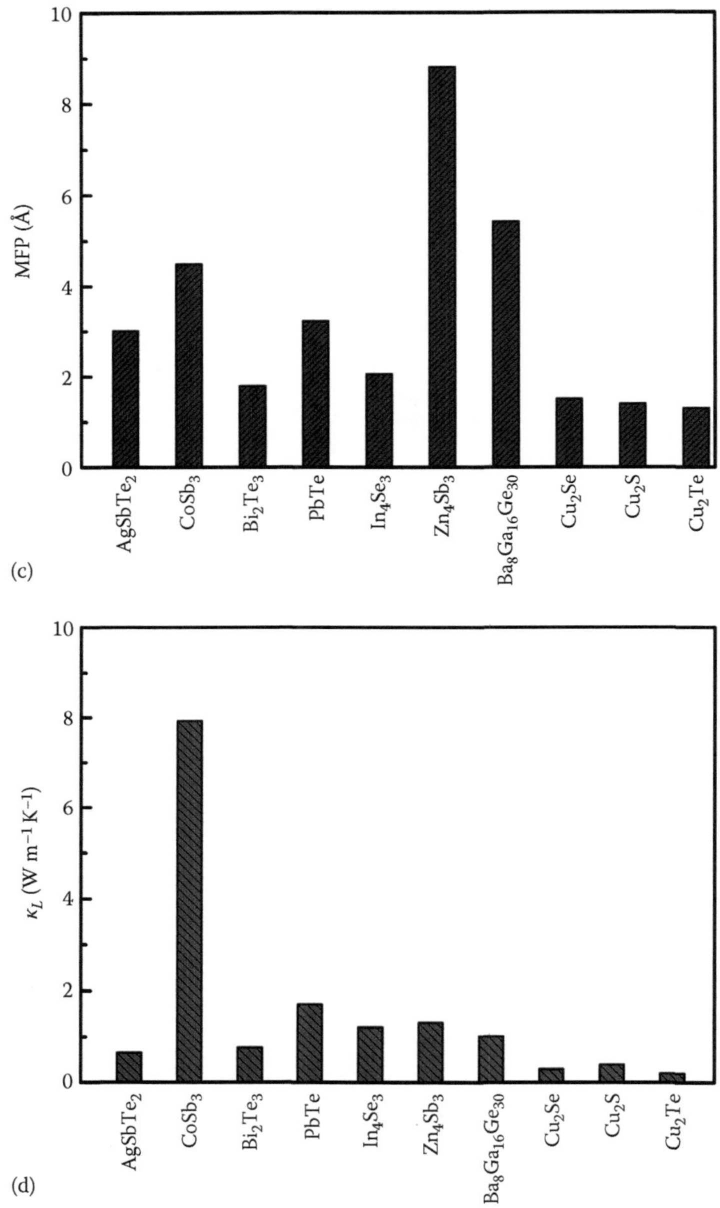

FIGURE 8.5 ***(Continued)***
Diagrams of (c) and (d) show the phonon mean free path and the lattice thermal conductivity for Cu_2Se and other typical TE materials. (From Lu, P., Liu, H.L., Yuan, X. et al., Multiformity and fluctuation of Cu ordering in Cu_2Se thermoelectric materials, *J. Mater. Chem. A*, 3, 6901–6908, 2015. Reproduced by permission of The Royal Society of Chemistry.)

8.4.2 Critical Phenomenon

The phase transition in stoichiometric Cu_2Se from the α-phase to the β-phase occurs near 400 K. In 1981, Vucic et al. first investigated the phase transition of Cu_2Se, indicating that the origin of the transition was the modulation of Cu distribution along the cubic [111] direction [36]. Using the ultrafast x-ray spectroscopy and scattering techniques, Miller et al. observed that the transformation from the Cu_2S α-phase to the superionic β-phase only takes less than 20 ps, suggesting that the transformation is determined by the intrinsic dynamics of the structure reorganization and not by the probability of a nucleation event [65]. In 2015, Lu et al. systematically investigated the phase transition of Cu_2Se by *in situ* TEM technique [39]. The ordered structure in the room-temperature Cu_2Se α-phase is found to severely fluctuate upon heating, especially when the temperature approaches the phase transition. The detailed process for the phase transition observed by the *in situ* TEM can be summarized as follows. Upon increasing temperature to the phase transition, the ordered structures of interlayer Cu atoms maintain multiformity and retain the in-plane tripled superlattices though atomic configurations vary continuously. When the temperature reaches close to the phase transition, the interlayer Cu atoms start to become disordered. At this stage, the layered structure still remains but parts of Cu atoms have moved to the neighboring interlayers. On further increasing the temperature, the layered structures completely disappear, indicating that the phase transition is completed and all Cu atoms randomly but evenly occupy the 8c and 32f sites in the fcc crystal structure. Meanwhile, these processes may be reversed instantly, showing an oscillation of structural changes at the nonenergy barrier critical states near the phase transition. These structural changes, including varying of ordered structures, disordering of Cu arrangements within the interlayers, and diffusion of Cu atoms across the interlayers, take place successively [39]. Figure 8.6a shows the calculated energies of transformation path during the phase transition in Cu_2Se. Very smooth changes in energy demonstrate that there are no energy barriers, a behavior typical of the second-order phase transition [16]. A slight oscillation in thermodynamics may result in switching between order and disorder or among different ordered structures. The gradual and continuous phase transition can also be confirmed by the high-temperature x-ray diffraction measurements shown in Figure 8.6b [16]. The peak of the α-phase at 12.9° steadily weakens and finally disappears, while the peak around 26.5° gradually shifts left and finally transits to the cubic peak.

The character of the second-order structural phase transition in stoichiometric Cu_2Se results in extremely large fluctuations in the sample density, concentration, structure, and correlated sample density concentration near the phase transition temperature [66]. These variations induced by thermal fluctuations lead to extremely large phonon critical scattering and significantly lowered lattice thermal conductivities during phase transitions as compared to the normal phases [16]. In addition, the thermal fluctuation also induces large electron critical scattering, which is evidenced by the dramatically decreased carrier mobility and enhanced temperature dependence of the carrier mobility during the phase transition [67]. Meanwhile, such strong electron critical scattering can also broaden the bandgap of Cu_2Se and lead to the rapid enhancement of the Seebeck coefficient (see Figure 8.6c) near the phase transition temperature [68]. Considering the greatly reduced lattice thermal conductivity, an extraordinary increase in the TE performance with the *ZT* value of 2.3 around 400 K has been achieved [16] in Cu_2Se (see Figure 8.6d). Notably, such ultrahigh *ZT* occurs within a range of dozens of degrees, which is quite meaningful for applications such as powerful solid-state heat pumps or power generators operating within a narrow temperature range around room temperature.

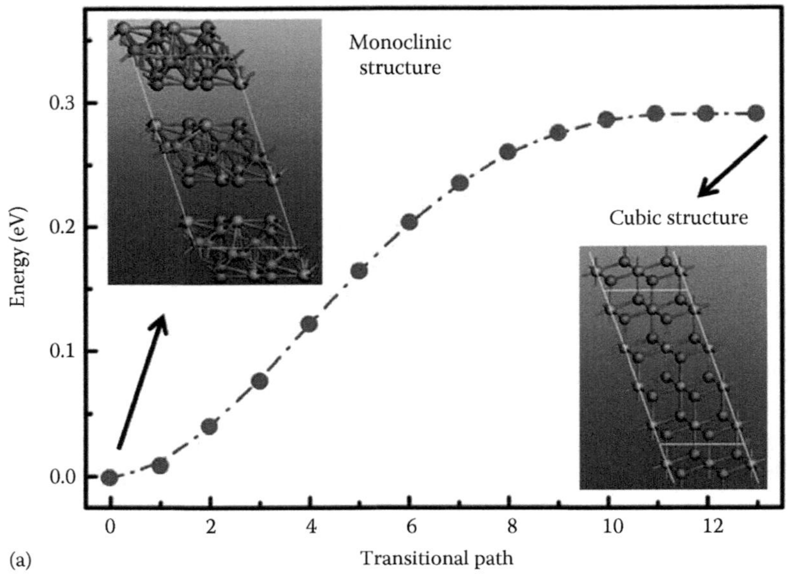

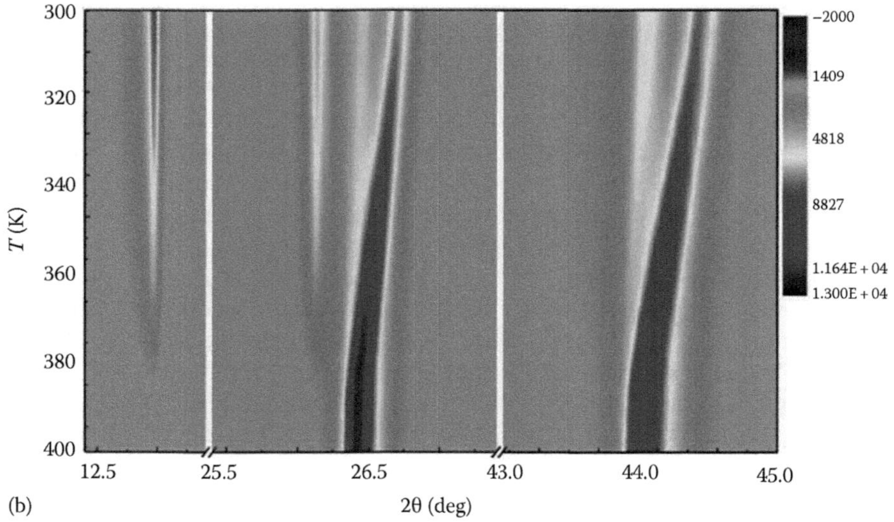

FIGURE 8.6
(a) Calculated energies of the transition states during the phase transition. (b) High-temperature x-ray diffraction data collected on heating of Cu_2Se powder from 300 K (top) to 410 K (bottom), showing the coalescence of the split peaks around 26.4°–26.7° and 44.0° and the disappearance of the extra peaks around 12.9° and 26.5° as the lower-symmetry α-phase transforms to the higher-symmetry cubic β-phase. *(Continued)*

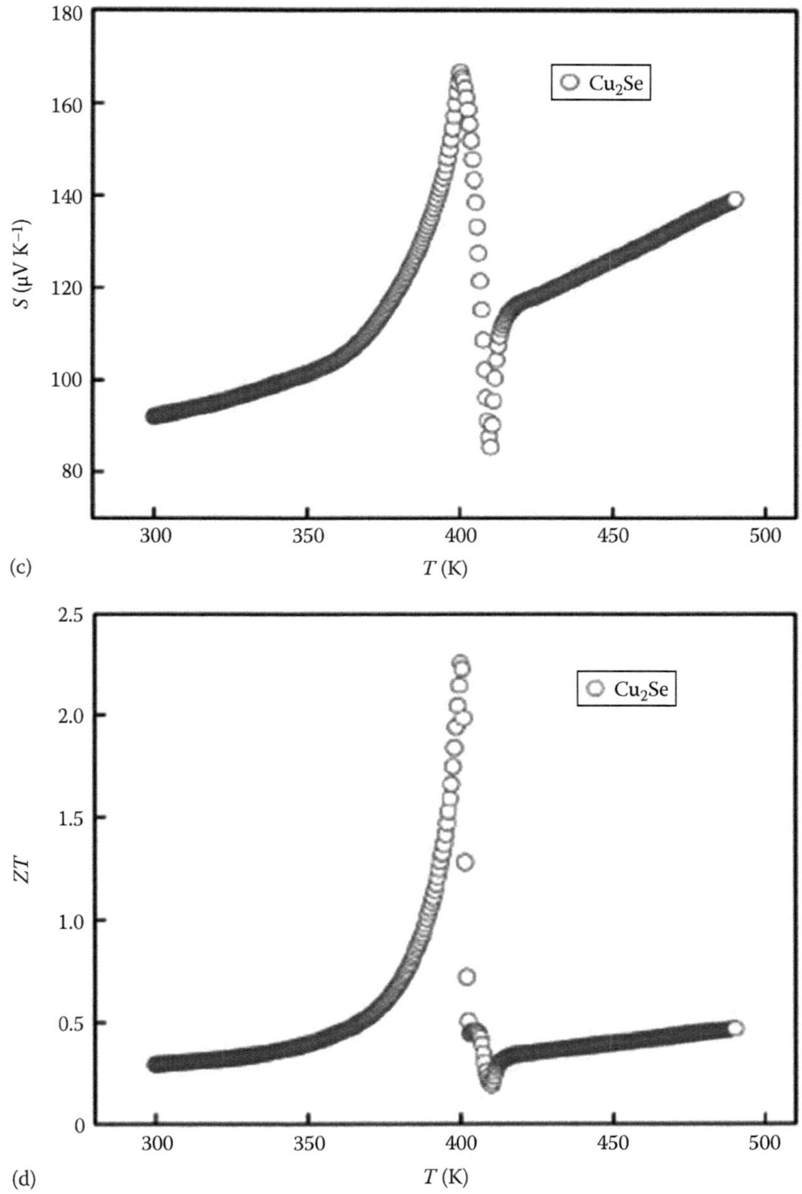

FIGURE 8.6 *(Continued)*
(c) and (d) are temperature dependences of the Seebeck coefficient and ZT of Cu_2Se, respectively. (From Liu, H.L., Yuan, X., Lu, P. et al.: Ultrahigh thermoelectric performance by electron and phonon critical scattering in $Cu_2Se_{1-x}I_x$. *Adv. Mater.* 2013. 25. 6607–6612. Copyright Wiley-VCH Verlag GmbH & Co. KGaA. Reproduced with permission.)

8.5 Investigation of Thermoelectric Performance of $Cu_{2-\delta}X$ (X = S, Se, and Te)

8.5.1 $Cu_{2-\delta}Se$

The investigation of Liu et al. in 2012 has reignited the attention of the TE community on $Cu_{2-\delta}Se$. Besides the phase transition study mentioned earlier, many works have been done to further understand and optimize the TE performance of $Cu_{2-\delta}Se$ [12–23]. Ballikaya et al. studied low-temperature transport properties of Cu_2Se [15]. They found the carrier-charged defect scattering rather than the carrier-acoustic phonon scattering as being the dominant scattering mechanism in Cu_2Se at 10–300 K. Specifically, the measurement of low-temperature thermal conductivity showed that there is no apparent peak typical of highly crystalline materials, suggesting the presence of severe disorder in Cu_2Se. Chi et al. cooled down Cu_2Se to liquid nitrogen temperatures and measured electrical transport properties and studied the crystal structure [55]. They found that upon temperature change, Cu_2Se might have another phase transition with a possible charge density wave character at 100–150 K as a result of the distortion/redistribution of the Se–Cu–Cu–Cu–Cu–Se layers. Around this temperature range, the electrical resistivity and the Hall effect of Cu_2Se exhibit a highly anomalous behavior, while the other TE properties, including the Seebeck coefficient, thermal conductivity, and specific heat, are quite normal.

In 2012, Liu et al. prepared Cu_2Se *via* the traditional melting–annealing method [12]. In the following years, many new methods have been adopted to prepare this compound. Yu et al. prepared Cu_2Se *via* ball milling and achieved a peak *ZT* of 1.6 at 973 K [14]. In 2015, Gahtori et al. also used the same ball milling method to prepare Cu_2Se and obtained a much higher *ZT* of 2.1 at 973 K [22]. The main reason for such high *ZT*s is believed to be the enhanced low-to-high wavelength phonon scattering by different kinds of defects, in a wide nano- to mesoscale spectrum generated during the ball milling process. In 2014, Li et al. successfully synthesized Cu_2Se nanoparticles *via* two simplified chemical processes, a coprecipitation and a reduction method [17]. Cu_2Se prepared by the coprecipitation method had better electronic transport properties and consequently higher *ZT* (1.35 at 1000 K) than Cu_2Se prepared by the reduction method. In 2014, Su et al. synthesized single phase Cu_2Se using a combustion process called self-propagating high-temperature synthesis [19]. Maximum *ZT*s of Cu_2Se synthesized *via* this synthesis technique are comparable to or better than *ZT*s of samples prepared by other methods. Importantly, this synthesis technique provides a new avenue for ultrafast, low-cost, large-scale production of Cu_2Se, which is meaningful for its real industrial applications. Based on the congruent melting character of Cu_2Se, Zhao et al. developed a fast melting–solidification method to prepare Cu_2Se in 2015 [20]. They found Cu_2Se to crystallize quickly with microscale grains during the quenching process and highly dense Cu_2Se bulk structures formed directly after the quenching process without the need for the traditional consolidation process. High *ZT*s around 1.7–1.8 at 973 K were achieved in Cu_2Se samples prepared by this fast melting–quenching method. The *ZT*s for Cu_2Se prepared by different synthesis methods are summarized and presented in Figure 8.7.

Zhong et al. doped Al into Cu_2Se and created highly aligned large lamellae in bulk $Cu_{1.94}Al_{0.02}Se$ using ball milling and the direct-current hot-pressing process [18]. The maximum *ZT*s of about 1.4 along the layer plane and 2.6 across the layer plane were obtained. Day et al. doped Ag at Cu sites and concluded that the introduction of Ag pushes the hole

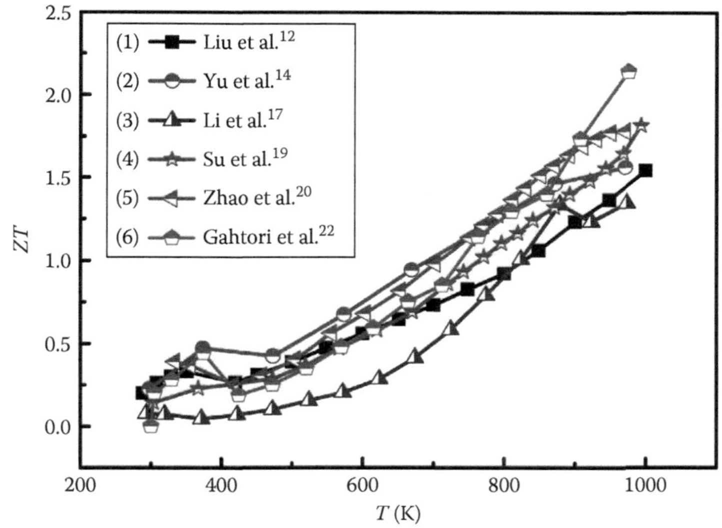

FIGURE 8.7
TE figure of merit (ZT) of $Cu_{2-\delta}Se$ prepared by different synthesis methods: (1) melting–annealing method (Liu et al. [12]), (2) ball milling method (Yu et al. [14]), (3) chemical coprecipitation method (Li et al. [17]), (4) self-propagating high-temperature synthesis technique (Su et al. [19]), (5) melting–solidification method (Zhao et al. [20]), and (6) ball milling method (Gahtori et al. [22]).

concentration close to the optimum value, leading to higher ZT for $Cu_{1.97}Ag_{0.03}Se$ than for Cu_2Se below 780 K [21]. Contrarily, creating Cu vacancies or adding excessive Se in Cu_2Se will push the hole concentration away from the optimum value and thereby will result in low ZTs [12,21]. However, when an overstoichiometric amount of Ag is doped into Cu_2Se, the ZT is greatly deteriorated due to bipolar conduction at elevated temperatures [15]. Zhao et al. doped I and Te into Cu_2Se and they found that all Te- and I-doped Cu_2Se samples have higher lattice thermal conductivity values compared to the undoped sample [23]. Thus, they concluded that I and Te are not helpful for optimizing the ZT of Cu_2Se.

8.5.2 $Cu_{2-\delta}S$

In 2014, He et al. have systematically studied the TE properties of Cu_2S from 300 to 1000 K [27]. Cu_2S is a p-type semiconductor with very low lattice thermal conductivity, which is even lower than its heavier counterpart Cu_2Se. On one hand, such low lattice thermal conductivity can be traced, in part, to the lower average speed of sound in Cu_2S. While the longitudinal speed of sound (v_l) and bulk modulus (B_T) in Cu_2S are actually higher than in Cu_2Se, as one might have expected from the smaller and more covalent sulfide compared to selenide, the transverse phonons travel much slower in Cu_2S than in Cu_2Se. The shear speed of sound (v_s) and shear modulus (G_T) are remarkably lower in Cu_2S than in Cu_2Se [27]. This leads to an overall reduction in the average speed of sound (both arithmetic and geometric averages) and a lower estimation for the minimum thermal conductivity (κ_{min}), which is the expected thermal conductivity in the diffusive limit of highly disordered and amorphous solids. On the other hand, at elevated temperatures, C_p of Cu_2S

is decreased to below the Dulong–Petit value derived for solids with shear modes that store elastic energy. Similar to Cu_2Se, the low and decreasing C_p would also contribute to the low thermal conductivity. At 1000 K, the peak ZT of 0.6 is obtained for Cu_2S [27]. The calculation based on the single parabolic band model suggested that the optimal hole concentration in the $Cu_{2-\delta}S$ system is around 1.3×10^{20} cm^{-3} at 750 K. *Via* introducing a slight

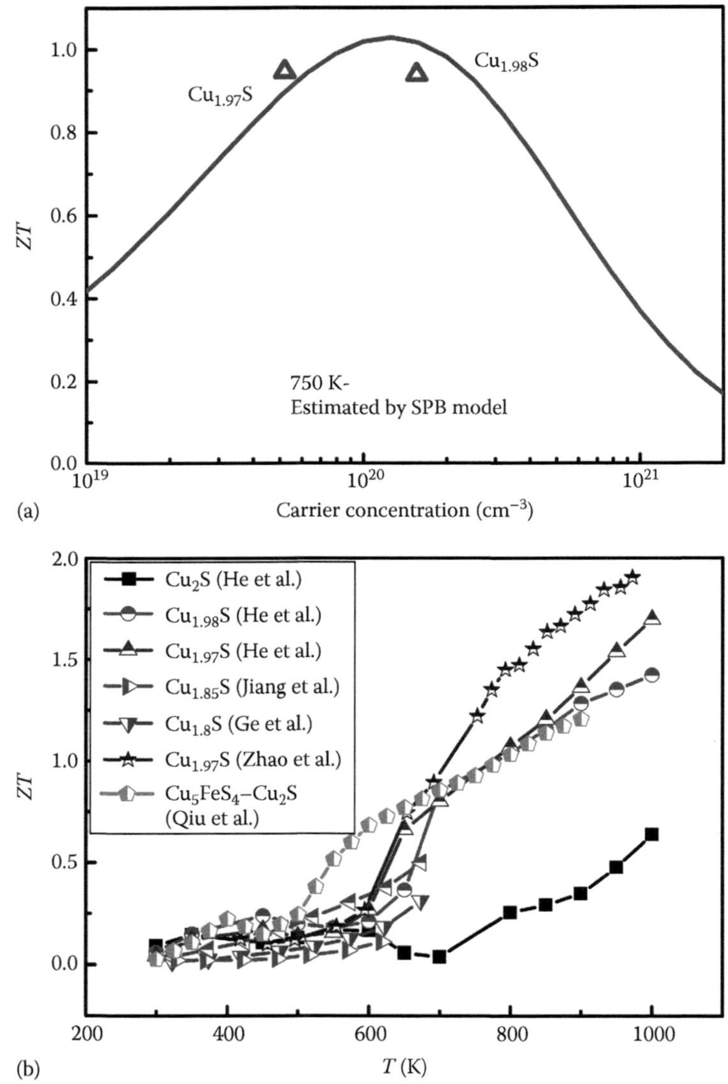

FIGURE 8.8
(a) TE figure of merit (ZT) as a function of Hall carrier concentration at 750 K for $Cu_{2-\delta}S$. Symbols are measurements and solid curves are predicted based on the single parabolic band model. (b) ZTs for different $Cu_{2-\delta}S$ compounds reported recently. (The data are taken from Ge, Z.H. et al., *Chem. Commun.*, 47, 12697, 2011; Jiang, Q.H. et al., *J. Mater. Chem. A*, 2, 9486, 2014; Qiu, P.F. et al., *Energy Environ. Sci.*, 7, 4000, 2014; He, Y. et al., *Adv. Mater.*, 26, 3974, 2014; Zhao, L.L. et al., *J. Mater. Chem. A*, 18, 9432, 2015.)

Cu deficiency, this optimal hole concentration can be achieved in $Cu_{1.97}S$. Correspondingly, higher *ZT* of around 1.7 at 1000 K has been obtained in $Cu_{1.97}S$ by He et al. [27]. In 2015, a similar high *ZT* of 1.9 at 970 K for $Cu_{1.97}S$ prepared by a melting–solidification technique was reported by Zhao et al. [28].

Besides the slight Cu-deficient $Cu_{2-\delta}S$ compounds, the TE properties of $Cu_{1.8}S$ and $Cu_{1.75}S$ have also been studied [24,25]. Although these compounds have different crystal structures at room temperature, they can convert to the superionic cubic structure after one or two phase transitions [33]. Due to the too high hole concentration, the Seebeck coefficients of these compounds are low and thus they usually show poor TE performance. For example, the maximum *ZT* for $Cu_{1.8}S$ and $Cu_{1.75}S$ is only 0.3 at 773 K and 0.08 at 623 K, respectively [24,25]. In 2014, Qiu et al. investigated TE properties of bornite Cu_5FeS_4, which is a widespread copper-sulfide mineral occurring in a variety of ore deposits formed under a wide range of geological conditions [26]. Cu_5FeS_4 has three temperature-dependent phases [69]. In the crystal structure of the high-temperature cubic phase (above 543 K), five Cu atoms, one Fe atom, and two vacancies (□) randomly occupy the center of eight sulfur tetrahedrons. Thus, it can be regarded as the Fe and vacancy-doped Cu_2S α-phase, and then the chemical formula can be written as $Cu_5Fe\square_2S_4$. Similar to Cu_2S, an ultralow lattice thermal conductivity (0.3–0.5 W m^{-1} K^{-1}) was also obtained for this compound. The maximum *ZT* for the stoichiometric Cu_5FeS_4 sample is 0.3 at 700 K. *Via* forming solid solutions with Cu_2S, the electrical conductivity can be significantly enhanced and, consequently, higher *ZT* of around 1.2 at 900 K was obtained for the composition $(Cu_2S)_{0.8}(Cu_5Fe\square_2S_4)_{0.2}$ [26]. The *ZT*s for $Cu_{2-\delta}S$ reported recently are summarized and presented in Figure 8.8.

8.5.3 $Cu_{2-\delta}Te$

In 2013, Ballikaya et al. reported on TE properties of Cu_2Te consolidated by the spark plasma sintering (SPS) technique [15]. Very high electrical conductivity, about 10^6 S m^{-1} at 300 K, was observed for Cu_2Te. Correspondingly, at this temperature, the Seebeck coefficient was only about 20 μV K^{-1} and the thermal conductivity was as high as 4 W m^{-1} K^{-1} due to too high carrier thermal conductivity. The maximum *ZT* for Cu_2Te reported by Ballikaya et al. was only 0.3 at 900 K. *Via* introducing an overstoichiometric content of Ag into Cu_2Te, a significantly enhanced Seebeck coefficient and suppressed thermal conductivity were observed, leading to an obviously improved *ZT* of 1.0 at 900 K [15]. According to optical studies, Cu_2Te has a bandgap of approximately 1.04 eV [53]. Thus, the ideal chemical stoichiometric ratio compound Cu_2Te should be an intrinsic semiconductor with a very low carrier concentration and electrical conductivity as well as a large Seebeck coefficient. This is inconsistent with Ballikaya's experimental results. In 2015, He et al. also prepared Cu_2Te by the SPS technique and observed similar experimental results [30]. It is concluded that there should be some compositional changes, such as the formation of Cu vacancies during the SPS sintering process, which induce the high hole concentrations (e.g., 5×10^{21} cm^{-3}) in the structure. He et al. cold-pressed their Cu_2Te powders into cylindrical pellets and then annealed them directly without the SPS sintering process [30]. High-density Cu_2Te bulks with low carrier concentrations, from 7.4×10^{20} to 1.5×10^{21} cm^{-3}, were obtained. Such greatly reduced carrier concentrations in the directly annealed Cu_2Te samples not only tune the electronic transport toward the optimal value but also significantly lower the carrier thermal conductivity as well as the total thermal conductivity. Thus, combining these two effects, significantly improved *ZT*s, with a maximum value of 1.1 at 1000 K,

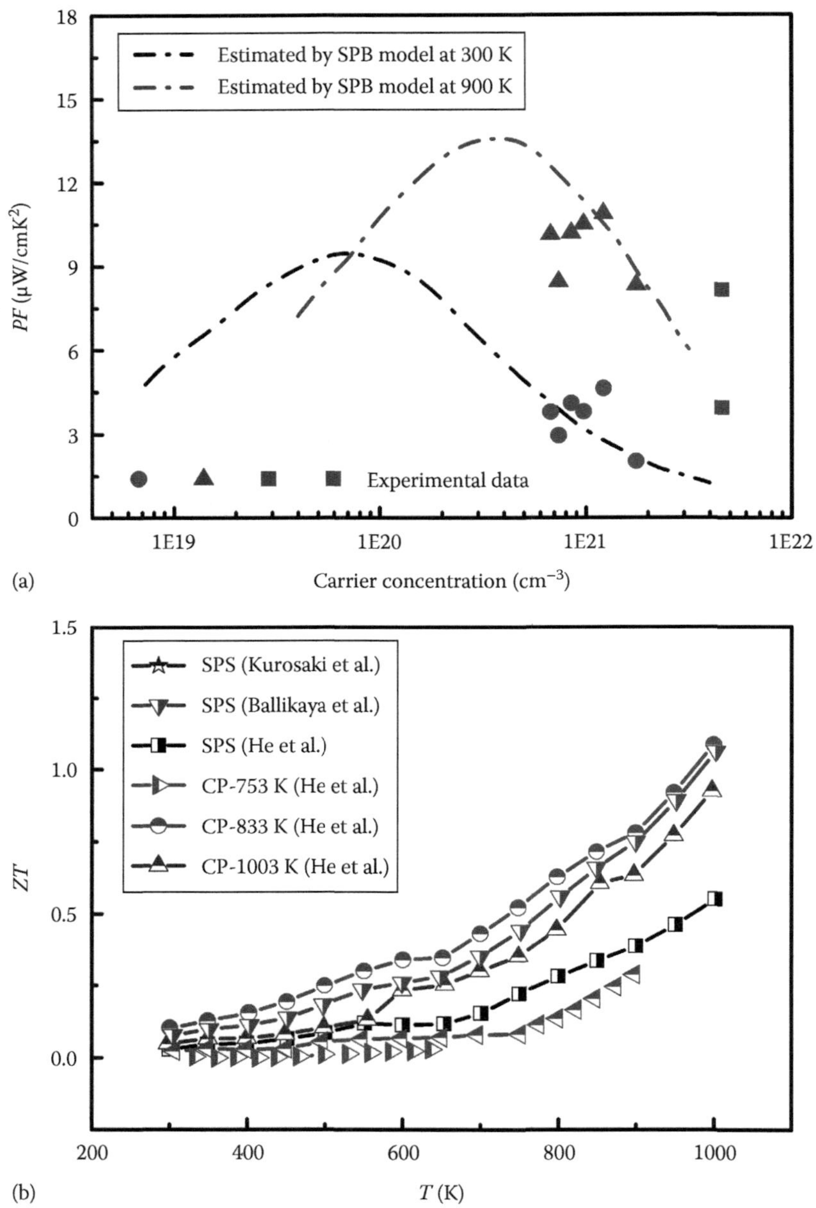

FIGURE 8.9
(a) Power factor ($PF = S^2\sigma$) as a function of carrier concentration in Cu_2Te. The dashed-dot lines were obtained assuming a single parabolic band model. (b) TE figure of merit for stoichiometric Cu_2Te consolidated by the spark plasma sintering technique and cold-pressing methods. ([a]: Reprinted by permission from Macmillan Publishers Ltd. *NPG Asia Mater.* He, Y., Zhang, T., Shi, X., Wei, S.H., and Chen, L.D., High thermoelectric performance in copper telluride, 7, e210, copyright 2015; [b]: The data are taken from Ballikaya, S et al., *J. Mater. Chem. A*, 1, 12478, 2013; Kurosaki, K. et al., *Mater. Trans.*, 47, 1432, 2006; He, Y. et al., *NPG Asia Mater.*, 7, e210, 2015.)

were obtained with these directly annealed Cu_2Te samples [30]. The *ZT*s for Cu_2Te prepared by the SPS technique and cold-pressing method are summarized and presented in Figure 8.9b.

8.5.4 $Cu_2(S,Te)$ Solid Solution

In 2015, He et al. reported an ultrahigh *ZT* of 2.1 at 1000 K for $Cu_2(S, Te)$ solid solutions [31]. They attributed such ultrahigh *ZT* to the mosaicity found in samples of $Cu_2(S, Te)$ solid solutions. A mosaic crystal is a very special state proposed by Darwin, involving high degree of perfection in the lattice translations throughout the crystal [70]. The crystal is composed of mosaic blocks with each block being a perfect crystal but tilted or rotated with respect to others by a fraction of an arc minute. Different from the random orientation in ordinary polycrystals or nanocrystal materials, blocks in mosaic crystals exhibit a nearly identical orientation whereby the bulk appears like a single crystal from the macroscale point of view but contains a number of small-angle boundaries (see Figure 8.10a) [70–72]. In general, flash cooling is usually used to form mosaic crystals. However, interestingly, He et al. successfully prepared $Cu_2(S, Te)$ mosaic crystals using the long-term annealing procedure. Figure 8.10b shows the high-resolution transmission electron microscopy analysis performed on the mosaic crystal of $Cu_2S_{0.5}Te_{0.5}$. The fast Fourier transform (FFT) of the diffractogram (the top right corner) from a region framed by the white square in Figure 8.10b shows a single crystal-type zone-axis pattern although it contains at least four nanocrystallites (outlined in white) of 15–20 nm. These four nanograins exhibit identical diffractograms except for some variation in the brightness of reflections. The intensity of reflections is proportional to the degree of favoring the Bragg condition for certain crystalline planes. Different distributions of reflection intensities (brightest reflections are highlighted using yellow circles in each diffractogram) suggest a slight misorientation in zone axes among these grains. The degree of misorientation is estimated as being within tens of milliradians. The reason for such unique microstructure is believed to be a large difference in atomic sizes of S and Te (about 70% mismatch), which induces the formation of subgrain boundaries inside the grain, which relax the lattice strain and lower the system's energy [31].

The strategy of using nanoscale mosaicity goes beyond the previous approaches by providing multiform effects to tune TE properties. In mosaic crystals, the very small mosaic grains or blocks suggest an ideal example of a lower-dimensional structure imported into bulk materials, which promotes excellent TE performance. In mosaic crystals, strong quantum confinement effects and perhaps even electron barrier filtering result in much enhanced electron effective masses and, thus, Seebeck coefficients. Mosaic crystals contain a number of small-angle tilt or rotation boundaries that introduce long-range disorder for thermal phonons, which can result in the glass-like low thermal conductivity. Thus, the unusual mosaicity composed of nanocrystals with tiny misorientations provides naturally independent pathways for electrons and phonons, that is, the electrons transfer along the nearly coherent crystalline framework, while the phonons are frequently blocked by the strained lattices due to a dense network of small-angle tilt or rotation boundaries. The former makes the material behave like a crystal from the perspective of electrons, while the latter makes it function as a glassy material as far as phonons are concerned. Consequently, as shown in Figure 8.10c, a record-high *ZT* value of ≈2.1 at 1000 K in Pb-free bulk nanomaterials has been achieved with $Cu_2S_{0.52}Te_{0.48}$ [31].

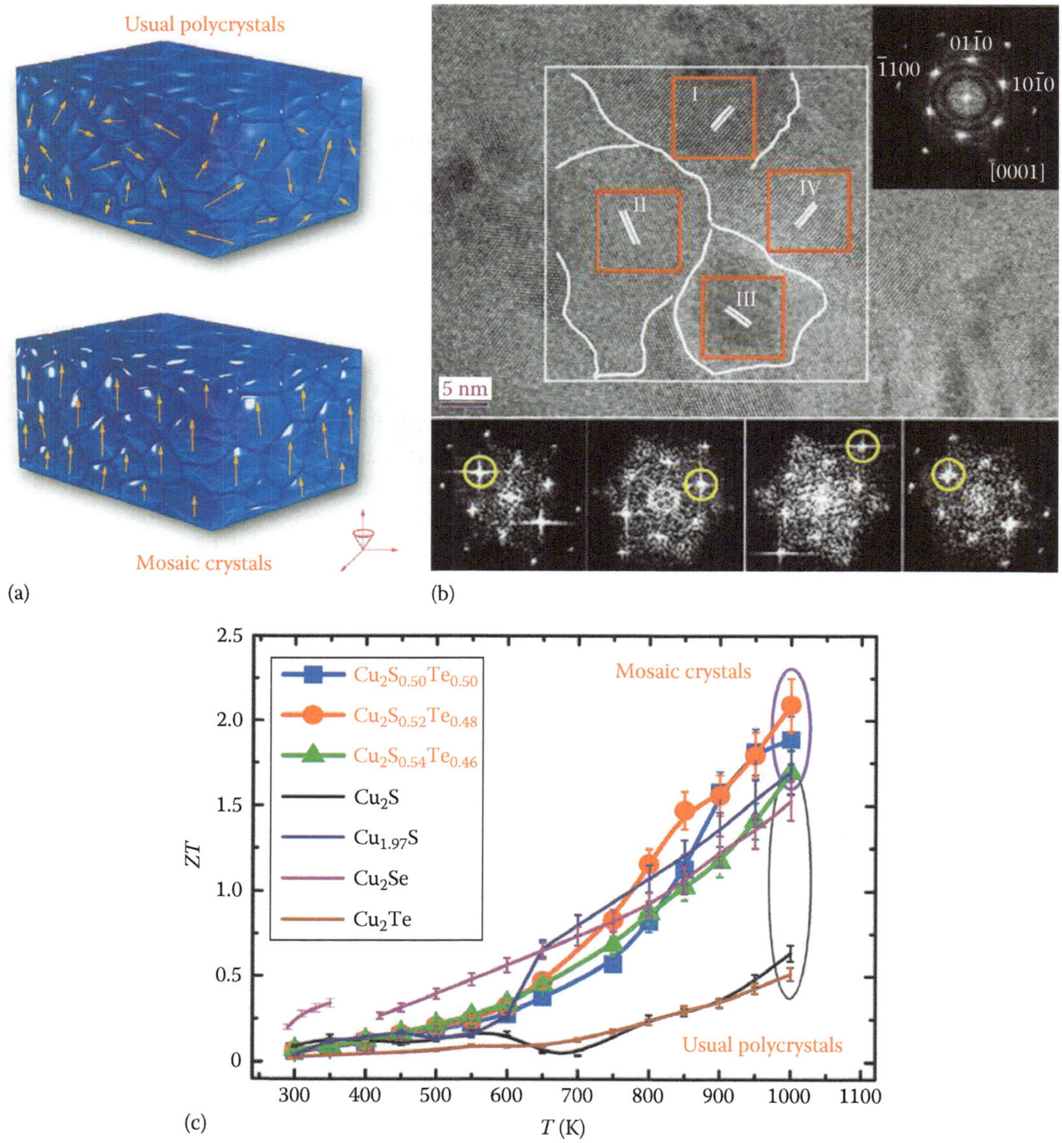

FIGURE 8.10
(a) Structural characteristics of a usual polycrystal and a mosaic crystal. (b) High-resolution transmission electron microscopy image and its fast Fourier transform diffractogram of selected areas I, II, III, and IV in Cu_2(S, Te) mosaic crystals, respectively. The nanocrystallites show nearly identical diffractograms except for variations in the brightness of different reflections that correspond to a slight misorientation with respect to each other (bright reflections are highlighted by yellow circles). (c) Temperature dependence of figure of merit (ZT) values for Cu_2(S, Te) mosaic crystals and usual polycrystals. The measurement uncertainty of ZT is about 15%. (From He, Y., Lu, P., Shi, X. et al.: Ultrahigh thermoelectric performance in mosaic crystals. *Adv. Mater.* 2015. 27. 3639–3644. Copyright Wiley-VCH Verlag GmbH & Co. KGaA. Reproduced with permission.)

8.6 Application Challenge for $Cu_{2-\delta}X$ (X = S, Se, and Te)

Although $Cu_{2-\delta}X$ (X = S, Se, and Te) compounds exhibit excellent *ZT*s, there are some key issues that must be solved before making TE devices for real applications. Among them, how to suppress the Cu-ion migration is one of the biggest challenges. Migration of Cu ions is quite obvious in $Cu_{2-\delta}X$ (X = S, Se, and Te) because of their large ionic conductivity, for example, 1.9 S cm^{-1} for Cu_2Se at 598 K [73] and 2.0 S cm^{-1} for Cu_2S at 628 K [74]. Under large currents or temperature gradients, the Cu-ion can migrate inside the superionic conductor from its initial location to the place where the electrical potential or chemical potential is lower. In 1953, Miyatani and Suzuki first reported on the severe Cu-ion migration phenomenon in Cu_2S [75]. An obvious copper precipitation at the cathode upon an application of the electrical field was observed. In copper selenide, a similarly severe Cu-ion migration phenomenon was also observed, which led directly to the termination of the program of Cu_2Se-based RTGs in the 1960s–1970s [9]. Recently, Brown et al. carried out a simple experiment to clarify the Cu-ion migration phenomenon in Cu_2Se [8]. As shown in Figure 8.11a, they located Cu_2Se between two conductive blocks under a temperature gradient of 275 K

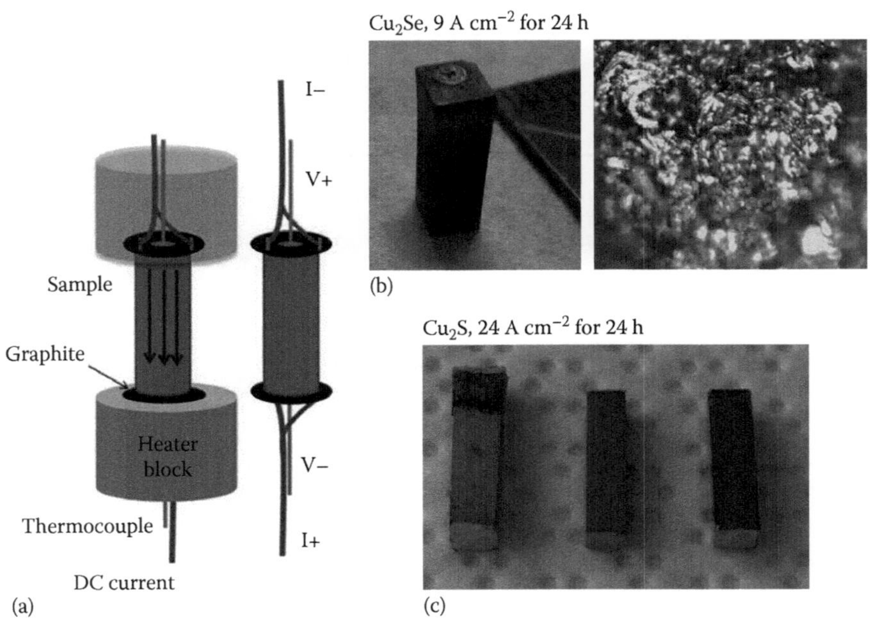

FIGURE 8.11
(a) Current stress test bench developed by Brown et al. Conductive graphite contacts are used for the current source and sink. (b) Left image: Cu_2Se sample after applying the current density (9 A cm^{-2}) for 24 h. Right image: optical microscopy image of the current-sink face of the Cu_2Se sample after applying a current for 24 h. Copper can be clearly seen precipitating on the surface. (c) From left to right: Cu_2S sample after applying the current (24 A cm^{-2}) for 24 h and $Cu_{1.8}S$ and CuS probes both after applying current (48 A cm^{-2}) for 72 h. ([b]: With kind permission from Springer Science+Business Media: *J. Electron. Mater.*, Chemical stability of (Ag, Cu)$_2$Se: A historical overview, 42, 2013, 2014–2019, Brown, D.R., Day, T., Caillat, T., and Snyder, G.J.; [c]: From Dennler, G., Chmielowski, R., Jacob, S. et al.: Are binary copper sulfides/selenides really new and promising thermoelectric materials?. *Adv. Energy Mater.* 2014. 4. 1301581. Copyright Wiley-VCH Verlag GmbH & Co. KGaA. Reproduced with permission.)

and then applied a large current (9 A cm^{-2}) for 24 h. When the sample was removed, visual inspection showed that it had undergone obvious deformation at the hot end. Figure 8.11b shows that the copper residue grew as bundles of nanowires and was found on the face attached to the hot-side heater and the current sink, proving that severe Cu-ion migration occurs during the passage of current. A similar experiment was also carried out on Cu_2S by Dennler et al., who found many cracks and copper whiskers on the Cu_2S surface after 24 h current stress with the current density of 24 A cm^{-2} (Figure 8.11c) [32]. Such Cu-ion migration would definitely change the initial chemical composition of the materials and, consequently, would greatly degrade the TE performance during service.

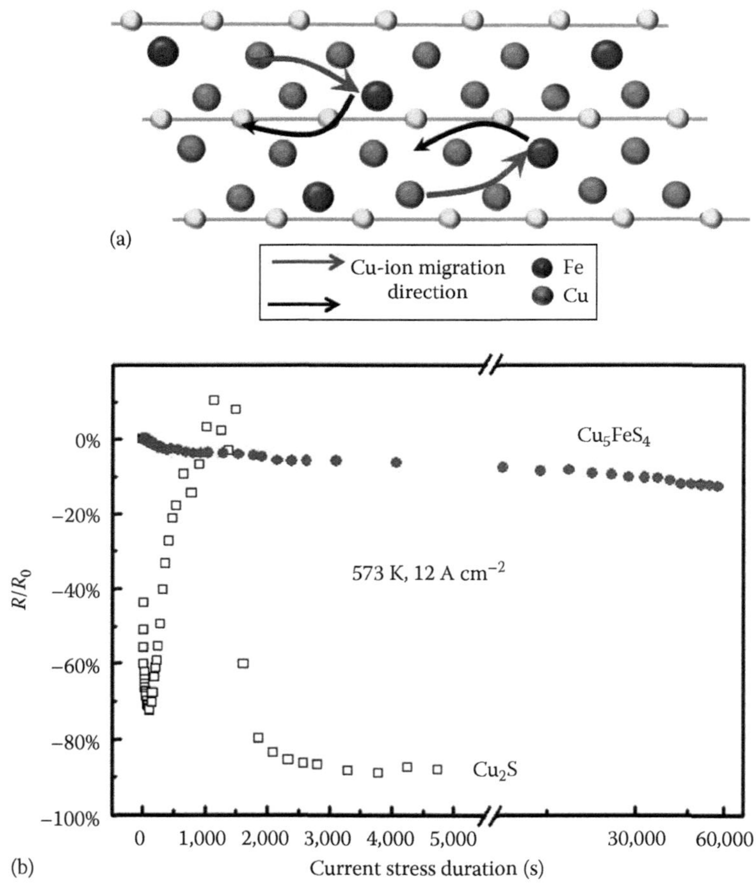

FIGURE 8.12
(a) Schematic map of using immobile Fe ions to suppress Cu-ion migration. The immobile Fe ions destroy the long-range Cu-ion migration channels and suppress Cu-ion migration. (b) Relative electrical resistance variation (R/R_0) of Cu_2S and Cu_5FeS_4 as a function of current stress duration at 573 K under a current density of 12 A cm^{-2}. Under large current, Cu-ion migration can create more Cu deficiencies in the matrix and significantly increase the hole concentration and reduce the electrical resistance. Therefore, the change of R/R_0 can well reflect on the degree of Cu-ion migration in these Cu-based materials. (From Qiu, P.F., Zhang, T.S., Qiu, Y.T., Shi, X., and Chen, L.D., Sulfide bornite thermoelectric material: A natural mineral with ultralow thermal conductivity, *Energy Environ. Sci.*, 7, 4000–4006, 2014. Reproduced by permission of The Royal Society of Chemistry.)

In order to promote the real application of Cu-based TE materials, the Cu-ion migration must be suppressed. However, as mentioned earlier, the excellent TE performance in Cu-based superionic conductors is mainly contributed to the *liquid-like* Cu-ion migration [12,64]. A possible solution for this dilemma between reliability and TE performance is to suppress the long-range Cu-ion migration while preserving short-range Cu-ion migration inside the material. This can be realized by introducing some immobile foreign ions at the initial Cu sites. In this case, the long-range Cu-ion migration channels can be significantly blocked or interrupted while most of the short-range Cu-ion migration channels are still well preserved. A schematic map for this solution is shown in Figure 8.12a. The practicability of this approach has been proved in the Cu_5FeS_4 cubic phase, in which the immobile Fe ions randomly located at Cu sites effectively suppressed the long-range Cu-ion migration and thereby resulted in a significantly enhanced stability under large currents as compared with the pure Cu_2S phase (see Figure 8.12b) [26]. Meanwhile, due to the well preserved short-range Cu-ion migration behavior, the Cu_5FeS_4 cubic phase still maintained very low lattice thermal conductivity comparable to Cu_2S. This approach is expected to be used in other Cu-based superionic conductors.

8.7 Conclusion

Following the discovery of classical TE materials (Bi_2Te_3, PbTe, and SiGe) in the mid-twentieth century, TE technology had received a great initial impetus but, in the subsequent 25 years, it stagnated. This lasted till the early 1990s when societal concerns regarding the global energy crisis led to the rejuvenation of interests in thermoelectricity and vigorous research efforts worldwide resulted in notable achievements. The performance of classical TE materials was significantly improved and many new families of efficient TE materials were discovered. Among them, $Cu_{2-\delta}X$ (X = S, Se, and Te) compounds have become a hot topic in TE community recently due to their intrinsic ultralow lattice thermal conductivity (see Figure 8.13a), which is lower than other TE materials. Significant developments and progress have been made with $Cu_{2-\delta}X$ (X = S, Se, and Te), which are also meaningful for the investigation of other kinds of TE materials. Some new TE phenomena and physical mechanisms were discovered in $Cu_{2-\delta}X$ (X = S, Se, and Te), and some new synthesis routes were proposed to fabricate them. As shown in Figure 8.13b, the $Cu_{2-\delta}X$ (X = S, Se, and Te) compounds have experimentally demonstrated excellent TE performances with the peak *ZT*s above 2, comparable to or exceeding many state-of-the-art TE materials reported before. Combining the low-cost and environmentally friendly features of copper, $Cu_{2-\delta}X$ (X = S, Se, and Te) show a great potential for future practical civil applications. The main challenge for $Cu_{2-\delta}X$ (X = S, Se, and Te) lies in the severe Cu-ion migration under large currents and temperature gradients, which greatly reduces the lifetime and reliability of the TE devices in real applications. Addressing this issue, a fundamental understanding of the mechanism of the Cu-ion migration in these materials, which is presently lacking, is required. Then, an effective method to suppress the long-range Cu-ion migration must be developed. We are optimistic that TE devices that use these $Cu_{2-\delta}X$ (X = S, Se, and Te) TE materials will appear soon when the long-range Cu-ion migration is greatly suppressed or even completely eliminated. We hope that further research in $Cu_{2-\delta}X$ (X = S, Se, and Te) TE materials will allow the TE technology to play a positive role in any future energy solution.

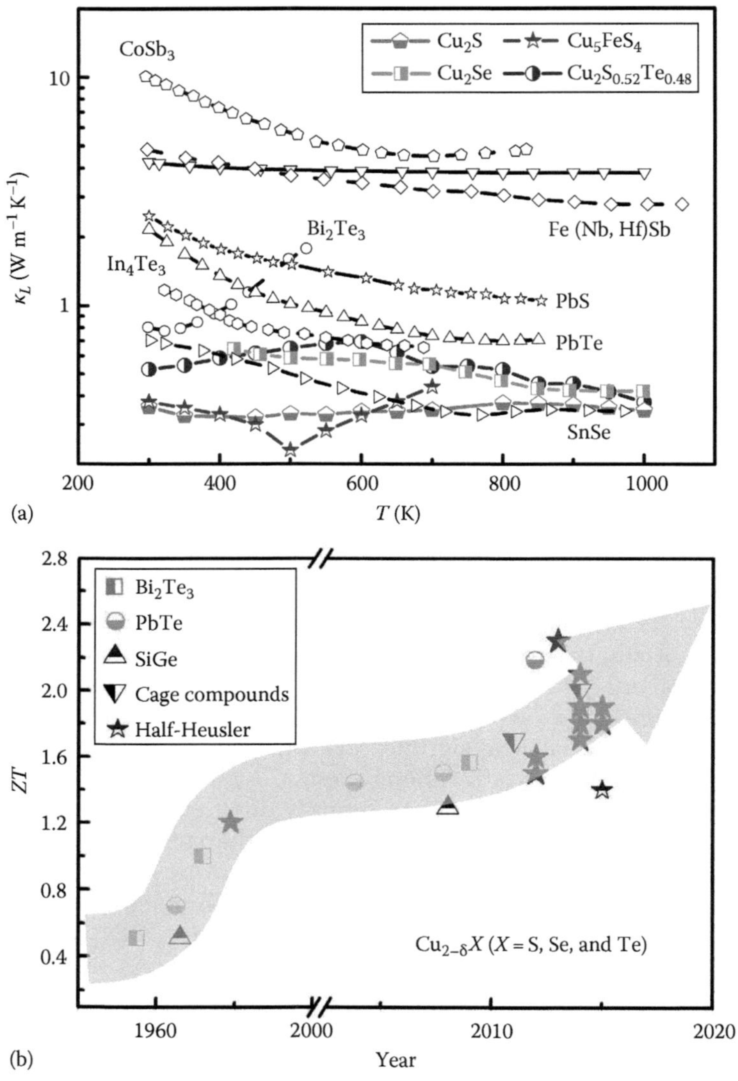

FIGURE 8.13
(a) Temperature dependence of lattice thermal conductivity for $Cu_{2-\delta}X$ (X = S, Se, and Te) and several kinds of representative TE materials. (b) Timeline underscoring the improvement in the TE figure of merit (ZT) of $Cu_{2-\delta}X$ (X = S, Se, and Te) and several kinds of representative TE materials. (The data are taken from Liu, H. et al., *Nat. Mater.*, 11, 422, 2012; Qiu, P.F. et al., *Energy Environ. Sci.*, 7, 4000, 2014; He, Y. et al., *Adv. Mater.*, 26, 3974, 2014; He, Y. et al., *Adv. Mater.*, 27, 3639, 2015; Poudel, B. et al., *Science*, 320, 634, 2008; Pei, Y. et al., *Nature*, 473, 66, 2011; Chen, L.D. et al., *J. Appl. Phys.*, 90, 1864, 2001; Rhyee, J.-S. et al., *Nature*, 459, 965, 2009; Wang, H. et al., *Adv. Energy Mater.*, 3, 488, 2013; Zhao, L.-D. et al., *Nature*, 508, 373, 2014; Fu, C. et al., *Nat. Commun.*, 6, 8144, 2015.)

References

1. Becquerel, A. C. 1827. *Ann. Chim. Phys.* 35: 328.
2. Becquerel, A. E. 1866. *Ann. Chim. Phys.* 48: 389.
3. Hirahara, E. 1947. Experimental studies on the electric conduction and heat capacity of cuprous sulfide semiconductors. *J. Phys. Soc. Jpn.* 2: 211–213.
4. Ogorelec, Z. and B. Celustka. 1966. Thermoelectric power and phase transitions of non-stoichiometric cuprous selenide. *J. Phys. Chem. Solids* 27: 615–617.
5. Sorokin, G. P. and A. P. Paradenko. 1966. Electrical properties of Cu_2S. *Sov. Phys. J.* 5: 91–95.
6. Epstein, A. and S. M. Kulifay. 1963. Monsanto Chemical Company. U.S. Patent 3,095,330.
7. Nolas, G. S., J. Sharp, and H. J. Goldsmid. 2001. *Thermoelectrics: Basic Principles and New Materials Developments.* Springer, New York.
8. Brown, D. R., T. Day, T. Caillat, and G. J. Snyder. 2013. Chemical stability of $(Ag,Cu)_2Se$: A historical overview. *J. Electron. Mater.* 42: 2014–2019.
9. Hinderman, J. D. 1979. Thermoelectric materials evaluation program. Annual technical report for fiscal year (Report No. MMM-2331-0642) (DOE, 1979), Minnesota Mining and Mfg. Co., St. Paul, MN. doi: 10.2172/5741759.
10. Hampl, E. F. J. 1972. Thermoelectric materials evaluation program. Quarterly technical task report No. 39 (Report No. MMM–2473-3421972) (DOE, 1972), Minnesota Mining and Mfg. Co., St. Paul, MN. doi: 10.2172/4562245.
11. Osmeyer, W. E. 1979. Selenide isotope generator for the Galileo Mission. Program final report (Report no. TES-33009-46) (DOE, 1979), Teledyne Energy Systems, Hunt Valley, Maryland. doi: 10.2172/5928964.
12. Liu, H., X. Shi, F. Xu et al. 2012. Copper ion liquid-like thermoelectrics. *Nat. Mater.* 11: 422–425.
13. Xiao, X. X., W. J. Xie, X. F. Tang, and Q. J. Zhang. 2011. Phase transition and high temperature thermoelectric properties of copper selenide $Cu_{2-x}Se$ ($0 \le x \le 0.25$). *Chin. Phys. B* 20: 087201.
14. Yu, B., W. S. Liu, S. Chen et al. 2012. Thermoelectric properties of copper selenide with ordered selenium layer and disordered copper layer. *Nano Energy* 1: 472–478.
15. Ballikaya, S., H. Chi, J. R. Salvador, and C. Uher. 2013. Thermoelectric properties of Ag-doped Cu_2Se and Cu_2Te. *J. Mater. Chem. A* 1: 12478–12484.
16. Liu, H. L., X. Yuan, P. Lu et al. 2013. Ultrahigh thermoelectric performance by electron and phonon critical scattering in $Cu_2Se_{1-x}I_x$. *Adv. Mater.* 25: 6607–6612.
17. Li, D., X. Y. Qin, Y. F. Liu et al. 2014. Chemical synthesis of nanostructured Cu_2Se with high thermoelectric performance. *RSC Adv.* 4: 8638–8644.
18. Zhong, B., Y. Zhang, W. Q. Li et al. 2014. High superionic conduction arising from aligned large lamellae and large figure of merit in bulk $Cu_{1.94}Al_{0.02}Se$. *Appl. Phys. Lett.* 105: 123902.
19. Su, X., F. Fu, Y. G. Yan et al. 2014. Self-propagating high-temperature synthesis for compound thermoelectrics and new criterion for combustion processing. *Nat. Commun.* 5: 4908.
20. Zhao, L. L., X. L. Wang, J. Y. Wang et al. 2014. Superior intrinsic thermoelectric performance with ZT of 1.8 in single-crystal and melt-quenched highly dense $Cu_{2-x}Se$ bulks. *Sci. Rep.* 5: 7671.
21. Day, T. W., K. A. Borup, T. S. Zhang et al. 2014. High-temperature thermoelectric properties of $Cu_{1.97}Ag_{0.03}Se_{1+y}$. *Mater. Renew. Sustain. Energy* 3: 26.
22. Gahtori, B., S. Bathul, K. Tyagi et al. 2015. Giant enhancement in thermoelectric performance of copper selenide by incorporation of different nanoscale dimensional defect features. *Nano Energy* 13: 36–46.
23. Zhao, L. L., X. L. Wang, F. F. Yun et al. 2015. The effects of Te^{2-} and I^- substitutions on the electronic structures, thermoelectric performance, and hardness in melt-quenched highly dense $Cu_{2-x}Se$. *Adv. Electron. Mater.* 1: 1400015.
24. Ge, Z. H., B. P. Zhang, Y. X. Chen, Z. X. Yu, Y. Liu, and J. F. Li. 2011. Synthesis and transport property of $Cu_{1.8}S$ as a promising thermoelectric compound. *Chem. Commun.* 47: 12697–12699.

25. Jiang, Q. H., H. X. Yan, J. R. Khaliq, Y. Shen, K. Simpson, and M. J. Reece. 2014. Enhancement of thermoelectric properties by atomic-scale percolation in digenite Cu_xS. *J. Mater. Chem. A* 2: 9486–9489.
26. Qiu, P. F., T. S. Zhang, Y. T. Qiu, X. Shi, and L. D. Chen. 2014. Sulfide bornite thermoelectric material: A natural mineral with ultralow thermal conductivity. *Energy Environ. Sci.* 7: 4000–4006.
27. He, Y., T. Day, T. S. Zhang et al. 2014. High thermoelectric performance in non-toxic earth-abundant copper sulfide. *Adv. Mater.* 26: 3974–3978.
28. Zhao, L. L., X. L. Wang, F. Y. Fei et al. 2015. High thermoelectric and mechanical performance in highly dense $Cu_{2-x}S$ bulks prepared by a melt-solidification technique. *J. Mater. Chem. A* 18: 9432–9437.
29. Kurosaki, K., K. Goto, A. Kosuga, H. Muta, and S. Yamanaka. 2006. Thermoelectric and thermophysical characteristics of Cu_2Te–Tl_2Te pseudo binary system. *Mater. Trans.* 47: 1432–1435.
30. He, Y., T. Zhang, X. Shi, S. H. Wei, and L. D. Chen. 2015. High thermoelectric performance in copper telluride. *NPG Asia Mater.* 7: e210.
31. He, Y., P. Lu, X. Shi et al. 2015. Ultrahigh thermoelectric performance in mosaic crystals. *Adv. Mater.* 27: 3639–3644.
32. Dennler, G., R. Chmielowski, S. Jacob et al. 2014. Are binary copper sulfides/selenides really new and promising thermoelectric materials? *Adv. Energy Mater.* 4: 1301581.
33. Chakrabarti, D. J. and D. E. Laughlin. 1983. The Cu–S (copper–sulfur) system. *Bull. Alloy Phase Diagr.* 4: 254–271.
34. Tonejc, A. 1980. Phase diagram and some properties of $Cu_{2-x}Se$ ($2.01 \geqq 2-x \geqq 1.75$). *J. Mater. Sci.* 15: 3090–3094.
35. Murray, R. M. and R. D. Heyding. 1975. Copper-selenium system at temperatures to 850 K and pressures to 50 Kbar. *Can. J. Chem.* 53: 878–887.
36. Vucic, Z., O. Milat, V. Horvatic, and Z. Ogorelec. 1981. Composition-induced phase-transition splitting in cuprous selenide. *Phys. Rev. B* 24: 5398–5402.
37. Kashida, S. and J. Akai. 1988. X-ray-diffraction and electron-microscopy studies of the room-temperature structure of Cu_2Se. *J. Phys. C: Solid State Phys.* 21: 5329–5336.
38. Nguyen, M. C., J. H. Choi, X. Zhao, C. Z. Wang, Z. Zhang, and K. M. Ho. 2013. New layered structures of cuprous chalcogenides as thin film solar cell materials: Cu_2Te and Cu_2Se. *Phys. Rev. Lett.* 111: 165502.
39. Lu, P., H. L. Liu, X. Yuan et al. 2015. Multiformity and fluctuation of Cu ordering in Cu_2Se thermoelectric materials. *J. Mater. Chem. A* 3: 6901–6908.
40. Buerger, M. J. and B. J. Wuensch. 1963. Distribution of atoms in high chalcocite. Cu_2S. *Science* 141: 276–277.
41. Wuensch, B. J. and M. J. Buerger. 1963. Crystal structure of chalcocite, Cu_2S. *Miner. Soc. Am.* Special Paper 1: 164.
42. Sadanaga, R., M. Ohmasa, and N. Morimoto. 1965. On the statistical distribution of copper ions in the structure of β-chalcocite. *Mineral. J.* 4: 275–290.
43. Will, G., E. Hinze, A. Rahman, and M. AbdelRahman. 2002. Crystal structure analysis and refinement of digenite $Cu_{1.8}S$ in the temperature range 20 to 500 degrees C under controlled sulfur partial pressure. *Eur. J. Mineral.* 14: 591–598.
44. Evans, H. T. 1979. Crystal-structure of low chalcocite and djurleite. *Z. Kristallogr.* 150: 299–320.
45. Xu, Q., B. Huang, Y. Zhao, Y. Yan, R. Noufi, and S. H. 2012. Crystal and electronic structures of Cu_xS solar cell absorbers. *Appl. Phys. Lett.* 100: 061906.
46. Cook, W. R. 1972. Phase changes in Cu_2S as a function of temperature. National Bureau of Standards Special Publication, Solid State Chemistry, 364, pp. 703–712.
47. Selivanov, E. N., R. I. Gulyaeva, and A. D. Vershinin. 2006. Thermal expansion and phase transformations of copper sulfides. *Inorg. Mater.* 43: 573–578.
48. Vouroutzis, N., N. Frangis, and C. Manolikas. 2005. The double modulation superstructure of the room temperature stable phase of stoichiometric Cu_2Te. *Phys. State Solidi* 202: 271–280.

49. Lukashev, P., W. L. Lambrecht, T. Kotani, and M. Schilfgaarde. 2007. Electronic and crystal structure of $Cu_{2-x}S$: Full-potential electronic structure calculations. *Phys. Rev. B* 76: 195202.
50. Zhang, Y., Y. Wang, L. Xi, R. Qiu, X. Shi, P. Zhang, and W. Zhang. 2014. Electronic structure of antifluorite Cu_2X (X = S, Se, Te) within the modified Becke-Johnson potential plus an on-site Coulomb U. *J. Chem. Phys.* 140:074702.
51. Silva, J. L., S. H. Wei, J. Zhou, and X. Wu. 2007. Stability and electronic structures of Cu_xTe. *Appl. Phys. Lett.* 91: 091902.
52. Kashida, S., W. Shimosaka, M. Mori, and D. Yoshimura. 2003. Valence band photoemission study of the copper chalcogenide compounds, Cu_2S, Cu_2Se and Cu_2Te. *J. Phys. Chem. Solids* 64: 2357–2363.
53. Sorokin, G. P., Yu. M. Papshev, and P. T. Oush. 1966. Photoconductivity of Cu_2S, Cu_2Se, and Cu_2Te. *Sov. Phys. Solid State* 7: 18101811.
54. Evans, H. T. 1971. Transmission of phytophthora pod rot of cocoa by invertebrates. *Nature* 232: 346.
55. Chi, H., H. Kim, J. C. Thomas et al. 2014. Low-temperature structural and transport anomalies in Cu_2Se. *Phys. Rev. B* 89: 195209.
56. Kim, H., S. Ballikaya, H. Chi et al. 2015. Ultralow thermal conductivity of β-Cu_2Se by atomic fluidity and structure distortion. *Acta Mater.* 86: 247–253.
57. Danilkin, S. A., M. Avdeev, M. Sale, and T. Sakuma. 2012. Neutron scattering study of ionic diffusion in Cu-Se superionic compounds. *Solid State Ionics* 225: 190–193.
58. Slack, G. A. 1995. New materials and performance limits for thermoelectrics. In *CRC Handbook of Thermoelectrics*, ed. Rowe, D. M. (CRC Press, Boca Raton, FL), pp. 407–440.
59. Trachenko, K. 2008. Heat capacity of liquids: An approach from the solid phase. *Phys. Rev. B* 78: 104021.
60. Pilgrim, W. C. and C. Morkel. 2006. State dependent particle dynamics in liquid alkali metals. *J. Phys. Condens. Matter* 18: 585–633.
61. Grimvall, G., 1975. The heat capacity of liquid metals. *Phys. Scr.* 11: 381–382.
62. Miyatani, S. Y. 1956. Point contact of Pt and gamma-Cu_2S. *J. Phys. Soc. Jpn.* 11: 1059–1063.
63. Wakamura, K. and I. Tsubota. 2000. Small band gap and high ionic conduction in Cu_2S. *Solid State Ionics* 130: 305–312.
64. Wang, L. W. 2012. High chalcocite Cu_2S: A solid-liquid hybrid phase. *Phys. Rev. Lett.* 108: 085703.
65. Miller, T. A., J. S. Wittenberg, H. Wen, S. Connor, Y. Cui, and A. M. Lindenberg. 2013. The mechanical of ultrafast structure switching in superionic copper (I) sulphide nanocrystals. *Nat. Commun.* 4: 1369.
66. Liu, H., X. Shi, M. Kirkham et al. 2013. Structure-transformation-induced abnormal thermoelectric properties in semiconductor copper selenide. *Mater. Lett.* 93: 121–124.
67. Thomas, G. A., K. Levin, and R. D. Parks. 1972. Thermoelectric anomaly near a critical-point. *Phys. Rev. Lett.* 29: 1321–1325.
68. Mahan, G. D. 2015. The Seebeck coefficient of superionic conductors. *J. Appl. Phys.* 117: 045101.
69. Grguric, B. A., A. Putnis, and R. J. Harrison. 1998. An investigation of the phase transitions in bornite (Cu_5FeS_4) using neutron diffraction and differential scanning calorimetry. *Am. Mineral.* 83: 1231–1239.
70. Darwin, C. G. 1922. Reflection of x-rays from imperfect crystals. *Philos. Mag.* 43: 800–829.
71. Geis, M. W., H. I. Smith, A. Argoitia et al. 1991. Large-area mosaic diamond films approaching single-crystal quality. *Appl. Phys. Lett.* 58: 2485–2487.
72. Er, A. O., J. Chen, J. Tang, and P. M. Rentzepis. 2012, Coherent acoustic wave oscillations and melting on Ag(111) surface by time resolved x-ray diffraction. *Appl. Phys. Lett.* 100: 151910.
73. Yakshibaev, R. A., V. N. Konev, and M. Kh. Balapanov. 1984. Ionic conductivity and diffusion in superionic conductor α-$Cu_{2-\delta}Se$. *Fiz. Tverd. Tela.* 26: 3641–3645.
74. Balapanov, M. Kh., I. G. Gafurov, U. Kh. Mukhamed'yanov, R. A. Yakshibaev, and R. Kh. Ishembetov. 2004. Ionic conductivity and chemical diffusion in superionic $Li_xCu_{2-x}S$ ($0 \le x \le 0.25$). *Phys. State Solidi (b)* 241: 114–119.

75. Miyatani, S.-Y. and Y. Suzuki. 1953. On the electric conductivity of cuprous sulfide: Experiment. *J. Phys. Soc. Jpn.* 8: 680–681.
76. Poudel, B., Q. Hao, Y. Ma et al. 2008. High-thermoelectric performance of nanostructured bismuth antimony telluride bulk alloys. *Science* 320: 634–638.
77. Pei, Y., X. Shi, A. Lalonde, H. Wang, L. D. Chen, and G. J. Snyder. 2011. Convergence of electronic bands for high performance bulk thermoelectrics. *Nature* 473: 66–69.
78. Chen, L. D., T. Kawahara, X. F. Tang, T. Goto, and T. Hirai. 2001. Anomalous barium filling fraction and n-type thermoelectric performance of $Ba_yCo_4Sb_{12}$. *J. Appl. Phys.* 90: 1864–1868.
79. Rhyee, J.-S., K.-H. Lee, S. M. Lee et al. 2009. Peierls distortion as a route to high thermoelectric performance in $In_4Se_{3-\delta}$ crystals. *Nature* 459: 965–968.
80. Wang, H., E. Schechtel, Y. Pei, and G. J. Snyder. 2013. High thermoelectric efficiency of n-type PbS. *Adv. Energy Mater.* 3: 488–495.
81. Zhao, L.-D., S.-H. Lo, Y. Zhang et al. 2014. Ultralow thermal conductivity and high thermoelectric figure of merit in SnSe crystals. *Nature* 508: 373–377.
82. Fu, C., S. Bai, Y. Liu et al. 2015. Realizing high figure of merit in heavy-band p-type half-Heusler thermoelectric materials. *Nat. Commun.* 6: 8144.

9

BiCuSeO: A Promising Thermoelectric Material

Li-Dong Zhao* and Jing-Feng Li**
**Beihang University*
***Tsinghua University*

CONTENTS

9.1 Introduction

Thermoelectric technology, harvesting electric power directly from heat, is a promising environmentally friendly means of energy saving and power generation. According to the optimal working temperature, thermoelectric materials can be sorted into three groups [1,2]. At low temperatures (<400 K), typical materials are Bi_2Te_3 based [3–5] and $CsBi_4Te_6$ [6];

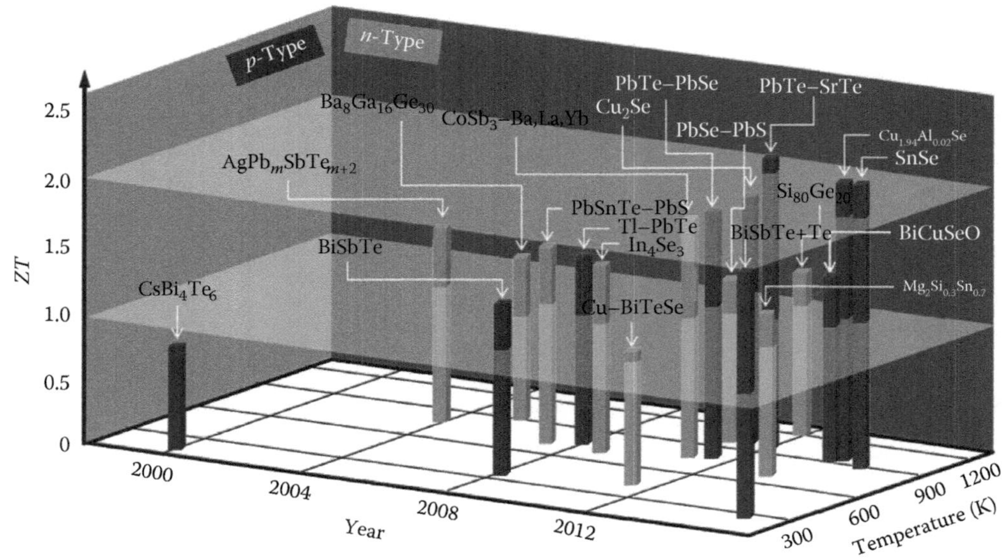

FIGURE 9.1
ZT peaks for the current state-of-the-art bulk thermoelectrics as a function of temperature and year: $CsBi_4Te_6$ [6], BiSbTe [5], Cu–BiTeSe [4], BiSbTe+Te [3], $AgPb_mSbTe_{m+2}$ [7], PbSnTe–PbS [8], PbTe–PbSe [10], PbTe–SrTe [9], Tl–PbTe [11], PbSe–PbS [12], $Ba_8Ga_{16}Ge_{30}$ [13], $CoSb_3$-(Ba, La, Yb) [14], $Mg_2Si_{0.3}Sn_{0.7}$ [15], SnSe [16, 17], In_4Se_3 [18], BiCuSeO [25,26], $Si_{80}Ge_{20}$ [19], Cu_2Se [20,21], and $Cu_{1.94}Al_{0.02}Se$ [22].

for mid-temperature ranges (600–900 K), typical materials are $AgPb_mSbTe_{m+2}$ [7], PbSnTe–PbS [8], PbTe–SrTe [9], PbTe–PbSe [10], Tl-PbTe [11], PbSe–PbS [12], $Ba_8Ga_{16}Ge_{30}$ [13], $CoSb_3$-(Ba, La, Yb) [14], $Mg_2Si_{0.3}Sn_{0.7}$ [15], SnSe [16,17], and In_4Se_3 [18], while for high temperatures (>900 K), the best materials are $Si_{80}Ge_{20}$ [19], Cu_2Se [20,21], and $Cu_{1.94}Al_{0.02}Se$ [22]. One can see that most of these thermoelectric materials (see Figure 9.1) are heavy-metal-based alloys, including rare-earth and low-abundance elements. Therefore, new prospective materials, composed of earth-abundant and cheap elements, are expected to play a vital role in extensive thermoelectric applications for energy conversion. One such novel material is BiCuSeO [23,24].

9.2 Historical Background

Figure 9.2 shows the historical development of layered compounds with the ZrCuSiAs structure, in which BiCuSeO is one of the family members [27]. LaAgSO, as an ancestry analogue of BiCuSeO, was initially studied as an ionic conductor [28]. LaAgSO crystallizes with conductive AgS layers and charge reservoir LaO layers that are mutually separated [29]. Due to the similarity of their layered crystal structures with that of superconducting cuprates, oxychalcogenides were thoroughly studied as new prospective superconductors during the 1990s [30–32]. Several isostructural compounds RMChO (R = Bi or Ce to Dy, M = Cu or Ag, and Ch = S, Se or Te) were also discovered [33–35]. However, no superconductivity was observed in any of these compounds, so the interest in this family gradually faded away. A few years later, apart from the superconductivity, an interest in oxychalcogenides was

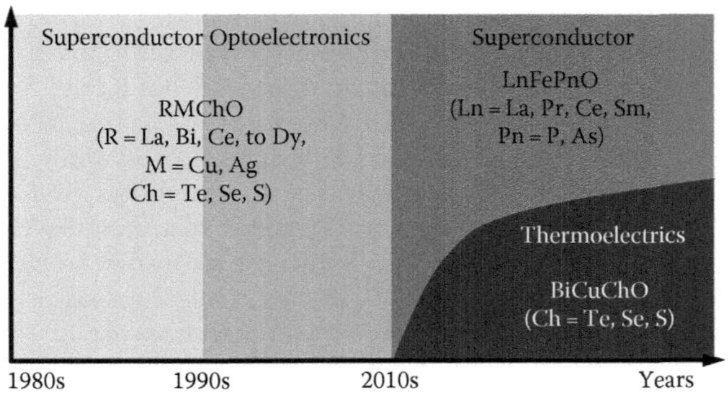

FIGURE 9.2
The historical development of layered compounds with the ZrCuSiAs structure.

rekindled because they were viewed as promising materials for optoelectronic applications [36] and, since then, numerous oxychalcogenides have been synthesized [37–39].

The search for superconductivity among oxychalcogenides started a new era, which was ignited by the discovery of superconductivity in two new layered compounds, LaFePO [40] and LaNiPO [41]. Due to the very low critical temperature T_c, this discovery did not lead to an enthusiastic response until 2008. The situation changed when the superconductivity with T_c ~26 K in the F-doped LaFeAsO was discovered, which initiated a tremendous research effort toward the high T_c iron-based superconductors [42] with the T_c rapidly increasing to 43 K [43,44] and 56 K [45]. The discovery of superconductivity in the tetragonal iron-based pnictide has stimulated a vigorous search for superconductivity in a wide class of materials with the similar structure. Berardan et al. [46] focused on superconductivity explorations in the BiCuSeO system, and the effect of various substitutions on the low temperature transport properties was also investigated. Whatever the substitution (Bi substituted by Sr, Cu by Fe, Se by Te, and O by F), no superconductivity was observed (see Figure 9.3a) and all samples revealed the p-type conduction mechanism,

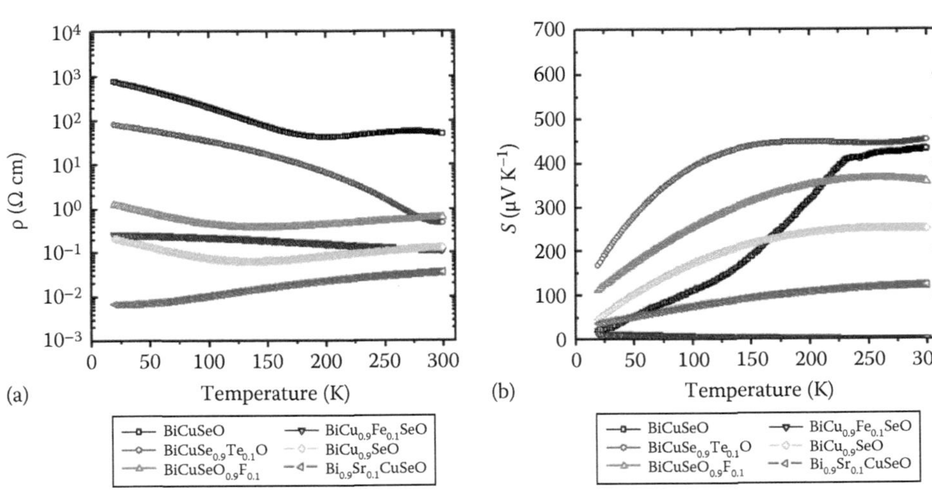

FIGURE 9.3
(a) The electrical resistivity and (b) Seebeck coefficient at 20–300 K for the BiCuSeO system.

Figure 9.3b [46]. Following this *Iron Rush* era, the oxypnictides became of interest, and just 3 months after T_c ~26 K was reported in 2008, LaFeAsO was found to exhibit a large Seebeck coefficient peak and a promising power factor at around 100 K, the features making the material interesting for thermoelectric cooling applications in the cryogenic temperature range [47,48]. Based on these observations, it made sense to reevaluate the potential of the oxychalcogenides for thermoelectric applications. It was found that the thermoelectric figure of merit *ZT* of these oxychalcogenides is mainly restricted by their high electrical resistivity, which, however, can be strongly reduced by tuning the carrier concentration through routine doping [49]. Indeed, in 2010, Zhao et al. [24] found Sr-doped BiCuSeO to exhibit a promising *ZT* of 0.76 at 873 K, which was then boosted to 1.4 at 923 K through sample texturing [50] and modulation doping (MD) [26]. To date, the BiCuSeO system has the highest reported *ZT* value of all oxygen containing thermoelectric materials [2].

9.3 Crystal Structure

BiCuSeO crystallizes in a layered ZrCuSiAs structure with the tetragonal unit cell $a = b = 3.9273$ Å, $c = 8.9293$ Å, $Z = 2$, and the space group *P*4/*nmm* (No. 129, PDF 45-0296). Figure 9.4a shows that, like every other LnCuChO (Ln = lanthanide, Ch = S, Se, Te), BiCuSeO exhibits a 2D layered structure containing fluorite like Bi_2O_2 layers alternately stacked with Cu_2Se_2 layers along the *c*-axis [38]. Figure 9.4b shows the top view along the *c*-axis. The Bi_2O_2 layers consist of slightly distorted Bi_4O tetrahedrons, Figure 9.4c, with two kinds of Bi–O–Bi bond angles, that is, ~106.95° and ~114.65° [38,51,52]. The coordination of Bi can also be considered as a distorted square antiprism with four O in one base

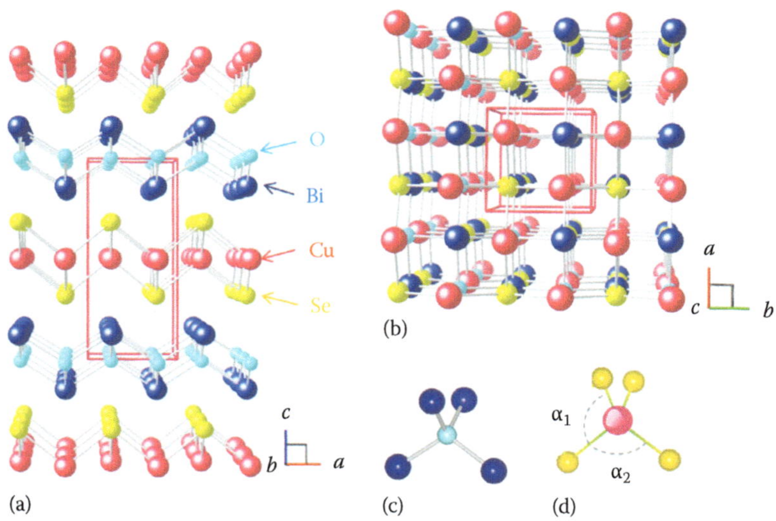

FIGURE 9.4
(a) Crystal structure of BiCuSeO along the *b*-axis, (b) structure along the *c*-axis, (c) Bi_4O tetrahedron, and (d) $CuSe_4$ tetrahedral.

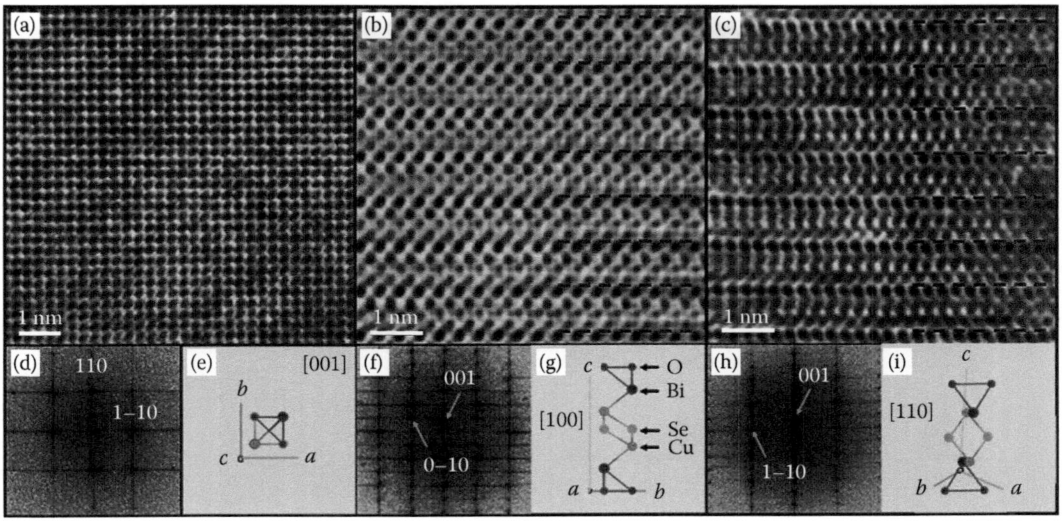

FIGURE 9.5
HRTEM images of undoped BiCuSeO sample: (a–c) lattice images along [001], [100], and [110] zone axes, respectively. (d, f, h) Fast Fourier-transformed images of (a) through (c), respectively. (e, g, i) Crystal structures of BiCuSeO along [001], [100], and [110] zone axes, respectively.

and four Se in the other. The Bi–Se bond length ~3.2 Å is longer than the Bi–O bond length ~2.33 Å. The Cu_2Se_2 layers consist of slightly distorted $CuSe_4$ tetrahedrons with Se–Se edges being shared, Figure 9.4d. Two kinds of Se–Cu–Se bond angles exist, with the small one being 102.6°–106.95° and the large one 113.1°–114.65°. The Cu_2Se_2 layer could be considered as a reversed version of the Bi_2O_2 layer because the position of Se atoms is the same as that of Bi atoms, while the position of Cu atoms is the same as that of O atoms in the Bi_2O_2 layer [52].

The layered structure of BiCuSeO has been investigated by transmission electron microscopy [23,53]. The layered feature, that is, conductive Cu–Se layers alternately stacked along the *c*-axis with insulating Bi–O layers, can only be seen along non-*c*-axes. As expected, the lattice image along the [001] zone axis (*c*-axis) shows homogeneous contrast (Figure 9.5a) and its fast Fourier transform image exhibits no spot splitting (Figure 9.5d), both of which reveal that the two sublattices (Cu–Se and Bi–O) are commensurate along the *c*-axis. On the contrary, lattice images along [100] and [110] zone axes in Figure 9.5b and c clearly show a six-period (according to the stacking sequence –Bi–O–Bi–Se–Cu–Se–) layered structure, and the respective distinct spot splitting is observed in fast Fourier transform images, as shown in Figure 9.5f and h.

9.4 Electronic Band Structure and DOS Effective Mass

Figure 9.6a shows the total and partial density of states (DOS) of BiCuSeO [23,53]. Bi 6s states are located deep in the valence band (–10 to –12 eV relative to the valence band maximum [VBM]) and so are the O 2p states (–5 to –7 eV). The bottom of the conduction

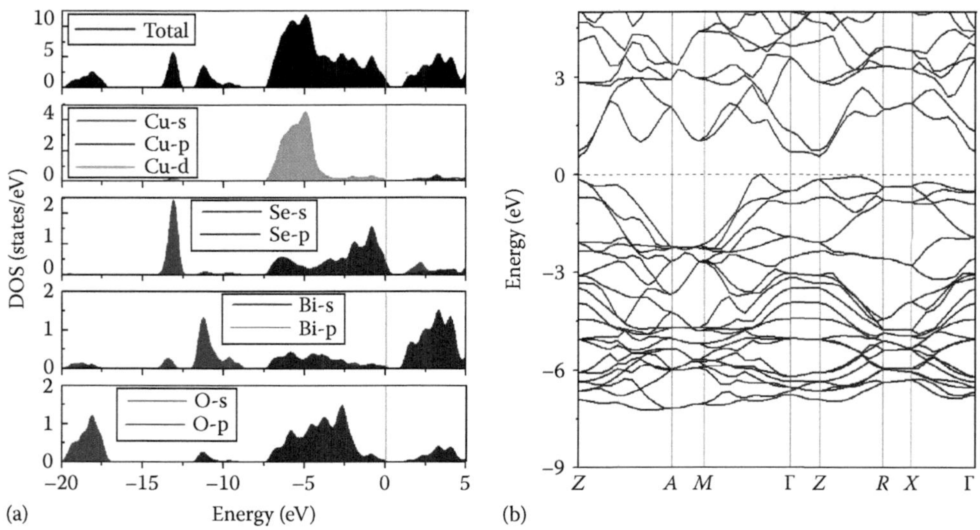

FIGURE 9.6
(a) The total and partial density of states of BiCuSeO and (b) the band structure of BiCuSeO.

band mainly consists of Bi 6p states. It is noteworthy that in the vicinity of the Fermi level (0 to −0.5 eV), Cu and Se contribute significantly to the DOS, with about 1 state eV^{-1} fu^{-1}, and synergistically determine the electronic conduction. The band structure indicates that BiCuSeO is a multiband semiconductor (Figure 9.6b). The top of the valence band consists of a hole pocket located on the Γ–*M* line. The other hole pockets can be observed on the Γ–*X* line and at the *Z* point, which are located about −0.15 eV below the Fermi level and participate in the electronic conduction at high hole concentration or at high temperature.

Liu et al. investigated the effects of band structure modifications on the thermoelectric transport properties of BiCuSeO through La doping [54]. Figure 9.7a indicates that the bandgaps of $Bi_{1-x}La_xCuSeO$ increase with the increasing La doping content. The top of the heavy valence band moves down as the bandgap increases, which leads to a decreasing band offset between the heavy and light valence bands (Figure 9.7b). The second band contributes to an increasing effective mass and Seebeck coefficient in $Bi_{1-x}La_xCuSeO$. The Pisarenko line, the dependence of the Seebeck coefficient on the carrier concentration, reflects the contribution of the second valence band at room temperature. As shown in Figure 9.7c and d, the Seebeck coefficient at high carrier concentrations deviates from the Pisarenko line based on the single parabolic band (SPB) model, indicating the contributions from the second valence band [55,56]. Despite the complex nonparabolic multiband structure in BiCuSeO, the SPB model can be adopted to give a rough estimation of the effective mass ~0.6 m_e, but one must keep in mind that this DOS effective mass is a mixed effective mass of multibands since there are at least two valence bands contributing to the electronic transport properties [53]. Lee et al. [57] determined the individual effective masses of the two valence bands of BiCuSeO. The origin of the Seebeck coefficient enhancement was discussed in terms of the two-band structure of the valence band of BiCuSeO, and the DOS effective masses of the heavy (~1.1 m_e) and light hole (~0.18 m_e) bands were estimated by using the Pisarenko relation.

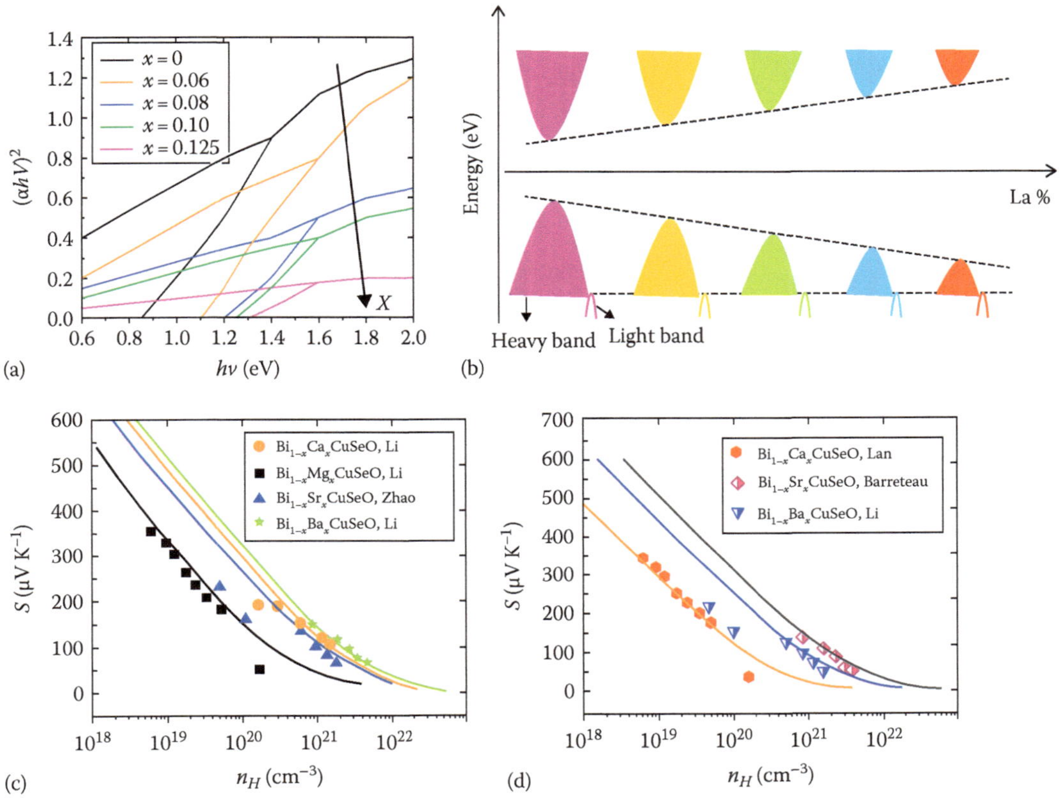

FIGURE 9.7
(a) The optical bandgaps and (b) proposed band model of $Bi_{1-x}La_xCuSeO$. (c, d) Pisarenko plot of the Seebeck coefficient vs. carrier concentration at room temperature. Mg-, Ca-, Sr-, and Ba-doped BiCuSeO are plotted for comparison.

9.5 Thermoelectric Transport Properties of Undoped BiCuSeO

As shown in Figure 9.8a, the electrical conductivity of BiCuSeO increases with rising temperature, indicating a semiconductor transport behavior [23,24]. The low electrical conductivity is related to the low carrier concentration of 1×10^{18} cm^{-3} and low carrier mobility of 22 cm^2 V^{-1} s^{-1}. The Seebeck coefficient of BiCuSeO is about +350 μV K^{-1} at room temperature and keeps increasing with rising temperature, reaching about +425 μV K^{-1} at 923 K. The combination of low electrical conductivity and large Seebeck coefficient results in a moderate power factor of about 2.5 μW cm^{-1} K^{-2} at 923 K (Figure 9.8b). As shown in Figure 9.8c, the total thermal conductivity is low at room temperature and keeps decreasing with rising temperature. The thermal conductivity is dominated by its lattice part, since the electronic thermal conductivity is low. The lattice thermal conductivity κ_{lat} was calculated by subtracting from the total thermal conductivity κ_{tot} the electronic thermal conductivity κ_{el} obtained from the Wiedemann–Franz law, $\kappa_{el} = L\sigma T$, where L is the Lorenz number that can be obtained by fitting the Seebeck coefficient to the reduced chemical potential [24]. Thus obtained, the lattice thermal conductivity is about 0.52 W m^{-1} K^{-1} at room temperature and decreases to 0.32 W m^{-1} K^{-1} at 923 K. The heat capacity C_p shows an increasing trend with

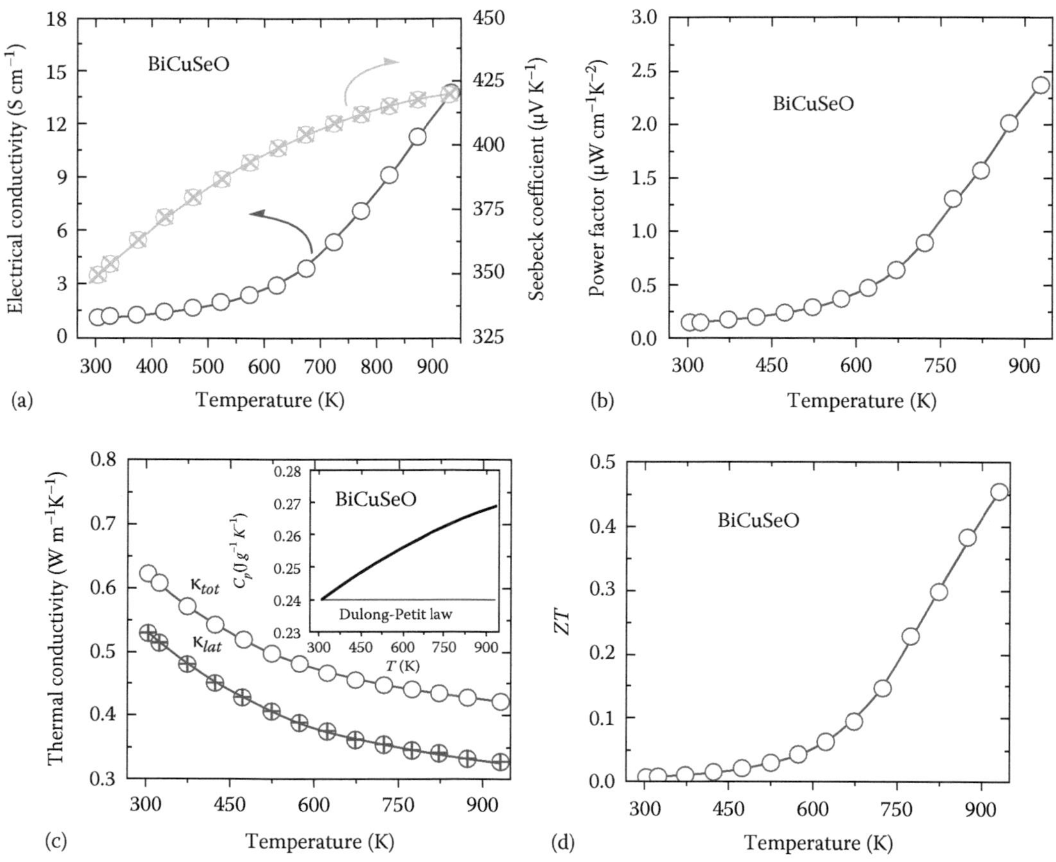

FIGURE 9.8
Thermoelectric properties of undoped BiCuSeO: (a) electrical conductivity and Seebeck coefficient. (b) Power factor. (c) Total thermal conductivity and lattice thermal conductivity; the inset shows the heat capacity. (d) Dimensionless figure of merit, *ZT*.

temperature (see the inset in Figure 9.8c). The heat capacity at room temperature is consistent with the Dulong–Petit law, which indicates that, unlike in Cu_2Se [20,21], the liquid-like behavior of the Cu ions, which might further reduce the thermal conductivity, is excluded. The relatively low power factor is compensated by the exceptionally low thermal conductivity, thus leading to a modest *ZT* ~0.5 at 923 K (see Figure 9.8d).

9.6 Origin of the Low Thermal Conductivity of BiCuSeO

The low thermal conductivity of BiCuSeO is impressive and outstanding. To clarify its origin, the elastic properties of polycrystalline BiCuSeO compounds have been experimentally analyzed [53]. Young's modulus is related to the stiffness of a compound's atomic bonds. Weaker bonds generally result in lower stiffness and lower Young's modulus. Materials with low Young's modulus are considered as having "soft" bonding, which

TABLE 9.1

Comparison of Room Temperature Sound Velocity (υ), Young's Modulus (E), and Lattice Thermal Conductivity (κ_{lat}) of BiCuSeO with Some Other Typical Oxides

	BiCuSeO	Ba_2YbAlO_5	$BaDyAlO_5$	$Gd_2Zr_2O_7$
υ (m s^{-1})	2107	2901	3078	3832
E (GPa)	76.5	109.4	116.5	234.3
κ_{lat} (W m^{-1} K^{-1})	0.9	1.4	1.8	2.2

results in a low speed of sound and thus in low lattice thermal conductivity [58]. Indeed, BiCuSeO exhibits a much lower speed of sound, Young's modulus, and lattice thermal conductivity at 300 K than other peer oxygen containing compounds [59,60] listed in Table 9.1.

Besides the relative low Young's modulus (E), the Grüneisen parameter (γ) is also relevant in evaluating the thermal conductivity. The Grüneisen parameter is often referred to as a temperature-dependent anharmonicity parameter that reflects the phonon vibrations in the crystal lattice that deviate from harmonic oscillations [61,62]. The Grüneisen parameter was calculated to be ~1.5 for BiCuSeO; this value is comparable or even larger than that of PbTe (1.45), which is an extensively studied conventional thermoelectric material [53,61,63]. It should be noted that γ ~1.5 is an average value for polycrystalline samples of BiCuSeO and it should be smaller than γ values along the a-, b-, or c-axes of a single crystal. Morelli et al. [61] reported that the intrinsically low thermal conductivity of $AgSbTe_2$ results from an extremely high Grüneisen parameter of 2.05. The high Grüneisen parameter, linked to the high bond anharmonicity, may originate from the presence of a lone pair of electrons in the Sb's sp-hybridized orbitals [61]. This lone electron pair gives rise to extra electron clouds surrounding Sb atoms and causes nonlinear repulsive forces that are manifested as bond anharmonicity. Similarly low lattice thermal conductivity should be observed in the Bi-based compounds, a sister group V element. The connection between the nature of bonding and the Grüneisen parameter has been explored in detail theoretically by Huang et al., who clearly showed the effect of large electron clouds on anharmonicity [64]. In addition, lone pair electrons of Bi probably lead to a more asymmetric electron density and thus more anharmonic lattice vibrations [65]. The very low thermal conductivity in BiCuSeO system is probably also due to other reasons, including the layered structure where phonons might be confined and scattered at the interfaces and the presence of heavy elements, which leads to lower phonon group velocities. All of these features can be summarily characterized as involving heavy atoms with "soft" bonding, the condition that meets the requirement for a low thermal conductivity material, as suggested by Clarke [66].

Saha [67] carried out first-principles density functional theory calculations and systematically investigated the lattice dynamics and related properties of BiCuSeO. From the marked differences in some of these properties of the two materials (LaCuSeO and BiCuSeO), the reasons for why BiCuSeO is a better thermal insulator than LaCuSeO were elucidated. Atomic displacement patterns were proposed for the respective lowest-frequency optic modes of BiCuSeO, as shown in Figure 9.9a where the arrows are proportional to the amplitude of the atomic motions. Bi atoms exhibit significantly larger displacements (in contrast to La, Figure 9.9c), indicating larger anharmonic effects. Assuming that the heat is conducted only by acoustic and quasiacoustic phonons and that Umklapp scattering processes dominate, one can obtain an average Grüneisen parameter of 2.9, Figure 9.9b. Additionally, the calculated thermal conductivity in the out-of-plane direction of BiCuSeO is about two times smaller than in the in-plane direction, indicating that BiCuSeO has an unprecedented

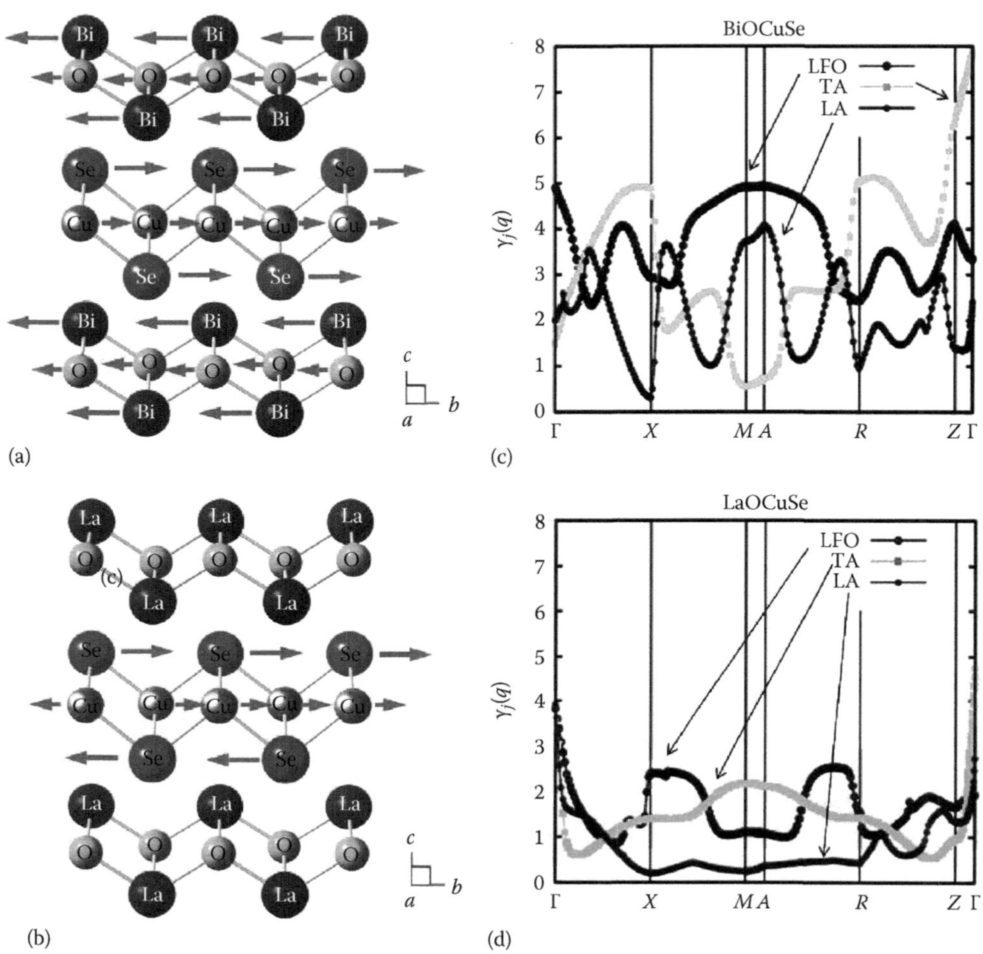

FIGURE 9.9
Atomic displacement patterns for the respective lowest-frequency optical mode in (a) BiCuSeO and in (b) LaCuSeO. Mode Grüneisen dispersion of (c) BiCuSeO and (d) LaCuSeO, respectively.

twofold anisotropy in its heat flow. In BiCuSeO, the lowest-frequency optic mode has the E_u symmetry, the same as its acoustic modes, and behaves as a quasiacoustic mode, whereas in LaCuSeO, the LFO mode has the E_g symmetry. The symmetry matching, as well as a lower value of the lowest-frequency optic mode in BiCuSeO than in LaCuSeO, clearly suggests that the acoustic modes can interact more strongly or can be hybridized more strongly in the former compound than in the latter (see Figure 9.9d), leading to a larger acoustic-optic phonon scattering and, hence, the lower thermal conductivity of BiCuSeO.

Ding et al. also calculated the phonon spectrum and mode Grüneisen parameters of BiCuSeO [68]. As shown in Figure 9.10a, the mode Grüneisen parameters vary between 0.08 and 6.74, with the average Grüneisen parameter being about 2.5. The total thermal conductivity of BiCuSeO was calculated using the lattice thermal conductivity with the obtained Grüneisen parameter and the electronic thermal conductivity determined from the Wiedemann–Franz relation. At low temperatures, the experimental data fall far below the calculated results (see Figure 9.10c) even after the addition of scattering contributions

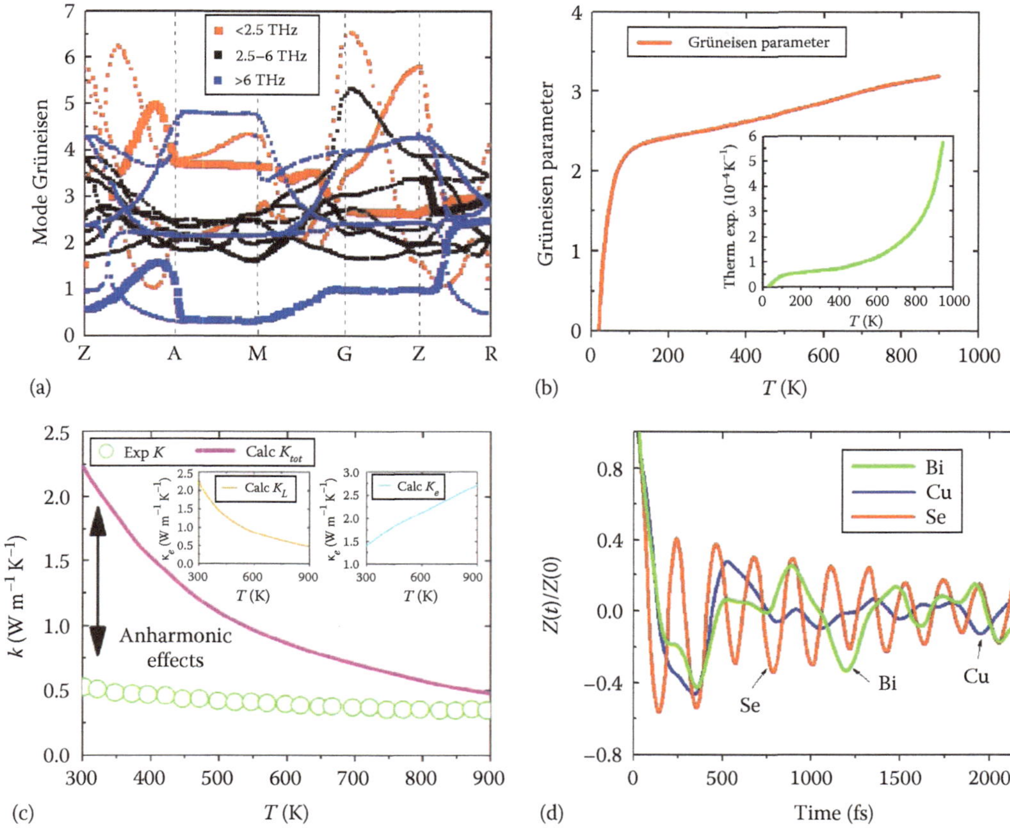

FIGURE 9.10
(a) Mode Grüneisen parameters of BiCuSeO. (b) Temperature-dependent average Grüneisen parameter of BiCuSeO. The inset shows the thermal expansion of BiCuSeO. (c) Calculated thermal conductivity. The solid line is the calculated total thermal conductivity, and the circles are the experimental data. The left-hand side inset shows the calculated electronic contribution, and the inset on the right is the lattice contribution. (d) Normalized self-velocity correlation functions as a function of time. The simulation time is greater than 2000 fs. Bi, Cu, and Se are shown in orange, dark green, and red, respectively.

from defects, which indicates that additional scattering mechanisms are likely behind the low thermal conductivity of BiCuSeO. As shown in Figure 9.10b, the Grüneisen parameter tends to increase with increasing temperature that causes an increase in the degree of thermal expansion. The anomalously high Grüneisen parameter of BiCuSeO reflects the crystal structure, which contains distorted edge-sharing $CuSe_4$ tetrahedra and a layered structure with zigzag geometry. This implies a soft lattice that can dissipate phonons easily. The presence of the heavy element Bi is also a contributing factor to the large Grüneisen parameter due to its large valence shell formed by lone pair electrons.

Ding et al. [68] also used the power spectrum in an innovative way, that is, to elucidate the vibration phase difference between atoms in a layered structure and, consequently, the interaction between them. The self-velocity correlation functions of Bi, Cu, and Se are shown in Figure 9.10d. In general, the correlation function and the variation in velocities of atoms have the same period of oscillations; thus, the decay of the correlation function can exactly reflect the decay in the correlation in atomic motions along the trajectories of the atoms. Moreover, a damped oscillating velocity autocorrelation function generally represents strong interactions between atoms, such as represented by Se atoms in our case, and their vibrations are very much localized. In contrast, the phonon spectrum for the liquid-like phase has a peak that is less pronounced than the peak for the solid-like phase, and the Bi and Cu elements behave differently. Although the typical solid-like features were not observed, the behavior actually observed was also far from that seen in liquids. These results reflect a highly anharmonic bonding that leads to a low thermal conductivity in the BiCuSeO system.

To unveil the origin of the unusual thermal transport properties of BiCuSeO, Vaqueiro et al. [69] carried out first-principles calculations of vibrational properties along with an analysis of in situ neutron diffraction. Results demonstrated that the anharmonicity and the large Grüneisen parameter in BiCuSeO are mainly related to the low-frequency Cu vibrations rather than to the Bi^{3+} lone electron pairs. Powder neutron diffraction data on BiCuQO (Q = Te, Se, S) revealed striking differences in the magnitude of the atomic displacement parameters (ADPs). All ADPs increase linearly with temperature, and the ADP of Cu increases at the fastest rate (see Figure 9.11a). Low-frequency optic modes are known

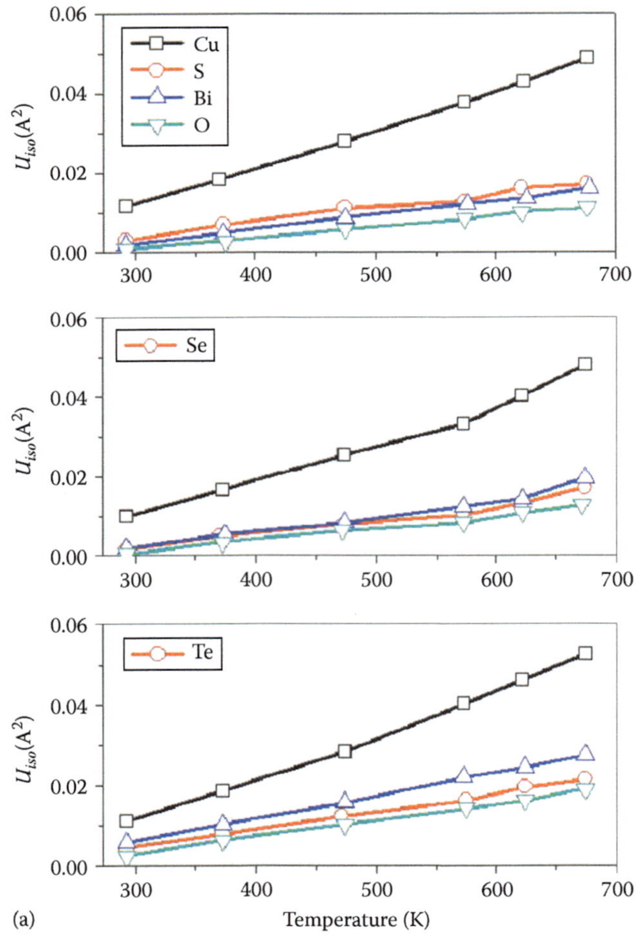

FIGURE 9.11
(a) Atomic displacement parameters for BiCuOQ (Q = Te, Se, S) as a function of temperature. (*Continued*)

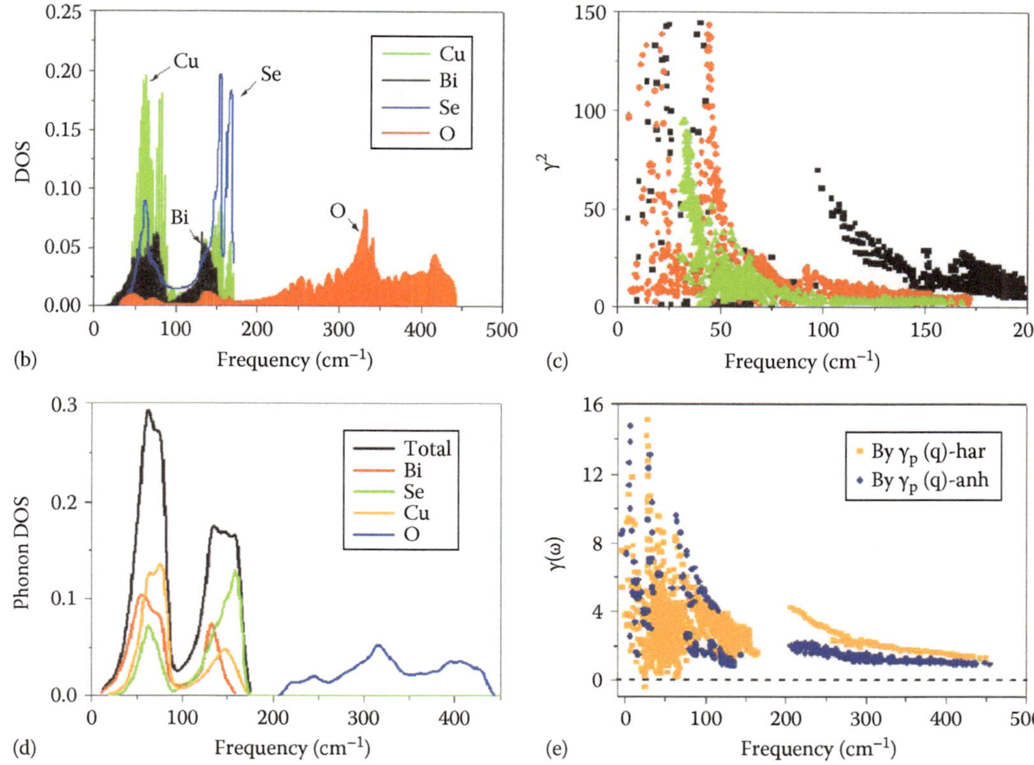

FIGURE 9.11 ***(Continued)***
(b) Total partial density of states (DOS) of BiCuSeO projected on each element. (c) Mode-resolved Grüneisen parameter vs. mode frequency for BiCuOQ with Q = S (black), Se (red), and Te (green). Phonon frequencies were computed using the experimentally determined structures at 293 K. (d) Total vibrational DOS of BiCuSeO projected on each element. (e) Frequency-dependent mode Grüneisen parameter of BiCuSeO.

to exist in low thermal conductivity materials containing heavy ions [69]. However, first-principles calculations (Figure 9.11b) indicate that the vibrations in BiCuQO have a large and unexpected Cu contribution to the phonon DOS, which outweighs that of the heavier Bi atom, as it is six times larger than that of Bi in BiCuSO, three times larger in BiCuSeO, and twice in BiCuTeO. The position of this low-frequency optic manifold is in excellent agreement with the neutron data analysis. In order to evaluate the role of Cu in reducing the lattice thermal conductivity, Vaqueiro et al. computed the mode-resolved Grüneisen parameter and projected it onto the atoms [69]. The calculated results indicate that the major contributors are Cu and chalcogen, while the Bi contribution is always small. The largest Grüneisen parameter at low frequencies is found in BiCuSO, BiCuSeO, and BiCuTeO (see Figure 9.11c).

The calculations of Shao et al. shed more light on the phonon transport properties in BiCuSeO [70]. Figure 9.11d shows that the high-frequency optic modes are separated from the low-frequency modes by a gap of 46 cm^{-1}. The high-frequency optic modes above 213 cm^{-1} are mainly from O vibrations, and these modes exhibit an obvious dispersion, indicating that they have relatively large group velocities. Acoustic and low-frequency optic modes are mainly from vibrations of Bi, Cu, and Se atoms. As shown in Figure 9.11e, throughout the Brillouin zone, the mode Grüneisen parameters of BiCuSeO are mostly positive. Moreover, the maxima of the mode Grüneisen parameters at low frequencies are very high (up to 14) near the Γ point. The calculated average Grüneisen parameter of BiCuSeO is larger than that of PbTe where it ranges from 1.96 to 2.18.

In sharp contrast to $Cu_{2-x}Se$ [20,21], where mobile Cu atoms are not constrained to their equilibrium positions within the structure and can move to neighboring sites, Vaqueiro et al. pointed out that the examination of the difference Fourier maps for BiCuQO at high temperatures indicate that Cu atoms remain at their crystallographic sites and there is no evidence of any structural disorder [69]. Measurements of the ionic conductivity in BiCuSeO confirm negligible ionic conduction and, therefore, the unusual thermal transport behavior of BiCuQO cannot be attributed to a liquid or partly liquid-like state of the Cu sublattice. This is very significant for thermoelectric applications, given the detrimental effect of ion migration that irreversibly damages the legs of the thermoelectric couples.

9.7 Approaches to Enhance the Thermoelectric Performance of BiCuSeO

9.7.1 Optimizing Carrier Concentration

The characteristic transport features of undoped BiCuSeO can be summarized as having a moderate power factor and impressively low lattice thermal conductivity. Therefore, the best way to enhance the thermoelectric *ZT* is to improve the electronic transport properties of BiCuSeO. The room temperature carrier concentration of BiCuSeO is $\sim 1 \times 10^{18}$ cm^{-3}. This is a much lower carrier concentration than the optimum values observed in the state-of-the-art thermoelectrics, which generally are of the order of $\sim 1 \times 10^{19}$ cm^{-3}. Figure 9.12 summarizes the thermoelectric *ZT*s as a function of temperature and year for BiCuSeO and its analogue systems developed since 2010 [24]. In most of these structures, high thermoelectric figures of merit were achieved through doping on the site of Bi.

In this section, we select and summarize the doping studies using alkaline earth metals M^{2+} (M = Mg, Ca, Sr, and Ba) on Bi sites [23,24,53,56,73,76]. Figure 9.13a shows the schematic crystal structure of BiCuSeO, which depicts the creation of holes generated by M^{2+} (M = Mg, Ca, Sr, and Ba) doping into Bi sites and their injection into $(Cu_2Se_2)^{2-}$ metallic layers. The carrier concentration can be enhanced through partial substitution of Bi^{3+} by M^{2+} in the insulating oxide $(Bi_2O_2)^{2+}$ layers, and the induced holes then transfer to the conductive $(Cu_2Se_2)^{2-}$ layers. This can be explained by a defect equation as follows:

$$2(Bi_{1-x}M_xCuSeO) = \left(Bi_{2(1-x)}M_{2x}O_2\right)^{2(1-x)+} + \left(Cu_2Se_2\right)^{2-} + 2xh^+ \tag{9.1}$$

The carrier concentration is significantly increased after M^{2+} (M = Mg, Ca, Sr, and Ba) doping, and a value as high as 1.3×10^{21} cm^{-3} can be reached at a 15% M^{2+} doping level. The mobility of charge carriers (holes) shows a pronounced decrease with the rising M^{2+} doping fraction, from 22 cm^2 V^{-1} s^{-1} for undoped BiCuSeO to about 1–2 cm^2 V^{-1} s^{-1} for samples with 0.15 M^{2+} (M = Mg, Ca, Sr, and Ba) doping. This dramatic decrease in the carrier mobility can be well explained by the increased ionized impurity scattering.

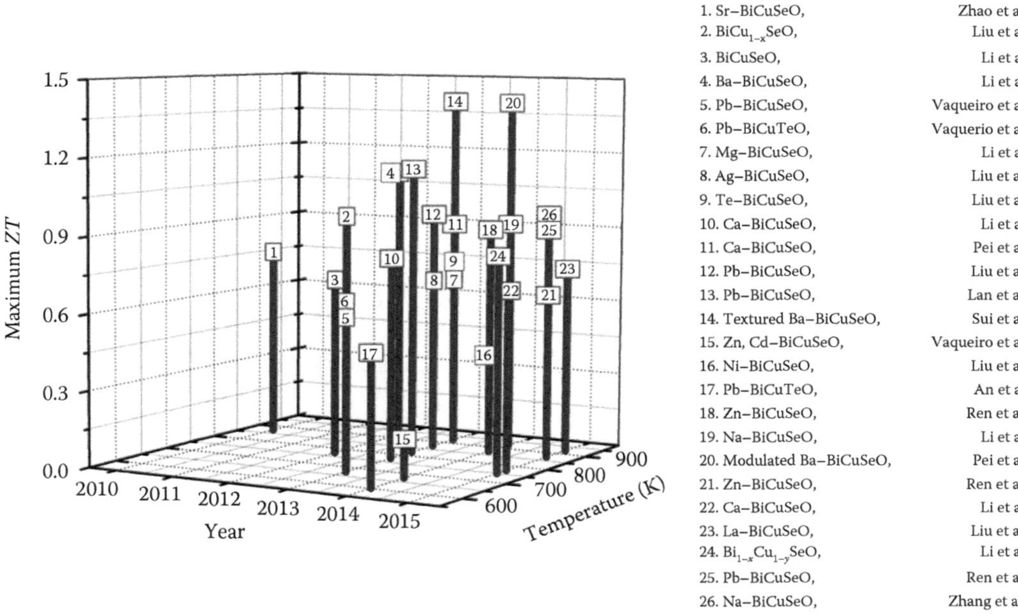

FIGURE 9.12
Thermoelectric figure of merit *ZT* as a function of temperature and year for BiCuSeO and its analogue systems developed from 2010: Sr–BiCuSeO [24], $BiCu_{1-x}SeO$ [71], ball-milled BiCuSeO [72], Ba–BiCuSeO [73], Pb–BiCuSeO [74], Pb–BiCuTeO [75], Mg–BiCuSeO [76], Ag–BiCuSeO [77], Te–BiCuSeO [78], Ca–BiCuSeO [56], Ca–BiCuSeO [53], Pb–BiCuSeO [79], Pb–BiCuSeO [74], textured Ba–BiCuSeO [50], (Zn, Cd)–BiCuSeO [80], Ni–BiCuSeO [81], Pb-BiCuTeO [82], Zn–BiCuSeO [83], Na–BiCuSeO [84], modulated Ba–BiCuSeO [26], Zn–BiCuSeO [85], Ca–BiCuSeO [86], La–BiCuSeO [54], $Bi_{1-x}Cu_{1-y}SeO$ [87], Pb–BiCuSeO [88], and Na–BiCuSeO [89].

Owing to the significantly increased carrier concentration, M^{2+} doping increases the electrical conductivity and decreases the Seebeck coefficient. Figure 9.13b shows that all power factor curves increase monotonously up to 923 K and can be enhanced with proper optimization of the carrier concentration. The lattice thermal conductivity of BiCuSeO is intrinsically low and ranges from 0.40 to 0.60 W m^{-1} K^{-1} at 300–923 K. Moreover, the lattice thermal conductivity can be further reduced by M^{2+} doping, as shown in Figure 9.13c. The reduction of the lattice thermal conductivity has been well explained by point defect scattering based on the Callaway model [59,60], in which point defect scattering in a solid solution originates from both the mass difference (mass fluctuations) and the size and the interatomic coupling force differences (strain field fluctuations) between the impurity atom and the host lattice. Figure 9.13d shows the temperature-dependent *ZT* for the $Bi_{1-x}M_xCuSeO$ systems (M = Mg, Ca, Sr, and Ba). Here, it should be noted that the thermoelectric properties of some BiCuSeO systems were measured up to 923 K, while some others not. The reason may be instrument limitations. To extend measurements to elevated temperatures, one can coat samples with a thin (0.1–0.2 mm) layer of boron nitride for the sake of protection. Specific details can be found in the literature [90–92].

It can be seen that the *ZT* has been significantly improved from 0.5 for an undoped sample of BiCuSeO to values of 0.67, 0.9, 0.76, and 1.1 for Mg-, Ca-, Sr-, and Ba-doped samples, respectively. These achievements are impressive for a new candidate thermoelectric material since the *ZT* of unity has been a threshold value for several decades until nanotechnology was improved in the 1990s [23].

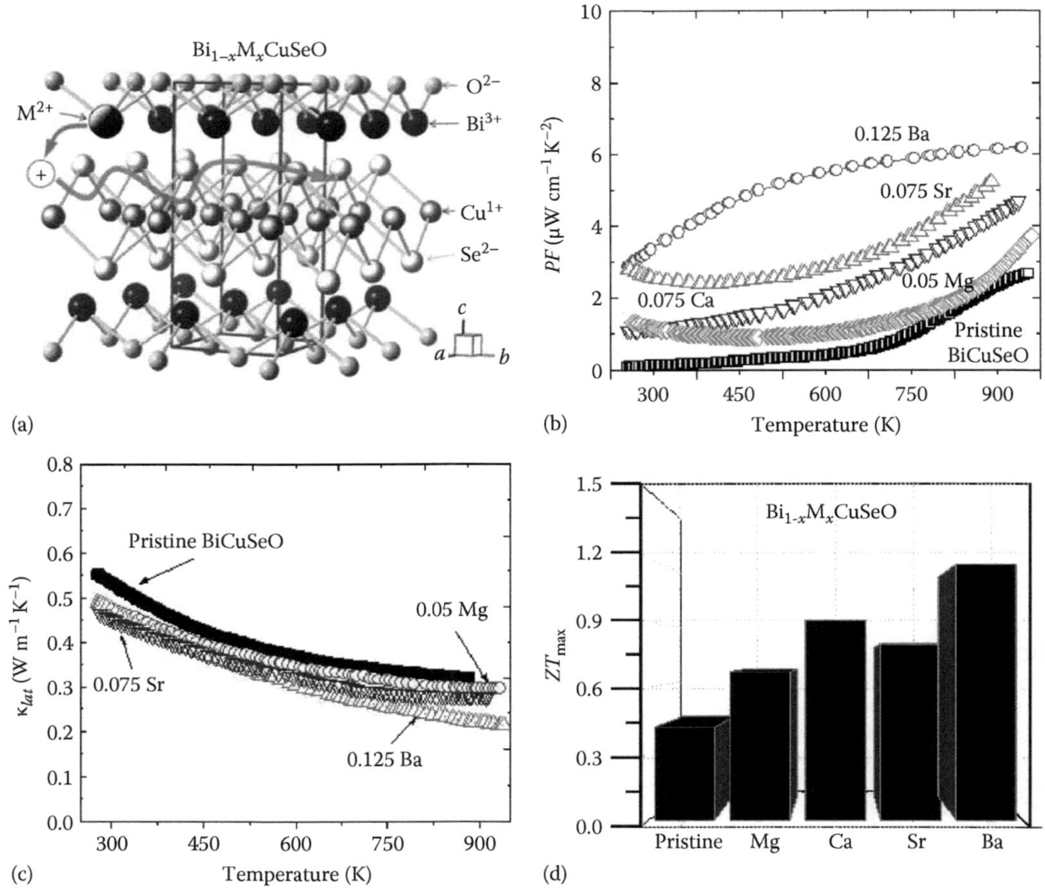

FIGURE 9.13
(a) Schematic crystal structure of BiCuSeO, in which hole carriers generated by M^{2+} (M = Mg, Ca, Sr, and Ba) doping into Bi sites are injected into CuSe metallic layers as a result of the large energy offset between these two layers. (b) Power factor. (c) Lattice thermal conductivity. (d) ZT for $Bi_{1-x}M_xCuSeO$.

9.7.2 Improving Performance via Simple and Facile Ball Milling

Li et al. [72] reported that a simple ball milling process could result in a remarkable enhancement of the thermoelectric properties of undoped BiCuSeO, with ZT values reaching up to ~0.70 at 773 K. The grain size of BiCuSeO was significantly reduced from 10–20 μm down to 1–2 μm by ball milling for 1500 min. Both the total and lattice thermal conductivity exhibited the same decreasing trend with increasing temperature, Figure 9.14a. The lattice thermal conductivity above 700 K showed a significant decrease after the ball milling process, stemming mainly from the enhanced grain boundary scattering due to the refined grain sizes by ball milling. Figures 9.14b shows that the carrier mobility was reduced by ball milling and, particularly so, with the prolonged ball milling time. The carrier concentration at room temperature was increased from 2.65×10^{17} to 5.22×10^{18} cm^{-3} after 500 min ball milling and further increased to 2.73×10^{19} cm^{-3} after 1500 min ball milling. The increased carrier concentration by ball milling probably comes

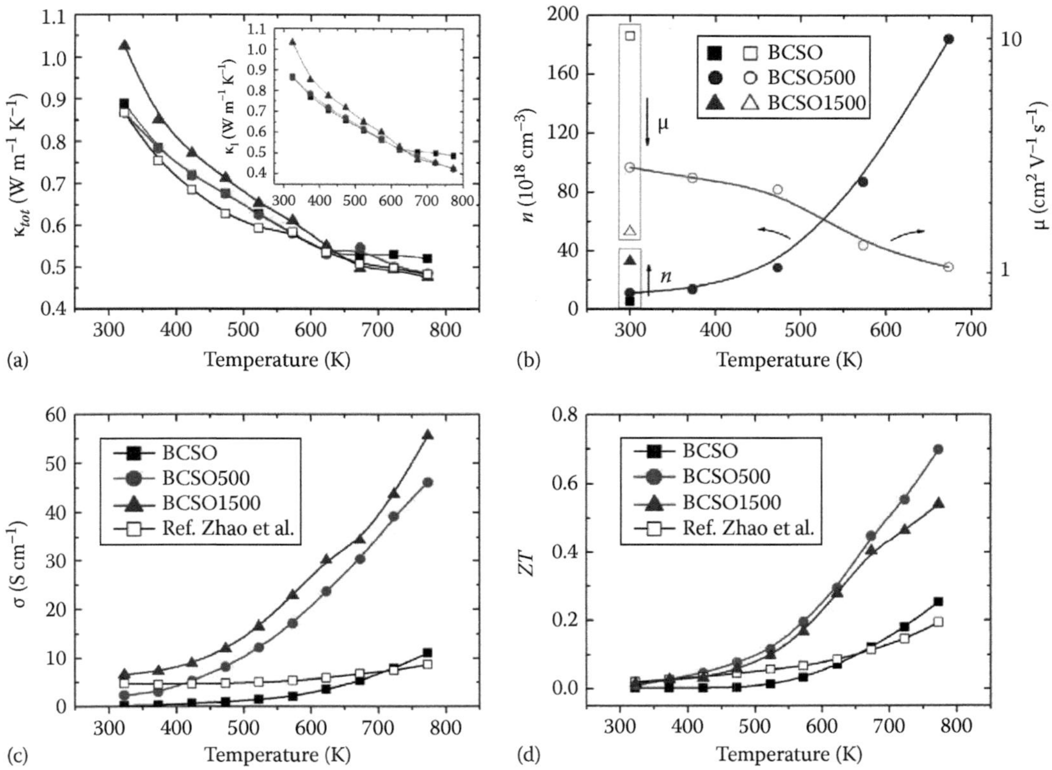

FIGURE 9.14
Temperature dependence of the thermoelectric transport properties of BiCuSeO with different milling times. Data reported by Zhao et al. are also given for comparison [24]. (a) Total thermal conductivity and (inset) the lattice thermal conductivity. (b) Carrier concentration and mobility. (c) Electrical conductivity. (d) *ZT*.

from the introduced Cu vacancies that form easily during the synthesis process. In addition, for the layer-structured compounds like Bi_2Te_3, the carrier concentration can also be raised by point defects induced in the ball milling process [72]. Figures 9.14c shows that the electrical conductivity of BiCuSeO was strongly improved after ball milling, especially at high temperatures. The electrical conductivity behavior is consistent with the observed carrier concentration variation, evidencing that ball milling is responsible for the electrical conductivity enhancement *via* unintentional doping.

Benefiting from the strongly enhanced electrical conductivity and lowered thermal conductivity, the *ZT* value was lifted up to ~0.70 at 773 K for BiCuSeO following the 500 min ball milling process, as shown in Figure 9.14d. This value is about 3.5 times higher than for the undoped BiCuSeO sample. It is quite astonishing that a simple ball milling process could cause such a remarkable enhancement in the thermoelectric performance of undoped BiCuSeO. The work reported by Li et al. also evidenced a good resilience to oxidation of BiCuSeO because the single-phase powders were ball milled in air without any protective atmosphere. Later, this antioxidation ability of BiCuSeO was further documented by a successful synthesis of single-phase BiCuSeO powders by mechanical alloying in air [93]. The resilience of BiCuSeO to oxidation outperforms most of the state-of-the-art thermoelectric materials [23,63].

9.7.3 Increasing Performance by Optimizing Cu Vacancies

It is generally believed that Cu vacancies form when BiCuSeO is synthesized by various processes. Liu et al. [71] optimized the carrier concentration and enhanced the thermoelectric performance of BiCuSeO through introducing Cu vacancies. In contrast to M^{2+} doping, the holes are directly generated in the conductive $(Cu_2Se_2)^{2-}$ layers due to Cu vacancies:

$$2BiCu_{1-x}SeO = \left(Bi_2O_2\right)^{2+} + \left(Cu_{2(1-x)}Se_2\right)^{2(1+x)-} + xV_{Cu}^{-} + 2xh^{+} \tag{9.2}$$

where V_{Cu}^{-} is the Cu vacancy, and h^+ is the generated hole carrier.

This equation implies that the deficiency of Cu can increase the concentration of holes just like a conventional acceptor.

As shown in Figure 9.15a, samples deficient in Cu exhibit significantly enhanced electrical conductivity. Interestingly, the Cu-deficient samples still maintain Seebeck coefficient values between 165 and 323 $\mu V\ K^{-1}$ over the entire investigated range of temperatures (see Figure 9.15b). Figure 9.15c shows that the thermal conductivity starts to increase as the Cu deficiency fraction x exceeds 0.025, due to an increasing contribution from the electronic thermal conductivity. The ZT increases with temperature and reaches its maximum value of about 0.81 at 923 K in $BiCu_{0.975}SeO$, as shown in Figure 9.15d. The data unambiguously show that the obtained ZT enhancement is mainly due to the improved electrical conductivity.

9.7.4 Enhancing Performance through Multifunctional Pb Doping

Besides alkaline earth anions M^{2+} (M = Mg, Ca, Sr, and Ba) doping on Bi^{3+} sites [24,53,73,76], Pb^{2+} doping has been extensively studied by several research teams [55,74,79,88,94] and was found to be an effective method to enhance the thermoelectric performance of BiCuSeO. Although the effect of Pb^{2+} doping resembles in most of its aspects doping by alkaline earth metals, it is nevertheless instructive to treat Pb^{2+} doping separately as there are some important differences.

The normalized carrier mobility, $\mu(m^*)^{3/2}$, was evaluated to investigate the origin of the enhancement of the thermoelectric performance. Lan et al. [74] found that the normalized mobility $\mu(m^*)^{3/2}$ of the Pb^{2+}-doped BiCuSeO is higher than the normalized mobility obtained with M^{2+} (M = Mg, Sr, and Ba) doping, Figure 9.16a. The increased normalized mobility $\mu(m^*)^{3/2}$ and the reduced thermal conductivity led to a maximum ZT of 1.14 at 823 K in $x = 0.06$ Pb-doped BiCuSeO, as shown in Figure 9.16b. The electronic structure calculations show that the enhancement of the hole mobility is very likely due to Pb-induced delocalized lone pair 6s orbitals in the electronic states close to the VBM, which are primarily responsible for hole transport. By using first-principles calculations, Pb substitution for a Bi atom gives the lowest defect formation energy in $Bi_{0.875}M_{0.125}CuSeO$ (M = Pb, Mg, Sr, and Ba) among all possible configurations. The calculation shows that Pb^{2+} doping introduces substantial 6s components into the states close to the VBM, while the s states from Mg, Sr, and Ba only contribute deep in the conduction band. Such delocalized 6s orbitals from the lone-pair electrons in Pb are believed to be helpful to enhance the hole mobility.

9.7.5 Enhancing Performance by Improving Carrier Mobility through Texturing

The low carrier mobility ~22 $cm^2\ V^{-1}\ s^{-1}$ at room temperature is the bottleneck for undoped BiCuSeO, and it greatly limits the improvement of the electronic transport properties and

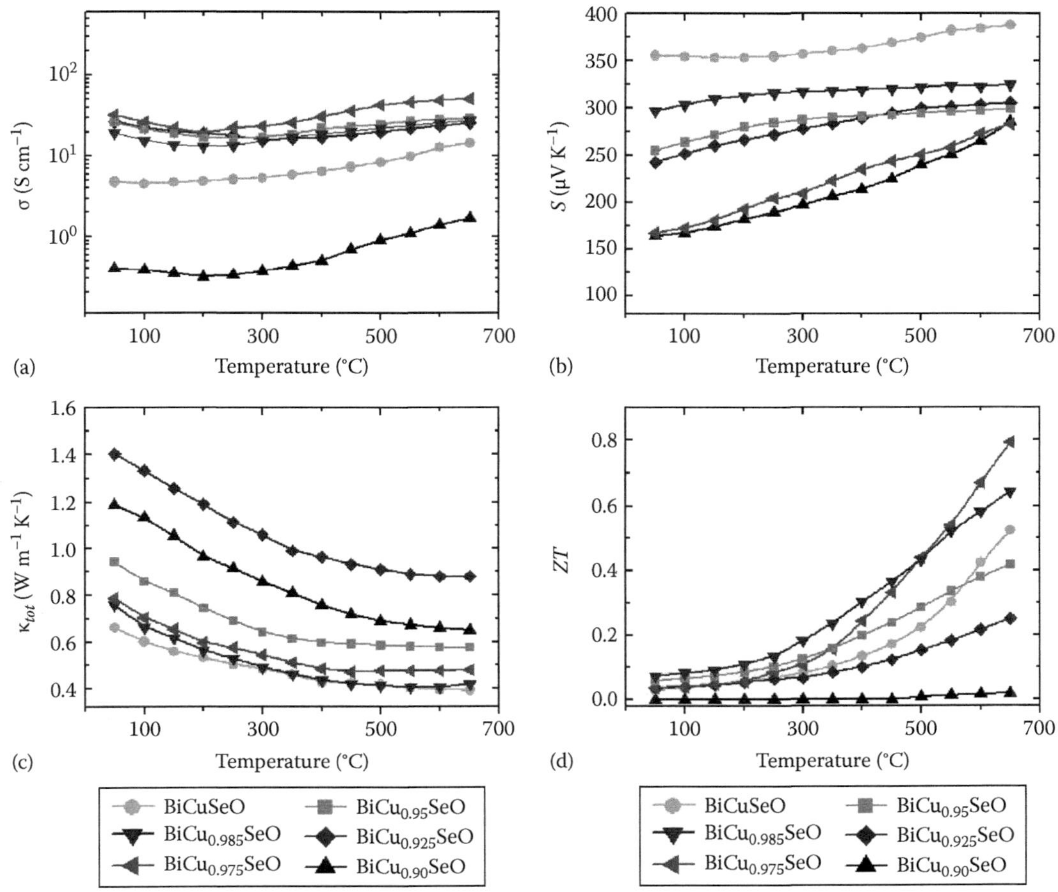

FIGURE 9.15
Temperature dependence of (a) electrical conductivity, (b) Seebeck coefficient, (c) total thermal conductivity, and (d) ZT of $BiCu_{1-x}SeO$.

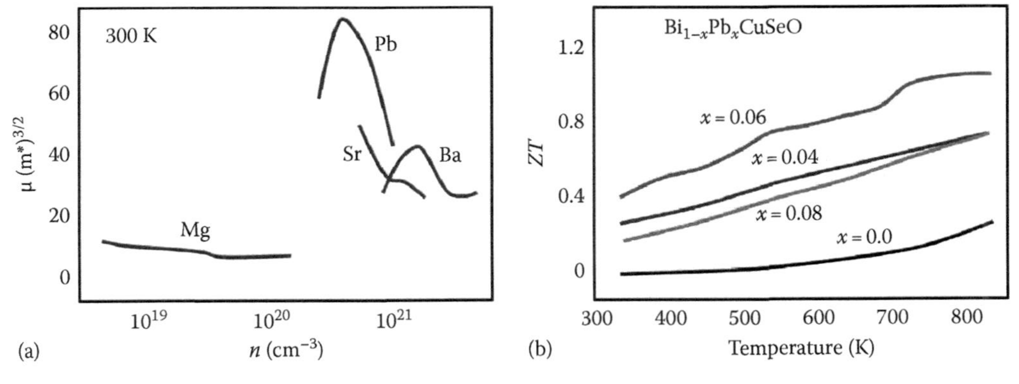

FIGURE 9.16
(a) Carrier concentration dependence of normalized mobility in $Bi_{0.875}M_{0.125}CuSeO$ [M = Pb, Mg, Sr, and Ba] systems and (b) ZT of the $Bi_{1-x}Pb_xCuSeO$.

the thermoelectric performance. Even worse, since most of the previous studies focused on the carrier concentration enhancement through various doping schemes, the carrier mobility was further deteriorated due to impurity scattering, even down to 1–2 $cm^2 V^{-1} s^{-1}$ as the carrier concentration increased up to the order of 1×10^{21} cm^{-3}. Because BiCuSeO oxyselenides crystallize in a layered structure, the electronic and thermal transport in the BiCuSeO system should be anisotropic in single crystals and a higher in-plane carrier mobility could be expected. Moreover, in polycrystalline materials, this intrinsic layered structure leads to the formation of platelet grains, with a preferential orientation due to extrinsic forces, and a strong anisotropy of the grain boundary orientation can be achieved between the *c*-axis and the in-plane direction. This anisotropy can be conveniently utilized to improve the carrier mobility.

Sui et al. [50] applied hot-forging processes to obtain textured samples, where the *a–b* planes of grains were preferentially oriented perpendicular to the pressing direction through sample deformation under externally applied pressure (hot forging). The highly textured microstructures after the hot-forging process were confirmed under scanning electron microscopy examination using an electron backscatter diffraction (EBSD) detector. Figure 9.17a shows that the platelet grains are oriented perpendicular to the pressing direction of $Bi_{0.875}Ba_{0.125}CuSeO$. The grains with different shades of gray indicate different orientations of grains; the lightest gray represents grains along [001], medium gray is for the [010] oriented grains, and the dark gray is for the [110] oriented grains. With increasing the number of hot-forging steps, the grains experience slight growth, and the preferential grain orientation along the [001] direction is more and more pronounced. The pole figures on the right clearly indicate that the orientation of grains evolves from that of absolute disorder before hot forging toward the [001] direction after hot forging. As expected, the carrier concentration was not influenced by the hot-forging process, while the carrier mobility was increased from 2 to ~3 , ~3.7, and ~4.2 $cm^2 V^{-1} s^{-1}$ after one, two, and three hot-forging steps, respectively. After the hot-forging process, the *ZT* measured parallel to the pressing direction was somewhat reduced, but it improved from 1.1 to 1.4 at 923 K when measured perpendicular to the pressing direction after three hot-forging steps (see the trend depicted in Figure 9.17b).

9.7.6 Enhancing Performance by Improving Carrier Mobility through Modulation Doping

As discussed in the previous section, the carrier mobility can be doubled by microstructure texturing through several steps of hot forging. This strategy is thus an effective way to enhance *ZT* values by taking the advantage of the anisotropic nature of BiCuSeO [50].

MD is a well-developed technique widely used in 2D electron gas thin-film devices in order to improve the carrier mobility and thus the electrical conductivity [95]. Pei et al. [26] applied the MD approach in the BiCuSeO system and the *ZT* was enhanced from 1.1 to 1.4.

The overall operating concept can be described as the following: the modulation-doped sample is actually a two-phase composite, that is, undoped BiCuSeO and heavily doped $Bi_{0.75}Ba_{0.25}CuSeO$. The undoped BiCuSeO phase shows a lower carrier concentration but higher carrier mobility, while the heavily doped $Bi_{0.75}Ba_{0.25}CuSeO$ phase exhibits a higher carrier concentration but lower carrier mobility. The Fermi level of undoped BiCuSeO is located in the middle between the conduction and valence bands and would move into

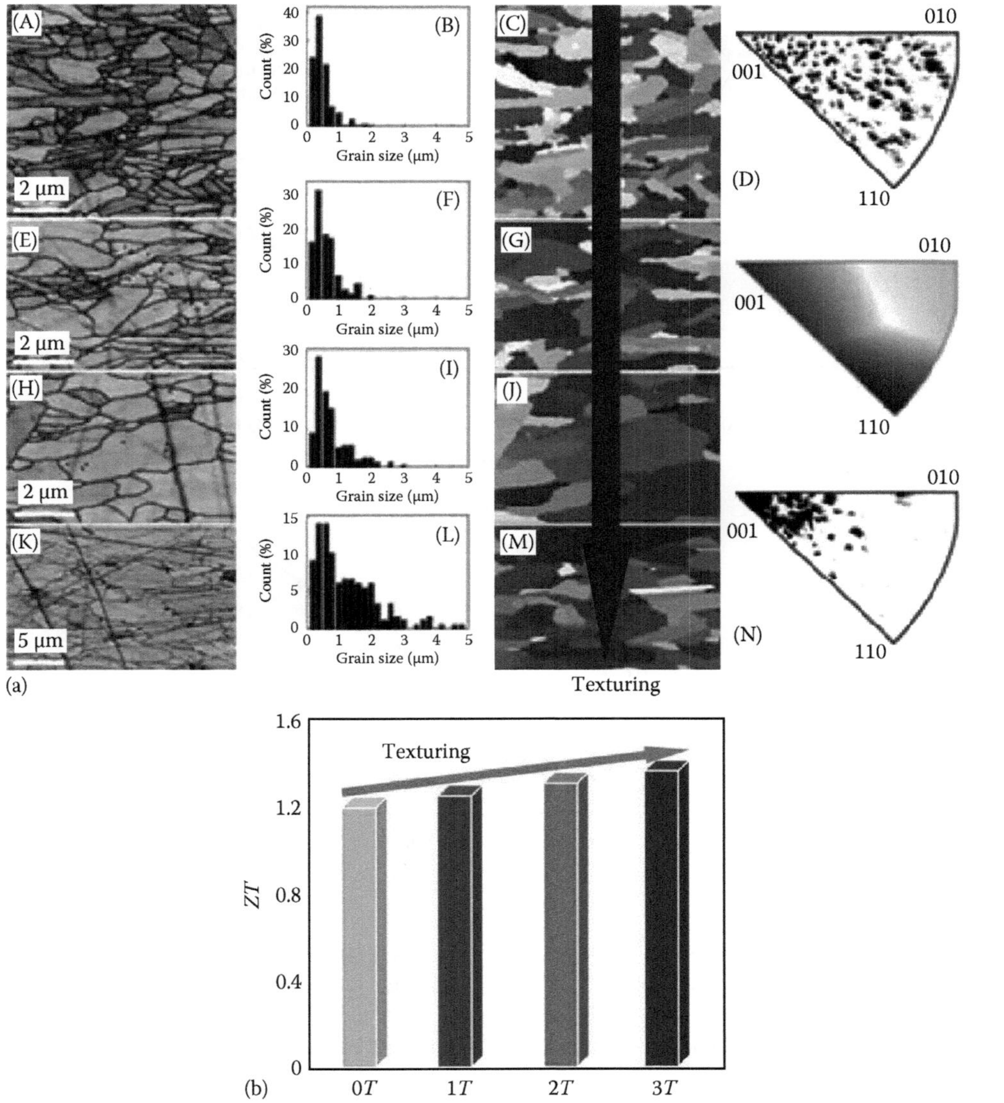

FIGURE 9.17
(a) Electron backscattering diffraction microstructures of textured $Bi_{0.875}Ba_{0.125}CuSeO$: (A, E, H, K) band contrast images, (B, F, I, L) grain size distribution histograms, (C, G, J, M) Z-Euler images, and (D, N) inverse pole figures of textured samples. (b) *ZT* of textured $Bi_{0.875}Ba_{0.125}CuSeO$. 0*T*, 1*T*, 2*T*, and 3*T* represent the samples without hot forging, after one-step, two-step, and three-step hot forging, respectively.

the valence band in the case of heavily doped $Bi_{0.75}Ba_{0.25}CuSeO$. As for the modulation-doped sample, $Bi_{0.875}Ba_{0.125}CuSeO$ (50% BiCuSeO + 50% $Bi_{0.75}Ba_{0.25}CuSeO$), the Fermi level is supposed to bridge the two end members. The Fermi level imbalance prompts the electrons to diffuse from the doped phase to the undoped phase, where the carrier mobility is higher owing to a smaller number of ionized scattering centers, thus resulting in the overall enhanced carrier mobility in comparison to a uniformly doped structure.

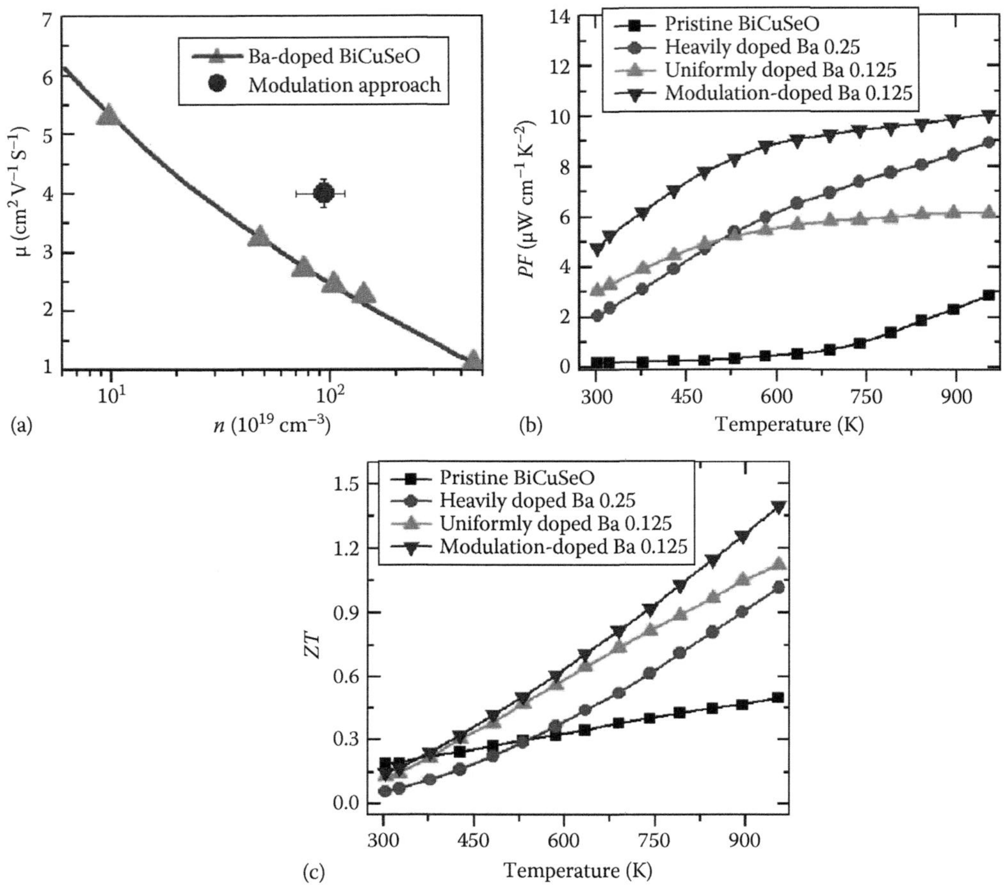

FIGURE 9.18
(a) Room temperature carrier mobility as a function of carrier concentration, (b) power factor, and (c) *ZT*, for the uniformly doped $Bi_{1-x}Ba_xCuSeO$ and modulation-doped $Bi_{0.875}Ba_{0.125}CuSeO$.

Figure 9.18a plots the room temperature carrier mobility as a function of carrier concentration for Ba-doped BiCuSeO, as well as for the modulation-doped structure. All Ba-doped samples follow a smooth curve that demonstrates an inversely proportional relationship between the carrier mobility and the carrier concentration. However, the carrier mobility of the modulation-doped sample deviates from the curve and shows a significant enhancement. It is worth noting that in the modulation-doped $Bi_{0.875}Ba_{0.125}CuSeO$ sample, the carrier mobility is ~4.1 $cm^2\ V^{-1}\ s^{-1}$, which is almost twice as high as the mobility in the uniformly doped $Bi_{0.875}Ba_{0.125}CuSeO$ sample having a similar carrier concentration. For the modulation-doped $Bi_{0.875}Ba_{0.125}CuSeO$, the improved electrical conductivity and the nearly unaffected Seebeck coefficient lead to a broad and high power factor ranging from 5 $\mu W\ cm^{-1}\ K^{-2}$ at 300 K to 10 $\mu W\ cm^{-1}\ K^{-2}$ at 923 K. Such power factors are even higher than the ones achieved with heavily doped $Bi_{0.75}Ba_{0.25}CuSeO$ (Figure 9.18b). The expressly high power factor of ~10 $\mu W\ cm^{-1}\ K^{-2}$ at 923 K, coupled with the extremely low thermal conductivity of ~0.25 $W\ m^{-1}\ K^{-1}$ at 923 K, yields the maximum *ZT* value of ~1.4 at 923 K for the modulation-doped $Bi_{0.875}Ba_{0.125}CuSeO$ [26], a nearly 30% increase in the figure of merit with respect to *ZT* ~1.1 for uniformly doped $Bi_{0.875}Ba_{0.125}CuSeO$ [73] (see Figure 9.18c).

9.7.7 Synergistically Optimizing the Electrical and Thermal Transport Properties in BiCuSeO through Dual Vacancies

Vacancy is a very important class of phonon scattering centers to reduce the thermal conductivity and develop high efficiency thermoelectric materials. Xiao et al. reported a successful synergistic optimization of electronic and thermal transport parameters through Bi/Cu dual vacancies in the BiCuSeO system [87]. As compared to undoped and monovacancy samples, these dual vacancies further increase phonon scattering and result in an ultralow thermal conductivity of 0.37 W m^{-1} K^{-1} at 750 K. Most importantly, the clear-cut evidence from positron annihilation measurements unambiguously confirms the interlayer charge transfer between these Bi/Cu dual vacancies, which results in a significant increase in the electrical conductivity while maintaining a relatively high Seebeck coefficient. As a result, BiCuSeO with dual vacancies shows a high ZT value of 0.84 at 750 K.

Figure 9.19a shows the temperature dependence of the electrical conductivity for $Bi_{1-x}Cu_{1-y}SeO$ samples. Undoped BiCuSeO exhibits quite low electrical conductivity values over the entire temperature range measured. In contrast, Bi- and/or Cu-deficient samples

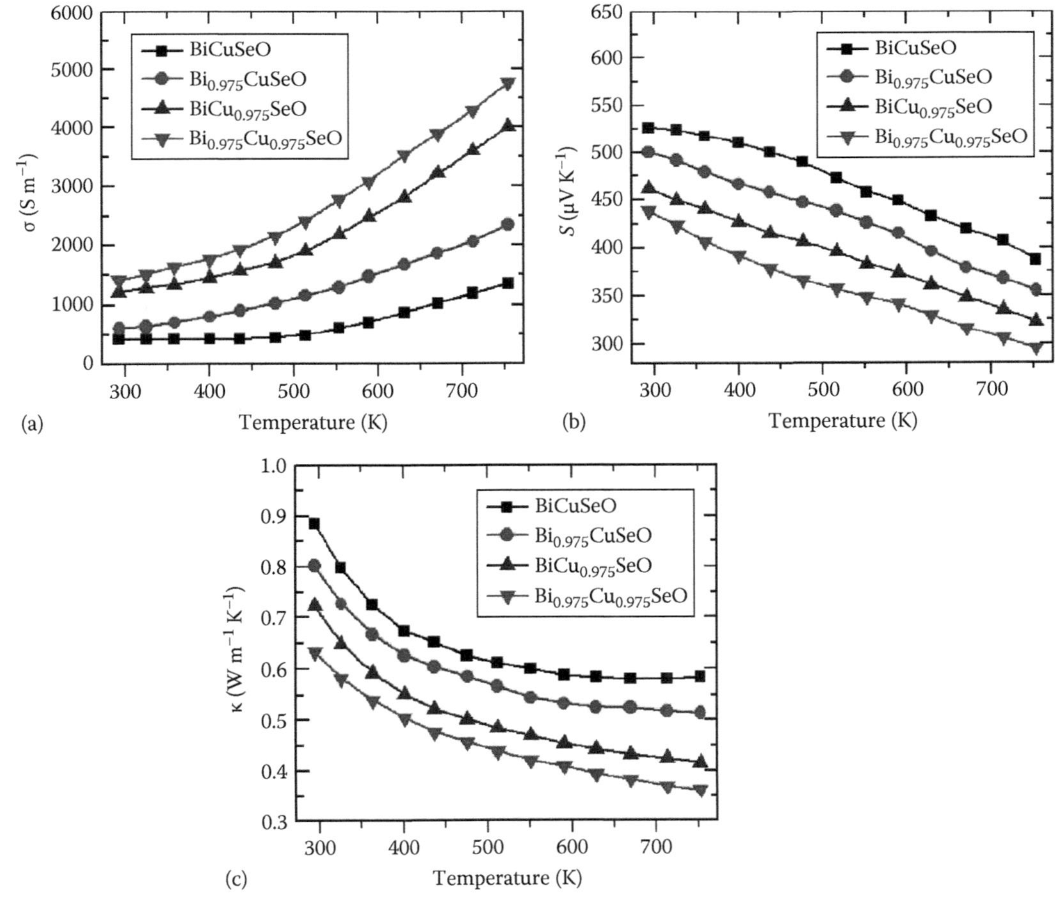

FIGURE 9.19
(a) Electrical conductivity, (b) Seebeck coefficient, and (c) total thermal conductivity as a function of temperature for $Bi_{1-x}Cu_{1-y}SeO$.

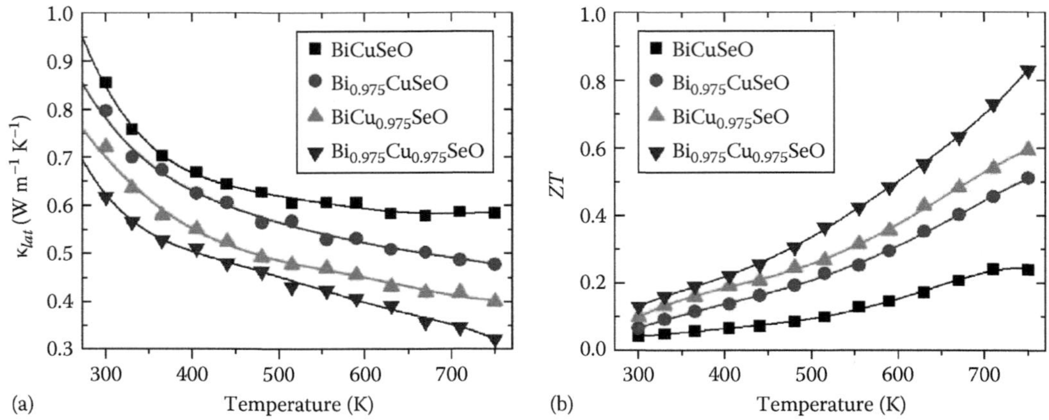

FIGURE 9.20
(a) Lattice thermal conductivity and (b) ZT for $Bi_{1-x}Cu_{1-y}SeO$.

exhibit significantly enhanced electrical conductivities. Upon introducing Bi and/or Cu vacancies, the Seebeck coefficient decreases due to the increased electrical conductivity (Figure 9.19b) but still maintains values of 300–433 μV K^{-1} for $Bi_{0.975}Cu_{0.975}SeO$ over the entire temperature range. Upon vacancy modulation, thermal conductivities are further reduced and reach the minimum of 0.63 W m^{-1} K^{-1} at 300 K in $Bi_{0.975}Cu_{0.975}SeO$ with Bi/Cu dual vacancies (Figure 9.19c).

Moreover, phonons with different wavelengths are more likely to be scattered by defects with comparable sizes. Thus, in vacancy-modulated $Bi_{1-x}Cu_{1-y}SeO$, while Bi/Cu vacancies and grain boundaries preferentially scatter short- and long-wavelength phonons, respectively, superlattice interfaces between adjacent $[Bi_2O_2]^{2+}$ and $[Cu_2Se_2]^{2-}$ layers strongly scatter mid-wavelength phonons, achieving the desirable all-length-scale phonon scattering. As shown in Figure 9.20a, the lattice thermal conductivity for vacancy-modulated samples is obviously lower than for the undoped counterpart. Figure 9.20b indicates that the ZT increases with temperature and reaches the maximum value of 0.84 at 750 K for the dual vacancies $Bi_{0.975}Cu_{0.975}SeO$ sample, which is thus superior to the undoped sample. The enhanced ZT in $Bi_{1-x}Cu_{1-y}SeO$ correlates with the relatively higher electrical conductivity and extremely low thermal conductivity, while a relatively high Seebeck coefficient is maintained over the whole temperature range.

9.8 Theoretical Calculations of the Thermoelectric Properties of BiCuSeO

9.8.1 Predicted Thermoelectric Properties as a Function of Doping Levels and Pressures for BiCuSeO by First-Principles Calculations

Theoretical calculations of thermoelectric properties were utilized to provide support and guidance for experiments. Zou et al. calculated the thermoelectric properties of BiCuSeO with different doping levels on the Bi site [96] and estimated the optimal doping concentration to achieve the maximum thermoelectric performance. Calculations of the electronic transport properties were based on a semiclassical Boltzmann transport equation in conjunction with the rigid band approximation and showed good agreement with experimental data.

As seen in Figure 9.21a, the calculated Seebeck coefficient of BiCuSeO agrees well with the experimental data, especially above 500 K. Figure 9.21b shows that the electrical conductivity at a given temperature increases with the increasing doping level. It is noteworthy that the deviation of the theoretical calculations from experimental observations in the low temperature regime can be ascribed to the oversimplified constant relaxation time approximation. To obtain the total thermal conductivity, the lattice thermal conductivity was assumed to obey the $1/T$ Umklapp phonon scattering behavior to fit the experimental data [73], and the electronic thermal conductivity was given by the Wiedemann–Franz law. The fitted thermal conductivity is plotted in Figure 9.21c and matches well with the experimental data. The calculations confirm experimental observations that the ZT can be enhanced in the high temperature range (above 600 K) *via* optimizing the doping level, Figure 9.21d. Theoretically, the ZT shows an increasing trend with temperature, and the maximum value of 1.32 is predicted at 1000 K for the 12.5% (doping level on the Bi site) doped BiCuSeO. Zou et al. also investigated the effect of high pressure on the electronic structure and thermoelectric properties of BiCuSeO [97]. They reported that the electronic bandgap increases with increasing pressure, and the band structure near the Fermi level of BiCuSeO is modified by applying hydrostatic pressure. The transport properties of BiCuSeO have been estimated based on the semi-classical Boltzmann transport theory, and the results show that the electrical conductivity of BiCuSeO can be enhanced by applying external pressure up to 10 GPa, and n-type BiCuSeO should be more suitable for improvement of thermoelectric properties under pressure than p-type doped BiCuSeO.

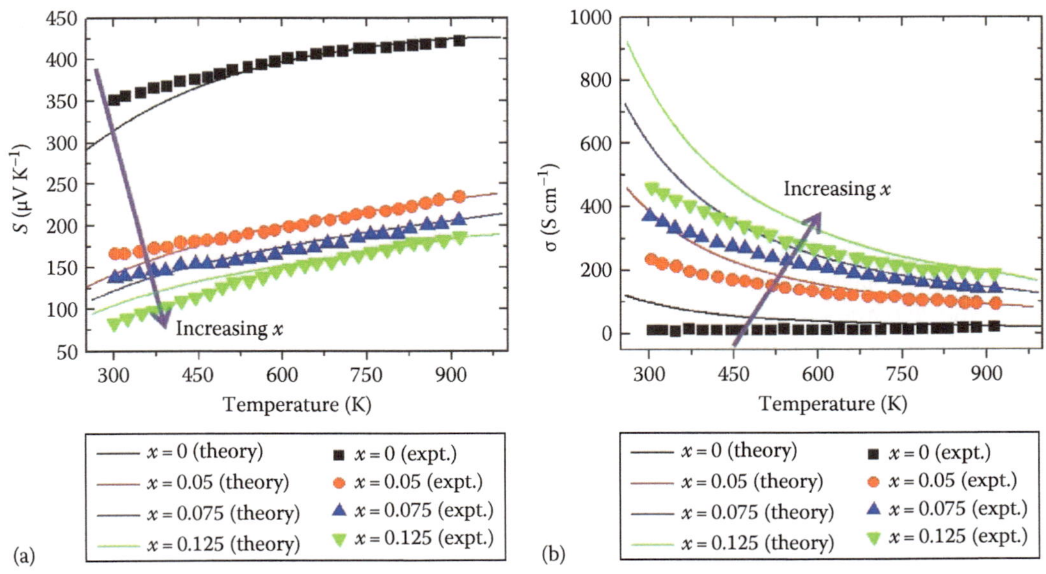

FIGURE 9.21
Calculated thermoelectric properties of BiCuSeO as a function of temperature with x the doping level on the Bi site: (a) Seebeck coefficient, (b) electrical conductivity. *(Continued)*

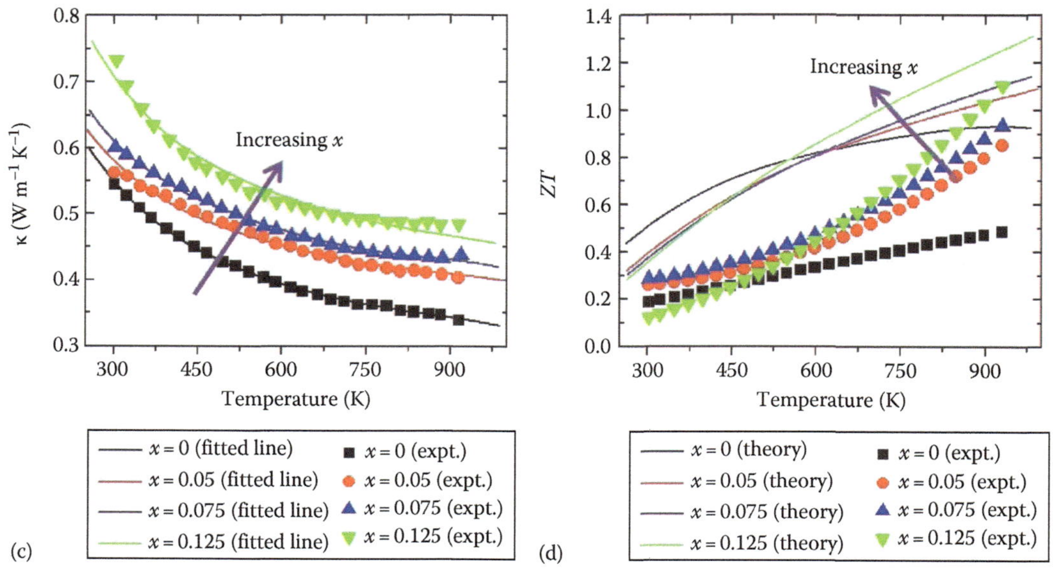

FIGURE 9.21 ***(Continued)***
Calculated thermoelectric properties of BiCuSeO as a function of temperature with x the doping level on the Bi site: (c) total thermal conductivity, and (d) ZT. The experimental data are from hole-doped BiCuSeO. (From Li, J. et al., *Energy Environ. Sci.*, 5, 8543, 2012.)

9.8.2 Low Effective Mass Leading to an Improved *ZT* by 32% for *n*-Type BiCuSeO: A First-Principles Study

To date, the highest ZT of 1.4 has been obtained in p-type BiCuSeO [26,50]. The development of high-performance n-type compounds based on the BiCuSeO system is very urgent. Experimentally, no n-type analogues have been reported so far, but calculations show that the maximum ZT value could be improved by 32% for n-type BiCuSeO compared with the prevailing p-type doping approach at 920 K [98].

The calculations indicate that the conductive pathway of p-type BiCuSeO is mainly determined by the interaction between Bi and Se atoms at the VBM. In contrast, at the conduction band minimum (CBM), the conductive pathway of n-type BiCuSeO is mainly determined by Bi atoms. Along the c direction, no charge density connects the $(Bi_2O_2)^{2+}$ and $(Cu_2Se_2)^{2-}$ layers at the VBM and CBM. It demonstrates that the electrical conductivity in the ab planes should be larger than that along the c direction for both n- and p-type BiCuSeO. When the isosurface level is set at a value of 0.0005, the Cu and Bi atoms are not yet connected by the charge density in p-type BiCuSeO; however, in n-type BiCuSeO, all Bi and Cu atoms are connected along the z direction. Therefore, the electrical conductivity of n-type BiCuSeO is most likely larger than that of p-type BiCuSeO along the z direction.

The calculation results also indicate that the electrical conductivity of n-type doped BiCuSeO is much larger than that of p-type doped BiCuSeO at the same temperature and carrier concentration. The electrical conductivity of n-type doped BiCuSeO is more than five times that of the p-type doped material. Even though the Seebeck coefficient of the p-type doped BiCuSeO is greater than that of the n-type doped counterpart when the carrier concentration is above 3.5×10^{19} cm^{-3}, the maximum values of the power factor of n-type doped BiCuSeO are much greater, by nearly a factor of 2 at 920 K, than that of the p-type doped material. The calculated maximum ZT values are 0.73 and 0.90 for p- and

n-type doped BiCuSeO at 780 K, respectively, corresponding to the carrier concentrations of 3.3×10^{20} and 6.8×10^{19} cm^{-3}, respectively. At 920 K, the calculated maximum ZT values are 0.91 and 1.2 for p- and n-type doped BiCuSeO, respectively, corresponding to the carrier concentrations of 3.2×10^{20} and 4.6×10^{19} cm^{-3}, respectively. Thus, the maximum ZT values could be improved by 32% using n-type doping in BiCuSeO compared with the prevailing p-type doping approach.

9.9 Studies of the Thermal Stability of BiCuSeO

BiCuSeO-based materials are expected to be suitable for applications in thermal to electrical power conversion modules in the 573–923 K temperature range. However, it is not clear whether this material is suitable for long-term operations at elevated temperatures in air. The phase structure and microstructure, as well as the electrical conductivity, of BiCuSeO-based samples after exposure to air at elevated temperatures were studied by Li et al. [86]. The authors used a PbTe-based thermoelectric alloy of $AgPb_{20}SbTe_{20}$ as a comparison. The results revealed that polycrystalline BiCuSeO possessed much better thermal stability in air than the PbTe-based alloy. The latter severely oxidized already at 573 K, and this resulted in a decreased carrier density and a significantly reduced electrical conductivity. An apparent decomposition occurred in BiCuSeO samples when heated at 773 K in air, although, at 573 K, the oxidation was limited to the surface and consisted of the formation of Bi oxide. As a result, the electrical conductivity decreased by 8% after an exposure to air at 573 K for 2 h, and it further decreased by 20% when prolonging the exposure to 6 h. When the sample was exposed to air at 773 K for 2 h, the electrical conductivity was decreased by several orders of magnitude, as shown in Figure 9.22a. Additionally, the

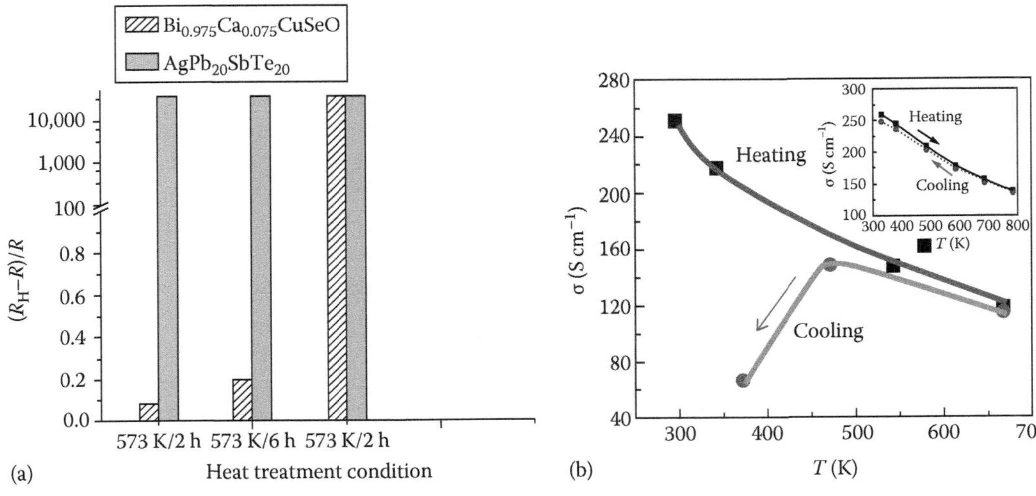

FIGURE 9.22
(a) Relative change of electrical resistivity for BiCuSeO- and $AgPb_{20}SbTe_{18}$-based materials after heat treatment in air, with R the electrical resistivity of the sample without heat treatment and R_H the electrical resistivity after heat treatment. (b) Electrical conductivity of the $Bi_{0.925}Ca_{0.075}CuSeO$ at 300–670 K with heating and cooling cycles in air. The inset shows the data obtained in inert atmosphere.

electrical conductivity of BiCuSeO samples repeatedly cycled in an inert atmosphere was not affected, while the data could not be reproduced when cycling in air, as depicted in Figure 9.22b. This further indicates that BiCuSeO-based materials are not suitable for long-term operation at high temperatures when exposed to air.

Barreteau et al. [99] also reported on the phase stability and thermal stability of these materials under inert and oxidizing atmosphere and discussed their actual potential for wide-scale thermoelectric applications. It was shown that, although they are stable in an inert atmosphere up to 923 K with a very slow volatilization rate, it is not the case under an oxidizing atmosphere because rapid oxidation starts even at moderate temperatures. On the basis of this result, isothermal measurements have been made on BiCuSeO powder. The isothermal oxidation curves recorded at 448 and 473 K do not evidence any significant mass gain, which is consistent with the x-ray diffraction results. Below 473 K, BiCuSeO samples do not oxidize in air. Above 473 K, the mass gain becomes more significant and it increases rapidly with temperature. Above 548 K, the Bragg peaks corresponding to BiCuSeO can hardly be detected, and only oxide phases are observed.

9.10 Summary and Outlook

Recently, BiCuSeO oxyselenides have been receiving an ever-increasing attention and have been extensively studied. Their thermoelectric performance, as measured by the dimensionless figure of merit ZT, was significantly increased from 0.5 for undoped BiCuSeO to 1.4 for chemically modified structures. The intrinsically low thermal conductivity of BiCuSeO implies that the most effective method to enhance ZT is *via* enhancing the electronic transport properties. In the past few years, many effective approaches have successfully been applied in the BiCuSeO system and they include optimizing the carrier concentration through doping and bandgap tuning, as well as increasing the carrier mobility through texturing and MD. If these effective power factor enhancing strategies were synergistically combined, they would likely further improve the performance of the BiCuSeO materials. Despite vast efforts devoted to improving the performance of BiCuSeO oxyselenides, the research on this family of promising thermoelectric materials is still at a relatively infant stage, and thus many fundamental and technical issues remain to be explored. First, novel synergistic approaches need to be developed to further enhance the power factor while maintaining the low thermal conductivity. For example, high carrier mobility can be expected in BiCuSeO single crystals, and possibly a high ZT of 2.0 could be achieved. However, growing single crystals in the BiCuSeO system is extremely challenging with the current synthesis technologies. Second, BiCuSeO belongs to the more general family of LnCuChO (Ln = lanthanide, Ch = S, Se, Te); thus, many analogues are worthy of exploration. Examples include BiCuTeO, BiCuSO, LaCuSeO, and CeCuSeO. Third, high-performance n-type compounds based on the BiCuSeO system are urgently needed. Finally, besides improving the thermoelectric performance, studies should focus on assessing and understanding the electrode materials, the contact resistivity, and the mechanical properties. The issues mentioned earlier are also critical for the fabrication of thermoelectric devices. Understanding the thermal stability and mechanical robustness of the new high-performance materials will help open the path to device assembly, testing and ultimately to commercialization.

Acknowledgments

This work was supported by the "Zhuoyue" program of Beihang University, the Recruitment Program for Young Professionals, and NSFC under grant no. 51571007 (LDZ) and NSFC under grant nos. 51172121 and 51221291 (JFL). This contribution was partly supported by the Triangle de la Physique under Project No. STP 2008-095T, the International Cooperation Department of Université Paris-Sud XI, and the ANR through the project OTHer (ANR2011JS0801201) (LDZ). The author (LDZ) is grateful to Professor Mercouri G. Kanatzidis for helpful discussions. We also thank Professors N. Dragoe, C. W. Nan, J. Q. He, Y. H. Lin, B. P. Zhang, D. Berardan, W. Cai, J. H. Sui, Y. L. Pei, E. Amzallag, F. Li, Y. Liu, and W. Xu for plentiful discussions and fruitful collaborations. Of course most of all, we are grateful to the numerous dedicated postdoctoral fellows (L. Pan et al.) and graduate students (J. Li and C. Barreteau et al.) who have contributed to our thermoelectric research. Their names appear in the various publications cited in this chapter.

References

1. Zhang, X. and L.-D. Zhao. 2015. Thermoelectric materials: Energy conversion between heat and electricity. *J. Materiomics* 1: 92–105.
2. Zhao, L.-D., V. P. Dravid, and M. G. Kanatzidis. 2014. The panoscopic approach to high performance thermoelectrics. *Energy Environ. Sci.* 7: 251–268.
3. Il Kim, S., K. H. Lee, H. A. Mun et al. 2015. Dense dislocation arrays embedded in grain boundaries for high-performance bulk thermoelectrics. *Science* 348: 109–114.
4. Liu, W.-S., Q. Zhang, Y. Lan et al. 2011.Thermoelectric property studies on Cu-doped n-type $Cu_xBi_2Te_{2.7}Se_{0.3}$ nanocomposites. *Adv. Energy Mater.* 1: 577–587.
5. Poudel, B., Q. Hao, Y. Ma et al. 2008. High-thermoelectric performance of nanostructured bismuth antimony telluride bulk alloys. *Science* 320: 634–638.
6. Chung, D.-Y., T. Hogan, P. Brazis et al. 2000. $CsBi_4Te_6$: A high-performance thermoelectric material for low-temperature applications. *Science* 287: 1024–1027.
7. Hsu, K. F., S. Loo, F. Guo et al. 2004. Cubic $AgPb_mSbTe_{2+m}$: Bulk thermoelectric materials with high figure of merit. *Science* 303: 818–821.
8. Androulakis, J., K. F. Hsu, R. Pcionek et al. 2006. Nanostructuring and high thermoelectric efficiency in p-type $Ag(Pb_{1-y}Sn_y)_mSbTe_{2+m}$. *Adv. Mater.* 18: 1170–1173.
9. Biswas, K., J. He, I. D. Blum et al. 2012. High-performance bulk thermoelectrics with all-scale hierarchical architectures. *Nature* 489: 414–418.
10. Pei, Y., X. Shi, A. LaLonde, H. Wang, L. Chen, and G. J. Snyder. 2011. Convergence of electronic bands for high performance bulk thermoelectrics. *Nature* 473: 66–69.
11. Heremans, J. P., V. Jovovic, E. S. Toberer et al. 2008. Enhancement of thermoelectric efficiency in PbTe by distortion of the electronic density of states. *Science* 321: 554–557.
12. Androulakis, J., I. Todorov, J. He, D.-Y. Chung, V. Dravid, and M. G. Kanatzidis. 2011. Thermoelectrics from abundant chemical elements: High-performance nanostructured PbSe-PbS. *J. Am. Chem. Soc.* 133: 10920–10927.
13. Saramat, A., G. Svensson, A. E. C. Palmqvist et al. 2006. Large thermoelectric figure of merit at high temperature in Czochralski-grown clathrate $Ba_8Ga_{16}Ge_{30}$. *J. Appl. Phys.* 99: 023708.
14. Shi, X., J. Yang, J. R. Salvador et al. 2011. Multiple-filled skutterudites: high thermoelectric figure of merit through separately optimizing electrical and thermal transports. *J. Am. Chem. Soc.* 133: 7837–7846.

15. Liu, W., X. Tan, K. Yin et al. 2012. Convergence of conduction bands as a means of enhancing thermoelectric performance of n-type $Mg_2Si_{1-x}Sn_x$ solid solutions. *Phys. Rev. Lett.* 108: 166601.
16. Zhao, L.-D., S.-H. Lo, Y. Zhang et al. 2014. Ultralow thermal conductivity and high thermoelectric figure of merit in SnSe crystals. *Nature* 508: 373–377.
17. Zhao, L.-D., G. Tan, S. Hao et al. 2015. Ultrahigh power factor and thermoelectric performance in hole-doped single-crystal SnSe. *Science* 351: 141–144.
18. Rhyee, J.-S., K. H. Lee, S. M. Lee et al. 2009. Peierls distortion as a route to high thermoelectric performance in $In_4Se_{3-\delta}$ crystals. *Nature* 459: 965–968.
19. Wang, X. W., H. Lee, Y. C. Lan et al. 2008. Enhanced thermoelectric figure of merit in nanostructured n-type silicon germanium bulk alloy. *Appl. Phys. Lett.* 93: 193121.
20. Liu, H., X. Shi, F. Xu et al. 2012. Copper ion liquid-like thermoelectrics. *Nat. Mater.* 11: 422–425.
21. Yu, B., W. Liu, S. Chen et al. 2012. Thermoelectric properties of copper selenide with ordered selenium layer and disordered copper layer. *Nano Energy* 1: 472–478.
22. Zhong, B., Y. Zhang, W. Li et al. 2014. High superionic conduction arising from aligned large lamellae and large figure of merit in bulk $Cu_{1.94}Al_{0.02}Se$. *Appl. Phys. Lett.* 105: 123902.
23. Zhao, L.-D., J. He, D. Berardan et al. 2014. BiCuSeO oxyselenides: New promising thermoelectric materials. *Energy Environ. Sci.* 7: 2900–2924.
24. Zhao, L.-D., D. Berardan, Y. L. Pei, C. Byl, L. Pinsard-Gaudart, and N. Dragoe. 2010. $Bi_{1-x}Sr_xCuSeO$ oxyselenides as promising thermoelectric materials. *Appl. Phys. Lett.* 97: 092118.
25. Sui, J., J. Li, J. He et al. 2013. Texturation boosts the thermoelectric performance of BiCuSeO oxyselenides. *Energy Environ. Sci.* 6: 2916–2920.
26. Pei, Y.-L., H. Wu, D. Wu, F. Zheng, and J. He. 2014. High thermoelectric performance realized in a BiCuSeO system by improving carrier mobility through 3D modulation doping. *J. Am. Chem. Soc.* 136: 13902–13908.
27. Barreteau, C., L. Pan, E. Amzallag, L.-D. Zhao, D. Berardan, and N. Dragoe. 2014. Layered oxychalcogenide in the Bi–Cu–O–Se system as good thermoelectric materials. *Semicond. Sci. Technol.* 29: 064001.
28. Palazzi, M., C. Carcaly, and J. A. Flahaut. 1980. A new ionic conductor (LaO)AgS. *J. Solid State Chem.* 35: 150–155.
29. Palazzi, M. and S. Jaulmes. 1981. Structure of the ionic conductor (LaO)AgS. *Acta Crystallogr. Sect. B Struct. Sci.* 37: 1337–1339.
30. Ohtani, T., M. Hirose, T. Sato, K. Nagaoka, and M. Iwabe. 1993. Synthesis and some physical-properties of a new series of layered selenides (LnO)CuSe (Ln = lanthanides). *Jpn. J. Appl. Phys. Part 1 Regular Papers Short Notes Rev. Papers* 32: 316–318.
31. Sekizawa, K., Y. Takano, K. Mori, and K. Yahagi. 1996. Magnetic and transport properties of layered oxysulfides $(La_{1-x}Ca_xO)Cu_{1-y}Ni_yS$ (y = 0 and y = x). *Czech. J. Phys.* 46: 1943–1944.
32. Takano, Y., C. Ogawa, Y. Miyahara, H. Ozaki, and K. Sekizawa. 1997. Single crystal growth of (LaO)CuS. *J. Alloys Compd.* 249: 221–223.
33. Ishikawa, K., S. Kinoshita, Y. Suzuki et al. 1991. Preparation and electrical-properties of (LaO) AgS and (LnO)CuS (Ln = La, Pr, or Nd). *J. Electrochem. Soc.* 138: 1166–1170.
34. Kusainova, A. M., P. S. Berdonosov, L. G. Akselrud, L. N. Kholodkovskaya, V. A. Dolgikh, and B. A. Popovkin. 1994. New layered compounds with the general composition (Mo)(CuSe), where M = Bi, Nd, Gd, Dy, and BiOCuS—Syntheses and crystal-structure. *J. Solid State Chem.* 112: 189–191.
35. Berdonosov, P. S., A. M. Kusainova, L. N. Kholodkovskaya, V. A. Dolgikh, L. G. Akselrud, and B. A. Popovkin. 1995. Powder x-ray and IR studies of the new oxyselenides MOCuSe (M = Bi, Gd, Dy). *J. Solid State Chem.* 118: 74–77.
36. Ueda, K., S. Inoue, S. Hirose, H. Kawazoe, and H. Hosono. 2000. Transparent p-type semiconductor: LaCuOS layered oxysulfide. *Appl. Phys. Lett.* 77: 2701–2703.
37. Inoue, S., K. Ueda, H. Hosono, and N. Hamada. 2001. Electronic structure of the transparent p-type semiconductor (LaO)CuS. *Phys. Rev. B* 64: 245211.

38. Hiramatsu, H., K. Ueda, H. Ohta, T. Kamiya, M. Hirano, and H. Hosono. 2005. Excitonic blue luminescence from p-LaCuOSe/n-InGaZn_5O_8 light-emitting diode at room temperature. *Appl. Phys. Lett.* 87: 211107.
39. Poettgen, R. and D. Johrend. 2008. Materials with ZrCuSiAs-type structure. *Zeitschrift fur Naturforschung Section B J. Chem. Sci.* 63: 1135–1148.
40. Kamihara, Y., H. Hiramatsu, M. Hirano et al. 2006. Iron-based layered superconductor: LaOFeP. *J. Am. Chem. Soc.* 128: 10012–10013.
41. Watanabe, T., H. Yanagi, T. Kamiya et al. 2007. Nickel-based oxyphosphide superconductor with a layered crystal structure, LaNiOP. *Inorg. Chem.* 46: 7719–7721.
42. Kamihara, Y., T. Watanabe, M. Hirano, and H. Hosono. 2008. Iron-based layered superconductor $LaO_{1-x}F_xFeAs$ (x = 0.05–0.12) with T_c = 26 K. *J. Am. Chem. Soc.* 130: 3296–3297.
43. Chen, X. H., T. Wu, G. Wu, R. H. Liu, H. Chen, and D. F. Fang. 2008. Superconductivity at 43 K in $SmFeAsO_{1-x}F_x$. *Nature* 453: 761–762.
44. Takahashi, H., K. Igawa, K. Arii, Y. Kamihara, M. Hirano, and H. Hosono. 2008. Superconductivity at 43 K in an iron-based layered compound $LaO_{1-x}F_xFeAs$. *Nature* 453: 376–378.
45. Wang, C., L. Li, S. Chi et al. 2008. Thorium-doping-induced superconductivity up to 56 K in $Gd_{1-x}Th_xFeAsO$. *Eur. Phys. Lett.* 83: 67006.
46. Berardan, D., L.-D. Zhao, C. Barreteau, and N. Dragoe. 2012. Low temperature transport properties of the BiCuSeO system. *Phys. Stat. Solidi A Appl. Mater.* 209: 2273–2276.
47. Pinsard-Gaudart, L., D. Berardan, J. Bobroff, and N. Dragoe. 2008. Large Seebeck coefficients in iron-oxypnictides: A new route towards n-type thermoelectric materials. *Phys. Stat. Solidi Rapid Res. Lett.* 2: 185–187.
48. Zhao, L.-D., D. Berardan, and N. Dragoe. 2010. Electrical transport properties of F-doped LaFeAsO oxypnictide. *J. Alloys Compd.* 508: 606–609.
49. Hiramatsu, H., H. Yanagi, T. Kamiya, K. Ueda, M. Hirano, and H. Hosono. 2008. Crystal structures, optoelectronic properties, and electronic structures of layered oxychalcogenides MCuOCh (M = Bi, La; Ch = S, Se, Te): Effects of electronic configurations of M^{3+} ions. *Chem. Mater.* 20: 326–334.
50. Sui, J. H., J. Li, J. Q. He et al. 2013. Texturation boosts the thermoelectric performance of BiCuSeO oxyselenides. *Energy Environ. Sci.* 6: 2916–2920.
51. Kamioka, H., H. Hiramatsu, M. Hirano, K. Ueda, T. Kamiya, and H. Hosono. 2005. Excitonic properties related to valence band levels split by spin-orbit interaction in layered oxychalcogenide LaCuOCh(Ch = S,Se). *J. Luminesc.* 112: 66–70.
52. Richard, A. P., J. A. Russell, A. Zakutayev, L. N. Zakharov, D. A. Keszler, and J. Tate. 2012. Synthesis, structure, and optical properties of BiCuOCh (Ch = S, Se, and Te). *J. Solid State Chem.* 187: 15–19.
53. Pei, Y. L., J. Q. He, J. F. Li et al. 2013. High thermoelectric performance of oxyselenides: intrinsically low thermal conductivity of Ca-doped BiCuSeO. *NPG Asia Mater.* 5: e47.
54. Liu, Y., J. Ding, B. Xu et al. 2015. Enhanced thermoelectric performance of La-doped BiCuSeO by tuning band structure. *Appl. Phys. Lett.* 106: 233903.
55. Lan, J.-L., B. Zhan, Y.-C. Liu et al. 2013. Doping for higher thermoelectric properties in p-type BiCuSeO oxyselenide. *Appl. Phys. Lett.* 102: 123905.
56. Li, F., T. R. Wei, F. Y. Kang, and J. F. Li. 2013. Enhanced thermoelectric performance of Ca-doped BiCuSeO in a wide temperature range. *J. Mater. Chem. A* 1: 11942–11949.
57. Lee, D. S., T. H. An, M. Jeong et al. 2013. Density of state effective mass and related charge transport properties in K-doped BiCuSeO. *Appl. Phys. Lett.* 103: 232110.
58. Kurosaki, K., A. Kosuga, H. Muta, M. Uno, and S. Yamanaka. 2005. Ag_9TlTe_5: A high-performance thermoelectric bulk material with extremely low thermal conductivity. *Appl. Phys. Lett.* 87(6): 061919.
59. Wan, C. L., W. Pan, Q. Xu et al. 2006. Effect of point defects on the thermal transport properties of $(La_xGd_{1-x})_2Zr_2O_7$: Experiment and theoretical model. *Phys. Rev. B* 74: 144109.
60. Wan, C. L., Z. X. Qu, Y. He, D. Luan, and W. Pan. 2008. Ultralow thermal conductivity in highly anion-defective aluminates. *Phys. Rev. Lett.* 101: 085901.

61. Morelli, D. T., V. Jovovic, and J. P. Heremans. 2008. Intrinsically minimal thermal conductivity in cubic I-V-VI(2) semiconductors. *Phys. Rev. Lett.* 101: 035901.
62. Nielsen, M. D., V. Ozolins, and J. P. Heremans. 2013. Lone pair electrons minimize lattice thermal conductivity. *Energy Environ. Sci.* 6: 570–578.
63. Zhao, L.-D., V. P. Dravid, and M. G. Kanatzidis. 2014. The panoscopic approach to high performance thermoelectrics. *Energy Environ. Sci.* 7: 251–268.
64. Huang, B. L. and M. Kaviany. 2006. Structural metrics of high-temperature lattice conductivity. *J. Appl. Phys.* 100: 123507.
65. Skoug, E. J. and D. T. Morelli. 2011. Role of lone-pair electrons in producing minimum thermal conductivity in nitrogen-group chalcogenide compounds. *Phys. Rev. Lett.* 107: 235901.
66. Clarke, D. R. 2003. Materials selection guidelines for low thermal conductivity thermal barrier coatings. *Surf. Coat Technol.* 163: 67–74.
67. Saha, S. K. 2015. Exploring the origin of ultralow thermal conductivity in layered BiOCuSe. *Phys. Rev. B* 92: 041202.
68. Ding, J., B. Xu, Y. Lin, C. Nan, and W. Liu. 2015. Lattice vibration modes of the layered material BiCuSeO and first principles study of its thermoelectric properties. *New J. Phys.* 17: 083012.
69. Vaqueiro, P., R. A. R. Al Orabi, S. D. N. Luu et al. 2015. The role of copper in the thermal conductivity of thermoelectric oxychalcogenides: Do lone pairs matter? *Phys. Chem. Chem. Phys.* 17: 31735–31740.
70. Shao, H., X. Tan, G.-Q. Liu, J. Jiang, and H. Jiang. 2015. A first-principles study on the phonon transport in layered BiCuOSe. *Sci. Rep.* 6: 21035.
71. Liu, Y., L.-D. Zhao, Y. C. Liu et al. 2011. Remarkable enhancement in thermoelectric performance of BiCuSeO by Cu deficiencies. *J. Am. Chem. Soc.* 133: 20112–20115.
72. Li, F., J. F. Li, L. D. Zhao et al. 2012. Polycrystalline BiCuSeO oxide as a potential thermoelectric material. *Energy Environ. Sci.* 5: 7188–7195.
73. Li, J., J. H. Sui, Y. L. Pei et al. 2012. A high thermoelectric figure of merit ZT > 1 in Ba heavily doped BiCuSeO oxyselenides. *Energy Environ. Sci.* 5: 8543–8547.
74. Lan, J.-L., Y.-C. Liu, B. Zhan et al. 2013. Enhanced thermoelectric properties of Pb-doped BiCuSeO ceramics. *Adv. Mater.* 25: 5086–5090.
75. Vaqueiro, P., G. Guelou, M. Stec, E. Guilmeau, and A. V. Powell. A copper-containing oxytelluride as a promising thermoelectric material for waste heat recovery. *J. Mater. Chem. A* 1: 520–523.
76. Li, J., J. H. Sui, C. Barreteau et al. 2013. Thermoelectric properties of Mg doped p-type BiCuSeO oxyselenides. *J. Alloys Compd.* 551: 649–653.
77. Liu, Y.-C., Y.-H. Zheng, B. Zhan et al. 2015. Influence of Ag doping on thermoelectric properties of BiCuSeO. *J. Eur. Ceram. Soc.* 35: 845–849.
78. Liu, Y., J. L. Lan, W. Xu et al. 2013. Enhanced thermoelectric performance of a BiCuSeO system via band gap tuning. *Chem. Commun.* 49: 8075–8077.
79. Liu, Y.-C., J.-L. Lan, B. Zhan et al. 2013. Thermoelectric properties of Pb-doped BiCuSeO ceramics. *J. Am. Ceram. Soc.* 96: 2710–2713.
80. Luu, S. D. N. and P. Vaqueiro. 2014. Thermoelectric properties of $BiOCu_{1-x}M_xSe$ (M = Cd and Zn). *Semicond. Sci. Technol.* 29: 064002.
81. Liu, Y., J. Liu, B. Zhang, and Y. Lin. 2014. Thermoelectric properties of Ni doped p-type BiCuSeO oxyselenides. In: Pan, W. and J. Gong (eds.). *High-Performance Ceramics* (Trans Tech Publications), Vol. 602–603, pp. 906–909.
82. An, T.-H., Y. S. Lim, H.-S. Choi et al. 2014. Point defect-assisted doping mechanism and related thermoelectric transport properties in Pb-doped BiCuOTe. *J. Mater. Chem. A* 2: 19759–19764.
83. Ren, G., S. Butt, C. Zeng et al. 2015. Electrical and thermal transport behavior in Zn-doped BiCuSeO oxyselenides. *J. Electron. Mater.* 44: 1627–1631.
84. Li, J., J. Sui, Y. Pei et al. 2014. The roles of Na doping in BiCuSeO oxyselenides as a thermoelectric material. *J. Mater. Chem. A* 2: 4903–4906.
85. Ren, G. K., S. Butt, Y. C. Liu et al. 2014. Enhanced thermoelectric performance of Zn-doped oxyselenides: $BiCu_{1-x}Zn_xSeO$. *Phys. Stat. Solidi A Appl. Mat.* 211: 2616–2620.

86. Li, F., T.-R. Wei, F. Kang, and J.-F. Li. 2014. Thermal stability and oxidation resistance of BiCuSeO based thermoelectric ceramics. *J. Alloys Compd.* 614: 394–400.
87. Li, Z., C. Xiao, S. Fan et al. 2015. Dual vacancies: An effective strategy realizing synergistic optimization of thermoelectric property in BiCuSeO. *J. Am. Chem. Soc.* 137: 6587–6593.
88. Ren, G.-K., J.-I Lan, S. Butt, K. J. Ventura, Y.-H. Lin, and C.-W. Nan. 2015. Enhanced thermoelectric properties in Pb-doped BiCuSeO oxyselenides prepared by ultrafast synthesis. *RSC Adv.* 5: 69878–69885.
89. Zhang, M., J. Yang, Q. Jiang et al. 2015. Multi-role of sodium doping in BiCuSeO on high thermoelectric performance. *J. Electron. Mater.* 44: 2849–2855.
90. Zhao, L.-D., S. Q. Hao, S. H. Lo et al. 2013. High thermoelectric performance via hierarchical compositionally alloyed nanostructures. *J. Am. Chem. Soc.* 135: 7364–7370.
91. Zhao, L.-D., J. Q. He, S. Q. Hao et al. 2012. Raising the thermoelectric performance of p-type PbS with endotaxial nanostructuring and valence-band offset engineering using CdS and ZnS. *J. Am. Chem. Soc.* 134: 16327–16336.
92. Zhao, L.-D. J. Q. He, C. I. Wu et al. 2012. Thermoelectrics with earth abundant elements: High performance p-type PbS nanostructured with SrS and CaS. *J. Am. Chem. Soc.* 134: 7902–7912.
93. Pele, V., C. Barreteau, D. Berardan, L.-D. Zhao, and N. Dragoe. 2013. Direct synthesis of BiCuChO-type oxychalcogenides by mechanical alloying. *J. Solid State Chem.* 203: 187–191.
94. Luu, S. D. N. and P. Vaqueiro. 2013. Synthesis, structural characterisation and thermoelectric properties of $Bi_{1-x}Pb_xOCuSe$. *J. Mater. Chem. A* 1: 12270–12275.
95. Tsukazaki, A., S. Akasaka, K. Nakahara et al. 2010. Observation of the fractional quantum Hall effect in an oxide. *Nat. Mater.* 9: 889–893.
96. Zou, D. F., S. H. Xie, Y.Y. Liu, J. G. Lin, and J. Y. Li. 2013. Electronic structures and thermoelectric properties of layered BiCuOCh oxychalcogenides (Ch = S, Se and Te): First-principles calculations. *J. Mater. Chem. A* 1: 8888–8896.
97. Zou, D. F., Y. Y. Liu, S. H. Xie, J. G. Lin, H. R. Zheng, and J. Y. Li. 2014. High pressure effect on the electronic structure and thermoelectric properties of BiCuSeO: First-principles calculations. *RSC Adv.* 4: 54819–54825.
98. Yang, J. M., G. Yang, G. B. Zhang, and Y. X. Wang. 2014. Low effective mass leading to an improved ZT value by 32% for n-type BiCuSeO: A first-principles study. *J. Mater. Chem. A* 2: 13923–13931.
99. Barreteau, C., D. Berardan, and N. Dragoe. 2015. Studies on the thermal stability of BiCuSeO. *J. Solid State Chem.* 222: 53–59.

10

Phase Diagram Study in n-*CoSb*$_3$ *Skutterudites*

Yinglu Tang, Chris Wolverton, and G. Jeffrey Snyder
Northwestern University

CONTENTS

10.1 Introduction

Phase diagrams are important in the study of thermoelectric materials in two essential aspects. First, thermoelectric performance is largely dependent on the carrier concentration, which is determined by the dopant solubility. To obtain a dopant solubility that properly optimizes carrier concentration, knowledge of the equilibrium phase diagram is required. Second, phase separation on different length scales [1–3] has been shown to have great ability in reducing the lattice thermal conductivity, thus enhancing thermoelectric performance. A variety of microstructures, as well as shape, size, and density of the different components, can be engineered to provide interfacial phonon scattering [4]. To control the solidification, precipitation, and decomposition reactions that provide these microstructures, knowledge of the phase diagram is indispensable.

Phase diagrams can be determined through either experiments or *ab initio* computations, each with different capabilities and accuracies. Experimental phase diagrams are more accurate at high temperature (where samples can be more easily equilibrated) and when significant differences in composition are compared. The ability of *ab initio* computations to predict defect stability can map small compositional changes in phase boundaries that would be next to impossible to measure experimentally. For instance, the phase diagram of Zn-Sb near Zn_4Sb_3 has been calculated using *ab initio* methods, which can explain the range of *p*-type compositions that can be deduced from carrier concentration measurements, specifically, $Zn_{3.9-\delta}Sb_3$ where $0.0025 < \delta < 0.0035$. The temperature dependence of the phase boundary explains the observation of nanometer-sized precipitates of Zn or ZnSb observed in some samples and even nanovoids due to retrograde solubility [5,6]. Such compositional differences of intrinsic defects might be impossible to measure but have profound effects on the electronic properties by controlling the carrier concentration.

Similar vacancy defects control the doping and therefore thermoelectric properties in the AZn_2Sb_2 Zintl phases [7] where A = Ca, Sr, Yb, and Eu. In this case, the small electronegativity difference of the A cation is enough to tune the vacancy concentration around the optimal carrier concentration needed for thermoelectric applications. Again, such compositional variation is exceedingly difficult to measure [8] but can be calculated using *ab initio* methods [7] that can explain the trends in electronegativity.

Small concentrations of impurities used as extrinsic dopants are possible but quite difficult to measure [9]. Here, again, *ab initio* calculations have proven to be exceedingly instructive by predicting fine details where experiments have large uncertainties [10]. *Ab initio* phase diagrams still have complications and limitations, particularly in quantitatively incorporating all sources of high-temperature entropic contributions (e.g., configurational and vibrational) [11]. Combining *ab initio* calculations with experimental values (such as combining Calphad and density functional theory [DFT] [12] or using experiments to guide the methods [13,14]) may give the best combination of predictive power and accuracy.

Single-phase and multiple-phase alloy compositions, such as IV–VI rock salt semiconductor alloy systems, are also of interest and much can be learned from the investigation of the coherent and incoherent phase stability by DFT calculations [15]. Nanostructures, or second-phase precipitates, formed by phase separation can improve the thermoelectric figure of merit in these alloys. The term "incoherent" indicates that the lattice parameters of coexisting phases are unconstrained and allowed to relax to their bulk, stress-free equilibrium values. Therefore, incoherent phase stability can be easily inferred from an equilibrium phase diagram. However, small nanoscale particles are often formed with lattice parameters coherent with the surrounding matrix, and the resulting large coherency strains in these alloys drastically alter the thermodynamics of phase stability. Owing to its metastable nature, the coherent phase stability is more difficult to assess experimentally. DFT calculations were thus used [15] to investigate the coherent and incoherent phase stability of the IV–VI rock salt semiconductor alloy systems. Though the thermodynamic ground state of these IV–VI pseudobinary systems is incoherent phase separation, the coherency strain energy represents a large fraction of the thermodynamic driving force for incoherent phase separation in all systems. The driving force for coherent phase separation is thus significantly reduced, and coherent nanostructures can only form at much lower temperatures [15].

In addition, DFT calculations can be used to analyze the various entropic contributions to the phase diagrams of these IV–VI rock salt systems. For instance, calculations of the thermodynamics of PbS-PbTe alloys [11] found that mixing energetics and configurational entropies alone were not sufficient to quantitatively explain the thermodynamic driving force for phase separation in these materials: first-principles calculations of the thermodynamics of phase separation overestimate the thermodynamic driving force for precipitation of PbS-rich nanostructures from PbS-PbTe alloys. DFT was recently used to reevaluate the thermodynamics of PbS-PbTe, including the effects of vibrational entropy in the free energy [11]. Incorporating the vibrational entropy of mixing reduces the calculated maximum miscibility gap temperature of PbS-PbTe, significantly reducing the error between calculated and experimental phase diagrams. Additionally, the vibrational spectra of PbS-PbTe solid solutions exhibit dynamic instabilities of S ions that corroborate reports of low-T ferroelectric-like phase transitions in solid solutions of PbS and PbTe, which are not present in either of the constituent compounds.

Although the determination of phase diagrams through experiments can be demanding, requiring the preparation of a large number of samples and longtime annealing in order to achieve the equilibrium state (e.g., in Ti-Ni-Sn and Ti-Ni-Sb ternary systems [16]), it still offers base knowledge in order to optimize thermoelectric performance. In this chapter, we focus on the experimental phase diagram study in n-$CoSb_3$ skutterudites, which demonstrates the added complexity due to filler addition that is also relevant to other systems beyond simple binary phase diagrams.

$CoSb_3$ skutterudites have body-centered cubic structure (space group $Im\bar{3}$), which allows one void per primitive cell Co_4Sb_{12} that can be filled by foreign atoms R, such as alkali, alkaline-earth, and rare-earth elements ($R_xCo_4Sb_{12}$, $0 < x < 1$) [17–23]. These filler atoms not only contribute to the reduction of the lattice thermal conductivity but also tune the carrier concentration by n-type doping. The solubility of each type of filler is usually considered as a single value and is often extracted by adding an excess amount of R into the $CoSb_3$ system, with impurity detection or deviation from Vegard's law as the experimental indication of maximum solubility [24,25]. Yet, as we show in this chapter here, by adding R into $CoSb_3$, a ternary system is under study, which leads to the breakdown of some common assumptions in determining the solubility limit in binary systems (e.g., impurity detection and Vegard's law), especially given the fact that the synthesized compositions are usually off the skutterudite stoichiometry due to either experimental uncertainty or excess Sb addition meant to compensate large Sb evaporation.

The complexity of R-Co-Sb ternary systems thus requires the following questions to be addressed. First, what is the solubility limit (equal to the commonly known filling fraction limit [FFL] when the void is the only soluble site) for each type of filler? Is it a single value? Second, when filler atom R is added to the $CoSb_3$ matrix, is the void the only soluble site? If not, what are the other possibilities? Third, how can we achieve precise doping control with large tolerance of synthesis uncertainty in a repeatable manner? This is important for large-scale commercialization of skutterudites in the future.

In Section 10.2, we describe briefly the methodology of how to apply phase diagram study in studying skutterudites. Sections 10.3 through 10.5 lay out the experimental results of several ternary systems (R-Co-Sb, with R = Yb, Ce, In, Ga), which answers the questions listed earlier. The results presented here are mainly based on the following published papers with copyright permission acquired: *Advanced Functional Materials* 23, 3194 (2013) [26]; *Energy and Environmental Science* 7, 812 (2014) [27]; *Journal of Materiomics* 1, 75 (2015) [28]; and *Nature Communications* 6, 7584 (2015) [29]. The readers are welcome to refer to these papers for more details. Section 10.6 discusses the influence of precipitation on the stability and lifetime performance of optimized compositions. Section 10.7 is a conclusion.

10.2 Methodology

The goal of phase diagram study of R-Co-Sb (R is the filler atom) ternary systems is to map out their isothermal sections at various temperatures lower than 874°C (peritectic temperature of $CoSb_3$ decomposition) [30]. From these isothermal sections, not only can phase regions near the skutterudite phase be revealed, but also the solubility of R, as well as its temperature dependence, can be extracted. To lower the experimental burden, only a subsection of the full isothermal section with the $CoSb_3$ phase included is studied, for example, $YbSb_2$-$CoSb_2$-Sb rather than Yb-Co-Sb. The methodology can be illustrated in detail with the Yb-Co-Sb ternary system as an example [28,31]. It can be summarized as follows: (1) the phase regions of an isothermal section can be estimated based on known binary phase diagram information; nominal compositions in different estimated phase regions are then synthesized. (2) Phase identification (x-ray diffraction [XRD] and scanning electron microprobe) and chemical composition characterization (electron probe microanalysis [EPMA]) are performed to determine phase region relations (shown in Figure 10.1). The solubility of a filler atom can thus be extracted from the determined phase diagram. (3) By varying the selected temperature of an isothermal section, the temperature dependence of the filler solubility can be obtained (shown in Figure 10.2), which gives another degree of

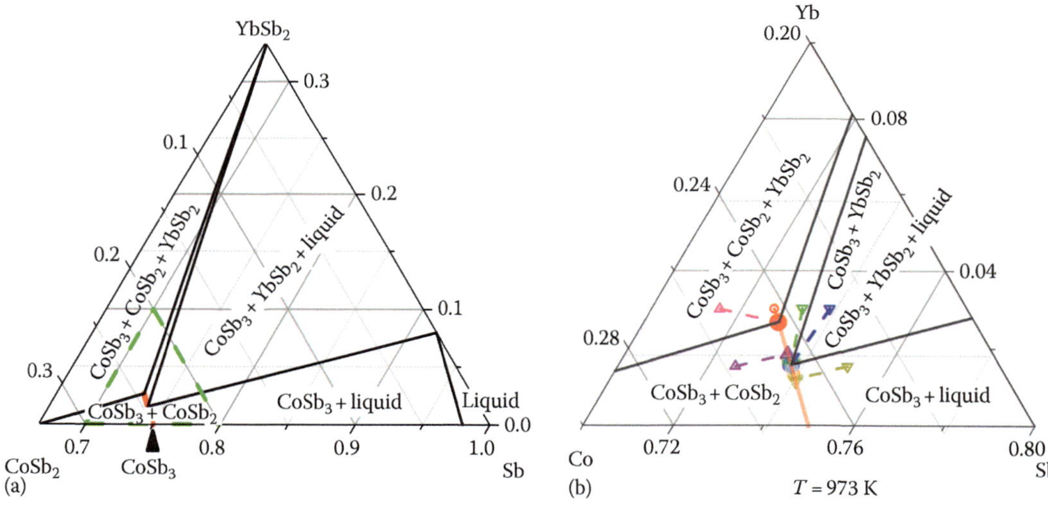

FIGURE 10.1
Phase regions near $CoSb_3$ in a section of the Yb-Co-Sb isothermal ternary phase diagram at 973 K [28]. (a) $YbSb_2$-$CoSb_2$-Sb where the red line represents the solubility region of $CoSb_3$. (b) Magnification of phase regions near $CoSb_3$. The solid red line indicates the void filling in the direction $(Yb_{VF})_xCo_4Sb_{12}$. The red point represents the stable composition (with maximum solubility $x = 0.44 \pm 0.01$) of the skutterudite phase obtained when the nominal composition is in the Co-rich three-phase region ($YbSb_2$, $CoSb_2$, and $Yb_xCo_4Sb_{12}$). The blue point represents the stable composition of the skutterudite phase obtained when the nominal composition is in the Sb-rich three-phase region ($YbSb_2$, liquid Sb, and $Yb_xCo_4Sb_{12}$). The nominal sample compositions are shown as open symbols; corresponding solid symbols represent the composition of the skutterudite majority phase using the experimental electron probe microanalysis value for the content of Yb and are positioned on the red line. (Reproduced from *J. Mater.*, 1, Tang, Y.L., Chen, S.W., and Snyder, G.J., Temperature dependent solubility of Yb in Yb-$CoSb_3$ skutterudite and its effect on preparation, optimization and lifetime of thermoelectrics, 74, Copyright 2015, with permission from Elsevier.)

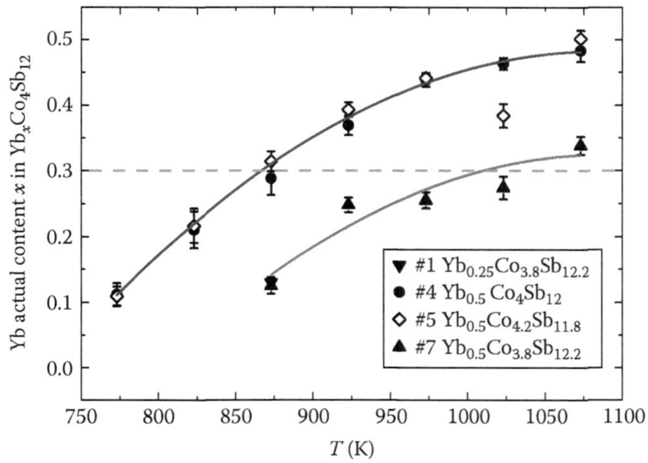

FIGURE 10.2
Temperature dependence of the solubility limit of Yb in Yb-doped skutterudites $Yb_xCo_4Sb_{12}$. The upper and lower lines indicate the temperature dependence of the actual Yb content in the stable skutterudite composition, represented as a red point (maximum solubility) and blue point, respectively, in Figure 10.1b. The dashed line is the actual Yb content ($x = 0.3$) targeted for optimized thermoelectric properties, which crosses the solubility lines at 873 and 1023 K, respectively. (Reproduced from *J. Mater.*, 1, Tang, Y.L., Chen, S.W., and Snyder, G.J., Temperature dependent solubility of Yb in Yb-$CoSb_3$ skutterudite and its effect on preparation, optimization and lifetime of thermoelectrics, 74, Copyright 2015, with permission from Elsevier.)

freedom to tune the carrier concentration. (4) Synthesis parameters are based on the determined phase diagram information and former knowledge of optimizing the material system under study (illustrated in Figure 10.3). (5) Optimized compositions are synthesized and thermoelectric performance of these samples is characterized (shown in Figure 10.4).

With compositions of the equilibrium phases determined for each nominal composition, the phase regions near $CoSb_3$ can be identified, as shown in Figure 10.1. The solubility limit of R in $CoSb_3$ can be found as the composition on the furthest point of the $CoSb_3$ phase region extending toward R (the red dot in Figure 10.1b). The solubility limit of Yb in $Yb_xCo_4Sb_{12}$ at 973 K is determined to be $x = 0.44 \pm 0.01$.

As temperature increases, Yb solubility increases by about a factor of five between 773 and 1073 K ($x = 0.11 \pm 0.02$ at 773 K and $x = 0.49 \pm 0.02$ at 1073 K). This large variance gives us another degree of freedom to tune the carrier concentration and thus allows us to optimize thermoelectric properties.

To determine synthesis parameters for optimized compositions, first an annealing temperature at which the solubility line crosses the optimized composition line needs to be determined, as shown in Figure 10.2. Then, the corresponding phase region is chosen such that the resulting stable compositions will produce optimized doping content. For example, if the annealing temperature in the Yb-Co-Sb system is chosen to be 873 K, then the nominal composition (red triangular and diamond samples in Figure 10.3a) should stay in the Co-rich three-phase region ($YbSb_2$, $CoSb_2$, and $Yb_xCo_4Sb_{12}$). If the annealing temperature is chosen to be 1023 K, then the nominal composition (blue circle sample in Figure 10.3b) should stay in the Sb-rich three-phase region ($YbSb_2$, liquid Sb, and $Yb_xCo_4Sb_{12}$).

Thermoelectric performance does not vary much among different optimized compositions. Thus, it enables the production of optimized skutterudites at an easily reproducible and commercial scale.

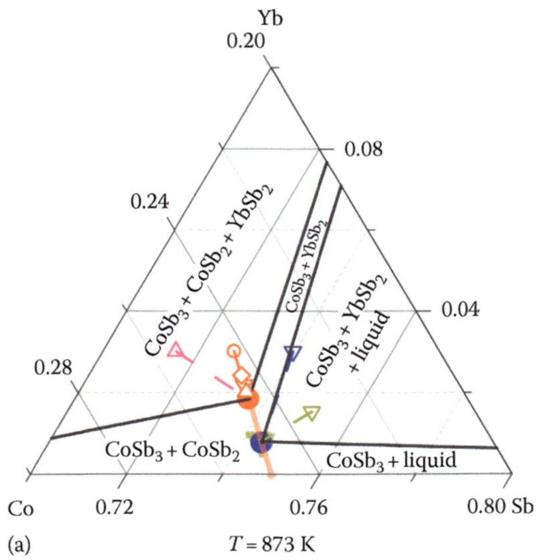

FIGURE 10.3
Magnification of phase regions near $CoSb_3$ of isothermal sections of the Yb-Co-Sb ternary phase diagram system at (a) 873 K and (b) 1023 K. (a) Samples with nominal compositions $Yb_xCo_4Sb_{12}$ (x = 0.33, 0.36, 0.40, empty red symbols on the solubility line) are in the Co-rich three-phase region and annealed at 873 K, which gives the desired optimum Yb content x = 0.30 (red dot). (b) Sample with nominal composition $Yb_xCo_4Sb_{12.2}$ (x = 0.33, blue circle) is in the Sb-rich three-phase region and annealed at 1023 K, which gives the desired optimum Yb content x = 0.30 (blue dot). (Reproduced from *J. Mater.*, 1, Tang, Y.L., Chen, S. W., and Snyder, G.J., Temperature dependent solubility of Yb in Yb-$CoSb_3$ skutterudite and its effect on preparation, optimization and lifetime of thermoelectrics, 74, Copyright 2015, with permission from Elsevier.)

10.3 Stable Compositions and Vegard's Law

In Section 10.2 we mentioned the concept of stable compositions. Now, we discuss in detail what stable compositions (invariant point) are and how they can be both misleading and powerful in guiding solubility design. We will take the Yb-Co-Sb system as an example for illustration.

10.3.1 Solubility Debate regarding Yb

Among various kinds of filled skutterudite materials, Yb-doped skutterudites have shown good thermoelectric properties due to the heavy mass and small radius of the Yb atom compared to other fillers and have been selected by many automotive companies, such as General Motors, as the most promising thermoelectric material for waste heat recovery [35]. Nolas et al. first reported a zT value of 1.0 for $Yb_{0.19}Co_4Sb_{12}$ at 600 K [23]. Composites containing Yb-filled $CoSb_3$ and well-distributed Yb_2O_3 particles were synthesized by Zhao et al. with a zT of 1.3 for $Yb_{0.25}Co_4Sb_{12}/Yb_2O_3$ at 850 K [32]. Later, a zT of 1.3 for $Yb_{0.285}Co_4Sb_{12.268}$ at 800 K was reported by Li et al. [36].

Despite the good thermoelectric performance of Yb-doped skutterudites, there has been a debate about the Yb solubility limit in the skutterudite phase. Because of the size and chemical difference between Yb and Co or Sb atoms, substitution of Yb for Co or Sb is not likely. Instead, the Yb atoms fit nicely into the void filling site, and thus, the solubility limit

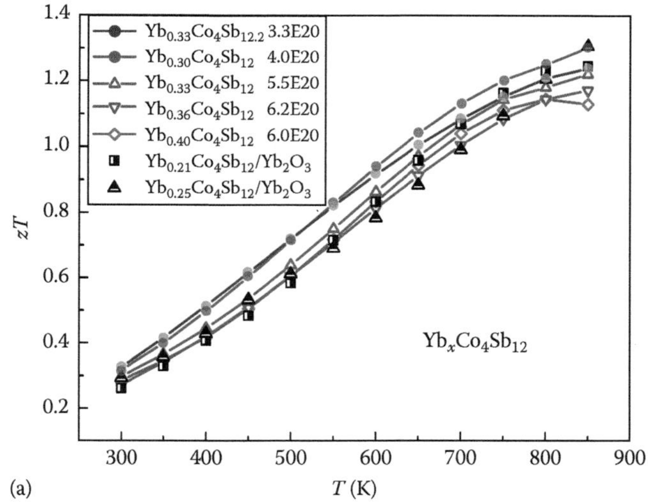

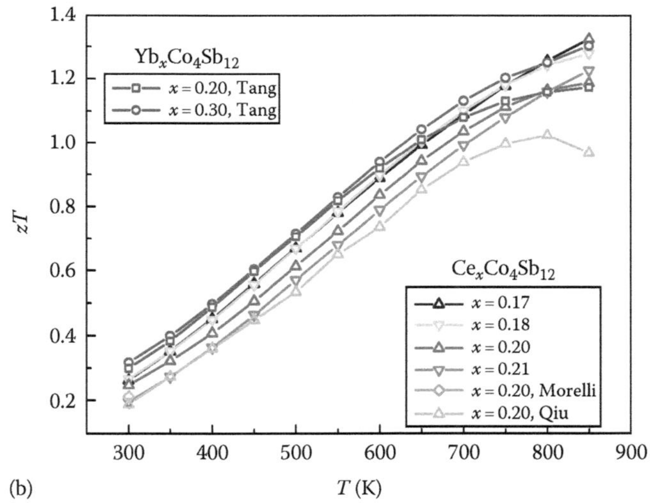

FIGURE 10.4
Temperature dependence of the thermoelectric figure of merit zT for optimized (a) Yb-$CoSb_3$ skutterudites and (b) Ce-$CoSb_3$ skutterudites [29]. (The reference data are from Morelli, D.T. et al., *Phys. Rev. B*, 56, 7376, 1997; Zhao, X.Y. et al., *Appl. Phys. Lett.*, 8, 092121, 2006; Qiu, P.F. et al., *Appl. Phys. Lett.*, 103, 062103, 2013; Tang, Y.L. et al., *Nat. Mater.*, 14, 1223, 2015; Reproduced from *J. Mater.*, 1, Tang, Y.L., Chen, S. W., and Snyder, G.J., Temperature dependent solubility of Yb in Yb-$CoSb_3$ skutterudite and its effect on preparation, optimization and lifetime of thermoelectrics, 74, Copyright 2015, with permission from Elsevier.)

discussed here corresponds to the FFL. Theoretically, the FFL was calculated by Mei et al. [37] to be 0.30 for Yb at 0 K; Shi et al. [38] studied the temperature dependence of FFLs of fillers in $CoSb_3$ skutterudites using a plane wave density functional method. By combining the formation energy of inserting an impurity R into the intrinsic void of $CoSb_3$ and that of secondary phases RSb_2 and $CoSb_2$, it was found that, theoretically, the FFL has a linear dependence on temperature, with $x = 0.2$ when R = Yb at 1000 K, which is close to the experimental value, but the temperature dependence was not demonstrated experimentally.

Dilley et al. synthesized $Yb_xCo_4Sb_{12}$ by induction heating samples at 1050°C for 12 min and then annealing at 600°C for 35 h [25]. Lattice constant values were obtained from the powder XRD data and plotted against nominal *x*, from which a Yb solubility limit of about $x = 0.22$ was extrapolated. Xia et al. synthesized $Yb_xCo_4Sb_{12}$ with nominal *x* up to 0.4 by melting samples at 1077°C for 10 h and subsequently annealing at 800°C for 7 days and observed no impurity phases [39]. The addition of oxygen to the system results in less filling due to the formation of Yb oxides [40]. In addition to traditional preparation methods (melting and longtime annealing), nonequilibrium synthesis methods were also adopted and gave different solubility limits, such as $x = 0.29$ by a combination of melt quenching with melt spinning (MMS) method by Li et al. [36] and $x = 0.5$ by ball milling by Yang et al. [41].

The equilibrium isothermal sections of the ternary phase diagram of the Yb-Co-Sb system at various temperatures were proposed by Tang et al. [28] based on the knowledge of the related binary phase diagrams and the microstructure of the samples [26,27,30,42,43]. The FFL is derived from the ternary phase diagram and, especially its temperature dependence, naturally explains the solubility debate and the confusion concerning the various reports of the maximum solubility of Yb in the skutterudite phase. It also leads to a strategy to synthesize optimized, high *zT* material.

10.3.2 Skutterudite Lattice Expansion due to Yb Filling

Lattice constants derived from the powder XRD data are plotted against Yb actual content determined from EPMA, in Figure 10.5a. There is a good linear relationship between the actual Yb content and the lattice constant of the skutterudite phase, which agrees with the

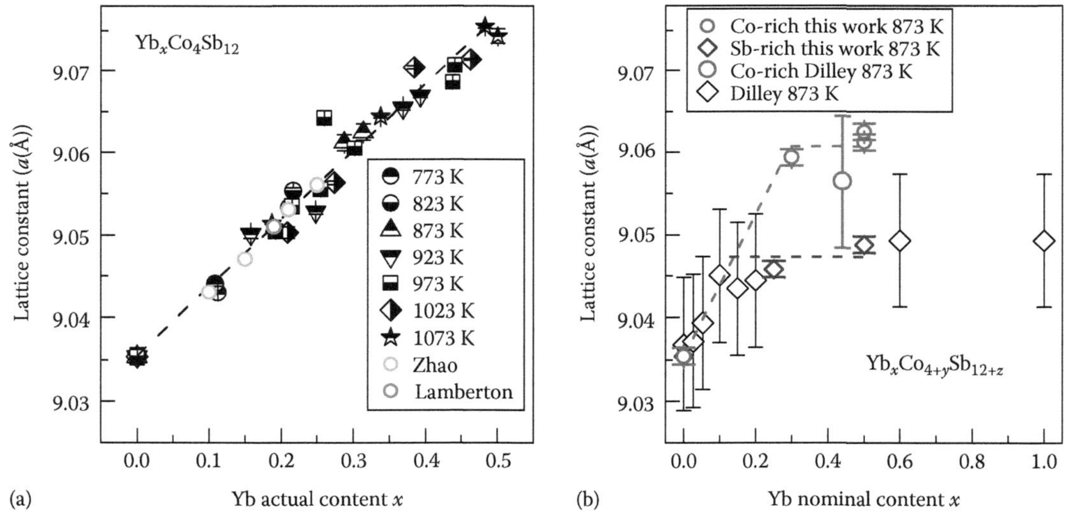

FIGURE 10.5
(a) Dependence of the lattice constant on the actual Yb content, *x*, in $Yb_xCo_4Sb_{12}$. (b) Comparison with samples annealed at 873 K by Dilley et al. [25]. The gray dashed line shows the increase of the lattice constant followed by a plateau when *x* is larger than 0.30 in Co-rich samples, which indicates the maximum filling fraction limit at 873 K (red point in Figure 10.3a). The dark dashed line shows another plateau of the lattice constant indicating a different stable composition $Yb_{0.13}Co_4Sb_{12}$ (blue point in Figure 10.3a) in Sb-rich samples. (Reproduced from *J. Mater.*, 1, Tang, Y.L., Chen, S.W., and Snyder, G.J., Temperature dependent solubility of Yb in Yb-$CoSb_3$ skutterudite and its effect on preparation, optimization and lifetime of thermoelectrics, 74, Copyright 2015, with permission from Elsevier.)

literature data and is consistent with Yb going into the same site across the whole range (up to $x = 0.5$), presumably the void site [32,44]. Lattice constants are also plotted against the nominal Yb content in order to compare with Reference 25 in Figure 10.5b. The two plateaus of lattice constants indicate that there are two stable skutterudite compositions at 873 K (explained in Section 10.3.3 from the Gibbs phase rule). The gray dashed line shows the increase of the lattice constant followed by a plateau when x is larger than 0.30 and when the nominal composition is in a Co-rich three-phase region. Here, the Yb content in the skutterudite phase reaches the maximum FFL at 873 K (red point in Figure 10.3a). The dark dashed line shows another plateau of the lattice constant when the nominal composition is in an Sb-rich three-phase region and where the skutterudite phase has a stable composition with lower Yb content ($Yb_{0.13}Co_4Sb_{12}$, blue point in Figure 10.3a). Moreover, for the $x = 0.44$ sample from Reference 25, which has a larger lattice constant than other samples, $YbSb_2$ and $CoSb_2$ impurities are present according to the XRD data, which agrees with our conclusion that the skutterudite phase with the maximum FFL coexists with $CoSb_2$ and $YbSb_2$ and is thus in a Co-rich three-phase region.

10.3.3 Stable Compositions in the Ternary Phase Diagram System

When the nominal composition contains higher Yb content than its solubility limit, it is in a three-phase region between $Yb_xCo_4Sb_{12}$, $YbSb_2$, and $CoSb_2$ (Figure 10.1). According to the phase rule, there are $F = C - P + 0 = 3 - 3 = 0$ degrees of freedom, where C is the number of components, P is the number of phases, and 0 indicates both temperature and pressure are fixed. Thus, the compositions of the three equilibrium phases in a three-phase region are fixed, and all the nominal compositions in this three-phase region (such as all the scattered empty circles in Figure 10.6a) will produce the same

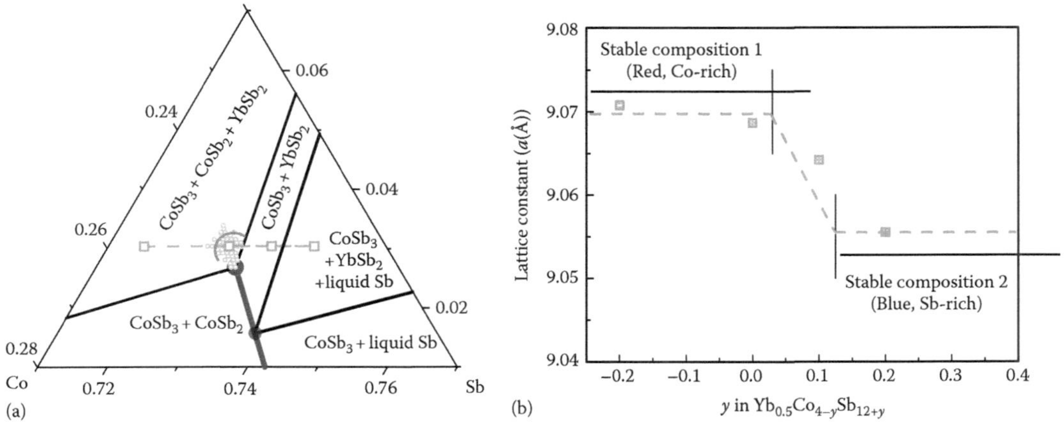

FIGURE 10.6
(a) Magnified region of the isothermal section at 973 K near $CoSb_3$ of the Yb-Co-Sb ternary phase diagram system with two stable skutterudite compositions (upper and lower points) and solubility line (thick gray line). Scattered points (empty circles) indicate possible synthesis error near the stable composition (upper point) in the three-phase region of $CoSb_3$, $CoSb_2$, and $YbSb_2$. (b) Dependence of the lattice constant derived from x-ray diffraction data on the Co/Sb ratio for nominal compositions with the same Yb content $x = 0.5$ marked as empty rectangles in (a). (Reproduced from *J. Mater.*, 1, Tang, Y. L., Chen, S. W., and Snyder, G. J., Temperature dependent solubility of Yb in Yb-CoSb$_3$ skutterudite and its effect on preparation, optimization and lifetime of thermoelectrics, 74, Copyright 2015, with permission from Elsevier.)

skutterudite composition with the same actual Yb content (upper point in Figure 10.6a). This is indicated by a plateau of the lattice constant marked as *stable composition* 1 from Figure 10.6b. At 973 K it was also confirmed by the EPMA measurements to be $x = 0.44 \pm 0.01$ ($Yb_{0.44}Co_4Sb_{12}$).

When the nominal composition is slightly Sb rich (e.g., $Yb_{0.5}Co_{3.9}Sb_{12.1}$ at 973 K), it moves into a two-phase region between $Yb_xCo_4Sb_{12}$ and $YbSb_2$. Due to the phase rule, the degrees of freedom $F = C - P + 0 = 3 - 2 = 1$, which means that the compositions of phases at equilibrium in this two-phase region are not fixed, but vary depending on the exact starting composition within this region. As a result, the Yb content in the skutterudite phase decreases as the nominal composition becomes more Sb rich, which is confirmed by the decrease of the lattice constant in Figure 10.6b.

However, when the nominal composition is even richer in Sb (e.g., $Yb_{0.5}Co_{3.8}Sb_{12.2}$ at 973 K), the nominal composition moves into another three-phase region between $Yb_xCo_4Sb_{12}$, $YbSb_2$, and liquid Sb. Thus, the compositions of phases at equilibrium are fixed again and all the nominal compositions in this three-phase region will produce the same skutterudite composition with the same actual Yb content ($Yb_{0.26}Co_4Sb_{12}$), albeit at a much lower Yb content (lower point in Figure 10.6a). This was confirmed by a lower plateau of lattice constant marked as *stable composition* 2 in Figure 10.6b, and the EPMA measurements verified this composition to be $x = 0.26 \pm 0.02$.

In the process of synthesizing samples, exact element control is not cost practical and thus the influence of impurity elements from low-cost starting material will lead to variations in the Co/Sb ratio and Yb content as well, shown as scattered empty circles in Figure 10.6a. If these scattered circles spill out of the three-phase region and enter into the two-phase region of $CoSb_3$ and $YbSb_2$, there will be a sharp decrease of Yb content in the skutterudite phase, resulting in a sharp decrease in the lattice constant as well, which is illustrated in Figure 10.6b.

In order to have precise control of the Yb content and thus the carrier concentration in the skutterudite phase, nominal compositions can be easily designed to remain in a three-phase region to make sure that the skutterudite phase has a stable composition. The existence of these stable compositions, as well as their temperature dependence, can thus be applied to facilitate the synthesis of optimized skutterudites for low-cost mass production.

10.3.4 Vegard's Law in the Ternary Phase Diagram System

In materials science, Vegard's law is the empirical rule that describes the relationship between the lattice parameter of a solid solution of two constituents and those of the pure constituents. According to Vegard, the lattice parameter of a solid solution should be approximately equal to a linear combination of the two constituents' lattice parameters at the same temperature given that both pure constituents have the same crystal structure [45]:

$$a_{A_{1-x}B_x} = (1-x)a_A + xa_B \tag{10.1}$$

where $a_{A_{1-x}B_x}$ is the lattice parameter of the solid solution, a_A and a_B are the lattice parameters of the pure constituents A and B, respectively, x is the molar fraction of B in the solution.

In our case, A is the unfilled $CoSb_3$, whereas B is the fully filled $YbCo_4Sb_{12}$, which does not exist in reality due to the solubility limit of Yb. For partially filled $Yb_xCo_4Sb_{12}$, it can be regarded as

$$Yb_xCo_4Sb_{12} = (1-x)Co_4Sb_{12} + x(YbCo_4Sb_{12}) \tag{10.2}$$

Thus, we should be able to extract a linear relationship between the lattice parameter of partially filled $Yb_xCo_4Sb_{12}$ and the filling fraction x, as shown in the following equation:

$$a_{Yb_xCo_4Sb_{12}} = (1-x)a_{Co_4Sb_{12}} + xa_{YbCo_4Sb_{12}} = a_{Co_4Sb_{12}} + x(a_{YbCo_4Sb_{12}} - a_{Co_4Sb_{12}}) \tag{10.3}$$

This equation is often applied to determine the FFL of fillers. From an experimental perspective, FFL is determined when the lattice parameter deviates from Vegard's law as the nominal filling amount x increases. Mostly, this deviation is observed as a plateau [46,47]; however, sometimes it is also observed as a linear behavior with a different slope [24,48,49]. This method is valid for binary phase diagram systems but is not always true for ternary phase diagram systems.

In a binary system, as shown in Figure 10.7, when increasing the amount x of B in A at a given temperature, the nominal composition experiences two-phase regions: first a single-phase region of the solid solution α and then a two-phase region of α and β when the amount of x exceeds the solubility limit (filled circle). In a single-phase region of α, due to the Gibbs phase rule, the degrees of freedom $F = C - P + 0 = 2 - 1 = 1$; the lattice parameter of α changes linearly with the addition of $B(x)$. In the two-phase region, the degree of freedom $F = C - P + 0 = 2 - 2 = 0$; the lattice parameter of phase α at equilibrium is fixed due to the fixed amount of B in phase α, indicated by the filled circle in Figure 10.7 (solubility limit of B in A). Thus, a linear relationship followed by a plateau will be observed between the lattice parameter of phase α and nominal content x of B.

However, in a ternary system, for example, R-Co-Sb, it is more complicated with an additional degree of freedom due to the thickness of the solubility region of $R_xCo_4Sb_{12}$, as shown in Figure 10.8. In a ternary system, the isothermal section is a surface rather

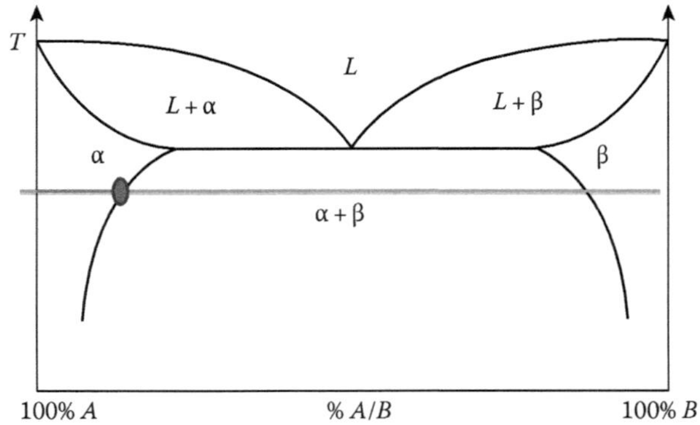

FIGURE 10.7
Illustration of a typical eutectic phase diagram of binary *A–B* system. The gray line represents the isothermal line. The section between 100% *A* and filled circle represents the solubility region of *B* in *A*. The filled circle represents the solubility limit of *B* in *A* at the selected temperature.

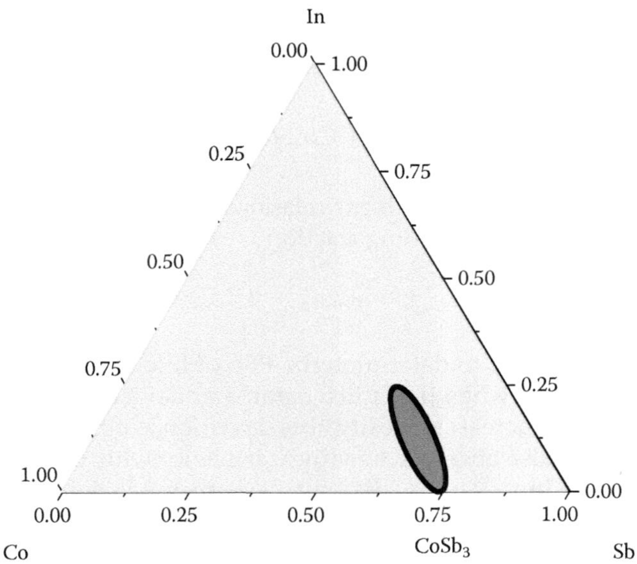

FIGURE 10.8
Illustration of an isothermal section of the ternary In-Co-Sb phase diagram. The light gray triangle represents the isothermal section. The dark gray area represents the solubility region of R-$CoSb_3$. The dark boundary represents the solubility limit of R in $CoSb_3$ at the selected temperature.

than a line in a binary system and so is the solubility region. The solubility limit also changes from a single value (represented by the filled circle in Figure 10.7) in a binary system to a boundary represented by the dark contour of the enclosed dark gray solubility region in Figure 10.8. The experimentally determined solubility limit is thus largely dependent on the phase regions adjacent to the skutterudite phase where the nominal compositions are, as illustrated in Figure 10.9. For Vegard's law to be applicable in a ternary system, the nominal compositions should be along the solubility direction, for example, $(In_{VF})_{2x/3}Co_4Sb_{12-x/3}(In_{Sb})_{x/3}$ – $CoSb_3$ (red solid line in Figure 10.17b), so it can be considered as in a pseudobinary system. If nominal compositions are off the solubility direction, for example, in the $(In_{VF})_xCo_4Sb_{12}$ – $CoSb_3$ for the In-Co-Sb system (purple solid line in Figure 10.17b) and the $(Yb_{VF})_xCo_4Sb_{12.2}$ – Co_4Sb_{12} for Yb-Co-Sb system (dashed line in Figure 10.10a), then deviations from Vegard's law can result from nominal compositions entering either two-phase (2-p in Figure 10.9) or three-phase regions (3-p′ in Figure 10.9) other than the one that will give the maximum solubility limit (x_{max}, Figure 10.9). As a result, the solubility limit observed will be smaller than the maximum solubility limit because its determination is misled by either the slope change or a false plateau in the relationship between the lattice parameter and the nominal filler content x.

Figure 10.10 serves as a good example to illustrate the inapplicability of Vegard's law when the nominal compositions are off the solubility direction in a ternary Yb-Co-Sb system. It demonstrates that reaching a plateau of the lattice constant could be because of the existence of stable compositions and thus does not necessarily imply reaching the solubility limit.

Excess of Sb is often added when synthesizing $CoSb_3$ skutterudites due to the high volatility of Sb. In samples with even a slight Sb excess (depicted in Figure 10.10a), increasing the nominal Yb content will lead to samples in the Sb-rich three-phase region of $Yb_xCo_4Sb_{12}$, $YbSb_2$, and liquid Sb first and then samples in the two-phase region of $Yb_xCo_4Sb_{12}$ and

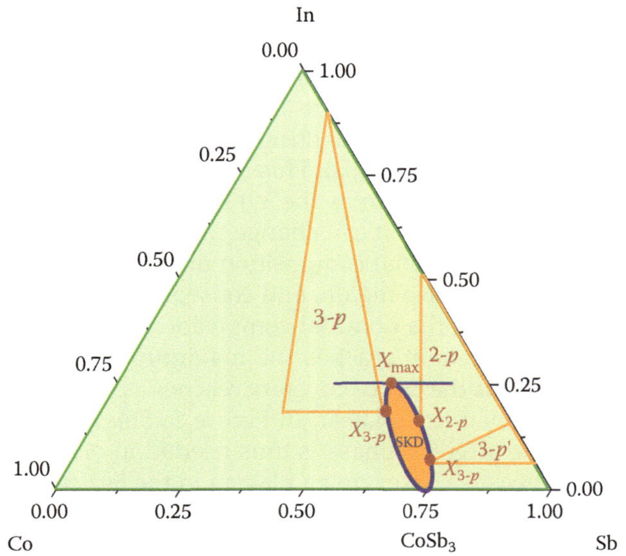

FIGURE 10.9
Dependence of the solubility limit on adjacent phase regions in a ternary phase diagram system. Triangles enclosed in yellow lines are three-phase regions (on the Co-rich side, 3-*p*; on the Sb-rich side, 3-*p*′), with $x_{3\text{-}p}$ and $x_{3\text{-}p'}$ indicating, respectively, the solubility limit of R in the equilibrium skutterudite phase when nominal compositions are in these phase regions. A two-phase region (2-*p*) is located right above the Sb-rich three-phase region. The solubility limit of R in the equilibrium skutterudite phase when nominal compositions are in this two-phase region is represented by the red dot marked at $x_{2\text{-}p}$. The maximum solubility limit x_{max} is represented by the red dot that is the furthest point along the elongated direction (solubility direction).

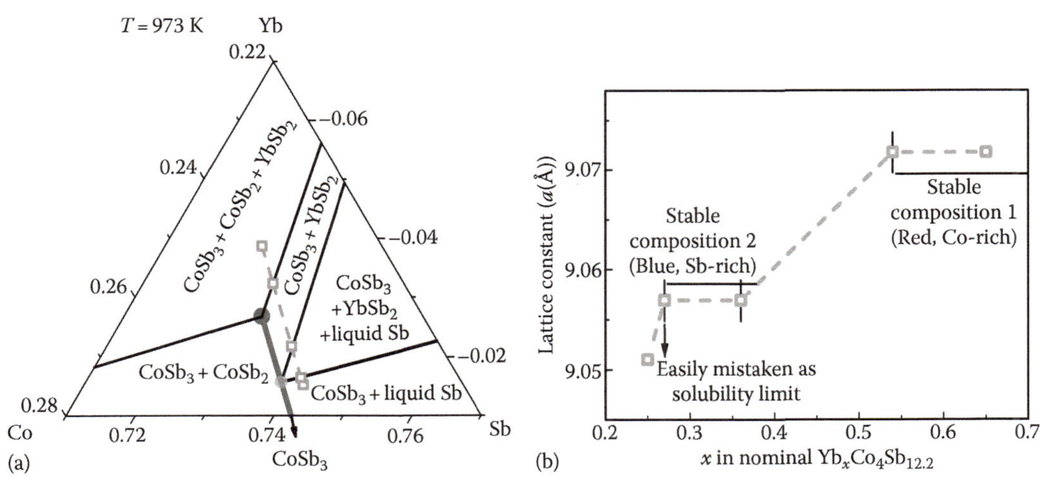

FIGURE 10.10
Samples with different nominal Yb content x in $Yb_xCo_4Sb_{12.2}$ (marked as empty rectangles) but the same Sb excess lead to a nonlinear dependence of lattice constant (b) due to the sample traversing different two- and three-phase regions of the phase diagram (a). (Reproduced from *J. Mater.*, 1, Tang, Y.L., Chen, S.W., and Snyder, G.J., Temperature dependent solubility of Yb in Yb-$CoSb_3$ skutterudite and its effect on preparation, optimization and lifetime of thermoelectrics, 74, Copyright 2015, with permission from Elsevier.)

$YbSb_2$ before samples reach the Co-rich three-phase region of $Yb_xCo_4Sb_{12}$, $YbSb_2$, and $CoSb_2$. Thus, the actual Yb content in Yb-filled $CoSb_3$, as measured by the lattice constant of the skutterudite phase, will stop increasing and be constant for a period (first plateau in Figure 10.10b or dark dashed line in Figure 10.5b), while the nominal composition is in the Sb-rich three-phase region and the resulting skutterudite phase has stable composition (lower light gray point in Figure 10.10a). This Yb content, at which the lattice constant stops increasing, can be easily mistaken as the Yb solubility limit [25]. However, as seen from Figure 10.10b, further Yb addition can change the Yb actual content in the skutterudite phase dramatically as the nominal composition moves away from the Sb-rich three-phase region and into the two-phase region, and consequently the lattice constant would resume the increasing trend. As the nominal composition moves into the Co-rich three-phase region, the actual Yb content reaches the maximum solubility (upper dark gray point in Figure 10.10a) and thus the lattice constant reaches a plateau again (second plateau in Figure 10.10b or gray dashed line plateau in Figure 10.5b). An overall investigation of the phase relations near the targeted phase is thus needed in order to make a conclusion about the solubility limit because observing a plateau in the lattice constant is insufficient in determining the solubility limit in ternary systems, nor does observing impurity phases help (the nominal composition could be in any two- or three-phase region).

10.4 Solubility Design Strategies

So far, we have discussed the difference of solubility in binary and ternary phase diagram systems (Section 10.3). It has to be noted that the dependence of the observed solubility on the phase regions adjacent to the skutterudite phase, as well as the temperature dependence of the solubility, might be the reason for many solubility debates.

Three solubility design strategies could then be proposed regarding high performance thermoelectric materials design. First of all, carefully determine the right solubility direction (see more in Section 10.5). Second, choose the phase region that will result in the maximum filler solubility limit. Third, vary the annealing temperature. Among the strategies mentioned, the large temperature dependence of the solubility especially expands our choice of optimized compositions for thermoelectric application, which can be helpful in changing thermoelectric materials with a dopant solubility barrier into promising ones instead. We will take Ce-doped $CoSb_3$ as an example for illustration in this section.

10.4.1 Solubility Barrier of Ce in Ce-$CoSb_3$

The intermediate temperature range (400°C–600°C) at which filled skutterudites have optimum thermoelectric performance makes this material a perfect candidate for automotive industry applications. As discussed previously, Yb-$CoSb_3$ skutterudites are so far the most promising thermoelectric materials for this application [50]. However, the deficiency of the rare-earth element Yb (its abundance in the earth's crust is only 3 ppm) could potentially be a problem in large-scale commercialization. Unlike Yb, Ce is the most earth abundant element of the rare earths (its abundance in the earth's crust is 68 ppm) [51]. In fact, there are huge resources of cerium because it is more abundant than other commonly known elements like lead (14 ppm), tin (2 ppm), silver (0.07 ppm), tellurium (0.005 ppm), and gold (0.001 ppm). The price of cerium is also more competitive compared to that of ytterbium. For large quantities, the difference is a factor larger than 10 (Ce, \$3,800/ton; Yb, \$53,200/ton [52]), placing Ce in the category of other commonly used elements such as Ni.

With its larger earth abundance and fairly low cost, the use of Ce would help the commercialization of *n*-type skutterudites for the automotive industry only if the thermoelectric properties of Ce-doped skutterudites can match those of Yb-doped skutterudites.

Interestingly, single-filled Ce-$CoSb_3$ skutterudites have been considered as unsuitable thermoelectric materials, and there has yet been little report of high-temperature thermoelectric properties of pure Ce-$CoSb_3$ skutterudites. Ce has been considered to be a good auxiliary filler in multiple-filled skutterudites, such as in (In, Ce)-$CoSb_3$ skutterudites (zT of 1.43 for $In_{0.2}Ce_{0.15}Co_4Sb_{12}$ at 800 K) [53] or (Ba, Ce)-$CoSb_3$ skutterudites (zT of 1.26 for $Ba_{0.18}Ce_{0.05}Co_4Sb_{12.02}$ at 850 K) [54]. Ce has not typically been considered as a single filler. This is most likely due to its reported small FFL $x = 0.09$ in $Ce_xCo_4Sb_{12}$ [23], which was concluded based on the observance of a plateau in the lattice constant against nominal Ce content, as well as the observation of impurity phases. These phenomena, as we have discussed in Section 10.3, are not a sufficient proof to conclude the solubility limit in a ternary system. A reinvestigation of the filling fraction of Ce was performed by solubility design using an equilibrium phase diagram approach [29]. This approach introduces both the annealing temperature and nominal composition as variables in determining the FFL.

To increase the FFL of Ce in $CoSb_3$ skutterudites, charge-compensational doping, such as with Fe or Mn, has been largely applied to dope the material from *n*-type to *p*-type [54–59]. Here, single-filling Ce-$CoSb_3$ skutterudites without charge-compensational doping are discussed [29].

10.4.2 Optimum Doping of Ce-$CoSb_3$ Skutterudites

Unfilled $CoSb_3$ is an intrinsic semiconductor with carrier concentration less than 10^{18} cm^{-3}. Rare-earth filler atoms in the void sites are nearly perfect *n*-type donors. The *n*-type carrier concentration is essentially determined by the number of filling atoms and the ionic charge state. In $CoSb_3$, a Yb filler has a valence charge +2 while Ce is +3, which is a lower oxidation state than often observed in oxides but common for Sb compounds [17]. Figure 10.11 demonstrates the trend of the carrier concentration versus the number of electrons per primitive cell, for single-, double-, or triple-filled skutterudites. To achieve high power factor in *n*-type skutterudites, 0.4–0.6 electrons per Co_4Sb_{12} unit is required, as indicated by the double arrow [63]. The carrier concentration of Ce-$CoSb_3$ samples in the previous literature study [24] is limited by the reported low FFL ($x = 0.09$ in $Ce_xCo_4Sb_{12}$ at 973 K); consequently, these samples are far from the high power factor region. In order to optimize the thermoelectric properties of Ce single-filled $CoSb_3$ skutterudites, a higher filling fraction of Ce is needed.

10.4.3 Ultrahigh Filling Fraction Limit of Ce in Ce-$CoSb_3$ Skutterudites

The lattice constant shows a simple linear dependence on the actual Ce content x in $Ce_xCo_4Sb_{12}$, which agrees with the literature data and is consistent with Ce going into the same site across the whole range (up to $x = 0.20$), presumably the void site. Samples annealed at higher temperatures show both higher Ce content by EPMA and higher lattice constants.

The partial filling of Ce in the void site increases the entropy due to the increase of disorder of the material, and thus, it can be expected that the equilibrium solubility should increase measurably with temperature. While this is known theoretically [38], the FFL was

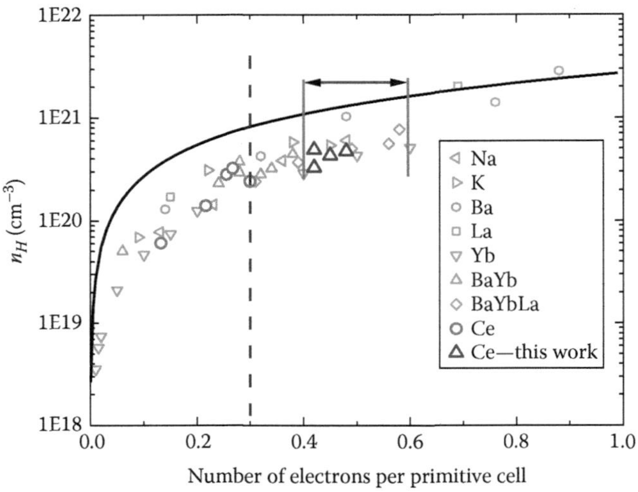

FIGURE 10.11
Hall carrier concentration versus number of electrons per primitive cell Co_4Sb_{12} [19,20,22,24,28,29,33,60–62]. The number of electrons per primitive cell is calculated as the sum of $q_i^* x_i$ for partially filled $(R_1)_{x1}(R_2)_{x2}\ldots(R_n)_{xn}Co_4Sb_{12}$, where q_i is the effective charge state and x_i is the filling fraction of the ith filler R in multiple-filled skutterudites. The solid line represents the theoretical curve calculated by using $n = 2x/a^3$ where x is the total number of electrons per primitive cell and a is the lattice constant of $CoSb_3$. The double arrow represents the estimated high power factor region (see discussions in Section 10.4.2) and the dashed line represents the number of electrons per primitive cell limited by the previously thought Ce filling fraction limit [24].

often considered to be a temperature-independent quantity. Tang et al. [28] showed that the Ce FFL (determined from the red point in Figure 10.12c) increases significantly with temperature (see Figure 10.12b). It increases to about 0.197 ± 0.007 at 1123 K and is more than twice the value reported from Morelli et al. ($x = 0.09$) [24].

Figure 10.12c shows a magnified region of an isothermal section at 973 K near $CoSb_3$ of the Ce-Co-Sb ternary phase diagram. There are three two-phase regions and three three-phase regions near $CoSb_3$. The stable compositions (marked as red and blue points in Figure 10.12c) were confirmed by the EPMA measurements to be $x = 0.120 \pm 0.005$ for the red stable composition ($Ce_{0.120}Co_4Sb_{12}$) and $x = 0.086 \pm 0.007$ for the blue stable composition ($Ce_{0.086}Co_4Sb_{12}$) at 973 K.

When the nominal composition is slightly Co rich (in addition to sufficient Ce, e.g., $Ce_{0.5}Co_{4.2}Sb_{11.8}$, red empty triangle), the sample is in the three-phase region between $CoSb_2$, $CeCoSb_3$ [64], and $Ce_xCo_4Sb_{12}$ with $x = 0.120 \pm 0.005$ (red point). Also if the nominal composition is slightly Sb rich (e.g., $Ce_{0.5}Co_{3.9}Sb_{12.1}$, orange empty triangle), the sample is in the three-phase region between $CeSb_2$, $CeCoSb_3$, and $Ce_xCo_4Sb_{12}$ with, coincidentally, the same $x = 0.120 \pm 0.005$ (red point) as found in the cobalt-rich three-phase region. In both cases, the equilibrium Ce content in the skutterudite phase will reach the same FFL $x = 0.120 \pm 0.005$ at 973 K (red point) as indicated by tie lines (red dashed lines). Note that these three-phase regions (one with $CeSb_2$, $CeCoSb_3$, and $Ce_xCo_4Sb_{12}$; the other with $CoSb_2$, $CeCoSb_3$, and $CexCo_4Sb_{12}$) are different from the previous phase diagram study of Ce-Co-Sb system at 400°C [64]. No similar ternary phase such as $CeCoSb_3$ is found in the Yb-Co-Sb or In-Co-Sb system, but the Ga-Co-Sb system has a ternary phase ($Co_3Ga_2Sb_4$) as well [65].

However, when the nominal composition is very antimony rich, such as used in reference [24], the samples are in another three-phase region between $Ce_xCo_4Sb_{12}$, $CeSb_2$, and liquid Sb (e.g., $Ce_{0.2}Co_{3.9}Sb_{12.1}$ and $Ce_{0.15}Co_{3.9}Sb_{12.1}$, marked as light and dark blue empty triangles, respectively). Here, the Ce content in the skutterudite phase remains constant with x = 0.087 ± 0.007 (blue point) as indicated by tie lines (blue dashed lines). Note that two nominal compositions ($Ce_{0.10}Co_4Sb_{12.2}$ and $Ce_{0.15}Co_4Sb_{12.2}$) from Reference 24 fall into this three-phase region of $Ce_xCo_4Sb_{12}$, $CeSb_2$, and liquid Sb, and one ($Ce_{0.20}Co_4Sb_{12.2}$) falls into the two-phase region of $Ce_xCo_4Sb_{12}$ and $CeSb_2$. The observation of the $CeSb_2$ impurity phase in those samples confirms that they are in a different field from the Co-rich three-phase region, which gives the higher Ce content. The blue stable composition point coincides with the reported FFL x = 0.09 [24], which is about 25% less of the FFL reported in Reference 29 for the same annealing temperature. As a result, an important solubility design strategy for enhancing the FFL of Ce in $CoSb_3$ is to focus on the Co-rich three-phase region rather than the lower Sb-rich three-phase region.

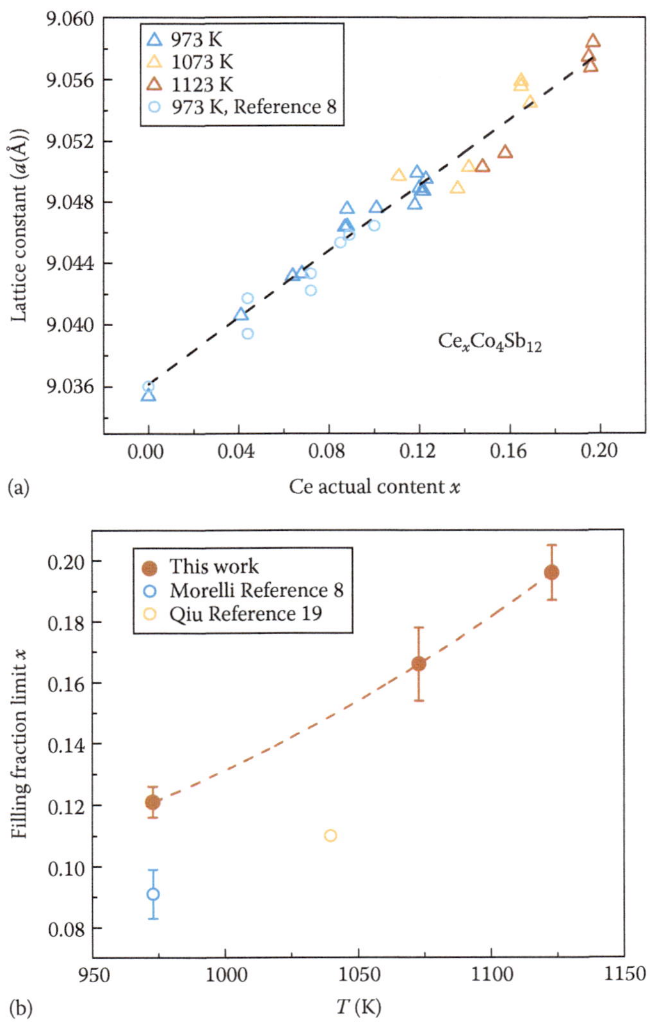

FIGURE 10.12
Temperature-dependent filling fraction limit of Ce in Ce-$CoSb_3$ skutterudites. (a) Skutterudite lattice expansion due to Ce filling. (b) Dependence of filling fraction limit (FFL) on the annealing temperature. The red dashed line is a guide for the eye.

(Continued)

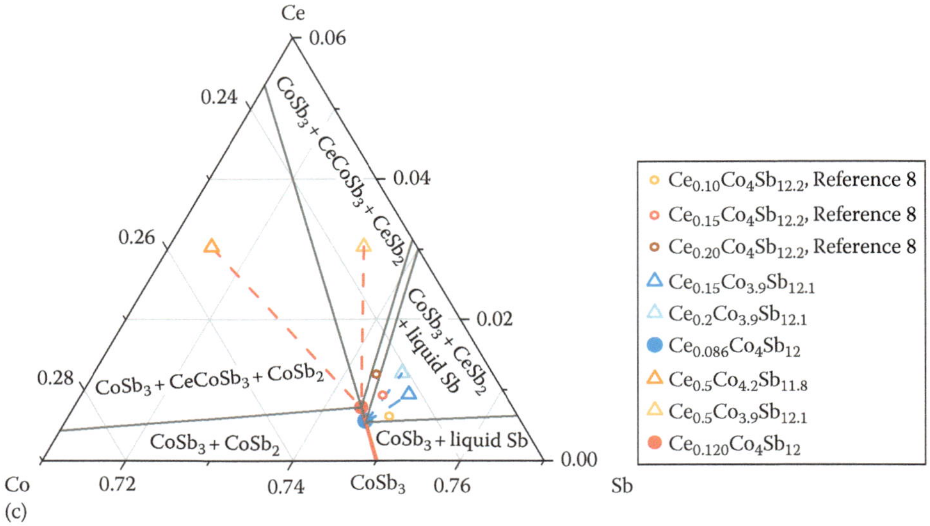

FIGURE 10.12 (*Continued*)
Temperature-dependent filling fraction limit of Ce in Ce-$CoSb_3$ skutterudites. (c) Dependence of FFL on nominal composition with annealing temperature 973 K. The phase region enclosed by red lines is a Co-rich three-phase region, which is related to the maximum solubility limit. The phase region enclosed by blue lines is an Sb-rich three-phase region, which is related to a lower solubility limit detected by Morelli et al. [24]. (Reproduced by permission from Macmillan Publishers Ltd., *Nat. Commun.*, Tang, Y.L., Hanus, R., Chen, S.W., and Snyder, G.J., Solubility design leading to high figure of merit in low-cost Ce-$CoSb_3$ skutterudites, 6, 7584, copyright 2015.)

From the analysis of the temperature-dependent FFL (Figure 10.12b), it is also seen that the Ce content in the skutterudite phase can be controlled through the annealing temperature when the nominal composition is in one of the three-phase regions, as demonstrated in the Yb-$CoSb_3$ system (Section 10.2) [28].

Atomic probe tomographic (APT) measurements of $Ce_xCo_4Sb_{12}$ skutterudites were also performed to confirm the ultrahigh filling fraction of $x = 0.20$. APT is particularly effective at chemically and structurally characterizing materials in 3D on the nanometer length scale. It provides accurate information about atomic concentrations and can identify even the smallest nanoparticles [66]. Small nanoparticles of Yb_2O_3 and InSb were observed in some Yb- and In-containing skutterudites, respectively [32,53,67]. Such small particles might go unnoticed in the SEM and would contribute to the EPMA signal. Here, APT was utilized to verify the Ce content and homogeneity in the $Ce_xCo_4Sb_{12}$ majority phase for the sample annealed at 1123 K with nominal composition $Ce_{0.5}Co_4Sb_{12}$. According to EPMA, the Ce content in the majority skutterudite phase should be $x = 0.20$ or 1.2 at%. Figure 10.13a contains a 3D reconstruction of the skutterudite phase at and around a grain boundary. As can be seen in Figure 10.13b, atomic concentrations were consistent and homogeneous away from the grain boundary. No nanoprecipitates of Ce-rich phases were observed. Atomic concentrations were averaged over a 30 nm diameter cylinder, from 20 to 60 nm on the scale bar in Figure 10.13a, and the Ce content was measured to be 1.4 ± 0.1 at%. This agrees with the values obtained by EPMA ($x = 0.20$ corresponding to Ce content of 1.2 at%) and supports our conclusion that such a high filling fraction of Ce is, indeed, uniformly distributed in the skutterudite phase. A uniform accumulation of Ce at the grain boundary was also observed, accompanied by a reduction in Co. Such atomic segregation is frequently observed at grain boundaries [66].

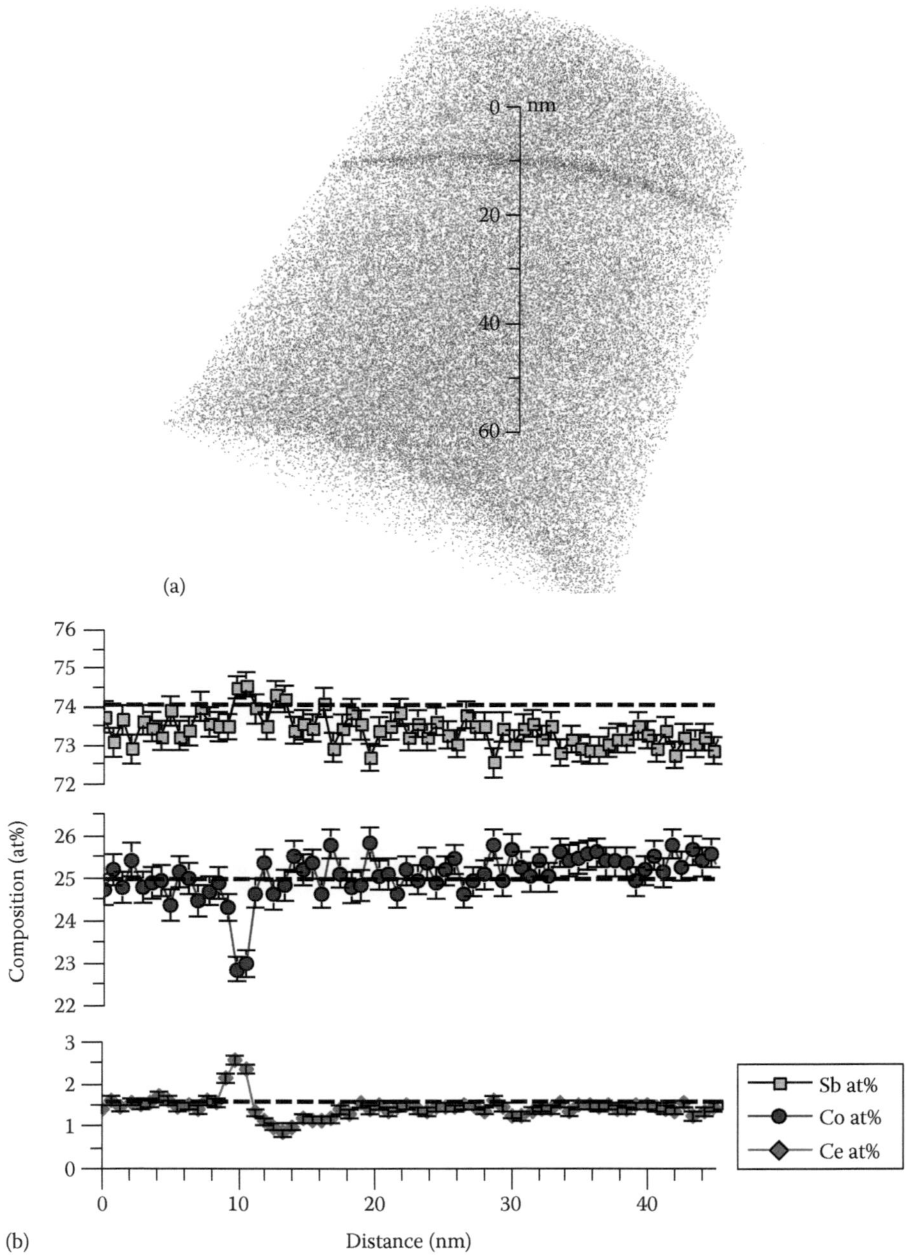

FIGURE 10.13
Atomic probe tomographic analysis of the most heavily doped sample $Ce_{0.2}Co_4Sb_{12}$ showing uniform Ce distribution within the grains and some impurity segregation to the grain boundary region. (a) 3D reconstruction of the microtip containing a grain boundary. Ce atoms are displayed in diamonds; Sb and Co atoms are omitted for clarity. (b) Concentration profile across the grain boundary and in the grain. The black dashed lines show values measured by electron probe microanalysis. (Reproduced by permission from Macmillan Publishers Ltd., *Nat. Commun.*, Tang, Y.L., Hanus, R., Chen, S.W., and Snyder, G.J., Solubility design leading to high figure of merit in low-cost Ce-$CoSb_3$ skutterudites, 6, 7584, copyright 2015.)

10.4.4 Thermoelectric Properties of Optimized Ce-$CoSb_3$ Skutterudites

The thermoelectric properties can be optimized when the filling fraction x of filler R (with valence q) in $R_xCo_4Sb_{12}$ satisfies $qx = 0.4–0.6$ [22,63]. From Figure 10.12, we can see that when nominal compositions are in the Ce-rich three-phase regions with an annealing temperature at 1073 and 1123 K, the Ce actual content in the skutterudite phase is about $x = 0.17$ and 0.20, respectively. With the Ce valence state being +3, qx for these materials would be in the optimum region. Thus, a set of samples with nominal compositions $Ce_xCo_4Sb_{12}$ ($x = 0.17$, 0.18, 0.20, and 0.21) were prepared for thermoelectric evaluation, of which the first two samples were annealed at 1073 K and the latter two were annealed at 1123 K. The temperature dependence of the transport properties of these samples is shown in Figure 10.14. Moreover, the transport data of two Yb-doped skutterudites with the doping level $x = 0.20$ and 0.30 are also plotted for the reason of comparison [28,34]. With Yb valence state being +2, qx for these two samples are right on the boundary of the optimum region.

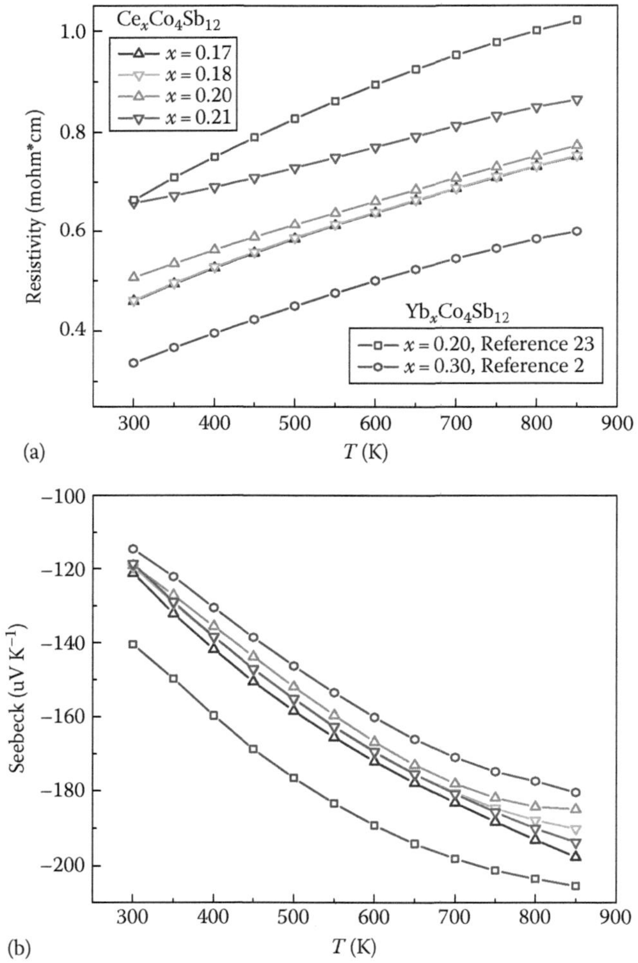

FIGURE 10.14
Transport properties of Ce- and Yb-doped skutterudites. The temperature dependence of (a) electrical resistivity, (b) Seebeck coefficient.

(Continued)

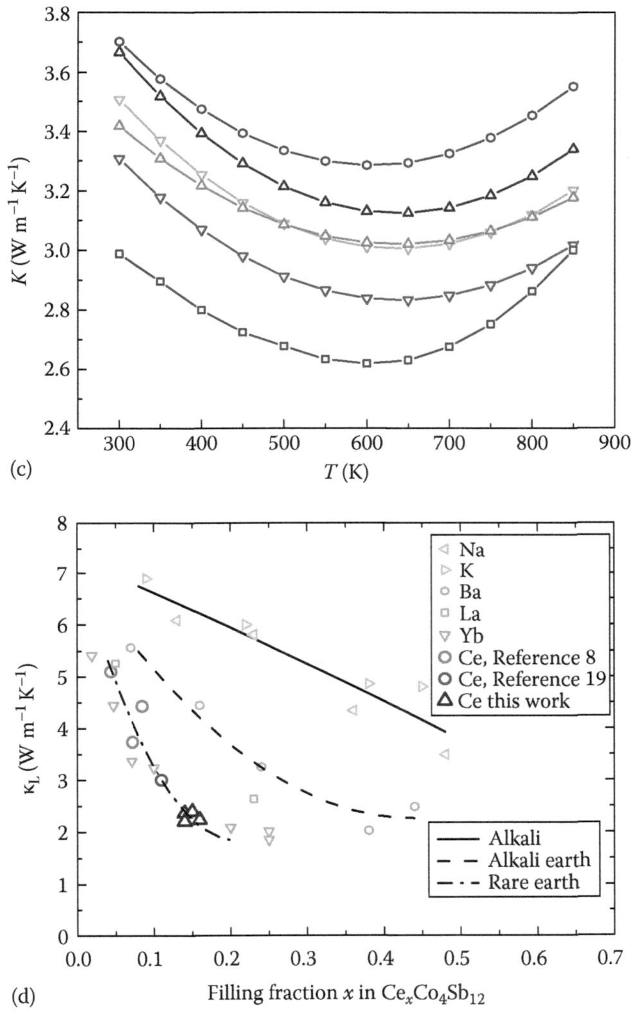

FIGURE 10.14 *(Continued)*
Transport properties of Ce- and Yb-doped skutterudites. The temperature dependence of (c) thermal conductivity is plotted in the temperature range of 300–850 K. In (d), the lattice thermal conductivity with a Lorenz number of $2.0 \times 10^{-8}\ V^2\ K^{-2}$ is plotted against the filling fraction for various types of fillers. x denotes the nominal doping level of fillers. (Reproduced from by permission from Macmillan Publishers Ltd., *Nat. Commun.*, Tang, Y.L., Hanus, R., Chen, S.W., and Snyder, G.J., Solubility design leading to high figure of merit in low-cost Ce-$CoSb_3$ skutterudites, 6, 7584, copyright 2015.)

Figure 10.4b shows that all Ce-$CoSb_3$ samples in this work have zT values higher than 1.0 at and above 750 K, with the maximum zT value in Ce-doped skutterudite samples reaching 1.3 at 850 K with the nominal Ce content $x = 0.17$ and $x = 0.18$ (with the composition $Ce_{0.14}Co_4Sb_{12}$ for both samples determined from EPMA). This zT value is 30% higher than that of the previously reported literature value for Ce single-filled skutterudites ($zT = 1.0$ for $Ce_{0.11}Co_4Sb_{12}$ at 850 K) [33]. The preceding zT values are similar to those for single-filled skutterudites [23,28,41,60], suggesting that Ce is a suitable earth abundant and low-cost replacement for Yb, for large-scale commercialization. The success of solubility design in Ce-$CoSb_3$ system is a good example that shows its potential in the optimization of thermoelectric properties in any other ternary systems with doping barriers, which allows wide applicability of this strategy.

10.5 Substitutional and Compound Defects

10.5.1 Solubility Debate regarding In and Ga

Unlike other filler atoms, the soluble sites as well as doping nature for both Ga and In generate much debate. Theoretically, a selection rule based on the electronegativity difference between Sb and R [68] was proposed by Shi et al. It was predicted that Ga and In could not be filled into the void site. Experimentally, it was assumed that group 13 elements occupy the void position exclusively and donate electrons, resulting in *n*-type ternary materials. Tl was expected to donate one electron in $CoSb_3$ [69], but for In and Ga, it is not clear how many electrons are donated from these void filling atoms [70–73]. Several studies have shown that GaSb- and InSb-containing composites were formed and concluded that Ga and In have virtually no solubility in $CoSb_3$ [53,74]. There was yet another report claiming that In could possibly substitute for Sb in $CoSb_3$ but hardly fill the void position [75]. Finally, since Ga^{3+} and In^{3+} are similar in size to Co^{3+}, it may also be possible for these atoms to substitute for the transition metals. The close electronegativity values among In (1.78), Ga (1.81), Sb (2.05), and Co (1.88) also support the possibility of both Co and Sb substitution.

Often lacking in this debate is the possibility that there is no dominant single defect type in skutterudites with group 13 impurities, but that compound defects form, as commonly found in wide-gap oxides. Amphoteric impurities were also observed in a few thermoelectric materials, in which one single-type impurity forms both donor and acceptor defects [76]. If both defects form in charge-balanced concentrations, the material will be nearly fully compensated with few free carriers, as expected for ionic compounds and many oxides. Charge-balanced defects were found in Ga-doped $CoSb_3$ skutterudites [71].

10.5.2 Density Functional Theory Calculations of In- and Ga-Doped $CoSb_3$ Systems

Before conducting an experimental phase diagram study of In-Co-Sb and Ga-Co-Sb systems, DFT calculations were applied to help navigate the research direction. Two types of calculations were performed: the first being the band structures of undoped $CoSb_3$ and different Ga-/In-doped systems and the second being the formation energies of possible defects as a function of both doping level and Fermi level [26,27].

The results from theoretical study can be summarized as follows [26,27]. First, unlike for other fillers, the void is *not* the only soluble site for both In and Ga. Second, Ga or In doping in $CoSb_3$-based skutterudites results in both void filling and Sb substituting, which leads to compound defects (charge compensated in the Ga case and not fully charge compensated in the In case). Third, the solubility limit is predicted to be 0.12 for Ga in a Co_4Sb_{12} formula unit at 923 K and 0.27 for In at 873 K.

The solubility region can then be estimated to be mainly directed along the direction of $(R_{VF})_{2x/3}Co_4Sb_{12-x/3}(R_{Sb})_{x/3}$ to $CoSb_3$ (R = Ga or In). This is different from the commonly known direction of $(R_{VF})_x$ Co_4Sb_{12} to $CoSb_3$ when the void is the only soluble site for alkali, alkaline-earth, and rare-earth element fillers. The experimental phase diagram study is accordingly designed, as shown in Section 10.5.3.

10.5.3 Phase Diagram of In-Co-Sb and Ga-Co-Sb Systems

The experimental phase diagrams were determined for Ga- or In-doped $CoSb_3$, both showing distinct characteristics due to Ga and In being in multiple sites [26,27,65]. The solubility direction $(R_{VF})_{2x/3}Co_4Sb_{12-x/3}(R_{Sb})_{x/3}$ to $CoSb_3$ for both In and Ga is different from the solubility direction $(R_{VF})_xCo_4Sb_{12}$ to $CoSb_3$ for other fillers, which leads to steep tie lines for nominal compositions in either Co- or Sb-rich phase regions in both In-Co-Sb and Ga-Co-Sb systems. As a result, any synthesis error could easily result in low In or Ga content skutterudites, which explains why these two fillers often generate many solubility debates. Moreover, the maximum solubility of In and Ga in the skutterudite phase for both In-Co-Sb and Ga-Co-Sb systems has to be achieved from a Co-rich three-phase region.

We will focus on the In-doped $CoSb_3$ system and point out the few differences with the Ga system. Samples with nominal compositions along or off the solubility direction $(R_{VF})_{2x/3}Co_4Sb_{12-x/3}(R_{Sb})_{x/3}$ to $CoSb_3$ were synthesized to confirm this as the maximum solubility direction.

Samples of $In_xCo_4Sb_{12-x/3}$ with the nominal indium content x higher than 0.30 contain a small amount of impurity phases (secondary phases other than the skutterudite phase), especially InSb and $CoSb_2$, which indicates that those samples are outside the single-phase region.

Figure 10.15 shows the dependence of the lattice parameter calculated from the XRD data on the actual indium content x. The dependence in the work of Qiu and Tang et al. [26] agrees with the data of Grytsiv et al. [77]. As x increases, the lattice parameter increases and becomes saturated when nominal $x \geq 0.30$, strongly supporting that the solubility limit

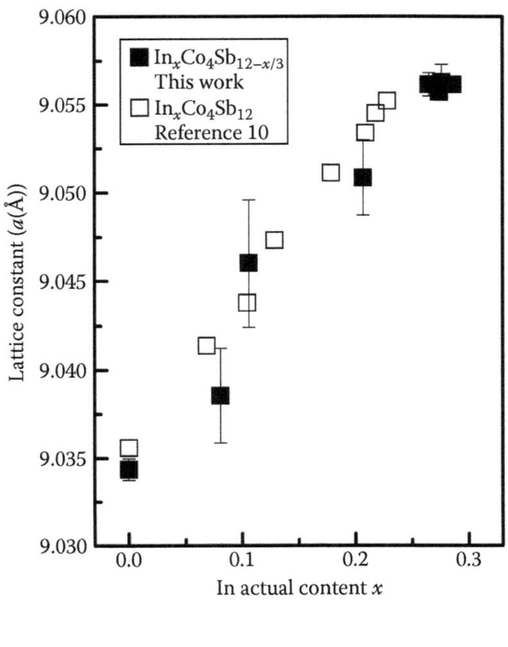

FIGURE 10.15
Dependence of the lattice parameter on actual indium content x for $In_xCo_4Sb_{12-x/3}$. (From Tang, Y.L., Qiu, Y.T., Xi, L.L. et al., Phase diagram of In-Co-Sb system and thermoelectric properties of In-containing skutterudites, *Energy Environ. Sci.*, 7, 812, 2014. Reproduced by permission of The Royal Society of Chemistry.)

of indium is around 0.30. This is close to the theoretical value $x = 0.27$ at 873 K. For Ga, the solubility is determined experimentally to be $x = 0.15$ at 923 K, which is close to the theoretical predicted value $x = 0.12$.

EPMA shows a significant volume fraction of In-rich impurity phases in the $In_{0.3}Co_{4-y}Sb_{11.9+y}$ samples with compositions that are off the Co/Sb ratio of the charge-compensated compound defect (CCCD) sample $(In_{VF})_{0.20}Co_4Sb_{11.90}(In_{Sb})_{0.10}$ (see Figure 10.16c and d). However, in CCCD samples, the indium is nearly completely homogeneously distributed (see Figure 10.16a and b) with few secondary phases, and the solubility limit of indium is estimated to be around $x = 0.27 \pm 0.01$. When the indium amount exceeds 0.27 in CCCD samples, a secondary phase, such as liquid and $CoSb_2$, is observed in SEM, which is consistent with XRD. The sensitivity of the formation of secondary phases to the ratio of Co/Sb in samples strongly supports the hypothesis that impurity indium atoms form compound defects in $CoSb_3$ by occupying both the void and Sb sites.

The proposed phase diagram of the In-containing skutterudites at 873 K is shown in Figure 10.17a. The related binary phases of In, Co, and Sb are shown on the axes including

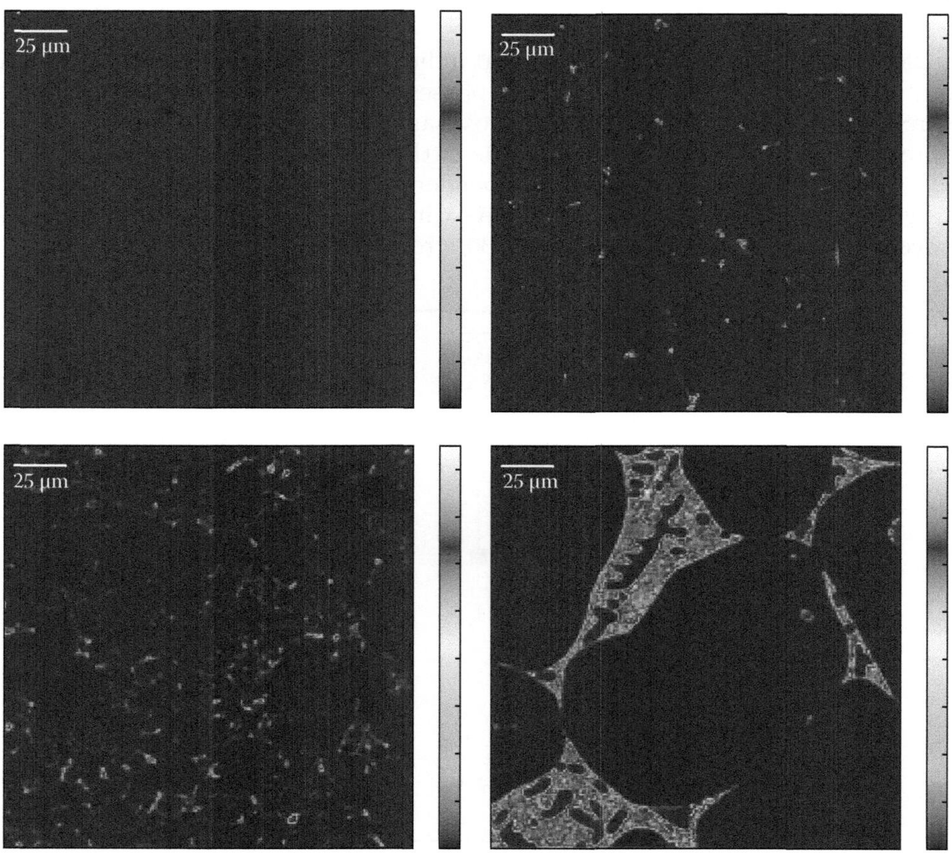

FIGURE 10.16
Electron probe microanalysis indium maps in samples with different nominal compositions [77]. (a) $In_{0.225}Co_4Sb_{11.925}$ (upper left), (b) $In_{0.30}Co_4Sb_{11.90}$ (upper right), (c) $In_{0.30}Co_{4.2}Sb_{11.7}$ (lower left), and (d) $In_{0.30}Co_{3.8}Sb_{12.1}$ (lower right). (From Tang, Y.L., Qiu, Y.T., Xi, L.L. et al., Phase diagram of In-Co-Sb system and thermoelectric properties of In-containing skutterudites, *Energy Environ. Sci.*, 7, 812, 2014. Reproduced by permission of The Royal Society of Chemistry.)

the known regions of solid or liquid solubility for the binary compounds and a rough estimate of the solubility of the ternary compounds. The exact regions of solid solubility of $CoSb_2$ and $CoSb_3$ are not known but they are enlarged for clarity in Figure 10.17a. The white regions of Figure 10.17a designate regions that are expected to contain two or three phases in equilibrium at 873 K.

The region near $CoSb_3$ is enlarged in Figure 10.17b. Experimental data suggest that the region of indium solubility extends toward the CCCD samples, $(In_{VF})_{2x/3}Co_4Sb_{12-x/3}(In_{Sb})_{x/3}$, with an estimated maximum x value around 0.27. When x exceeds 0.27, the CCCD samples contain an In-dual-site-doped skutterudite phase with indium doping level around $x = 0.27$ and impurity phases InSb and $CoSb_2$. When the nominal composition is off the Co/Sb ratio of CCCD $(In_{VF})_{2x/3}Co_4Sb_{12-x/3}(In_{Sb})_{x/3}$ samples, significant impurity phases form from $x = 0.30$ samples.

When the nominal composition is slightly Sb rich ($In_{0.3}Co_{4-y}Sb_{11.9+y}$, $y > 0$), the nominal composition is in a two-phase region between In-$CoSb_3$ and liquid In-Sb, where both phases have variable indium content in this region. This two-phase region has an additional degree of freedom to describe the tie line between which phases with variable In content form. We observe that the tie line connecting the nominal composition and the skutterudite phase is quite steep, which leads to a much lower In content in the skutterudite phase and an In-rich liquid In-Sb impurity phase. It is reasonable that In prefers the high entropy of the liquid rather than the solid skutterudite phase.

In contrast to the Sb-rich side, when the nominal composition is slightly Co rich ($In_{0.3}Co_{4-y}Sb_{11.9+y}$, $y < 0$), it is in a three-phase region between three phases with fixed

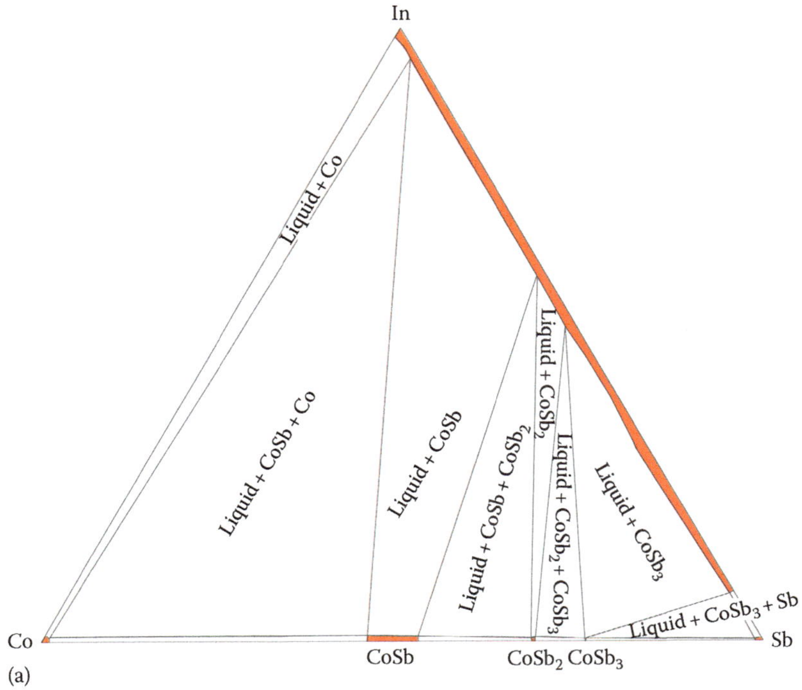

FIGURE 10.17
Proposed phase diagram for In-containing skutterudites at 873 K. (a) Full diagram with related binary phases and approximate regions of solubility indicated by red regions. *(Continued)*

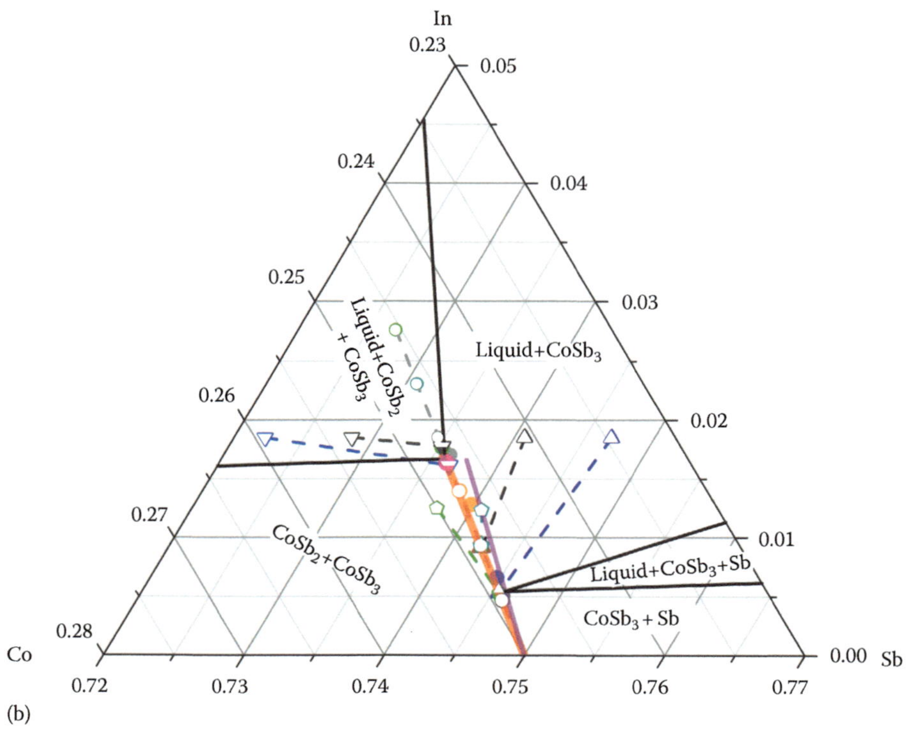

FIGURE 10.17 (*Continued*)
Proposed phase diagram for In-containing skutterudites at 873 K. (b) Phase regions enlarged near $CoSb_3$. Because of the stability of indium in dual sites (charge-compensated compound defect [CCCD] compositions), the region of indium solubility is extended as a red line in the direction of $(In_{VF})_{2x/3}Co_4Sb_{12-x/3}(In_{Sb})_{x/3}$ up to $x = 0.27$. The solid purple line indicates the direction of pure void filling only $(In_{VF})_xCo_4Sb_{12}$ up to $x = 0.27$. The magenta circular symbol represents the composition of the skutterudite phase $In_{0.27}Co_4Sb_{11.9}$ obtained when the nominal composition is in the three-phase region (liquid, $CoSb_2$, and In-$CoSb_3$). The nominal sample compositions are shown as open symbols. Solid symbols represent the composition of the skutterudite majority phase using the experimental electron probe microanalysis value for indium content, which is below 0.27 for all samples and positioned on the red CCCD line. The samples with nominal CCCD compositions when $x < 0.30$ appear to be within the solubility limit, the closed symbols of which are shifted away from the red solubility line according to the observed carrier concentration, as explained in Section 10.5.4. (From Tang, Y.L., Qiu, Y.T., Xi, L.L. et al., Phase diagram of In-Co-Sb system and thermoelectric properties of In-containing skutterudites, *Energy Environ. Sci.*, 7, 812, 2014. Reproduced by permission of The Royal Society of Chemistry.)

compositions: liquid InSb, $In_{0.27}Co_4Sb_{11.9}$, and $CoSb_2$. The composition of the liquid InSb ($In_{0.53}Sb_{0.47}$) is fixed near the composition of the InSb phase that is stable up to 799 K. Because the phase boundary between the two-phase region and three-phase region must be a straight line, the point of contact (magenta circular symbol in Figure 10.17b) between the skutterudite phase and the phase triangle must have relatively high indium content. Because of the phase rule, all the compositions in this three-phase region will produce the same skutterudite composition with high actual indium content ($In_{0.27}Co_4Sb_{11.9}$).

The steep slope of the tie line on the Sb-rich side of the stable phase indicates that, being slightly Co rich instead of Sb rich, is more favorable in order to achieve high indium content in the skutterudite phase. Moreover, it also sheds light on why there is a solubility difference when different impurity phases appear [77]. As the phase width appears to be quite narrow, small variations of composition will lead to different two- or three-phase equilibrium regions on the phase diagram, which explains why different attempts to produce the same material may in fact produce very different skutterudite materials with different filling fraction, and therefore different thermoelectric properties. In addition, the Co-rich and Sb-rich sides of the single-phase region (red line in Figure 10.17b) will have different temperature-dependent characteristics and could lead to different precipitation products as found in Zn-Sb [6]. For example, solid solubility is typically temperature dependent and can be used to form nanoscale precipitates [78], which may explain the appearance of InSb in some samples [53,79,80] but not others [81–84].

10.5.4 Thermoelectric Properties of In-$CoSb_3$

10.5.4.1 Thermoelectric Figure of Merit

Figure 10.18 shows the temperature dependence of the dimensionless thermoelectric figure of merit (zT) for In-doped skutterudites [27,85]. The maximum zT value in $In_xCo_4Sb_{12-x/3}$ samples reaches 1.2 (at 750 K) with indium content $x = 0.276$, which is among the highest zT values for single-filled skutterudites [19,41].

10.5.4.2 Not Fully Charge-Compensated Compound Defects

Compared with $(Ga_{VF})_{2x/3}Co_4Sb_{12-x/3}(Ga_{Sb})_{x/3}$ samples, the compound defect samples of the indium system have relatively larger carrier concentrations, which may be due to the defects

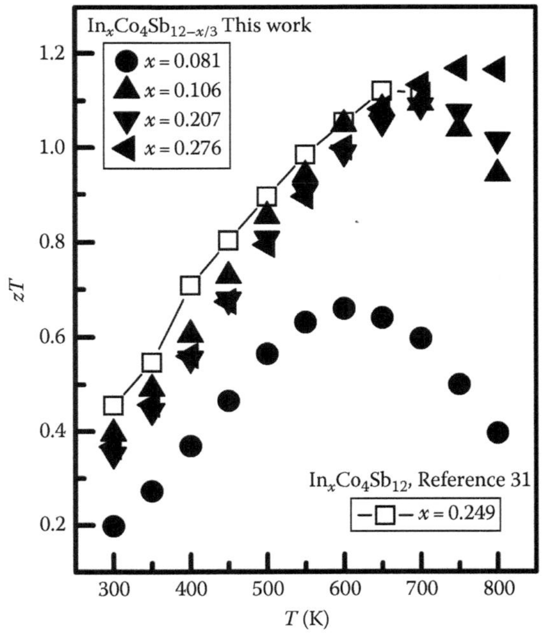

FIGURE 10.18
Temperature dependence of the thermoelectric figure of merit zT for In-containing skutterudites with complex compound defect. The x values are the same as the ones in Figure 10.15. (From Tang, Y.L., Qiu, Y.T., Xi, L.L. et al., Phase diagram of In-Co-Sb system and thermoelectric properties of In-containing skutterudites, *Energy Environ. Sci.*, 7, 812, 2014. Reproduced by permission of The Royal Society of Chemistry.)

not being fully charge compensated. If we write the compositions in a way containing both single-filling defects ($(\text{In}_{VF})_\delta$) and dual-site compound defects ($(\text{In}_{VF})_{2(x-\delta)/3}$ and $(\text{In}_{Sb})_{(x-\delta)/3}$), such as $(\text{In}_{VF})_{(2x+\delta)/3}\text{Co}_4\text{Sb}_{12-(x-\delta)/3}(\text{In}_{Sb})_{(x-\delta)/3}$, where δ can be calculated using the measured Hall carrier concentration n_H and lattice constant a: $n_H = 2\delta/a^3$, then we can get the value of filling fraction $(2x + \delta)/3$ with x being the content of indium determined by EPMA.

If we plot the compositions $(\text{In}_{VF})_{(2x+\delta)/3}\text{Co}_4\text{Sb}_{12-(x-\delta)/3}(\text{In}_{Sb})_{(x-\delta)/3}$ in the enlarged ternary phase diagram in Figure 10.17b, we can see that the maximal solubility line will slightly change its direction to somewhere between the direction of $(\text{In}_{VF})_{2x/3}\text{Co}_4\text{Sb}_{12-x/3}(\text{In}_{Sb})_{x/3}$ – $CoSb_3$ and the direction of $(\text{In}_{VF})_x\text{Co}_4\text{Sb}_{12}$ – $CoSb_3$, which corresponds to a certain thickness of the solubility region and a ratio of void filling/Sb substituting of indium larger than two (with complete charge compensation, this ratio is two).

10.6 Stability of Optimized Compositions

The temperature-dependent solubility limit predicts the precipitation of impurity phases at intermediate temperatures, as shown schematically in Figure 10.19. Such temperature-dependent solubility and precipitation has been observed in other ternary systems, such as Na-doped PbTe as well [9]. In heavily doped SiGe, the precipitation and redissolution of dopants during high-temperature transport measurements was also reported [86]. The rate and size of the precipitates will depend on the temperature and time. As the optimum carrier concentration corresponds to approximately x = 0.3, $\text{Yb}_{0.3}\text{Co}_4\text{Sb}_{12}$ material above

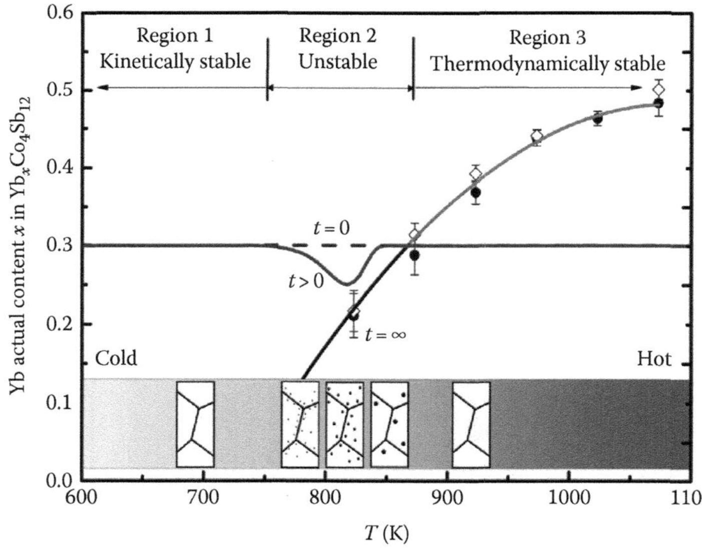

FIGURE 10.19
Stability of $\text{Yb}_x\text{Co}_4\text{Sb}_{12}$ skutterudites with optimum dopant composition x = 0.3. During operation (time $t > 0$), dopant precipitation is predicted in region 2 where the skutterudite phase is neither thermodynamically stable (as in region 3) nor kinetically stable (as in region 1). (Reproduced from *J. Mater.*, 1, Tang, Y.L., Chen, S.W., and Snyder, G.J., Temperature dependent solubility of Yb in Yb-CoSb$_3$ skutterudite and its effect on preparation, optimization and lifetime of thermoelectrics, 74, Copyright 2015, with permission from Elsevier.)

873 K will be stable indefinitely. The $Yb_{0.3}Co_4Sb_{12}$ material at low temperatures will be thermodynamically unstable but kinetically stable because precipitation will require the Yb diffusion, which will be small at low temperatures. In the intermediate temperature range, some precipitation of impurity phases is expected. Reduction of the Yb content will also reduce the carrier concentration, which will raise the electrical resistivity of the material. Despite the formation of impurity phases, the maximum performance of the material may not be dramatically reduced because a lower carrier concentration at low temperature may actually increase the zT at that temperature [87]. Such a temperature-dependent doping effect was found in Ag-doped *n*-type PbTe-Ag_2Te nanocomposites to improve the thermoelectric performance [88]. The large difference in solubility limit at different temperatures could also be used to form nanoscale precipitates with controlled size [4,78].

10.7 Conclusion

In ternary R-Co-Sb phase diagram systems, some common concepts were clarified and new strategies were developed for optimizing thermoelectrics. Observing impurity phases is not sufficient in determining the solubility limit because the nominal composition could be in any two- or three-phase region. A similar caveat applies for the departure of Vegard's law. A changing slope of the lattice constant as a function of nominal filling content x can simply indicate crossing a phase boundary, while a plateau can be due to the existence of only one of several stable compositions. Three solubility design strategies are proposed based on the study of various ternary systems (Yb-Co-Sb in Section 10.3, Ce-Co-Sb in Section 10.4, In-Co-Sb and Ga-Co-Sb in Section 10.5): first, carefully determine the correct solubility direction; second, aim at the right phase region; third, vary annealing temperature. These strategies can be generalized to any ternary system with a solubility barrier of doping elements.

Acknowledgments

Y.T. acknowledges support by the Swiss Federal Office of Energy under contract SI/501310-01. C.W. acknowledges support by the U.S. Department of Energy, Office of Science and Office of Basic Energy Sciences, under award no. DE-SC0014520. G.J.S. acknowledges support by the Solid-State Solar-Thermal Energy Conversion Center (S3TEC), an Energy Frontier Research Center funded by the U.S. Department of Energy, Office of Science, Basic Energy Sciences under award no. DE-SC0001299.

References

1. Pei, Y. Z., J. Lensch-Falk, E. S. Toberer, D. L. Medlin, and G. J. Snyder. 2011. High Thermoelectric performance in PbTe due to large nanoscale Ag_2Te precipitates and La doping. *Adv. Funct. Mater.* 21: 241.
2. Biswas, K., J. Q. He, I. D. Blum et al. 2012. High-performance bulk thermoelectrics with all-scale hierarchical architectures. *Nature* 489: 414.

3. Schwall, M. and B. Balke. 2013. Phase separation as a key to a thermoelectric high efficiency. *Phys. Chem. Chem. Phys.* 15: 1868.
4. Heinz, N. A., T. Ikeda, Y. Z. Pei, and G. J. Snyder. 2014. Applying quantitative microstructure control in advanced functional composites. *Adv. Funct. Mater.* 24: 2135.
5. Toberer, E. S., P. Rauwel, S. Gariel, J. Taftø, and G. J. Snyder. 2010. Composition and the thermoelectric performance of β-Zn_4Sb_3. *J. Mater. Chem.* 20: 9877.
6. Pomrehn, G. S., E. S. Toberer, G. J. Snyder, and A. van de Walle. 2011. Entropic stabilization and retrograde solubility in Zn_4Sb_3. *Phys. Rev. B* 83: 094106.
7. Pomrehn, G. S., A. Zevalkink, W. G. Zeier, A. van de Walle, and G. J. Snyder. 2014. Defect-controlled electronic properties in AZn_2Sb_2 Zintl phases. *Angew. Chem. Int. Ed.* 53: 3422.
8. Bux, S. K., A. Zevalkink, D. Uhl et al. 2015. Enhancing the performance of Zintl phases via defect chemistry. Presentation at a meeting on *Nuclear and Emerging Technologies for Space*, NETS-2015, Albuquerque, NM, February 2015 (unpublished).
9. Aminorroaya-Yamini, S., T. Ikeda, A. Lalonde, Y. Z. Pei, S. X. Dou, and G. J. Snyder. 2013. Rational design of *p*-type thermoelectric PbTe: Temperature dependent sodium solubility. *J. Mater. Chem. A* 1: 8725.
10. Doak, J. W., K. J. Michel, and C. Wolverton. 2015. Determining dilute-limit solvus boundaries in multi-component systems using defect energetics: Na in PbTe and PbS. *J. Mater. Chem. C* 3: 10630.
11. Doak, J. W., C. Wolverton, and V. Ozoliņš. 2015. Vibrational contributions to the phase stability of PbS-PbTe alloys. *Phys. Rev. B* 92: 17.
12. Wolverton, C., X. Y. Yan, R. Vijayaraghavan, and V. Ozolins. 2012. Incorporating first-principles energetics in computational thermodynamics approaches. *Acta Mater.* 50: 2187.
13. Bajaj, S., H. Wang, J. W. Doak, C. Wolverton, and G. J. Snyder. 2016. Calculation of dopant solubilities and phase diagrams of X-Pb-Se (X = Br, Na) limited to defects with localized charge. *J. Mater. Chem. C* 4: 1769.
14. Bajaj, S., G. S. Pomrehn, J. W. Doak et al. 2015. *Ab Initio* study of intrinsic point defects in PbTe: An insight into phase stability. *Acta Mater.* 92: 72.
15. Doak, J. W. and C. Wolverton. 2012. Coherent and incoherent phase stabilities of thermoelectric rocksalt IV-VI semiconductor alloys. *Phys. Rev. B* 86: 14.
16. Romaka, V. V., P. Rogl, L. Romaka et al. 2013. Phase equilibria, formation, crystal and electronic structure of ternary compounds in Ti–Ni–Sn and Ti–Ni–Sb ternary systems. *J. Solid State Chem.* 197: 103.
17. Uher, C. 2000. Skutterudites: Prospective novel thermoelectrics. In *Semiconductors and Semimetals*, Vol. 69, T. M. Tritt (ed.), Academic Press, New York, 2000, pp. 139–253.
18. Zhao, X. Y., X. Shi, L. D. Chen, W. Q. Zhang, W. B. Zhang, and Y. Z. Pei. 2006. Synthesis and thermoelectric properties of Sr-filled skutterudite $Sr_yCo_4Sb_{12}$. *J. Appl. Phys.* 99: 053711.
19. Pei, Y. Z., J. Yang, L. D. Chen, W. Zhang, J. R. Salvador, and J. Yang. 2009. Improving thermoelectric performance of caged compounds through light-element filling. *Appl. Phys. Lett.* 95: 042101.
20. Pei, Y. Z., L. D. Chen, W. Zhang et al. 2006. Synthesis and thermoelectric properties of $K_yCo_4Sb_{12}$. *Appl. Phys. Lett.* 89: 221107.
21. Pei, Y. Z., S. Q. Bai, X. Y. Zhao, W. Zhang, and L. D. Chen. 2008. Thermoelectric properties of $Eu_yCo_4Sb_{12}$ filled skutterudites. *Solid State Sci.* 10: 1422.
22. Shi, X., J. Yang, J. R. Salvador et al. 2011. Multiple-filled skutterudites: High thermoelectric figure of merit through separately optimizing electrical and thermal transports. *J. Am. Chem. Soc.* 133: 7837.
23. Nolas, G. S., M. Kaeser, R. T. Littleton, and T. M. Tritt. 2000. High figure of merit in partially filled ytterbium skutterudite materials. *Appl. Phys. Lett.* 77: 1855.
24. Morelli, D. T., G. P. Meisner, B. Chen, S. Hu, and C. Uher. 1997. Cerium filling and doping of cobalt triantimonide. *Phys. Rev. B* 56: 7376.

25. Dilley, N. R., E. D. Bauer, M. B. Maple, and B. C. Sales. 2000. Thermoelectric properties of chemically substituted skutterudites $Yb_yCo_4\ Sn_xSb_{12-x}$. *J. Appl. Phys.* 88: 1948.
26. Qiu, Y. T., L. L. Xi, X. Shi et al. 2013. Charge-compensated compound defects in Ga-containing thermoelectric skutterudites. *Adv. Funct. Mater.* 23: 3194.
27. Tang, Y. L., Y. T. Qiu, L. L. Xi et al. 2014. Phase diagram of In-Co-Sb system and thermoelectric properties of In-containing skutterudites. *Energy Environ. Sci.* 7: 812.
28. Tang, Y. L., S. W. Chen, and G. J. Snyder. 2015. Temperature dependent solubility of Yb in Yb-$CoSb_3$ skutterudite and its effect on preparation, optimization and lifetime of thermoelectrics. *J. Materiomics* 1: 74.
29. Tang, Y. L., R. Hanus, S. W. Chen, and G. J. Snyder. 2015. Solubility design leading to high figure of merit in low-cost Ce-CoSb3 skutterudites. *Nat. Commun.* 6: 7584.
30. Okamoto, H. 1991. Co-Sb (Cobalt-antimony). *J. Phase Equilib.* 12: 244.
31. Tang, Y. L. 2015. Phase diagram studies in *n*-$CoSb_3$ skutterudites. PhD thesis, The California Institute of Technology (Caltech), Pasadena, CA.
32. Zhao, X. Y., X. Shi, L. D. Chen et al. 2006. Synthesis of $Yb_yCo_4Sb_{12}/Yb_2O_3$ composites and their thermoelectric properties. *Appl. Phys. Lett.* 8: 092121.
33. Qiu, P. F., X. Shi, Y. T. Qiu et al. 2013. Enhancement of thermoelectric performance in slightly charge-compensated $Ce_yCo_4Sb_{12}$ skutterudites. *Appl. Phys. Lett.* 103: 062103.
34. Tang, Y. L., Z. M. Gibbs, L. A. Agapito et al. 2015. Convergence of multi-valley bands as the electronic origin of high thermoelectric performance in $CoSb_3$ skutterudites. *Nat. Mater.* 14: 1223.
35. Meisner, G. P. 2011. Advanced thermoelectric materials and generator technology for automotive waste heat at GM. Presented at *Directions in Engine-Efficiency and Emissions Research*, Detroit, MI (unpublished).
36. Li, H., X. F. Tang, Q. J. Zhang, and C. Uher. 2008. Rapid preparation method of bulk nanostructured $Yb_{0.3}Co_4Sb_{12+y}$ compounds and their improved thermoelectric performance. *Appl. Phys. Lett.* 93: 252109.
37. Mei, Z. G., W. Q. Zhang, L. D. Chen, and J. H. Yang. 2006. Filling fraction limits for rare-earth atoms in $CoSb_3$: An *Ab Initio* approach. *Phys. Rev. B* 74: 153202.
38. Shi, X., W. Q. Zhang, L. D. Chen, J. H. Yang, and C. Uher. 2008. Thermodynamic analysis of the filling fraction limits for impurities in $CoSb_3$ based on Ab initio calculations. *Acta Mater.* 56: 1733.
39. Xia, X. G., P. F. Qiu, X. Shi, X. Y. Li, X. Y. Huang, and L. D. Chen. 2012. High-temperature oxidation behavior of filled skutterudites $Yb_yCo_4Sb_{12}$. *J. Electron. Mater.* 41: 2225.
40. Salvador, J. R., J. Yang, X. Shi et al. 2009. Transport and mechanical properties of Yb-filled skutterudites. *Philos. Mag.* 89: 1517.
41. Yang, J., Q. Hao, H. Wang et al. 2009. Solubility study of Yb in *n*-type skutterudites $Yb_xCo_4Sb_{12}$ and their enhanced thermoelectric properties. *Phys. Rev. B* 80: 115329.
42. Okamoto, H. 1990. Co-Yb (Cobalt-Ytterbium). In *Binary Alloy Phase Diagrams*, Vol. 2, T. B. Massalski (ed.), ASM International, Materials Park, OH, p. 2.
43. Okamoto, H. 1990. Sb-Yb (Antimony-Ytterbium). In *Binary Alloy Phase Diagrams*, Vol. 3, T. B. Massalski (ed.), ASM International, Materials Park, OH, p. 3317.
44. Lamberton, G. A., R. H. Tedstrom, T. M. Tritt, and G. S. Nolas. 2005. Thermoelectric properties of Yb-filled Ge-compensated $CoSb_3$ skutterudite materials. *J. Appl. Phys.* 97: 113715.
45. Vegard, L. 1921. Die Konstitution der Mischkristalle und die Raumfüllung der Atome. *Zeitschrift für Physik* 5: 17.
46. He, T., J. Chen, H. David Rosenfeld, and M. A. Subramanian. 2006. Thermoelectric properties of indium-filled skutterudites. *Chem. Mater.* 18: 759.
47. Li, X. Y., L. D. Chen, J. F. Fan, W. B. Zhang, T. Kawahara, and T. Hirai. 2005. Thermoelectric properties of Te-doped $CoSb_3$ by spark plasma sintering. *J. Appl. Phys.* 98: 083702.

48. Wojciechowski, K. T., J. Toboła, and J. Leszczyński. 2003. Thermoelectric properties and electronic structure of $CoSb_3$ doped with Se and Te. *J. Alloy. Compd.* 361: 19.
49. Liu, W. S., B. P. Zhang, J. F. Li, H. L. Zhang, and L. D. Zhao. 2007. Enhanced thermoelectric properties in $CoSb_{3-x}Te_x$ alloys prepared by mechanical alloying and spark plasma sintering. *J. Appl. Phys.* 102: 103717.
50. Fairbanks, J. W. 2010. Vehicular thermoelectrics: A new green technology. Presented at *Directions in Engine-Efficiency and Emissions Research*, Detroit, MI (unpublished).
51. Emsley, J. 2011. *Nature's Building Blocks: An A-Z Guide to the Elements*, New Edition Oxford University Press, Inc., New York, p. 699.
52. The Chinese Rare Earth Website. 2015. Rare earth average unit prices for 2009, 2010, 2011, 2013 and 2016. http://www.pcreml.com/rare-earth-pricing, July 2016.
53. Li, H., X. Tang, Q. Zhang, and C. Uher. 2009. High performance $In_xCe_yCo_4Sb_{12}$ Thermoelectric materials with *in situ* forming nanostructured InSb phase. *Appl. Phys. Lett.* 94: 102114.
54. Bai, S., Y. Pei, L. Chen, W. Zhang, X. Zhao, and J. Yang. 2009. Enhanced thermoelectric performance of dual-element-filled skutterudites $Ba_xCe_yCo_4Sb_{12}$. *Acta Mater.* 57: 3135.
55. Zhang, L., A. Grytsiv, M. Kerber, P. Rogl, E. Bauer, and M. Zehetbauer. 2010. Thermoelectric performance of mischmetal skutterudites $Mm_yFe_{4-x}Co_xSb_{12}$ at elevated temperatures. *J. Alloy. Compd.* 490: 19.
56. Liu, R., P. Qiu, X. Chen, X. Huang, and L. Chen. 2011. Composition optimization of *p*-type skutterudites $Ce_yFe_xCo_{4-x}Sb_{12}$ and $Yb_yFe_xCo_{4-x}Sb_{12}$. *J. Mater. Res.* 26: 1813.
57. Tang, X., Q. Zhang, L. Chen, T. Goto, and T. Hirai. 2005. Synthesis and thermoelectric properties of *p*-type- and *n*-type-filled skutterudite $R_yM_xCo_{4-x}Sb_{12}$ (R: Ce, Ba, Y; M: Fe, Ni). *J. Appl. Phys.* 97: 093712.
58. Chen, B., J. H. Xu, C. Uher et al. 1997. Low-temperature transport properties of the filled skutterudites $CeFe_{4-x}Co_xSb_{12}$. *Phys. Rev. B* 55: 1476.
59. Meisner, G., D. Morelli, S. Hu, J. Yang, and C. Uher. 1998. Structure and lattice thermal conductivity of fractionally filled skutterudites: Solid solutions of fully filled and unfilled end members. *Phys. Rev. Lett.* 80: 3551.
60. Chen, L. D., T. Kawahara, X. F. Tang et al. 2001. Anomalous barium filling fraction and *n*-type thermoelectric performance of $Ba_yCo_4Sb_{12}$. *J. Appl. Phys.* 90: 1864.
61. Nolas, G. S., J. L. Cohn, and G. A. Slack. 1998. Effect of partial void filling on the lattice thermal conductivity of skutterudites. *Phys. Rev. B* 58: 164.
62. Shi, X., H. Kong, C. P. Li et al. 2008. Low thermal conductivity and high thermoelectric figure of merit in *n*-type $Ba_xYb_yCo_4Sb_{12}$ double-filled skutterudites. *Appl. Phys. Lett.* 92: 182101.
63. Yang, J., L. L. Xi, W. Q. Zhang, L. D. Chen, and J. H. Yang. 2009. Electrical transport properties of filled $CoSb_3$ skutterudites: A theoretical study. *J. Electron. Mater.* 38: 1397.
64. Luo, R. M., F. S. Liu, J. Q. Li, and X. W. Feng. 2009. The isothermal section of the Ce–Co–Sb ternary system at 400°C. *J. Alloy. Compd.* 471: 60.
65. Gierlotka, W., S. W. Chen, W. A. Chen, J. S. Chang, G. J. Snyder, and Y. L. Tang. 2015. The Co-Sb-Ga system: Isoplethal section and thermodynamic modeling. *Metall. Mater. Trans. A* 46A: 1488.
66. Kim, Y. J., I. D. Blum, J. Q. He, M. G. Kanatzidis, V. P. Dravid, and D. N. Seidman. 2014. Three-dimensional atom-probe tomographic analyses of lead-telluride based thermoelectric materials. *J. Miner. Met. Mater. Soc. (TMS)* 66: 2288.
67. Ding, J., H. Gu, P. F. Qiu et al. 2013. Creation of Yb_2O_3 nanoprecipitates through an oxidation process in bulk Yb-filled skutterudites. *J. Electron. Mater.* 42: 382.
68. Shi, X., W. Zhang, L. Chen, and J. Yang. 2005. Filling fraction limit for intrinsic voids in crystals: Doping in skutterudites. *Phys. Rev. Lett.* 95: 18.
69. Harnwunggmoung, A., K. Kurosaki, Y. Ohishi, H. Muta, and S. Yamanaka. 2011. Effects of Tl-filling into the voids and Rh substitution for Co on the thermoelectric properties of $CoSb_3$. *J. Alloy. Compd.* 509: 1084.

70. Mallik, R. C., C. Stiewe, G. Karpinski, R. Hassdorf, and E. Muller. 2009. Thermoelectric properties of Co_4Sb_{12} skutterudite materials with partial in filling and excess in additions. *J. Electron. Mater.* 38: 1337.
71. Harnwunggmoung, A., K. Kurosaki, T. Plirdpring et al. 2011. Thermoelectric properties of Ga-added $CoSb_3$ based skutterudites. *J. Appl. Phys.* 110: 5.
72. Visnow, E., C. P. Heinrich, A. Schmitz et al. 2015. On the true Indium content of in-filled skutterudites. *Inorg. Chem.* 54: 7818.
73. Zhao, W., P. Wei, Q. Zhang et al. 2015. Multi-localization transport behaviour in bulk thermoelectric materials. *Nat. Commun.* 6: 6197.
74. Xiong, Z., X. H. Chen, X. Y. Huang, S. Q. Bai, and L. D. Chen. 2010. High thermoelectric performance of $Yb_{0.26}Co_4Sb_{12}/_yGaSb$ nanocomposites originating from scattering electrons of low energy. *Acta Mater.* 58: 3995.
75. Sesselmann, A., T. Dasgupta, K. Kelm, and E. Müller. 2011. Transport properties and microstructure of indium-added cobalt–antimony-based skutterudites. *J. Mater. Res.* 26: 1820.
76. Jaworski, C. M., J. Tobola, E. M. Levin, K. Schmidt-Rohr, and J. P. Heremans. 2009. Antimony as an amphoteric dopant in lead telluride. *Phys. Rev. B* 80: 12.
77. Grytsiv, A., P. Rogl, H. Michor, E. Bauer, and G. Giester. 2013. $In_yCo_4Sb_{12}$ skutterudite: Phase equilibria and crystal structure. *J. Electron. Mater.* 42: 2940.
78. Ikeda, T., V. Ravi, and G. J. Snyder. 2009. Formation of Sb_2Te_3 widmanstätten precipitates in thermoelectric PbTe. *Acta Mater.* 57: 666.
79. Graff, J., S. Zhu, T. Holgate, J. Y. Peng, J. He, and T. M. Tritt. 2011. High-temperature thermoelectric properties of Co_4Sb_{12}-based skutterudites with multiple filler atoms: $Ce_{0.1}In_xYb_yCo_4Sb_{12}$. *J. Electron. Mater.* 40: 696.
80. Ballikaya, S., N. Uzar, S. Yildirim, J. R. Salvador, and C. Uher. 2012. High thermoelectric performance of In, Yb, Ce multiple filled $CoSb_3$ based skutterudite compounds. *J. Solid State Chem.* 193: 31.
81. Zhao, W., P. Wei, Q. J. Zhang, C. L. Dong, L. S. Liu, and X. F. Tang. 2009. Enhanced thermoelectric performance in barium and indium double-filled skutterudite bulk materials via orbital hybridization induced by indium filler. *J. Am. Chem. Soc.* 131: 3713.
82. Zhao, W. Y., C. L. Dong, P. Wei et al. 2007. Synthesis and high temperature transport properties of barium and indium double-filled skutterudites $Ba_xIn_yCo_4Sb_{12-z}$. *J. Appl. Phys.* 102: 6.
83. Peng, J. Y., P. N. Alboni, J. He et al. 2008. Thermoelectric properties of (In,Yb) double-filled $CoSb_3$ skutterudite. *J. Appl. Phys.* 104: 5.
84. Peng, J. Y., J. He, Z. Su, P. N. Alboni, S. Zhu, and T. M. Tritt. 2009. High temperature thermoelectric properties of double-filled $In_xYb_yCo_4Sb_{12}$ skutterudites. *J. Appl. Phys.* 105: 5.
85. Mallik, R. C., J. Y. Jung, S. C. Ur, and I. H. Kim. 2008. Transport properties of Sn-filled and Te-doped $CoSb_3$ skutterudites. *Met. Mater. Int.* 14: 223.
86. Vining, C. B. 1991. A model for the high-temperature transport properties of heavily doped *n*-type silicon-germanium alloys. *J. Appl. Phys.* 69: 331.
87. Snyder, G. J. and E. S. Toberer. 2008. Complex thermoelectric materials. *Nat. Mater.* 7: 105.
88. Pei, Y. Z., A. F. May, and G. J. Snyder. 2011. Self-tuning the carrier concentration of PbTe/Ag_2Te composites with excess Ag for high thermoelectric performance. *Adv. Energy Mater.* 1: 291.

11

Chain-Forming A_3MPn_3 and $A_5M_2Pn_6$ Zintl Phases

Alex Zevalkink, Umut Aydemir, and G. Jeffrey Snyder
Northwestern University

CONTENTS

11.1 Introduction

Significant improvements in thermoelectric efficiency have been achieved in the past decade owing to both innovations in controlling thermal and electronic transport properties and to the discovery of a large number of entirely new thermoelectric materials. Zintl phases represent an important and incredibly diverse new class of thermoelectrics [1–3], with promising novel structure types and compounds being reported on a frequent basis [4,5]. Zintl phases are a subset of intermetallic compounds that are characterized by covalently bonded anionic "substructures," surrounded by highly electropositive cations [6,7]. Their unique combination of covalent and ionic bonding, together with a high degree of structural complexity, leads to excellent thermoelectric properties, as evidenced by initial reports of $ZT \sim 1$ in compounds such as $YbZn_2Sb_2$, Zn_4Sb_3, and $Yb_{14}MnSb_{11}$ [8–10].

In Zintl phases, the electropositive cation can be treated like a fully ionized element that provides valence electrons to the more electronegative elements that, in turn, either use those electrons to form a closed shell or, if necessary, to form bonds in order to achieve a full octet. Owing to this well-defined relationship between the chemical structure and the total number of valence electrons, Zintl phases are classically considered to be line compounds with very narrow homogeneity ranges. Of course, if all Zintl phases were strict line compounds, they would be uninteresting as thermoelectric materials, since doping to control the electronic properties would be impossible. Fortunately, many Zintl phases tolerate quite drastic deviations from the Zintl formalism. For example, $Yb_{14}AlSb_{11}$ is a

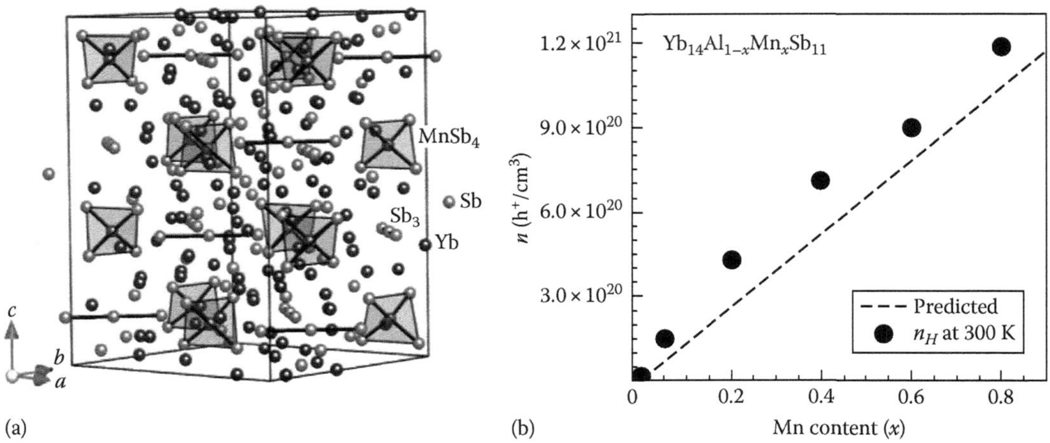

FIGURE 11.1
(a) Structure of the Zintl phase $Yb_{14}AlSb_{11}$ along the [101] direction. (b) The carrier concentration and electronic properties can be tuned by substituting Mn^{2+} on the Al^{3+} site, leading to a transition from intrinsic (nondegenerate) to extrinsic (degenerate) semiconducting behavior. (Data from Toberer, E.S. et al., *Adv. Funct. Mater.*, 18, 2795, 2008.)

well-behaved Zintl phase in which the $14Yb^{2+}$ cations donate 28 valence electrons to support the formation of $4Pn^{3-}$ anions, an MPn_4^{9-} tetrahedron, and a Pn_3^{7-} linear anion, leading to overall charge balance (shown in Figure 11.1a). As such, $Yb_{14}AlSb_{11}$ behaves as an intrinsic semiconductor with low carrier concentrations and very high resistivity. However, a variety of different substitutions can be used to tune the electronic properties by shifting the system away from a charge-balanced state, the most effective of which is the substitution of Mn^{2+} on the Al^{3+} site [11,12]. Figure 11.1b shows excellent agreement between the predicted and experimental p-type carrier concentration as a function of x in $Yb_{14}Al_{1-x}Mn_xSb_{11}$. The optimum ZT is achieved at intermediate Mn concentrations. Beyond Mn, alternative doping possibilities include several possible Ln^{3+} species on the Yb site, Zn or Mg on the Al site, and Te on the Sb site [13–18]. Isoelectronic substitutions, which influence the formation of intrinsic point defects [19], can also be used to fine-tune the carrier concentration [20].

The success of $Yb_{14}MnSb_{11}$ and its subsequent development and integration into modules by the Jet Propulsion Laboratory [21] has inspired widespread interest in Zintl phases [2,3,22]. During the past several years, as the focus has turned toward materials containing inexpensive, nontoxic elements, A_3MPn_3 and $A_5M_2Pn_6$ Zintl compounds have emerged as a new class of thermoelectric materials with promising thermoelectric performance [23–28]. The rich chemistry of this class of compounds allows for extensive possibilities to tune bonding character and transport properties: the cation, A, can be an alkaline earth or divalent rare-earth metal (Ca, Sr, Ba Eu, Yb), M is either a triel element (Al, Ga, In) or Sn, and Pn is a pnictogen species (P, As, Sb, Bi). A_3MPn_3 and $A_5M_2Pn_6$ compounds form eight distinct structure types that are unified by a common structural motif: MPn_4 tetrahedra that are either corner or edge linked to form either infinite chains or isolated pairs.

The majority of A_3MPn_3 and $A_5M_2Pn_6$ Zintl phases (those in which A = alkaline earth metal) were discovered between 1980 and 1990 by Schäfer and Cordier, who described the new phases as valence-precise compounds consistent with the Zintl–Klemm formalism [29–32,40]. More than a decade later, the search for air-stable Zintl phases by Kanatzidis and others led to the discovery of several additional compounds containing divalent rare-earth metals [34–38], and, in 2002, Park et al. reported semiconducting behavior in $Eu_5In_2Sb_6$ [38], hinting at their potential for thermoelectric applications. In 2010, the high temperature thermoelectric properties of Na-doped $Ca_5Al_2Sb_6$ ($ZT \sim 0.6$) were reported by Toberer et al. [26], spurring a much wider investigation into the thermoelectric properties of these phases.

The chain-forming Zintl phases discussed in this chapter—despite the fact that they form several different structure types—have chemical and structural similarities that lead to very similar electronic and thermal transport behaviors. By exploring the transport properties of these closely related structure types, we aim to identify trends that can potentially be used to guide future work on Zintl thermoelectrics.

11.2 Crystal Structure

As shown in Figure 11.2, all A_3MPn_3 and $A_5M_2Pn_6$ Zintl compounds form anionic substructures based on MPn_4 tetrahedra. To reach the 1:3 ratio of M to Pn atoms required by stoichiometry, each MPn_4 tetrahedron must either share corners with neighboring tetrahedra, yielding infinite chains of tetrahedra, or share edges, yielding tetrahedral dumbbells (pairs). A combination of this corner and edge sharing is a third possibility.

The three different types of tetrahedral substructures found in A_3MPn_3 compounds—linear chains, spiral chains, and dumbbells—are shown in Figure 11.2a through c. We can readily show that these structures are valence precise in the limit of ionic bonding as each compound contains three A^{2+}, one M^{3+}, and three Pn^{3-} ions. Alternatively, Zintl electron counting in the covalent limit yields $(A^{2+})_3(4b\ M^-)(2b\ Pn^-)(1b\ Pn^{2-})_2$, where 4b, 2b, and 1b indicate the number of covalent M–Pn bonds. In either limit, M and Pn combine to form a polyanion, $[MPn_3]^{6-}$, which is balanced by the charge of the three A^{2+} cations.

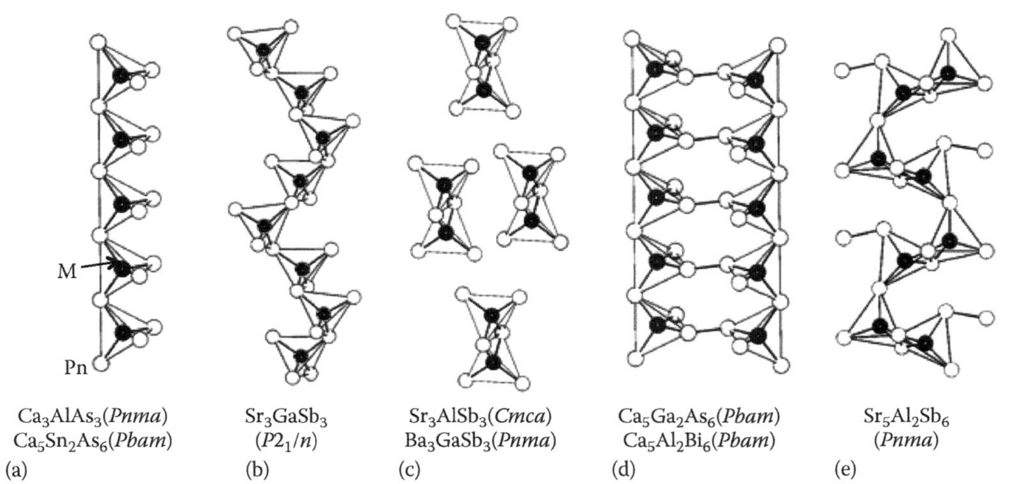

FIGURE 11.2
(a–e) Five types of anionic substructures and eight unique structure types occur among A_3MPn_3 and $A_5M_2Pn_6$ compounds with A = Ca, Sr, Ba, Eu, Yb, M = Al, Ga, In, Sn, and Pn = P, As, Sb, Bi. The MPn_4 tetrahedra building blocks are either corner or edge linked to form chain dumbbells. M and Pn are shown as black and white spheres, respectively.

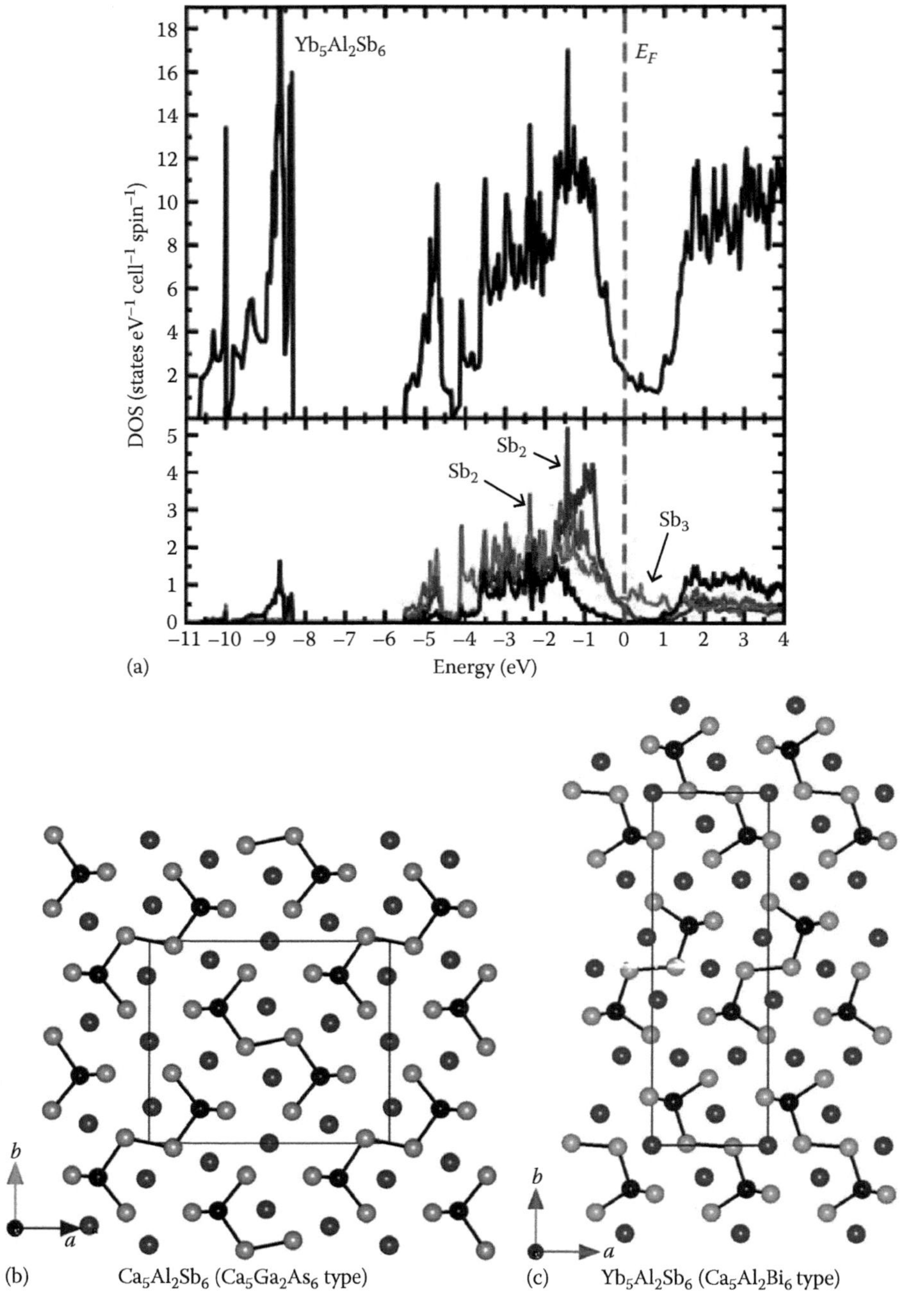

FIGURE 11.3
(a) The density of states of $Yb_5Al_2Sb_6$ shows electronic states arising from the antibonding Sb–Sb interactions (labeled as Sb_3) at the Fermi level. In contrast, states associated with Sb–Sb antibonding in $Ca_5M_2Sb_6$ compounds lie in the conduction band minimum. (Details of the calculations can be found in Zevalkink, A. et al., *Chem. Mater.*, 24, 2091, 2012; Aydemir, U. et al., *Dalton Trans.*, 44, 6767, 2015.) (b) The $Ca_5Al_2Bi_6$ and (c) $Ca_5Ga_2As_6$ structure types are characterized by differing arrangements double-chain MPn_4 tetrahedra, which are bridged by Pn–Pn dumbbells. White, gray, and black atoms represent Pn, A, and M, respectively.

Within A_3MPn_3 compounds, which structure type is favored is mainly a question of geometry. Compounds with smaller cations (A = Ca), smaller anions (Pn = P, As), or both, tend to adopt the Ca_3AlAs_3 structure with infinite linear chains. Compounds with large cations and large anions, in contrast, form one of the two possible structure types containing dumbbells (Ba_3AlSb_3 or Ba_3GaSb_3 types). Finally, Sr_3GaSb_3 forms a completely unique structure that has been described by Schäfer et al. as a compromise between the former structure types [31].

If the chemical formula A_3MPn_3 is rewritten as "$A_6M_2Pn_6$," the relationship to $A_5M_2Pn_6$ compounds becomes clearer, and we can readily see that $A_5M_2Pn_6$ compounds are cation deficient relative to the "$A_6M_2Pn_6$" type and thus lack two electrons per formula unit. To account for this electron deficiency, $A_5M_2Pn_6$ compounds form two covalent Pn–Pn bonds per formula unit. For example, the $Ca_5Ga_2As_6$ structure type contains linear, corner-linked chains of tetrahedra (Figure 11.2d), similar to the chains in Ca_3GaAs_3. Here, however, neighboring chains are oxidatively coupled via Pn–Pn bonds, leading to ladderlike anionic structures. The Zintl electron counting scheme in this case is satisfied by $(A^{2+})_5(4b\ M^-)_2(2b\ Pn^-)_4(1b\ Pn^{2-})_2$. The $Ca_5Al_2Bi_6$ structure type contains the same double chains but differs in the way the chains are packed (see Figure 11.3b). This cannot be readily explained by geometrical differences, as Ca^{2+} and Yb^{2+}, for example, are nearly identical in radius. Instead, the difference appears to stem from increased covalency in the A–Pn interactions in the latter structure type [35,37]. The anionic substructure shown in Figure 11.2e is found only in one compound: $Sr_5Al_2Sb_6$. In this unique structure type, corner- and edge-linked tetrahedra form infinite, curvy chains, and additional Sb–Sb bonds extend from every second tetrahedron. Zintl electron counting in the covalent limit yields $(Sr^{2+})_5(4b\ Al^-)_2(2b\ Pn^-)_4(1b\ Pn^{2-})_2$.

Finally, it is worth including in our discussion Sn-containing $A_5M_2Pn_6$ compounds, which form the $Ca_5Sn_2As_6$ structure type. In these, the additional electron provided by Sn allows for the formation of an anionic structure identical to that found in the Ca_3AlAs_3 structure: simple corner-linked chains. The structures differ only in the packing of the chains and the cation coordination. Table 11.1 lists all known A_3MPn_3 and $A_5M_2Pn_6$ compounds that crystallize in the eight structure types discussed in this chapter.

11.3 Electronic Structure

Since the dominant structural features in all $A_5M_2Sb_6$ and A_3MSb_3 compounds are related, it is not surprising that the electronic density of states also show some similar characteristics across all structure types. As expected, the valence band edge is dominated by M and Sb states, while the conduction band contains primarily cation electronic states, consistent with the salt-like nature of Zintl phases [24,33,35,48,50–52]. In the investigated compounds, M–Pn bonding and antibonding interactions lead to distinctive peaks in the valence and conduction bands, respectively. Peaks associated with M–Pn bonding interactions are found deep in the valence band (between −4 and −6 eV), as evidenced by the large fraction of M-character, while the corresponding M–Pn antibonding states have been shown to play an important role near the band gap edge [50].

In $A_5M_2Pn_6$ compounds (except for M = Sn), the Pn–Pn bonds are the most critical interaction in determining the size of the band gap. In the $Ca_5Al_2Sb_6$ structure type, Sb–Sb antibonding states form the conduction band minimum, while M–Sb states are found at

TABLE 11.1

All Known A_3MPn_3 and $A_5M_2Pn_6$ Compounds That Crystallize in the Eight Structure Types

Ca_3AlAs_3	Ba_3AlSb_3	Ba_3GaSb_3	Sr_3GaSb_3
Pnma	*Cmca*	*Pnma*	*P*2_1/*n*
Ca_3InP_3 [39]	Sr_3AlSb_3 [40]	Ba_3GaSb_3 [39]	Sr_3GaSb_3 [31]
Sr_3InP_3 [31]	Ba_3AlSb_3 [43]		
Eu_3InP_3 [34]	Eu_3GaP_3 [42]		
Ca_3AlAs_3 [41,43]			
Ca_3GaAs_3 [39]			
Ca_3AlSb_3 [32]			
$Ca_5Ga_2As_6$	**$Ca_5Al_2Bi_6$**	**$Sr_5Al_2Sb_6$**	**$Ca_5Sn_2As_6$**
Pbam	*Pbam*	*Pnma*	*Pbam*
$Ca_5Ga_2As_6$ [44]	$Ca_5Al_2Bi_6$ [32]	$Sr_5Al_2Sb_6$ [45]	$Ca_5Sn_2As_6$ [46]
$Ca_5Al_2Sb_6$ [29]	$Sr_5In_2Bi_6$ [47]		$Sr_5Sn_2As_6$ [48]
$Ca_5Ga_2Sb_6$ [30]	$Yb_5Al_2Sb_6$ [35]		$Eu_5Sn_2As_6$ [48]
$Ca_5In_2Sb_6$ [30]	$Yb_5Ga_2Sb_6$ [36]		$Sr_5Sn_2P_6$ [49]
$Sr_5In_2Sb_6$ [30]	$Yb_5In_2Sb_6$ [37]		
$Ba_5In_2Sb_6$ [45]			
$Eu_5In_2Sb_6$ [38]			

The relevant crystallographic studies are cited.

slightly higher energies and thus have only an indirect role in determining the size of the band gap [50]. In contrast, in Yb-based compounds ($Ca_5Al_2Bi_6$ structure type), the Pn–Pn bonds lead to antibonding states at the Fermi level, closing the band gap entirely, as shown in Figure 11.3a for $Yb_5Al_2Sb_6$. $Yb_5M_2Sb_6$ compounds were confirmed to behave as semimetals experimentally, with extremely low Seebeck coefficients and low *ZT* values [35–37,53].

Table 11.2 lists the electronic band gaps (E_g) of selected $A_5M_2Sb_6$ and A_3MSb_3 compounds estimated using several independent methods: density functional theory (DFT), the slope of the high temperature resistivity (ρ), the maximum Seebeck value (α), and from optical absorption measurements. Note that the interpretation of the optical absorption edge was unambiguous. Details of the analysis can be found in References 50 and 54. Among compounds with the $Ca_5M_2Sb_6$ structure type, a general trend between composition and the

TABLE 11.2

Band Gap Estimates from DFT Calculations and from Experimental Resistivity, Seebeck Coefficients, and Optical Absorption Measurements

	DFT (eV)	From ρ (eV)	From α (eV)	Optical (eV)	Reference
$Ca_5Al_2Sb_6$	0.5	0.7	0.5	0.65	[50]
$Ca_5Ga_2Sb_6$	0.29	0.6	0.3	0.43	[50]
$Ca_5In_2Sb_6$	0.48	0.7	0.4	0.64	[50]
$Sr_5In_2Sb_6$	0.44	0.6	0.4	0.55	[54]
$Sr_5Al_2Sb_6$	0.78	0.5	0.4	—	[55]
Ca_3AlSb_3	0.25	—	0.6	—	[25, 41]
Sr_3GaSb_3	0.77	—	0.5	—	[24]
Sr_3AlSb_3	1.05	0.7	0.6	—	[33]

TABLE 11.3

$N_{v,\,edge}$ Is the Valence Band Degeneracy at $E_F = 0$, and $N_{v,\,add}$ Is the Number of Additional Bands within 0.1 eV of the Band Edge

	$N_{v,edge}$	$N_{v,add}$	m_1^* (m_e)	m_2^* (m_e)	m_3^* (m_e)	m_{DOS}^* (m_e)	Reference
$Ca_5Al_2Sb_6$	2	0	0.20	6.49	1.36	1.92	[50]
$Ca_5Ga_2Sb_6$	2	0	0.18	3.74	1.19	1.46	[50]
$Ca_5In_2Sb_6$	2	0	0.21	3.76	1.53	1.80	[50]
$Sr_5Al_2Sb_6$	—	—	—	—	—	—	[55]
Ca_3AlSb_3	2	1	0.15, 0.75	0.15, 0.75	0.75	0.85–1.11	[25]
Sr_3GaSb_3	1	2	0.18	0.57	0.66	0.41–0.85	[24]
Sr_3AlSb_3	1	2	1.80	1.25	0.21	0.78–1.62	[33]

The single-band effective mass, m^*, is given in three perpendicular directions, where m_1^* is parallel to the tetrahedral chains or pairs. m_{DOS}^* is the estimated range for the density of states effective mass.

size of the band gap was noted by Yan et al. [52]: $Ca_5Ga_2As_6 > Ca_5Al_2Sb_6 > Ca_5In_2Sb_6 > Sr_5In_2Sb_6 > Ba_5In_2Sb_6 > Ca_5Ga_2Sb_6$. This trend is in good agreement with the available experimental data and the DFT results in Table 11.2.

11.3.1 Band Degeneracy

The key features of the electronic band dispersions (the band degeneracy, mass, and anisotropy) are presented in Table 11.3 for selected compounds. Only the valence band is considered here since, experimentally, only *p*-type doping has been achieved. The number of bands that contribute to transport in the valence band is difficult to specify precisely due to multiple bands within a few k_BT of the band edge. We define the number of bands at the valence band maximum as $N_{v,edge}$ and the number of additional bands lying within 0.1 eV of the band edge is given by $N_{v,add}$. Whether or not the additional bands are involved in transport depends on several factors: the energy of the band maximum relative to $E_F = 0$, the degree of Fermi spreading at a given temperature (nearly 0.1 eV at very high temperatures), and the location of the Fermi level, which varies significantly with doping. In general, A_3MPn_3 and $A_5M_2Pn_6$ compounds have relatively few bands contributing to transport, especially when compared with high-efficiency thermoelectric materials such as PbTe [56,57]. Low band degeneracy, resulting from a band gap at the Γ point, is one of the factors limiting the maximum performance of Zintl thermoelectrics investigated to date [58].

11.3.2 Electronic Band Mass

The structure types discussed in this chapter are highly anisotropic, leading to a large degree of anisotropy in the electronic band structures. Table 11.3 lists the single-band effective mass (m^*) in three perpendicular directions. As might be expected, the orientation of the anionic substructure influences m^*; in compounds with chain-forming structures, the band mass parallel to the chains (m_1^*) is generally lightest. However, this simple picture is complicated by the presence of multiple bands with different masses, such as in Ca_3AlSb_3. While the topmost bands ($N_{v,edge}$) in the chain-forming structures show preference for transport along the chains, the additional bands ($N_{v,add}$) are essentially isotropic. In contrast, in Sr_3AlSb_3, which contains pairs of tetrahedra, the topmost band is the lightest

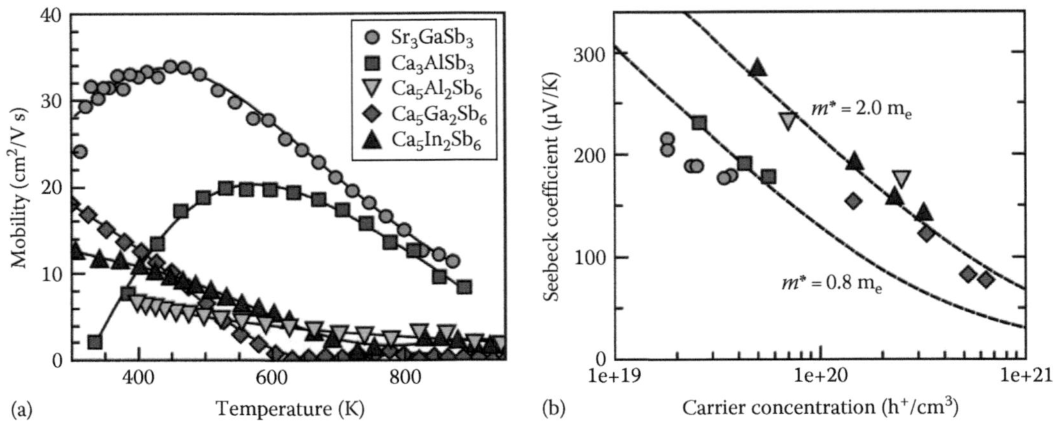

FIGURE 11.4
(a) The Hall mobility of undoped samples and (b) Seebeck coefficients of Zn-doped samples are consistent with higher effective mass in $Ca_5M_2Sb_6$ compounds (M = Al, Ga, In) compared with Sr_3AlSb_3 and Ca_3AlSb_3. (Data from Johnson, S.I. et al., *J. Mater. Chem. A*, 1, 4244, 2013; Zevalkink, A. et al., *Energy Environ. Sci.*, 5, 9121, 2012; Zevalkink, A. et al., *Energy Environ. Sci.*, 4, 510, 2011; Zevalkink, A. et al., *J. Appl. Phys.*, 110, 013721, 2011; Zevalkink, A. et al., *Dalton Trans.*, 42, 9713, 2013; Zevalkink, A. et al., *Chem. Mater.*, 24, 2091, 2012.)

in the direction perpendicular to the tetrahedral pairs. The density of states effective mass, m^*_{DOS}, was estimated using the following equation:

$$m^*_{DOS} = (m^*_1 m^*_2 m^*_3)^{1/3} N_v^{2/3} \tag{11.1}$$

To express the possible range in m^*_{DOS} that stems from differences in the total number of contributing bands, $N_v = N_{v,edge}$ and $N_v = N_{v,add} + N_{v,edge}$ were used as lower and upper bounds. Although these estimates are imprecise, they highlight the lower mass found in Sr_3GaSb_3 and Ca_3AlSb_3 relative to the $Ca_5M_2Sb_6$ compounds.

Larger band mass implies lower carrier mobility, larger Seebeck coefficients, and the need for carrier concentrations to optimize *ZT*. The calculated differences in effective mass were largely consistent with experimental results. $Ca_5M_2Sb_6$ compounds (M = Al, Ga, In) were found to have lower Hall mobility in undoped samples and higher Seebeck coefficients (Figure 11.4a and b, respectively) compared with Sr_3AlSb_3 and Ca_3AlSb_3. The dashed curves in Figure 11.4b were calculated using a single parabolic band (SPB) model, yielding band masses of 2.0 and 0.8 m_e, respectively, for the upper and lower curves. These values are in approximate agreement with the m^*_{DOS} values predicted by DFT.

11.4 Tunable Electronic Transport Properties

A_3MPn_3 and $A_5M_2Pn_6$ Zintl phases, consistent with Zintl electron counting and DFT calculations, behave as intrinsic semiconductors, displaying low carrier concentrations ($n_H < 10^{19}$ h$^+$/cm^3), high resistivity, ρ, and high Seebeck coefficients, α. The only reported exceptions are compounds with the $Ca_5Al_2Bi_6$ structure type, which lack band gaps and exhibit semimetallic behavior [53]. To optimize the *ZT* of a semiconducting material, the carrier concentration, *n*, must be tuned by either electronic doping or by altering the stoichiometry. Doping with aliovalent elements results in a valence imbalance and an increased concentration of free carriers, leading to a transition from nondegenerate to degenerate semiconducting behavior. Compared with undoped Zintl phases, which tend to be poor thermoelectric materials, the peak *ZT* can be significantly improved by tuning the electronic properties by doping [11,60].

In particular, $A_5M_2Sb_6$ compounds (A = Ca, Sr, or Eu and M = Al, Ga, or In) with the $Ca_5Ga_2As_6$ structure type have proved to be quite tolerant to doping. The aforementioned $A_5M_2Sb_6$ compounds are naturally slightly *p*-type, and their intrinsic semiconducting behavior leads to high resistivity and thus low *ZT* values of approximately 0.1 [50]. Motivated by the effective doping of divalent Zn or Mn on the trivalent Al site in $Yb_{14}AlSb_{11}$, similar substitutions were attempted and found to be effective in $A_5M_2Sb_6$ compounds. The hole carrier concentration resulting from Zn^{2+} replacing In^{3+} in $Ca_5In_2Sb_6$ can be estimated as a function of Zn concentration (x) by assuming that each Zn substitution yields a single free hole, shown as a dashed line in Figure 11.5a. Experimentally, the Hall carrier concentration of $Ca_5In_{2-x}Zn_xSb_6$ samples increases as predicted at low Zn concentration

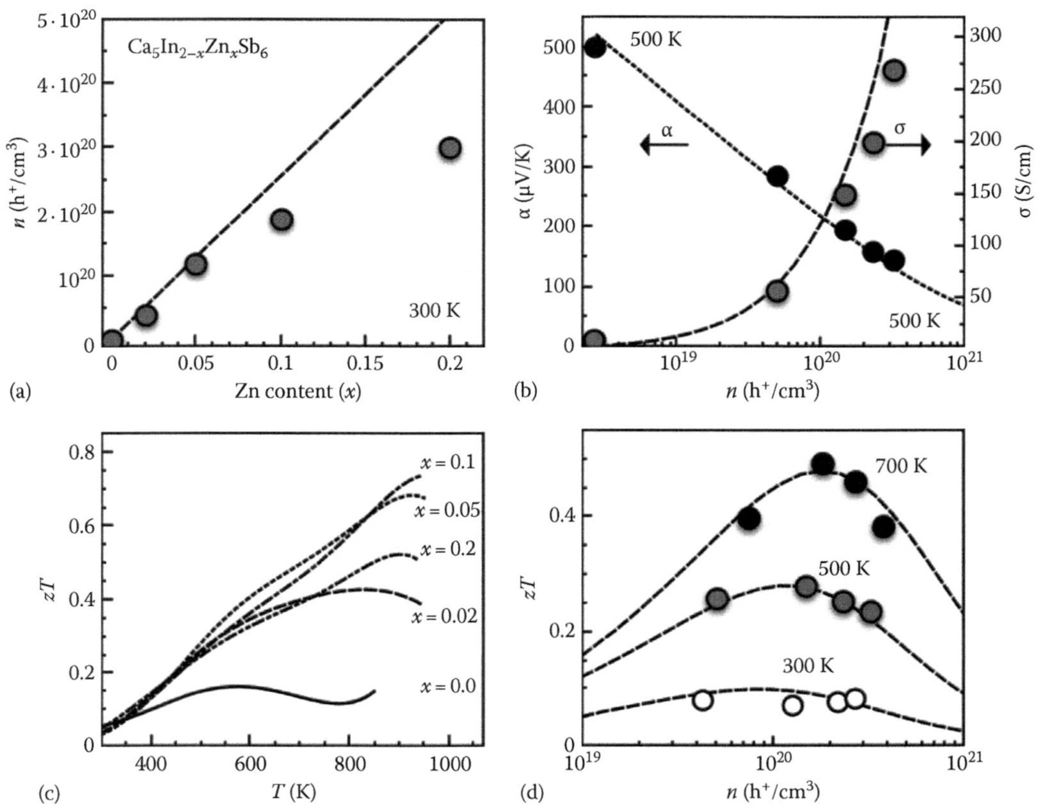

FIGURE 11.5
(a) Substituting Zn^{2+} on the In^{3+} site increases the *p*-type carrier concentration in $Ca_5In_{2-x}Zn_xSb_6$ and (b) causes a transition from intrinsic to extrinsic semiconducting behavior. An SPB model was used to calculate the dotted curves with $m^* = 2\ m_e$ and $\mu_o = 6.9\ cm^2/Vs$ at 500 K. (c) With increasing Zn doping, the maximum *ZT* increased from ~0.1 to 0.7. (d) The optimum carrier concentration predicted from the curves in part (b), assuming *n*-independent κ_L, corresponds to $x = 0.05$. (Data from Zevalkink, A. et al., *Dalton Trans.*, 42, 9713, 2013.)

(gray symbols in Figure 11.5a) but begins to deviate for $x > 0.05$ [28]. Figure 11.5b shows the decreasing Seebeck coefficient and increasing conductivity with increasing Zn content (x). The peak ZT increases from 0.1 to 0.7 with increasing Zn concentration (Figure 11.5c). Zn-doped $Ca_5Al_2Sb_6$, $Ca_5Ga_2Sb_6$, $Sr_5In_2Sb_6$, and $Eu_5In_2Sb_6$ were found to behave similarly, allowing for optimized ZT in each case [23,27,54,61].

The SPB model is a simple tool for approximating the dependence of α and σ on carrier concentration (see dashed curves in Figure 11.5b), which can, in turn, be used to estimate ZT as a function of n. In this model, we assume that the band mass, m^*, does not change with doping (often referred to as a "rigid band" approximation) and that the dominant scattering results from interaction of electronic carriers with acoustic phonons. ZT can then be predicted as a function of n by considering κ_L as independent of doping, thus ruling out the need to synthesize large numbers of samples. Figure 11.5d shows the success of this model for $Ca_5In_{2-x}Zn_xSb_6$ at multiple temperatures. Additional details regarding the SPB model and its limitations are discussed in Reference 61.

The doping efficiency (i.e., the ratio of experimental to predicted carrier concentration) was not high in all of the compounds discussed in this chapter. Figure 11.6 compares the doping efficiency for a few selected cases. In the three $Ca_5M_2Sb_6$ compounds, the carrier concentrations resulting from doping were very close to the predicted values. However, in Zn-doped Sr_3GaSb_3 and Na-doped Ca_3AlSb_3, the doping efficiency is very low; n_H increases, but only at a fraction of the expected rate [24,25]. Similarly, substitution of Na on the divalent cation site in $Ca_5Al_2Sb_6$ also leads to an increase of the hole carrier concentration but with lower doping efficiency. Finally, doping $Sr_5Al_2Sb_6$ and Sr_3AlSb_3 with Zn had no effect on n_H [33,35]. This disparity in doping efficiency can stem from multiple factors,

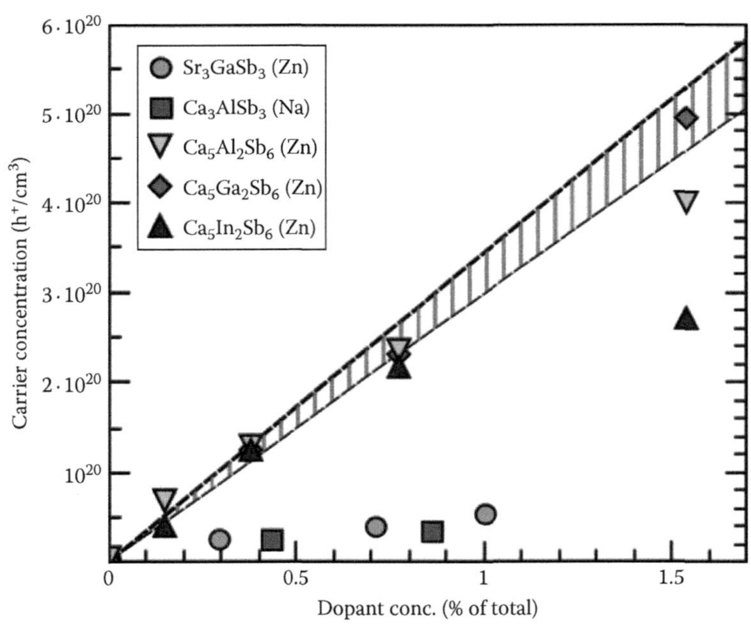

FIGURE 11.6
Dependence of the experimental Hall carrier concentration on dopant concentration [23–25,27,28]. Dashed lines show the predicted carrier concentration. *Note*: The difference between the upper and lower dashed lines is a result of slight differences in the average volume per atom.

including geometric and electronic differences, both of which influence the solubility of a dopant (and whether it precipitates as a secondary phase) and the tendency of the primary phase to form compensating site defects. As observed for most Zintl phases, the A_3MPn_3 and $A_5M_2Pn_6$ Zintl phases tolerate only *p*-type doping, likely due to a lower energy barrier to the formation of *p*-type lattice defects, compared with *n*-type defects [19,62].

11.5 Glass-Like Lattice Thermal Conductivity

Shown in Figure 11.7a are the lattice thermal conductivities, κ_L, of Zn-doped Sr_3GaSb_3, Ca_3AlSb_3, and $Ca_5M_2Sb_6$ and undoped $Sr_5Al_2Sb_6$ and Sr_3AlSb_3 as a function of temperature. Remarkably, the temperature dependence of κ_L is similar in all compounds, exhibiting a $1/T$ decay indicative of Umklapp scattering. In all of the compounds shown here at high temperatures, κ_L approaches the so-called glass limit or minimum lattice thermal conductivity [64], given by Equation 11.2 for $T > T_{Debye}$:

$$\kappa_{\min} = \frac{1}{2}\left(\frac{\pi}{6}\right)^{1/3} k_B (V)^{-2/3} (2v_T + v_L) \tag{11.2}$$

The minimum lattice thermal conductivity, $\kappa_{\min}$, is a function of the longitudinal and transverse sound velocities (ν_L and ν_T) and volume per atom, *V*. For A_3MPn_3 and $A_5M_2Pn_6$ phases, ν_L and ν_T were determined experimentally. From these values, the elastic moduli

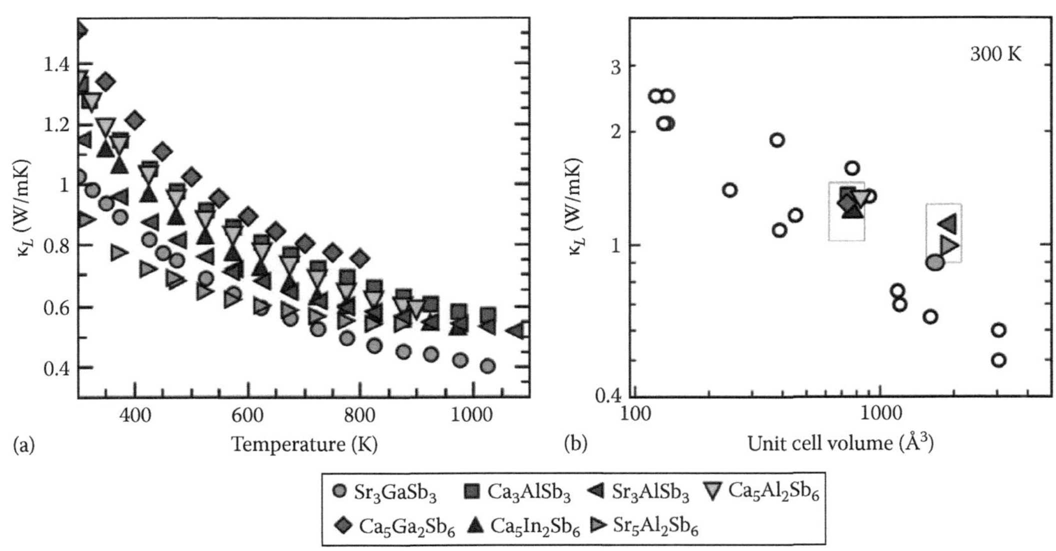

FIGURE 11.7
(a) The lattice thermal conductivity is shown for selected compounds as a function of temperature. (b) The room temperature lattice thermal conductivity of Zintl antimonides decreases with increasing unit cell volume. $A_5M_2Sb_6$ and A_3MSb_3 compounds fit the expected trend. (Data for unfilled circles are from Cahill, D. et al., *Phys. Rev. B*, 46, 6131, 1992; data for filled symbols are from Zevalkink, A. et al., *Energy Environ. Sci.*, 4, 510, 2011; Zevalkink, A. et al., *J. Appl. Phys.*, 110, 013721, 2011; Zevalkink, A. et al., *Dalton Trans.*, 42, 9713, 2013; Zevalkink, A. et al., *Chem. Sus. Chem.*, 6, 2316, 2013; Zevalkink, A. et al., *Dalton Trans.*, 43, 14720, 2014.)

TABLE 11.4

The Number of Atoms Per Unit Cell (N), Theoretical Density (d), Shear and Bulk Moduli (G and K), and Longitudinal and Transverse Speeds of Sound (ν_L and ν_T)

	N	d (g/cm³)	G (GPa)	K (GPa)	ν_l (m/s)	ν_t (m/s)	Reference
$Ca_5Al_2Sb_6$	26	4.31	24.8	39.4	2400	4100	[50]
$Ca_5Ga_2Sb_6$	26	4.52	23.5	40.9	2280	3970	[50]
$Ca_5In_2Sb_6$	26	4.90	21.9	38.2	2120	3710	[50]
$Sr_5In_2Sb_6$	26	5.32	21.1	28.6	1994	3268	[59]
$Eu_5Ga_2Sb_6$	26	6.59	23.4	32.6	1750	2920	N/A
$Eu_5In_2Sb_6$	26	6.77	20.6	30.1	1880	3110	[60]
$Yb_5Al_2Sb_6$	26	7.47	26.5	37.0	1610	2650	[53]
$Yb_5Ga_2Sb_6$	26	7.68	20.0	27.4	1860	3030	[53]
$Yb_5In_2Sb_6$	26	7.89	27.3	35.9	1790	2930	[53]
$Sr_5Al_2Sb_6$	52	4.50	18.9	37.2	2050	3720	[55]
Ca_3AlSb_3	28	4.14	24.6	39.1	2440	4170	[25]
Sr_3GaSb_3	56	4.92	22.3	39.4	2130	3750	[24]
Sr_3AlSb_3	56	4.58	196	34.2	2070	3630	[33]

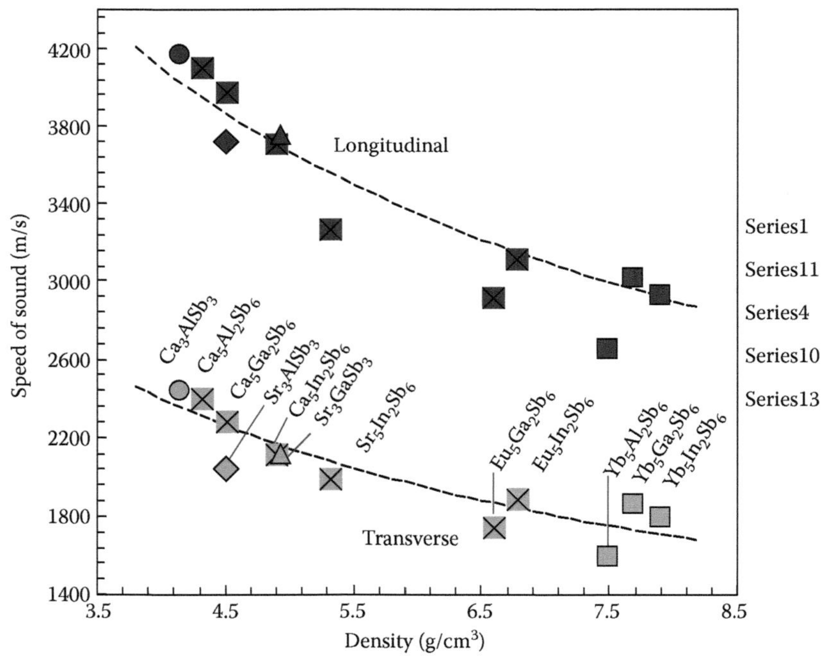

FIGURE 11.8
The speed of sound decreases with increasing density. The dashed lines represent a $\sqrt{1/d}$ dependence, deviations from which are due to differences in elastic moduli. The data and sources are listed in Table 11.4.

were determined according to Equation 11.3, where d is the theoretical density and K and G are the bulk and shear moduli, respectively. Table 11.4 lists the experimental sound velocities and elastic moduli for selected compounds. The decrease in the experimental sound velocity with increasing theoretical density is illustrated in Figure 11.8:

$$v_T = \sqrt{\frac{G}{d}} \quad v_L = \sqrt{\frac{K + \frac{4}{3}G}{d}} \tag{11.3}$$

In complex materials, the size of the unit cell is an additional predictor of lattice thermal conductivity [63], especially when considering materials with otherwise similar bonding characteristics and density. Figure 11.7b illustrates the decrease in κ_L with increasing unit cell volume in Zintl antimonides. This trend also applies to the specific compounds described here; the lowest values of κ_L were found in Sr_3GaSb_3, Sr_3AlSb_3, and $Sr_5Al_2Sb_6$, all of which have unit cells containing at least 50 atoms. It is instructive to compare Sr_3GaSb_3 and $Ca_5In_2Sb_6$, which have nearly identical sound velocity, but unit cells that differ in size by a factor of two. This leads to approximately 20% lower κ_L in Sr_3GaSb_3 across the entire temperature range. Assuming that the Grüneisen parameter does not vary greatly between these compounds, the difference in κ_L can most likely be explained by a larger unit cell of Sr_3GaSb_3.

11.6 Point Defect Scattering

Alloying is a classic strategy for the reduction of the lattice thermal conductivity of thermoelectric materials [66,67]. Among Zintl phases, it has been applied most successfully to AM_2Sb_2 compounds, leading to large improvements in ZT [8,65,68,69]. Due to the large number of isoelectronic substitutions possible in compounds with the $Ca_5Ga_2As_6$ structure type, it is natural to consider alloying as a possible route to increase thermoelectric efficiency. However, this strategy can only work when the reduction of thermal conductivity surpasses reductions in the carrier mobility [70].

The results of alloying of Al and In on the M site in undoped and Zn-doped $Ca_5M_2Sb_6$ are shown in Figure 11.9a through d. In agreement with simple electron counting, the Al:In ratio does not influence the carrier concentration or Seebeck coefficients. All of the undoped samples behave as intrinsic semiconductors, while all Zn-doped samples have optimized carrier concentrations of approximately 2×10^{20} h^+/cm^3. The main effect of alloying is to decrease both the mobility, μ, and the lattice thermal conductivity, as shown in Figure 11.9a and b, respectively. At high temperatures (773 K), scattering of electrons and phonons by point defects is outweighed by the competing influence of acoustic and Umklapp phonon scattering processes, so that the effect of alloying on κ_L and μ becomes negligible. The combined effect of alloying thus has very little influence on ZT. Wang et al. [70] have shown that alloying is most effective at low temperatures and in compounds that possess high lattice thermal conductivity and mobility. For example, $Si_{1-x}Ge_x$ is an ideal alloy system, achieving a 95% reduction in κ_L and enhanced ZT. Considering both $Ca_5Al_2Sb_6$ and $Ca_5In_2Sb_6$ have very intrinsically low glass-like lattice thermal conductivity, there is limited room for further improvement via alloying. We anticipate that alloying for the sole purpose of decreasing thermal conductivity will be similarly ineffective in other

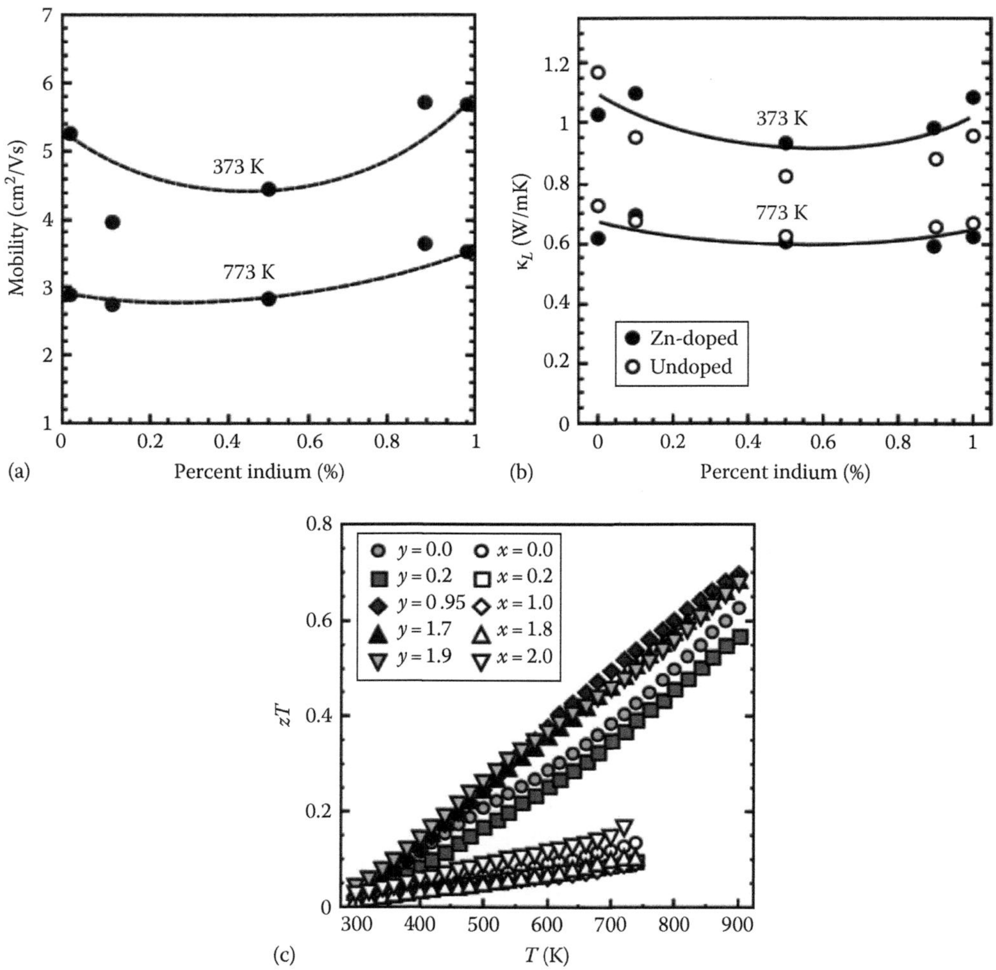

FIGURE 11.9
(a) In $Ca_5Al_{2-x}In_xSb_6$ samples, alloying reduces the Hall mobility and (b) κ_L relative to the pure end members. This reduction is less significant at higher temperatures. *Note*: The *x*-axis is given as percent In on the Al site. (c) The figure of merit of doped (higher *ZT*) and undoped (low *ZT*) $Ca_5Al_{2-x}In_xSb_6$ is not a strong function of alloying. (Data from Zevalkink, A. et al., *Dalton Trans.*, 43, 15872, 2014.)

A_3MPn_3 and $A_5M_2Pn_6$ Zintl compounds. However, alloying to control the electronic band structure may be a promising strategy to achieve band convergence or to fine-tune the band gap, for example.

11.7 Figure of Merit

The maximum figures of merit achieved to date in A_3MSb_3 and $A_5M_2Sb_6$ compounds are shown in Figure 11.10. Zn^{2+} on the M^{3+} site was used in all cases except Ca_3AlSb_3 and $Ca_5Al_2Sb_6$, in which slightly higher peak *ZT* values were achieved with Na^{1+} on the Ca^{2+}

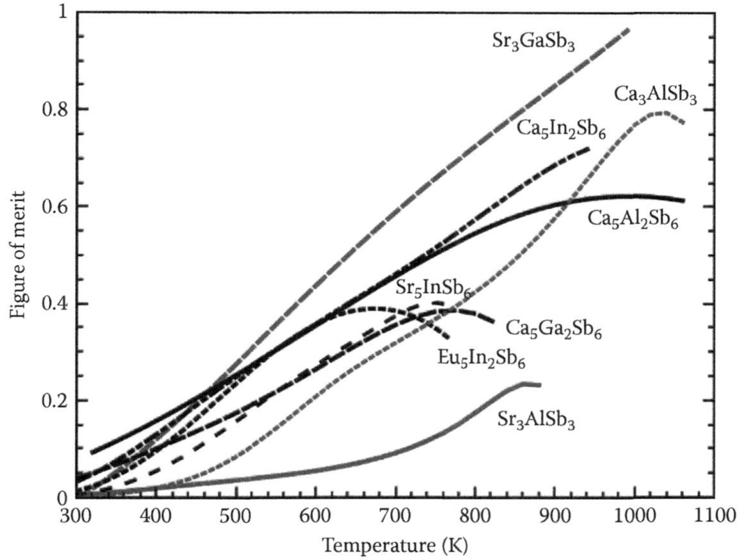

FIGURE 11.10
The figures of merit of optimally doped Ca_3AlSb_3 and Sr_3GaSb_3 and $Ca_5In_2Sb_6$ are highest due to their higher electronic mobility and lower κ_L. (Data taken from Zevalkink, A. et al., *Energy Environ. Sci.*, 4, 510, 2011; Zevalkink, A. et al., *J. Appl. Phys.*, 110, 013721, 2011; Zevalkink, A. et al., *Dalton Trans.*, 42, 9713, 2013; Zevalkink, A. et al., *Chem. Sus. Chem.*, 6, 2316, 2013; Chanakian, S. et al., *J. Mater. Chem. A*, 19, 10289, 2015; Zevalkink, A. et al., *Dalton Trans.*, 43, 14720, 2014; Chanakian, S. et al., *J. Mater. Chem. C*, 3, 10518, 2015.)

site. Due to the difference in effective mass, the optimal carrier concentration for $A_5M_2Sb_6$ compounds was approximately $n_H = 10^{20}$ h^+/cm^3, while Ca_3AlSb_3 and Sr_3GaSb_3 are optimized when $n_H = 5 \times 10^{19}$ h^+/cm^3. Relative to the other chain-forming Zintl compounds, Sr_3GaSb_3 exhibits a higher figure of merit across the entire measured temperature range, which can be attributed almost entirely to its lower κ_L. However, in all optimally doped samples, promising values were obtained for the peak *ZT*, highlighting the value of continued investigations of new Zintl phases. Future improvements to the thermoelectric performance of these compounds may stem from optimization of the synthetic techniques and from further tuning of the band structure through isoelectronic substitutions.

11.8 Conclusion

Zintl phases are efficient thermoelectric materials with complex structures and bonding, highly tunable properties, and, at times, exceptionally low, glass-like lattice thermal conductivity. Several Zintl compounds have been reported with *ZT* values in excess of unity, including $Yb_{14}MnSb_{11}$, YbM_2Sb_2 (M = Zn, Cd), and many others have exciting properties that need only to be optimized. However, the vast majority of Zintl compounds have not been studied beyond the initial crystallographic characterization, suggesting that many potential thermoelectric materials remain "undiscovered." Recognized only in past several years, the chain-forming A_3MPn_3 and $A_5M_2Pn_6$ compounds represent an entirely new class of Zintl thermoelectrics. Like $Yb_{14}MnSb_{11}$, they are stable at relatively high temperatures, they exhibit intrinsically low lattice thermal conductivity, and doping can be used to tune their electronic properties. This combination has led to promising *ZT* values in the range of 0.5–1 in initial studies. In the future, focus should be placed on strategies such as band structure engineering through chemical substitutions to improve the band degeneracy and electronic carrier mobility, which may lead to improved thermoelectric performance.

Acknowledgment

This research was carried out in part at the Jet Propulsion Laboratory, California Institute of Technology, under a contract with the National Aeronautics and Space Administration and was supported by the NASA Science Missions Directorate's Radioisotope Power Systems Technology Advancement Program. A. Zevalkink and U. Aydemir acknowledge the financial assistance, respectively, of the National Science Foundation and the Scientific and Technological Research Council of Turkey.

References

1. Kauzlarich, S. M. 1996. *Chemistry, Structure, and Bonding of Zintl Phases and Ions.* New York: Wiley-VCH.
2. Kauzlarich, S. M., S. R. Brown, and G. J. Snyder. 2007. Zintl phases for thermoelectric devices. *Dalton Trans.* 21: 2099–2107.
3. Toberer, E. S., A. F. May, and G. J. Snyder. 2010. Zintl chemistry for designing high efficiency thermoelectric materials. *Chem. Mater.* 22: 624–634.
4. Wang, J., S. Q. Xia, and X. T. Tao. 2013. $A_{10}LaCdSb_9$ (A = Ca, Yb): A highly complex Zintl system and the thermoelectric properties. *Chem. Asian J.* 8: 251–257.
5. He, H., A. Zevalkink, Z. M. Gibbs, G. J. Snyder, and S. Bobev. 2012. Synthesis, structural characterization, and physical properties of the type-I clathrates $A_8Zn_{18}As_{28}$ (A = K, Rb, Cs) and $Cs_8Cd_{18}As_{28}$. *Chem. Mater.* 24: 3596–3603.
6. Schaefer, H. 1985. On the problem of polar intermetallic compounds: The stimulation of E. Zintl's work for the modern chemistry of intermetallics. *Ann. Rev. Mater. Sci.* 15: 1–41.
7. von Schnering, H. G. 1981. Homoatomic bonding of main group elements. *Angew. Chem. Int. Ed.* 20: 33–51.
8. Gascoin, F., S. Ottensmann, D. Stark., M. S. Haile, and G. Snyder. 2005. Zintl phases as thermoelectric materials: Tuned transport properties of the compounds $Ca_xYb_{1-x}Zn_2Sb_2$. *Adv. Funct. Mater.* 15: 1860–1864.
9. Snyder, G. J., M. Christensen, E. Nishibori, T. Caillat, and B. B. Iversen. 2004. Disordered zinc in Zn_4Sb_3 with phonon-glass and electron-crystal thermoelectric properties. *Nat. Mater.* 3: 458–463.
10. Brown, S., S. Kauzlarich, F. Gascoin, and G. J. Snyder. 2006. $Yb_{14}MnSb_{11}$: New high efficiency thermoelectric material for power generation. *Chem. Mater.* 18: 1873–1877.
11. Toberer, E. S., C. A. Cox, S. R. Brown et al. 2008. Traversing the metal-insulator transition in a Zintl phase: Rational enhancement of thermoelectric efficiency in $Yb_{14}Mn_{1-x}Al_xSb_{11}$. *Adv. Funct. Mater.* 18: 2795–2800.

12. Cox, C. A., E. S. Toberer, A. A. Levchenko et al. 2009. Structure, heat capacity, and high-temperature thermal properties of $Yb_{14}Mn_{1-x}Al_xSb_{11}$. *Chem. Mater.* 21: 1354–1360.
13. Yi, T., M. N. Abdusalyamova, F. Makhmudov, and S. M. Kauzlarich. 2012. Magnetic and transport properties of Te doped $Yb_{14}MnSb_{11}$. *J. Mater. Chem.* 22: 14378.
14. Toberer, E. S., S. R. Brown, T. Ikeda, S. M. Kauzlarich, and G. J. Snyder. 2008. High thermoelectric efficiency in lanthanum doped $Yb_{14}MnSb_{11}$. *Appl. Phys. Lett.* 93: 062110.
15. Brown, S., E. S. Toberer, T. Ikeda et al. 2008. Improved thermoelectric performance in $Yb_{14}Mn_{1-x}Zn_xSb_{11}$ by the reduction of spin-disorder scattering. *Chem. Mater.* 20: 3412–3419.
16. Roudebush, J. H., J. Grebenkemper, Y. Hu, N. Kazem, and M. N. Abdusalyamova, and S. M. Kauzlarich. 2014. $Yb_{14-x}Tm_xMnSb_{11}$ ($0 < x < 0.5$): Structure and magnetic properties. *J. Solid State Chem.* 211: 206–211.
17. Hu, Y., J. Wang, A. Kawamura, K. Kovnir, and S. Kauzlarich. 2014. $Yb_{14}MgSb_{11}$ and $Ca_{14}MgSb_{11}$: New Mg containing Zintl compounds and their structures, bonding, and thermoelectric properties. *Chem. Mater.* 27: 343–351.
18. Hu, Y., S. K. Bux, J. H. Grebenkemper, and S. M. Kauzlarich. 2015. The effect of light rare earth element substitution in $Yb_{14}MnSb_{11}$ on thermoelectric properties. *J. Mater. Chem. C* 3: 10566–10573.
19. Pomrehn, G., A. Zevalkink, W. G. Zeier, A. van de Walle, and G. J. Snyder. 2014. Defect-controlled electronic properties in AZn_2Sb_2 Zintl phases. *Angew. Chem. Int. Ed.* 53: 3422–3426.
20. Uvarov, C. A., F. Ortega-Alvarez, and S. M. Kauzlarich. 2012. Enhanced high-temperature thermoelectric performance of $Yb_{14-x}Ca_xMnSb_{11}$. *Inorg. Chem.* 51: 7617–7624.
21. Paik, J.-A., E. Brandon, T. Calliat, R. Ewell, and J. P. Fleurial. 2011. Life testing of $Yb_{14}MnSb_{11}$ for high performance thermoelectric couples. *Proceedings of Nuclear and Emerging Technologies for Space.*
22. Toberer, E. S., A. F. May, B. C. Melot, E. Flage-Larsen, and G. J. Snyder. 2010. Electronic structure and transport in thermoelectric compounds AZn_2Sb_2 (A = Sr, Ca, Yb, Eu). *Dalton Trans.* 39: 1046–1054.
23. Johnson, S. I., A. Zevalkink, and G. J. Snyder. 2013. Improved thermoelectric properties in Zn-doped $Ca_5Ga_2Sb_6$. *J. Mater. Chem. A* 1: 4244.
24. Zevalkink, A., W. G. Zeier, G. Pomrehn, E. Schechtel, W. Tremel, and G. J. Snyder. 2012. Thermoelectric properties of Sr_3GaSb_3, a chain-forming Zintl compound. *Energy Environ. Sci.* 5: 9121.
25. Zevalkink, A., E. S. Toberer, W. G. Zeier, E. Flage-Larsen, and G. J. Snyder. 2011. Ca_3AlSb_3: An inexpensive, non-toxic thermoelectric material for waste heat recovery. *Energy Environ. Sci.* 4: 510.
26. Toberer, E. S., A. Zevalkink, and G. J. Snyder. 2010. The Zintl compound $Ca_5Al_2Sb_6$ for low cost thermoelectric power generation. *Adv. Funct. Mater.* 20: 4375–4380.
27. Zevalkink, A., E. S. Toberer, T. Bleith, E. Flage-Larsen, and G. J. Snyder. 2011. Improved carrier concentration control in Zn-doped $Ca_5Al_2Sb_6$. *J. Appl. Phys.* 110: 013721.
28. Zevalkink, A., J. Swallow, and G. J. Snyder. 2013. Thermoelectric properties of Zn-doped $Ca_5In_2Sb_6$. *Dalton Trans.* 42: 9713–9719.
29. Cordier, G., E. Czech, M. Jakowski, and H. Schaefer. 1981. Zintlphasen mit komplexen anionen: Zur kenntnis von $Ca_5Al_2Sb_6$ und $Ca_3Al_2As_4$. *Rev. Chem. Miner.* 18: 9–18.
30. Cordier, G., H. Schafer, and M. Stelter. 1986. Perantimonidogallate und indate: Zur kenntnis von $Ca_5Ga_2Sb_6$, $Ca_5In_2Sb_6$, $Sr_5In_2Sb_6$. *Z. Naturforsch. B* 33: 5–8.
31. Cordier, G., H. Schaefer, and M. Stelter. 1987. Sr_3GaSb_3 und Sr_3InP_3 zwei neue Zintlphasen mit komplexen anionen. *Z. Naturforsch.* 42b: 1268–1272.
32. Cordier, G., H. Schaefer, and M. Stelter. 1984. Ca_3AlSb_3 und $Ca_5Al_2Bi_6$, zwie neue Zintlphasen mit kettenförmigen Anionen. *Z. Naturforsch.* 39b: 727–732.
33. Zevalkink, A., G. Pomrehn, J. Swallow, Y. Takagiwa, and G. J. Snyder. 2013. Thermoelectric properties and electronic structure of the Zintl-phase Sr_3AlSb_3. *Chem. Sus. Chem.* 6: 2316–2321.
34. Jiang, J., A. C. Payne, M. M. Olmstead et al. 2005. Complex magnetic ordering in Eu_3InP_3: A new rare earth metal zintl compound. *Inorg. Chem.* 44: 2189–2197.

35. Todorov, I., D. Y. Chung, L. Ye, A. J. Freeman, and M. G. Kanatzidis. 2009. Synthesis, structure and charge transport properties of $Yb_5Al_2Sb_6$: A Zintl phase with incomplete electron transfer. *Inorg. Chem.* 48: 4768–4776.
36. Subbarao, U., S. Sarkar, V. K. Gudelli, V. Kanchana, G. Vaitheeswaran, and S. C. Peter. 2000. $Yb_5Ga_2Sb_6$: A mixed valent and narrow-band gap material in the $RE_5M_2X_6$ family. *J. Solid State Chem.* 155: 55–61.
37. Kim, S.-J., J. R. Ireland, C. R. Kannewurf, and M. G. Kanatzidis. 2000. $Yb_5In_2Sb_6$: A new rare earth zintl phase with a narrow band gap. *J. Solid State Chem.* 155: 55–61.
38. Park, S.-M., E. S. Choi, W. Kang, and S.-J. Kim. 2002. $Eu_5In_2Sb_6$, $Eu_5In_{2-x}Zn_xSb_6$: Rare earth zintl phases with narrow band gaps. *J. Mater. Chem.* 12: 1839–1843.
39. Cordier, G., H. Schaefer, and M. Stelter. 1985. Neue Zintlphasen: Ba_3GaSb_3, Ca_3GaAs_3 und Ca_3InP_3. *Z. Naturforsch. B* 40: 1100–1104.
40. Cordier, G., M. Stelter, and H. Schaefer. 1984. Zintlphasen mit komplexen Anionen: Zur kenntnis von $Sr_6Al_2Sb_6$. *J. Less-Common Metals* 98: 285–290.
41. Cordier, G. and H. Schaefer. 1981. Ca_3AlAs_3—Ein intermetallisches Analogon zu den Kettensilicaten. *Angew. Chem.* 93: 474.
42. Somer, M., W. Carrillo Cabrera, K. Peters, von Schnering H. G., and G. Cordier. 1996. Crystal structure of Eu_6GaP_6. *Z. Kristal.* 211: 257.
43. Cordier, G., M. Stelter, and H. Schaefer. 1982. Zintl phases with complex anions—Data on Ca_3AlAs_3 and $Ba_6Al_2Sb_6$. *Z. Naturforsch.* 37: 975–980.
44. Verdier, P. P., P. L'Haridon, M. Maunaye, and Y. Laurent. 1976. Etude structurale de $Ca_5Ga_2As_6$. *Acta Cryst. B* 32: 726.
45. Cordier, G. and M. Stelter. 1988. $Sr_5Al_2Sb_6$ und $Ba_5In_2Sb_6$: Zwei neue Zintlphasen mit unterschiedlichen Baenderanionen. *Z. Naturforsch.* 43b: 463–466.
46. Eisenmann, B., H. Jordan, and H. Schaefer. 1985. $Ca_5Sn_2As_6$, das erste Inoarsenidostannat. *Z. Anorg. Allg. Chem.* 530: 74–78.
47. Zhongming, S., M. J. Gao, and P. Dachun. 2006. Synthesis and crystal structure of a new Zintl phases $Sr_5In_2Bi_6$. *J. Alloys Compd.* 421: 190–194.
48. Wang, J., S. Q. Xia, and X. T. Tao. 2012. $A_5Sn_2As_6$ (A = Sr, Eu). Synthesis, crystal and electronic structure, and thermoelectric properties. *Inorg. Chem.* 51: 5771–5778.
49. Eisenmann, B., H. Jordan, and H. Schaefer. 1986. Phasen im Grenzbereich der Zintlkonzeption: Zur Kenntnis von $Sr_5Sn_2P_6$ und SrSnP. *J. Less-Common Metals* 116: 251–258.
50. Zevalkink, A., G. S. Pomrehn, S. Johnson, J. Swallow, Z. M. Gibbs, and G. J. Snyder. 2012. Influence of the trial elements (M = Al, Ga, In) on the transport properties of $Ca_5M_2Sb_6$ Zintl compounds. *Chem. Mater.* 24: 2091–2098.
51. Yan, Y. L. and Y. X. Wang. 2011. High thermoelectric performance of $Ca_5Ga_2As_6$ as a Zintl compound. *J. Mater. Chem.* 21: 12497.
52. Yan, Y. L., Y. X. Wang, and G. B. Zhang. 2014. A key factor improving the thermoelectric properties of Zintl compounds $A_5M_2Pn_6$ (A = Ca, Sr, Ba; M = Ga, Al, In; Pn = As, Sb). *Comp. Mater. Sci.* 85: 88–93.
53. Aydemir, U., A. Zevalkink, A. Ormeci et al. 2015. Thermoelectric properties of the Zintl phases $Yb_5M_2Sb_6$ (M = Al, Ga, In). *Dalton Trans.* 44: 6767–6774.
54. Chanakian, S., A. Zevalkink, U. Aydemir et al. 2015. Enhanced thermoelectric properties of $Sr_5In_2Sb_6$ via Zn-doping. *J. Mater. Chem. A* 19: 10289–10295.
55. Zevalkink, A., Y. Takagiwa, K. Kitahara, K. Kimura, and G. J. Snyder. 2014. Thermoelectric properties and electronic structure of the Zintl phase $Sr_5Al_2Sb_6$. *Dalton Trans.* 43: 14720–14725.
56. Pei, Y., X. Shi, A. LaLonde, H. Wang, L. Chen, and G. J. Snyder. 2011. Convergence of electronic bands for high performance bulk thermoelectrics. *Nature* 473: 66–69.
57. DiSalvo, F. J. 1999. Thermoelectric cooling and power generation. *Science* 285: 703–706.
58. Zeier, W., A. Zevalkink, A. Gibbs, G. Hautier, M. Kanatzidis, and G. J. Snyder, 2016. Thinking like a chemist: Intuition in thermoelectric materials. *Angew. Chem. Int. Ed.* 55: 6826–6841.
59. Aydemir, U., A. Zevalkink, A. Ormeci, Z. M. Gibbs, S. Bux, and G. J. Snyder. 2015. Thermoelectric enhancement in $BaGa_2Sb_2$ by Zn doping. *Chem. Mater.* 27: 1622–1630.

60. Chanakian, S., U. Aydemir, A. Zevalkink et al. 2015. High temperature thermoelectric properties of Zn-doped $Eu_5In_2Sb_6$. *J. Mater. Chem. C* 3: 10518–10524.
61. May, A. F., E. S. Toberer, A. Saramat, and G. J. Snyder. 2009. Characterization and analysis of thermoelectric transport in n-type $Ba_8Ga_{16-x}Ge_{30+x}$. *Phys. Rev. B* 80: 125205.
62. Zunger, A. 2003. Practical doping principles. *Appl. Phys. Lett.* 83: 57.
63. Toberer, E. S., A. Zevalkink, and G. J. Snyder. 2011. Phonon engineering through crystal chemistry. *J. Mater. Chem.* 21: 15843.
64. Cahill, D., S. Watson, and R. Pohl. 1992. Lower limit to the thermal conductivity of disordered crystals. *Phys. Rev. B* 46: 6131–6140.
65. Zevalkink, A., S. Chanakian, U. Aydemir et al. 2015. Thermoelectric properties and electronic structure of the Zintl phase $Sr_5In_2Sb_6$ and the $Ca_{5-x}Sr_xIn_2Sb_6$ solid solution. *J. Phys. Condens. Matter.* 27: 015801.
66. Abeles, B., D. S. Beers, G. D. Cody, and J. P. Dismukes. 1962. Thermal conductivity of Ge–Si alloys at high temperature. *Phys. Rev.* 125: 44–46.
67. Klemens, P. G. 1955. The scattering of low-frequency lattice waves by static imperfections. *Proc. Phys. Soc. A* 68: 1113–1127.
68. Cao, Q.-G., H. Zhang, M.-B. Tang et al. 2010. Zintl phase $Yb_{1-x}Ca_xCd_2Sb_2$ with tunable thermoelectric properties induced by cation substitution. *J. Appl. Phys.* 107: 053714.
69. Zhang, H., M. Baitinger, M. B. Tang et al. 2010. Thermoelectric properties of $Eu(Zn_{1-x}Cd_x)_2Sb_2$. *Dalton Trans.* 39: 1101–1104.
70. Wang, H., A. D. LaLonde, Y. Z. Pei, and G. J. Snyder. 2013. The criteria for disorder in thermoelectric solid solutions. *Mater. Views* 23: 1586–1596.
71. Zevalkink, A., J. Swallow, S. Ohno, U. Aydemir, S. Bux, and G. J. Snyder. 2014. Thermoelectric properties of the $Ca_5Al_{2-x}In_xSb_6$ solid solution. *Dalton Trans.* 43: 15872–15878.

12

Thallium-Based Chalcogenides as Thermoelectrics

Quansheng Guo, Abdeljalil Assoud, and Holger Kleinke
University of Waterloo

CONTENTS

12.1 Introduction

Thermoelectric materials are capable of converting a temperature gradient into useful electricity (Seebeck effect) and creating a temperature gradient from electricity (Peltier effect) [1,2]. The Seebeck effect has been used for power generation in spacecrafts such as the Voyager deep space probes for decades [3]. Moreover, thermoelectrics are investigated for electricity generation from the waste heat in automobiles and trucks to ultimately enhance the gas mileage by reducing the load on the alternator [4–7]. The Peltier effect is used for cooling in portable coolers and laptops and investigated for temperature control in car seats [8].

The thermoelectric performance depends on the figure of merit, $ZT = T\alpha^2\sigma\,\kappa^{-1}$, with α being the Seebeck coefficient, σ the electrical conductivity, κ the thermal conductivity, and T the average absolute temperature. A thermoelectric device contains alternating n- and p-doped semiconducting legs, with the electrons of the n-type legs moving in the same direction, that is, with the heat flow, as the holes of the p-type legs. The device ZT depends on the average ZT values of all materials in both the n- and the p-legs. The efficiency of the power generation η is related to the device's average figure of merit ZT, written as $\overline{ZT}$ in Equation 12.1 (with T_H is the hot temperature, T_C is the cold temperature):

$$\eta = \frac{T_H - T_C}{T_H}\,\frac{\sqrt{1+\overline{ZT}}-1}{\sqrt{1+\overline{ZT}}+(T_C/T_H)} \tag{12.1}$$

The physical properties in the equation of the figure of merit all depend on the charge carrier concentration, n; the higher the n is, the smaller is the Seebeck coefficient and the larger is the electrical conductivity. The thermal conductivity has two different contributions, one being independent of the charge carriers, namely, the lattice (phonon) thermal conductivity, κ_L, and the other being the electronic contribution, κ_e, which is directly related to the electrical conductivity *via* the Wiedemann–Franz law, $\kappa_e = L_0\sigma T$, with L_0 = Lorenz number. With the total thermal conductivity being the sum of these two contributions, thus $\kappa = \kappa_L + \kappa_e = \kappa_L + L_0\sigma T$, κ also increases with increasing carrier concentration. The interdependence of the above transport parameters constitutes a major obstacle in optimizing the thermoelectric performance, for no parameter—with the possible exception of κ_L—can be enhanced without negatively affecting the other parameters. The best compromise is often a carrier concentration of the order of 10^{19}–10^{20} cm^{-3}, corresponding to that of a heavily doped semiconductor. κ_L can be minimized by working with heavy elements, such as thallium, lead, and bismuth, and complex crystal structures. It should be noted, however, that complex structures, in turn, often occur with low mobility of the charge carriers, detrimental to the electrical conductivity.

Advanced thermoelectric materials are thus heavily doped narrow gap semiconductors comprising heavy elements, typical example materials being Bi_2Te_3 for room temperature applications and PbTe for high-temperature applications. State-of-the-art bulk materials have demonstrated maximal ZT values of unity or even above, for example, heavily doped PbTe:Na with ZT_{max} = 1.4 at 750 K [9] and $Na_{0.02}Pb_{0.98}Se_{0.15}Te_{0.85}$ with ZT_{max} = 1.8 at 850 K [10], the triple-filled skutterudite n-type $Ba_{0.08}La_{0.05}Yb_{0.04}Co_4Sb_{12}$ with ZT_{max} = 1.7 at 850 K [11] and the p-type $LaFe_3CoSb_{12}$ with an extrapolated ZT_{max} = 1.4 at 1000 K [12], the clathrate n-type $Ba_8Ni_{0.31}Zn_{0.52}Ga_{13.06}Ge_{32.2}$ with ZT_{max} = 1.2 at 1000 K [13], and the Zintl phases p-type Zn_4Sb_3 and $La_{0.4}Yb_{13.6}MnSb_{11}$ with ZT_{max} = 1.2 at 1150 K [14]. Various methods of nanostructuring were reported to lead to significant improvements, in some cases beyond ZT_{max} = 2. Examples are thin films made of superlattices of Sb_2Te_3/Bi_2Te_3 with ZT_{max} = 2.4 at room temperature [15], n-type $AgPb_{18}SbTe_{20}$ with Sb-rich and Sb-poor nanodomains with ZT_{max} = 2.2 at 800 K [16], and $Na_xPb_{1-x}Te$ with SrTe nanocrystals with ZT_{max} = 2.2 at 915 K [17].

This chapter focuses on Tl-based thermoelectric materials in bulk form. Thallium has been used to enhance the performance of known thermoelectrics such as PbTe, which, after doping with Tl, achieves a ZT_{max} = 1.5 at 773K [18], and also skutterudites. Tl-filled n-type $Tl_{0.25}Co_4Sb_{12}$ exhibits a ZT_{max} = 0.90 at 600 K, and p-type $Tl_{0.2}FeCo_3Sb_{12}$ has ZT_{max} = 0.36 at 723 K [19]. In these cases, however, other dopants were used more successfully, with the aforementioned results. Research into thermoelectric properties of thallium chalcogenides began in 1965, where liquid alloys from the composition $Tl_{31}Te_{69}$ to $Tl_{68}Te_{32}$ were studied above 500 K. These liquids mostly exhibited low thermal conductivity below $\kappa = 0.4$ W m^{-1} K^{-1}, and the best figure of merit occurred in $Tl_{31}Te_{69}$ at 500 K (ZT = 0.8) [20]. Thereafter, several crystalline thallium tellurides were characterized and in part optimized *via* doping, as discussed in this chapter. In 2013, Kurosaki et al. [21] published a review on the thermoelectric performance of group 13 chalcogenides, which shed light on the low thermal conductivity of several thallium tellurides. Generally speaking, the latter is a consequence of both the distorted coordination polyhedra around the thallium atoms, giving rise to complex crystal structures, and also the presence of the heavy constituent elements thallium and tellurium.

In the following, we first introduce the structures and thermoelectric properties of binary thallium tellurides and then discuss the currently best known thallium tellurides, which all attain figure of merit values above 0.4 and sometimes beyond unity.

12.2 Binary Thallium Tellurides

Of the known binary thallium tellurides, only Tl_5Te_3 melts congruently around 723 K, while Tl_2Te, TlTe, and Tl_2Te_3 all exhibit incongruent melting points at lower temperatures [22,23]. They all adopt relatively complex crystal structures with irregular coordination environments around the Tl atoms, in contrast to PbTe and Bi_2Te_3 with perfect $PbTe_6$ octahedra and very regular $BiTe_6$ octahedra, respectively. The Tl_5Te_3 structure (In_5Bi_3 type) is composed of tetragonally compressed $TlTe_6$ octahedra (centered by Tl1) with two Tl–Te bonds of 3.15 Å and four of 3.35 Å, which form linear chains along the *c*-axis *via* corner sharing. These chains are interconnected in the *a*, *b* plane *via* common corners, resulting in a truly three-dimensional network of $TlTe_6$ octahedra (see Figure 12.1). The second Tl atom (Tl2) is bonded to five Te atoms at distances varying from 3.16 to 3.6 Å and participates in secondary Tl–Tl contacts between 3.46 and 3.72 Å [24], which are typical for other Tl^+ chalcogenides as well [25–27]. With no Te–Te contacts present, this telluride is typically described as being mixed valent, that is, with $Tl2^{1+}$ and $Tl1^{2+}$, according to the charge-balanced formula $(Tl2^+)_4Tl1^{2+}(Te^{2-})_3$. A Tl^{2+} contains one unpaired electron, in agreement with the metallic properties of Tl_5Te_3 [28].

Despite the formally closed shell configuration of Tl^+ ($6s^26p^0$), Tl^+ is known to often participate in (weakly) bonding Tl–Tl interactions [29,30]. This driving force is likely the reason for the different structures of, for example, $Tl_6Si_2Te_6$ and $K_6Si_2Te_6$, with the former structure having numerous Tl–Tl contacts <3.6 Å, and the latter zero such K–K interactions because of the repulsive nature of K^+–K^+ contacts [31]. We will encounter several structures

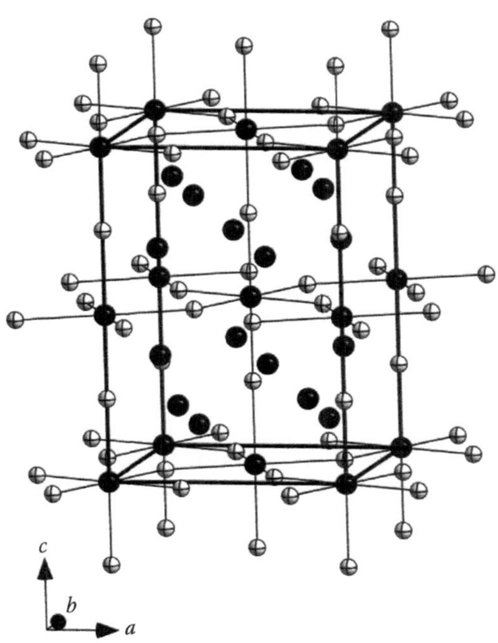

FIGURE 12.1
Crystal structure of Tl_5Te_3, emphasizing the network of $TlTe_6$ octahedra. Black circles, Tl atoms; crossed, Te atoms.

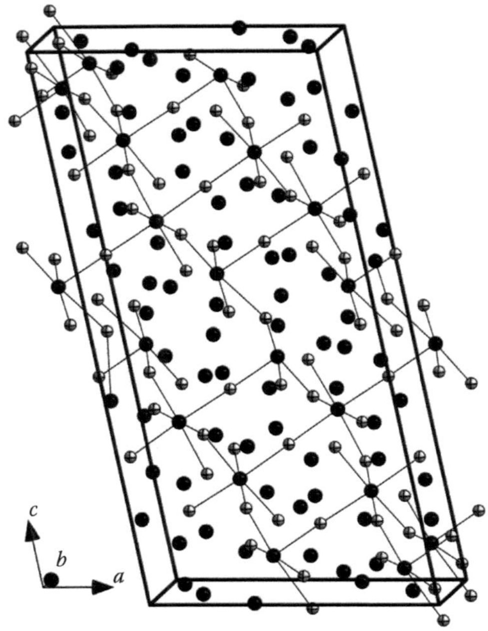

FIGURE 12.2
Crystal structure of Tl_2Te, emphasizing the network of $TlTe_6$ and $TlTe_5$ polyhedra. Black circles, Tl atoms; crossed, Te atoms.

in this chapter that exhibit these Tl–Tl contacts, which further contribute to the complexity of the crystal structures.

The Tl_2Te structure contains a strongly related network of corner-sharing $TlTe_6$ octahedra—with the linear chains running parallel to the [101] instead of the [001] direction—and $TlTe_5$ square pyramids shown in Figure 12.2, which encompasses more Tl atoms with more Tl–Tl contacts, compared to Tl_5Te_3. Thus, the Tl_2Te structure was discussed as an intergrowth of elemental thallium and Tl_5Te_3 [32]. As such, it contains several secondary Tl–Tl contacts of 3.3–3.6 Å.

TlTe undergoes a phase transition around 172 K. Above 172 K, the structure of TlTe contains an equidistant linear $^{1}_{\infty}[Te^{-}]$ chain with Te–Te distances of 3.08 Å. A second linear equidistant Te chain also runs parallel to the *c*-axis, which is more complex because each chain atom is bonded to two additional Te atoms *via* hypervalent bonds of 3.01 Å [33]. This material is charge-balanced according to $(Tl^{+})_4Te_3^{3-}Te^{-}$. Upon cooling below 172 K, the chains in part further distort [34]. Tl_2Te_3 contains comparable hypervalently bonded Te_3 units, which are interconnected *via* short Te–Te bonds to zigzag chains. Here too, Tl is in the +1 valence state, and the material is charge-balanced according to $(Tl^{+})_2Te_3^{2-}$ [35].

The thermoelectric properties of these materials were published in 2009 by Matsumoto et al. after hot-pressing, albeit without any attempts to optimize the carrier concentration [36]. All of these binary tellurides have comparably low Debye temperatures of θ = 107–124 K, which should be indicative of low lattice thermal conductivity, κ_L. Matsumoto et al. found values of $\kappa_L \approx 1$ W m^{-1} K^{-1} for Tl_5Te_3 and $\kappa_L \approx 0.5$ W m^{-1} K^{-1} for Tl_2Te and Tl_2Te_3, but around 2.8 W m^{-1} K^{-1} for TlTe. For comparison, κ_L of PbTe—composed of similarly heavy elements but adopting the simple NaCl structure—is typically listed as $\kappa_L \approx 2$ W m^{-1} K^{-1} [37].

Because of the poor electrical performance, expressed *via* the power factor $PF = \alpha^2\sigma$, none of the binary Tl tellurides exhibits good thermoelectric performance. Tl_5Te_3 and TlTe are metallic, in agreement with calculations of the electronic structures [28,38], and thus have a too small Seebeck coefficient α. Correspondingly, we determined $\alpha = 18\ \mu V\ K^{-1}$, $\sigma = 2700\ \Omega^{-1}\ cm^{-1}$, and $\kappa = 2.3\ W\ m^{-1}\ K^{-1}$ at 320 K for Tl_5Te_3 (unpublished data). Tl_2Te_3 is insulating (too small electrical conductivity), and Tl_2Te exhibits only a moderate Seebeck coefficient. In the end, Tl_2Te has the highest figure of merit values of the binary Tl tellurides, culminating in $ZT_{max} = 0.2$ at 586 K [36].

12.3 Ternary Thallium Tellurides

12.3.1 TlAgTe and $TlAg_9Te_5$

Although TlAgTe is an equiatomic ternary telluride, that is, has a very simple formula, and adopts a small unit cell, its structure is unique thus far and shows relatively low symmetry. Therein, corner-sharing $AgTe_4$ tetrahedra form a three-dimensional network, which encompasses the Tl atoms in linear channels. Each Tl atom is surrounded by a highly irregular polyhedron—as, for example, in the case of Tl_5Te_3 of six Te atoms (see Figure 12.3) [39]. This material thus possesses some characteristics of a low thermal conductivity material, namely, distorted coordination polyhedra and heavy elements; however, the unit cell is rather small. Nevertheless, its thermal conductivity was determined to be as low as $0.25\ W\ m^{-1}\ K^{-1}$, although this was measured on a hot-pressed pellet of only 90% density [40].

To correct for the relatively high porosity of 10%, we calculated the hypothetical corrected densities of the fully dense samples *via* the Maxwell–Eucken correction (Equation 12.2),

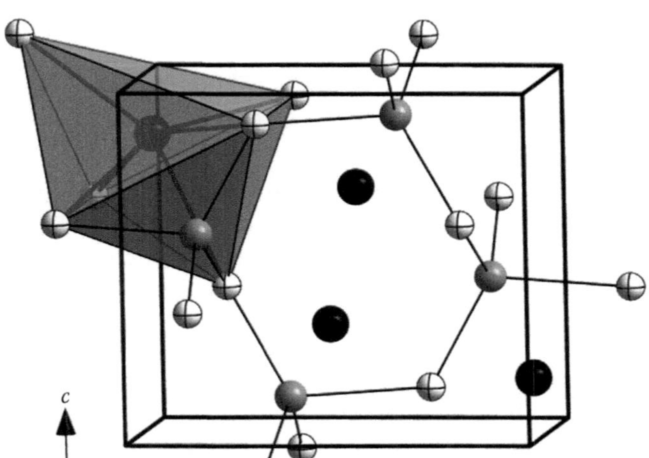

FIGURE 12.3
Crystal structure of TlAgTe, emphasizing the irregular $TlTe_6$ octahedron. Black circles, Tl atoms; gray, Ag atoms; crossed, Te atoms.

with κ_P = measured thermal conductivity with a porosity P and κ_0 = theoretical thermal conductivity at full density, that is, $P = 0$ [41,42]:

$$\kappa_P = \kappa_0 \frac{1-P}{1+\beta P} \tag{12.2}$$

As is common practice, we set the correction factor $\beta = 2$ as for spherical pores. This results in a pore-corrected thermal conductivity of $\kappa_0 = 0.33$ W m^{-1} K^{-1}, comparable to the liquid binary thallium tellurides [20]. The maximum figure of merit, determined to be ZT_{max} = 0.61 at 580 K, originates from a high Seebeck coefficient but low electrical conductivity, indicative of low carrier concentration (and relatively low density).

In contrast to TlAgTe, $TlAg_9Te_5$ adopts a very complex crystal structure. According to a single-crystal structure study [43], its formula is actually $Tl_{12}Ag_{96.5}Te_{66} \approx Tl_2Ag_{16}Te_{11}$, with some uncertainties because of five deficient Ag sites, exhibiting refined occupancies between 46% and 88%. Its huge unit cell with a volume of 4760 Å^3 is composed of hexagonal channels formed by the Ag and Te atoms that include the Tl atoms. Such high complexity, together with the Ag deficiency, should cause a very low mean-free phonon path length, and thus extremely low thermal conductivity. This was experimentally confirmed on a hot-pressed sample with a 98% density [44], which exhibited a low speed of sound of 1200 ms^{-1} and low Debye temperature of θ = 120 K, comparable to the aforementioned binary thallium tellurides. Accordingly, a basically temperature-independent thermal conductivity of κ = 0.23 W m^{-1} K^{-1} was found, despite the rather low Tl content of only 7 at.%, compared to 33 at.% in TlAgTe. That these two Tl-Ag tellurides have the same low thermal conductivity can be understood as a trade-off between a high average mass per atom and a highly complex crystal structure.

In addition, $TlAg_9Te_5$ possesses a high Seebeck coefficient, which decreases with increasing temperature, reaching α = 320 µV K^{-1} at 700 K, and a low electrical conductivity increasing with increasing temperature to σ = 38 Ω^{-1} cm^{-1} at 700 K. These values and temperature dependencies point toward a low charge carrier concentration; nevertheless, a high thermoelectric figure of merit of ZT_{max} = 1.23 at 700 K is obtained at the highest temperature investigated. As the melting point is 772 K, one cannot safely expand the temperature range to look for higher ZT values.

12.3.2 $TlETe_2$ with E = Ga, In, Sb, Bi, and Lanthanides

$TlETe_2$ can crystallize in two different structure types, one being truly three-dimensional (when E = Ga, In) and the other one being comprised of layers of $TlTe_6$ and ETe_6 octahedra, respectively, with E = Sb, Bi, or a lanthanide Ln. The former consists of a network of ETe_4 tetrahedra and $TlTe_8$ square antiprisms, the aristotype being the TlSe ($Tl^+Tl^{3+}Se_2$) structure, with E = Ga, In replacing the Tl^{3+} atoms. These materials are charge-balanced according to the formula $Tl^+E^{3+}(Te^{2-})_2$ and, indeed, exhibit semiconducting properties, for example, low electrical conductivity increasing with increasing temperature. Like the previously discussed representatives, these Tl tellurides also possess low thermal conductivity and Debye temperatures. With κ = 0.5 W m^{-1} K^{-1}, and θ = 155 K for $TlGaTe_2$ and 148 K for $TlInTe_2$, these values are somewhat higher because of the simpler structure compared to $TlAg_9Te_5$ and because of the lower Tl content compared to TlAgTe [45]. Overall, their thermoelectric performance is less competitive, with ZT_{max} = 0.44 for $TlGaTe_2$ and $TlInTe_2$ (both at 852 K).

$TlETe_2$ adopts the layered α-$NaFeO_2$ type with E = Sb, Bi, Ln, as these larger E^{3+} cations are octahedrally coordinated in the α-$NaFeO_2$ structure. This structure may be described based on close packing of the anions, with the two different cations occupying all octahedral holes in such a way that the Tl and E atoms occur in separate layers. Along the *c*-axis, the stacking order is Te–Tl–Te–E–Te–Tl–Te–*E*–.... $TlBiTe_2$ is dimorphic, transforming at 695 K to the NaCl type, which is basically the disordered variant of α-$NaFeO_2$, with all metal atoms being symmetry equivalent [46].

$TlSbTe_2$ and $TlBiTe_2$ were both calculated to be narrow gap semiconductors, according to the formula $Tl^+E^{3+}(Te^{2-})_2$, with the Sb variant exhibiting a direct gap and the Bi variant an indirect gap [47]. Again, the lattice thermal conductivity values are significantly below 1 W m^{-1} K^{-1}, that is, κ_L = 0.6 W m^{-1} K^{-1} in the case of $TlSbTe_2$ [48] and κ_L = 0.3 W m^{-1} K^{-1} in the case of $TlGdTe_2$. For comparison, $TlGdSe_2$ was also analyzed and showed a significantly lower electrical conductivity because of the expected larger band gap and a higher lattice thermal conductivity of 1.3 W m^{-1} K^{-1} because of the substitution of the Te atoms by the lighter Se atoms [49]. $TlBiTe_2$ is the only representative of this family that exhibits *n*-type conduction [50]. Overall, $TlSbTe_2$ has the largest ZT_{max} = 0.87 at 715 K of the $TlETe_2$ materials.

12.3.3 Tl_2ETe_3 with E = Ge, Sn

Tl_2GeTe_3 [51] and Tl_2SnTe_3 [52] both adopt the Tl_2SnTe_3 structure type, which is composed of pairs of edge-sharing ETe_4 tetrahedra. In Tl_2GeTe_3, the two crystallographically different Tl atoms are located in a severely distorted $TlTe_6$ "octahedron" with Tl–Te distances between 3.23 and 3.77 Å and a $TlTe_8$ square antiprism with Tl–Te distances between 3.34 and 3.71 Å, respectively (see Figure 12.4). In addition, three secondary Tl–Tl contacts of 3.50, 3.53, and 3.75 Å complete the coordination spheres. It is thus not surprising that, again, a low Debye

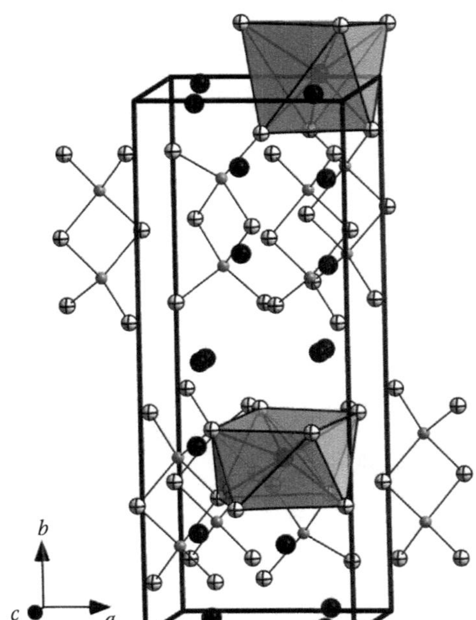

FIGURE 12.4
Crystal structure of Tl_2GeTe_3, emphasizing the irregular $TlTe_6$ octahedron and the $TlTe_8$ square antiprism. Black circles, Tl atoms; gray, Ge atoms; crossed, Te atoms.

temperature (θ = 85 K for Tl_2GeTe_3) and low thermal conductivity were observed, namely, κ_L = 0.29 W m^{-1} K^{-1} for Tl_2GeTe_3 at room temperature [53] and 0.24 W m^{-1} K^{-1} for Tl_2SnTe_3 at 420 K [54]. In combination with relatively low power factors, despite reasonable carrier concentrations of several 10^{19} cm^{-3}, the figure of merit values obtained are moderate, culminating in ZT_{max} = 0.29 at 473 K for Tl_2GeTe_3 and ZT_{max} = 0.37 at 530 K for Tl_2SnTe_3. Here, again, the possible temperature range is limited because of the low incongruent melting point of 637 K of Tl_2SnTe_3 [55].

12.3.4 Tl_2ETe_5 with E = Ge, Sn

Tl_2GeTe_5 [56] and Tl_2SnTe_5 [57] crystallize in different but strongly related structures. As demonstrated in Figure 12.5, both structures exhibit $TlTe_8$ cubes and square antiprisms, as well as ETe_4 tetrahedra and unique Te-centered Te_4 squares, reminiscent of the square planar XeF_4 molecule. In contrast to Tl_2SnTe_5, the $TeTe_4$ squares form pairs *via* edge condensation in the case of Tl_2GeTe_5.

These materials have even lower (incongruent) melting points than the Tl_2ETe_3 tellurides discussed earlier, namely, about 540 K in the case of Tl_2GeTe_5 and 560 K in the case of Tl_2SnTe_5, which limits the range of their applications. Experimental band gaps were derived from the shape of the Seebeck curves [58] to be 0.17 eV for Tl_2GeTe_5 and 0.25 eV for Tl_2SnTe_5, which are useful sizes for low temperature thermoelectric applications. Sharp et al. [59] compared the thermal displacement parameters for various Tl atoms, as determined both by x-ray and neutron diffraction. In both cases, the Tl atoms in the cubes experience larger Tl–Te distances than the Tl atoms in the square antiprisms, as well as

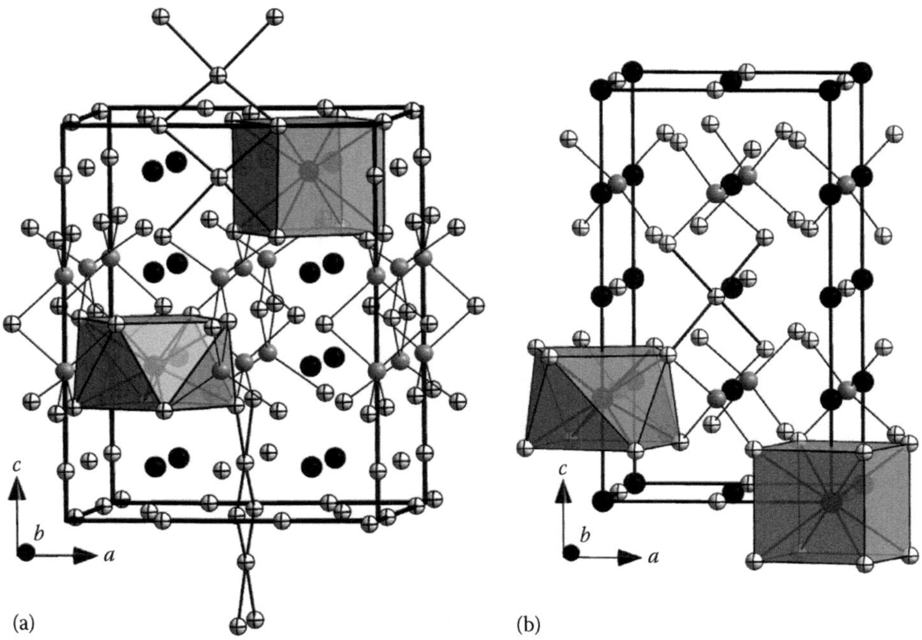

FIGURE 12.5
Crystal structures of (a) Tl_2GeTe_5 and (b) Tl_2SnTe_5, emphasizing $TlTe_8$ cubes and square antiprisms as well as $TeTe_4$ squares. Black circles, Tl atoms; gray, Ge/Sn atoms; crossed, Te atoms.

doubled or even tripled displacement parameters. Enlarged displacement parameters are indicative of more thermal vibration, which, in turn, may cause flatter phonon bands and thus lower phonon velocity, ultimately lowering the lattice thermal conductivity [60]. Such effects were observed to a larger extent in the case of the filled skutterudites and clathrates, where the filler atom may exhibit up to six times higher thermal displacement parameters than the host atoms, as, for example, found in $LaFe_4Sb_{12}$ [61], $LaFe_3CoSb_{12}$ [62], and $Sr_8Ga_{16}Ge_{30}$ [63]. Correspondingly, the lattice thermal conductivity of both Tl_2GeTe_5 and Tl_2SnTe_5 is approximately $\kappa_L = 0.5$ W m^{-1} K^{-1}. Again, with moderate power factors, *ZT* values of up to 0.6 were achieved in case of Tl_2SnTe_5.

12.3.5 $Tl_{10-x}E_xTe_6$ with E = Sn, Pb, Sb, Bi, and Lanthanides

Electronic structure calculations of Tl_5Te_3 point toward the existence of a band gap well above the Fermi level [60], which may be reached by adding one valence electron per formula unit, such as replacing one Tl with one Sn or Pb atom. Alternatively, one may replace half a Tl atom with one Sb, Bi, or La atom. Correspondingly, both $(Tl^+)_4(Sn^{2+})(Te^{2-})_3$ and $(Tl^+)_4(Pb^{2+})(Te^{2-})_3$ would be semiconducting, like $(Tl^+)_{4.5}(Sb^{3+})_{0.5}(Te^{2-})_3 \equiv (Tl^+)_9(Sb^{3+})(Te^{2-})_6$ in agreement with electronic structure calculations on Tl_9SbTe_6 and Tl_9BiTe_6 [64]. These substitutions occur foremost on the octahedrally coordinated Tl1 site in Wyckoff position *4c*, not on the Tl2 site, and Wyckoff position *16l*, with its irregular Te coordination and multiple Tl–Tl contacts [65,66]. Moreover, a stepwise replacement of Tl1 with the spherically shaped La^{3+} atom leads to an increasingly regular Te_6 polyhedron, for example, from Tl1–Te distances of 2 × 3.15 and 4 × 3.36 Å in the case of Tl_5Te_3 to 2 × 3.22 and 4 × 3.36 Å in the case of $Tl_{9.4}La_{0.6}Te_6$ with 30% La on the *4c* site [67]. The same is true for replacing Tl1 with Pb, although Pb^{2+} also has a free electron pair, which is apparently less sterically active: the Pb–Te distances in Tl_4PbTe_3 are 2 × 3.26 and 4 × 3.31 Å, corresponding to an increase in the apical and a decrease in the equatorial distances, compared to $Tl1Te_6$ in Tl_5Te_3 (see Figure 12.6). Since the apical distances in the $Tl1Te_6$ polyhedron are parallel to the *c*-axis, and thus the equatorial ones parallel to the *a*, *b* plane of the tetragonal unit cell, these substitutions barely change—or even shorten—the *a*- and *b*-axes, but elongate the *c*-axis, for example, from $a = 8.92$ Å and $c = 12.61$ Å for Tl_5Te_3 to $a = 8.84$ Å and $c = 13.06$ Å for Tl_4PbTe_3 [65].

Furthermore, the Tl2 atom exhibits larger thermal displacement parameters than the Pb and Te sites, namely, $U_{eq} = 0.031$ Å^2 for Tl2 versus 0.016–0.022 Å^2 for the other sites in

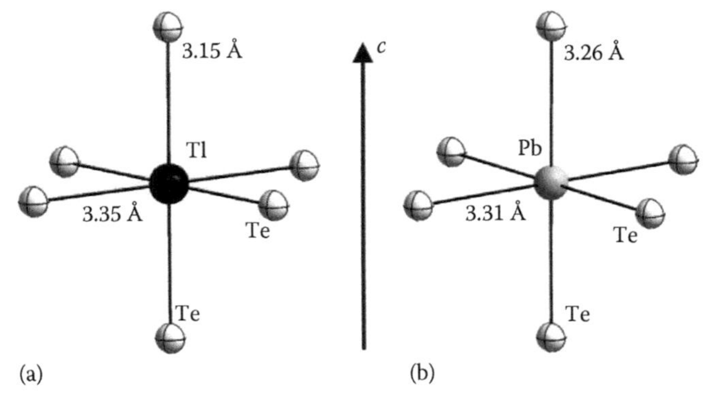

FIGURE 12.6
(a) $TlTe_6$ octahedron in Tl_5Te_3 and (b) $PbTe_6$ octahedron in Tl_4PbTe_3. The difference between the apical and equatorial bond lengths is 0.15 Å in Tl_5Te_3 and 0.05 Å in Tl_4PbTe_3.

our investigation [68] and U_{eq} = 0.024 Å^2 for Tl_2 versus 0.010–0.015 Å^2 in Böttcher's work [65]. Similarly, the values in $Tl_{9.4}La_{0.6}Te_6$ are U_{eq} = 0.037 Å^2 for T12 and U_{eq} = 0.028 Å^2 for the mixed Tl/La *4c* site. As mentioned earlier, these enlargements should cause a lowered lattice thermal conductivity.

A further potential advantage of these substitution variants stems from the higher melting points, compared to the 723 K of Tl_5Te_3, which increases the usable temperature range of these materials. For example, Tl_4SnTe_3 melts at 827 K, while Tl_4PbTe_3 decomposes at 848 K [65], and Tl_9BiTe_6 melts at 813 K [69], Tl_9LaTe_6 at 800 K [67], and Tl_9NdTe_6 at 820 K [70].

The first report on outstanding thermoelectric properties in this series was published in 2001, dealing with Tl_9BiTe_6 that, after zone refining, was extrapolated to attain a ZT_{max} = 1.2 at 500 K [71]. Like basically all of the Tl tellurides discussed earlier, this material is bestowed with a very low thermal conductivity of κ_L = 0.39 W m^{-1} K^{-1} around room temperature. Its rather low electrical conductivity of σ = 140 Ω^{-1} cm^{-1} at 325 K is the major drawback, while the Seebeck coefficient is quite high with α = +280 μV K^{-1} at 325 K. These values are indicative of a small number of carriers, likely a consequence of the zone refining that typically leads to almost defect-free samples. Hot-pressing of Tl_9BiTe_6 results in significantly higher electrical conductivity and lower Seebeck coefficient (and similar thermal conductivity), culminating in a significantly lower, albeit still impressive, figure of merit curve, with ZT = 0.78 at 500 K and ZT_{max} = 0.86 at 590 K [72]. Subsequently, we were able to improve on the latter after optimizing the processing procedure, which resulted in ZT = 0.98 at 500 K, and ZT = 1.1 after introducing 2% vacancies on the Bi site. These defects increase the *p*-type charge carriers, reflected in higher electrical conductivity and lower Seebeck coefficient [73].

The isoelectronic Tl_9SbTe_6, first investigated in the year 2003, after cold-pressing and sintering [48], exhibits higher electrical and thermal conductivity than Tl_9BiTe_6, possibly because of a higher charge carrier concentration. The sintered pellet attains ZT = 0.34 at 500 K, and the hot-pressed pellet has ZT = 0.84 at 500 K. The latter was improved upon *via* increasing the Sb content, as expressed in the nominal formula $Tl_{8.98}Sb_{1.02}Te_6$, to ZT = 0.90 [73]. Sn- and Pb-doping experiments on Tl_9BiTe_6 and Tl_9SbTe_6 failed to yield improvements in the figure of merit [74].

Moreover, none of the variants with lanthanides Ln, first discovered by Babanly et al. [70,75,76] for Ln = Ce, Nd, Sm, Gd, achieve comparable ZT values despite comparable thermal conductivity, regardless of the Ln content. When the electrical conductivity is of the order of the values of Tl_9BiTe_6, the Seebeck coefficient is significantly lower in the $Tl_{10-x}Ln_xTe_6$ series. The best performance was found when Ln = La and x = 1, that is, in Tl_9LaTe_6, with ZT = 0.51 at 500 K and 0.57 at 600 K [77].

Thermoelectric property investigations of hot-pressed pellets of Tl_4SnTe_3 and Tl_4PbTe_3 demonstrated again very low lattice thermal conductivities with κ_L = 0.29 W m^{-1} K^{-1} and κ_L = 0.58 W m^{-1} K^{-1} and Debye temperatures of θ = 118 and 120 K, respectively [53]. In contrast to the Sb and Bi case, where the heavier analogue showed the higher Seebeck coefficient α and lower electrical conductivity σ, Tl_4SnTe_3 had higher α and lower σ than Tl_4PbTe_3 below 600 K and quite comparable values above 600 K. In the end, both materials achieved practically equivalent figure of merit values of ZT_{max} = 0.74 (Tl_4SnTe_3) and 0.71 (Tl_4PbTe_3) at 673 K. As in the case of Tl_9BiTe_6, we were able to enhance the performance, demonstrating ZT_{max} = 1.02 (Tl_4SnTe_3) and 1.16 (Tl_4PbTe_3) at 685 K [68]. In addition, adjusting x in $Tl_{10-x}Sn_xTe_6$ and $Tl_{10-x}Pb_xTe_6$ yielded even higher figure of merit values of ZT_{max} = 1.26 for 1.95 Sn and ZT_{max} = 1.42 for 1.90 Pb.

In Figure 12.7, we compare the thermoelectric properties of all hot-pressed, stoichiometric Tl_9ETe_6 and Tl_4ETe_3 samples, which are either from Kurosaki and Yamanaka et al.

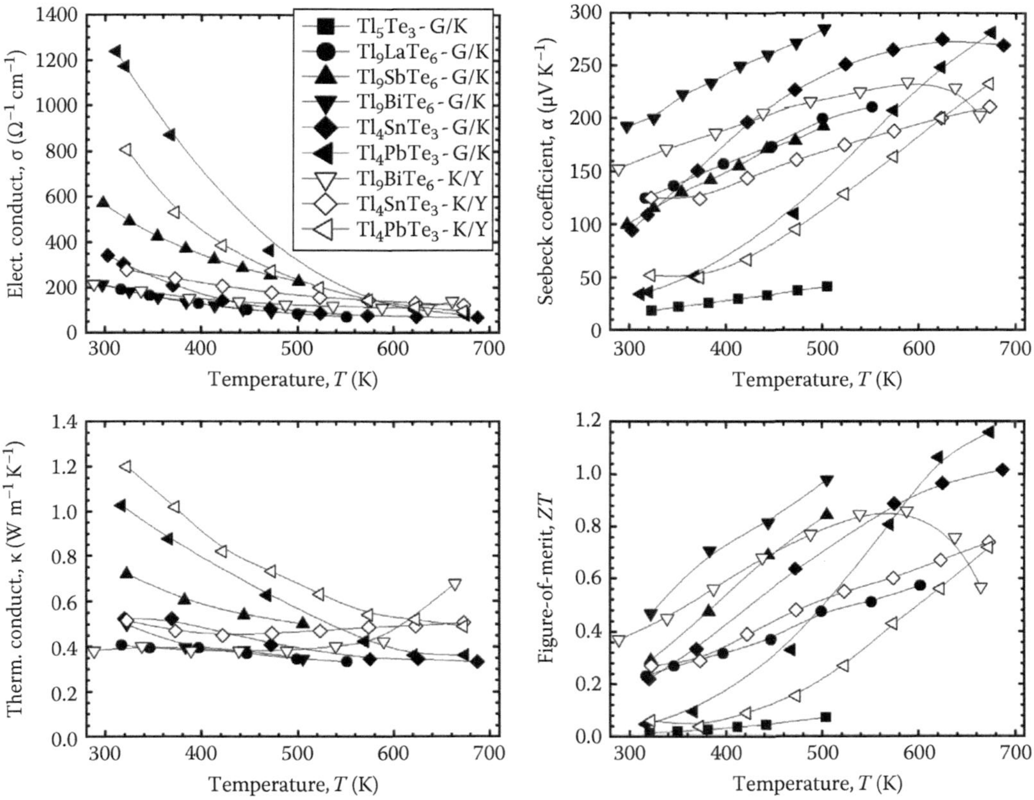

FIGURE 12.7
Thermoelectric properties of stoichiometric, hot-pressed Tl_5Te_3, Tl_9ETe_6, and Tl_4ETe_3. The electrical and thermal conductivity data of Tl_5Te_3 are omitted, because they are much higher than for the other materials. (Solid black symbols indicate data from Guo/Kleinke [68,73,77] and open symbols from Kurosaki/Yamanaka [53].)

(open symbols) or from our group (Guo et al., black symbols). All of these materials exhibit the typical temperature dependencies of doped semiconductors despite attempts to prepare the electron-precise variants. Evidently, these samples are usually prepared with a large amount of *p*-type defects, resulting in room temperature carrier concentrations of the order of 3×10^{19} to 10×10^{19} cm^{-3} (unpublished Hall data). The samples of the same nominal formula from the two different groups exhibit mostly parallel running curves with comparable values, which implies that the differences between the samples with different E atoms are systematic for these samples. Aside from Tl_5Te_3, the two most different materials are Tl_9BiTe_6 (smallest σ, largest α, highest *ZT* around room temperature) and Tl_4PbTe_3 (largest σ, smallest α, lowest *ZT* around room temperature), which also exhibits the steepest slopes for those properties. In the end, Tl_4PbTe_3 has the highest *ZT* values above 600 K but the lowest below 500 K. At this point, we are trying to analyze the reasons for the different behaviors *via* electronic structure calculations and low- and high-temperature Hall measurements in collaboration with the Felser group (MPI Dresden) and the Snyder group (Northwestern University, Evanston, IL).

In Figure 12.8, we demonstrate the changes after deliberately doping Tl_9SbTe_6, Tl_9BiTe_6, Tl_4SnTe_3, and Tl_4PbTe_3. In the latter three cases, we decreased the content of E (Bi, Sn, and

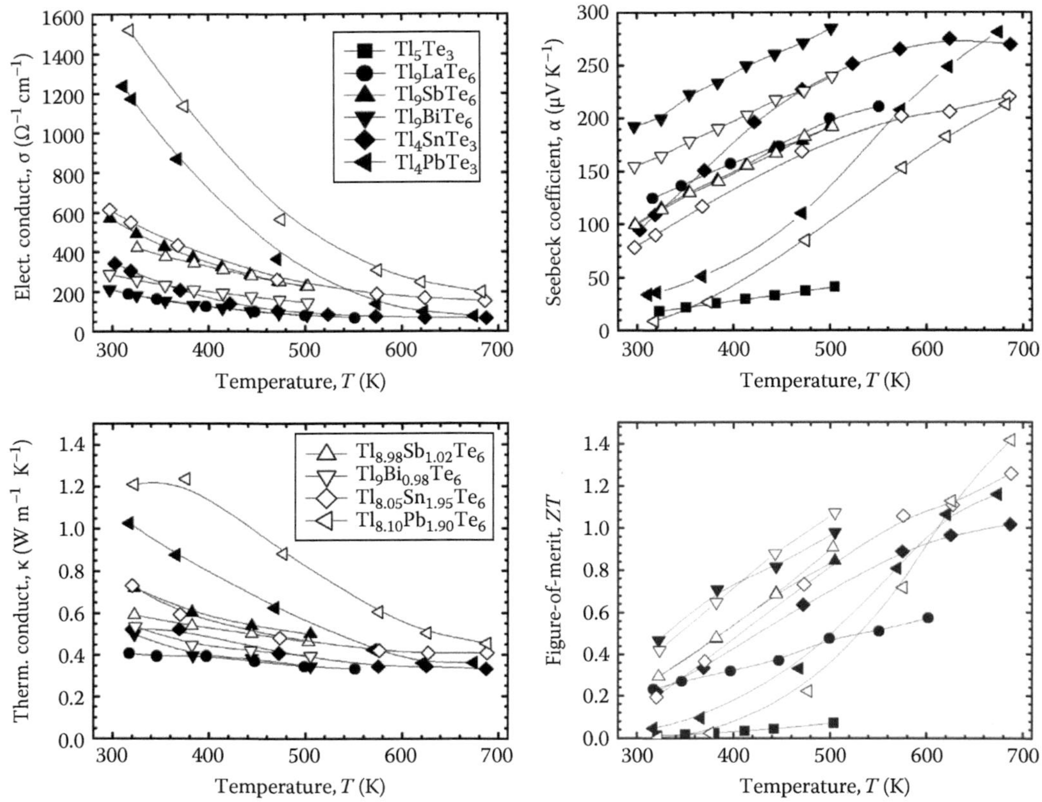

FIGURE 12.8
Thermoelectric properties of stoichiometric, hot-pressed Tl_5Te_3, Tl_9ETe_6, and Tl_4ETe_3 in comparison to the optimized stoichiometries. The electrical and thermal conductivity data of Tl_5Te_3 are omitted, because they are much higher than for the other materials. (All data are from Guo/Kleinke [68,73,77].

Pb) to increase the concentration of the holes, which enhanced electrical and thermal conductivities while decreasing the Seebeck effect as expected. The improvements in the figure of merit occur only at elevated temperatures, with $Tl_9Bi_{0.98}Te_6$ outperforming Tl_9BiTe_6 above 420 K, $Tl_{8.05}Sn_{1.95}Te_6$ outperforming Tl_4SnTe_3 above 350 K, and $Tl_{8.10}Pb_{1.90}Te_6$ outperforming Tl_4PbTe_3 above 600 K. In case of E = Sb, on the other hand, only insignificant improvements were found after increasing the Sb concentration by 2%, according to the formula $Tl_{8.98}Sb_{1.02}Te_6$.

12.4 Conclusions

The structures of many thallium chalcogenides are particularly complex, that is, have large unit cells of low symmetry. This is a consequence of the Tl atoms in the +1 oxidation state—its preferred state in chalcogenides—possessing a stereochemically active electron pair and their tendency toward secondary (closed shell) Tl–Tl interactions.

This is advantageous for thermoelectric materials, for it causes low thermal conductivity. Another reason for the low thermal conductivity is that thallium is similar in weight to the heavy elements lead and bismuth, which neither commonly adopt such distorted coordination spheres nor participate in closed shell Pb–Pb or Bi–Bi interactions. Comparing the structures of PbTe and Bi_2Te_3 with Tl_2Te and TlTe clearly illustrates these differences.

In this chapter, we have described the structures and properties of a number of such low thermal conductivity thallium tellurides, while the selenides are typically more ionic and exhibit higher thermal conductivity because of their lighter molar masses. Materials such as $TlAg_9Te_5$, Tl_9SbTe_6, Tl_9BiTe_6, Tl_4SnTe_3, and Tl_4PbTe_3 all exhibit high thermoelectric figure of merit values around or above the traditional benchmark of unity (at moderate temperatures of 500–700 K) as-prepared, that is, without any of the typical optimization methods such as doping or grain boundary engineering or nanostructuring. In part, doping was attempted and led to further improvements, such as in Tl_4PbTe_3 by adjusting the Tl/Pb ratio from $ZT_{max} = 1.16$ to $ZT_{max} = 1.42$. More enhancements are likely to follow, as this research began only about 16 years ago.

The high toxicity of thallium constitutes a major hindrance on the way to commercial applications of these materials. The permissible exposure limit of thallium is 100 $\mu g\ m^{-3}$ in the United States, comparable to lead (50 $\mu g\ m^{-3}$) and tellurium (100 $\mu g\ m^{-3}$), while bismuth is environmentally benign. Therefore, bismuth tellurides are preferred over thallium or lead tellurides, but they too are not sustainable because of their large tellurium contents. In conclusion, while exciting performances of various thallium tellurides are evident, there is a clear need for more sustainable thermoelectric materials without thallium and tellurium. Finding materials with the benefits of thallium, but without its toxicity, remains as a huge challenge for the future.

Acknowledgment

Financial support from the Natural Sciences and Engineering Research Council of Canada (NSERC) is acknowledged.

References

1. Rowe, D. M. 1995. *CRC Handbook of Thermoelectrics*. Boca Raton, FL: CRC Press.
2. Rowe, D. M. 2006. *Thermoelectrics Handbook: Macro to Nano*. Boca Raton, FL: CRC Press/Taylor & Francis Group.
3. Furlong, R. R. and E. J. Wahlquist. 1999. US space missions using radioisotope power systems. *Nucl. News* 42: 26–35.
4. Yang, J. and T. Caillat. 2006. Thermoelectric materials for space and automotive power generation. *MRS Bull.* 31: 224–229.
5. Yang, J. and F. R. Stabler. 2009. Automotive applications of thermoelectric materials. *J. Electron. Mater.* 38: 1245–1251.
6. Bell L. E. 2008. Cooling, heating, generating power, and recovering waste heat with thermoelectric systems. *Science* 321: 1457–1461.

7. Matsumoto, M., M. Mori, T. Haraguchi et al. 2015. Development of state of the art compact and lightweight thermoelectric generator using vacuum space structure. *SAE Int. J. Engines* 8: 1815–1825.
8. Choi, H.-S., S. Yun, and K.-I. Whang. 2007. Development of a temperature-controlled car-seat system utilizing thermoelectric device. *Appl. Thermal Eng.* 27: 2841–2849.
9. Pei, Y., A. LaLonde, S. Iwanaga, and G. J. Snyder. 2011. High thermoelectric figure of merit in heavy hole dominated PbTe. *Energy Environ. Sci.* 4: 2085–2089.
10. Pei, Y., X. Shi, A. LaLonde, H. Wang, L. Chen, and G. J. Snyder. 2011. Convergence of electronic bands for high performance bulk thermoelectrics. *Nature* 473: 66–69.
11. Shi, X., J. Yang, J. R. Salvador et al. 2011. Multiple-filled skutterudites: High thermoelectric figure of merit through separately optimizing electrical and thermal transports. *J. Am. Chem. Soc.* 133: 7837–7846.
12. Sales, B. C., D. Mandrus, and R. K. Williams. 1996. Filled skutterudite antimonides: A new class of thermoelectric materials. *Science* 272: 1325–1328.
13. Shi, X., J. Yang, S. Bai et al. 2010. On the design of high-efficiency thermoelectric clathrates through a systematic cross-substitution of framework elements. *Adv. Funct. Mater.* 20: 755–763.
14. Toberer, E. S., S. R. Brown, T. Ikeda, S. M. Kauzlarich, and G. J. Snyder. 2008. High thermoelectric efficiency in lanthanum doped $Yb_{14}MnSb_{11}$. *Appl. Phys. Lett.* 93: 062110.
15. Venkatasubramanian, R., E. Slivola, T. Colpitts, and B. O'Quinn. 2001. Thin-film thermoelectric devices with high room-temperature figures of merit. *Nature* 413: 597–602.
16. Hsu, K. F., S. Loo, F. Guo et al. 2004. Cubic $AgPb_mSbTe_{2+m}$: Bulk thermoelectric materials with high figure of merit. *Science* 303: 818–821.
17. Biswas, K., J. He, I. D. Blum et al. 2012. High-performance bulk thermoelectrics with all-scale hierarchical architectures. *Nature* 489: 414–418.
18. Heremans, J. P., V. Jovovic, E. S. Toberer et al. 2008. Enhancement of thermoelectric efficiency in PbTe by distortion of the electronic density of states. *Science* 321: 554–557.
19. Kim, D., K. Kurosaki, Y. Ohishi, H. Muta, and S. Yamanaka. 2013. How thermoelectric properties of p-type Tl-filled skutterudites are improved. *APL Mater.* 1: 032115.
20. Cutler, M. and C. E. Mallon. 1965. Thermoelectric properties of liquid semiconductor solutions of thallium and tellurium. *J. Appl. Phys.* 36: 201–205.
21. Kurosaki, K. and S. Yamanaka. 2013. Low-thermal-conductivity group 13 chalcogenides as high-efficiency thermoelectric materials. *Phys. Status Solidi* 210: 82–88.
22. Massalski, T. B. 1986. *Binary Alloy Phase Diagrams*. Metals Park, OH: American Society for Metals.
23. Record, M.-C., Y. Feutelais, and H. L. Lukas. 1997. Phase diagram investigation and thermodynamic evaluation of the thallium-tellurium system. *Z. Metallk.* 88: 45–54.
24. Schewe, I., P. Böttcher, and H. G. von Schnering. 1989. The crystal structure of Tl_5Te_3 and its relationship to the Cr_5B_3 type. *Z. Kristallogr.* 188: 287–298.
25. Sankar, C. R., S. Bangarigadu-Sanasy, A. Assoud, and H. Kleinke. 2010. Syntheses, crystal structures and thermoelectric properties of two new thallium tellurides: Tl_4ZrTe_4 and Tl_4HfTe_4. *J. Mater. Chem.* 20: 7485–7490.
26. Sankar, C. R., S. Bangarigadu-Sanasy, A. Assoud, and H. Kleinke. 2011. Crystal structures and physical properties of Tl_4MQ_4 (M = Zr, Hf, Q = S, Se). *Inorg. Chem.* 50: 245–249.
27. Sankar, C. R., A. Becker, A. Assoud, and H. Kleinke. 2013. New quaternary chalcogenides, $Tl_{18}Pb_2M_7Q_{25}$ (M = Ti, Zr, and Hf; Q = S and Se): Crystal structure, electronic structure, and electrical transport properties. *Inorg. Chem.* 52: 1895–1900.
28. Nordell, K. J. and G. J. Miller. 1996. Electronic structure, superconductivity, and substitution patterns in Tl_5Te_3. *J. Alloys Compd.* 241: 51–62.
29. Janiak, C. and R. Hoffmann. 1990. Thallium(I)-thallium(I) and indium(I)-indium(I) interactions: From the molecular to the solid state. *J. Am. Chem. Soc.* 112: 5924–5946.
30. Pyykkö, P. 1997. Strong closed shell interactions. *Chem. Rev.* 97: 597–636.

31. Assoud, A., N. Soheilnia, and H. Kleinke. 2006. Crystal structure, electronic structure and physical properties of the new low-valent thallium silicon telluride $Tl_6Si_2Te_6$ in comparison to $Tl_6Ge_2Te_6$. *J. Solid State Chem.* 179: 2707–2713.
32. Cerny, R., J. M. Joubert, Y. Filinchuk, and Y. Feutelais. 2002. Tl_2Te and its relationship with Tl_5Te_3. *Acta Crystallogr. C* 58: i63–i65.
33. Graf, C., A. Assoud, O. Mayasree, and H. Kleinke. 2009. Solid state polyselenides and polytellurides: A large variety of Se–Se and Te–Te interactions. *Molecules* 14: 3115–3131.
34. Stöwe, K. 2000. The phase transition of TlTe: Crystal structure. *J. Solid State Chem.* 149: 123–132.
35. Doert, T., R. H. Cardoso Gil, and P. Böttcher. 1999. The crystal structure of Tl_2Te_3—A reinvestigation. *Z. Anorg. Allg. Chem.* 625: 2160–2163.
36. Matsumoto, H., K. Kurosaki, H. Muta, and S. Yamanaka. 2009. Thermoelectric properties of the thallium-tellurium binary compounds. *Mater. Trans.* 50: 1582–1585.
37. Ahn, K., C. P. Li, C. Uher, and M. G. Kanatzidis. 2009. Improvement in the thermoelectric figure of merit by La/Ag cosubstitution in PbTe. *Chem. Mater.* 21: 1361–1367.
38. Wagner, F. R. and K. Stöwe. 2001. Electronic band structure of TlTe. *J. Solid State Chem.* 157: 193–205.
39. Ayral-Marin, R. M., B. Liautard, M. Maurin, J. C. Tedenac, and G. Brun. 1988. AgTlTe—Etude de la zone de solution solide et de ses modifications structurales. *J. Phys. Chem. Solids* 49: 939–944.
40. Kurosaki, K., H. Uneda, H. Muta, and S. Yamanaka. 2005. Extremely low thermal conductivity of AgTlTe. *J. Alloys Compd.* 395: 304–306.
41. Datta, A. and G. S. Nolas. 2012. Solution-based synthesis and low-temperature transport properties of $CsBi_4Te_6$. *ACS Appl. Mater. Interfaces* 4: 772–776.
42. Nandihalli, N., S. Gorsse, and H. Kleinke. 2015. Effects of additions of carbon nanotubes on the thermoelectric properties of $Ni_{0.05}Mo_3Sb_{5.4}Te_{1.6}$. *J. Solid State Chem.* 226: 164–169.
43. Paccard, D., L. Paccard, G. Brun, and J. C. Tedenac. 1992. A new phase in the Tl–Ag–Te system: Crystal structure of $Tl_2Ag_{16}Te_{11}$. *J. Alloys Compd.* 184: 337–342.
44. Kurosaki, K., A. Kosuga, H. Muta, M. Uno, and S. Yamanaka. 2005. Ag_9TlTe_5: A high-performance thermoelectric bulk material with extremely low thermal conductivity. *Appl. Phys. Lett.* 87: 061919.
45. Matsumoto, H., K. Kurosaki, H. Muta, and S. Yamanaka. 2008. Systematic investigation of the thermoelectric properties of $TlMTe_2$ (M = Ga, In, or Tl). *J. Appl. Phys.* 104: 073705.
46. Chrissafis, K., E. S. Vinga, K. M. Paraskevopoulos, and E. K. Polychroniadis. 2003. On the study of the solid–solid phase transformation of $TlBiTe_2$. *Phys. Status Solidi A* 196: 515–522.
47. Hoang, K. and S. D. Mahanti. 2008. Atomic and electronic structures of thallium-based III-V-VI_2 ternary chalcogenides: Ab initio calculations. *Phys. Rev. B* 77: 205107.
48. Kurosaki, K., H. Uneda, H. Muta, and S. Yamanaka. 2004. Thermoelectric properties of thallium antimony telluride. *J. Alloys Compd.* 376: 43–48.
49. Sankar, C. R., S. Bangarigadu-Sanasy, and H. Kleinke. 2012. Thermoelectric properties of $TlGdQ_2$ (Q = Se, Te) and Tl_9GdTe_6. *J. Electron. Mater.* 41: 1662–1666.
50. Kurosaki, K., A. Kosuga, and S. Yamanaka. 2003. Thermoelectric properties of $TlBiTe_2$. *J. Alloys Compd.* 351: 279–282.
51. Klepp, K. O. and P. Ecker. 1995. The crystal structure of Tl_2GeTe_3: A tellurogermanate(IV) with bitetrahedral anions. *J. Solid State Chem.* 117: 351–355.
52. Agafonov, P. V., B. Legendre, N. Rodier, J. M. Cense, E. Dichi, and G. Kra. 1991. Structure de Tl_2SnTe_3. *Acta Crystallogr. C* 47: 1300–1301.
53. Kosuga, A., K. Kurosaki, H. Muta, and S. Yamanaka. 2006. Thermoelectric properties of Tl–X–Te (X = Ge, Sn, and Pb) compounds with low lattice thermal conductivity. *J. Appl. Phys.* 99: 063705.
54. Sankar, C. R., M. Guch, A. Assoud, and H. Kleinke. 2011. Structural, thermal and physical properties of the thallium zirconium telluride Tl_2ZrTe_3. *Chem. Mater.* 23: 3886–3891.
55. Dichi, E., M. Sghaier, and G. Kra. 2008. Thermoelectric properties of ternary phases of thallium–tin–tellurium system. *J. Alloys Compd.* 458: 109–114.

56. Marsh, R. E. 1990. On the structure of thallium germanium telluride (Tl_2GeTe_5). *J. Solid State Chem.* 87: 467–468.
57. Agafonov, V., B. Legendre, N. Rodier, J. Cense, E. Dichi, and G. Kra. 1991. Structure of Tl_2SnTe_5. *Acta Crystallogr. C* 47: 850–852.
58. Goldsmid, H. J. and J. W. Sharp. 1999. Estimation of the thermal band gap of a semiconductor from Seebeck measurements. *J. Electron. Mater.* 28: 869–872.
59. Sharp, J. W., B. C. Sales, D. G. Mandrus, and B. C. Chakoumakos. 1999. Thermoelectric properties of Tl_2SnTe_5 and Tl_2GeTe_5. *Appl. Phys. Lett.* 74: 3794–3796.
60. Christensen, M., A. B. Abrahamsen, N. B. Christensen et al. 2008. Avoided crossing of rattler modes in thermoelectric materials. *Nat. Mater.* 7: 811–815.
61. Braun, D. J. and W. Jeitschko. 1980. Preparation and structural investigations of antimonides with the lanthanum iron phosphide ($LaFe_4P_{12}$) structure. *J. Less-Common Met.* 72: 147–156.
62. Chakoumakos, B. C., B. C. Sales, D. Mandrus, and V. Keppens. 1999. Disparate atomic displacements in skutterudite-type $LaFe_3CoSb_{12}$, a model for thermoelectric behavior. *Acta Crystallogr. B* 55: 341–347.
63. Chakoumakos, B. C., B. C. Sales, D. G. Mandrus, and G. S. Nolas. 2000. Structural disorder and thermal conductivity of the semiconducting clathrate $Sr_8Ga_{16}Ge_{30}$. *J. Alloys Compd.* 146: 80–86.
64. Tao, X., P. Jund, R. Viennois, and J.-C. Tédenac. 2011. Physical properties of thallium–tellurium based thermoelectric compounds using first-principles simulations. *J. Phys. Chem. A* 115: 8761–8766.
65. Bradtmöller, S. and P. Böttcher. 1993. Darstellung und Kristallstruktur von $SnTl_4Te_3$ und $PbTl_4Te_3$. *Z. Anorg. Allg. Chem.* 619: 1155–1160.
66. Böttcher, P., T. Doert, C. Druska, and S. Bradtmöller. 1997. Investigations on compounds with Cr_5Bi_3 and In_5Bi_3 structure types. *J. Alloys Compd.* 246: 209–215.
67. Bangarigadu-Sanasy, S., C. R. Sankar, A. Assoud, and H. Kleinke. 2011. Crystal structures and thermoelectric properties of the series $Tl_{10-x}La_xTe_6$ with $0.2 \le x \le 1.15$. *Dalton Trans.* 40: 862–867.
68. Guo, Q., A. Assoud, and H. Kleinke. 2014. Improved bulk materials with thermoelectric figure-of-merit > 1: $Tl_{10-x}Sn_xTe_6$ and $Tl_{10-x}Pb_xTe_6$. *Adv. Energy Mater.* 4: 1400348.
69. Kurosaki, K., A. Kosuga, and S. Yamanaka. 2003. Thermophysical properties of Tl_9BiTe_6 and $TlBiTe_2$. *J. Alloys Compd.* 351: 14–17.
70. Imamalieva, S. Z., F. M. Sadygov, and M. B. Babanly. 2008. New thallium neodymium tellurides. *Inorg. Mater.* 44: 935–938.
71. Wölfing, B., C. Kloc, J. Teubner, and E. Bucher. 2001. High performance thermoelectric Tl_9BiTe_6 with an extremely low thermal conductivity. *Phys. Rev. Lett.* 86: 4350–4353.
72. Yamanaka, S., A. Kosuga, and K. Kurosaki. 2003. Thermoelectric properties of Tl_9BiTe_6. *J. Alloys Compd.* 352: 275–278.
73. Guo, Q., M. Chan, B. A. Kuropatwa, and H. Kleinke. 2013. Enhanced thermoelectric properties of variants of Tl_9SbTe_6 and Tl_9BiTe_6. *Chem. Mater.* 25: 4097–4104.
74. Guo, Q., M. Chan, B. A. Kuropatwa, and H. Kleinke. 2014. Thermoelectric properties of Sn- and Pb-doped Tl_9BiTe_6 and Tl_9SbTe_6. *J. Appl. Phys.* 116: 183702.
75. Babanly, M. B., S. Z. Imamaliyeva, D. M. Babanly, and F. M. Sadigov. 2009. Tl_9LnTe_6 (Ln–Ce, Sm, Gd) compounds—The new structural analogies of Tl_5Te_3. *Azerb. Chem. J.* 2: 122–125.
76. Babanly, M. B., J.-C. Tedenac, S. Z. Imamalieva, F. N. Guseynov, and G. B. Dashdieva. 2010. Phase equilibria study in systems Tl–Pb(Nd)–Bi–Te new phases of variable composition on the base of Tl_9BiTe_6. *J. Alloys Compd.* 491: 230–236.
77. Guo, Q. and H. Kleinke. 2015. Thermoelectric properties of hot-pressed Tl_9LnTe_6 (Ln = La, Ce, Pr, Nd, Sm, Gd, Tb) and $Tl_{10-x}La_xTe_6$ ($0.90 \le x \le 1.05$). *J. Alloys Compd.* 630: 37–42.

13

Higher Manganese Silicides

Yuzuru Miyazaki
Tohoku University

CONTENTS

13.1 Introduction

Si-rich compounds in the Mn–Si binary system are called "higher manganese silicides" (HMSs) and have been known as potential thermoelectric (TE) materials for more than 50 years [1–3]. Although the figure of merit, *ZT*, of the material is less than unity, HMSs have recently attracted renewed interest as they consist of naturally abundant and less-toxic elements and are chemically and thermally stable in the atmosphere. By adopting first-principles calculations to design alloys and cutting-edge techniques to modify microstructures, the TE performance has recently been gradually improved. Now, it is no longer a dream that the TE performance exceeds $ZT > 1$. In this chapter, we will review key milestones in the long history of HMS research of the stoichiometry, crystal structure, TE properties, thermal expansion, and so on. We will also give some insights on how to obtain HMS-based materials with superior TE performance.

13.2 Chemical Composition

Reports on HMSs used as heat-resistant semiconducting materials can be found as far back as the 1950s [4]. At the early stages, HMSs were already known to exhibit high electrical conductivity, as well as a high Seebeck coefficient and low thermal conductivity. However, the chemical formula was not revealed for a long time. Initially, the formula was considered to be simple stoichiometric compound of $MnSi_2$ [5], but subsequent detailed metallurgical studies clarified that the Si composition is in the range of 46–47 wt%, corresponding to formulas between $MnSi_{1.67}$–$MnSi_{1.73}$ [6]. Subsequently, at least four phases were identified as stable HMSs, as shown in Table 13.1: Mn_4Si_7 ($MnSi_{1.750}$, 63.6 at% Si) [7,8], $Mn_{11}Si_{19}$ ($MnSi_{1.727}$, 63.3 at% Si) [9], $Mn_{15}Si_{26}$ ($MnSi_{1.733}$, 63.4 at% Si) [10,11], and $Mn_{27}Si_{47}$ ($MnSi_{1.741}$, 63.5 at% Si) [12]. According to Nowotny's classification, all of these HMSs are represented as a series of the formula $Mn_nSi_{(2n-m)}$ [13]. There are an infinite number of integers m and n that satisfy the condition $(2n-m)/n$ to be approximately 1.73, yet only the above four phases have been structurally determined and their space group, lattice parameters, and atomic coordinates reported. The existence of other compositions, such as Mn_7Si_{12} ($MnSi_{1.712}$, 63.1 at% Si) [14], has been reported (also shown in Table 13.1) but has only been observed in electron diffraction (ED) patterns and not yet isolated as bulk compounds. All these HMSs structurally resemble each other and are called Nowotny chimney ladder phases, as the Mn atoms form a rectangular chimney along the c-axis and the Si atoms form a spiral staircase in each chimney [13].

In general, the word "commensurate" represents a simple ratio of integers. Hence, the formula Mn_4Si_7 can be regarded as commensurate but more complicated formulas such as $Mn_{15}Si_{26}$ or $Mn_{27}Si_{47}$ may not be commensurate, although there is no essential change in the actual Mn:Si ratio. It is quite strange why such a tiny change in the chemical formula in the order of ~10^{-3} causes a huge change in the c-axis length and space group from a primitive tetragonal to a body-centered tetragonal unit cell. From an academic point of view, it would be more elegant and correct to represent such unique compounds as incommensurate $MnSi_\gamma$ with a small compositional range of γ.

The first equilibrium phase diagram of the Mn–Si binary system to report the existence of HMS was probably made by Korshunov et al. [15]. They summarized a phase relation in the MnSi–Si pseudobinary system and claimed the chemical formula of the HMSs to be $MnSi_{1.67}$–$MnSi_{1.73}$. They also reported that the HMS phase is formed by a peritectic reaction (peritectic temperature of 1150°C) upon cooling from the liquid. Later, Dudkin and Kuznetsova [16] reported a similar phase relation and the formation of the HMS (denoted as "Mn_3Si_5") phase as a peritectic reaction from the liquid at a peritectic temperature of 1159°C. Figure 13.1 depicts the currently accepted phase relation around HMS, represented in the figure as $Mn_{11}Si_{19}$. When a liquid phase with the composition of ~$Mn_{11}Si_{19}$ starts to cool slowly, the MnSi (monosilicide) phase solidifies as a primary crystal. To grow single crystals of $Mn_{11}Si_{19}$, the composition of the liquid phase must be quite strict at around 65 at% Si, corresponding to a narrow two-phase region of L + $Mn_{11}Si_{19}$. Several reports also claim that HMS is not a strict line compound but has a compositional range of 63–64 at%, corresponding to $MnSi_{1.70}$–$MnSi_{1.75}$. Most of the reports agree well with that of Korshunov et al. [15], claiming that the HMSs are formed as a peritectic reaction from the liquid and a two-phase region of L + HMS is incredibly narrow to solidify the HMS as a primary crystal. Exceptionally, Fujino et al. [17] claimed that the HMS phase melts congruently upon heating at 1145°C because single crystals can be successfully prepared from the liquid with the composition of 46.80 wt% Si.

13.3 Crystal Structure

Through a series of extensive studies using transmission electron microscopy (TEM) [18–20], HMSs are now recognized as a single compound consisting of two sublattices of

TABLE 13.1

Variation of HMSs with the Representation of $Mn_nSi_{(2n-m)}$

Formula $Mn_nSi_{(2n-m)}$	n	m	$(2n-m)/n$	a (nm)[a]	c (nm)[a]	Space Group	c_{Mn} (nm)	c_{Si} (nm)	No. of Mn Sites	No. of Si Sites	References
Mn_4Si_7	4	1	1.750	0.5525(1)	1.7463(3)	$P\bar{4}c2$	0.4366	0.2494	5	4	[7]
Mn_4Si_7	4	1	1.750	0.55259(5)	1.75156(8)	$P\bar{4}c2$	0.4379	0.2502	5	4	[8]
Mn_7Si_{12}	7	2	1.714								[14]
$Mn_{11}Si_{19}$	11	3	1.727	0.5518	4.8136	$P\bar{4}c2$	0.4376	0.2533	12	10	[9]
$Mn_{15}Si_{26}$	15	4	1.733	0.5531	6.5311	$I\bar{4}2d$	0.4354	0.2512	8	7	[10]
$Mn_{15}Si_{26}$	15	4	1.733	0.5525	6.555	$I\bar{4}2d$	0.4370	0.2521			[11]
$Mn_{19}Si_{33}$	19	5	1.737								[14]
$Mn_{26}Si_{45}$	26	7	1.731	0.5515	11.336		0.4360	0.2519			[11]
$Mn_{27}Si_{47}$	27	7	1.741	0.5530	11.794	$P\bar{4}n2$	0.4368	0.2509	28	24	[12]
$Mn_{39}Si_{68}$	39	10	1.744								[14]
$MnSi_\gamma$			1.7326(1)	0.55242(3)		$I4_1/amd(00\gamma)00ss$	0.43665(3)	0.25202(3)	1	1	[22]

[a] A number in parentheses represents the estimated standard deviation for the last significant digit.

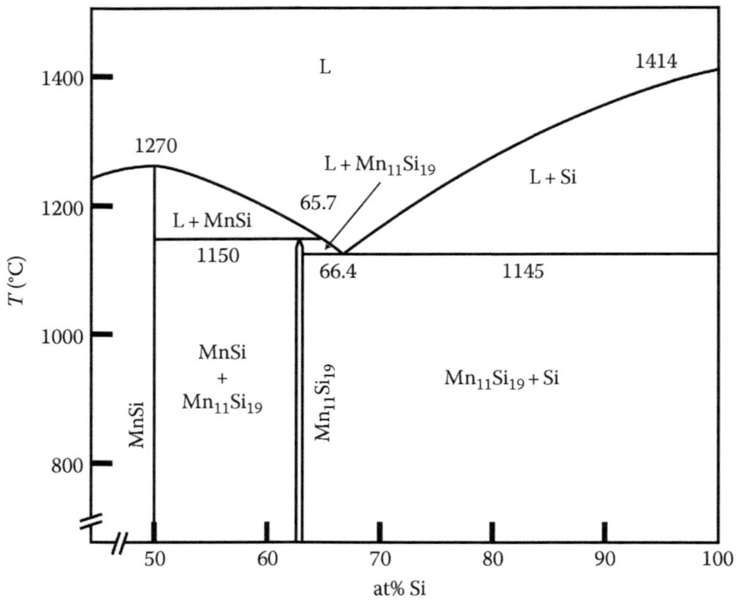

FIGURE 13.1
A partial phase diagram of the Mn–Si binary system. (Redrawn with data from Korshunov, V.A. et al., *Fiz. Metal. Metalloved.*, 12, 277, 1961; Dudkin, L.D. and Kuznetsova, E. S., *Dokl. Akad. Nauk SSSR*, 141, 97, 1961.)

Mn and Si atoms with different stacking lengths along the *c*-axis. This single compound is usually not uniform throughout the crystals but allows a variety of ED patterns and lattice images owing to the combination of a "spacing anomaly" and "orientation anomaly." Ye and Amelickxs [14] made further progress with TEM studies and revealed that the spacing anomalies originate from the difference in stacking periods between the two sublattices and are commonly observed in HMSs, while orientation anomalies originate from the difference in the three dimensional (3D) arrangement of consecutive Si sublattices relative to the 3D arrangement of Mn sublattices. A variety of moiré-like fringes are formed depending on the relative angles between the 3D arrangement of Mn and Si sublattices along the *c*-axis. Figure 13.2a shows a typical high-resolution image with and without orientation anomalies in a single domain. It is easy to recognize the orientation difference between the bright dots (Mn sublattices) and the bright-dark modulated dots (Si sublattices) in Figure 13.2b. In contrast, only parallel rows are observed in Figure 13.2c.

Based on the ED patterns, Yamamoto [21], a specialist of higher dimensional crystallography, has pointed out that the HMSs belong to incommensurate composite crystals consisting of two tetragonal subsystems (equivalent to sublattices), [Mn] and [Si], with different stacking periods along the *c*-axis. He also pointed out that the crystal structure is accurately represented by using a (3 + 1)-dimensional superspace approach. Composite crystal is a special term in crystallography consisting of two or more subsystems with different periods and sometimes different symmetries. Owing to such a periodic difference, composite crystals usually involve displacive modulations, wherein every atomic site is cooperatively shifted. In the present HMSs, [Mn] and [Si] have identical tetragonal *a*- and *b*-axes but different *c*-axis of c_{Mn} and c_{Si}, being parallel but incommensurate with each other. To accurately describe the crystal structure, a (3 + 1)-dimensional superspace description of $I4_1/amd(00\gamma)00ss$ is necessary. This superspace group represents the [Mn]

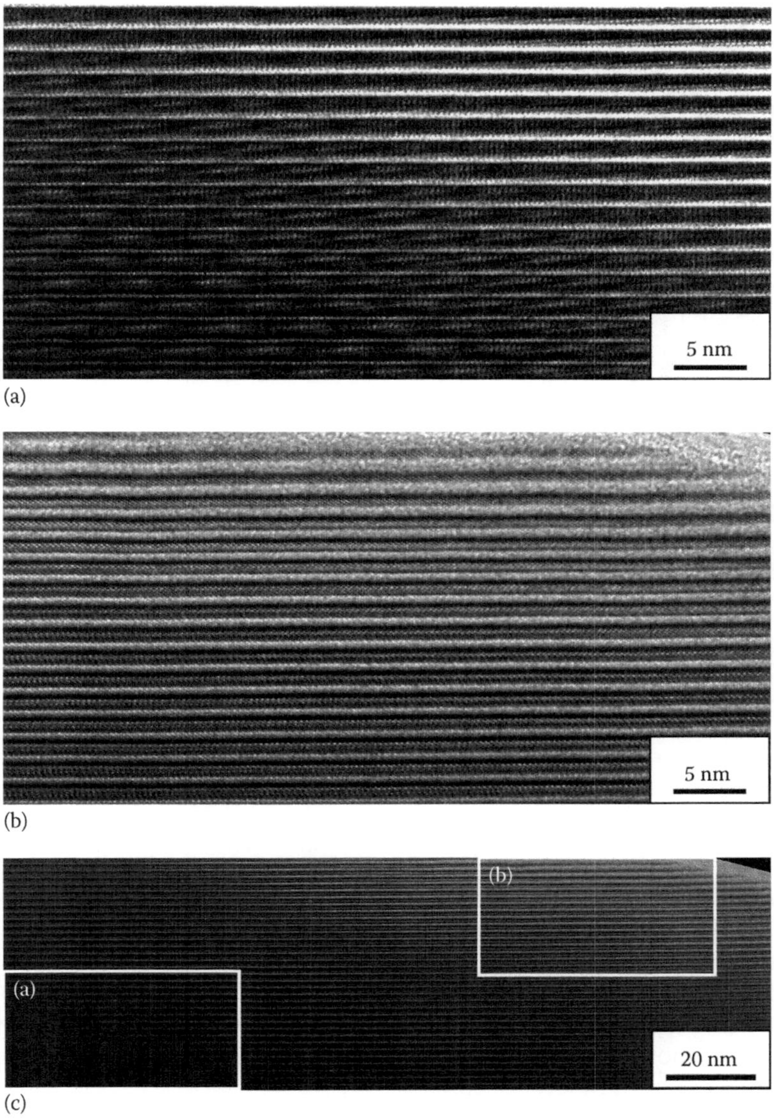

FIGURE 13.2
Typical high-resolution images with (a) or without (b) orientation anomalies in a single domain of HMS (c). (Images provided by the courtesy of Prof. K. Yubuta, IMR, Tohoku University, Sendai, Japan.)

has the 3D space group $I4_1/amd$ and the [Si] has the 3D space group $P4/nnc$ with an origin shift between the subsystems.

Recently, Miyazaki et al. [22] successfully determined the modulated structure of an HMS by means of Rietveld refinement of neutron diffraction data based on the identical superspace group. By defining the c-axis ratio $\gamma = c_{Mn}/c_{Si}$, the structure formula should be represented as $MnSi_\gamma$ and γ of their sample is determined to be 1.7364(1); the number in the parenthesis represents the estimated standard deviation for the last significant digit. The parameter γ also represents the modulation vector component along c^* as $\mathbf{k} = (00\gamma)$.

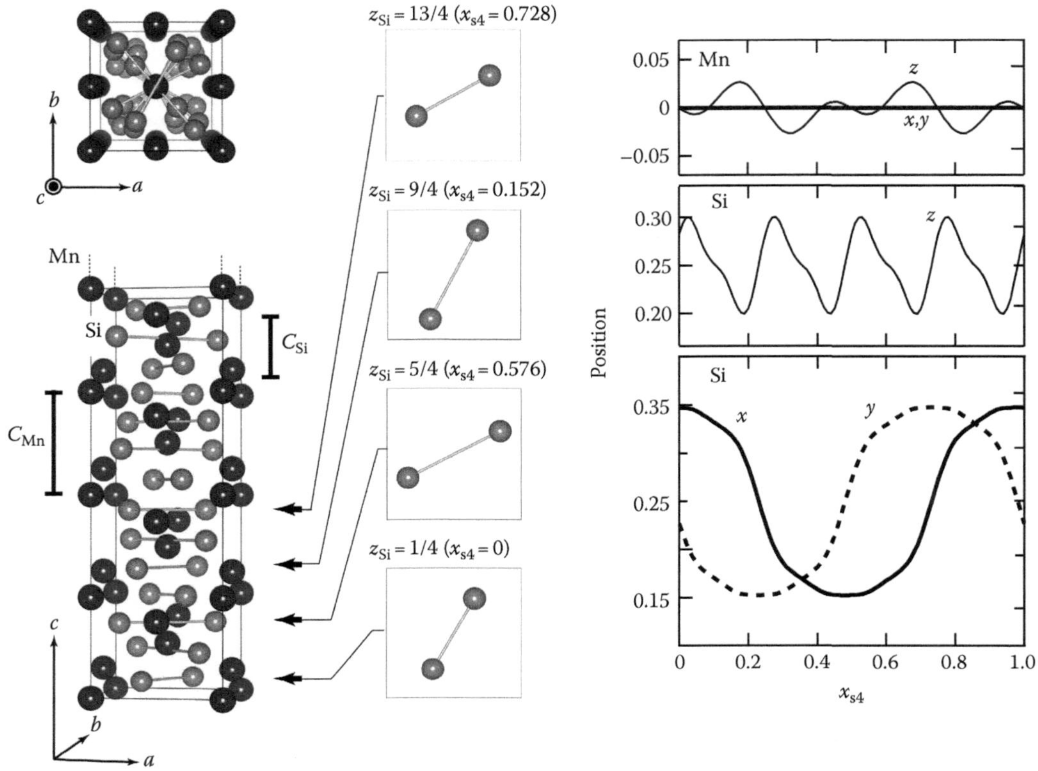

FIGURE 13.3
Modulated arrangement of atoms in $MnSi_{1.7364}$ determined by the Rietveld analysis of neutron diffraction data of References 22 and 23. (Adapted with permission from Miyazaki, Y., Igarashi, D., Hayashi, K., Kajitani, T., and Yubuta, K., *Phys. Rev. B*, 78, 214104, 2008. Copyright 2008 by the American Physical Society.)

The modulated arrangement of atoms in $MnSi_{1.7364}$ is depicted in Figure 13.3. The precise (x, y, z) coordinates can be obtained by putting the γ value either into $x_{s4} = (1/\gamma)(z_{Si} - 1/4)$ for the atoms in [Si], or $x_{s4} = \gamma z_{Mn}$ for the atoms in [Mn], and the six universal curves of modulation functions are shown in the right panel. A more detailed description of the modulated structure can be found in Reference 23.

13.4 Electronic Structure

To date, there are no experimental studies on the electronic structure of HMSs by means of x-ray photoelectron spectroscopy. Therefore, we will review the electronic structure and density of states (DOS) by means of first-principles calculations. Unfortunately, we have no software to compute the electronic structure of incommensurate composite crystals because a finite length of a 3D unit cell cannot be defined.

Imai et al. [24] first computed DOSs of Mn_4Si_7 and $Mn_{11}Si_{19}$ based on the first-principles pseudopotential method. They demonstrated that the overall shape of both DOSs is similar; a broad band ranging from −4 to −1 eV and a sharp DOS at −1 to 0 eV. Another similarity is that the conduction band (CB) is located 0.7 eV higher from the top of the valence band (VB) for both HMSs. One essential difference is the position of the Fermi level (E_F), which is located in the middle of the band gap for Mn_4Si_7 and just below the VB top for $Mn_{11}Si_{19}$. Hence, Mn_4Si_7 is a band semiconductor, whereas $Mn_{11}Si_{19}$ is a metal or a degenerate semiconductor. Later, Imai and Watanabe [25] demonstrated that both the VB top and CB are mainly the 3d character of the Mn atoms in the $Mn_{11}Si_{19}$ phase.

Migas et al. [26] computed the electronic structures of four crystalline phases, Mn_4Si_7, $Mn_{11}Si_{19}$, $Mn_{15}Si_{26}$, and $Mn_{27}Si_{47}$, based on the first-principles pseudopotential method. They also computed the effect of stacking faults of Si atoms on the electronic structure. Although the characteristic features of the DOS computed for Mn_4Si_7 show a close resemblance to that reported by Imai, the contribution of the Mn-d and Si-p character was precisely determined and the magnitude of the DOS at the CB was comparably larger than that of the VB. Similar to the results reported in [25], Mn_4Si_7 is revealed to be a band semiconductor, while the other three phases are metals, wherein the E_F crosses the k-points between the Γ- and Z-points in the first Brillouin zone. The Fermi surface has a typical cylindrical shape characteristic of a 2D conductor, suggesting that the HMSs would exhibit anisotropic transport and TE properties. The effective mass is determined to be $m_{xx} = 1.64\ m_0$ and $m_{zz} = 5.72\ m_0$ for holes and $m_{xx} = 8.17\ m_0$ and $m_{zz} = 3.40\ m_0$ for electrons, respectively; xx and zz represent the principle-axis components and m_0 the free-electron mass [26].

The electronic structure of the four phases reported by Migas et al. [26] shows essentially identical features yet differs in minor details; the only significant difference is the position of E_F. Hence, the electronic structure of all the incommensurate HMSs can be treated semiquantitatively based on the rigid band model. To predict the TE properties and to avoid time-consuming calculations, it is enough to calculate the electronic structure based on the simplest commensurate Mn_4Si_7 structure and apply the rigid band model. For such a purpose, it is popular to determine the TE properties using BoltzTraP code [27] based on the Boltzmann transport theory. Here, we present the expected TE properties of HMSs at 800 K, at which the material exhibits the best TE performance.

In Figure 13.4, we depict the Seebeck coefficient S, the electrical conductivity σ, and the TE power factor P expected at 800 K, based on the Mn_4Si_7 model, plotted against the chemical potential μ. The solid and dotted curves in each panel represent parameters parallel to the a- and c-axes to represent anisotropy. The position at $\mu = 0$ eV corresponds to the nondoped state, and a positive (negative) shift from $\mu = 0$ represents electron (hole) doping. As it is impossible to evaluate the relaxation time of carriers, τ, with this method, σ and P contain the term τ. At 800 K, the Seebeck coefficient approaches ~220 $\mu V\ K^{-1}$ for both directions (S_a is 5 $\mu V\ K^{-1}$ higher than S_c). A small amount of electron doping effectively increases S and further electron doping turns S into negative, yielding n-type materials and major carriers of electrons. Such a transition from p-type to n-type has been experimentally confirmed by the partial substitution of Fe in the Mn sites, as we will describe in Section 13.7. The electrical conductivity shows a distinct anisotropy, $\sigma_a > \sigma_c$, although both components are in the order of $10^{18}\ S\ m^{-1}\ s^{-1}$. By putting a typical relaxation time of carriers, $\tau = 1 \times 10^{-14}$ s, we obtain σ_a ~$5 \times 10^4\ S\ m^{-1}$. The electrical conductivity at 800 K increases with hole doping but decreases with electron doping and reaches a minimum value at the p to n transition. With further electron doping, the electrical conductivity increases in a symmetrical manner against μ in the n-type regime. The power factor, $P = S^2\sigma$, is expected to increase with a small amount of hole doping and more than doubled performance is expected for the n-type samples under an appropriate doping level, as long as the rigid band model is still applicable. In the actual case, the shift of μ is achieved by partial substitution of the Mn sites, as the Si sites allow little accommodation for foreign atoms. At the same time,

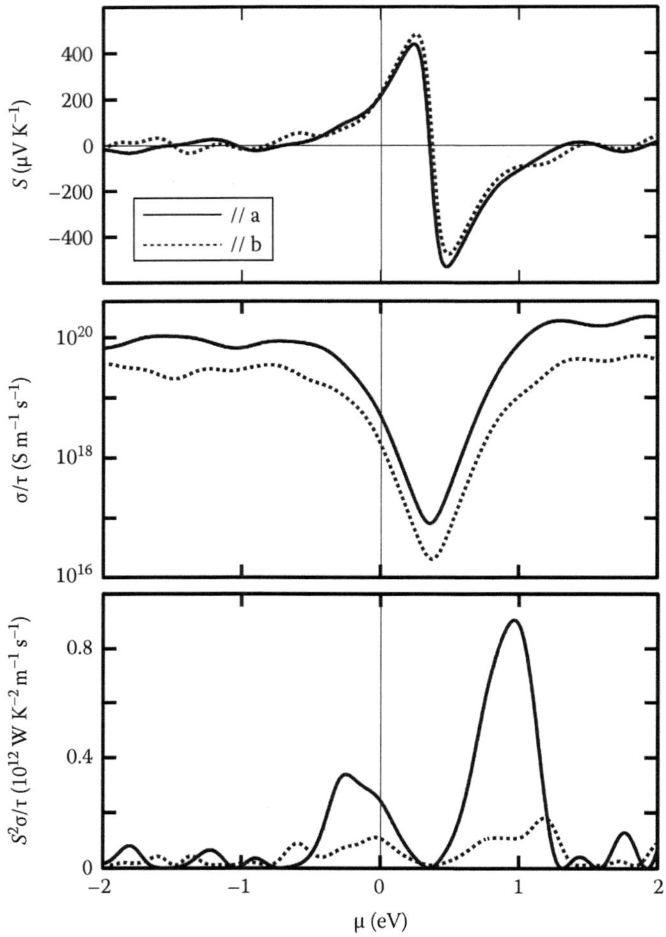

FIGURE 13.4
Calculated Seebeck coefficient S, electrical conductivity σ, and thermoelectric power factor P expected at 800 K based on the Mn_4Si_7 model, plotted against chemical potential μ.

the partial substitution will cause a certain decrease in the relaxation time of the carriers because of impurity scattering and, hence, deteriorates the power factors. Special care must be taken to design and prepare alloys based on the rigid band model.

13.5 TE Properties of HMSs

Substitution-free HMSs are p-type degenerate semiconductors with a typical hole concentration of 2.0–2.5 × 10^{21} cm^{-3} at room temperature [28]. Kawasumi et al. [29] demonstrated that the hole concentration remains unchanged up to ~500 K and then starts to increase exponentially above this temperature in accordance with a simultaneous increase in electron concentration as a result of the bipolar diffusion effect (BPE). At ~1000 K, the electron concentration of ca. 5.0 × 10^{21} cm^{-3} equals that of hole carriers. A typical band gap ranges between 0.7 and 0.8 eV at room temperature, and the hole mobility is reported to be on the order of 10 cm^2 V^{-1} s^{-1}.

Typical TE properties measured on a single crystal of $MnSi_{1.75}$ summarized by Fedorov and Zaitsev [2] are shown in Figure 13.5. As expected from the electronic structure, transport is more metallic parallel to the *a*-axis. The Seebeck coefficient parallel to *c*, S_c, exhibits a negative value below 50 K but gradually increases with temperature *T*, reaching S_c = 200 μV K^{-1} at 300 K and a maximum of 280 μV K^{-1} at 600–650 K. In contrast, the Seebeck coefficient parallel to *a*, S_a, exhibits a broad maximum at 50 K because of the phonon drag effect and increases with *T* toward another broad maximum of 190 μV K^{-1} at 800 K. A typical polycrystalline sample exhibits *S* values between S_a and S_c over the entire *T* range. The anisotropy of the Seebeck coefficient, S_c/S_a, decreases from 2 at lower temperatures

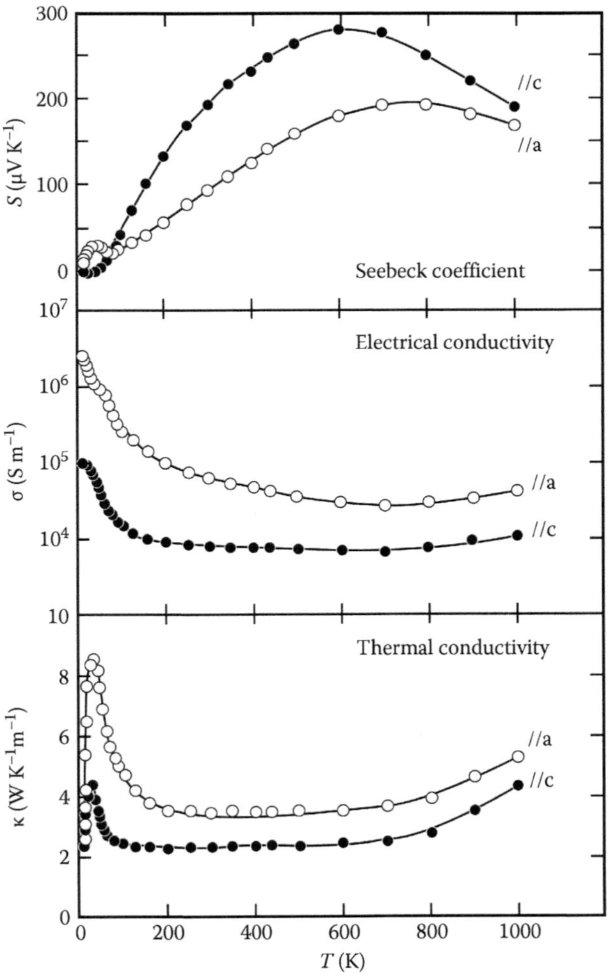

FIGURE 13.5
Typical thermoelectric properties measured on a single crystal of $MnSi_{1.75}$. (Redrawn with the data from Fedorov, M.I., Thermoelectrics of transition metal silicides, in *Thermoelectrics Handbook: Macro to Nano*, ed. D. M. Rowe, CRC Press, Boca Raton, FL, 2006.)

to 1.6 at 600 K and still decreases with T but never reaches 1, contrary to the calculations. Anisotropy remains with S_c = 250 μV K^{-1} and S_c = 180 μV K^{-1} at 800 K.

The electrical conductivity parallel to a, σ_a, is approximately one order of magnitude higher than that parallel to c, σ_c. This observation qualitatively agrees with calculations. Both σ_a and σ_c show metallic behavior from the lowest temperature up to 600–700 K, at which the BPE takes place and gradually increases with T above this temperature. At 800 K, σ_a becomes ~3.5 × 10^4 S m^{-1}, quite close to the calculated value if we assume carrier conduction is mainly governed by majority hole carriers even at this temperature. Along c, σ_c ~ 8.5 × 10^5 S m^{-1} is also comparable to the calculated value if a slightly lower relaxation time is used.

Experimental data of thermal conductivity show typical sharp peaks at ~30 K in both directions and increase in a similar manner to those of electrical conductivity. Thermal conductivity above room temperature is quite low κ = 2–4 W m^{-1} K^{-1}, as a result of the displacive modulation mentioned earlier. The contribution of carriers κ_c and lattice κ_L can be estimated by applying the Wiedemann–Franz law, $\kappa_c = L\sigma T$, to the total thermal conductivity $\kappa = \kappa_c + \kappa_L$. The derived values are κ_c = 0.65 W m^{-1} K^{-1} and κ_L = 3.25 W m^{-1} K^{-1} parallel to a, while κ_c = 0.17 W m^{-1} K^{-1} and κ_L = 2.53 W m^{-1} K^{-1} parallel to c; the lattice contribution still dominates for both directions. This is advantageous for further enhancement of ZT, because there is still room to reduce κ_L by introducing phonon scattering centers.

13.6 Effect of Partial Substitution on Crystal Structure

The TE power factor can be enhanced through partial substitution of atoms as presented in Section 13.4. Partial substitution of Mn with Cr or Fe was first reported by Tarasov and coworkers [30–32]. They reported a diagram of S and σ similar to Figure 13.4, plotted against the amount of substitution x in $(Mn_{1-x}M_x)Si_{\sim 1.7}$ (M = Cr or Fe) measured at room temperature. The number of valence electrons in Cr (Fe) is one less (more) than that of Mn and thus corresponds to hole (electron) doping. In the case of Fe substitution, a transition from p-type to n-type is observed when the amount of Fe is around x_{Fe} = 0.30. However, the absolute value of the Seebeck coefficient in the n-type samples is limited to approximately 60 μV K^{-1}. At the time, it was already known that the Si content slightly increases (decreases) with increasing x_{Cr} (x_{Fe}) and that it is necessary to change the nominal Si compositions [33].

In Figure 13.6, the determined Si content γ (original description is $(2n-m)/n$) reported in [33] is plotted as open symbols for Cr-, Fe-, and Co-substituted solid solutions. In the figure, our data obtained using the superspace approach from x-ray Rietveld analysis are also superimposed as filled symbols [34–36]. Data from both groups agree well for all the solid solutions. The substitution-free HMSs typically have a γ value of approximately 1.732, closest to that of quasicommensurate $Mn_{15}Si_{26}$ ($MnSi_{1.733}$). With Cr substitution, the γ value increases linearly according to Vegard's law up to 1.76 at the solubility limit of x_{Cr} = 0.20. With Fe or Co substitution, both data are overlapping and the γ value decreases to 1.67 at the solubility limit of x_{Fe} = 0.35 and to 1.71 at the solubility limit of x_{Co} = 0.10. Such a change in γ is not attributed to the difference in atomic radii, but as a consequence of hole (electron) doping. From the results, it can be considered that the solid solutions are a group of incommensurate compounds, because the γ value changes continuously.

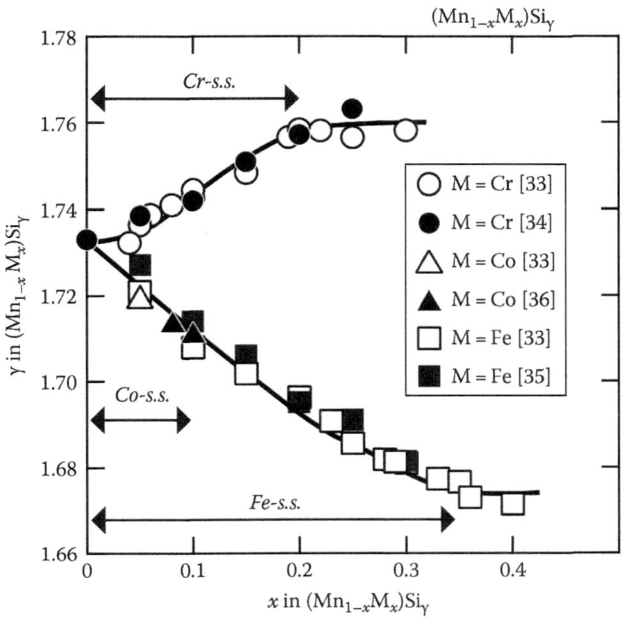

FIGURE 13.6
Determined Si content γ (original description is $(2n-m)/n$) reported in Reference 33 and plotted as open symbols for the Cr-, Fe-, and Co-substituted solid solutions. The data obtained from the superspace approach using x-ray Rietveld analysis in References 34–36 are also superimposed as filled symbols.

The Mn sites can also allow partial substitutions of other atoms such as V and Re, up to several at% [37,38]. A complete solid solution is formed between Ru_2Si_3 and $MnSi_\gamma$ [39]. A unique microstructure consisting of an inhomogeneous distribution of Mn-rich and Mn-poor dendritic domains is reported to occur in Cr- and Ru-substituted solid solutions [34,39]. A discernible reduction of κ_L is reported in such samples and, hence, further reduction in κ_L can be expected by achieving a well-controlled microstructure with homogeneous distribution of such domains.

By contrast, the Si sites cannot be effectively substituted and only a small amount of Ge, B, and Al can occupy the Si sites [40–42]. The solubility limit of Ge at the Si sites is reported as $MnSi_{1.715}Ge_{0.015}$ and that of Al as 0.13–0.15 wt% in the formula of $Mn_{11}Si_{19}$. In the case of B atoms, no solubility occurs and added boron will form the compound MnB as a secondary phase. No significant change in the lattice parameters is reported as the solubility limit of Ge and Al is also quite small.

13.7 Effect of Partial Substitution on TE Properties

In this section, we will review the effect of partial substitution on TE properties. As described, a small amount of hole carriers should be introduced to enhance the power factor of p-type materials. In Figure 13.7, we show the results of Cr substitution performed on polycrystalline samples [34]. With increasing Cr content x up to 0.20 in $(Mn_{1-x}Cr_x)Si_\gamma$, the

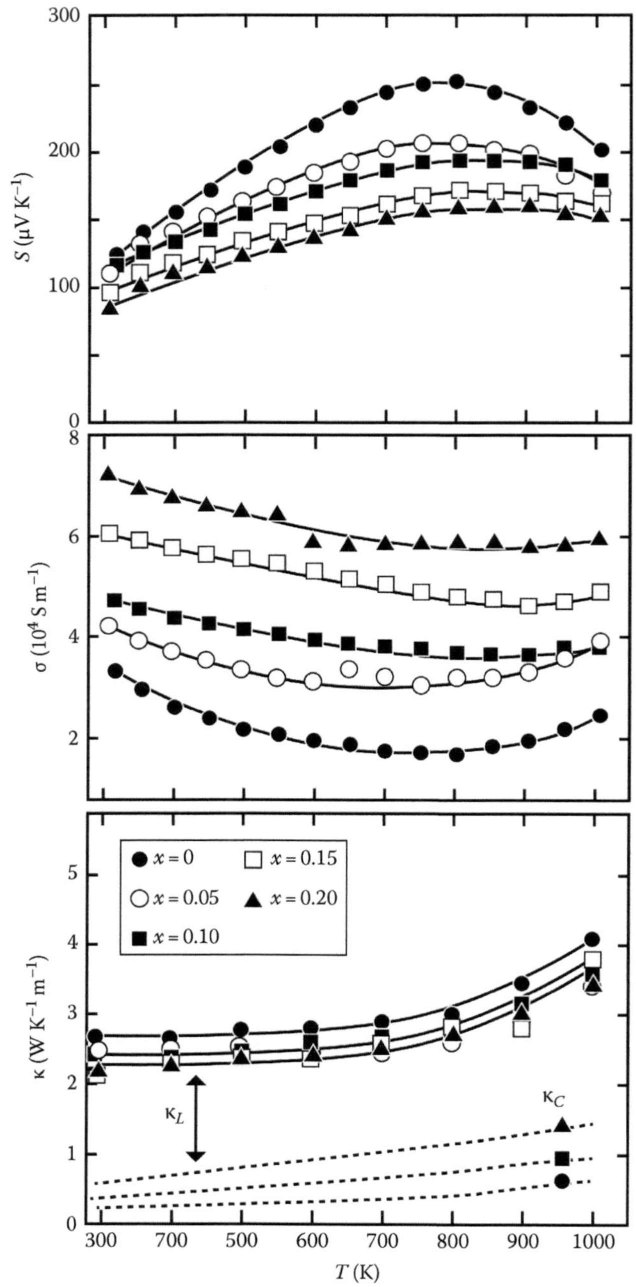

FIGURE 13.7
Thermoelectric properties of Cr-substituted $(Mn_{1-x}Cr_x)Si_\gamma$ polycrystalline samples. (Redrawn with the data from Kikuchi, Y., *Jpn. J. Appl. Phys.*, 51, 085801, 2012. Copyright 2012 The Japan Society of Applied Physics.)

Seebeck coefficient gradually decreases and the electrical conductivity increases due to hole doping. As a result, the maximum power factor is estimated to increase from $P = 1.1 \times 10^{-3}$ W m^{-1} K^{-2} for $x = 0$ (at 850 K) to 1.5×10^{-3} W m^{-1} K^{-2} for $x = 0.20$ (at 900 K) samples. Despite the increase in σ, the total thermal conductivity gradually decreases with increasing x. This decrease in κ originates from a certain reduction of κ_L resulting from the microdomain separation mentioned earlier. Consequently, a substantial increase in ZT from 0.28 for $x = 0$ (at 800 K) to 0.45 for $x = 0.20$ (at 900 K) is achieved. Another benefit is that Cr substitution raises the starting temperature of BPE, increasing the highest operating temperature to 900 K.

The performance of p-type is further enhanced by partial substitution of Ru [39,43]. In Figure 13.8, we show traced data of the power factor, thermal conductivity, and dimensionless figure of merit of Ru-substituted samples reported by Okamoto et al. [39]. The chemical formula has been modified to represent Mn-based solid solution as $(Mn_{1-x}Ru_x)Si_\gamma$. A partial substitution with Ru for Mn sites effectively increases the power factor up to $P = 1.5 \times 10^{-3}$ W m^{-1} K^{-2} for $x = 0.10$ (at 800 K), as in the case of Cr substitution. A significant reduction of the total thermal conductivity can be achieved with the heavier Ru, lowering the κ down to 1.5 W m^{-1} K^{-1} at 800 K. As a result, the highest ZT of 0.76 (at 874 K) was realized. Such an increase in P cannot be expected from the calculation shown in Figure 13.4, as the Ru substitution corresponds to electron doping and the P value monotonically decreases with electron doping in the p-type regime. n-Type conduction is only realized for the samples with $x \geq 0.45$.

Partial substitution with Fe [36,44] also induces the transition from p-type to n-type at $x \sim 0.28$, but the electron concentration cannot be increased as the solubility limit of Fe at $x \sim 0.35$ is close to the transition point. The highest ZT is thus limited to ~0.07 for $x = 0.30$ samples (at 644 K).

Substitution at the Si sites, although not very effective because of the low solubility limit, has also been reported. Si site substitution may not change the electronic structure very much, but one remarkable finding is that the substitution of ~1 at% Ge for Si effectively improves the microstructure by dissipating the layered striations and such samples exhibit much higher electrical conductivity to enhance ZT [41,42].

Double doping of both the Mn and Si sites simultaneously has also been performed. Several combinations are possible: isovalent substitution, for example, Mn with Re and Si with Ge, and aliovalent substitution but maintaining the formal charge as in Mn with Fe and Si with Al. A lot of effort has been devoted, and some successful results have been reported [43,45–47].

13.8 Striation Problem and Thermal Expansion

Ivanova et al. [48] first observed the microstructure of HMS single crystals prepared by the Czochralski method. They found that "bands" are arranged perpendicular to the c-axis of HMS, within 3°, regardless of the growth direction. Kojima et al. [49] confirmed that the bands are layered precipitates (striations) of MnSi monosilicide by means of X-ray diffraction (XRD) and Electron Probe Microanalysis (EPMA). A typical thickness of the striations ranges from 1.6 to 3.0 μm, and the striations are located with an interval of 80–150 μm perpendicular to the c-axis of HMS. Such striations can often be observed for the samples prepared from the liquid state, even at a Si-rich composition where the eutectic HMS + Si region is thermodynamically stable [50]. However, the samples prepared by chemical transport [51] or flux method [52] prepared far below the peritectic temperature never contain such striations.

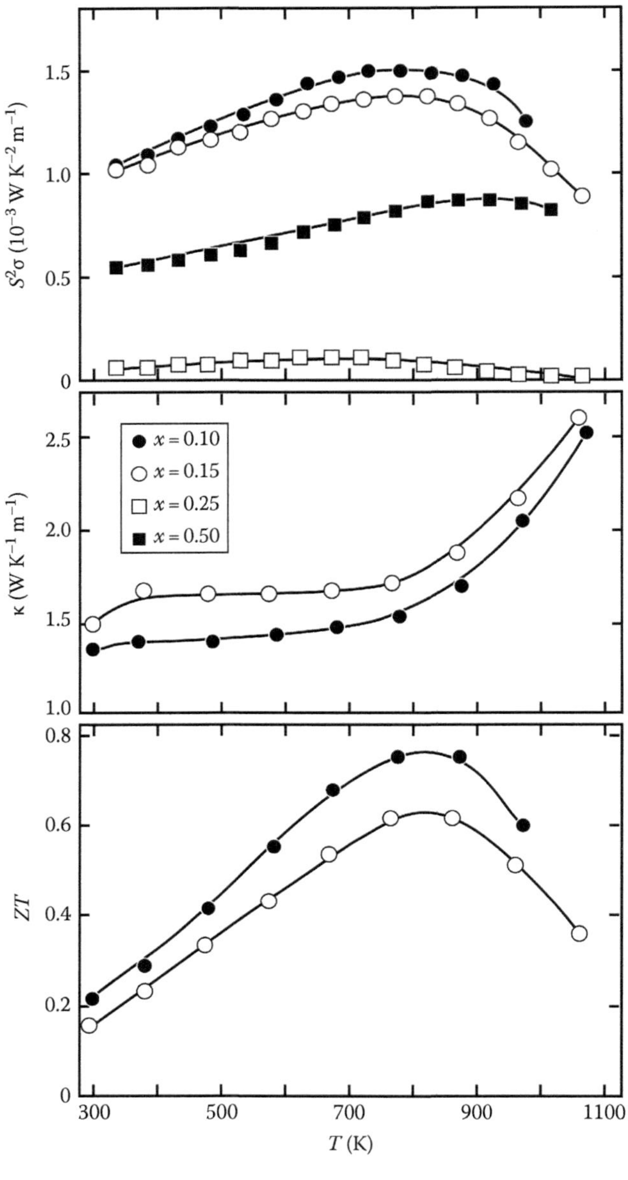

FIGURE 13.8
Thermoelectric properties of Ru-substituted polycrystalline samples. The chemical formula has been modified to represent Mn-based solid solution as $(Mn_{1-x}Ru_x)Si_\gamma$. (Redrawn with the data from, *Acta Mater.*, 57, Okamoto, N.L., Crystal structure and thermoelectric properties of chimney–ladder compounds in the Ru_2Si_3–Mn_4Si_7 pseudobinary system, 5036, Copyright 2009, with permission from Elsevier.)

There have been several facts that question whether all the striations are in fact MnSi precipitates. According to the Mn–Si binary phase diagram [15,16], MnSi appears as a *primary crystal* of a peritectic reaction at a composition of ~65 at% Si during freezing. The MnSi phase has a primitive cubic structure with the space group of $P2_13$ and the lattice parameter of a = 0.456 nm [53]. Generally, primary crystals with a cubic structure, that is, no

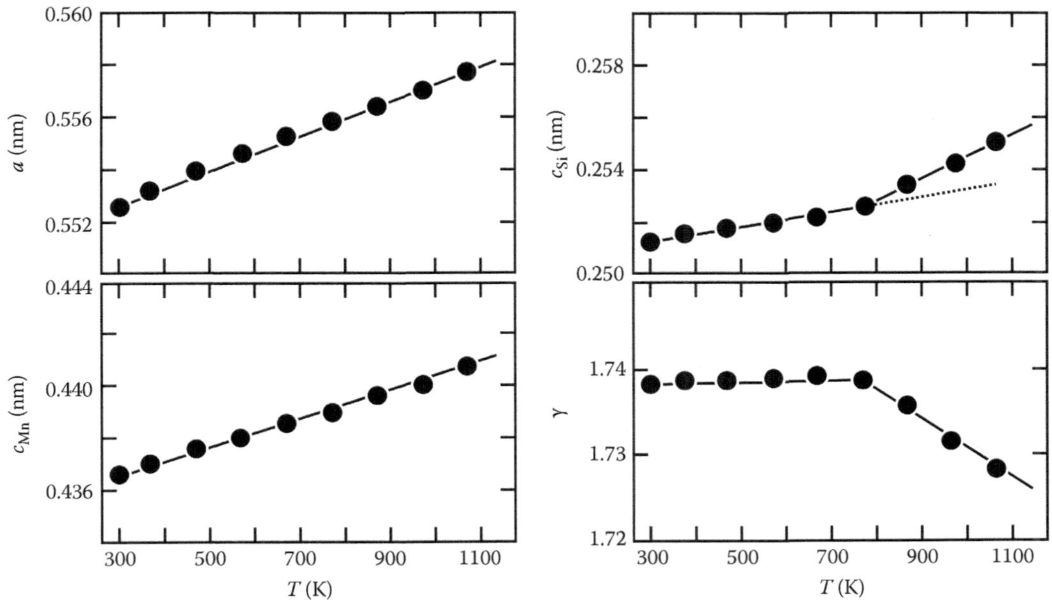

FIGURE 13.9
Temperature dependence of the lattice parameters, *a*, c_{Mn}, c_{Si}, and γ of an $MnSi_{1.7364}$ sample. (Redrawn with the data from, *J. Alloys Compd.*, 616, Kikuchi, Y., High temperature X-ray diffraction study on incommensurate composite crystal MnSiγ – (3+1)-dimensional superspace approach, 263, Copyright 2014, with permission from Elsevier.)

preferable growth orientation, grow isotropically to form globules. A subsequent peritectic reaction takes place and HMSs will nucleate at the boundaries of MnSi and the liquid. The HMSs polycrystals can grow in *any* direction. The *a*-axis length of HMSs is ~0.552 nm [22], and no consistent epitaxial relationship between MnSi and HMSs is found. Hence, there is no reason to regularly form striations of MnSi with a highly anisotropic shape always perpendicular to the *c*-axis of HMS crystals during cooling. Moreover, experiments have shown that the electrical conductivity becomes lower when such striations exist. Why does electrical conductivity of samples decrease despite the presence of a good electrical conductor of MnSi? The authors have performed high-temperature (HT) XRD to investigate the structural change of HMSs. Figure 13.9 represents the temperature dependence of the lattice parameters, *a*, c_{Mn}, c_{Si}, and γ [54]. The *a*- and c_{Mn}-axis lengths linearly increase with increasing temperature because of thermal expansion. The c_{Si}-axis also expands with *T*, but a sudden change in the trend is discovered at 773 K, above which the temperature coefficient of $c_{Si}(T)$ almost tripled. As a result, the γ value starts to deviate from a constant value above $T = 773$ K, at which BPE becomes significant. This fact means that *HMSs change their composition* and a small amount of Si must go out from the structure at temperatures higher than ~800 K. If samples were cooled down from $T > 800$ K, microcracks perpendicular to the *c*-axis would be formed, and although HMSs have the capability to allow structure modulation, stacking faults and orientation disorder would arise to some extent.

The coefficient of thermal expansion (CTE) can be evaluated using HTXRD [54]. HMSs are found to possess the linear CTEs of $\alpha_{cMn} = 12.6(2) \times 10^{-6}$ K^{-1} and $\alpha_{cSi} = 10.7(5) \times 10^{-6}$ K^{-1} at 298 K. CTEs of HMSs are comparable to those of typical metals, for example, Ti of 11.1 × 10^{-6} K^{-1}, and hence Ti can be used as electrodes for HMS-based modules. However, at

1173 K, linear CTEs become α_{cMn} = 12.5(2) × 10^{-6} K^{-1} and α_{cSi} = 32.8(4) × 10^{-6} K^{-1}; only the CTE of [Si] triples. Such a large change in CTE could be an origin of microcracks always formed perpendicular to the *c*-axis of HMSs.

The existence of striations is a serious drawback and must be resolved if TE materials are to be operated at HTs. The substitution of 0.5 at% Ge for Si, in $Mn(Si_{1-x}Ge_x)_{1.7}$, is quite effective at dissipating such striations, and the electrical conductivity is improved from 2 × 10^4 S m^{-1} to 5 × 10^4 S m^{-1} at room temperature [45]. A tiny amount of isovalent substitution of Ge for Si would not introduce carriers, but the improvement of the microstructure will increase the mobility of the carriers, yielding a large increase in the power factor of HMS-based alloys.

13.9 Concluding Remarks: Micro-/Nanostructuring to Improve TE Properties

To decrease the lattice thermal conductivity to a minimum and hence increase *ZT*, the grain size needs to be controlled so that it is larger than the mean free path (MFP) of carriers but smaller than the MFP of phonons [55]. Usually, planetary ball milling is used to reduce the grain sizes of powder samples. However, promising results [56] have not been easily obtained by a simple dry ball milling process. Ball milling with excess Si can be effective, and two groups [57,58] have indeed observed an enhancement of *ZT*, as free Si particles in the sintered body behave as phonon scattering centers to reduce the lattice thermal conductivity. Gas-phase atomizing (GPA) is another route to obtain fine powders with uniform sizes. Bernard-Granger et al. [59] obtained *ZT* ~0.7 at 800 K, using gas GPA and subsequent Spark Plasma Sintering (SPS) technique. Recently, Chen et al. [60] performed neutron scattering experiments to evaluate the phonon MFP of single crystalline samples. They concluded that the effective MFP is ~10 nm to reduce 50% of the total thermal conductivity along *c*. Obtaining such small grain sizes may not be possible using conventional ball milling as the typical achievable grain size is an order of ~100 nm; a longer ball milling time often results in the decomposition of HMSs into MnSi [61,62].

A melt-spinning and subsequent SPS process is quite effective to obtain matrix with a nanometer order and finely dispersed precipitates to effectively block phonon propagations. Luo et al. [63] have prepared HMS samples, with *in situ* formed nano-MnSi, showing *ZT* ~0.62 at 800 K. Quite recently, Yamamoto et al. [64] have also employed this technique and succeeded in preparing samples with *ZT* > 1 in partially Re-substituted alloys. The solubility limit of Re to Mn is very low using a conventional technique such as arc melting. However, a melt-spinning and subsequent SPS process is quite effective to obtain supersaturated solid solutions. Thus prepared samples show a remarkable reduction in lattice thermal conductivity and, hence, enhanced *ZT* values. Such a new preparation method would open a new era for the old but prospective TE materials like HMSs.

Acknowledgments

The author is indebted to Dr. Y. Kikuchi, Y. Saito, and H. Hamada for their cooperation. He is also thankful to Professor K. Hayashi and Professor K. Yubuta for fruitful discussion.

References

1. Zaitsev, V. K. 1995. Thermoelectric properties of anisotropic $MnSi_{1.75}$. In *CRC Handbook of Thermoelectrics*, ed. D. M. Rowe, Sect. 25. New York: CRC Press.
2. Fedorov, M. I. and V. K. Zaitsev. 2006. Thermoelectrics of transition metal silicides. In *Thermoelectrics Handbook: Macro to Nano*, ed. D. M. Rowe, Sect. 31. Boca Raton, FL: CRC Press.
3. Miyazaki, Y. and Y. Kikuchi. 2013. Higher manganese silicide, $MnSi_\gamma$. In *Thermoelectric Nanomaterials*, Springer Series in Materials Science, eds. K. Koumoto and T. Mori, Chap. 7. Heidelberg, Germany: Springer.
4. Nikitin, E. N. 1958. Study of temperature dependencies of electrical conductivity and thermopower of silicides. *Zh. Tekh. Fiz.* 8: 20–23.
5. Boren, B. 1933. Röntgenuntersuchung der legierungen von silicium mit chrom, mangan, kobalt und nickel. *Arkiv Kemi. Min. Geol.* 11A: 1–28.
6. Korshunov, V. A. and P. V. Gel'd. 1961. The electrical properties of the higher silicide of manganese. *Fiz. Metal. Metalloved.* 11: 945–947.
7. Karpinskii, O. G. and B. A. Evseev. 1969. The crystal structure of the compound Mn_4Si_7. *Izv. Akad. Nauk SSSR Neorg. Mater.* 5: 525–529.
8. Gottlieb, U., A. Sulpice, B. Lambert-Andron, and O. Laborde. 2003. Magnetic properties of single crystalline Mn_4Si_7. *J. Alloys Compd.* 361: 13–18.
9. Schwomma, O., A. Preisinger, H. Nowotny, and A. Wittmann. 1964. Die kristallstruktur von $Mn_{11}Si_{19}$ und deren zusammenhang mit disilicid-typen. *Monatch. Chem.* 95: 1527–1537.
10. Knott, H. W., M. H. Mueller, and L. Heaton. 1967. The crystal structure of $Mn_{15}Si_{26}$. *Acta Cryst.* 23: 549–555.
11. Flieher, G., H. Völlenkle, and H. Nowotny. 1967. Die Kristallstruktur von $Mn_{15}Si_{26}$. *Monatch. Chem.* 98: 2173–2179.
12. Zwilling, G. and H. Nowotny. 1973. Zur struktur der defekt-mangansilicide. *Monatch. Chem.* 104: 668–675.
13. Nowotny, H. 1970. Crystal chemistry of transition element defect silicides and related compounds. In *The Chemistry of Extended Defects in Non-Metallic Solids*, eds. L. Eyring and M. O'Keeffe, pp. 223–237. Amsterdam, the Netherlands: North Holland Pub.
14. Ye, H. Q. and S. Amelinckx. 1986. High-resolution electron microscopic study of manganese silicides $MnSi_{2-x}$. *J. Solid State Chem.* 61: 8–39.
15. Korshunov, V. A., F. A. Sidorenko, P. V. Gel'd, and K. N. Davydov. 1961. Phase components of the system MnSi-Si. *Fiz. Metal. Metalloved.* 12: 277–284.
16. Dudkin, L. D. and E. S. Kuznetsova. 1961. Study of Mn-Si system in the Si-rich region (in Russian). *Dokl. Akad. Nauk SSSR* 141: 94–97.
17. Fujino, Y., D. Shinoda, S. Asanabe, and Y. Sasaki. 1964. Phase diagram of the partial system of MnSi-Si. *Jpn. J. Appl. Phys.* 3: 431–435.
18. De Ridder, R. and S. Amelinckx. 1971. The structure of defect manganese silicides. *Mater. Res. Bull.* 6: 1223–1234.
19. De Ridder, R., G. Van Tendeloo, and S. Amelinckx. 1975. Incommensurate superstructure in $MnSi_{2-x}$. *Phys. Stat. Sol. (a)* 30: K99–K101.
20. De Ridder, R., G. Van Tendeloo, and S. Amelinckx. 1976. Electron microscopic study of the chimney ladder structures $MnSi_{2-x}$ and $MoGe_{2-x}$. *Phys. Stat. Sol. (a)* 33: 383–393.
21. Yamamoto, A. 1993. Determination of composite crystal structures and superspace groups. *Acta Cryst.* A49: 831–846.
22. Miyazaki, Y., D. Igarashi, K. Hayashi, T. Kajitani, and K. Yubuta. 2008. Modulated crystal structure of chimney-ladder higher manganese silicides $MnSi_\gamma$ (γ~1.74). *Phys. Rev. B* 78: 214104.
23. Miyazaki, Y. 2011. Superspace group approach to the crystal structure of thermoelectric higher manganese silicides $MnSi_\gamma$. In *Neutron Diffraction*, ed. I. Khidirov, Chap. 11. Rijeka, Croatia: Intech Open Access Book.

24. Imai, Y., M. Mukaida, and T. Tsunoda. 2000. Comparison of density of states of transition metal silicides and their related compounds systematically calculated by a first-principle pseudopotential method using plane-wave basis. *Intermetallics* 8: 381–390.
25. Imai, Y. and A. Watanabe. 2005. Consideration of the validity of the 14 valence electron rule for semiconducting chimney-ladder phase compounds. *Intermetallics* 13: 233–241.
26. Migas, D. B., V. L. Shaposhnikov, A. B. Filonov, V. E. Borisenko, and N. N. Dorozhkin. 2008. Ab initio study of the band structures of different phases of higher manganese silicides. *Phys. Rev. B* 77: 075205.
27. Madsen, G. K. H. and D. J. Singh. 2006. BoltzTraP: A code for calculating band-structure dependent quantities. *Compt. Phys. Commun.* 175: 67–71.
28. Nishida, I. 1972. Semiconducting properties of nonstoichiometric manganese silicides. *J. Mater. Sci.* 7: 435–440.
29. Kawasumi, I., M. Sakata, I. Nishida, and K. Masumoto. 1981. Crystal growth of manganese silicide, $MnSi_{\sim 1.73}$, and semiconducting properties of $Mn_{15}Si_{26}$. *J. Mater. Sci.* 16: 355–366.
30. Nikitin, E. N. and V. I. Tarasov. 1971. Electrical properties of solid solutions based on higher silicides of manganese. *Fiz. Tverd. Tela.* 13: 3473–3475.
31. Zaitsev, V. K., V. I. Tarasov, and A. A. Adilbekov. 1975. Metal-nonmetal transition in compensated higher manganese silicide. *Fiz. Tverd. Tela.* 17: 581–584.
32. Tarasov, V. I., E. N. Nikitin, and L. N. Shumilova. 1975. Conductivity of the solid solution system $Mn_{1-x}Fe_xSi_{1.7}$. *Izv. Akad. Nauk SSSR, Neorg. Mater.* 11: 1038–1040.
33. Flieher, G., H. Völlenkle, and H. Nowotny. 1968. Neue abkömmlinge der $TiSi_2$-struktur. *Monatch. Chem.* 99: 2408–2415.
34. Kikuchi, Y., Y. Miyazaki, Y. Saito, K. Hayashi, K. Yubuta, and T. Kajitani. 2012. Enhanced thermoelectric performance of a chimney-ladder $(Mn_{1-x}Cr_x)Si_\gamma$ ($\gamma \sim 1.7$). *Jpn. J. Appl. Phys.* 51: 085801.
35. Miyazaki, Y., Y. Saito, K. Hayashi, K. Yubuta, and T. Kajitani. 2010. Effect of cobalt-substitution on the structure and thermoelectric properties of chimney-ladder solid solution $(Mn_{1-x}Co_x)Si_\gamma$ ($\gamma \sim 1.7$). *Adv. Sci. Tech.* 74: 22–25.
36. Miyazaki, Y., Y. Saito, K. Hayashi, K. Yubuta, and T. Kajitani. 2011. Preparation and thermoelectric properties of a chimney-ladder $(Mn_{1-x}Fe_x)Si_\gamma$ ($\gamma \sim 1.7$) solid solution. *Jpn. J. Appl. Phys.* 50: 035804.
37. Setz, S., H. Nowotny, and F. Benesovsky. 1968. Untersuchungen in den systemen: Mangan-{vanadin, rhenium, eisen}-silicium. *Monatch. Chem.* 99: 2004–2015.
38. Chen, X., S. N. Girard, F. Meng et al. 2014. Approaching the minimum thermal conductivity in rhenium-substituted higher manganese silicides. *Adv. Energy Mater.* 4: 1400452.
39. Okamoto, N. L., T. Koyama, K. Kishida, K. Tanaka, and H. Inui. 2009. Crystal structure and thermoelectric properties of chimney–ladder compounds in the Ru_2Si_3–Mn_4Si_7 pseudobinary system. *Acta Mater.* 57: 5036–5045.
40. Nikitin, E. N., A. F. Sidorov, V. I. Tarasov, and A. I. Zaslavskii. 1970. Alloying of higher manganese silicide on the basis of microprobe-analysis results. *Izv. Acad. Nauk SSSR Neorg. Mater.* 6: 604–605.
41. Aoyama, I., M. I. Fedorov, V. A. Zaitsev et al. 2005. Effects of Ge doping on micromorphology of MnSi in $MnSi_{\sim 1.7}$ and on their thermoelectric transport properties. *Jpn. J. Appl. Phys.* 44: 8562–8570.
42. Zhou, A. J., T. J. Zhu, X. B. Zhao et al. 2010. Improved thermoelectric performance of higher manganese silicides with Ge additions. *J. Electron. Mater.* 39: 2002–2007.
43. Ponnambalam, V., D. T. Morelli, S. Bhattacharya, and T. M. Tritt. 2013. The role of simultaneous substitution of Cr and Ru on the thermoelectric properties of defect manganese silicides $MnSi_\delta$ ($1.73 < \delta < 1.75$). *J. Alloys Compd.* 580: 598–603.
44. Ponnambalam, V. and D. T. Morelli. 2012. Effect of Cr and Fe substitution on the transport properties of the Nowotny chimney-ladder $MnSi_\delta$ ($1.73 < \delta < 1.75$) compounds. *J. Electron. Mater.* 41: 1389–1394.

45. Aoyama, I., H. Kaibe, L. Rauscher et al. 2005. Doping effects on thermoelectric properties of higher manganese silicides (HMSs, $MnSi_{1.74}$) and characterization of thermoelectric generating module using p-type (Al, Ge and Mo)-doped HMSs and n-type $Mg_2Si_{0.4}Sn_{0.6}$ Legs. *Jpn. J. Appl. Phys.* 44: 4275–4281.
46. Barczak, S. A., R. A. Downie, S. R. Popuri, R. Decourt, M. Pollet, and J. W. G. Bos. 2015. Thermoelectric properties of Fe and Al double substituted $MnSi_\gamma$ ($\gamma \sim 1.73$). *J. Solid State Chem.* 227: 55–59.
47. Truong, D. Y. N., D. Berthebaud, F. Gascoin, and H. Kleinke. 2015. Molybdenum, tungsten, and aluminium substitution for enhancement of the thermoelectric performance of higher manganese silicides. *J. Electron. Mater.* 44: 3603–3611.
48. Ivanova, L. D., N. K. Abrikosov, E. I. Elagina, and V. D. Khvostikova. 1969. Production and investigation of the properties of single crystals of the higher manganese silicide. *Izv. Akad. Nauk SSSR Neorg. Mater.* 5: 1933–1937.
49. Kojima, T., I. Nishida, and T. Sakata. 1979. Crystal growth of $Mn_{15}Si_{26}$. *J. Cryst. Growth* 47: 589–592.
50. Kawasumi, I., M. Sakata, I. Nishida, and K. Masumoto. 1980. Crystal growth and characterization of $MnSi_{\sim 1.73}$. *J. Cryst. Growth* 49: 651–658.
51. Kojima, T. and I. Nishida. 1975. Crystal growth of $MnSi_{1.73}$ by chemical transport method. *Jpn. J. Appl. Phys.* 14: 141–142.
52. Udono, H., K. Nakamori, Y. Takahashi, Y. Ujiie, I. J. Ohsugi, and T. Iida. 2011. Solution growth and thermoelectric properties of single-phase $MnSi_{1.75-x}$. *J. Electron. Mater.* 40: 1165–1170.
53. Jeong, T. and W. E. Pickett. 2004. Implications of the B20 crystal structure for the magnetoelectronic structure of MnSi. *Phys. Rev. B* 70: 075114.
54. Kikuchi, Y., T. Nakajo, K. Hayashi, and Y. Miyazaki. 2014. High temperature X-ray diffraction study on incommensurate composite crystal $MnSi_\gamma$ – (3+1)-dimensional superspace approach. *J. Alloys Compd.* 616: 263–267.
55. Norouzzadeh, P., Z. Zamanipour, J. S. Krasinski, and D. Vashaee. 2012. The effect of nanostructuring on thermoelectric transport properties of p-type higher manganese silicide $MnSi_{1.73}$. *J. Appl. Phys.* 112: 124308.
56. Truong, D. Y. N., H. Kleinke, and F. Gascoin. 2015. Preparation of pure higher manganese silicides through wet ball milling and reactive sintering with enhanced thermoelectric properties. *Intermetallics* 66: 127–132.
57. Yoshikura, M. and T. Itoh. 2010. Thermoelectric properties of higher manganese silicide compounds synthesized by MG-PDS method. *J. Jpn. Soc. Powder Powder Metall.* 57: 242–246 (in Japanese).
58. Sadia, Y. and Y. Gelbstein. 2012. Silicon-rich higher manganese silicides for thermoelectric applications. *J. Electron. Mater.* 41: 1504–1508.
59. Bernard-Granger, G., M. Soulier, H. Ihou-Mouko et al. 2015. Microstructure investigations and thermoelectrical properties of a p-type polycrystalline higher manganese silicide material sintered from a gas-phase atomized power. *J. Alloys Compd.* 618: 403–412.
60. Chen, X., A. Weathers, J. Carrete et al. 2015. Twisting phonons in complex crystals with quasi-one-dimensional substructures. *Nat. Commun.* 6: 7723.
61. Zhou, A. J., X. B. Zhao, T. J. Zhu et al. 2010. Mechanochemical decomposition of higher manganese silicides in the ball milling process. *Intermetallics* 18: 2051–2056.
62. Chen, X., L. Shi, J. Zhou, and J. B. Goodenough. 2015. Effects of ball milling on microstructures and thermoelectric properties of higher manganese silicides. *J. Alloys Compd.* 641: 30–36.
63. Luo, W., H. Li, Y. Yan et al. 2011. Rapid synthesis of high thermoelectric performance higher manganese silicide with in-site formed nano-phase of MnSi. *Intermetallics* 19: 404–408.
64. Yamamoto, A., S. Ghodke, H. Miyazaki et al. 2016. Thermoelectric properties of supersaturated Re solid solution of higher manganese silicides. *Jpn. J. Appl. Phys.* 55: 020301.

14

Boron-Based Materials

Takao Mori
University of Tsukuba

CONTENTS

14.1 Introduction

The need to develop high-temperature thermoelectric materials has increased with the demonstration of viable application examples, such as the utilization of high-temperature waste heat in thermal power plants [1,2], steelworks, factories, and incinerators, and also the conversion of concentrated solar power [3]. Topping cycles to enhance output in thermal power plants may have an especially large impact considering the amounts of energy involved. For example, in the United States, there are times when over 50% of the electricity consumed is generated by midscale coal-fired power plants [1,2]. Yazawa and Shakouri have performed calculations showing a 6% increase in the overall efficiency of a 520 MW class Rankine cycle power plant enabled by a topping cycle of high-temperature thermoelectric material with ZT ~0.7 [2].

Boron cluster compounds are attractive candidates as high-temperature thermoelectric materials. First of all, they possess high melting points, which is the first basic requirement. In regard to thermoelectric properties, generally large Seebeck coefficients are observed [4–6], and the compounds have been found to exhibit intrinsic low thermal conductivity [7–9], which are both inherent advantages for thermoelectrics. In comparison to similarly strongly covalently bonded solids with high sound velocities on both neighboring sides of the periodical table like beryllium and carbon-based materials, the thermal conductivities of boron cluster compounds are anomalously low. Several mechanisms to explain this, such as a symmetry mismatch effect or degrees of freedom of boron dumbbells, in addition to the well-known crystal complexity first demonstrated by Slack [10], have been proposed [6,10,11]. Layered borides, such as the AlB_2-type analogue $TmAlB_4$, have also been found to have "tiling" building defects [12–14] that strongly affect the thermal conductivity [15].

Another advantage of the boron cluster compounds at high temperatures is that they typically follow the variable range hopping (VRH) transport mechanism [16] where the electrical conductivity and Seebeck coefficient both increase with temperature [17,18].

Another attractive feature of such boron cluster compounds for material science development in general is that the network structures and physical properties have been found to be controllable to some degree through the incorporation of metal atoms in the voids of clusters and also by the addition of third elements like C, N, and Si that can act as bridging sites of the boron network framework [19].

Up to now, active investigations have been carried out into the thermoelectric properties of boron-based materials such as boron carbide [7,8,20–23], rare earth borocarbonitrides $RB_{15.5}CN$, $RB_{22}C_2N$, and $RB_{28.5}C_4$ (R = rare earth) [24–28], doped β-boron [29–34], B_6S_{1-x} [35], RB_{66} [18,35–38], $RB_{44}Si_2$ [5,17,39,40], YAl_xB_{14} [41–44], $MgAlB_{14}$ [45,46], YB_{25} [43], $YB_{18}Si_5$ [47], $B_{12}As_2$ [48,49], $B_{12}P_2$ [50], and $B_{12}O_2$ (B_6O) [51,52]. The p, n characteristics are summed up in Table 14.1. The boron icosahedral compounds are predominantly p-type with the exception of Fe-doped and V-doped β-boron (albeit this is "metallization" with the absolute value of the Seebeck coefficient becoming smaller toward the important high-temperature region) [29,31], the rare earth borocarbonitrides [24–28], and YAl_xB_{14}-type compounds in the Al-rich region (Section 14.2) [41–44]).

Also, as different classes of materials compared to the boron icosahedral cluster compounds with intrinsic low thermal conductivity, AB_6 [53–60] (A = alkaline rare earth), which is a dirty metal (semimetal/small gap semiconductor) with high thermal conductivity and some difficulty in reliable control of properties, and RMB_4 (M = transition metal, Al) [15,61,62] layered compounds, for example, have also been considered as thermoelectric materials.

TABLE 14.1

p, n Characteristics of Boron Cluster Compounds

p-Type Compounds	References	n-Type Compounds	References
β-boron Ce, Ni, Co, Cu, Cr, Mn, Al, Ti, Ta, Nb, Si, Ge, Mg, Zr, Hf, Sc-doped	[29–34]	β-boron Fe, V-doped	[29,31]
Boron carbide $B_{4.1}C \sim B_{10.5}C$	[7,8,20–23]	$RB_{15.5}CN$, $RB_{22}C_2N$, $RB_{28.5}C_4$	[24–28]
RB_{66}	[18,35–38]	YAl_xB_{14} ($0.57 \leqq x \leqq 0.63$ determined from Rietveld refinement)	[41–44]
$MgAlB_{14}$	[45,46]	$MgAlB_{14}$ (Mg:Al:B = 1.04:1.06:14 sinter mixing ratio)	[46]
$RB_{44}Si_2$	[5,17,39,40]		
YB_{25}	[43]		
$YB_{18}Si_5$	[47]		
$B_{12}As_2$	[48,49]		
$B_{12}P_2$	[50]		
$B_{12}O_2$	[51,52]		
B_6S_{1-x} ($0.37 < x < 0.40$)	[35]		

In this chapter, the following recent striking developments for boron-based thermoelectrics are discussed:

- Renaissance of the RB_{66} system
- Excellent p, n control in aluminoborides
- Hybrid effect
- Reinvestigation/consideration by Slack of boron-based materials with rhombohedral α-boron-type structure, B_6O

14.2 Renaissance of the RB_{66} System

14.2.1 Discovery of a Metal-Rich Composition

RB_{66} has a complex structure with more than 1600 atoms in the unit cell, and a view of the crystal structure is given in Figure 14.1a. Despite the large number of atoms, a simplified view of the crystal structure can be obtained by just three structural units: the $(B_{12})_{13}$ supericosahedra, the B_{80} cluster, and the rare earth atoms (Figure 14.1b). Previously, Slack and coworkers discovered RB_{66} to be a crystalline material exhibiting glass-like thermal conductivity [7,8], which can be considered as a precursor to his famous phonon glass electron crystal concept [62].

Regarding the composition of this compound, it had previously been reported that the homogeneity region of YB_{66} is quite wide, extending from YB_{56} to YB_{68} [63,64]. During the crystal growth of $YB_{44}Si_2$, an anomalously metal-rich YB_{47} crystal with the YB_{66}-type structure was discovered by chance. Subsequently, the YB_{48} crystal could be grown purposely [37]. RB_{66} had not been considered to be a good thermoelectric material since the maximum ZT among all the known compounds, including doped samples, had been only 4×10^{-3} [18]. YB_{48} showed a dramatic improvement, attaining a value around 0.1 at 1000 K [37]. This is still not a high enough value, but the working temperature of the material has the potential to be much higher (melting point ~2600 K), and the ZT shows a sharp increase toward higher temperatures. Furthermore, this represents just the basic properties of this simple binary compound, and doping and/or fabricating composites (as discussed in Section 14.3) are potential routes for further improvement.

The detailed mechanism of the large overall improvement in thermoelectric properties remains unclear. In comparison to conventional YB_{66}, YB_{48} showed a very large reduction in the electrical resistivity with minor detrimental decreases in the Seebeck coefficient and an increase in the thermal conductivity, leading to approximately 30 times enhancement of ZT [37]. One striking difference in the crystal structure is that YB_{48} has an additional Y site, Y3, where Y replaces the boron dumbbell in the B_{80} cluster (Figure 14.2). Another difference is that while YB_{66} and YB_{48} are both p-type materials with dominant hole conduction, the metal-rich YB_{48} can be expected to have additional electrons injected into the system, although Hall measurements of RB_{66} samples to investigate the carrier concentration in detail have not been successful up to now.

It has been demonstrated in Reference 11 that the localization length in VRH systems can give insights into the disorder of similar systems. This is reasonable, since the disorder

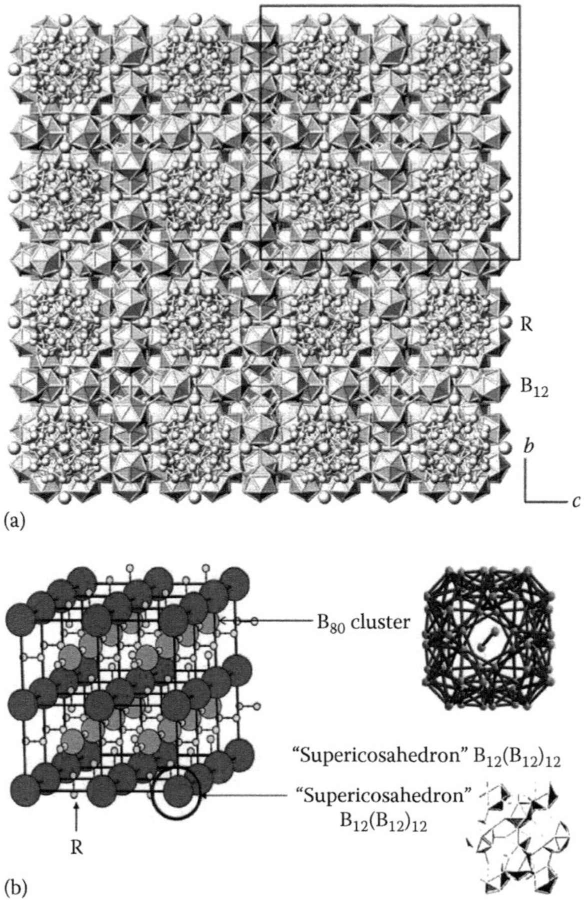

FIGURE 14.1
Crystal structure of RB_{62} (RB_{66}). (Redrawn from Sussardi, A. et al., *J. Materiomics*, 1, 196, 2015.)

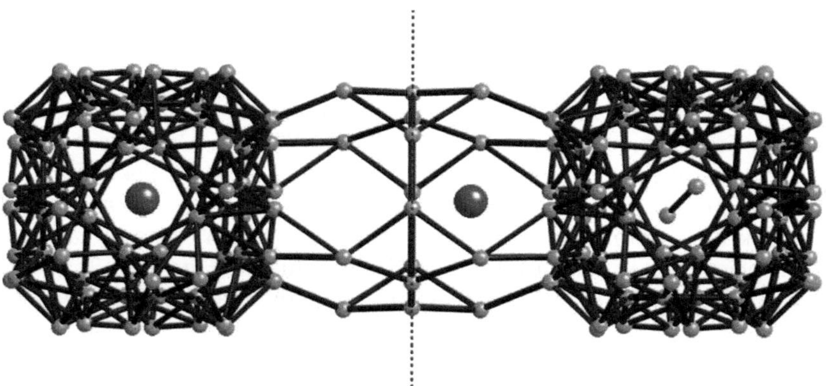

FIGURE 14.2
View of the structure of YB_{48} showing the B_{80} cluster and new Y3 site at (1/4, 1/4, 1/4) replacing the B-B dumbbell pair.

is the main driving force for the carrier wave functions to be localized to begin with [16,65–67]. The VRH resistivity follows the temperature dependence of

$$\rho = \rho_0 \exp\left[\left(\frac{T_0}{T}\right)^{0.25}\right] \tag{14.1}$$

where ρ_0 is the prefactor of resistivity, and T_0 is the characteristic temperature.

The localization length at the Fermi level can be expressed as

$$\xi = \left[\frac{1.81}{k_B T_0 D(E_F)}\right]^{1/3} \tag{14.2}$$

where k_B, $D(E_F)$, and ξ are the Boltzmann constant, the density of states at the Fermi level energy, and the localization length, respectively.

$D(E_F)$ can be estimated from the electronic γ coefficient of the specific heat in a first approximation where the electron–phonon coupling constant is set at zero:

$$\gamma = 1/3\pi^2 k_B^2 D(E_F) \tag{14.3}$$

and, therefore, the localization length can be estimated.

In the case of YB_{48}, it is revealed that the localization length ξ = 2.14 Å [37], compared to ξ = 0.56 Å for YB_{62} [16]. The significantly larger ξ indicates that YB_{48} is less disordered than YB_{62}. The single crystal refinement data of YB_{48} [37] compared with YB_{62} [68] show that the partially occupied B-site occupancies are similar. Therefore, the difference in the disorder cannot be explained by them. A concrete difference is that the boron dumbbell B13 sites are replaced by the Y3 site, and this could mean that the boron dumbbell is the origin of the large disorder. This is a fascinating possibility because it has been proposed before that this dumbbell-like B pair inside the B_{80} cluster possesses some strong phonon scattering mechanism to contribute to the especially low glass-like thermal conductivity of the RB_{66} systems [6,69]. The thermal conductivity of YB_{48}, indeed, has a value of 2.3 W m^{-1} K^{-1} at 300 K, which is slightly higher than 2.0 W m^{-1} K^{-1} of YB_{66}, but the boron dumbbell may have a much larger effect on the electrical conductivity. The effect of the dumbbell B pair on the physical properties should be studied further, hopefully also with theoretical methods. Such dumbbell crystal structures may play a larger role than expected in the thermoelectric properties of other systems also.

14.2.2 Enhancement Discovered in the Sm Phase

In a series of rare earth compounds, the samarium and ytterbium phases have sometimes yielded anomalous properties [70]. This is usually related to the mixed valency where these two rare earth atoms can sometimes take the divalent state instead of the typical trivalent state. Mixed valency has previously been reported for SmB_6, for example [71–73], which has attracted increasing interest recently from claims that it is a topological insulator [74–81]. There have been various theoretical works and reviews reporting extremely enhanced thermoelectric properties of topological insulators [82].

With these developments in mind, the samarium phase of RB_{66} was revisited [38]. Single crystals of SmB_{62} were grown using the floating zone method. As shown in Figure 14.3a,

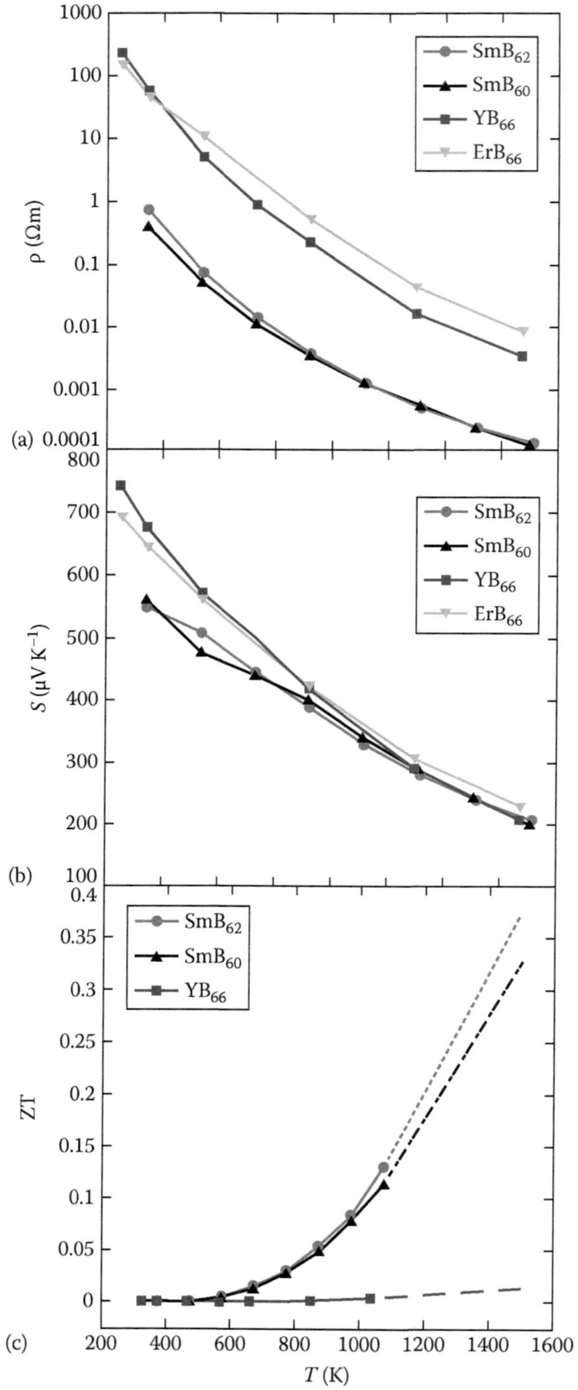

FIGURE 14.3
(a) Electrical resistivity ρ and (b) Seebeck coefficient S of SmB_{60} and SmB_{62} crystals with YB_{66} and ErB_{62} plotted for comparison. (c) ZT of SmB_{62} with YB_{66} plotted for comparison. (Redrawn from Sussardi, A. et al., *J. Materiomics*, 1, 196, 2015.)

TABLE 14.2

Relevant Parameters of the VRH, Electronic Specific Heat, and ZT for YB_{66}, and New Discoveries in YB_{48} and SmB_{62}

Compound	T_0 (K)	γ (mJ mol^{-1} K^{-2})	ξ (Å)	ZT_{max}	Reference
YB_{48}	8.1×10^6	3.2	2.14	0.1 at 990 K	[37]
YB_{66}	13×10^8	1.9	0.56	0.0035 at 1035 K	[18]
SmB_{62}	7.4×10^8	24.7	0.33	0.13 at 1050 K (extrapolated ~0.4 at 1500 K)	[38]

the electrical resistivity of SmB_{62} shows a huge reduction of approximately two orders of magnitude compared to YB_{66} and ErB_{66}. This is somewhat understandable considering that from a straightforward analysis of the magnetic measurements of SmB_{62}, a mixed valency of Sm^{2+}:Sm^{3+} = 1:1 was indicated. This points to a large increase in the carrier (hole) concentration that could lead to the reduction in resistivity. In such a case, the Seebeck coefficient would normally be expected to have an accompanying large reduction. What is striking is that only a very small decrease is actually observed (Figure 14.3b). As a result, since the thermal conductivity of SmB_{62} is also slightly decreased compared to the yttrium phase with a lighter rare earth atom, the ZT of SmB_{62} is approximately 40 times enhanced compared to YB_{66}. ZT ~0.13 at 1050 K and, with a very sharp increase toward higher temperatures, an extrapolated value of ~0.4 at 1500 K is expected. The melting point of SmB_{62} is around 2850 K [60–61]. The analysis of the VRH electrical conductivity and specific heat shows that the localization length ξ = 0.33 Å for SmB_{62} [38] (we note that a numerical error was made in the ξ value in Reference 38 and we correct it here), which is similar to ξ = 0.56 Å for YB_{62} [16], indicating that delocalization does not appear to be a factor.

In the case of enhanced thermoelectric properties for SmB_{62} [38] or YB_{48} [37], it appears that, respectively, "delocalization" in YB_{48} may be a key, while the γ coefficient of the electronic specific heat (the density of states at the Fermi level) of SmB_{62} is about an order of magnitude larger than in YB_{66} and YB_{48} (Table 14.2). In the case of YB_{48}, it appears that the replacement of the boron dumbbell by the additional yttrium site Y3 may also be important, while an increase in carriers may lead to delocalization in the weak Anderson localization [65–67]. The density of states at the Fermi level is also enhanced for SmB_{62}. The thermoelectric enhancement mechanisms need to be further carefully considered to reach the full potential of these compounds. However, as a starting point, these relatively good performances for simple binary borides at high temperatures, where few materials can operate, are a promising development.

14.3 Excellent p, n Control in Aluminoborides

Finding good p, n matching materials has been one of the large obstacles for TE application. In the case of boron-based materials in particular, boron carbide was found at an early stage to be an excellent p-type high-temperature thermoelectric material (ZT ~0.6 for composites [83]) and was even used in commercial modules at one point by the Hi-Z company, but the lack of a good n-type counterpart has been a long-standing problem (the Hi-Z module used SiGe as the n-type element, but it went out of production). As briefly noted earlier, the rare earth borocarbonitrides $RB_{15.5}CN$, $RB_{22}C_2N$, and $RB_{28.5}C_4$ were found to be promising n-type counterparts [24–28], but there are still difficulties such as the complete removal of boron carbide inclusions that are reported to degrade the n-type performance [28].

A new promising development was discovered in the aluminoborides. Through particular synthesis processes, yttrium aluminoboride, YAl_xB_{14}, can be prepared with different occupancies of the aluminum site. As a result (Figure 14.3), excellent control of p-n characteristics (400 to −200 μV K^{-1}) was achieved in a system with the same crystal structure and consisting of the same elements [41–43].

This discovery of control of the occupancy is quite surprising, because in metal borides, such as MB_2, MB_4, MB_6, MB_{12}, MB_{66}, or the novel compounds recently discovered like $RB_{15.5}CN$, $RB_{22}C_2N$, and $RB_{28.5}C_4$, $RB_{44}Si_2$, or even other aluminoborides like $RAlB_4$ [12,84] or previous reports on $RAlB_{14}$ ($R_{\sim0.6}Al_{\sim0.7}B_{14}$) itself [85,86], the homogeneity region of the metal site occupancy has been found to be very small. This is also related to the fact that boron-based materials are electron deficient (in contrast to carbon-based materials) and fundamentally require the electrons from the metal atoms to form these particular crystal structures [19]. First-principles calculations based on the present x-ray diffraction data of YAl_xB_{14} reveal the stable configurations of the atomic sites with different occupancies, indicating that such control of the occupancy of the metal site is possible [42]. Determination of the density of states for the different configurations also clearly illustrates the shift between positive to zero to negative Seebeck coefficients. However, recent neutron diffraction experiments indicate that there may be local features of the boron network structure undergoing changes that are still not fully elucidated and investigations are ongoing [87]. The crystal structure of boron icosahedral cluster compounds is fascinating [19], and also very robust, with "self-healing" occurring even under heavy irradiative bombardment, with interstitial boron atoms migrating to repair the icosahedra, for example [88] (Figure 14.4).

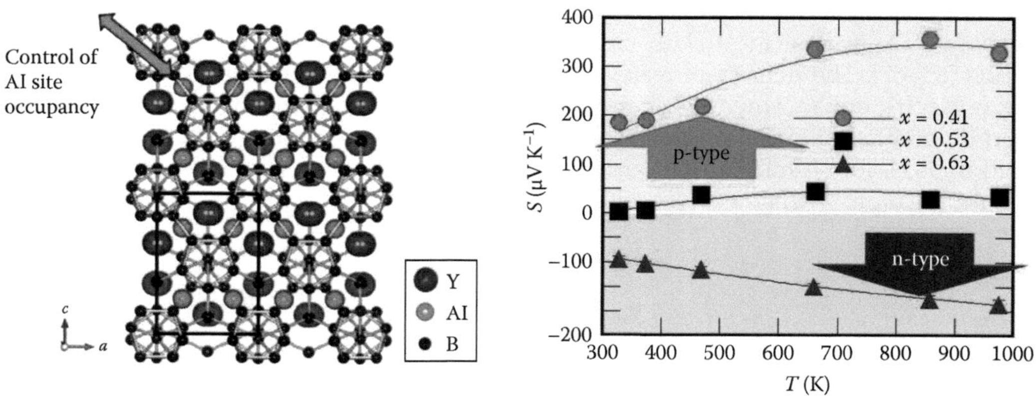

FIGURE 14.4
Image of the Al occupancy control and Seebeck coefficients of YAl_xB_{14}. (From Maruyama, S. et al., *Appl. Phys. Lett.*, 101, 152101, 2012; Sahara, R. et al., *Sci. Technol. Adv. Mater.*, 15, 035012, 2014; Maruyama, S. et al., *J. Appl. Phys.*, 115, 123702, 2014; Maruyama, S. et al., *Mater. Renew. Sustain. Energy*, 3, 31, 2014.)

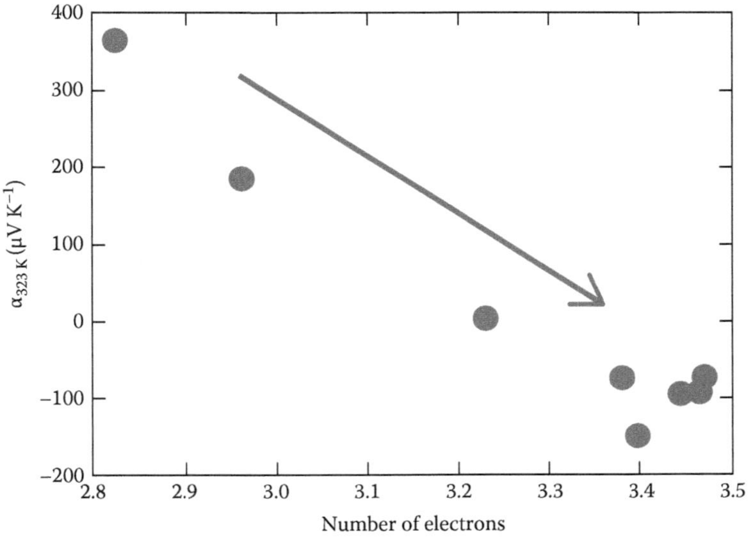

FIGURE 14.5
Seebeck coefficient at 323 K of YAl_xB_{14} plotted versus estimated electron contribution from Y and Al (the Y occupancy of these compounds is ~0.56). (The data are taken from Maruyama, S. et al., *Appl. Phys. Lett.*, 101, 152101, 2012; Sahara, R. et al., *Sci. Technol. Adv. Mater.*, 15, 035012, 2014; Maruyama, S. et al., *J. Appl. Phys.*, 115, 123702, 2014; Maruyama, S. et al., *Mater. Renew. Sustain. Energy*, 3, 31, 2014.)

The n-type aluminoboride composition compounds are particularly interesting considering the scarcity of n-type higher borides. p-Type boron carbide composites (Section 14.3) are already close to the ZT ~0.7 used in the topping cycle calculations mentioned in the "Introduction" [2]. A problem is the lack of n-type similar counterparts and, therefore, the n-type YAl_xB_{14} is a good candidate. The maximum ZT for n-type YAl_xB_{14} is around 0.09, which is not high, but is actually similar to boron carbide when considered as a starting point of boron-based composites for enhanced performance (Section 14.3). There is also interest in whether the absolute value of the negative Seebeck coefficient can be further enhanced. In Figure 14.5 is plotted the Seebeck coefficient of YAl_xB_{14} at 323 K versus a plot of total expected electron contributions from Y and Al assuming trivalent states (it should be noted that the Y site has a basically constant partial occupancy around 0.56 in this compound). Attempts to increase the Al occupancy in YAl_xB_{14} beyond 0.63 were not successful, although fortuitously, excess Al acted as a sintering aid to improve the density of the material [43,44]. Another attempt to enhance the n-type behavior was made by substitutionally doping boron with carbon. This was attempted also because previously the carbon-doped YB_{66} was found to have the beneficial effect of lowering the thermal conductivity [18]. Although substitutional doping of B by C was successful, accompanying the increase of C, the Al occupancy was found to decrease. Namely, despite the fact that control of the Al occupancy could be carried out in YAl_xB_{14} as described previously, above the Al-rich boundary of $x = 0.63$, a charge compensation set in, and the n-type characteristics could not be tuned further [89].

Therefore, fabrication of composites should be the next step for these n-type borides. And despite the present relatively low ZT, it should be stressed that the boron-based materials do not depend on nanostructuring for their good thermoelectric performance, which is a distinct advantage when considering applications at high temperatures.

14.4 Hybrid Effect

Previously, enhanced thermoelectric properties were realized in several systems where composites composed of semiconducting borides and highly metallic borides were formed (with the metallic boride being the minority phase).

For example, the highest performance in boron carbide materials was attained in boron carbide/TiB_2 composites. The enhancement was reported by multiple groups [83,90], with a maximum ZT ~0.6, which is significantly higher than the ZT of pristine boron carbide [21,22]. Other boron carbide and metallic material composites have also shown improved properties [91,92]. In the case of the rare earth borocarbonitrides, two examples were found. RB_4 and/or RB_6 seeding into $RB_{22}C_2N$ or $RB_{17}CN$ resulted in a similar substantial reduction of resistivity with just a slight degradation of the Seebeck coefficients [25]. A more dramatic hybrid effect was realized with Co or VB_2 seeding into $YB_{22}C_2N$. A very large reduction of the electrical resistivity by more than 2 orders of magnitude, that is, ~10,000%, was obtained, while simultaneously an enhancement of the maximum absolute value of the Seebeck coefficient by up to 220% was achieved [93,94].

An EDS mapping of the vanadium distribution in VB_2-seeded $YB_{22}CN$ is shown in Figure 14.6, and the VB_2 can be seen in many places to be forming a thin web covering some surfaces of the bulk host [95]. Since VB_2 has a very high electrical conductivity, the large reduction in resistivity can be explained by this partial web-facilitating

FIGURE 14.6
EDS mapping of V of VB_2 doped/seeded $YB_{22}CN$. (Reprinted from *Scr. Mater.*, 111, Mori, T. and Hara, T., Hybrid effect to possibly overcome the trade-off between Seebeck coefficient and electrical conductivity, 44–48. Copyright 2016, with permission from Elsevier.)

conduction. These metallic nanoinclusions are relatively small in the total bulk and apparently do not achieve full percolation throughout the material, considering that significant decreases of the bulk Seebeck coefficient values are not observed, although this may also be due to some interface effect that is not completely clear yet. Furthermore, in the particular case of VB_2-seeded $YB_{22}CN$, a hybrid effect was achieved, where the Seebeck coefficient has a simultaneous enhancement, presumably from intrinsic V doping into the material.

The hybrid effect is fortuitous, but the enhancement by the incorporation of a metallic nano/microstructure is possibly a method that can be broadly utilized for prospective thermoelectric systems having large Seebeck coefficients but relatively low conductivity [95]. Compared to many proposed state-of-the-art methods (i.e., confinement effects and band structure engineering), it is relatively simple and straightforward with a large benefit experimentally realized for several cases already.

14.5 Reinvestigation/Consideration by Slack of Boron-Based Materials with Rhombohedral α-Boron-Type Structure, B_6O

Recently, Slack et al. made a very comprehensive review and analysis of boron-based materials with the rhombohedral α-boron-type structure [52]. They reviewed the crystal structures and properties (thermodynamic, mechanical, optical bandgap, carrier mobility, thermal neutron capture) of $B_{12}O_2$ (B_6O), $B_{12}N_2Be$, $B_{12}N_2$, $B_{13}O$, $B_{13}C_2$, $B_{13}S$, $B_{13}P$, $B_{13}Se$, $B_{12}S_2$, $B_{12}P_2$, and $B_{12}As_2$. It was pointed out that there have been instances of scattering and ambiguity of reported compositions of these compounds over the past half century and that a continuous series of crystals formed with an expanding unit cell volume from $B_{12}O_2$ to $B_{12}As_2$. They demonstrated that if the lattice positions of the interstitial atoms (O, As, etc.) were known, the volume increases would be directly related to the radii of those atoms. Therefore, if the unit cell volume was known, the accurate chemical composition could be determined.

In relation to thermoelectric properties, they noted that the electrical properties of many of the compounds like $B_{12}S_2$ and $B_{12}N_2$ are unknown and pointed to the relatively high carrier mobility of $B_{12}P_2$ and $B_{12}As_2$ of 80–100 $cm^2 V^{-1} s^{-1}$. All known boron-based materials with the rhombohedral α-boron-type structure have been p-type. Importantly, $B_{12}O_2$ (B_6O) is also expected to have high mobility, but the reported value is 1000 times smaller, and Slack et al. were able to determine that the sample was actually oxygen deficient with 20% of the O sites occupied by B and that a stoichiometric $B_{12}O_2$ could be expected to have high mobility. They further provided a recipe for the synthesis of stoichiometric $B_{12}O_2$ that is not trivial. Stoichiometric $B_{12}O_2$ was therefore revealed as a promising thermoelectric material to be targeted [52].

One recent experimental advancement in the rhombohedral α-boron-type structure compounds was made by the synthesis, crystal structure determination, and thermoelectric measurements of B_6S_{1-x} ($0.37 < x < 0.40$) [35]. The synthesized samples had very low densities below 50% and, therefore, electrical resistivities were very high, but the Seebeck coefficients showed large p-type values above 200 μV K^{-1} increasing toward high temperatures, similar in behavior to boron carbide in the attractive composition range (Figure 14.7) [35].

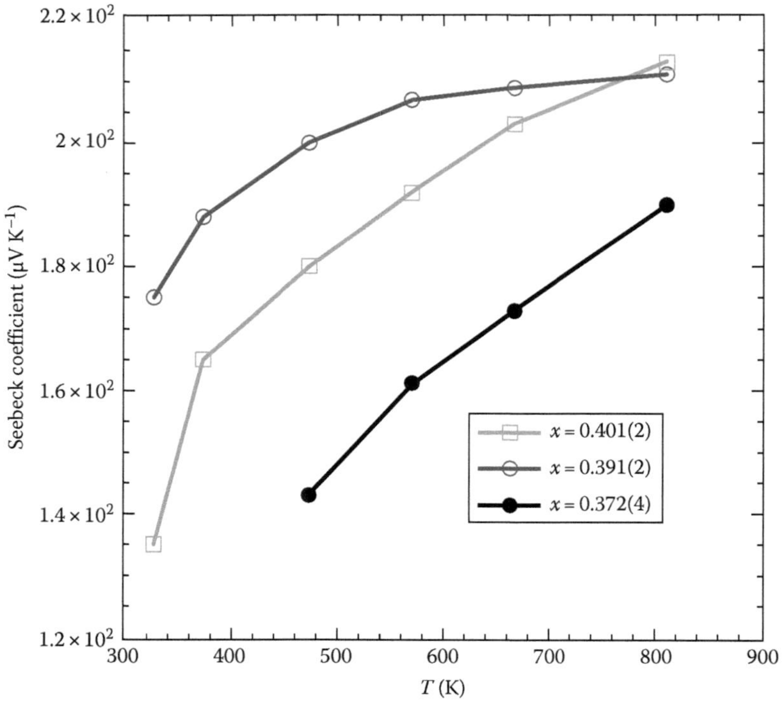

FIGURE 14.7
Seebeck coefficients of B_6S_{1-x}. (Redrawn from Sologub, O. et al., *Scr. Mater.*, 68, 289, 2013.)

14.6 Conclusions

Several attractive high-temperature thermoelectric applications, in particular the topping cycle to enhance the efficiency of thermal power plants, have increased the impetus to develop high-temperature thermoelectric materials. Boron-based materials are an attractive target because of their high thermal stability, large Seebeck coefficients, and intrinsically low thermal conductivity for the icosahedral cluster compounds. Indeed, boron carbide has attracted much attention as a good p-type material. I have reviewed several hot topics, such as the recent discoveries in the RB_{66} system, in which Slack et al. first discovered glass-like thermal conductivity in a crystalline compound, and where recently 30–40 times thermoelectric performance enhancement was discovered. An excellent p, n control in aluminoborides yields a new n-type possibility in addition to the borocarbonitrides, which is an imperative need for boride thermoelectrics. The hybrid effect discovered in nanocomposites of semiconducting borides and highly metallic borides is possibly a thermoelectric enhancement method that can be readily applied to other non-boride systems also. And, finally, the rhombohedral α-boron-type structure compounds have been systematically reinvestigated with new insights into the crystal structure and composition, and stoichiometric $B_{12}O_2$, which has not been synthesized yet, is highlighted as a promising target for high-temperature thermoelectrics.

Acknowledgments

The author acknowledges support from CREST, Japan Science and Technology Agency (JST). The author also thanks S. Maruyama with assistance in drawing Figs. 14.4 and 14.5.

References

1. Yazawa, K., Y. R. Koh, and A. Shakouri. 2013. Optimization of thermoelectric topping combined steam turbine cycles for energy economy. *Appl. Energy* 109: 1–9.
2. Yazawa, K. and A. Shakouri. 2016. Thermoelectric topping cycles with scalable design and temperature dependent material properties. *Scr. Mater.* 111: 58–63.
3. Weidenkaff, A., M. Trottmann, P. Tomes, C. Suter, A. Steinfeld, and A. Veziridis. 2013. In *Thermoelectric Nanomaterials*, eds. K. Koumoto and T. Mori (Springer, Heidelberg, Germany), pp. 365–382.
4. Wood, C. and D. Emin. 1984. Conduction mechanism in boron carbide. *Phys. Rev. B* 29: 4582–4587.
5. Mori, T., D. Berthebaud, T. Nishimura, A. Nomura, T. Shishido, and K. Nakajima. 2010. Effect of Zn doping on improving crystal quality and thermoelectric properties of borosilicides. *Dalton Trans.* 39: 1027–1030.
6. Mori, T. 2012. In *Modules, Systems, and Applications in Thermoelectrics*, Boride Thermoelectrics: High-Temperature Thermoelectric Materials, ed. D. M. Rowe (CRC Press/Taylor & Francis, London, U.K.), Ch. 14.
7. Slack, G. A., D. W. Oliver, and F. H. Horn. 1971. Thermal conductivity of boron and some boron compounds. *Phys. Rev. B* 4: 1714–1720.
8. Cahill, D. G., H. E. Fischer, S. K. Watson, R. O. Pohl, and G. A. Slack. 1989. Thermal properties of boron and borides. *Phys. Rev. B* 40: 3254–3260.
9. Mori, T. 2006. Thermal conductivity of rare-earth B_{12}-icosaherdal compound. *Phys. B* 383: 120–121.
10. Slack, G. A. 1979. In *Semiconductors and Semimetals*, The Thermal Conductivity of Nonmetallic Crystals, eds. F. Seitz, D. Turnbull, and H. Ehrenreich (Academic Press, New York), Vol. 34, p. 1.
11. Mori, T., J. Martin, and G. Nolas. 2007. Thermal conductivity of $YbB_{44}Si_2$. *J. Appl. Phys.* 102: 073510.
12. Mori, T., H. Borrmann, S. Okada et al. 2007. Crystal structure, chemical bonding, electrical transport, and magnetic behaviour of $TmAlB_4$. *Phys. Rev. B* 76: 064404.
13. Mori, T., I. Kuzmych-Ianchuk, K. Yubuta et al. 2012. Direct elucidation of the effect of building defects on the physical properties of alpha-$TmAlB_4$: An AlB_2-type analogous "tiling" compound. *J. Appl. Phys.* 111: 07E127.
14. Yubuta, K., T. Mori, S. Okada et al. 2013. High-resolution electron microscopy and x-ray diffraction study of intergrowth structures in α- and β-type $YbAlB_4$ single crystals. *Philos. Mag.* 93: 1054–1064.
15. Wang, X. J., T. Mori, I. Kuzmych-Ianchuk et al. 2014. Thermal conductivity of layered borides: The effect of building defects on the thermal conductivity of $TmAlB_4$ and the anisotropic thermal conductivity of AlB_2. *APL Mater.* 2: 046113.
16. Mott, N. F. 1968. Conduction in non-crystalline materials. *J. Non-Cryst. Solids* 1: 1–18.
17. Mori, T. 2005. High temperature thermoelectric properties of $B_{12}B_{12}$ icosahedral cluster-containing rare earth boride crystals. *J. Appl. Phys.* 97: 093703.
18. Mori, T. and T. Tanaka. 2006. Effect of transition metal doping and carbon doping on thermoelectric properties of YB_{66} single crystals. *J. Solid State Chem.* 179: 2889–2894.
19. Mori, T. 2008. Higher borides. In *Handbook on the Physics and Chemistry of Rare Earths*, eds. K. A. Gschneidner Jr., J.-C. Bunzli, and V. Pecharsky (North-Holland, Amsterdam, the Netherlands), Vol. 38, pp. 105–173.

20. Aselage, T. L., D. Emin, S. S. McCready, and R. V. Duncan. 1998. Large enhancement of boron carbides' Seebeck coefficients through vibrational softening. *Phys. Rev. Lett.* 81: 2316–2319.
21. Bouchacourt, M. and F. Thevenot. 1985. The correlation between the thermoelectric properties and stoichiometry in the boron carbide phase B_4C–$B_{10.5}C$. *J. Mater. Sci.* 20: 1237–1247.
22. Cai, K. F., C. W. Nan, and X. M. Min. 1998. The influence of silicon dopant and processing on thermoelectric properties of B_4C ceramics. In *Symposium Z—Thermoelectric Materials, 1998—The Next Generation*; MRS Online Proceedings Library; Vol. 545, pp. 131–136.
23. Caruso, A. N., S. Balaz, B. Xu et al. 2004. Surface photovoltage effects on the isomeric semiconductors of boron-carbide. *Appl. Phys. Lett.* 84: 1302–1304.
24. Mori, T. and T. Nishimura. 2006. Thermoelectric properties of homologous p- and n-type boron-rich borides. *J. Solid State Chem.* 179: 2908–2915.
25. Mori, T., T. Nishimura, K. Yamaura, and E. Takayama-Muromachi. 2007. High temperature thermoelectric properties of a homologous series of nn-type boron icosahedra compounds: A possible counterpart to pp-type boron carbide. *J. Appl. Phys.* 101: 093714.
26. Belokoneva, E. L. and T. Mori. 2009. Topology and symmetry analysis of rare earth borocarbides structural family, analogy to hexaferrites and relation to properties. *Cryst. Res. Technol.* 44: 19–24.
27. Berthebaud, D., T. Nishimura, and T. Mori. 2010. Thermoelectric properties and spark plasma sintering of doped $YB_{22}C_2N$. *J. Mater. Res.* 25: 665–669.
28. Mori, T., T. Nishimura, W. Schnelle, U. Burkhardt, and Y. Grin. 2014. The origin of the n-type behavior in rare earth borocarbide $Y_{1-x}B_{28.5}C_4$. *Dalton Trans.* 43: 15048–15054.
29. Slack, G. A., J. H. Rosolowski, C. Hejna, M. Garbauskas, and J. S. Kasper. 1987. Semiconducting properties of boron. In *Proceedings of the Ninth International Symposium Boron, Borides and Related Compounds*, Duisburg, Germany, 1987, p. 132.
30. Slack, G. A., C. I. Hejna, M. F. Garbauskas, and J. S. Kasper. 1988. X-ray study of transition-metal dopants in β-boron. *J. Solid State Chem.* 76: 64–86.
31. Werheit, H., K. DeGroot, and W. Malkovich. 1981. On n-type β rhombohedral boron. *J. Less Common Metals* 82: 163–168.
32. Werheit, H., R. Schmechel, V. Kueffel, and T. Lundström. 1997. On the electronic properties of β-rhombohedral boron interstitially doped with 3d transition metal atoms. *J. Alloys Compd.* 262–263: 372–380.
33. Nakayama, T., K. Tanaka, J. Shimizu, and K. Kimura. 1998. Thermoelectric properties of metal-doped β-rhombohedral boron. In *Proceedings of the 17th International Conference on Thermoelectrics*, IEEE Catalog Number 98TH8365, pp. 570–573.
34. Takizawa, H., N. Haze, K. Okamoto, K. Uheda, and T. Endo. 2002. Microwave synthesis of Fe-doped β-rhombohedral boron. *Mater. Res. Bull.* 37: 113–121.
35. Sologub, O., Y. Matsushita, and T. Mori. 2013. An α-rhombohedral boron-related compound with sulfur: Synthesis, structure and thermoelectric properties. *Scr. Mater.* 68: 289–292.
36. Golikova, O. A., N. Amandzhanov, M. M. Kazanin, G. M. Klimashin, and V.V. Kutasov. 1990. Electrical activity of impurities in β-rhombohedral boron. *Phys. Stat. Solidi A* 121: 579–586.
37. Hossain, M. A., I. Tanaka, T. Tanaka, A. U. Khan, and T. Mori. 2015. YB_{48} the metal rich boundary of YB_{66}; crystal growth and thermoelectric properties. *J. Phys. Chem. Solids* 87: 221–227.
38. Sussardi, A., T. Tanaka, A. U. Khan, L. Schlapbach, and T. Mori. 2015. Enhanced thermoelectric properties of samarium boride. *J. Materiomics* 1: 196–204.
39. Berthebaud, D., A. Sato, Y. Michiue et al. 2011. Effect of transition element doping on crystal structure of rare earth borosilicides $REB_{44}Si_2$. *J. Solid State Chem.* 184: 1682–1687.
40. Hossain, M. A., I. Tanaka, T. Tanaka, A. U. Khan, and T. Mori. 2016. Crystal growth and anisotropy of high temperature thermoelectric properties of yttrium borosilicide single crystals. *J. Solid State Chem.* 233: 1–7.
41. Maruyama, S., Y. Miyazaki, K. Hayashi, T. Kajitani, and T. Mori. 2012. Excellent p-n control in a high temperature thermoelectric boride. *Appl. Phys. Lett.* 101: 152101.
42. Sahara, R., T. Mori, S. Maruyama, Y. Miyazaki, K. Hayashi, and T. Kajitani. 2014. Theoretical and experimental investigation of the excellent p–n control in yttrium aluminoborides. *Sci. Technol. Adv. Mater.* 15: 035012.

43. Maruyama, S., A. Prytuliak, Y. Miyazaki, K. Hayashi, T. Kajitani, and T. Mori. 2014. Al insertion and additive effects on the thermoelectric properties of yttrium boride. *J. Appl. Phys.* 115: 123702.
44. Maruyama, S., T. Nishimura, Y. Miyazaki, K. Hayashi, T. Kajitani, and T. Mori. 2014. Microstructure and thermoelectric properties of $Y_xAl_yB_{14}$ samples fabricated through the spark plasma sintering. *Mater. Renew. Sustain. Energy* 3: 31.
45. Wan, L. F. and S. P. Beckman. 2014. Lattice instability in the $AlMgB_{14}$ structure. *Phys. B Condens. Matter* 438: 9–12.
46. Miura, S., H. Sasaki, K. Takagi, and T. Fujima. 2014. Effect of varying mixture ratio of raw material powders on the thermoelectric properties of $AlMgB_{14}$-based materials prepared by spark plasma sintering. *J. Phys. Chem. Solids* 75: 951–953.
47. Mori, T. and H. Nishijima. unpublished.
48. Gong, Y., Y. Zhang, M. Dudley et al. 2010. Thermal conductivity and Seebeck coefficients of icosahedral boron arsenide films on silicon carbide. *J. Appl. Phys.* 108: 084906.
49. Frye, C. D., J. H. Edgar, I. Ohkubo, and T. Mori. 2013. Seebeck coefficient and electrical resistivity of single crystal $B_{12}As_2$ at high temperatures. *J. Phys. Soc. Jpn.* 82: 095001.
50. Y. Kumashiro, T. Yokoyama, K. Sato, Y. Ando, S. Nagatani, and K. Kajiyama. 2000. Electrical and thermal properties of $B_{12}P_2$ wafers. *J. Solid State Chem.* 154, 33–38.
51. Akashi, T., T. Itoh, I. Gunjishima, H. Masumoto, and T. Goto. 2002. Thermoelectric properties of hot-pressed boron suboxide (B_6O). *Mater. Trans.* 43: 1719–1723.
52. Slack, G. A. and K. E. Morgan. 2014. Some crystallography, chemistry, physics, and thermodynamics of $B_{12}O_2$, $B_{12}P_2$, $B_{12}As_2$, and related alpha-boron type crystals. *J. Phys. Chem. Solids* 75: 1054–1074.
53. Paderno, Yu. B., V. I. Novikov, and E. S. Garf. 1970. Electrical properties of hexaborides of alkali- and rare-earth metals. *Poroshkovaja Metallurgija* 11: 70–73.
54. Etourneau, J., J. P. Mercurio, R. Naslain, and P. Hagenmuller. 1970. Structure electronique de quelques hexaborures de type CaB_6. *J. Solid State Chem.* 2: 332–342.
55. Gianno, K., A. V. Sologubenko, H. R. Ott, A. D. Bianchi, and Z. Fisk. 2002. Low-temperature thermoelectric power of CaB_6. *J. Phys. Condens. Matter* 14: 1035–1043.
56. Takeda, M., F. Domingo, T. Miura and T. Fukuda. 2002. Resistivity and thermopower of CaB_6 single crystal. *Mater. Res. Soc. Symp. Proc.* 691: G8.7.1–G8.7.6.
57. Yagasaki, K., S. Notsu, Y. Shimoji et al. 2003. Resistivity and thermopower of CaB_6single crystal. *Physica B* 329–333: 1259–1260.
58. Gürsoy, M., M. Takeda, and B. Albert. 2015. High-pressure densified solid solutions of alkaline earth hexaborides (Ca/Sr, Ca/Ba, Sr/Ba) and their high-temperature thermoelectric properties. *J. Solid State Chem.* 221: 191–195.
59. Tynell, T., T. Aizawa, I. Ohkubo, K. Nakamura, and T. Mori, unpublished.
60. Lassoued, S., R. Gautier, and J. F. Halet. 2011. In *Boron Rich Solids, Sensors, Ultra High Temperature Ceramics, Thermoelectrics, Armor,* The Electronic Properties of Metal Borides and Borocarbides: Differences and Similarities, eds. N. Orlovskaya and M. Lugovy (Springer, New York), pp. 95–114.
61. Simonson, J. W. and S. J. Poon. 2010. Applying an electron counting rule to screen prospective thermoelectric alloys: The thermoelectric properties of $YCrB_4$ and Er_3CrB_7-type phases. *J. Alloys Compd.* 504: 265–272.
62. Slack, G. A. and V. G. Tsoukala. 1994. Some properties of semiconducting $IrSb_3$. *J. Appl. Phys.* 76: 1665–1671.
63. Tanaka, T., Z. U. Rek, J. Wong, and M. Rowen. 1998. FZ crystal growth of monochromator-grade YB_{66} single crystals as guided by topographic and double-crystal diffraction characteristics. *J. Cryst. Growth* 192: 141–151.
64. Slack, G. A., D. W. Oliver, G. D. Brower, and J. D. Young. 1977. Properties of melt-grown single crystals of YB_{68}. *J. Phys. Chem. Solids* 38: 45–49.
65. Efros, A. L. and M. Pollak, eds. 1985. *Electron-Electron Interactions in Disordered Systems* (North-Holland, Amsterdam, the Netherlands), p. 409.
66. Mori, T., Y. Shimazu, and S. Ikehata. 1993. NMR Study on compensated Si:P. *J. Phys. Soc. Jpn.* 62: 3172–3180.

67. Mori, T., Y. Shimazu, and S. Ikehata. 1994. Magnetic susceptibility of compensated Si:P. *Solid State Commun.* 91: 13–15.
68. Higashi, I., K. Kobayashi, T. Tanaka, and Y. Ishizawa. 1997. Structure refinement of YB_{62} and YB_{56} of the YB_{66}-type structure. *J. Solid State Chem.* 133: 16–20.
69. Tanaka T., K. Kamiya, T. Numazawa, A. Sato, and S. Takenouchi. 2006. The effect of transition metal doping on thermal conductivity of YB_{66}. *Z. Kristallogr.* 221: 472–476.
70. Wachter, P. 1994. Intermediate valence and heavy fermions. In *Handbook on the Physics and Chemistry of Rare Earths*, Ch. 132, Vol. 19, pp. 177–382.
71. Nickerson, J., R. White, K. Lee, R. Bachmann, T. Geballe, and G. Hull. 1971. Physical properties of SmB_6. *Phys. Rev. B* 3: 2030–2042.
72. Gabáni, S., E. Bauer, S. Berger et al. 2001. Pressure-induced Fermi-liquid behaviour in the Kondo insulator SmB_6: Possible transition through a quantum critical point. *Phys. Rev. B* 67: 172406.
73. Roman, J., V. Pavlík, K. Flachbart et al. 1997. Transport and magnetic properties of mixed valent SmB_6. *Phys. B Condens. Matter* 230–232: 715–717.
74. Dzero, M. and V. Galitski. 2013. A new exotic state in an old material: A tale of SmB_6. *J. Exp. Theor. Phys.* 117: 499–507.
75. Phelan, W. A., S. M. Koohpayeh, P. Cottingham et al. 2014. Correlation between bulk thermodynamic measurements and the low-temperature-resistance plateau in SmB_6. *Phys. Rev. X* 4: 031012.
76. Xu, N., P. K. Biswas, J. H. Dil et al. 2014. Direct observation of the spin texture in SmB_6 as evidence of the topological Kondo insulator. *Nat. Commun.* 5: 4566.
77. Wolgast, S., Ç. Kurdak, K. Sun, J. W. Allen, D. J. Kim, and Z. Fisk. 2013. Low-temperature surface conduction in the Kondo insulator SmB6. *Phys. Rev. B* 88: 180405(R).
78. Jiang, J., S. Li, T. Zhang et al. 2013. Observation of possible topological in-gap surface states in the Kondo insulator SmB_6 by photoemission. *Nat. Commun.* 4: 3010.
79. Zhu, Z. H., A. Nicolaou, G. Levy et al. 2013. Polarity-driven surface metallicity in SmB6. *Phys. Rev. Lett.* 111: 216402.
80. Takimoto, T. 2011. SmB_6: A promising candidate for a topological insulator. *J. Phys. Soc. Jpn.* 80: 123710.
81. Müchler, L., B. Yan, F. Casper, S. Chadov, and C. Felser. 2013. Topological insulators. In *Thermoelectric Nanomaterials*, eds. K. Koumoto and T. Mori (Springer, Heidelberg, Germany), Ch. 6, pp. 123–139.
82. Goto T. 1998. *Kinzoku* (in Japanese) 68: 1086.
83. Okada, S., T. Shishido, T. Mori, K. Kudou, K. Iizumi, T. Lundstrom, and K. Nakajima. 2006. Properties of $REAlB_4$ and Lu_2AlB_6 crystals grown from RE-AL-B (RE = Tm, Yb, Lu) melts. *J. Alloys Compd.* 408–412: 547–550.
84. Korsukova, M. M., V. N. Gurin, Yu. B. Kuzma et al. 1989. Crystal structure, electrical, and magnetic properties of the new ternary compounds $LnAlB_{14}$. *Phys. Stat. Solidi A* 114: 265–272.
85. Korsukova, M. M., T. Lundström, L.-E. Tergenius, and V. N. Gurin. 1992. The crystal structure of defective $YAlB_{14}$ and $ErAlB_{14}$. *J. Alloys Compd.* 187: 39–48.
86. Maruyama, S., T. Kamiyama, Y. Miyazaki, K. Hayashi, T. Kajitani, and T. Mori, unpublished.
87. Emin, D. 2006. Unusual properties of icosahedral boron-rich solids. *J. Solid State Chem.* 179: 2791–2798.
88. Maruyama, S. and T. Mori. unpublished.
89. Bing, F., H. P. Martin, and A. Michaelis. 2013. In situ preparation and thermoelectric properties of B_4C_1_x–TiB_2 composites. *J. Electron. Mater.* 42: 2314–2319.
90. Cai, K. F. and C. W. Nan. 2000. The influence of W_2B_5 addition on microstructure and thermoelectric properties of B_4C ceramic. *Ceram. Int.* 26: 523–527.
91. Liu, C. H. 2000. Structure and properties of boron carbide with aluminum incorporation. *Mater. Sci. Eng. B* 72: 23–26.

92. Prytuliak, A. and T. Mori. 2011. Effect of transition-metal additives on thermoelectric properties of $YB_{22}C_2N$. *J. Electron. Mater.* 40: 920–925.
93. Prytuliak, A., S. Maruyama, and T. Mori. 2013. Anomalous effect of vanadium boride seeding on thermoelectric properties of $YB_{22}C_2N$. *Mater. Res. Bull.* 48: 1972–1977.
94. Mori, T. and T. Hara. 2016. Hybrid effect to possibly overcome the trade-off between Seebeck coefficient and electrical conductivity. *Scr. Mater.* 111: 44–48.

15

Complex Chalcogenides: Pseudo-Hollandites, Structures and Properties

Franck Gascoin
University of Normandy

CONTENTS

15.1 Introduction

This chapter is dedicated to a newer group of prospective thermoelectric materials, the so-called pseudo-hollandite compounds of the general formula $A_yM_5X_8$. Although their thermoelectric figure of merit *ZT* is less impressive than the *ZT*s of some other classical and newly discovered materials, this family of chalcogenides exemplifies very well how one might think about using solid-state chemistry to discover new ways to achieve the desired properties.

Our quest to discover always better performing thermoelectric materials basically utilizes the following three main strategies: (1) choose an already known thermoelectric compound and try to improve its properties using different approaches, such as nanostructuration, band engineering, and texturation [1–3]; (2) recognizing that the properties of many known compounds have never been reported or measured, take the crystal database and select, utilizing different criteria, the best target materials; and (3) start from scratch and try, using educated guesses and sometime pure luck or even a good hunch, to find the next best material. This third method might, at first, seem a bit fishy, but there are two kinds of fishermen: (1) those who know the fishes, their habits and playgrounds, and the techniques to catch them (let alone the influence of the tides and weather!) and (2) occasional fishermen … who, I confess, can still get lucky from time to time! That said, embarking on systematic studies of unknown intermetallic systems, especially if the ultimate goal is to find the next best (thermoelectric) material, needs and takes time. On the other hand, such a fundamental task is essential and is often rewarded by interesting findings in terms of structural chemistry and physical properties, although not always the ones we were aiming for. But, just like Professor John Corbett wrote in his article "Exploratory synthesis in the solid state. Endless wonders," [4] "it is always difficult to predict the unimaginable."

In the past 10–15 years, there have been very extensive studies carried out on antimonides and chalcogenides showing that, indeed, these classes of materials have superior thermoelectric properties. In fact, the propensity of these elements to form heavily doped semiconductors must be recognized and must be taken into account in our current research.

Pseudo-hollandites are a family of compounds with the general formula $A_xT_5X_8$, where A is a monovalent or divalent cation, T is a transition metal, and X is a chalcogen. These compounds are named after the "Hollandites," a large family of transition metal oxides of formula $A_xT_8O_{16}$, as their structure is somewhat related, with both types of compounds composed of a 3D framework of octahedrally coordinated transition metal atoms that contain infinite channels in which sit cations A. As evidenced by the difference of stoichiometry and, in particular, by the ratio T/X or T/O, the pseudo-hollandite structure is more condensed than the structure of the classical hollandite. Indeed, it only contains one type of channel formed around the arrangement of 10 TX_6 octahedra, while the hollandite displays two types of channels, thus leading to a visibly more open structure. The larger of the two channels contains the A cations and is formed by only 8 TO_6 octahedra. Another way to describe the pseudo-hollandite structure is to recognize the CdI_2-type layers that are connected together via pillars made of double chains of edge-sharing TX_6 octahedra (see Figure 15.1).

Interestingly, these three structure types are characterized by their chemical flexibility, as they can be formed by numerous combinations of elements, thus offering an ideal playground for tuning the transport properties. Also interesting is the fact that these structures have all been recently investigated for their thermoelectric properties (among others). The best results were obtained for pseudo-hollandite chromium selenides $A_xCr_5Se_8$ and for CdI_2-type titanium sulfoselenides Ti(S, Se)$_2$, with the maximum *ZT*s just above 0.5, so far [5].

FIGURE 15.1
(a) The "classical" hollandite of formula AT_8O_{16}. (b) The intermetallic pseudo-hollandite of formula AT_5X_8. (c) The CdI_2-type layered structure.

15.2 Synthesis of Pseudo-Hollandites

Thermoelectricity, with its practitioners called "thermoelectricians," brings together researchers from very diverse fields spanning from "hardcore" physicists to sales managers, but overall, the community's interest and focus are cutting-edge materials science. Since the renewed interest in thermoelectric materials in the early 2000s, it seems that many newcomers have not yet appreciated that to make a material with a well-defined chemical composition and free of impurities is not that simple! This is even more true when dealing with large batches of hundreds of grams (or even kilograms if the large-scale production is the goal). Moreover, when using selenium, the degree of complexity is even greater as the evaporation of Se may dramatically affect the stoichiometry, homogeneity, and purity of the product.

Whenever doable, the opportunity to work with single crystals is always welcome, rewarding, and satisfying. Indeed, it allows the precise determination of the crystal structure, including the actual stoichiometry, and if the crystals are large enough, physical properties can be measured accurately. However, obtaining single crystals is not a task that is always trivial or even possible.

Concerning the pseudo-hollandites, it is tricky to obtain big crystals (again, "big" meaning large enough so that physical properties can be measured by contacting the crystals with at least four electrical leads), but good results have been obtained by growing single crystals in sealed tubes using a carbonate as the source of the A cation and/or a slight excess of selenium [6]. This observation seems to indicate that either an increased pressure inside the ampoule helps the crystal growth or that a phenomenon of self-transport takes place resulting in the formation of sizable crystals. Single crystals of pseudo-hollandite tend to form as long and extremely fine needles, difficult to handle and often twinned, as they like to form bundles rather than staying isolated.

When pure polycrystalline samples are the target, it is evidently not recommended to use precursors in excess, as that will leave behind unwanted impurities. Thus, larger batches of pseudo-hollandite can be obtained by using a stoichiometric mixture of precursors (raw elements or binary compounds) and an adequate heat treatment (that depends on the composition) in a sealed container.

15.3 Properties of Pseudo-Hollandites

Recognizing that the structure of pseudo-hollandite is rather complex, possesses two distinct subnetworks, and can be made from a variety of element combinations was the first step toward the investigation of their transport properties, the assumption made being that such a structure "should" have a relatively low lattice thermal conductivity. Actually, this is the only assumption one can make with some degree of confidence when looking at crystallographic structures (going through a database for instance), and even this is merely a hypothesis that must be verified.

15.3.1 $TlCr_5Se_8$: The First Target Material

If the choice of selenium as the chalcogen might seem obvious (sulfides are often very resistive and tellurides are often metallic and somewhat trickier to make), chromium was

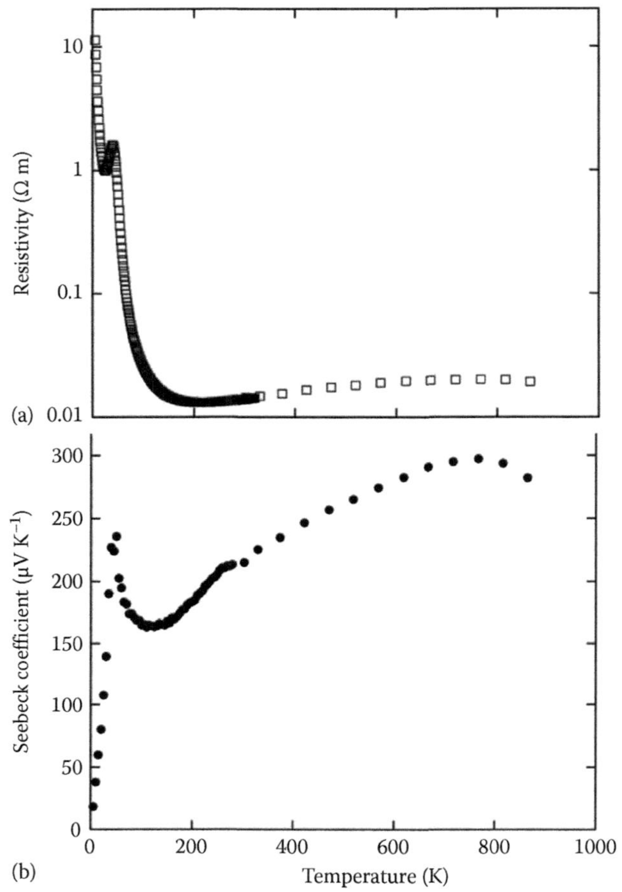

FIGURE 15.2
(a) Variation of the electrical resistivity with temperature. (b) Variation of the Seebeck coefficient with temperature.

chosen because our previous work on $AgCrSe_2$ showed promising results [7]. However, selecting Tl as the A cation might be surprising as Tl is, indeed, very toxic. Nevertheless, the rationale behind this choice is the fact that Tl is the least electropositive cation that is known to exist in the channel of the pseudo-hollandite. Therefore, $TlCr_5Se_8$ is very likely to be less resistive than if an alkali metal is used instead. This is evidently true only in a *p*-type compound as the A cation provides electrons. The more electropositive it is, the more electrons are transferred and, hence, a lower hole concentration results. Just the opposite effect would arise in the case of an *n*-type compound.

The transport properties of $TlCr_5Se_8$ were measured from 4 K to about 900 K on polycrystalline samples densified by spark plasma sintering. The temperature-dependent electrical resistivity data (Figure 15.2a) indicated a semiconducting behavior, as one would expect since the compound can be rationalized as an electron precise Zintl phase, for example, Tl^+ + $5Cr^{3+}$ + $6Se^{2-}$. Accordingly, a band gap of about 50 meV was calculated by fitting the data in the range of 40–200 K using the relationship $\rho = \exp(-\Delta E/2k_BT)$. Above room temperature, the electrical resistivity stood between 5 and 20 mΩ cm, much lower values compared to

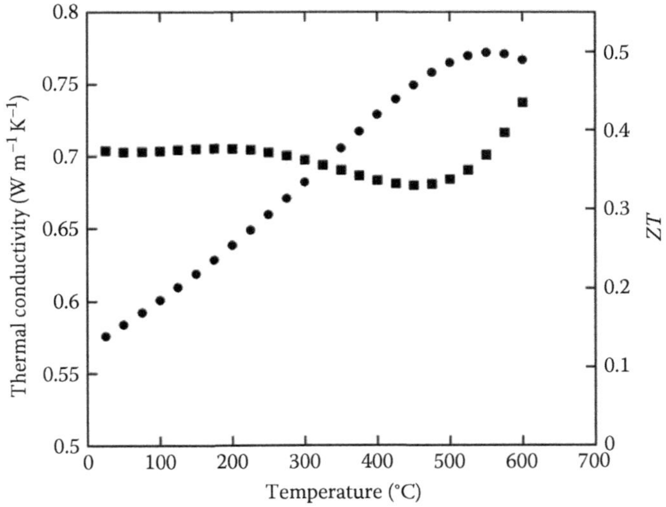

FIGURE 15.3
Variation with temperature of the total thermal conductivity (square symbols) and of the thermoelectric figure of merit (circular symbols) of $TlCr_5Se_8$.

the values reported by Bensch et al. [8]. As for the Seebeck coefficient, it reached a very large value of 300 μV K^{-1} at 800 K (Figure 15.2b) before dropping because of the thermal excitation of the minority carriers. Both transport coefficients showed anomalies in the vicinity of the magnetic transition occurring at around 40 K. As it is out of the scope of this contribution, more details concerning this transition can be found elsewhere [5].

The total thermal conductivity (Figure 15.3) was very low and remained below 1 W m^{-1} K^{-1} over the whole temperature range, with a slight increase at high temperatures due to the onset of the bipolar conduction at about 800 K. This total thermal conductivity virtually corresponded to the lattice thermal conductivity, as the electronic thermal conductivity is, indeed, negligible because the electrical resistivity was high (above 20 mΩ cm over this temperature range). This result, of course, validated our assumption concerning the relationship between the structure and its resulting thermal conductivity. Overall, the calculated thermoelectric figure of merit ZT (Figure 15.3), far from being ridiculously small, reached a value of 0.5 at 800 K. This is rather unexpected when looking at the electrical resistivity, as the resistivity values of several tens of mΩ cm typically lead to a much lower ZT. This unusual combination of a very high Seebeck coefficient, an extremely low thermal conductivity, and a somewhat high electrical resistivity (compared to the usual thermoelectric materials) might open the door for new classes of thermoelectric materials that, till now, were mostly disregarded.

15.3.2 Attempts to Improve $TlCr_5Se_8$

Since the first investigation of the transport properties of $TlCr_5Se_8$ revealed the very low lattice thermal conductivity, it seemed appropriate to devote more efforts to optimize the electronic transport properties in order to attempt to improve the figure of merit ZT of such a compound. With $TlCr_5Se_8$, being a p-type semiconductor with a relatively high electrical resistivity, the first goal is to increase its hole concentration. There are two obvious ways to do so: reducing the thallium content and thus supplying fewer electrons, consequently, increasing the hole concentration, and by partially substituting selenium by tellurium.

Reducing the thallium content is possible, as various older studies have shown [9,10]. However, we have found that the direct synthesis (from the elements) of $Tl_xCr_5Se_8$ is limited to values of x of about 0.90. Below this limit, a pure phase cannot be synthesized because impurities of Cr_2Se_3 crystallize as a secondary phase. As expected, a decrease in the content of thallium led to a decrease of the electrical resistivity by a considerable extent, as seen in Figure 15.4. Unfortunately, this decrease was concomitant with a decrease of the Seebeck coefficient (Figure 15.5), thus resulting in a decreased power factor. Moreover, as the resistivity decreased, the electronic component of the thermal conductivity became nonnegligible and, consequently, the total thermal conductivity increased. Overall, reducing the thallium content did not lead to the improvement of the thermoelectric figure of merit.

The substitution of selenium by tellurium has been realized up to the composition $TlCr_5Se_5Te_3$ without having to modify our synthesis conditions. As expected, this substitution, indeed, results in a drastic change of the electrical resistivity and of the Seebeck coefficient, as shown in Figures 15.6 and 15.7. Again, the overall thermoelectric performance was not improved.

This first study of the thermoelectric properties of the pseudo-hollandite $Tl_xCr_5Se_8$ revealed that the crystallographic features of this structure lead to a very low (below 1 W m^{-1} K^{-1}) lattice thermal conductivity, thus verifying our first assumption. Evidently, this has encouraged us to pursue our investigations within this structure type.

15.3.3 Modifications of $Tl_xCr_5Se_8$

Starting with $TlCr_5Se_8$, the first evident modification to make is to get rid of thallium, for obvious reasons of its toxicity. Acting as a monovalent cation, thallium can be often

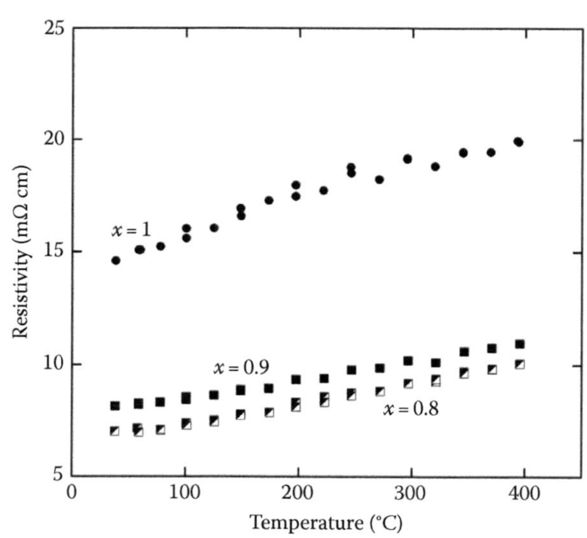

FIGURE 15.4
Variation of the electrical resistivity of $Tl_xCr_5Se_8$ with temperature.

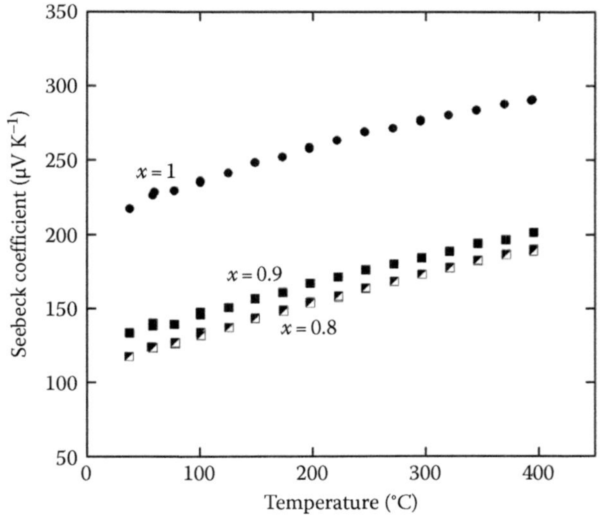

FIGURE 15.5
Variation of the Seebeck coefficient of $Tl_xCr_5Se_8$ with temperature.

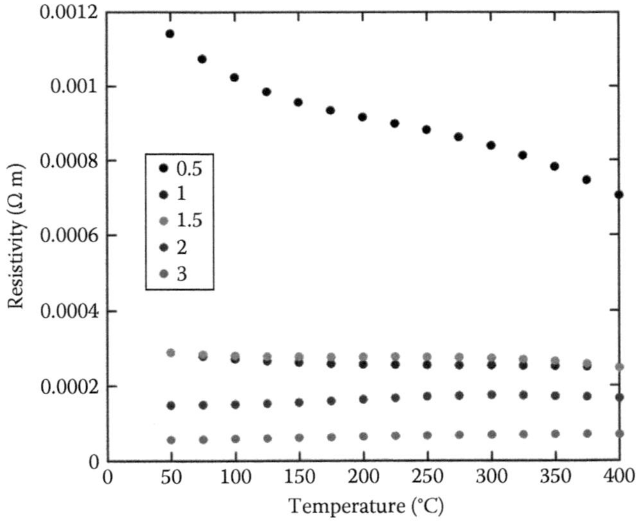

FIGURE 15.6
Variation of the electrical resistivity of $TlCr_5Se_{8-x}Te_x$ with temperature for different Te content x.

replaced by alkali, alkaline-earth, or even rare-earth metals, with the charge and the size taken into consideration. Of course, substituting Tl by any of the aforementioned metals might not help decrease the electrical resistivity, as these substituting elements are more electropositive than Tl, thus leading to an increased ionicity of bonding between them and the Cr_5Se_8 framework. This should, therefore, result in an opening of the band gap and, evidently, to even more dominant semiconducting properties. Nevertheless, by varying the occupancy of the cation, the electronic properties might still be tunable.

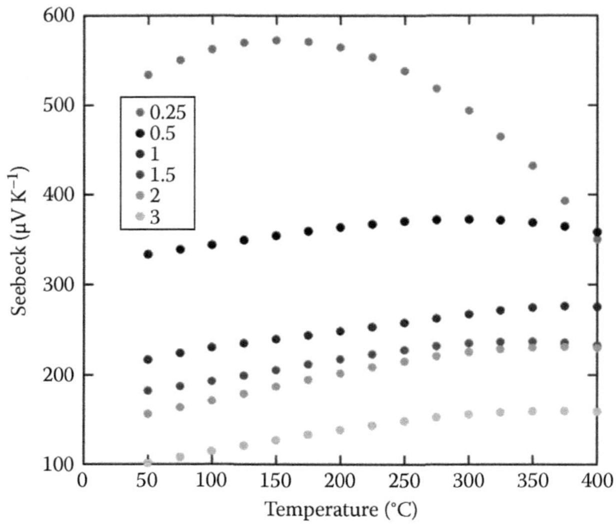

FIGURE 15.7
Variation of the Seebeck coefficient of $TlCr_5Se_{8-x}Te_x$ with temperature; the higher the Te content, the lower the Seebeck.

We have thus conducted investigations with $Ba_xCr_5Se_8$. Barium is relatively easy to handle (compared to alkali metals) and its metallic radius (2.24 Å) seems suitable for the Cr_5Se_8 framework. Evidently, in order to keep the electron count balanced, the target stoichiometry is now $Ba_{0.5}Cr_5Se_8$, as the divalent barium replaces the monovalent thallium.

The synthesis of $Ba_{0.5}Cr_5Se_8$ is somewhat less straightforward than that of $TlCr_5Se_8$, mainly because of the propensity of Ba to react with the silica tubes. Hence, it was decided to make first the precursor BaSe via mechanical alloying. Thus, high-temperature reactions successfully led to the formation of $Ba_{0.5}Cr_5Se_8$ when BaSe, Se, and Cr were used as starting materials. Following densification using spark plasma sintering, the x-ray diffraction powder patterns always showed unindexed Bragg peaks that were first thought to originate due to a superstructure (Figure 15.8), most likely the result of the Ba disorder within the channels.

It was thus decided to try to grow single crystals. This was possible using an excess of elemental selenium during the high-temperature synthesis. The single-crystal x-ray analysis revealed that the compound $Ba_{0.5}Cr_5Se_8$ does not crystallize in the usual monoclinic cell of the pseudo-hollandite structure type but, rather, in a new triclinic setting. Consequently, our data could be perfectly fit by a Rietveld refinement using this new setting (see Figure 15.9). More details concerning this crystallographic study can be found elsewhere [11]. It has to be stressed that this kind of thorough crystallographic study, although very time consuming, is and should be mandatory in all material studies, especially when also dealing with transport properties.

The transport properties of $Ba_{0.5}Cr_5Se_8$ have been measured in the range of 4–875 K. As expected and already noted before, the electrical resistivity of $Ba_{0.5}Cr_5Se_8$ was typical of a semiconductor and was higher than that of $TlCr_5Se_8$. The Seebeck coefficient was very large and reached values over 300 μV K^{-1} at temperatures between 550 and 750 K (see Figure 15.10). Also expected was the extremely low thermal conductivity shown in

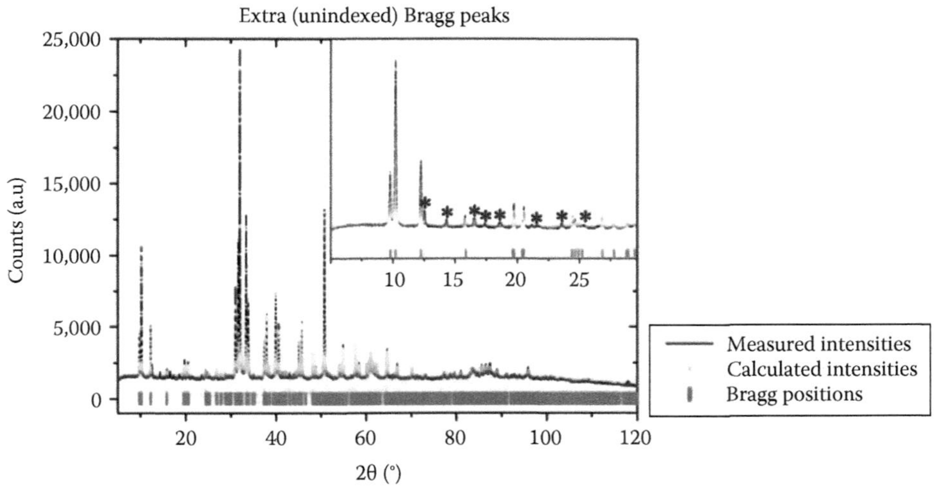

FIGURE 15.8
Rietveld refinement of $Ba_{0.5}Cr_5Se_8$ in the "classical" pseudo-hollandite monoclinic cell. Extra, unindexed peaks are visible in the inset.

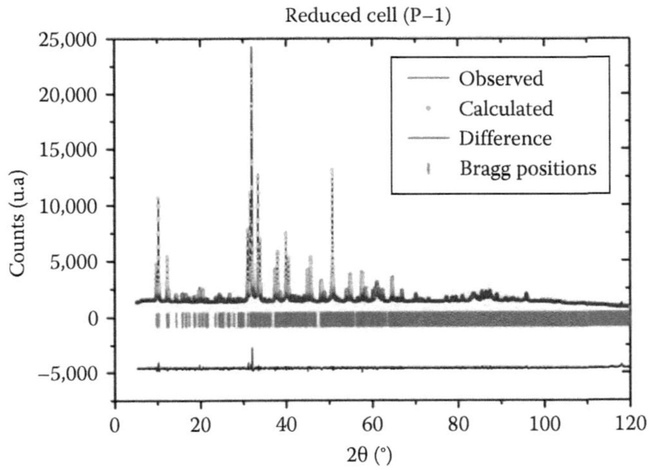

FIGURE 15.9
Rietveld refinement of $Ba_{0.5}Cr_5Se_8$ after lowering the symmetry, proving that the compound crystallizes in a new structure type.

Figure 15.11, well below 1 W m^{-1} K^{-1} over the entire range, except for a peak at low temperatures that is most likely related to a magnetic transition. Investigations of this anomaly are underway.

Naturally, the next step to try was to modify the barium content in order to tune the transport properties. Thus, the study of solid solutions of $Ba_xCr_5Se_8$ was performed. Bulk samples with barium sub- and overstoichiometry were synthesized, corresponding to x = (0.2, 0.3, 0.4, 0.45, 0.48) and x = (0.51, 0.52, 0.55, 0.58, 0.6, 0.7, 0.8). For selected stoichiometries, x-ray diffraction patterns are shown in Figure 15.12.

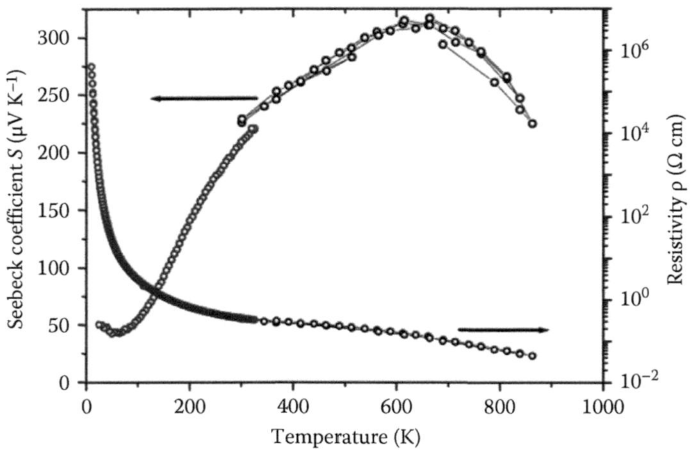

FIGURE 15.10
Variation with temperature of the Seebeck coefficient and of the electrical resistivity of $Ba_{0.5}Cr_5Se_8$.

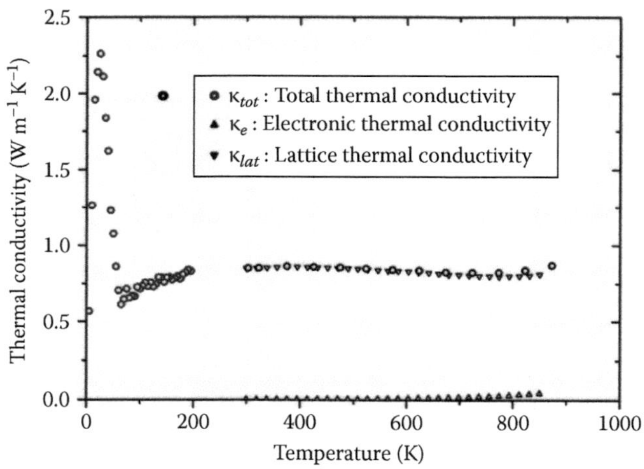

FIGURE 15.11
Variation of the thermal conductivity of $Ba_{0.5}Cr_5Se_8$ with temperature. The electronic and lattice parts of the thermal conductivity are also plotted.

For compositions with $x < 0.5$, the presence of Cr_2Se_3 as an impurity was always observed, and this prevented any further reliable transport property measurements. However, and this is more surprising, it appeared that $Ba_xCr_5Se_8$ can be made as a pure phase with the value of x up to 0.55. Above this value, impurities of $Ba_{1-p}Cr_2Se_{4-p}$ always appeared. This trend was surprising as the deficiency or excess of Ba might be considered as being partially compensated by the oxidation or the reduction of the transition metal. Indeed, just like it was reported for the thallium homologue, decreasing the cation content led to the oxidation of Cr^{III} to Cr^{IV}. The opposite trend should therefore result in the reduction of Cr^{III} to Cr^{II}. Here in $Ba_xCr_5Se_8$ with the content of x between 0.5 and 0.55, magnetic studies clearly indicated an increase of the magnetic moment, in good agreement with the

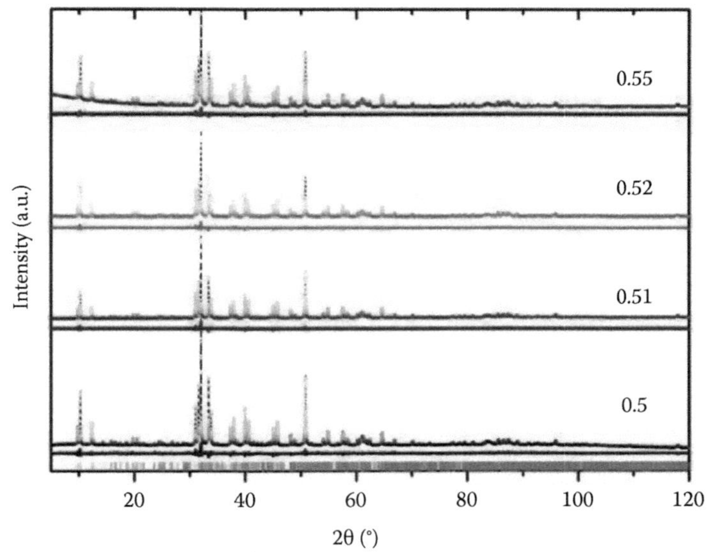

FIGURE 15.12
X-ray diffraction powder pattern of $Ba_xCr_5Se_8$. There was no trace of any impurity for x between 0.5 and 0.55.

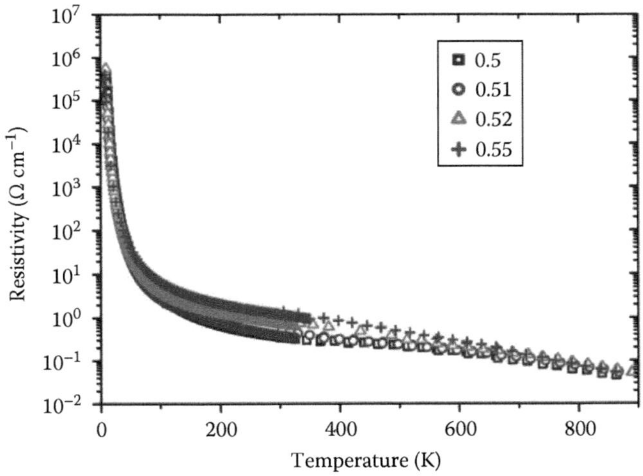

FIGURE 15.13
Variation of the electrical resistivity of $Ba_xCr_5Se_8$ with temperature. Different symbols indicate different Ba content x.

reduction of chromium. However, the magnetic moment reached extremely high values that are still under investigation, especially since the same values have been reported elsewhere in other chromium selenide [11].

Concerning the transport properties of these overstoichiometric compounds, they all exhibited semiconducting electrical resistivity, as shown in Figure 15.13. However, there was a clear decrease of the Seebeck coefficient with the increasing Ba content

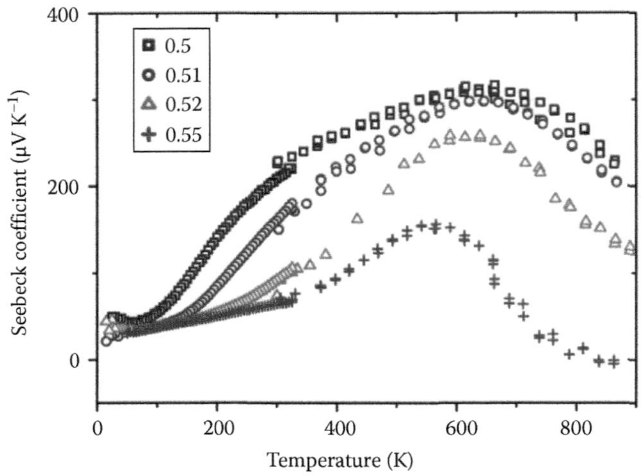

FIGURE 15.14
Variation of the Seebeck coefficient of $Ba_xCr_5Se_8$ with temperature. Different symbols indicate different Ba content x.

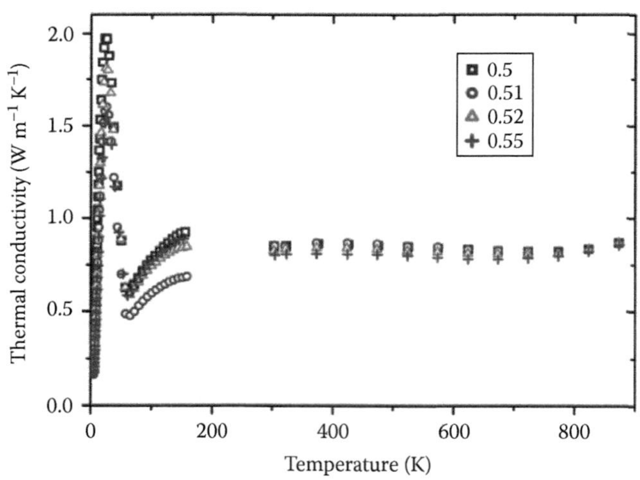

FIGURE 15.15
Variation of the total thermal conductivity of $Ba_{0.5}Cr_5Se_8$ with temperature. Different symbols indicate different Ba content x.

(see Figure 15.14). Furthermore, the temperature at which the maximum value of the Seebeck coefficient was attained decreases with the increasing content of Ba. Such a behavior might be the result of the presence of both types of carriers, thus leading to an increased carrier concentration and therefore a decreased Seebeck coefficient.

As expected, the changing content of Ba does not modify the thermal conductivity, which remained below 1 W m^{-1} K^{-1} above room temperature (Figure 15.15). Also interesting is the peak of the thermal conductivity observed at low temperatures, which might be associated with a magnetic transition. Refinement of the magnetic structure is needed to clarify the situation, and this obviously requires further investigation.

15.4 Conclusion

Transition metal chalcogenides have been and remain fascinating structures that stimulate the interest and attention of many research groups, on account of their unique physical properties and their intriguing structural features. In this contribution, compounds adopting the pseudo-hollandite structure type have demonstrated their significance and versatility. Indeed, starting from a simple picture of a rather complex structure that leads to an extremely low lattice thermal conductivity, this structure type offers broad opportunities to modify its stoichiometry and explore what effect such modifications have on the transport properties. Of course, this statement could be made for any family of compounds. However, in this particular case, such a structure type combines many attractive features that should help to shed light on how to engineer outstanding thermoelectric properties. Furthermore, and this should not be underestimated, the exploratory research, even if centered around a particular structure type, can often lead to the discovery of new compounds whose properties are by default unknown and therefore worth the time and effort to explore in detail, despite the lengthy and challenging task.

Acknowledgments

This work was benefitted from the efforts of past and present graduate students, R. Nunna, R. Lefèvre, and S. Maier. The author is also indebted to his colleagues D. Berthebaud, S. Hébert, O. Perez, and D. Pelloquin for their help and commitment.

References

1. Pei, Y. Z., X. Y. Shi, A. Lalonde, H. Wang, L. Chen, and G. J. Snyder. 2011. Convergence of electronic bands for high performance bulk thermoelectrics. *Nature* 474: 66–69.
2. Bux, K. S., J.-P. Fleurial, and R. Kaner. 2010. Nanostructured materials for thermoelectric applications. *Chem. Commun.* 46: 8311–8324.
3. Lognoné, Q., F. Gascoin, O. I. Lebedev, L. Lutterotti, S. Gascoin, and D. Chateigner. 2014. Qualitative texture analysis of spark plasma textured n-Bi_2Te_3. *J. Am. Ceram. Soc.* 97: 2038–2045.
4. Corbett, J. D. 2000. Exploratory synthesis in the solid state. Endless Wonders. *Inorg. Chem.* 39: 5178–5191.
5. Takahashi, H., N. Raghavendra, F. Gascoin, D. Pelloquin, S. Hébert, and E. Guilmeau. 2013. Transport properties of an intermetallic with pseudo-hollandite structure as a potential thermoelectric materials, the example of $Tl_xCr_5Se_8$. *Chem. Mater.* 25: 1809–1815.
6. Bronger, H., C. Herudek, J. Huster, and D. Schmitz. 1993. Neue Alkalimetallchromalkogenide und ihre Struktursystematik. *Zeit. Anorg. Allgem. Chemie* 619: 243–252.
7. Gascoin, F. and A. Maignan. 2011. Order-disorder transition in $AgCrSe_2$: A new route to efficient thermoelectrics. *Chem. Mater.* 23: 2510–2513.
8. Bensch, W. and B. Sander. 2002. The influence of sulphur and selenium onto the physicochemical properties of the quasi-ternary chromium chalcogenides $TlCr_5S_{8-z}Se_z$ ($z = 0 - 8$). *Solid State Sci.* 4: 701–708.

9. Bensch, W., E. Wörner, M. Muhler, and U. Ruschewitz. 1994. Single crystal structure, magnetic properties, and electronic structure of $Tl_xCr_5S_8$ ($x = 1.0$ and 0.7). *J. Solid State Chem.* 110: 234–242.
10. Ohtani, T., Y. Sano, K. Kodama, and S. Onoue. 1993. Synthesis of new ternary tunnel chalcogenides by ion exchange reactions and deintercalation of the ternary chromium. *Mater. Res. Bull.* 28: 501–508.
11. Lefèvre, R., D. Berthebaud, O. Perez, D. Pelloquin, S. Hébert, and F. Gascoin. 2015. Polar transition metal chalcogenides: Structure and properties of the new pseudo-Hollandite $Bi_{0.5}Cr_5Se_8$. *Chem. Mater.* 27: 7110–7118.

16

Tetrahedrites: Earth-Abundant Thermoelectric Materials with Intrinsically Low Thermal Conductivity

Xu Lu and Donald T. Morelli
Michigan State University

CONTENTS

16.1 Introduction

Thermoelectric (TE) materials can directly convert heat into electricity through the Seebeck effect and thereby have potential application in waste heat recovery in transportation, industrial, commercial, and residential processes. The efficiency of conversion of heat into electricity for a TE device is determined by the figure of merit $ZT = (S^2\sigma/k)T$, where S is the Seebeck coefficient, σ is the electrical conductivity, and k is the total thermal conductivity. A good TE material should possess high Seebeck coefficient, high electrical conductivity, and low thermal conductivity. Since the electronic part of thermal conductivity is governed by the Wiedemann–Franz law and thus is coupled to the electrical conductivity, minimizing the lattice thermal conductivity is crucial to improve the figure of merit of TE materials. Although progress has been made in the design and synthesis of new TE materials during the last decade by reducing lattice thermal conductivity through solid solution formation or nanostructuring, thereby increasing the value of figure of merit to over two (almost 20% efficiency at typical operating temperatures), TE power generation is still far from large-scale application due to the facts that most high-performance TE materials contain expensive/toxic elements, require complicated synthesis procedures, or have poor mechanical properties. The need for earth-abundant and environment-friendly TE materials with high performance is ever more pressing. Some success has been realized, for example, in *n*-type skutterudites $CoSb_3$ [1,2] and Mg_2Si-based compounds [3,4], with ZT values above 1.5; finding *p*-type counterparts with comparable performance, however, has been challenging.

Copper-containing TE materials often behave as *p*-type semiconductors and some have great potential for TE applications. For example, binary copper-containing compounds like Cu_2Se and Cu_2S show high ZT values despite the fact that they are ionic conductors [5]. Most recently, ternary copper-containing compounds such as $Cu_3SbSe_{4-x}S_x$ solid solutions [6], Cu_2SnSe_3 [7], $CuInTe_2$ [8], $CuGaTe_2$ [9], and Cu_5FeS_4 [10] have been shown to be good TE materials with ZT values close to or exceeding unity.

Tetrahedrites are ternary copper-containing compounds with the chemical formula of $Cu_{12-x}Tr_xSb(As)_4S_{13}$, where Tr are transition metal elements including Cu, Zn, Mn, and Ni [11]. The tetrahedrite–tennantite (As bearing) series is an important family of copper natural ore mineral and also one of most widespread sulfosalts on the earth [12]. They are capable of incorporating various atoms on different sites. For example, Ag, Zn, Ni, Fe, or Co may all reside on the Cu site, Bi, As, and Te on the Sb site, and Se on S site, which gives rise to the strong potential for tuning their electronic structure by controlling chemical composition. Overall, these compounds display intrinsic low lattice thermal conductivity and tunable electronic transport properties, making them suitable for TE power generation. In addition, tetrahedrites display mechanical properties comparable to or better than a number of other mainstream TE materials.

16.2 Intrinsic Low Thermal Conductivity and Crystal Structure of Tetrahedrites

As mentioned earlier, TE materials with high figure of merit should have both low lattice thermal conductivity and high electrical conductivity. Thus, they are simultaneously required to behave with respect to phonons as a glassy or disordered material and with respect to electrons as a good crystal. This concept, known as "phonon-glass, electron-crystal," was introduced by Slack in 1995 [13]. In a phonon-glass compound, the phonon mean free path approaches the interatomic spacing, reducing the lattice thermal conductivity to a minimum value described by the relation [14]

$$\kappa_L = \frac{1}{3} C_V v \lambda \tag{16.1}$$

where C_V is the specific heat, v the phonon group velocity, and λ the phonon mean free path.

For phonon-glass compounds, the heat-carrying acoustic phonons are scattered via phonon–phonon interactions governed by lattice vibrations characteristic of a given crystal structure. Some TE materials show extremely low (<1 W m^{-1} K^{-1}) lattice thermal conductivities without nanostructuring or alloying, and these values are almost independent of grain size and temperature. The crystal structures in such circumstances possess the following characteristics:

16.2.1 Large Primitive Unit Cell Volume

A large primitive unit cell for a specific compound implies a large number of atoms in the cell. Based on the classic heat transport theory, the number of phonon modes is determined by the number of atoms (N) in the primitive unit cell. There will be $3(N - 1)$ optical phonon modes and three acoustic phonon modes induced by lattice vibrations. Although the increased number of optical phonon modes will, to first order, not affect heat transfer due to the low group velocity of these modes, the contribution to total specific heat from the heat-carrying acoustic phonon modes, given by $C_{V,acoustic} = C_v/N$, will decrease. The total lattice specific heat at a high temperature can be estimated by the Dulong–Petit value $C_v = 3k_BN/V$, where k_B is the Boltzmann constant and V is the unit cell volume. Therefore, Equation 16.1 becomes $\kappa_L = k_B v\,\lambda/V$, indicating that a larger primitive cell volume leads to lower lattice thermal conductivity, which explains the low thermal conductivities usually found in solids with complex crystal structures.

16.2.2 "Rattling" Atoms

Slack initiated the concept that the rattling atoms in cage-like compounds with oversized atomic sites, such as clathrates and partially filled skutterudites, will result in low thermal conductivity by providing more phonon scattering centers [13]. The large atomic displacement of these rattling atoms in the voids of host structures has been demonstrated by x-ray diffraction analysis [15] and neutron scattering [1]. From another point of view, the coupling between rattling atoms and host framework leads to low phonon group velocity and thus low lattice thermal conductivity.

16.2.3 Strong Anharmonicity

Anharmonicity can be considered as the anharmonic terms in the expansion of the ionic interaction potential energy in solids. Lattice thermal resistivity in a solid originates with these anharmonic terms, as a perfect harmonic crystal without grain boundaries or other types of defects or impurities would have an infinite thermal conductivity. Anharmonicity can arise due to specific atomic arrangements and chemical bonding in certain compounds. Based on an old idea of Zhuze [16], Morelli et al. [17] attributed the low lattice thermal conductivity in $AgSbTe_2$ to the lone pair–driven anharmonicity. Skoug and Morelli [18] first identified the role of lone pair electrons in the low thermal conductivity of nitrogen-group chalcogenide compounds ($CuSbSe_2$ and Cu_3SbSe_3). In these compounds, lone pair electrons from the V group atom (Sb) are believed to give rise to minimum lattice thermal conductivity due to anharmonicity caused by electrostatic repulsion between the lone pair electrons and neighboring chalcogen ions (Se). The strength of anharmonicity was shown to correlate with the bond length, the angle between the group V atom and the chalcogen atom, and the coordination number of the group V atom. More recently, a similar phenomenon was reported in I–V–VI_2 compounds based on the rock salt structure [19]. Marginally stable I–V–VI_2 compounds with lone pair electrons from group V atoms have anharmonic bonds that result in strong phonon–phonon interactions, which let their thermal conductivity reach to the amorphous limit.

A direct measure of anharmonic strength is the Grüneisen parameter, which is defined as the negative of the logarithmic derivative of each phonon mode's frequency ω_i with respect to the crystal volume V:

$$\gamma_i = -\frac{\partial \ln(\omega_i)}{\partial \ln V} \tag{16.2}$$

The definition of the Grüneisen parameter reflects how much the chemical bond stiffness in solids is affected by changes in the distances between atoms. If we take into account only the acoustic phonon modes for the heat conduction process, the lattice thermal conductivity can be expressed as [20]

$$\kappa_L = A\frac{\overline{M}\theta^3\delta}{\gamma^2 N^{2/3} T} \tag{16.3}$$

where, M is the average mass of the atoms in the compound, N is the number of atoms in the primitive unit cell, δ^3 is the volume per atom, θ is the Debye temperature, γ is the average of the mode Gruneisen parameters.

A large value of the Grüneisen parameter, that is, strong anharmonicity, favors low lattice thermal conductivity, and if the anharmonic phonon scattering is strong enough, the thermal conductivity can approach the theoretical minimum value.

The preceding arguments can be applied specifically to the crystal structure and the associated intrinsic low lattice thermal conductivity of tetrahedrites, which was found to have a value below 0.5 W m^{-1} K^{-1} over the entire measured temperature range. Experimentally, the average Grüneisen parameter in solids can be determined by

$$\gamma = \frac{9\alpha B V_m}{C_V} \tag{16.4}$$

where, α is the linear thermal expansion coefficient, B is the bulk modulus, V_m is the molar volume, C_V is the specific heat.

For $Cu_{10}Zn_2Sb_4S_{13}$, α is 10.35×10^{-6} K^{-1} [21], B is 44.05 GPa [22], and C_V is around 4.06 J g^{-1} K^{-1} at 300 K [23], so the calculated Grüneisen parameter is ≈2.06 at 300 K, which is a typical large value for solids with strong anharmonicity (for instance, $AgSbTe_2$ has $\gamma = 2.05$ and $AgBiSe_2$ has $\gamma = 2.9$ [17]).

The composition and crystal structure of tetrahedrite compounds were historically studied mostly by geologists. There are 29 atoms in the primitive unit cell with a cell volume of 1100 Å^3, which is comparable to the value of Zintl antimonides with similarly low thermal conductivity. It is a cubic structure with group symmetry of $I\bar{4}3m$. It is useful to consider that the structural subunits comprise the tetrahedrite crystal structure; see Figure 16.1 [24]. First, there are two distinct Cu sites in the lattice. Six of the twelve Cu atoms are distributed on tetrahedral 12d Cu(1) sites, while the remaining Cu atoms occupy trigonal planar 12e Cu(2) sites. In terms of a simple crystal-chemical formula, four of the six tetrahedral sites are thought to be occupied by monovalent Cu, while the other two are occupied by Cu^{2+} ions; the trigonal planar sites are occupied solely by monovalent Cu. Additionally, there are two distinct sulfur sites: the S(1) site, tetrahedrally coordinated by

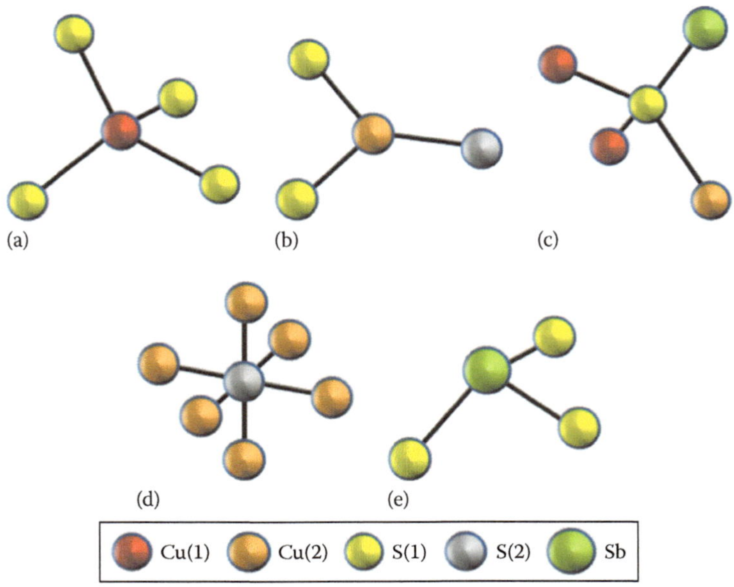

FIGURE 16.1
Structural subunits in tetrahedrites: (a) tetrahedrally coordinated Cu(1), (b) trigonally coordinated Cu(2), (c) tetrahedrally coordinated S(1), (d) octahedrally coordinated S(2), and (e) trigonally coordinated Sb.

two Cu(1), one Cu(2), and Sb; and the S(2) site, octahedrally coordinated by Cu(2). Finally, there is the Sb site, trigonally coordinated by S(1). Of these structural subunits, our attention is drawn to two in particular that have importance in the context of anharmonicity and low intrinsic thermal conductivity. First, because Sb is trigonally coordinated in tetrahedrite, it is nominally in the trivalent Sb^{3+} state and exhibits the classic chemical lone pair of electrons. Second, because these lone pair–containing Sb atoms reside above and below the CuS_3 plane, the entire ensemble of $Sb[CuS_3]Sb$ forms a trigonal bipyramid, which can be considered a cage-like structure similar to clathrates and partially filled skutterudites; see Figure 16.2.

The vibration of Cu atoms normal to the S trigonal plane was first studied by Suekuni through an atomic displacement parameter analysis [25], and these Cu vibrations were believed to be the origin of low lattice thermal conductivity. It was claimed that such low-energy vibrations will interact with the heat-carrying phonons, as rattlers do in other cage compounds. A subsequent study on low-temperature heat capacity of tetrahedrites [23] revealed that the experimental specific heat data can be well reproduced only if the data were fit by using a Debye lattice and three additional Einstein oscillators, associated with three localized modes from trigonally coordinated Cu atoms. The Einstein oscillator with an energy of 8.4 meV is believed to arise from the vibration of the trigonally coordinated Cu atoms normal to the plane of the sulfur triangle. The large amplitude vibration of this oscillator is a clear sign of strong anharmonicity resulting from the unstable overlap of Cu d-orbitals with S p-orbitals. Two low-energy Einstein oscillators result from the motion of trigonally coordinated Cu atoms perpendicular to the trigonal plane. The high-energy Einstein oscillation peak was found in measured density of states (DOS) of phonons collected by inelastic neutron scattering at the exact energy of 8.4 meV as extrapolated from

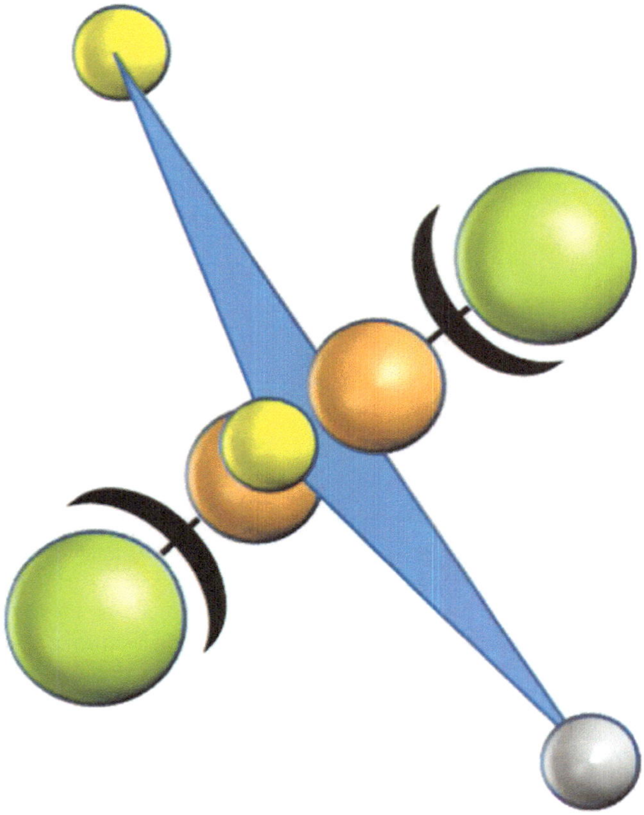

FIGURE 16.2
Trigonal bipyramidal "cage" formed by the trigonal Cu–S plane and lone pair–containing Sb atoms above and below it. The black crescents indicate the lone pairs, and the two copper atoms indicate the two possible positions of the Cu atom above or below the sulfur trigonal plane.

specific heat, which proves that the strong anharmonicity in phonon–phonon interactions results from the motion of Cu atoms above and below the sulfur trigonal planes. Furthermore, lattice dynamics calculations [26] found unstable vibration modes with imaginary frequencies; these were believed to be associated with the trigonally coordinated Cu atoms. A double-well potential related to motion of the trigonally coordinated Cu was also found by calculation.

A detailed study on chemical bonding in tetrahedrites was recently performed by Lai et al. [27] through synchrotron x-ray diffraction and theoretical calculation. The results highlight the important role of lone pair electrons from Sb atoms in tetrahedrites. The interaction between trigonally coordinated Cu and lone pair electrons gives rise to the oversized five-atom atomic cage, which is required in the context of phonon glass electron crystal (PGEC) and rattlers. Along the direction normal to the S trigonal plane, a weak bonding configuration exists between Cu atoms and Sb atoms that varies from lone pair to covalent bonding. Additionally, for the trigonally coordinated Cu atoms, bond strength among the three S atoms is different. The strong local bonding asymmetry is believed to be the origin of anharmonicity.

16.3 Electronic Structure and Electrical Transport Properties of Tetrahedrites

Early experimental study performed by Johnson and Jeanloz [28] showed that the electrical resistivities of tetrahedrites can vary over seven orders of magnitude with different compositions, which can be explained by a Brillouin zone model. In this model, the number of valence electrons per unit cell was taken into account for the electronic structure of tetrahedrites. All the Cu atoms in tetrahedrites are assumed to be monovalent and provide one electron. Divalent atoms such as Zn contribute two electrons, while Sb and S atoms provide three and six electrons, respectively. Therefore, from $Cu_{12}Sb_4S_{13}$ to $Cu_{10}Zn_2Sb_4S_{13}$, the valence electron count increases from 204 to 208 per unit cell. A total of 204 electrons completely fill the 51st Brillouin zone and 208 valence electrons completely fill the 52nd Brillouin zone, since each Brillouin zone can hold four electrons in the unit cell. The completely filled Brillouin zone indicates insulating or semiconducting nature of compounds, while for compounds with the partially filled Brillouin zone, there are available states for electrons, rendering those compounds metallic. In fact, isolated divalent Cu atoms were found to exist in synthetic $Cu_{12}Sb_4S_{13}$ by magnetic and electron paramagnetic resonance (EPR) measurements [29], which indicates that not all Cu atoms are in the monovalent state in the real case. For convenience, 2 of 12 atoms are considered to be in the mixed valence or pseudo-divalent state between 1+ and 2+. Therefore, $Cu_{12}Sb_4S_{13}$ behaves as highly degenerate semiconductor or metallic compound, which is consistent with experiment. In addition, the Brillouin zone model for tetrahedrites also indicates that the composition limit of tetrahedrites is the arrangement with 208 valence electrons per unit cell and that the stability increases as compositions approach this limit. An optical spectra analysis of a natural mineral tetrahedrite sample with 208 valence electrons per unit cell showed that the absorption edge is centered at 720 nm, suggesting a large gap semiconductor with the bandgap of 1.7 ± 0.2 eV. The Cu atoms in the pseudo-divalent state are found to be mobile in the tetrahedrite structure and may take part in charge transfer producing an ionic component of the conductivity [30], which is not acceptable in TE applications. However, replacing these mobile Cu atoms with other transition metals can effectively eliminate the ionic conductivity [31]. For example, for natural mineral samples, no ionic component of conductivity is observed.

The results of electronic structure obtained by density functional theory (DFT) calculations were first reported by Lu et al. [26]. The Fermi level of $Cu_{12}Sb_4S_{13}$ was found to fall near a sharp peak in the DOS at the top of the valence band, representing a metallic character. The flat characteristic band feature near the Fermi level indicates favorable thermopower (Seebeck coefficient) for TE applications. A large semiconducting gap (1.2 eV) forms between the valence bands and the conduction bands. The hybridized sulfur 3p and copper 3d orbitals give rise to the top of the valence bands, resulting in two unoccupied states per formula unit. Therefore, synthetic $Cu_{12}Sb_4S_{13}$ has low resistivity due to the absence of divalent Cu atoms. In contrast, the divalent atoms in tetrahedrites, for example, replacing Cu with Zn, will fill the unoccupied states, pushing the Fermi level into the large bandgap and rendering the material insulating. Besides Zn, the effect of other transition metal element (Fe, Ni, Mn, Co) substitutions on Cu sites in tetrahedrites has been studied both theoretically and experimentally [26,32–35]. The 3d bands for these transition metals exhibit exchange splitting above 2 eV between the majority spin states and minority spin states indicated by DFT calculations. Makovicky et al. [36]

have shown that Fe in synthetic $Cu_{12-x}Fe_xSb_4S_{13}$ is trivalent for $0 < x < 1$ and divalent for $1 \leq x \leq 2$. This implies that each Fe atom can provide an extra electron to fill holes in the valence band compared to that of each Zn atom. Recent DFT calculations confirm this interpretation, showing that iron in $Cu_{11}FeSb_4S_{13}$ is indeed in the Fe^{3+} s^0d^5 configuration due to the formation of an exchange gap between the Fe majority and minority spin, suggesting the high-spin d^5 configuration of Fe^{3+} [26].

The behavior with Ni doping is quite different from Zn doping [32]. The majority spin states are fully occupied. In the minority spin channel, however, three partially empty acceptor-like bands exist above the Fermi level and thus give rise to a finite DOS at the Fermi level. This means that Ni substitution introduces an additional hole state near the top of the valence band by hybridizing Ni 3d states with the p-d hybridized valence bands of the host. Therefore, the Ni-substituted tetrahedrite remains metallic even with two Cu atoms replaced by Ni atoms. Mn substitution for Cu in tetrahedrites also results in an occupied majority spin state and an unoccupied minority spin state for 3d orbitals. However, these unoccupied orbitals reside in the conduction band of the host, separated by large bandgap from the valence band. Similar to Mn substitution, for Co substitution, the majority spin 3d orbitals are in occupied valence band, while the minority spin states are found in the bandgap. Therefore, both Mn substitution and Co substitution can be considered to have the same effect as Zn substitution, pushing the Fermi level into bandgap and driving the compound insulating.

Experimentally, the effect of Zn and Fe doping on Cu sites of tetrahedrites $Cu_{12-x}Zn(Fe)_xSb_4S_{13}$ was first systematically studied by Lu et al. [26]. It was found that with increasing Zn concentration, the Seebeck coefficient rises considerably, exceeding 200 μV K^{-1} at 700 K for the $x = 1.5$ sample. This is consistent with the theoretical calculation that the carrier concentration is reduced as Zn fills the hole states in the valence band. For all the samples with $x \leq 1.5$, the resistivities were in the range of 10^{-3} Ω cm and the Seebeck coefficients in the range of 100–200 μV K^{-1}; these are values that give rise to TE power factors comparable to some of the best TE materials in this temperature range. The pure $Cu_{12}Sb_4S_{13}$ has the highest power factor of 12 W cm^{-1} K^{-2} at 700 K among all the samples. As predicted by calculation, the $x = 2$ Zn-substituted sample is insulating with no TE efficiency. Fe substitution plays the same role as Zn in transport properties except Fe with 3+ valence state will fill the hole states more effectively. The highest ZT values obtained were 0.9 and 0.8 for $x = 1$ Zn substitution and $x = 0.5$ at 700 K for Fe substitution, respectively. The maximum ZT values of different element substitutions are displayed in Figure 16.3. Ni substitution was reported by Suekuni et al. [32], and those results showed that all of the Ni-substituted samples were conducting, with resistivities on the order of 10^{-3} Ω cm. The highest ZT found in this study was 0.7 at 670 K for the $x = 1.5$ Ni substitution; this sample had a power factor of approximately 9 W cm^{-1} K^{-2}. Experiments on Mn substitution conducted by Heo et al. [33] showed that the ZT value could be raised to 1.1 by $x = 1$ Mn substitution, due to the extremely low thermal conductivity measured in this study. However, a subsequent study on the same composition [34] reported that Mn substitution had no advantage over other dopants, and the reported low thermal conductivity in the previous study arose from the low sample density.

Te substitution for Sb in tetrahedrites $Cu_{12}Sb_{4-x}Te_xS_{13}$ was also carried out by Lu and Morelli [37]. It was concluded that Te atoms with a 4+ valence state replacing Sb atoms with 3+ valence state will introduce extra electrons to fill the unoccupied states in the valence band of pure tetrahedrites. Therefore, Te substitution on Sb sites has the similar effect as

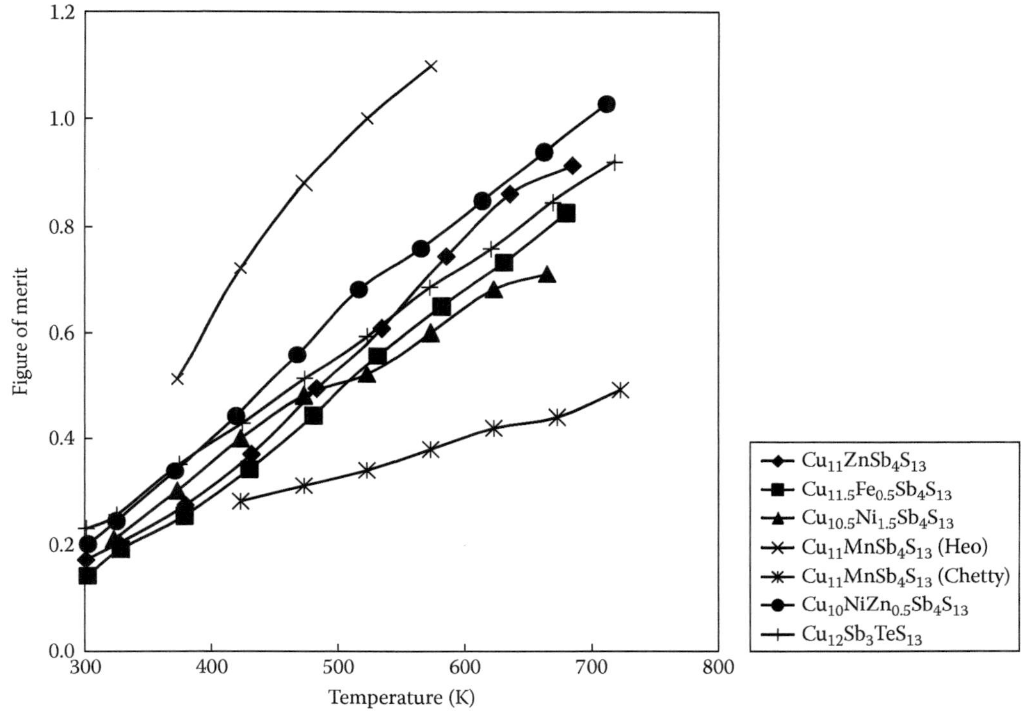

FIGURE 16.3
TE figure of merit for various transition metal–substituted tetrahedrites.

Zn substitution on Cu sites. The highest ZT in this study was found to be 0.9 at 700 K for $x = 1$ Te substitution. A selection of ZT values for various transition metal substitutions in tetrahedrite is given in Figure 16.3.

In summary, for all the substitution elements mentioned earlier, most of them behave as electron donors, which fill the hole states and cause the increase in both electrical resistivity and the Seebeck coefficient. In all these cases, there is a balance between the power factor and the figure of merit. Unlike most other TE materials, the maximum power factor does not necessarily result in the maximum ZT value for tetrahedrites. Because the lattice thermal conductivity of these compounds is approaching the amorphous limit, the electronic thermal conductivity assumes the main role in the total thermal conductivity and thus governs their TE properties. With increasing substitutional element content, even though the power factor is reduced, caused by increased resistivity, the overall figure of merit can be enhanced due to the reduction in the electronic thermal conductivity. To understand the relationship between the composition of substituted tetrahedrites and their ZT values, the notion of the fraction (f) of hole states filled in the valence band upon substitution was used: $f = xN_e/2$, where N_e is the number of excess electrons contributed by each Tr ion. For example, for $x = 0.5$ Fe substitution ($N_e = 2$), the fraction is 0.5, while for $x = 0.5$ Zn substitution ($N_e = 1$), the fraction is 0.25. All the Cu atoms are assumed to be monovalent in this model. The maximum ZT values are found at $f = 0.5$ and ZT begins to decrease for higher filling fraction for both substitutions. ZT values above 0.8 can be maintained over a large range of substitution compositions with a hole filling fraction up to 0.8.

16.4 Codoping

Codoping, which means adding two elements into the parent compound, is usually applied in TE materials to improve their performance by tuning both the electronic structure and the phonon transport at the same time. Usually, the added two doping elements have different effects on the properties of the parent compound. In Section 16.3 it was concluded that Zn substitution fills the hole states in the valence band, while Ni substitution introduces additional unoccupied states at the top of the valence band. Motivated by this conclusion, Ni and Zn codoping was performed by making $Cu_{10.5}Ni_{1.5-x}Zn_xSb_4S_{13}$, assuming that Zn doping may drive the Fermi level close to the top edge of the valence band and fill the holes introduced by Ni substitution [38].

The ZT value for the $x = 0.5$ sample in this study exceeds unity at 700 K; in this sample, the optimum Fermi energy level position and the minimum lattice thermal conductivity are achieved simultaneously by Zn and Ni codoping. Another advantage of codoping with Ni and Zn is that it minimizes the number of mobile Cu atoms in tetrahedrites, largely reducing the risk of ionic conductivity for TE applications.

16.5 Direct Use of Natural Mineral Tetrahedrites and Other Rapid Synthesis Methods

In traditional methods of synthesizing TE materials, pure elements are extracted from natural minerals and combined into compounds with accurate chemical stoichiometry by heating or mechanical alloying in vacuum or inert atmosphere. The process can be costly and time consuming. Although tetrahedrites are a fairly new field of investigation, several approaches have already been investigated, aimed at producing a low-cost material on a large scale. In these materials, the fact that the figure of merit is not sensitive to the impurity concentration but rather is related to the hole filling fraction in the valence band makes it possible to directly use natural minerals as a raw source for TE material synthesis, since the wide range of hole filling fraction also allows for a wide range of impurity concentrations that is often found in natural minerals. The systematic investigation of the TE properties of mixtures of synthetic $Cu_{12}Sb_4S_{13}$ and natural mineral specimens ($Cu_{10.5}Fe_{1.5}As_{3.6}Sb_{0.4}S_{13}$ and $Cu_{9.7}Zn_{1.9}Fe_{0.4}As_4S_{13}$) showed that complete solid solutions can be formed in these mixtures after 30 min ball milling and simple hot pressing [39]. For the composition $Cu_{9.7}Zn_{1.9}Fe_{0.4}As_4S_{13}$, which mainly contains Zn as an impurity substitution element, by 50% weight ratio mixing with synthetic $Cu_{12}Sb_4S_{13}$, a ZT value close to unity was achieved. A subsequent study by Lu and Morelli [40] reported a further simplified method of synthesis directly from natural mineral; in this method, the natural mineral was mixed with the pure elements, Cu, Sb, and S, in the ratio of 12:4:3 by high-energy ball milling. Surprisingly, a single-phase tetrahedrite solid solution was easily

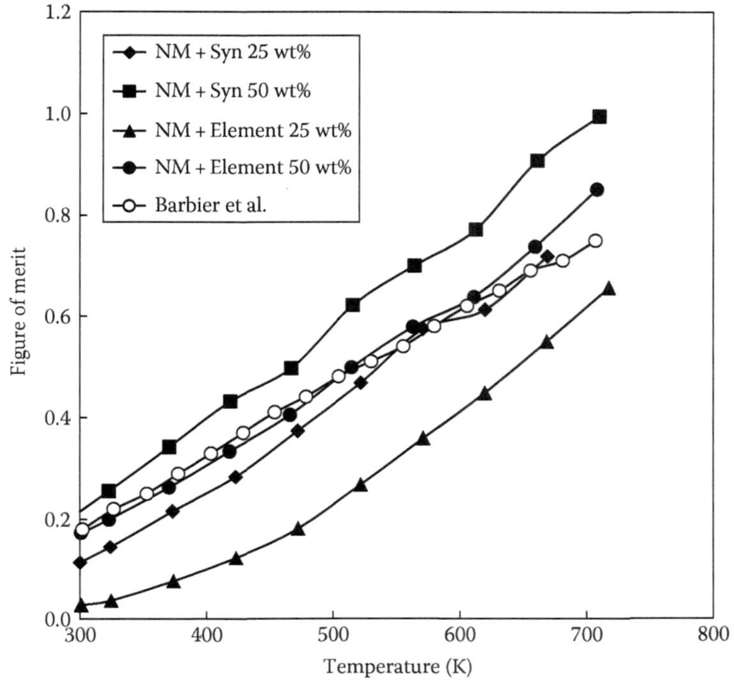

FIGURE 16.4
TE figure of merit for tetrahedrites synthesized using "manufacturable" methods: "NM + Syn," mixture of natural mineral and synthetic tetrahedrite [39], and "NM + element," mixture of natural mineral and Cu, Sb, and S elements [40]. Also shown are the results by Barbier et al. [41] for samples synthesized using high-energy ball milling of elemental precursors.

obtained without any heat treatment in 3–6 h. In this process, the natural mineral serves as a "seed matrix" and incorporates pure elements into a parent compound. The highest *ZT* found in this study was 0.9 with natural mineral (Zn rich) mixing with pure elements in a mass ratio of 1:1. Barbier et al. [41] combined a high-energy ball milling procedure using elemental precursors with spark plasma sintering and were able to reduce the total processing time compared to previous methods by almost an order of magnitude. The *ZT* values of samples made from some of these new methods of synthesis are summarized in Figure 16.4.

In addition, high-performance TE materials can be synthesized directly from natural minerals by mixing with pure synthetic $Cu_{12}Sb_4S_{13}$ or pure elements and also using high-energy ball milling combined with spark plasma sintering. Considering their potential low cost and convenience, these methods have great prospects in real-market applications.

16.6 Mechanical Properties and Structure Stability of Tetrahedrites

For real TE applications, such as waste heat recovery and power generation, in addition to the TE figure of merit, the mechanical properties of the materials play an important role.

Thermal stress induced by a temperature gradient between cold and hot sides of TE modules is an important parameter. The thermal stress can be described as [22]

$$\sigma_T = \frac{E(T)\alpha(T)\Delta T}{(1-\upsilon(T))} \tag{16.5}$$

where, $E(T)$ is the temperature-dependent Young's modulus, $\alpha(T)$ is the temperature-dependent linear thermal expansion coefficient, ΔT is the temperature gradient, $\upsilon(T)$ is the temperature-dependent Poisson's ratio.

In order to minimize thermal stress, small values of Young's modulus, linear thermal expansion coefficient, and Poisson's ratio are required at a fixed temperature gradient. The relevant parameters for tetrahedrites are listed in Table 16.1, which also contains reported mechanical properties for other TE materials including the Si–Ge alloy $Si_{80}Ge_{20}$, the $CoSb_3$-based skutterudites, Mg_2Si, and PbTe-based compounds. Little difference in Poisson's ratio is exhibited between tetrahedrites and other TE materials. Young's modulus of tetrahedrites is comparable to that of PbTe-based materials but lower than others. The linear thermal expansion coefficient of tetrahedrites is comparable to Mg_2Si and $CoSb_3$, which is smaller than that of PbTe-based compounds. By combining three parameters, tetrahedrites have an advantage over other state-of-the-art TE materials in terms of thermal stress, indicating TE modules made from tetrahedrites may have a higher likelihood of surviving after long-term thermal cycling. The fracture toughness K_c is a measure of resistance to crack propagation in materials. Unfortunately, the fracture toughness value of 0.47 MPa for tetrahedrites is only comparable to PbTe-based materials but lower than others due to its anharmonic bonding nature. The brittle index (BI) is usually used as an evaluation of the machinability of materials, which is a rough gauge of the chipping induced by cutting of a material, with the tendency to chip increasing as the magnitude of BI increases. It can be calculated as

$$\mathrm{BI} = \frac{E * H}{K_c^2} \upsilon \tag{16.6}$$

where, E is Young's modulus, H is the hardness, K_c is the fracture toughness.

The calculated BI value of tetrahedrites indicates that it has comparable machinability to Mg_2Si and PbTe and is better than Si–Ge alloys. Therefore, from the mechanical point

TABLE 16.1

Mechanical Properties of Tetrahedrites

Materials	E (GPa)	α (10^{-6} K^{-1})	υ	H (GPa)	K_c (MPa*$m^{0.5}$)	BI (μm^{-1})	Reference
Synthetic $Cu_{12}Sb_4S_{13}$	34	10.5	0.34	1.7	0.47	240	[22]
Synthetic $Cu_{12}Sb_4S_{13}$ + natural mineral (50% weight)	55	10.5	0.31	3.7	0.47	520	[22]
$Si_{80}Ge_{20}$	135–147	3.1	0.24	12.4–14.7	0.98–1.03	1820	[42]
$Pb_{0.95}Sn_{0.05}Te$–PbS	53	19.8	0.24	1.1–1.3	0.31–0.44	398	[43]
Mg_2Si	117	7.5	0.18	4.8–5.6	0.7–1.3	585	[44]
$CoSb_3$	127	13.5	0.24	5.6–7.0	0.82	900	[45]

E stands for Young's modulus, α is the linear coefficient of thermal expansion, υ is Poisson's ratio, H is the hardness, K_c is the fracture toughness, and BI is the brittle index.

of view, tetrahedrites are promising candidates for large-scale TE applications with great capability to resist thermal stress and desirable machinability, in spite of the fact that the fracture toughness must be improved. It was also found that by mixing with synthetic $Cu_{12}Sb_4S_{13}$, both Young's modulus and the hardness are enhanced, while the fracture toughness is independent of composition variations.

The high-temperature structure stability of $Cu_{12}Sb_4S_{13}$ and $Cu_{10.4}Ni_{1.6}Sb_4S_{13}$ was studied by Barbier et al. [41] by using DSC (differential scanning calorimetry)/TGA (thermogravimetric analysis) measurements. It was found that $Cu_{12}Sb_4S_{13}$ starts to decompose into Cu_3SbS_3 at 791 K. However, in contrast, the DSC curve of $Cu_{10.4}Ni_{1.6}Sb_4S_{13}$ shows that the decomposition is suppressed by Ni substitution, and there is no endothermic peak until the temperature reaches 880 K, which is the melting point of tetrahedrites. No phase transformation was observed, supported by the absence of weight loss in the TGA curve. Therefore, impurity substitution in tetrahedrites helps stabilize the structure and raise the operation temperature for TE applications. For instance, Ni substitution in this case increases the operation temperature from 673 to 783 K.

16.7 Conclusions and Future Work

In this review, we have highlighted the unique crystal structure and chemical bonding in tetrahedrites, which possess intrinsic low lattice thermal conductivity. The effects of transition metal substitution on electrical transport properties are summarized. The conclusion that the TE figure of merit of tetrahedrites is quite robust against impurity substitution makes it plausible to directly use natural minerals as a raw source for TE materials synthesis. Additionally, tetrahedrites are found to have favorable capability to resist thermal stress and possess good machinability properties. Transition metal substitutions are proved to not only tune the electronic structure but also enhance the crystal structure stability of tetrahedrites in preliminary studies. All the factors listed earlier make tetrahedrites very promising candidates for large-scale waste heat recovery.

It should be noted that like many other copper-containing compounds, pure synthetic $Cu_{12}Sb_4S_{13}$ has at least a small portion of ionic conductivity. Electromigration of Cu ions in TE materials can deteriorate their performance and irreversibly alter chemical composition. Further study of the relationship between impurity substitution and ion migration in tetrahedrites is thus required, as are investigations of thermal stability at high operating temperatures. If these problems can be addressed successfully, there are good prospects for the use of these materials in large-scale power generation applications.

Acknowledgments

This work was supported as part of the Center for Revolutionary Materials for Solid State Energy Conversion, an Energy Frontier Research Center funded by the U.S. Department of Energy, Office of Science, Office of Basic Energy Sciences under Award Number DE-SC0001054.

References

1. Shi, X., J. Yang, J. Salvador et al. 2011. Multiple-filled skutterudites: High thermoelectric figure of merit through separately optimizing electrical and thermal transports. *J. Am. Chem. Soc.* 133: 7837–7846.
2. Shi, X., H. Kong, C. P. Li et al. 2008. Low thermal conductivity and high thermoelectric figure of merit in n-type $Ba_xYb_yCo_4Sb_{12}$ double-filled skutterudites. *Appl. Phys. Lett.* 92: 182101.
3. Liu, W., X. Tang, H. Li, J. Sharp, X. Zhou, and C. Uher. 2011. Optimized thermoelectric properties of Sb-doped $Mg_{2(1+z)}Si_{0.5-y}Sn_{0.5}Sb_y$ through adjustment of the Mg content. *Chem. Mater.* 23: 5256–5263.
4. Liu, W., Q.J. Zhang, K. Yin et al. 2013. High figure of merit and thermoelectric properties of Bi-doped $Mg_2Si_{0.4}Sn_{0.6}$ solid solutions. *J. Solid State Chem.* 203: 333–339.
5. G. Dennler, R. Chmielowski, S. Jacob et al. 2014. Are binary copper sulfide/selenides really new thermoelectric materials? *Adv. Energy Mater.* 4: 1301581.
6. Skoug, E. J., J. D. Cain, and D. T. Morelli. 2011. High thermoelectric figure of merit in the Cu_3SbSe_4-Cu_3SbS_4 solid solution. *Appl. Phys. Lett.* 98: 261911.
7. Skoug, E. J., J. D. Cain, and D. T. Morelli. 2010. Thermoelectric properties of the Cu_2SnSe_3–Cu_2GeSe_3 solid solution. *J. Alloys Compd.* 506: 18–21.
8. Liu, R., L. Xi, H. Liu, X. Shi, W. Zhang, and L. Chen. 2012. Ternary compound $CuInTe_2$: A promising thermoelectric material with diamond-like structure. *Chem. Commun.* 48: 3818–3820.
9. Plirdpring, T., K. Kurosaki, A. Kosuga et al. 2012. Chalcopyrite $CuGaTe_2$: A high-efficiency bulk thermoelectric material. *Adv. Mater.* 24: 3622–3626.
10. Qiu, P., T. Zhang, Y. Qiu, X. Shi, and L. Chen. 2014. Sulfide bornite thermoelectric material: A natural mineral with ultralow thermal conductivity. *Energy Environ. Sci.* 7: 4000–4006.
11. Makovicky, E. and S. K. Moller. 1994. Exploratory studies on substitution of minor elements in synthetic tetrahedrite. *Neues Jahrb. Miner. Abh.* 167: 89–123.
12. Pfitzner, A., M. Evain, and V. Petricek. 1997. $Cu_{12}Sb_4S_{13}$: A temperature-dependent structure investigation. *Acta Cryst. B* 53: 337–345.
13. Slack, G. A. 1995. New materials and performance limits for thermoelectric cooling. In *CRC Handbook of Thermoelectrics* (ed. D. M. Rowe), CRC Press, Boca Raton, FL, Chap. 34. p. 407.
14. Cahill, D. G. and R. O. Pohl. 1987. Thermal conductivity of amorphous solids above the plateau. *Phys. Rev. B* 35: 4067–4073.
15. Braun, D. J. and W. J. Jeitschko. 1980. Preparation and structural investigations of antimonides with the $LaFe_4P_{12}$ structure. *J. Less Common Met.* 72: 147–156.
16. Zhuze, V. P., V. M. Sergeeva, and E. L. Shtrum. 1958. Semiconducting compounds with the general formula ABX_2. *Sov. Phys. Tech. Phys.* 3: 1925.
17. Morelli, D. T., V. Jovovic, and J. P. Heremans. 2008. Intrinsically minimal thermal conductivity in cubic I-V-VI2 semiconductors. *Phys. Rev. Lett.* 101: 035901.
18. Skoug, E. J. and D. T. Morelli. 2011. Role of lone-pair electrons in producing minimum thermal conductivity in nitrogen-group chalcogenide compounds. *Phys. Rev. Lett.* 107: 235901.
19. Nielsen, M. D., V. Ozolins, and J. P. Heremans.2013. Lone pair electrons minimize lattice thermal conductivity. *Energy Environ. Sci.* 6: 570–578.
20. Slack, G. A. 1979. The thermal conductivity of nonmetallic crystals. In *Solid State Physics* (eds. H. Ehrenreich, F. Weitz, and D. Turnbull, Academic Press, New York), pp. 1–71.
21. Lara-Curzio, E. Unpublished data. Oak Ridge National Lab, Oak Ridge, TN.
22. Fan, X., E. D. Case, X. Lu, and D. T. Morelli. 2013. Room temperature mechanical properties of natural-mineral-based thermoelectrics. *J. Mater. Sci.* 48: 7540–7550.
23. Lara-Curzio, E., A. F. May, O. Delaire et al. 2014. $Cu_2ZnSnS_xO_{4-x}$ and $Cu_2ZnSnS_xSe_{4-x}$: First principles simulations of optimal alloy configurations and their energies. *J. Appl. Phys.* 115: 193515.

24. Wuensch, B. J. 1964. The crystal structure of tetrahedrite $Cu_{12}Sb_4S_{18}$. *Zeit. für Kristallogr.* 119: 437–442.
25. Suekuni, K., K. Tsuruta, T. Ariga, and M. Koyano. 2012. Thermoelectric properties of mineral tetrahedrites $Cu_{10}Tr_2Sb_4S_{13}$ with low thermal conductivity. *Appl. Phys. Exp.* 5: 051201.
26. Lu, X., D. T. Morelli, Y. Xia et al. 2013. High performance thermoelectricity in earth-abundant compounds based on natural mineral tetrahedrites. *Adv. Energy Mater.* 3: 342–348.
27. Lai, W., Y. Wang, D. T. Morelli, and X. Lu. 2015. From bonding asymmetry to anharmonic rattling in $Cu_{12}Sb_4S_{13}$ tetrahedrites: When lone-pair electrons are not so lonely. *Adv. Funct. Mater.* 25: 3648–3657.
28. Johnson, M. L. and R. Jeanloz. 1983. A Brillouin-zone model for compositional variation in tetrahedrite. *Am. Mineral.* 68: 220–226.
29. Di Benedetto, F., G. P. Bernardini, C. Cipriani, C. Emiliani, D. Gatteschi, and M. Romanelli. 2005. The distribution of Cu(II) and the magnetic properties of the synthetic analogue of tetrahedrite: $Cu_{12}Sb_4S_{13}$. *Phys. Chem. Minerals* 32: 155–164.
30. Makovicky, E. and B. J. Skinner. 1979. Studies of the sulfosalts of copper; VII, Crystal structures of the exsolution products $Cu_{12.3}Sb_4S_{13}$ and $Cu_{13.8}Sb_4S_{13}$ of unsubstituted synthetic tetrahedrite. *Can. Mineral.* 17: 619–634.
31. Mozgova, N., V. Mikucionis, V. I. Valiukenas, A. Tsepin, and A. Orliukas. 1987. Some electrical properties of fahlore $Cu_{10}(Zn, Fe)_2(As,Sb)_4S_{13}$. *Phys. Chem. Minerals* 15: 171–172.
32. Suekuni, K., K. Tsuruta, M. Kunii et al. 2013. High-performance thermoelectric mineral $Cu_{12-x}Ni_xSb_4S_{13}$ tetrahedrite. *J. Appl. Phys.* 113: 043712.
33. Heo, J., G. Laurita, S. Muir, M. A. Subramanian, and D. A. Keszler. 2014. Enhanced thermoelectric performance of synthetic tetrahedrites. *Chem. Mater.* 26: 2047–2051.
34. Chetty, R., P. Kumar, G. Rogl et al. 2015. Thermoelectric properties of a Mn substituted synthetic tetrahedrite *Phys. Chem. Chem. Phys.* 17: 1716–1727.
35. Suekuni, K., Y. Tomizawa, T. Ozaki and M. Koyano. 2014. Systematic study of electronic and magnetic properties for $Cu_{12-x}TM_xSb_4S_{13}$ (TM = Mn, Fe, Co, Ni, and Zn) tetrahedrite. *J. Appl. Phys.* 115: 143702.
36. Makovicky, E., K. Forcher, W. Lottermoser, and G. Amthauer. 1990. The role of Fe^{2+} and Fe^{3+} in synthetic Fe-substituted tetrahedrite. *Mineral. Petrol.* 43: 73–81.
37. Lu, X. and D. T. Morelli. 2014. The effect of Te substitution for Sb on thermoelectric properties of tetrahedrite. *J. Electron. Mater.* 43: 1983–1987.
38. Lu, X., D. Morelli, Y. Xia, and V. Ozolins. 2015. Increasing the thermoelectric figure of merit of tetrahedrites by co-doping with nickel and zinc. *Chem. Mater.* 27: 408–413.
39. Lu, X. and D. T. Morelli. 2013. Natural mineral tetrahedrite as a direct source of thermoelectric materials. *Phys. Chem. Chem. Phys.* 15: 5762–5766.
40. Lu, X. and D. T. Morelli 2013. Rapid synthesis of high-performance thermoelectric materials directly from natural mineral tetrahedrite. *MRS Commun.* 3: 129–133.
41. Barbier, T., P. Lemoine, S. Gascoin et al. 2015. Structural stability of the synthetic thermoelectric ternary and nickel-substituted tetrahedrite phases. *J. Alloys Compd.* 634: 253–262.
42. Kallel, A. C., G. Roux, and C. L. Martin. 2013. Thermoelectric and mechanical properties of a hot pressed nanostructured n-type $Si_{80}Ge_{20}$ alloy. *Mater. Sci. Eng. A* 564: 65–70.
43. Ni, J. E., E. D. Case, K. N. Khabir et al. 2010. Room temperature Young's modulus, shear modulus, Poisson's ratio and hardness of PbTe–PbS thermoelectric materials. *Mater. Sci. Eng. B* 170: 58–66.
44. Schmidt, R. D., E. D. Case, J. Giles, J. E. Ni, and T. P. Hogan. 2012. Room-temperature mechanical properties and slow crack growth behavior of Mg_2Si thermoelectric materials. *J. Electron. Mater.* 41: 1210–1216.
45. Schmidt, R. D. 2010. Mechanical Properties of Thermoelectric Skutterudite Compounds, Master thesis, UMI No. 1493371, Michigan State University, East Lansing, MI.

17

Organic Thermoelectric Materials

Gun-Ho Kim and Kevin P. Pipe
University of Michigan

CONTENTS

17.1 Introduction

More than the half of the energy utilized in the United States is immediately rejected as waste heat [1], a significant amount of which exists at low temperatures over large areas. At such temperatures, the conversion efficiency of thermoelectric materials is low, making cost and capability for large-area deposition prime concerns in implementing TE generation. Organic semiconductors (OSCs) satisfy these requirements, given that their base elements are abundant in the earth's crust (e.g., C, H, O) and they offer low manufacturing costs and capability for large-area deposition.

The thermoelectric performance of OSCs has dramatically improved recently, particularly in poly(3,4-ethylenedioxythiophene) (PEDOT) [2–4], increasing their promise for waste heat recovery and solid-state refrigeration. For example, optimization of oxidation level increased ZT (= $S^2\sigma T/\kappa$, where S, σ, κ, and T are the Seebeck coefficient, electrical conductivity, thermal conductivity, and absolute temperature, respectively) to 0.25 in PEDOT doped by tosylate (Tos) at room temperature [2]. Various redox agents [5] or electrochemical transistors [6,7] have been used to optimize carrier concentration, demonstrating large values of $S^2\sigma$ (e.g., 1270 μW m^{-1} K^{-2} [7]). Composites of polymers with materials having large S (e.g., Te nanowires [8] or Bi_2Te_3 powders [9]) or large σ (e.g., carbon nanotubes [10] or graphene [11]) have shown a large enhancement in $S^2\sigma$ with miniscule increase in κ.

Several parameters have been found to be particularly important for improving ZT in OSCs. For example, minimizing dopant volume was found to be critical to maximizing thermoelectric performance in heavily doped polymers [3] and ambient humidity was found to greatly increase S in PEDOT doped by poly(sulfonate) (PSS) [12,13]. In this chapter, fundamental transport parameters and important engineering considerations for maximizing ZT in OSCs are discussed.

17.2 Carrier Localization Length

While OSCs have low κ in general (~0.5 W m^{-1} K^{-1}), their charge carrier mobility μ is low, limiting $S^2\sigma$ and hence ZT. The reason for low μ in OSCs is the localization of charge carrier wave functions caused by disorder that is either static (i.e., impurities or amorphous morphology) or dynamic [14] (i.e., lattice vibrations). The degree of localization, which is quantified by the carrier localization length α, strongly affects both μ and $S^2\sigma$, as shown in Figure 17.1a. For example, PEDOT and polyacetylene, which have $\alpha = 2.1L$ and $4.7L$, respectively (where L is the molecular spacing) [15], have exhibited the highest values of $S^2\sigma$ among OSCs (~1000 μW m^{-1} K^{-2}) [7,16].

Electron paramagnetic resonance is a common method to measure α [19,20] but requires large sample thickness, since its sensitivity is not typically sufficient to detect spins in a thin organic film [21]. Another approach is to utilize the sensitivity of α to S (Figure 17.1b) and derive it from S and σ data over a wide range of carrier concentrations [15].

17.3 Optimization of Carrier Concentration

While ionized impurity scattering typically causes μ to decrease with increasing doping for traditional inorganic thermoelectric materials with extended carrier wave functions, μ increases with n for typical carrier concentrations when carriers are localized. Figure 17.2 shows the Gaussian density of states (DOS) typically observed in OSCs [22], in which the

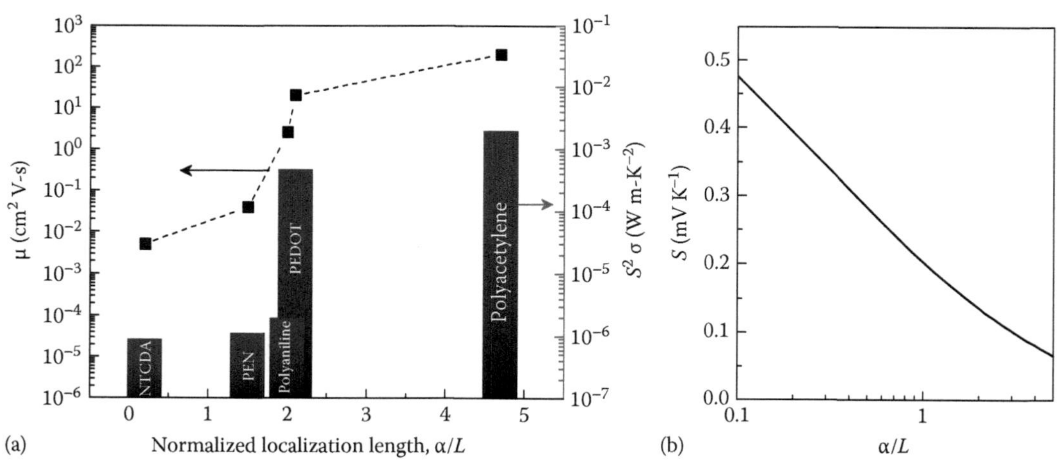

FIGURE 17.1
Carrier mobility (μ), thermoelectric power factor ($S^2\sigma$), and Seebeck coefficient (S) versus carrier localization length (α) normalized by the molecular spacing (L). (a) μ and $S^2\sigma$ in various organic semiconductors: naphthalenetetracarboxylic dianhydride [17], pentacene [18], polyaniline [16,19], poly(3,4-ethylenedioxythiophene) [3], and polyacetylene [16] as a function of α/L. α was extracted by fitting a model to measured S data [15]. Both μ and $S^2\sigma$ rapidly increase with α/L, demonstrating α to be a critical indicator when evaluating the potential of organic thermoelectric materials. (b) S as a function of α/L, calculated for constant n/N_0 [15], showing its sensitivity to α.

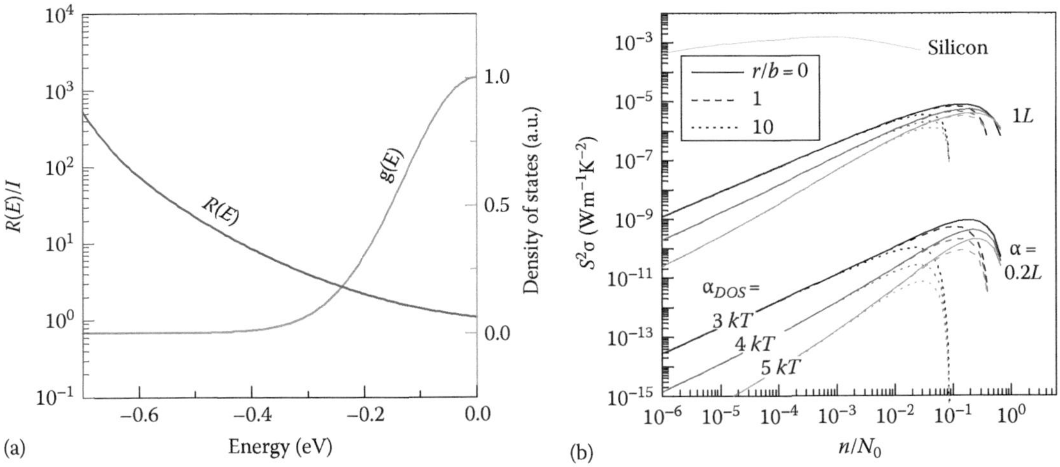

FIGURE 17.2
Gaussian density of states $g(E)$, averaged hopping distance $R(E)$, and $S^2\sigma$. (a) At high energy, the number of available states exponentially increases, and hence the averaged hopping distance decreases (Equation 17.2). (b) The mobility enhancement at large n leads to a very large optimum n that maximizes $S^2\sigma$, which is much greater than that of inorganic materials [15].

number of available states rapidly increases at high energies. As carrier concentration increases, the Fermi energy moves to a higher energy; since the number of available states is then larger for the majority of carriers that participate in transport, the average hopping distance between these (localized) states is reduced, making μ increase. This enhancement in μ for increasing n results in $S^2\sigma$ being maximized at a large value of n (close to 10% of the total DOS). Reaching such a high n requires a large number of ionized dopants; practically, dopants must be as efficient as possible (in terms of the number of contributed carriers per dopant molecule) since nonionized dopants present a steric hindrance to dopant ionization. PSS, Tos, and iodine are relatively efficient dopants for PEDOT (PSS and Tos) and polyacetylene (iodine); together with their large α (Section 17.1), this large efficiency has resulted in large values of $S^2\sigma$.

The small dielectric constants of OSCs make charge carriers associated with dopants more bound to the dopant species, limiting dopant efficiency. For this reason, the ratio of the number of mobile carriers to the number of dopants is small and given by [23]

$$\frac{n}{n_d} = b = \exp(-E_a/kT) = \exp\left[-\left(E_{a,\max} - \beta n_d^{1/3}\right)/kT\right] \quad (17.1)$$

where $E_{a,\max}$ is the maximum activation energy of the dopant at dilute concentrations (approximately equal to its exciton binding energy), and β represents the dependence of E_a on dopant concentration.

Given a typical exciton binding energy (~0.5 eV) for OSCs, a typical fraction of mobile carriers per dopant (b) is 10^{-3} at low concentrations. For example, heavily doped pentacene contains 3.5 times more iodine dopants than host pentacene molecules [24], and highly conductive commercial PEDOT contains 2.5, 6, or even 20 times as many PSS dopant molecules as PEDOT molecules. The large volume of dopants (most of which are nonionized) significantly increases hopping distance between conductive PEDOT molecules and hence reduces μ.

17.4 Minimization of Dopant Volume

Models for electrical conduction in OSCs have included the mechanisms of hopping transport, band transport [25], transport through trap states [26], and charge transfer between small aggregates [27]. Under the hopping transport description, which has successfully described experimental data for various conducting polymer and small molecule samples [15,28,29], the probability of hopping exponentially decays with the distance between hopping sites, which is given by

$$R(E) = \left(\frac{4\pi}{3B} \int_{-\infty}^{E} g(\varepsilon)\left(1 - f(\varepsilon)\right) d\varepsilon \right)^{-1/3} \tag{17.2}$$

where B is the percolation threshold [30], $f(\varepsilon)$ is Fermi–Dirac distribution, and $g(\varepsilon)$ is the Gaussian DOS, given by.

$$g(\varepsilon) = \frac{N_t(r\chi)}{\sqrt{2\pi a_{\text{DOS}}^2}} \exp\left(-\frac{\varepsilon^2}{2a_{\text{DOS}}^2} \right) \tag{17.3}$$

where $N_t(r\chi)$ is the total DOS as a function of dopant volume (= $r\chi$, where r is the ratio of dopant molecule volume to host molecule volume and χ is their number ratio). In OSCs, $r\chi$ is not negligible, and the dopant volume dependent total DOS is given by [15]

$$N_t(r\chi) = \frac{N_0}{1 + r\chi} - \eta(r\chi)N_0 \tag{17.4}$$

where N_0 is the total DOS of the host molecules (without dopants). Small b (Equation 17.1) and large optimum n (~$0.1N_0$) make the dopant volume large in OSCs, which limits $S^2\sigma$. Figure 17.3 illustrates the effect of dopant volume on $S^2\sigma$ in OSCs with different α [15]. The circular shape of the color contour indicates that the dependence of $S^2\sigma$ on $r\chi$ is just as strong as it is on n in OSCs, making $r\chi$ a key metric in maximizing $S^2\sigma$ together with the traditional optimization of n.

PEDOT:PSS is a commercially available conducting polymer blend, and has been used for transparent electrodes [31] and solar cell hole injection layers [32]. It has a large α (= 2.1L) [15] and good stability in air, making it a promising thermoelectric material [9]. For PEDOT:PSS mass fractions typical of commercially available highly conductive PEDOT:PSS, the material falls within the upper right of Figure 17.3d. Since PSS is soluble in a hydrophilic solvent and PEDOT is not, PSS can be selectively removed (dedoped) in spin-cast PEDOT:PSS thin films by immersion in a hydrophilic solvent such as ethylene glycol (EG) [3] for a given time. Figure 17.4a and b shows the decrease in film thickness and

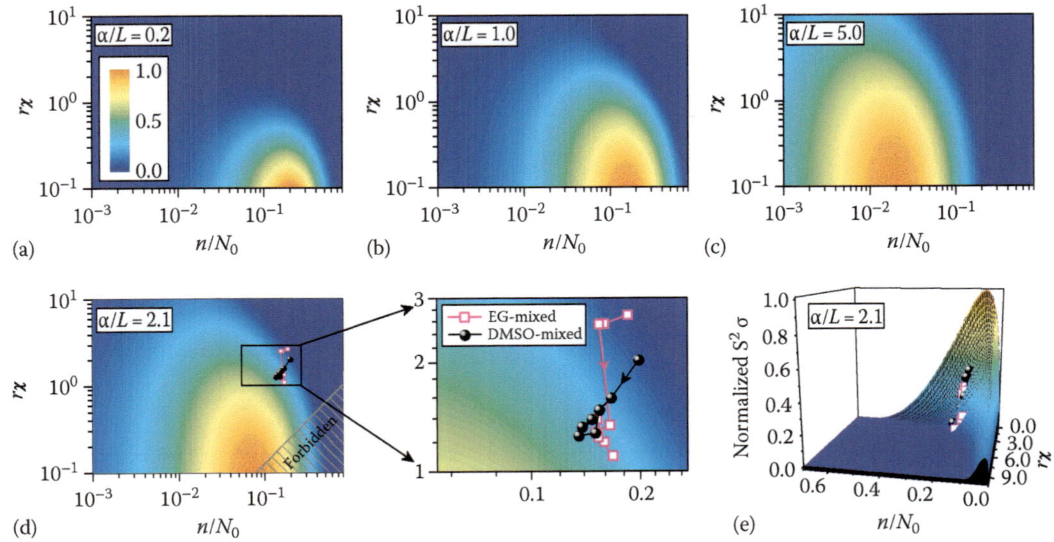

FIGURE 17.3
Two-dimensional trajectory of doping (or dedoping) for efficient maximization of thermoelectric power factor in OSCs. (a–c) Calculations for various values of α. (d) Dedoping trajectory of PEDOT:PSS during EG treatment [3], determined by χ as measured by x-ray photoelectron spectroscopy data and n/N_0 as derived by fitting measured data for *S* to a numerical model [15]. The forbidden area represents $n/N_0 > \chi$ for $r = 1.3$. The arrows in the magnified image represent the time evolution of the data during the dedoping process. (e) A steep increase in $S^2\sigma$ along the measured dedoping trajectory that was realized by reducing both *n* and $r\chi$ [15]. (Reprinted by permission from Macmillan Publishers Ltd., *Nat. Mater.*, Kim, G.-H., Shao, L., Zhang, K., and Pipe, K.P., Engineered doping of organic semiconductors for enhanced thermoelectric efficiency, 12, 719, copyright 2013.)

x-ray photoelectron spectroscopy (XPS) intensity of S_{2p} in PSS during EG treatment, both of which indicate the selective removal of PSS. The monomer number ratio of PEDOT to PSS as measured by XPS and *n* derived by fitting the measured *S* to a model [15] were used to plot the dedoping trajectory of PEDOT:PSS during EG treatment. As can be seen in Figure 17.3d, the dedoping trajectory aims directly at the point of the maximum $S^2\sigma$, leading to a steep increase in $S^2\sigma$ (Figure 17.3e and d) [3]. The dedoping process ends at $\chi = 0.96$ and $n/N_0 = 0.13$, at which $S^2\sigma$ reaches 469 μW m^{-1} K^{-2}. Further optimization of *n* ($= 0.07N_0$) and minimization of dopant volume (*b* close to unity) are predicted to increase $S^2\sigma$ to as high as 1100 μW m^{-1} K^{-2}, which is close to $S^2\sigma = 1270$ μW m^{-1} K^{-2} measured in electrochemical PEDOT transistors [7].

In addition to simultaneously enhancing *S* and σ, selectively removing PSS reduces the cross-plane thermal conductivity of PEDOT:PSS from 0.31 to 0.22 W m^{-1} K^{-1} (Figure 17.4c). This reduction in κ is likely due to an increase in the density of weak van der Waals bonds (which are relatively inefficient at transferring heat [33]) as large PSS molecules (molecular weight of 400,000) are removed and the sample becomes rich in PEDOT (molecular weight of 2,000). Because of the practical challenges of measuring in-plane thermal conductivity in a thin, dedoped polymer film, a 1.4 μm thick PEDOT:PSS film was prepared by multiple spin castings with the same spin conditions as the thin films and exhibited a similar in-plane σ. The in-plane κ of this thick film was measured using the 3ω technique; calculating *ZT* of dedoped PEDOT:PSS using a thermal conductivity anisotropy ratio derived from this value yields a *ZT* as high as 0.42 at room temperature,

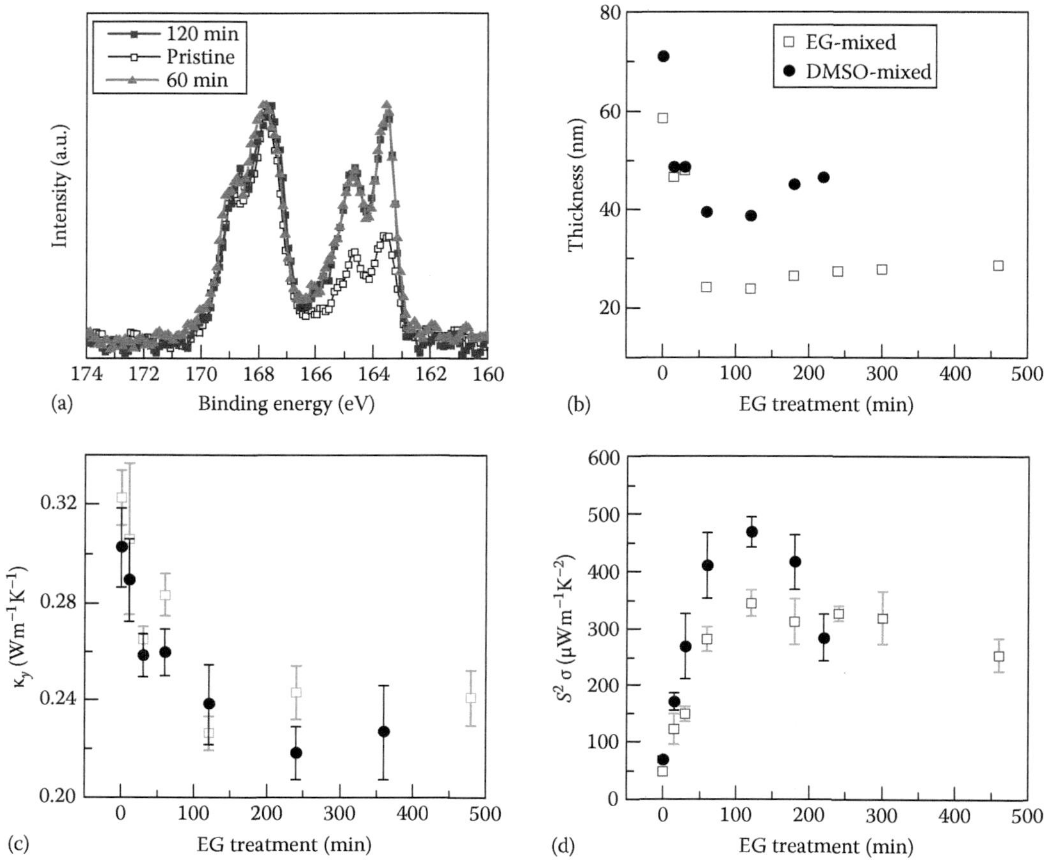

FIGURE 17.4
Effects of the selective removal of poly(sulfonate) (PSS) on the thermoelectric properties of PEDOT:PSS. (a) X-ray photoelectron spectroscopy and (b) thickness data demonstrating the selective removal of PSS by EG treatment. (c) κ and (d) $S^2\sigma$ of PEDOT:PSS at various dedoping times. (From Kim, G.-H. et al., *Nat. Mater.*, 12, 719, 2013.).

which is the highest value yet reported among OSCs. It should be noted that recent studies have observed a large electronic contribution to the thermal conductivity in 20 μm thick PEDOT:PSS [34] and 100 nm thick PEDOT:Tos freestanding films [35], as discussed in Section 17.6.

17.5 Effect of Water on Thermoelectric Properties in PEDOT

In contrast to inorganic crystalline thermoelectric materials, in which dopants replace host atoms and form covalent bonds, dopants in organic thermoelectric materials reside between host molecules and are bound to them through relatively weak electrostatic interactions. Highly polar molecules such as water can further weaken the electrostatic interaction between molecular dopants and host molecules and hence lower the energetic barrier for a redox reaction. Furthermore, water molecules can act as a reducing agent for $PEDOT^+$, given the redox potential of $PEDOT^+:PSS^-$ (~2 V) [6] is large enough to make the oxidation of water ($2H_2O \rightarrow H_2O_2 + 2H^+ + 2e^-$: 1.76 V) thermodynamically favorable [36]. Since *n* in commercially available PEDOT:PSS is typically higher than the optimum concentration (Figure 17.3d), this redox reaction by water increases its thermoelectric performance. Several studies have observed a large increase in *S* under high humidity conditions, suggesting a decrease in *n*. For example, highly conductive PEDOT:PSS (~1000 S cm^{-1}) exhibited an increase from 17 μV K^{-1} at relative humidity (RH) = 30% to 65 μV K^{-1} at RH > 90% [12], which resulted in a large value of $S^2\sigma$ (~370 μW m^{-1} K^{-2}). This increase in *S* was found to be even more profound in PEDOT:PSS samples with relatively low σ, from 10 μV K^{-1} at RH = 10% to 215 μV K^{-1} at RH = 80% [13], leading to four orders of magnitude enhancement in $S^2\sigma$.

Given the fact that PSS is hydrophilic and PEDOT is not, the amount of PSS in PEDOT:PSS is expected to strongly affect the extent to which high humidity enhances $S^2\sigma$. Furthermore, reduction in the elastic modulus of PEDOT:PSS is expected at high humidity [37], with consequent reduction in κ. Figure 17.5 summarizes the humidity dependences of thermoelectric properties in PEDOT:PSS with different amounts of PSS. At RH = 35%, *S* and σ were measured to be 26 μV K^{-1} and 301.5 S cm^{-1} for pristine PEDOT:PSS (PH1000 as purchased from H. C. Starck; 1:3.3 weight ratio as measured by XPS) and were found to increase to 35 μV K^{-1} and 1040 S cm^{-1} after 120 min of dedoping

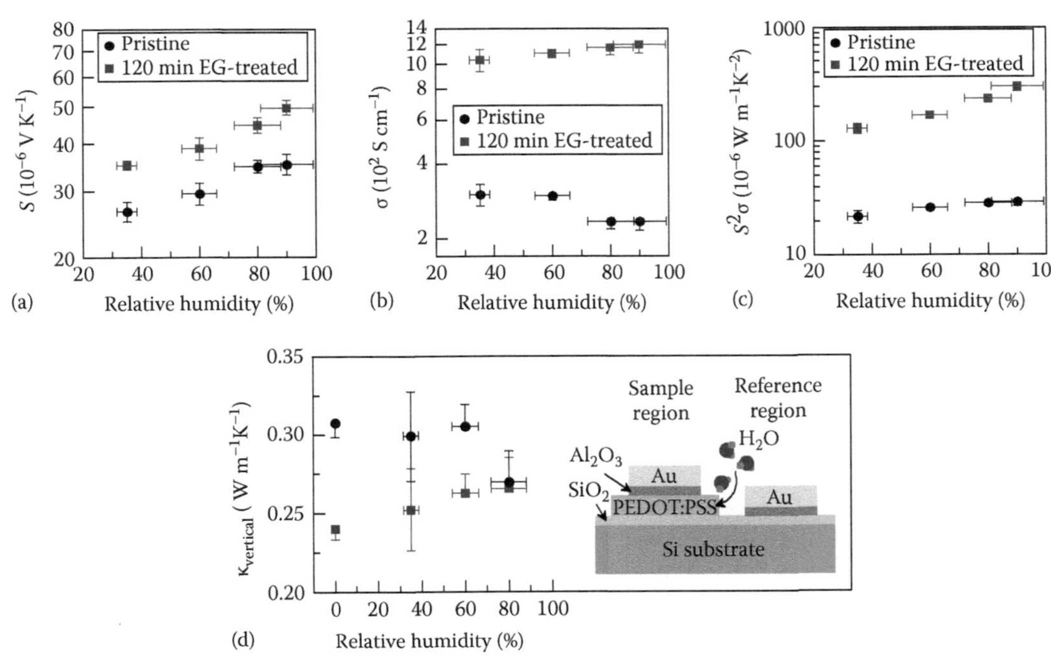

FIGURE 17.5
Humidity-dependent thermoelectric properties of PEDOT:PSS with different amounts of poly(sulfonate). (a) *S*, (b) σ, (c) $S^2\sigma$, and (d) κ measured for pristine and 120 min EG-treated (dedoped) PEDOT:PSS at various humidity conditions. For data measured at RH ≠ 35%, relative changes in the film thickness due to water-induced swelling (assumed to be proportional to previously measured weight increase at high humidity [12]) were used to calculate the relative error bars for σ and κ with respect to the data measured at RH = 35%. Inset in (d) illustrates the differential 3ω technique used for κ measurements.

by EG treatment (which reduced the weight ratio to 1:1.6 as measured by XPS). At higher humidity conditions, both pristine and dedoped PEDOT:PSS exhibited large increases in S (to 35 and 49 $\mu V\ K^{-1}$, respectively), which indicates a reduction in n at high humidity. On the other hand, σ exhibited different trends depending on the amount of PSS: σ in pristine PEDOT:PSS decreased from 300 S cm^{-1} at RH = 35% to 235 S cm^{-1} at RH = 90%, while it increased in dedoped PEDOT:PSS from 1040 S cm^{-1} at RH = 35% to 1203 S cm^{-1} at RH = 90%.

Several competing effects arise in the properties that make up σ (= $n\mu q$, where q is the unit charge) upon water absorption, including a reduction in n (as evidenced by increase in S) by a redox reaction between $PEDOT^+$ and water molecules, a decrease in μ due to an increase in hopping distance caused by water-induced swelling, an increase in μ due to a reduction in the binding energy between PSS and PEDOT caused by highly polar water molecules, and an increase in μ caused by realignment of PEDOT chains upon solvent annealing [38].

For PEDOT:PSS with a large amount of PSS ($r\chi$ = 3.3), analysis of S and σ data using a thermoelectric transport model [15] indicates that the decrease in σ is primarily due to a reduction in n (which is apparent due to the increase in S). μ is found to increase slightly, indicating that water-induced swelling effects that reduce μ by increasing hopping distance are outweighed by other effects (e.g., reduction of binding energy) that increase it. For PEDOT:PSS with less PSS ($r\chi$ = 1.6), the increase in σ with increasing humidity suggests that the reduction in n (as evidenced by an increase in S) is overwhelmed by an increase in μ. This increase in μ is expected to derive from a reduction in binding energy and/or realignment of PEDOT molecules; reduction in μ due to water-induced swelling is expected to be relatively small due to the smaller amount of (hydrophilic) PSS.

While measuring μ in heavily doped PEDOT:PSS over a wide range of humidity is difficult, interpreting S and σ data using a model [15] can provide meaningful insights regarding n and μ. The dielectric constant of PEDOT:PSS can be used to estimate α, which is expected to be on a similar order as the Bohr radius [39]. The average dielectric constant ($\varepsilon_{average}$) in PEDOT:PSS with water is given by $(\varepsilon_{PEDOT})^{\phi PEDOT}(\varepsilon_{water})^{\phi water}$ [40], where ϕ = mass fraction = volume fraction, since the mass density of PEDOT:PSS (~1 g cm^{-3}) is similar to that of water. The relative mass of water with respect to PEDOT:PSS has been measured to be 0.3 at high humidity [12]; $\varepsilon_{average}$ is then calculated to be 3.9. Assuming the increase in hopping distance due to water-induced swelling is isotropic (= $1.3^{1/3}$) and taking the Bohr radius (= $\varepsilon h^2/4\pi^2 q\mu_0$, where h is Planck's constant and μ_0 is the reduced mass of charge carriers) estimated for $\varepsilon_{average}$ = 3.9 as α [39], the model [15] predicts a factor of 4 increase in μ and a factor of 2.5 reduction in n under high humidity conditions, with respect to their values at RH = 35%. The fact that μ is enhanced more than n is reduced explains the measured simultaneous increase of S and σ.

Like σ, the dependence of κ on humidity depends on the amount of PSS. For PEDOT:PSS with less PSS (i.e., pristine), κ was observed to decrease at high humidity, which can be attributed to a decrease in the elastic constant [37] since κ is proportional to its square root [41]. For EG-treated (dedoped) PEDOT:PSS, the amount of (hydrophilic) PSS is small, and κ was observed to increase slightly at high humidity. Given the increase in σ at high humidity, this increase in κ may be caused by an increase in heat transport by charge carriers. Further study is required to quantitatively determine the electronic contribution to κ, as available techniques for measuring in-plane κ are not applicable at present to 20 nm thick films under controlled humidity.

17.6 Thermal Conductivity

The thermal conductivities of conducting polymers are typically low, which is favorable to thermoelectric efficiency. Enhancements in *ZT* for PEDOT have mainly arisen due to methods for increasing $S^2\sigma$ [2,3]. Reported values of κ in these studies remain small (~0.3 W m^{-1} K^{-1}) as σ increases, consistent with other experimental studies [42] and models predicting a violation of the Wiedemann–Franz law in such material systems [43]. In-plane κ has been measured to be small (~0.2 W m^{-1} K^{-1}) even in highly conductive PEDOT:PSS samples [44]. In contrast, other recent studies have found that the Lorenz number is close to the Sommerfeld value in 20 μm thick PEDOT:PSS freestanding films [34] or even larger than the Sommerfeld value in 100 nm thick PEDOT:Tos freestanding films [35], which leads to a large value of the in-plane κ (greater than 1 W m^{-1} K^{-1}). Given the uncertainty in current understanding regarding this topic at present, ideally one should measure all three thermoelectric parameters (*S*, σ, and κ) in the same sample and in the same direction to accurately determine *ZT*.

17.7 Conclusions

The thermoelectric energy conversion efficiencies of OSCs have recently exhibited great improvements, with further gains possible through emerging strategies for molecular design, synthesis, and deposition. Previously we have discussed thermoelectric transport effects related to doping and humidity that are particular to OSCs and strongly impact their conversion efficiency. Low cost and nontoxicity, other properties which many OSCs possess, could prove to be at least as important as energy conversion efficiency for thermoelectric device applications.

Acknowledgments

This work was supported by the U.S. Department of Energy, Office of Basic Energy Sciences, as part of the Center for Solar and Thermal Energy Conversion in Complex Materials, an Energy Frontier Research Center (DE-SC0000957). It was also supported by the Samsung Global Research Outreach program under Agreement No. 86091294. Gun-Ho Kim acknowledges a PISET fellowship from the University of Michigan Energy Institute. Both authors acknowledge the Lurie Nanofabrication Facility and Electron Microbeam Analysis Laboratory at the University of Michigan for sample preparation and characterization.

References

1. LLNL. 2014. Estimated U.S. energy use in 2014: ~98.3 Quads. Lawrence Livermore National Laboratory, Livermore, CA. https://flowcharts.llnl.gov. Accessed on January 17, 2016.
2. Bubnova, O., Z. U. Khan, A. Malti et al. 2011. Optimization of the thermoelectric figure of merit in the conducting polymer poly(3,4-ethylenedioxythiophene). *Nat. Mater.* 10: 429–433.

3. Kim, G.-H., L. Shao, K. Zhang, and K. P. Pipe. 2013. Engineered doping of organic semiconductors for enhanced thermoelectric efficiency. *Nat. Mater.* 12: 719–723.
4. Zhang, Q., Y. Sun, W. Xu, and D. Zhu. 2014. Organic thermoelectric materials: Emerging green energy materials converting heat to electricity directly and efficiently. *Adv. Mater.* 26: 6829–6851.
5. Park, H., S. H. Lee, F. S. Kim, H. H. Choi, I. W. Cheong, and J. H. Kim. 2014. Enhanced thermoelectric properties of PEDOT: PSS nanofilms by a chemical dedoping process. *J. Mater. Chem. A* 2: 6532–6539.
6. Bubnova, O., M. Berggren, and X. Crispin. 2012. Tuning the thermoelectric properties of conducting polymers in an electrochemical transistor. *J. Am. Chem. Soc.* 134: 16456–16459.
7. Park, T., C. Park, B. Kim, H. Shin, and E. Kim. 2013. Flexible PEDOT electrodes with large thermoelectric power factors to generate electricity by the touch of fingertips. *Energy Environ. Sci.* 6: 788–792.
8. See, K. C., J. P. Feser, C. E. Chen, A. Majumdar, J. J. Urban, and R. A. Segalman. 2010. Water-processable polymer-nanocrystal hybrids for thermoelectrics. *Nano Lett.* 10: 4664–4667.
9. Zhang, B., J. Sun, H. E. Katz, F. Fang, and R. L. Opila. 2010. Promising thermoelectric properties of commercial PEDOT:PSS materials and their Bi_2Te_3 powder composites. *ACS Appl. Mater. Interfaces* 2: 3170–3178.
10. Moriarty, G. P., K. Briggs, B. Stevens, Ch. Yu, and J. C. Grunlan. 2013. Fully organic nanocomposites with high thermoelectric power factors by using a dual-stabilizer preparation. *Energy Technol.* 1: 265–272.
11. Zhang, K., Y. Zhang, and S. R. Wang. 2013. Enhancing thermoelectric properties of organic composites through hierarchical nanostructures. *Sci. Rep.* 3: 3448.
12. Wei, Q., M. Mukaida, K. Kirihara, Y. Naitoh, and T. Ishida. 2014. Thermoelectric power enhancement of PEDOT:PSS in high-humidity conditions. *Appl. Phys. Express* 7: 031601.
13. Wang, H., U. Ail, R. Gabrielsson, M. Berggren, and X. Crispin. 2015. Ionic Seebeck effect in conducting polymers. *Adv. Energy Mater.* 5: 1500044.
14. Bohlin, J., M. Linares, and S. Stafstrom. 2011. Effect of dynamic disorder on charge transport along a pentacene chain. *Phys. Rev. B* 83: 085209.
15. Kim, G.-H. and K. P. Pipe. 2012. Thermoelectric model to characterize carrier transport in organic semiconductors. *Phys. Rev. B* 86: 085208.
16. Shakouri, A. and S. Li. 1999. Thermoelectric power factor for electrically conductive polymers. *Proceedings of ICT 1999*, Piscataway, NJ, IEEE Catalog Number 99TH8407, pp. 402–406.
17. Nollau, A., M. Pfeiffer, T. Fritz, and K. Leo. 2000. Controlled n-type doping of a molecular organic semiconductor: Naphthalenetetracarboxylic dianhydride (NTCDA) doped with bis(ethylenedithio)-tetrathiafulvalene (BEDT-TTF). *J. Appl. Phys.* 87: 4340–4343.
18. Kim, G.-H., M. Shtein, and K. P. Pipe. 2011. Thermoelectric and bulk mobility measurements in pentacene thin films. *Appl. Phys. Lett.* 98: 093303.
19. Kahol, P. K., K. K. S. Kumar, S. Geetha, and D. C. Trivedi. 2003. Effect of dopants on electron localization length in polyaniline. *Synth. Met.* 139: 191–200.
20. Marumoto, K., S. Kuroda, T. Takenobu, and Y. Iwasa. 2006. Spatial extent of wave functions of gate-induced hole carriers in pentacene field-effect devices as investigated by electron spin resonance. *Phys. Rev. Lett.* 97: 256603.
21. Vardeny, Z. V. 2010. *Organic Spintronics*. Boca Raton, FL: CRC Press.
22. Yogev, S., E. Halpern, R. Matsubara, M. Nakamura, and Y. Rosenwaks. 2011. Direct measurement of density of states in pentacene thin film transistors. *Phys. Rev. B* 84: 165124.
23. Gregg, B. A., S. G. Chen, and R. A. Cormier. 2004. Coulomb forces and doping in organic semiconductors. *Chem. Mater.* 16: 4586–4599.
24. Brinkmann, M., V. S. Videva, A. Bieber et al. 2004. Electronic and structural evidences for charge transfer and localization in iodine-doped pentacene. *J. Phys. Chem. A* 108: 8170–8179.
25. Sakanoue, T. and H. Sirringhaus. 2010. Band-like temperature dependence of mobility in a solution-processed organic semiconductor. *Nat. Mater.* 9: 736–740.

26. Nicolai, H. T., M. Kuik, G. A. H. Wetzelaer et al. 2012. Unification of trap-limited electron transport in semiconducting polymers. *Nat. Mater.* 11: 882–887.
27. Noriega, R., J. Rivnay, K. Vandewal et al. 2013. A general relationship between disorder, aggregation and charge transport in conjugated polymers. *Nat. Mater.* 12: 1037–1043.
28. Pasveer, W. F., J. Cottaar, C. Tanase et al. 2005. Unified description of charge-carrier mobilities in disordered semiconducting polymers. *Phys. Rev. Lett.* 94: 206601.
29. Fishchuk, I. I., A. Kadashchuk, S. T. Hoffmann et al. 2013. Unified description for hopping transport in organic semiconductors including both energetic disorder and polaronic contributions. *Phys. Rev. B* 88: 125202.
30. Stauffer, D. and A. Aharony. 1992. *Introduction to Percolation Theory*, 2nd ed. London, U.K.: Taylor & Francis.
31. Xia, Y. J., K. Sun, and J. Y. Ouyang. 2012. Solution-processed metallic conducting polymer films as transparent electrode of optoelectronic devices. *Adv. Mater.* 24: 2436–2440.
32. Kim, Y. H., C. Sachse, M. L. Machala, C. May, L. Muller-Meskamp, and K. Leo. 2011. Highly conductive PEDOT:PSS electrode with optimized solvent and thermal post-treatment for ITO-free organic solar cells. *Adv. Funct. Mater.* 21: 1076–1081.
33. Kim, G.-H., D. Lee, A. Shanker et al. 2015. Significant electronic thermal conductivity in amorphous polymer blends by engineered interchain interactions. *Nat. Mater.* 14: 295–300.
34. Liu, J., X. J. Wang, D. Y. Li, N. E. Coates, R. A. Segalman, and D. G. Cahill. 2015. Thermal conductivity and elastic constants of PEDOT:PSS with high electrical conductivity. *Macromolecules* 48: 585–591.
35. Weathers, A., Z. U. Khan, R. Brooke et al. 2015. Significant electronic thermal transport in the conducting polymer poly(3,4-ethylenedioxythiophene). *Adv. Mater.* 27: 2101–2106.
36. Bard, A. J., R. Parsons, J. Jordan, and International Union of Pure and Applied Chemistry. 1985. *Standard Potentials in Aqueous Solution*, 1st edn., Monographs in Electroanalytical Chemistry and Electrochemistry. New York: M. Dekker.
37. Lang, U., N. Naujoks, and J. Dual. 2009. Mechanical characterization of PEDOT:PSS thin films. *Synth. Met.* 159: 473–479.
38. Yeo, J. S., J. M. Yun, D. Y. Kim et al. 2012. Significant vertical phase separation in solvent-vapor-annealed poly(3,4-ethylenedioxythiophene):poly(styrene sulfonate) composite films leading to better conductivity and work function for high-performance indium tin oxide-free optoelectronics. *ACS Appl. Mater. Interfaces* 4: 2551–2560.
39. Pollak, M. and B. I. Shklovskii. 1991. *Hopping Transport in Solids, Modern Problems in Condensed Matter Sciences*. Amsterdam, the Netherlands: Elsevier Science Publishers.
40. Hartshorn, L., N. J. L. Megson, and E. Rushton. 1937. The structure and electrical properties of protective films. *J. Soc. Chem. Ind.* 56: 266–270.
41. Hsieh, W. P., M. D. Losego, P. V. Braun, S. Shenogin, P. Keblinski, and D. G. Cahill. 2011. Testing the minimum thermal conductivity model for amorphous polymers using high pressure. *Phys. Rev. B* 83: 174205.
42. Yan, H., N. Sada, and N. Toshima. 2002. Thermal transporting properties of electrically conductive polyaniline films as organic thermoelectric materials. *J. Therm. Anal. Calorim.* 69: 881–887.
43. Casian, A. 2010. Violation of the Wiedemann-Franz law in quasi-one-dimensional organic crystals. *Phys. Rev. B* 81: 155415.
44. Hagino, H., M. Hokazono, H. Anno, N. Toshima, and K. Miyazaki. 2014. Thermal transport properties of pedot-pss thin films. Paper read at the *15th International Heat Transfer Conference*, Kyoto, Japan.

18

Inorganic/Organic Hybrid Superlattice Materials

Kunihito Koumoto*, Ruoming Tian*, Ronggui Yang, and Chunlei Wan*****
**Toyota Physical and Chemical Research*
***University of Colorado*
****Tsinghua University*

CONTENTS

18.1 Introduction

Thermoelectric materials have been mostly inorganics based on metals, alloys, and inorganic compounds for a long time, but recently we have witnessed a rapid growth of organics for near room-temperature applications [1–5]. High thermoelectric performance of organic materials appear to be mainly due to their low thermal conductivity, and the guiding principle to get high ZT (= $S^2\sigma T/\kappa$, S: Seebeck coefficient, σ: electrical conductivity, κ: thermal conductivity, T: absolute temperature) is to enhance the power factor, $S^2\sigma$, of low thermal conductivity organics. This is completely opposite to that for exploiting inorganic thermoelectric materials, which usually possess high power factor but also high thermal conductivity.

Accordingly, nanoscale hybridization (not a mere composite) of inorganic and organic compounds would be promising to realize high power factor and low thermal conductivity simultaneously to get high ZT. We have recently found that TiS_2/organic hybrid superlattices demonstrate good n-type TE properties at 300–400 K with ZT ~0.3, and we proposed that this material can be combined with p-type organic materials to fabricate flexible devices and modules for energy harvesting applications [6,7].

18.2 Transition Metal Dichalcogenide: TiS_2

Transition metal dichalcogenides (TMDCs) with 2D layer structures are well known to show various electromagnetic and optical properties derived from their specific band structures. Applications of these compounds to nanomaterials and devices have been widely researched recently [8] because a material that is just a few atoms thick can have very different fundamental properties from a material made of the same molecules in solid form [9]. Among these TMDCs, TiS_2 is a very interesting compound for a thermoelectric material due to its unique band structure, and we have investigated this compound for more than 5 years from the point of view of reducing its high thermal conductivity while maintaining its high power factor [10–14].

TiS_2 crystallizes into a CdI_2-type layer structure (trigonal system, space group) with TiS_2 slabs stacking alternately in the *c*-axis direction. The band structure of TiS_2 with a stoichiometric composition, that is, Ti/S = 1/2, possesses an indirect band gap, as shown in Figure 18.1, indicating it is a semimetal [15]. However, a crystal synthesized by a conventional method has a nonstoichiometric composition, $Ti_{1+x}S_2$, in which excess Ti atoms are

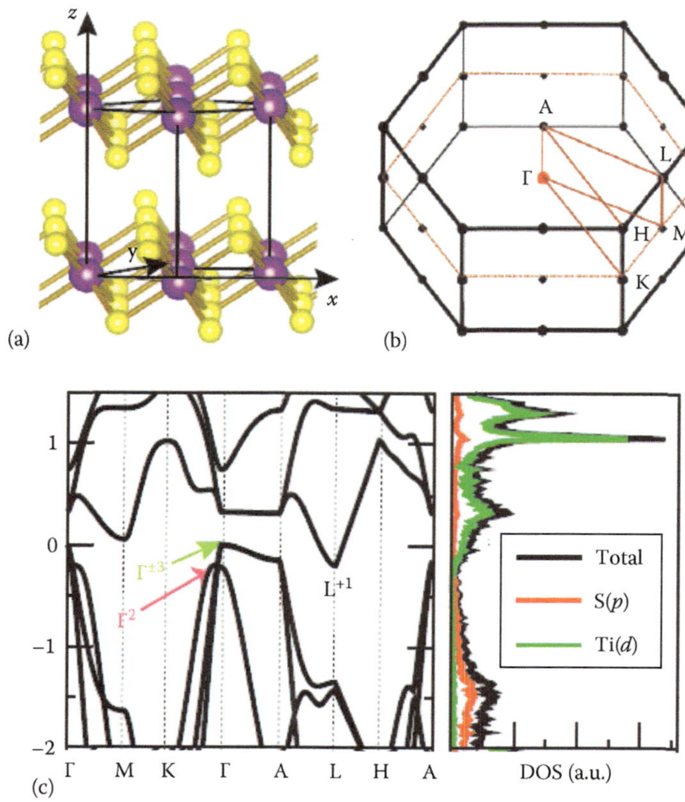

FIGURE 18.1
(a) Unit cell of CdI_2-type TiS_2 crystal structure, (b) corresponding Brillouin zone, and (c) band structure and density of states. (Reprinted with permission from Samanta, A., Pandey, T., and Singh, A.K., *Phys. Rev. B*, 90, 174301. Copyright 2014 by the American Physical Society.)

self-intercalated into the van der Waals gap between TiS_2 layers. Excess Ti atoms become ionized and the released electrons become excited to the conduction band together with the elevation of the Fermi level, and hence it becomes an *n*-type degenerate semiconductor showing metallic conduction. The bottom of the conduction band is composed of triple degenerate Ti3d (t_{2g}) orbitals with high density of states, so that electrons excited to the conduction band would become heavy (effective mass ~4 m_0) and this gives rise to a large Seebeck coefficient. Accordingly, TiS_2 is a good candidate TE material with high power factor even at room temperature. However, its *ZT* is rather low, ~0.16 at 300 K [16], as its thermal conductivity (4–5 W m^{-1} K^{-1} at room temperature) is too high compared with the conventional TE materials with thermal conductivity below 1–2 W m^{-1} K^{-1}.

As the TiS_2 layers are weakly bonded by van der Waals forces, a crystal can be cleaved easily to nanosheets, just like a graphite crystal is easily cleaved to form graphene. The resulting monolayer is called Ti sulphene [17]. When the crystal is reduced to a single layer of TiS_2, the density of states near the conduction band minimum sharply increases, two-dimensionality is very much enhanced (Figure 18.2a) and the Seebeck coefficient increases dramatically with a decreasing number of Ti sulphene layers (Figure 18.2b). So, on the one hand, the 2D confinement of electrons in very thin nanosheets gives rise to a large in-plane

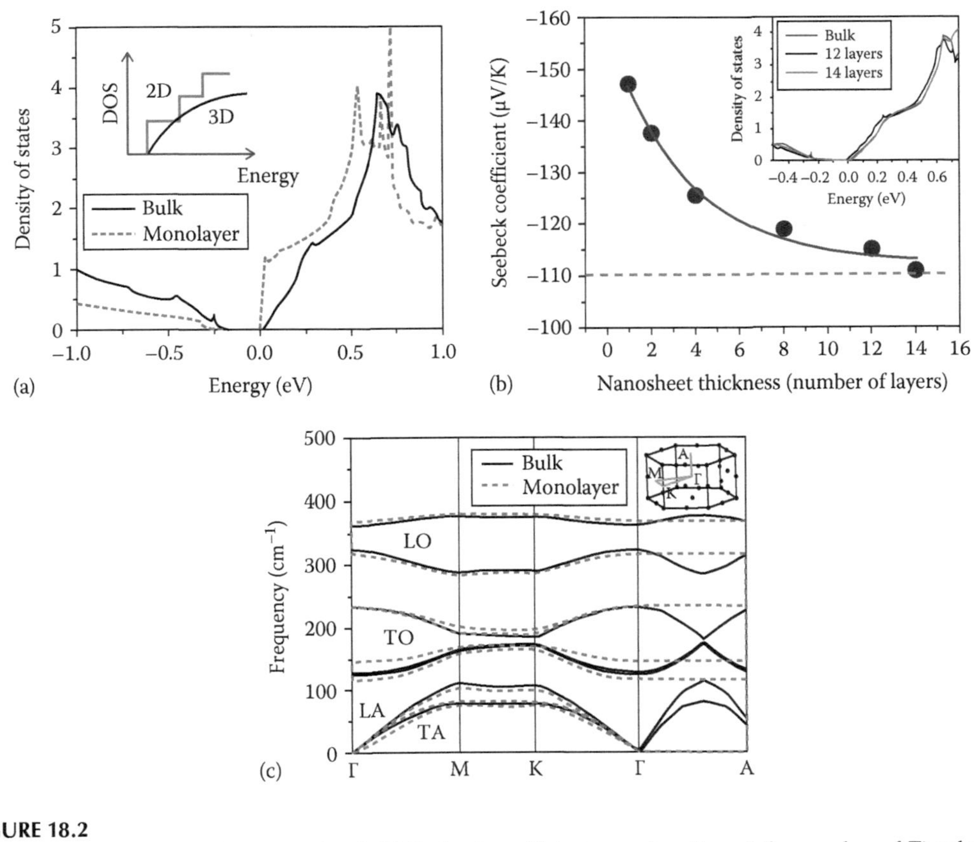

FIGURE 18.2
(a) Density of states near the Fermi level, (b) Seebeck coefficient as a function of the number of Ti sulphene layers, and (c) calculated phonon dispersion for bulk TiS2 (solid line) and TiS2 monolayer (dashed line). (From Zhang, R.Z., Wan, C.L., Wang, Y.F., and Koumoto, K., Titanium sulphene: Two-dimensional confinement of electrons and phonons giving rise to improved thermoelectric performance, *Phys. Chem. Chem. Phys.*, 14, 15641, 2012. Reproduced by permission of The Royal Society of Chemistry.)

Seebeck coefficient, and on the other hand, the 2D confinement of phonons reduces the thermal conductivity as the phonon dispersion is suppressed in thin nanosheets (Figure 18.2c).

Accordingly, titanium sulphene nanosheets can be regarded as very nice functional nanoblocks, and integrating them with other different nanoblocks in a periodic manner would lead to the construction of superlattice structures. For example, metal atoms or metal sulfide nanoblocks can be intercalated into the van der Waals gap between the sulphene layers to form a natural superlattice. This has been proven efficient in enhancing the thermoelectric performance of TiS_2 [10–14]. Another variation is inorganic/organic hybrid superlattices. This is exactly what we have proposed as a new strategy for the next-generation TE materials design.

18.3 Advantages of Inorganic/Organic Hybrid Superlattices

Construction of inorganic/organic hybrid superlattices would provide us with a lot of advantages in enhancing thermoelectric performance through interlayer as well as intralayer interactions:

- *Quantum confinement effect*: Electronic carriers confined in an inorganic layer sandwiched by two insulating organic layers would become heavy and generate a large Seebeck coefficient. Phonons are also confined within an inorganic layer and their two-dimensionally restricted movement would lead to lower thermal conductivity.
- *Chemical interactions*: Inorganic layers and organic layers in a hybrid superlattice would form certain chemical bonds that are different from weak van der Waals bond existing in a pristine inorganic layer compound. This bond formation would affect both electron and phonon transport.
- *Physical interactions*: Electrostatic interactions between carrier electrons (holes) in an inorganic layer and cations (anions) in organic layers would affect the electron transport. Long-range interactions between phonons in an inorganic layer and the dipole moment of polar molecules in organic layers would affect the phonon transport.

It would be possible to enhance *ZT* by deliberately controlling these effects and/or interactions, though they are correlated in a complicated manner. In the following sections, synthesis, composition and structure, and thermoelectric properties of TiS_2-based hybrid superlattices will be described.

18.4 TiS_2/Organic Hybrid Superlattices

18.4.1 Synthesis and Characterization

We synthesized an inorganic/organic superlattice with facile two-step chemical processes, electrochemical intercalation and solvent exchange, and the procedure is shown

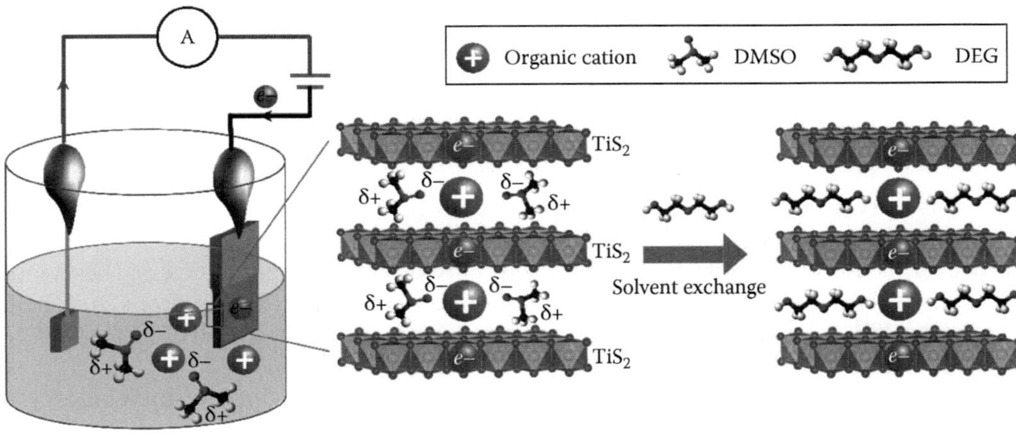

FIGURE 18.3
Electrochemical intercalation of solvated organic cations and hybrid superlattice formation by solvent exchange. (Reprinted with permission from Wan, C.L., Kodama, Y., Kondo, M. et al., Dielectric mismatch mediates carrier mobility in organic-intercalated layered TiS_2, *Nano Lett.*, 15, 6302. Copyright 2015 American Chemical Society.)

in Figure 18.3. TiS_2 single crystals, 3–5 mm in length and 0.1–0.2 mm in thickness, were synthesized in a silica ampoule using the chemical vapor transport method [18] and they were used as the host material. The TiS_2 crystals were used as the cathode in electrochemical reaction cells with the organic salt dissolved in a solvent acting as the electrolyte. When an electrical potential was applied, part of the Ti^{4+} in TiS_2 was reduced to Ti^{3+} and the TiS_2 layers were negatively charged with additional carriers. The organic cations in the electrolyte were intercalated into the van der Waals gap, driven by the Coulomb force, and an inorganic/organic superlattice was formed. Owing to the cation–dipole effect, the solvent molecules are cointercalated with the organic cations, which can be changed into other polar solvent molecules desired by a solvent exchange reaction [19].

Hexylammonium (HA) chloride dissolved in dimethyl sulfoxide (DMSO) was chosen here as the electrolyte. The electrochemical intercalation process was performed under a constant voltage of 1.8 V for 20 min, which resulted in a hybrid superlattice of $TiS_2(HA)_x(DMSO)_w$ (Figure 18.3). The hybrid superlattice was then immersed in a solution of polar molecules to form $TiS_2(HA)_x(PM)_y(DMSO)_z$ (PM = polar molecules) after most of the DMSO molecules were exchanged with polar molecules. Thermogravimetric differential thermal analysis was performed on the samples to study their thermal stability. $TiS_2(HA)_x(DMSO)_w$ was found to have low thermal stability and continuously lost weight when it was heated above room temperature. In contrast, $TiS_2(HA)_x(PM)_y(DMSO)_z$ maintained its weight and was stable up to over 100°C, which is suitable for thermoelectric applications around room temperature.

18.4.2 Carrier Mobility Tuning by the Dielectric Screening Effect

Charge carriers, electrons in this case, move around in the inorganic TiS_2 layers and organic cations (HA) are situated between the inorganic layers, so that the movement of negatively charged electrons is always limited by Coulomb attractive force (Equation 18.1) exerted by positively charged organic cations:

$$F = \frac{1}{4\pi r^2}\frac{1}{\varepsilon_0}q_1 q_2 \qquad (18.1)$$

where F is the Coulomb attractive force, ε_0 is the static dielectric constant of the medium, and r is the distance between two charges (q_1 and q_2).

However, if the polar molecules are present between the inorganic layers, they would affect the electrostatic interactions between organic cations and carrier electrons. According to Equation 18.1, a larger static dielectric constant of polar molecules would weaken the Coulomb attractive force and hence enhance carrier mobility.

To systematically vary the dielectric constants, a series of organic molecules with polar hydroxyl groups including ethylene glycol (EG), diethylene glycol (DEG), and poly(ethylene glycol) of average molecular weight with 200, 400, and 1000 g·mol^{-1} (denoted as polyethylene glycol (PEG) 200, PEG 400, and PEG 1000, respectively) were chosen. The static dielectric constant decreases with an increasing degree of polymerization (molecular weight) as shown in Table 18.1 [20]. The chemical resemblance of these molecules could result in similar structures in terms of the interlayer distance of the inorganic/organic hybrid superlattices (Figure 18.4a), enabling a systematic comparison of their transport properties.

The nuclear magnetic resonance (NMR) analysis confirmed that the organic molecules, including HA and DMSO, had been incorporated into the TiS_2 crystal with a composition of $TiS_2(HA)_{0.09}(DMSO)_{0.15}$ after the electrochemical intercalation process. The interlayer distance between TiS_2 expanded from 5.69 to 13.9 Å. The as-prepared $TiS_2(HA)_{0.09}(DMSO)_{0.15}$ was then used as a precursor by being immersed in the solutions of other polar molecules and heated to 50°C for 24 h to prepare a series of hybrid inorganic/organic superlattices through solvent exchange reactions. The NMR results showed that the target molecules had been successfully incorporated and there was no DMSO residue (see Table 18.1). There were some deviations in HA ion content, which might have been caused by the differences in the sizes and shapes of the original TiS_2 single crystals. The X-ray diffraction (XRD) result showed that the interlayer distance for all the hybrid materials were almost identical, around 13.9–14.2 Å. The similar XRD patterns indicated that all hybrid inorganic/

TABLE 18.1

Static Dielectric Constants (ε_l) of the Glycols in the Inorganic/Organic Hybrid Superlattices $TiS_2(HA)_x(Glycol)_y$ and the Composition Parameters (x, y) Derived from the Nuclear Magnetic Resonance Analysis

Composition	ε_l	x	y
$TiS_2(HA)_x(EG)_y$	41.2	0.096	0.31
$TiS_2(HA)_x(DEG)_y$	31.5	0.092	0.24
$TiS_2(HA)_x(PEG200)_y$	22.1	0.072	0.18
$TiS_2(HA)_x(PEG400)_y$	17.3	0.087	0.11
$TiS_2(HA)_x(PEG1000)_y$	10.7	0.097	0.028

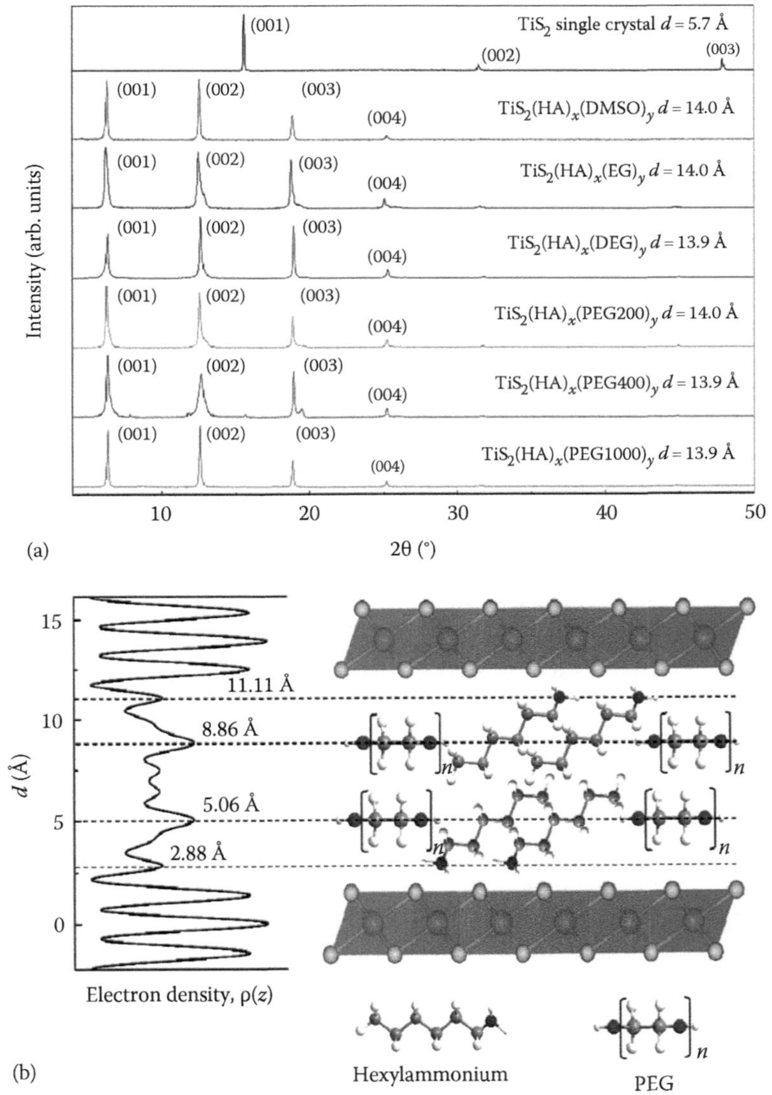

FIGURE 18.4
(a) XRD patterns and interlayer distances of inorganic/organic hybrid superlattices and (b) 1D electron density map along the *c*-axis showing molecular arrangement. (Reprinted with permission from Wan, C.L., Kodama, Y., Kondo, M. et al., Dielectric mismatch mediates carrier mobility in organic-intercalated layered TiS_2, *Nano Lett.*, 15, 6302. Copyright 2015 American Chemical Society.)

organic materials have similar structures. High angle annular dark field scanning transmission electron microscopy (HAADF-STEM) observation of $TiS_2(HA)_{0.097}(PEG1000)_{0.028}$, for example, showed the layered structure with alternating bright and dark areas, where the bright areas represent the inorganic TiS_2 layers and the dark areas are the organic layers. The image clearly indicated that every TiS_2 layer is sandwiched between two organic layers, forming a stage-1 structure [7].

To determine the geometric structures (the size and the steric arrangement) of the intercalated organic molecules, the hybrid material was ultrasonically pulverized and a thin film was prepared by dropping the dispersion solution onto the surface of a CaF_2 substrate that was then analyzed by polarized Fourier transform infrared spectroscopy [7]. It was found that the HA ions and the TiS_2 layers formed an angle of 64° with respect to the normal of the TiS_2 basal plane. Long HA molecules were always well aligned and formed paraffin-like bilayer structures in the intercalation layered compounds due to the hydrophobic interactions [20]. The 1D electron density map along the *c*-axis of the hybrid inorganic/organic material (derived from the Fourier transform of the XRD pattern) showed three sharp peaks appearing at −1.43, 0, and 1.43 Å. These peaks correspond to the triple atomic planes in TiS_2, including a layer of Ti atoms (at 0 Å) sandwiched by two S atom planes (at ∓1.43 Å). The small peaks located between the neighboring strong triple peaks for TiS_2 should be due to the ordered organic molecules. The shape of the electron density map clearly suggested that the layered structure was centrosymmetric, which is consistent with the bilayer model for the HA cations. The small peaks at 2.88 and 11.11 Å corresponded to the ammonium head groups. There were also two relatively strong symmetric peaks at 5.06 and 8.86 Å, which were assigned to the PEG molecules. The PEG molecules that were confined with an interlayer distance of 8.2 Å between two TiS_2 layers could have double layers with planar zigzag conformation or could have adopted a coil conformation. However, if the latter case was true, the electron density map would have had a broadened peak, dispersed halfway between the TiS_2 peaks. The appearance of the two strong symmetric peaks (3.8 Å apart at 5.06 and 8.86 Å) instead clearly indicated that the PEG molecules had a double-layer planar zigzag conformation and the molecular planes were parallel to the TiS_2 layers, as shown in Figure 18.4b.

Figure 18.5a and b shows the carrier concentration and the in-plane Hall mobility, respectively, as a function of the static dielectric constant of the intercalated polar molecule. The Hall mobility was systematically enhanced due to the increase of the dielectric constants of the polar molecules, while the carrier concentration almost remained constant. The change of the carrier mobility is likely dominated by the relaxation time (τ) of the quasi-2D electrons inside the inorganic TiS_2 layers. Pristine TiS_2 singe crystals have a carrier mobility of 7.5 cm^2 V^{-1} s^{-1}, where the electrons are mainly scattered by acoustic phonons [21]. Ionic impurity scattering was introduced after incorporating organic cations (HA ions), due to the Coulomb force between the cations and the electrons, which would decrease the electron mobility. However, the overall electrical conductivity of hybrid materials could be increased due to the significant increase in carrier concentration satisfying the electrical neutrality.

Using Matthiessen's rule, the effective mobility can be expressed as

$$\mu = \left(\frac{1}{\mu_l} + \frac{1}{\mu_i}\right)^{-1} = \left(\frac{1}{\mu_l} + \frac{1}{f\varepsilon^2}\right)^{-1} \tag{18.2}$$

where, μ_l and μ_i denote the mobility by acoustic phonon scattering and impurity scattering, respectively, ε is the average dielectric constant of our hybrid material.

The coefficient *f* is a functional parameter that relates to effective mass, effective charge of impurities, and concentration of impurities.

In our hybrid materials, it can be treated as a constant value [7]. To calculate the effective mobility μ, μ_l was defined as the mobility of pristine TiS_2 single crystal (μ_l = 7.5 cm^2 V^{-1} s^{-1}) [21],

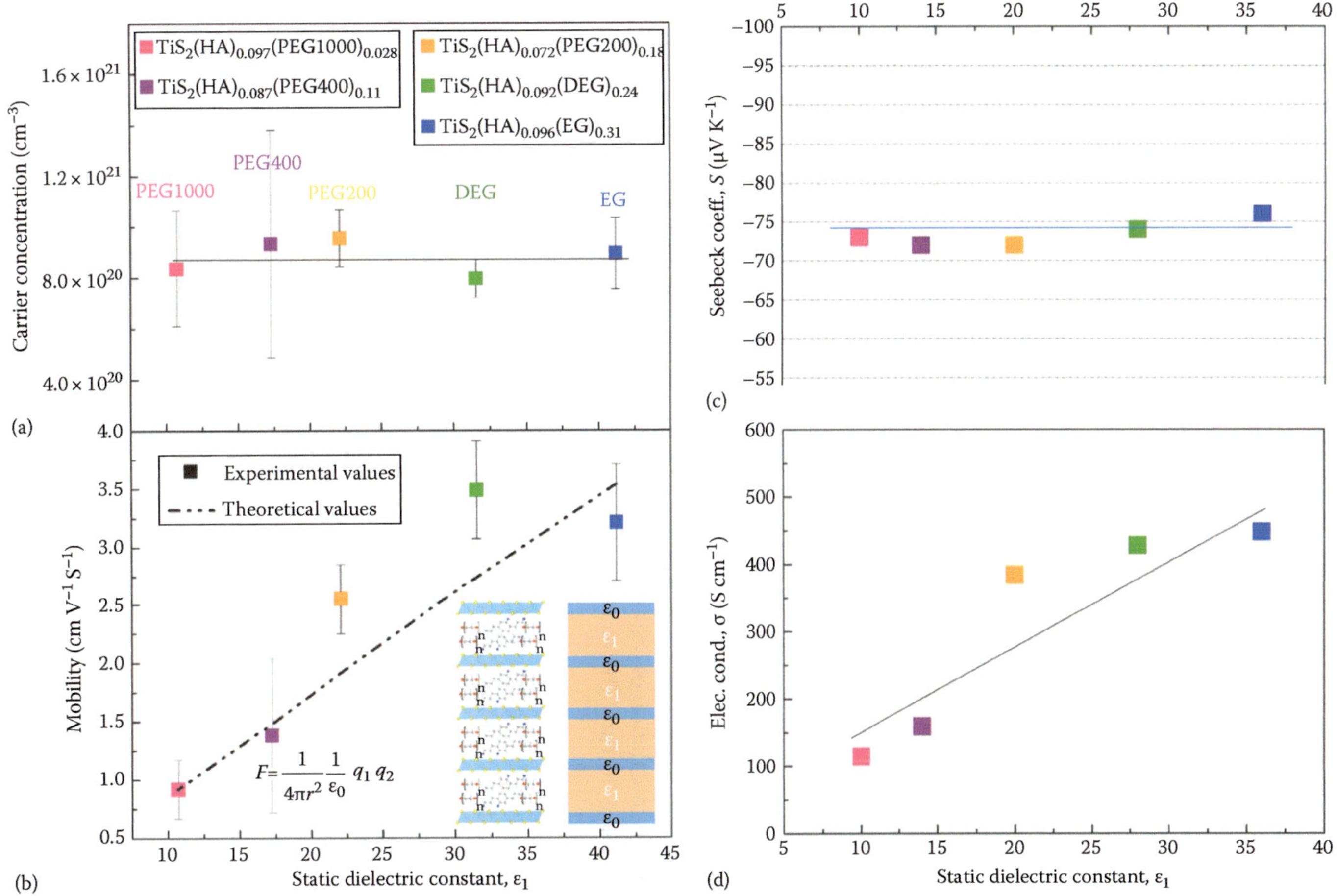

FIGURE 18.5
Transport properties of $TiS_2(HA)_x(PM)_y$ hybrid superlattices as a function of static dielectric constant of polar molecule (PM, ethylene glycol series) at 300 K: (a) carrier concentration, (b) carrier mobility, (c) Seebeck coefficient, and (d) electrical conductivity. (Reprinted with permission from Wan, C.L., Kodama, Y., Kondo, M. et al., Dielectric mismatch mediates carrier mobility in organic-intercalated layered TiS_2, *Nano Lett.*, 15, 6302. Copyright 2015 American Chemical Society.)

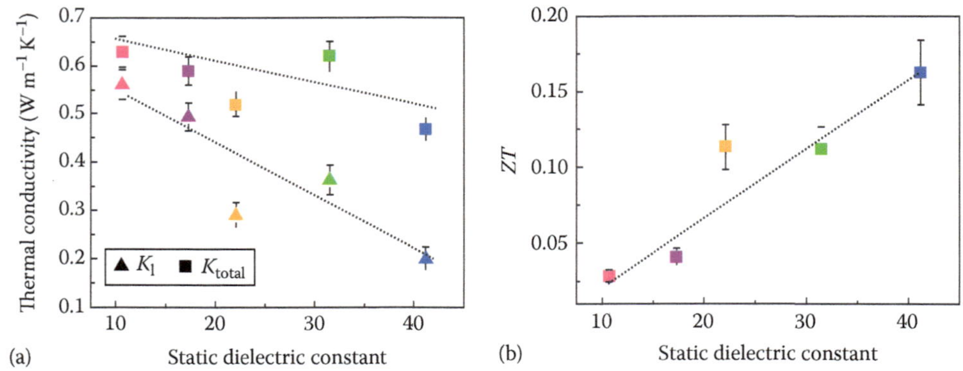

FIGURE 18.6
(a) Thermal conductivity as a function of static dielectric constant of the polar molecule and (b) *ZT* versus static dielectric constant at 300 K. (Reprinted with permission from Wan, C.L., Kodama, Y., Kondo, M. et al., Dielectric mismatch mediates carrier mobility in organic-intercalated layered TiS_2, *Nano Lett.*, 15, 6302. Copyright 2015 American Chemical Society.)

and the coefficient *f* was derived from the mobility of intercalated $TiS_2(HA)_{0.096}(PEG1000)_{0.028}$ crystal, ($f = 0.00737$ cm^2 V^{-1} s^{-1}). The calculated curve of effective mobility (theoretical) is close to a straight line for $10 < \varepsilon_l < 42$ and agrees well with the experimental values of all hybrid inorganic/organic materials, as shown in Figure 18.5b. Such agreement of the mobility from the simple theoretical model and experimental results, indeed, proved that the electronic band structures of these hybrid materials might have only been slightly modified by the intercalation of organic molecules, as articulated earlier, and the dielectric mismatch is the key to tune the electron mobility in these hybrid materials by screening the Coulomb potentials.

Accordingly, the Seebeck coefficient stayed almost constant but the electrical conductivity increased with increasing static dielectric constant of polar molecules (Figure 18.5c and d). This is a very important finding, we believe, because it suggests that we can enhance the power factor of a hybrid superlattice by incorporating polar molecules with high static dielectric constant. It is also interesting to see that the lattice thermal conductivity was very low and decreased with increasing static dielectric constant (Figure 18.6a). This ultralow lattice thermal conductivity would indicate that long-range interactions between the dipole moment in polar molecules and the phonons in TiS_2 layers should affect the phonon scattering process. Consequently, *ZT* was increased with increasing static dielectric constant of polar molecules, and 0.17 at room temperature was obtained for the case of EG, which has the largest static dielectric constant (Figure 18.6b). So, we can conclude that the intercalation of polar molecules with high static dielectric constant enhances carrier mobility and electrical conductivity and greatly reduces lattice thermal conductivity, leading to high *ZT*.

18.4.3 Enhancement of *ZT* by Intercalation of H_2O

Based on the findings described in the previous section, we further tried to get higher *ZT* by intercalating polar molecules with higher static dielectric constant, that is, water molecule ($\varepsilon_0 = 80$). Figure 18.7 shows the in-plane thermoelectric transport properties of

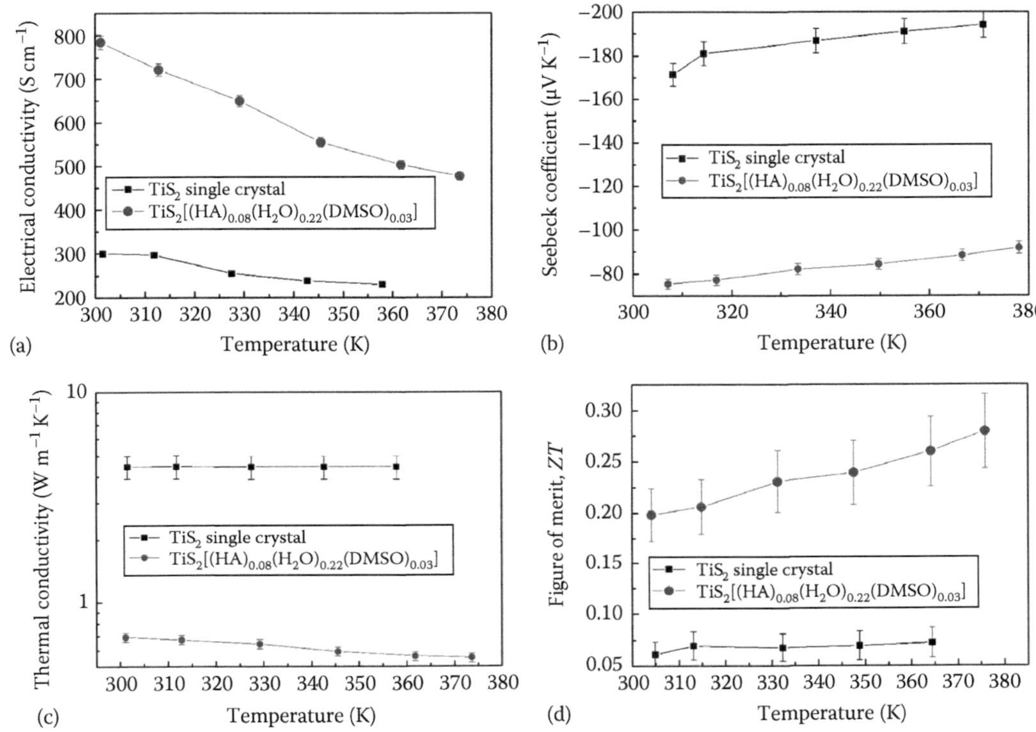

FIGURE 18.7
In-plane thermoelectric transport properties of TiS_2 single crystal and $TiS_2(HA)_{0.08}(H_2O)_{0.22}(DMSO)_{0.03}$ hybrid superlattice: (a) electrical conductivity, (b) Seebeck coefficient, (c) thermal conductivity, and (d) thermoelectric figure of merit, *ZT*. (Reprinted by permission from Macmillan Publishers Ltd., Wan, C.L., Gu, X.K., Dang, F. et al., Flexible *n*-type thermoelectric materials by organic intercalation of layered transition metal dichalcogenide TiS_2, 14, 622, copyright 2015.)

$TiS_2(HA)_{0.08}(H_2O)_{0.22}(DMSO)_{0.03}$ in comparison with those of the single-crystal TiS_2 host material before organic intercalation. During the single-crystal growth, the introduction of interstitial Ti atoms readily occurred in the van der Waals gap in the TiS_2 layers and induced nonstoichiometry, resulting in additional electron carriers in the single-crystal TiS_2 [22]. A carrier concentration of 3.4×10^{20} cm^{-3} was obtained by the Hall measurement for the single crystal. However, even at such high carrier concentration, the Seebeck coefficient is still very high, −171 μV K^{-1} at room temperature. The electrical conductivity decreased and the absolute Seebeck coefficient increased with increasing temperature, exhibiting the behavior of a typical degenerate semiconductor. After the intercalation of the HA ions, the in-plane Seebeck coefficient decreased to −78 μV K^{-1} and the electrical conductivity increased to 790 S cm^{-1} at room temperature. During the electrochemical process, the TiS_2 layers were electrochemically reduced [23], giving rise to Ti^{3+}, and the negative charge was balanced by the intercalated HA ions (i.e., the organic cations) according to the requirement for electrical neutrality. The total carrier density in the hybrid material increased to 7.59×10^{20} cm^{-3} after the electrochemical intercalation. This corresponds to an effective carrier concentration of 1.29×10^{21} cm^{-3} for the TiS_2 layers, if we assume that electrons are confined inside them. The in-plane mobility is calculated to be 6.41 cm^2 V^{-1} s^{-1}, using $\mu = \sigma/ne$, where μ, σ, n, and e are the mobility, electrical conductivity, carrier concentration, and unit electron charge, respectively. This value is comparable to that of the TiS_2 single crystal (7.24 $cm^2\ V^{-1}\ s^{-1}$), which indicates that the electron transport is still confined along the TiS_2 layers in the hybrid superlattice. The positive HA ions could potentially scatter the electrons inside the TiS_2 layers by Coulomb interaction. However, because of the presence of the polar water molecules with a high static dielectric constant that surround the positive HA ions, the Coulomb force can be effectively screened [7,24], and the electron mobility inside the TiS_2 layers can remain almost unaffected. Overall, the in-plane electrical transport properties of the hybrid superlattice exhibited degenerate semiconducting behavior similar to the host single-crystal TiS_2; the electrical conductivity decreases and the absolute Seebeck coefficient increases with increasing temperature. The in-plane power factor of $TiS_2(HA)_{0.08}(H_2O)_{0.22}(DMSO)_{0.03}$ was 0.45 $mW\ m^{-1}\ K^{-2}$ at room temperature, lower than that of TiS_2, 1.05 $mW\ m^{-1}\ K^{-2}$.

The in-plane thermal conductivities of the TiS_2 single crystal and $TiS_2(HA)_{0.08}(H_2O)_{0.22}(DMSO)_{0.03}$ were measured to be 4.45 and 0.69 $W\ m^{-1}\ K^{-1}$ using the parallel thermal conductance method developed for small samples [25], respectively. The experimental value of the TiS_2 single crystal is close to the literature value [16]. As a result of the significant reduction in the thermal conductivity, the in-plane ZT value has been tripled in $TiS_2(HA)_{0.08}(H_2O)_{0.22}(DMSO)_{0.03}$ and reached 0.28 at 373 K, which is the highest value for n-type flexible thermoelectrics and is close to that of the most promising p-type organic material, PEDOT-PSS [26].

18.4.4 Reduction of Thermal Conductivity by Hybridization

It is very interesting that such a low lattice thermal conductivity can be realized in a hybrid superlattice. Both phonons and electrons contribute to the thermal conductivity, and the electronic contribution can be comparable to the phonon contribution in high-efficiency thermoelectric materials. After subtracting the electronic thermal conductivity, which was estimated according to the Wiedemann–Franz law using a calibrated Lorenz constant ($2.40 \times 10^{-8}\ V^2\ K^{-2}$), the in-plane lattice thermal conductivity was estimated to be 4.2 $W\ m^{-1}\ K^{-1}$ for the TiS_2 single crystal and an ultralow value of 0.12 $W\ m^{-1}\ K^{-1}$ was obtained for $TiS_2(HA)_{0.08}(H_2O)_{0.22}(DMSO)_{0.03}$ at 300 K. The ultralow thermal conductivities across inorganic/organic interfaces have been studied for planar self-assembled monolayers [27] and nanocrystal arrays [28]. Low cross-plane thermal conductivity has also been reported for inorganic/organic superlattices [29,30]. However, the in-plane thermal conductivity of inorganic/organic superlattices has not been reported. Assuming that the TiS_2 layers and the organic molecules formed a parallel thermal circuit along the in-plane direction of the hybrid superlattice, the thermal conductivity was estimated to be 2.59 $W\ m^{-1}\ K^{-1}$, almost half of the thermal conductivity of bulk TiS_2, which is much higher than the measured value.

To explain such ultralow thermal conductivity of $TiS_2(HA)_{0.08}(H_2O)_{0.22}(DMSO)_{0.03}$, we conducted equilibrium molecular dynamics simulations with the Green–Kubo method to understand the fundamental phonon transport mechanisms in both bulk TiS_2 and the hybrid superlattice material with HA molecules. In our simulations, we observed an eightfold reduction of the lattice thermal conductivity when the TiS_2 bulk is intercalated by organic components. This reduction is significantly higher than the simple volumetric averaging, indicating that the coupling between the inorganic and organic parts plays an important role in phonon transport in the in-plane direction. As the thermal conduction in the inorganic part is much better than the organic part, the low thermal conductivity of the hybrid material could be attributed to the reduction in the thermal conductivity of the inorganic layer after the intercalation. To qualitatively explain the difference in the phonon dynamics between bulk TiS_2 and the hybrid superlattice, we calculated the spectral energy density [31], $\phi(\mathbf{k},\omega)$, with phonon wavevector $\mathbf{k}$ and frequency ω of the bulk TiS_2 and the TiS_2 layer from the hybrid superlattice, as shown in Figure 18.8. The positions of peaks, ω_0, in the spectral energy density distribution for a given wavevector, $\mathbf{k}_0$, correspond to the phonon dispersion that the material supports, and the broadening of the peak in the frequency domain is related to the phonon relaxation time. The smaller the broadening is, the longer the relaxation time is. In both materials, the three acoustic branches are clearly distinguishable. For all acoustic dispersions, larger broadening can be observed for the hybrid case, indicating stronger scattering in the hybrid superlattice, which can be understood as a result of the inorganic TiS_2 layers being ionically combined with the dangling HA ions. The electrostatic interaction is much stronger than weak van der Waals interaction, which is dominant in the bulk TiS_2.

It is noted that other scattering mechanisms, besides the interfacial coupling, can potentially further reduce the thermal conductivity, which is worthy of further investigation. One possibility is the wavy structure of the layers observed for $TiS_2(HA)_{0.08}(H_2O)_{0.22}(DMSO)_{0.03}$ (Figure 18.9a). As a reference, we measured the thermal conductivity of $TiS_2(HA)_{0.097}(PEG1000)_{0.028}$, where the layers were aligned straight owing to the long chain structure of PEG (Figure 18.9b). The lattice thermal conductivity of the flat hybrid superlattice $TiS_2(HA)_{0.097}(PEG1000)_{0.028}$ was measured to be 0.56 W m^{-1} K^{-1}, four to five times higher than the wavy $TiS_2(HA)_{0.08}(H_2O)_{0.22}(DMSO)_{0.03}$. This can be further explained using the thermal conductivity accumulation as a function of the mean

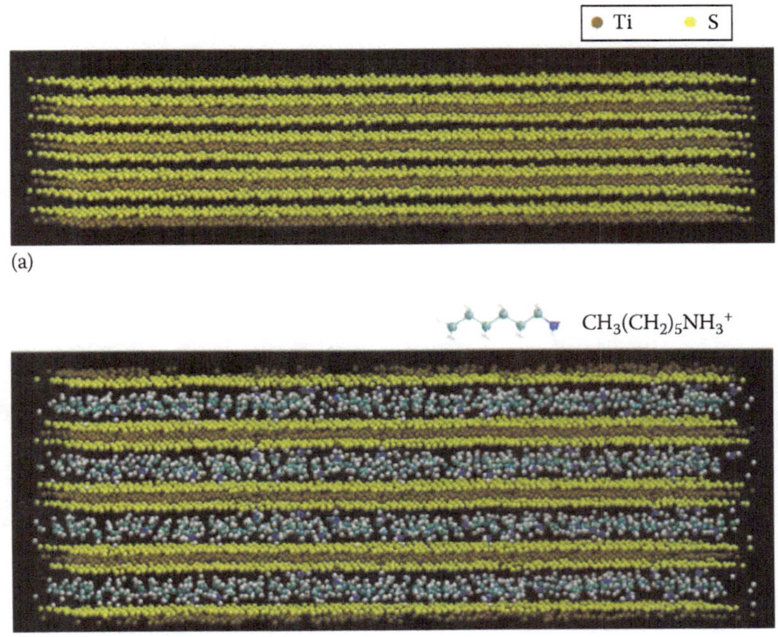

FIGURE 18.8
Molecular dynamics simulations of the thermal conductivities: structural models of (a) TiS_2 and (b) $TiS_2(HA)_x(H_2O)_y(DMSO)_z$. *(Continued)*

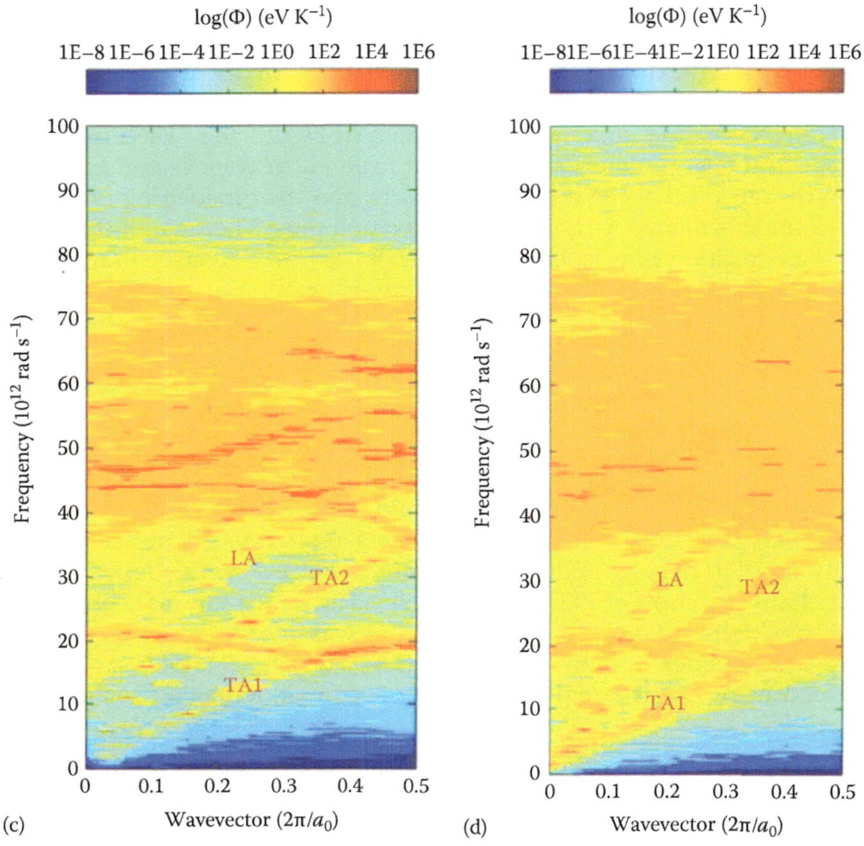

FIGURE 18.8 (*Continued*)
Molecular dynamics simulations of the thermal conductivities: spectral energy densities of (c) TiS_2 and (d) $TiS_2(HA)_x(H_2O)_y(DMSO)_z$). (Reprinted by permission from Macmillan Publishers Ltd., *Nat. Mater.*, Wan, C.L., Gu, X.K., Dang, F. et al., Flexible *n*-type thermoelectric materials by organic intercalation of layered transition metal dichalcogenide TiS_2, 14, 622, copyright 2015.)

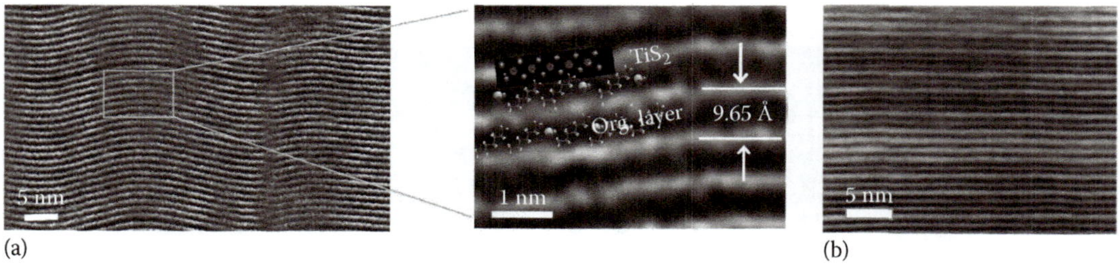

FIGURE 18.9
HAADF-STEM images of (a) $TiS_2(HA)_{0.08}(H_2O)_{0.22}(DMSO)_{0.03}$ and (b) $TiS_2(HA)_{0.097}(PEG1000)_{0.028}$. ([a]: Reprinted by permission from Macmillan Publishers Ltd., *Nat. Mater.*, Wan, C.L., Gu, X.K., Dang, F. et al., Flexible *n*-type thermoelectric materials by organic intercalation of layered transition metal dichalcogenide TiS_2, 14, 622, copyright 2015. [b]: Reprinted with permission from Wan, C.L., Kodama, Y., Kondo, M. et al., Dielectric mismatch mediates carrier mobility in organic-intercalated layered TiS_2, *Nano Lett.*, 15, 6302. Copyright 2015 American Chemical Society.)

free path. The phonons with a mean free path above 10 nm (the half-wavelength of the wavy structure) that contribute to 82% of the total thermal conductivity would be strongly scattered by the wavy structure. In addition, local atomic disordering of the TiS_2 layers in the hybrid superlattice in the wavy structure could be additional phonon scattering mechanism. Compared with the sharp peaks of XRD for TiS_2 single crystals, the peaks of $TiS_2(HA)_{0.08}(H_2O)_{0.22}(DMSO)_{0.03}$ become broader, which is a sign of structural disordering, as further confirmed by HAADF-STEM (Figure 18.9a).

18.4.5 Enhancement of *ZT* by Optimizing Carrier Concentration

The *ZT* values so far obtained for the hybrid superlattices are lower than the estimated maximum value, and there are several issues to overcome in order to further enhance *ZT*. The most important issue to overcome is to gain control over the carrier concentration. It is always too high when the organic cations are fully intercalated together with the electron injection into TiS_2 layers, and hence the Seebeck coefficient is greatly suppressed. It is desirable to reduce somehow the carrier concentration so as to achieve higher Seebeck coefficients and higher power factors.

There are at least two ways to reduce the carrier concentration during processing. One way is to remove the organic cations by heating in vacuum. This seems to work well. The carrier concentration does decrease and the Seebeck coefficient increases. Even though the thermal conductivity greatly increases, possibly due to destruction of the superlattice structure (Figure 18.10a), *ZT* increases from 0.07 (for $TiS_2(HA)_{0.08}(DMSO)_{0.15}$) to 0.2 (for $TiS_2(HA)_{0.025}$) at room temperature (Figure 18.10b).

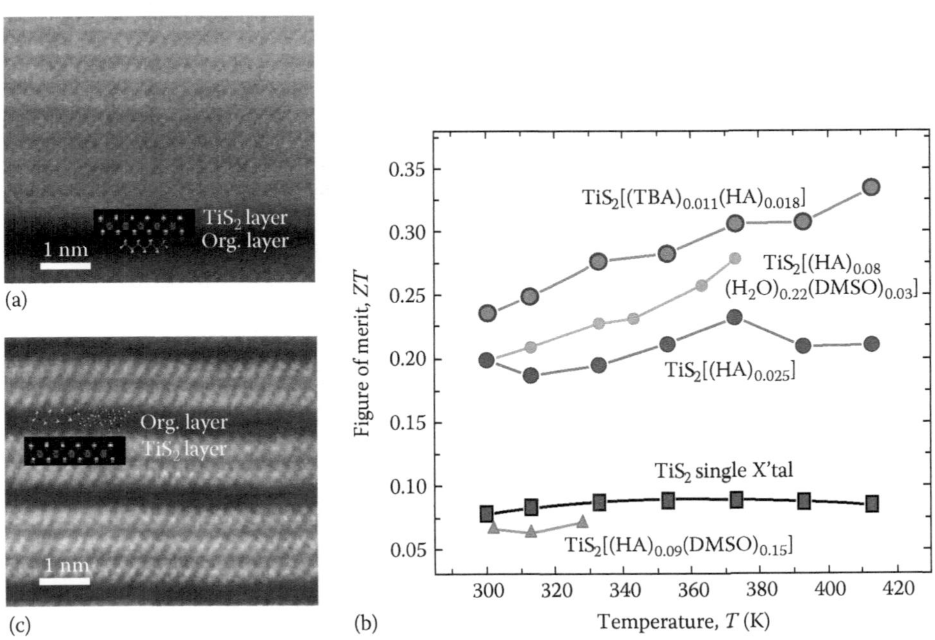

FIGURE 18.10
(a) HAADF-STEM image of the cross section of the $TiS_2(HA)_{0.025}$ hybrid superlattice, (b) temperature dependence of *ZT* for hybrid superlattices with various compositions, (c) HAADF-STEM image of the cross section of the $TiS_2(TBA)_{0.011}(HA)_{0.018}$ hybrid superlattice.

The other way is to intercalate two kinds of organic cations that possess different thermal stability. Tetrabutylammonium (TBA) and HA are first cointercalated into the van der Waals gap of TiS_2, and then HA, which is more thermally unstable, is removed preferentially by heating in vacuum while maintaining a slightly modified superlattice structure (Figure 18.10c). Carrier concentration decreases, the Seebeck coefficient increases, and the electrical conductivity increases as a result. Even though the thermal conductivity appears to be rather high due to certain structural modifications, the *ZT* is greatly enhanced and reaches about 0.33 (for $TiS_2\,(TBA)_{0.011}(HA)_{0.018}$) at 413 K in an ambient atmosphere, as shown in Figure 18.10b.

Although post-annealing of the hybrid superlattice in vacuum can reduce the carrier concentration, it is always accompanied by the modification or destruction of a superlattice structure, often leading to an increased thermal conductivity and, hence, suppressing the enhancement of *ZT*. It is also difficult to adjust the composition to optimize the carrier concentration without destroying the superlattice structure. Accordingly, we should find other possible ways to control (i.e., reduce) the carrier concentration. One way would be to dope electron acceptors with TiS_2 layers, but it should be done together with incorporating polar molecules with high static dielectric constants to enhance the carrier mobility. It might be a worthwhile challenge in the future.

18.5 Flexible *n*-Type Thermoelectrics and Their Novel Applications

Apart from the low thermal conductivity, the hybrid superlattice became mechanically flexible compared with a TiS_2 single crystal. The flexural modulus was measured to be ~145 MPa at 300 K, which is two orders of magnitude lower than that of a TiS_2 single crystal and even lower than a typical plastic material such as Teflon [32]. In the TiS_2 single crystal, the TiS_2 layers are connected by van der Waals forces, but the interlayer interaction is significantly reduced by the expansion of the interlayers, as a result of the intercalation of organic molecules. Moreover, the organic molecules located between neighboring TiS_2 layers could also accommodate deformations and contribute to the flexibility. Therefore, the 2D TiS_2 monolayers could be freely deformed and would be highly flexible.

Organic semiconductors attract increasing interest as flexible thermoelectric materials due to material abundance, easy processing, and low thermal conductivity. Though exciting progress in *p*-type polymers and composites has been reported [33,34], their *n*-type counterpart has fallen behind due to difficulties in *n*-type doping of organic semiconductors. Here, we propose a TiS_2-based inorganic/organic hybrid superlattice for *n*-type flexible thermoelectric materials.

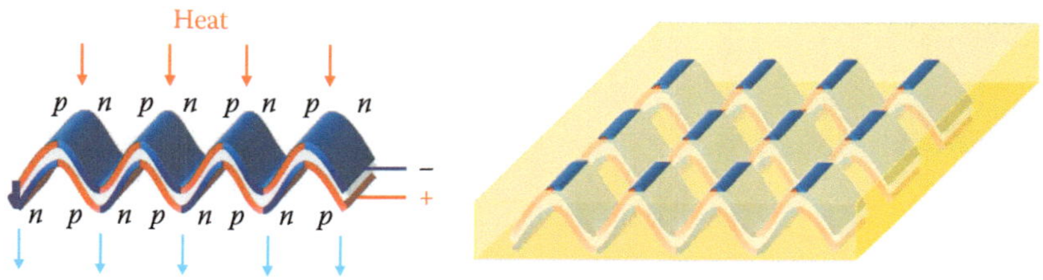

FIGURE 18.11
Example of a new type of device/module made of flexible thermoelectric materials.

Taking advantage of the flexibility of the thermoelectric materials, it would be possible to design new types of devices and modules outside of the domain of conventional solid metal or hard inorganic materials (Figure 18.11). Such new flexible devices would enable us to explore novel applications. High-power flexible solar cells, for example, will be realized hopefully by combining new thin-film solar cells and flexible thermoelectric modules.

Acknowledgments

K. K. acknowledges financial support from JSPS KAKENHI Grant Number 25289226 and TherMAT. R.Y. acknowledges AFOSR (FA9550-11-1-0109). The simulation work used the Janus supercomputer, supported by NSF (0821794). C.W. acknowledges financial support from a Murata Science Foundation Research Grant, a Thermal and Electrical Energy Technology Foundation Research Grant, Takahashi Industrial and Economic Research Foundation and JSPS KAKENHI Grant Number 26820295.

References

1. Leclerc, M. and A. Najari. 2011. Organic thermoelectrics: Green energy from a blue polymer. *Nat. Mater.* 10: 409–410.
2. Nonoguchi, Y., K. Ohashi, R. Kanazawa et al. 2013. Systematic conversion of single walled carbon nanotubes into n-type thermoelectric materials by molecular dopants. *Sci. Rep.* 3: 3344.
3. Suemori, K., S. Hoshino, and T. Kamata. 2013. Flexible and lightweight thermoelectric generators composed of carbon nanotube–polystyrene composites printed on film substrate. *Appl. Phys. Lett.* 103: 153902.
4. Wei, Q. S., M. Mukaida, K. Kirihara, Y. Naitoh, and T. Ishida. 2014. Thermoelectric power enhancement of PEDOT:PSS in high-humidity conditions. *Appl. Phys. Express* 7: 031601.
5. Yan, H. and N. Toshima. 1999. Thermoelectric properties of alternatively layered films of polyaniline and (+/−)-10-camphorsulfonic acid-doped polyaniline. *Chem. Lett.* 28: 1217–1218.
6. Wan, C. L., X. K. Gu, F. Dang et al. 2015. Flexible *n*-type thermoelectric materials by organic intercalation of layered transition metal dichalcogenide TiS_2. *Nat. Mater.* 14: 622–627.
7. Wan, C. L., Y. Kodama, M. Kondo et al. 2015. Dielectric mismatch mediates carrier mobility in organic-intercalated layered TiS_2. *Nano Lett.* 15: 6302–6308.
8. Chhowalla, M., H. S. Shin, G. Eda, L.-J. Li, K. P. Loh, and H. Zhang. 2013. The chemistry of two-dimensional layered transition metal dichalcogenide nanosheets. *Nat. Chem.* 5: 263–75.
9. Gibney, E. 2015. 2D or not 2D. *Nature* 522: 274–276.
10. Guilmeau, E., A. Maignan, C. Wan, and K. Koumoto. 2015. On the effects of substitution, intercalation, non-stoichiometry and block layer concept in TiS_2 based thermoelectrics. *Phys. Chem. Chem. Phys.* 17: 24541–24555.
11. Putri, Y. E., C. L. Wan, Y. F. Wang, W. Norimatsu, M. Kusunoki, and K. Koumoto. 2012. Effects of alkaline earth doping on the thermoelectric properties of misfit layer sulfides. *Scr. Mater.* 66: 895–898.
12. Wan, C. L., Y. F. Wang, W. Norimatsu, M. Kusunoki, and K. Koumoto. 2012. Nanoscale stacking faults induced low thermal conductivity in thermoelectric layered metal sulfides. *Appl. Phys. Lett.* 100: 101913.

13. Wan, C. L., Y. F. Wang, N. Wang, and K. Koumoto. 2010. Low-thermal-conductivity $(MS)_{1+x}(TiS_2)_2$ (M = Pb, Bi, Sn) misfit layer compounds for bulk thermoelectric materials. *Materials* 3: 2606–2617.
14. Wan, C. L., Y. F. Wang, N. Wang, W. Norimatsu, M. Kusunoki, and K. Koumoto. 2010. Development of novel thermoelectric materials by reduction of lattice thermal conductivity. *Sci. Technol. Adv. Mater.* 11: 044306.
15. Samanta, A., T. Pandey, and A. K. Singh. 2014. Strain-induced electronic phase transition and strong enhancement of thermopower of TiS_2. *Phys. Rev. B* 90: 174301.
16. Imai, H., Y. Shimakawa, and Y. Kubo. 2001. Large thermoelectric power factor in TiS_2 crystal with nearly stoichiometric composition. *Phys. Rev. B* 64: 241104.
17. Zhang, R. Z., C. L. Wan, Y. F. Wang, and K. Koumoto. 2012. Titanium sulphene: Two-dimensional confinement of electrons and phonons giving rise to improved thermoelectric performance. *Phys. Chem. Chem. Phys.* 14: 15641–15644.
18. Barry, J. J., H. P. Hughes, P. C. Klipstein, and R. H. Friend. 1983. Stoichiometry effects in angle-resolved photoemission and transport studies of $Ti_{1+x}S_2$. *J. Phys. C: Solid State Phys.* 16: 393–402.
19. Schöllhorn, R. and A. Weiss. 1974. Cation exchange reactions and layer solvate complexes of ternary phases M_xMoS_2. *J. Less Common Met.* 36: 229–236.
20. Sengwa, R. J., K. Kaur, and R. Chaudhary. 2000. Dielectric properties of low molecular weight poly(ethylene glycol)s. *Polym. Int.* 49: 599–608.
21. Wan, C. L., Y. F. Wang, N. Wang, and K. Koumoto. 2011. Intercalation: Building a natural superlattice for better thermoelectric performance in layered chalcogenides. *J. Electron. Mater.* 40: 1271–1280.
22. Kukkonen, C. A., W. J. Kaiser, E. M. Logothetis et al. 1981. Transport and optical properties of $Ti_{1+x}S_2$. *Phys. Rev. B* 24: 1691–709.
23. Schöllhorn, R. and H. Meyer. 1974. Cathodic reduction of layered transition metal chalcogenides. *Mater. Res. Bull.* 9: 1237–1245.
24. Jena, D. and A. Konar. 2007. Enhancement of carrier mobility in semiconductor nanostructures by dielectric engineering. *Phys. Rev. Lett.* 98: 136805.
25. Zawilski, B. M., R. T. Littleton, and T. M. Tritt. 2001. Description of the parallel thermal conductance technique for the measurement of the thermal conductivity of small diameter samples. *Rev. Sci. Instrum.* 72: 1770–1774.
26. Kim, G. H., L. Shao, K. Zhang, and K. P. Pipe. 2013. Engineered doping of organic semiconductors for enhanced thermoelectric efficiency. *Nat. Mater.* 12: 719–723.
27. Losego, M. D., M. E. Grady, N. R. Sottos, D. G. Cahill, and P. V. Braun. 2012. Effects of chemical bonding on heat transport across interfaces. *Nat. Mater.* 11: 502–506.
28. Ong, W.-L., S. M. Rupich, D. V. Talapin, A. J. H. McGaughey, and J. A. Malen. 2013. Surface chemistry mediates thermal transport in three-dimensional nanocrystal arrays. *Nat. Mater.* 12: 410–415.
29. Liu, J., B. Yoon, E. Kuhlmann et al. 2013. Ultralow thermal conductivity of atomic/molecular layer-deposited hybrid organic–inorganic zincone thin films. *Nano Lett.* 13: 5594–5599.
30. Losego, M. D., I. P. Blitz, R. A. Vaia, D. G. Cahill, and P. V. Braun. 2013. Ultralow thermal conductivity in organoclay nanolaminates synthesized via simple self-assembly. *Nano Lett.* 13: 2215–2219.
31. Thomas, J. A., J. E. Turney, R. M. Iutzi, C. H. Amon, and A. J. H. McGaughey. 2010. Predicting phonon dispersion relations and lifetimes from the spectral energy density. *Phys. Rev. B* 81: 081411.
32. Ouyang, J., C.-W. Chu, C. R. Szmanda, L. Ma, and Y. Yang. 2004. Programmable polymer thin film and non-volatile memory device. *Nat. Mater.* 3: 918–922.
33. Bubnova, O., Z. U. Khan, A. Malti et al. 2011. Optimization of the thermoelectric figure of merit in the conducting polymer poly(3,4-ethylenedioxythiophene). *Nat. Mater.* 10: 429–33.
34. Sariciftci, N. S., L. Smilowitz, A. J. Heeger, and F. Wudl. 1992. Photoinduced electron transfer from a conducting polymer to buckminsterfullerene. *Science* 258: 1474–1476.

19

Recent Progress in Skutterudites

G. Jeffrey Snyder*, Yinglu Tang*, Caitlin M. Crawford, and Eric S. Toberer****
**Northwestern University*
***Colorado School of Mines*

CONTENTS

19.1 Introduction

Since the skutterudite mineral (Co, Ni, Fe)As_3 was first discovered in 1845, this structure has proven to be one of the most diverse chemical families with a broad range of electronic, magnetic, and thermal phenomena. The potential of skutterudite materials for thermoelectric applications has been extensively studied and has led to some of the best thermoelectric performance to date [1–3]. On a more fundamental level, the skutterudite family has been fascinating. Electronic transport in these materials spans uncorrelated to highly correlated behavior and phonon transport is unconventional due to the incorporation of guest atoms in large interstitial sites. In this chapter, we focus on recent progress (since 2010) in skutterudites to update prior reviews, including in this book series [4–7]. In this short time period, more than 600 papers have been published; as such, we approach this literature in terms of case examples to highlight particularly intriguing developments rather than an extensive review. In an effort to focus this review, the cobalt antimonide ($CoSb_3$) skutterudite serves as the primary emphasis of this chapter.

We begin with a discussion of chemical advances in skutterudites, including both framework and interstitial chemistry. Due to the differences in charge transport between *n*- and *p*-type compositions, the impact of these chemical modifications in electronic properties is discussed separately. Thermal transport is separated into atomic-scale and microstructural design strategies due to the different governing principles. We finish with a discussion of generator-related advances, including both performance and thermal/mechanical reliability.

19.2 Chemistry

The binary skutterudite structure takes the general form of MX_3 where M represents one of the group IX transition metals (Co, Rh, Ir) and X is of the pnictogen family (P, As, Sb). Skutterudite crystallizes in a body-centered-cubic (BCC) structure in the $Im\bar{3}$ space group. This structure may also be described as a distorted ReO_3 structure where the M–X framework introduces large interstitial voids within the system, as seen in Figure 19.1 by the polyhedron. For the binary $CoSb_3$ compound, this void radius was calculated to be about 1.892 Å [4]. Filler atoms are often placed within these voids to reduce the lattice thermal conductivity *via* phonon scattering and perturbations to the phonon dispersion. The following sections begin with recent advances in filler chemistry and then discuss fundamental studies of the framework and modifications therein.

19.2.1 Filler Chemistry

Filler chemistry and the associated impact on transport properties have been a major focus of skutterudite research since their initial application in thermoelectrics [8]. New ternary compositions continue to be explored and the incorporation of multiple fillers within a single framework creates an infinite search space. As the field matures, we see an increased emphasis on the synthesis environment (e.g., temperature and pressure) and composition. These serve to both enable new chemistry and shed light on the interior dynamics of these materials.

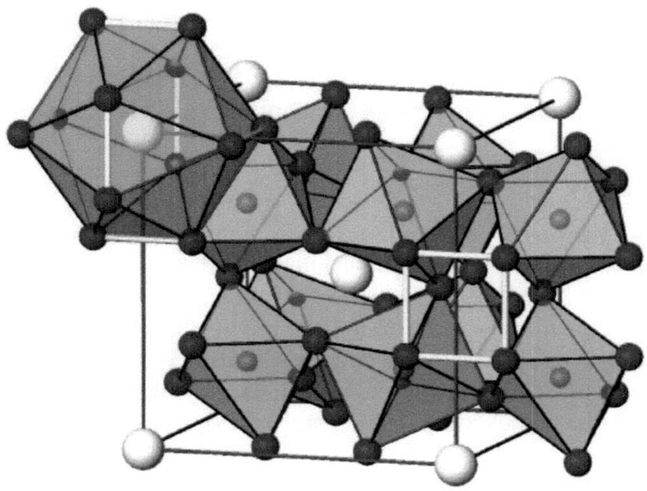

FIGURE 19.1
Structure of skutterudite $A_xCo_4Sb_{12}$ showing the $CoSb_6$ octahedra, the Sb_4 rings, and the icosahedral coordination of the filling atom. The atoms are denoted by colors grey, black, and white for cobalt, antimony, and the filler, A, respectively.

19.2.1.1 Chemical Diversity and Multiple Fillers

The last 5 years have seen an increased emphasis on expanding and understanding the filler chemistry. Historically, the focus has been on electropositive fillers that donate electrons to the framework, such as Na [9], K [10], Sr [11], Ba [12], Yb [13,14], La [15], Ce [16], and Eu [17]; recent filler studies have expanded choices among alkali, alkali earth, halogen, and rare earth elements, such as Li [18], Mg [19], I [20], Sm [21], and Gd [22]. The potential for multiple filler elements to further reduce thermal conductivity was initially pioneered by Yang et al. in 2007 ($Ba_xA_yCo_4Sb_{12}$, A = La, Ce, and Sr) [23]; this field has blossomed with studies of double- [24] and triple-filled [1] compositions. Much of this work has been focused on achieving high *ZT* through the suppression of lattice thermal conductivity. Significant opportunities remain to understand the chemistry of multiple fillers, including local ordering and filler–filler interactions.

19.2.1.2 High Pressure

According to theoretical calculations [25,26], elements with small ionic radii such as Li cannot fill the voids at ambient pressure to any significant concentration. Here, the filling fraction limit (FFL) depends on the competition between the formation of an *A*-filled skutterudite and that of the secondary phase A_mSb with the lowest formation enthalpy (i.e., Li_3Sb in the case of A = Li [18]). However, since enthalpies vary with pressure, high-pressure (3–4 GPa) synthesis could induce Li filling into the skutterudite structure [18]. On the other extreme, high pressures were required to insert iodine into the framework due to its large atomic size [20]. Here, the filler site is quite strained by the large I ion, as observed by the significant lattice expansion. The incorporation of a halogen into the filler site provides a new avenue for electronic and thermal control and is discussed in the following.

Self-insertion of Sb was found in the compound Sb_xCoSb_{3-x} synthesized at high pressure and high temperature (7.7 GPa and 550°C, respectively) [27]. This self-insertion reaction is characterized by the collapse of some framework Sb atoms into the cages formed by the

Co and Sb atoms in the skutterudite structure, as was further confirmed by Bader's analysis of maximum entropy charge density maps obtained from synchrotron radiation x-ray powder diffraction data. The opposite reaction (i.e., $Sb_xCoSb_{3-x} \rightarrow CoSb_3$) occurs, however, when Sb_xCoSb_{3-x} is heated above 180°C at ambient pressure.

19.2.1.3 Filler Solubility

The solubility of each type of filler (A) is usually considered as a single value and often extracted by adding A into the $CoSb_3$ system until a secondary phase precipitates. However, by adding A into $CoSb_3$, a ternary system is under study, which leads to the breakdown of some common assumptions in determining solubility limits in binary systems due to the additional degree of freedom. Solubility thus depends strongly on the phase region of the nominal composition as well as the annealing temperature [28,29]. In other words, the intentional excess or deficiency of Co or Sb has a profound effect on the filling fraction.

To highlight how filler solubility is more complex than initially appreciated, we consider two case examples involving Ce and Yb fillers. For the Ce–Co–Sb ternary system (Figure 19.2), the Sb-rich three-phase region (blue triangle: $Ce_xCo_4Sb_{12}$, $CeSb_2$, and liquid Sb) gives a Ce solubility $x = 0.09$ (similar to previous reported value [16]), which is about 25% less compared to that from the Co-rich three-phase region (red triangle: $CoSb_2$, $CeCoSb_3$,

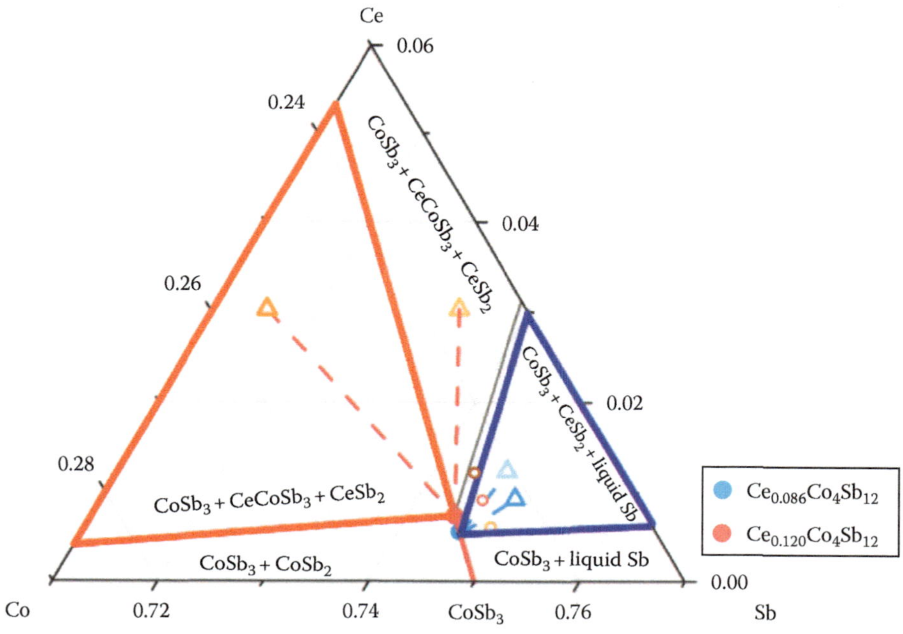

FIGURE 19.2
Isothermal section of Ce–Co–Sb ternary phase diagram at 973 K. In ternary skutterudites, the filling fraction limit can depend on the nominal composition, as shown for Ce–Co–Sb in this figure. The single phase $Ce_xCo_4Sb_{12}$ region is represented as a red solid line extending from $CoSb_3$. The Co-rich three-phase region (red triangle) has a higher filling fraction ($x = 0.12$) than the Sb-rich three-phase region (blue triangle, $x = 0.09$).

and $Ce_xCo_4Sb_{12}$, $x = 0.12$) for the same annealing temperature. As a result, an important solubility design strategy for enhancing the FFL of Ce in $CoSb_3$ is to focus on the Co-rich three-phase region rather than the lower Sb-rich three-phase region. This example also serves to highlight a potential cause of disagreement in the literature concerning filling fractions of skutterudites with other fillers. The temperature dependence of filler solubility was also demonstrated experimentally in Yb-doped $CoSb_3$-based skutterudites [28]. As temperature increases, Yb solubility increases by about a factor of five between 773 and 1073 K ($x = 0.11 \pm 0.02$ at 773 K and $x = 0.49 \pm 0.02$ at 1073 K) [28]. This large variance gives another degree of freedom to tune the carrier concentration and thus allows optimizing thermoelectric properties. Likewise, it can serve to cause disagreements in the literature when samples are quenched.

19.2.1.4 Ordering in Partially Filled Compositions

The traditional view of partially filled skutterudites is that the occupation of the voids is random; however, recent work suggests that entropy may not always dominate over filler–filler interactions. In Ba-filled $CoSb_3$, theoretical calculations indicate that there are two ordered phases (α and γ) [30]. System energetics included the configurational entropy associated with all possible ways of distributing Ba and vacancies over the BCC sites of the filler sublattice. The α phase has a high symmetry with the filled Ba sites forming a diamond network. The γ phase is highly anisotropic with Ba–Ba nearest neighbors along a [111] axis, and Ba vacancy nearest neighbors pair along the other ⟨111⟩ directions. The observance of ordered phases from calculations is not particularly surprising on its own, as calculations struggle to simulate true disorder within finite cells.

Positron annihilation spectroscopy was used to further investigate the potential for filling order in Ba-filled $CoSb_3$ [31]. It was found that for low Ba concentrations, the Ba fillers randomly occupy the structural voids. From $x = 0.16$, the γ phase forms locally and with increasing concentration, a fully ordered γ phase appears at $x = 0.25$. If the Ba composition is further increased in the sample, Ba atoms start to take the diamond positions and the α phase forms at some local places and hence a two-phase mixture of α and γ phases appears.

This evidence for ordering in one partially filled composition has potential implications for both electronic and thermal transport. In partially filled skutterudites where the rigid electronic band approximation breaks down upon incorporation (e.g., $Yb_{0.5}Co_4Sb_{12}$), ordering and associated zone folding may affect the band curvature and degeneracy. For thermal transport, the presumption has been that partial filling can be viewed as a scattering center. However, ordered supercells lead to zone folding and long-range vibrational modes involving the fillers. In such a case, phonon transport may be more akin to fully filled skutterudites. For mixed compositions, molecular dynamics simulations find strong interface scattering at domains of the α and γ phases [31].

19.2.2 Framework Chemistry

While the filler site has received the majority of recent interest, there remains significant potential to control transport in skutterudites *via* the $CoSb_3$ framework. We begin with a review of recent work on the native defects and bonding within $CoSb_3$. Developments in framework alloying are then considered, with an emphasis on aliovalent pairs (e.g., replacing Co with Fe and Ni).

19.2.2.1 Native Defects and Bonding

Understanding the native defects in pure $CoSb_3$ is critical to understanding how the Fermi level varies with synthesis conditions. Recent calculations have estimated defect populations from the internal formation energetics and the chemical potential of the elemental constituents [32]. For Sb-rich conditions, calculations found V_{Co} to be the most common defect, with Sb_i the next highest in energy. Charge neutrality pins the Fermi level to a p-type composition, which is consistent with experiment. Experimental studies of undoped Co-rich $CoSb_3$ have found n-type behavior at room temperature that transitions to p-type at higher temperatures. This unexpected result can be rationalized by defect calculations. Considering isolated defects, as can be expected at high temperature where thermal energy dissociates defect pairs, the dominant defect is Co_i and the Fermi level is pinned near the valence band edge. However, considering defect pairs of Co_i, a slight reduction in defect formation energy is observed, and the Fermi level is shifted close to the conduction band. This work highlights how rich the defect chemistry is in binary $CoSb_3$; much remains to be understood about defects in the more complex skutterudites with high ZT.

The thermoelectric community commonly describes bonding in $CoSb_3$ in a Zintl-like picture, where four Co^{3+} donate electrons to three Sb_4^{4-} rings [33,34]. Recent work has called this simple picture into question through a combination of high-resolution single-crystal diffraction and density functional theory calculations [35]. Refinement of single-crystal data allows the electron density to be resolved and compared with calculations. Common to the Zintl description is a charge buildup between Sb and Sb in the Sb_4 rings. However, the Zintl description fails in several ways: (1) the integrated charge associated with Co is found to be slightly negative, (2) the rings are slightly positively charged, and (3) the Co–Sb interaction shows significant covalent character. These findings are significantly more consistent with the only slight electronegativity contrast (Co: 1.88, Sb: 2.05). In the context of the filler atoms' potential well and associated "rattling" behavior, these findings are particularly significant. Rather than viewing the filler as surrounded by negatively charged Sb-rings, this picture suggests that a more neutral framework surrounds the filler atoms.

19.2.2.2 Framework Alloying

Alterations to the $CoSb_3$ framework allow carrier concentration tuning, decrease the lattice thermal conductivity through phonon scattering, and alter the electronic structure and scattering. The antimony in binary skutterudite can be substituted with various levels of group 14 or group 16 elements [36]; when substituted in equal measure, the carrier concentration is nominally unaffected. Similar strategies have been adopted with the cobalt through substitution with equal portions of iron and nickel [37,38]. Zevalkink et al. showed that stoichiometric equal amounts of Sn and Te on the antimony site produce an ordered ring structure that replace the Sb_4 rings in pure skutterudite as depicted in Figure 19.3. Substituting antimony with various pnictogen atoms (e.g., $ASn_{1.5}Te_{1.5}$ where A = Co, Ir, Rh) was shown to increase density of states near the band edge [39]. While such electron counting schemes have been used in skutterudites since their inception, high throughput systematic approaches have recently leveraged this to examine hypothetical skutterudites. Here, adjusting the filler and framework chemistry reveals compounds that may otherwise be overlooked (e.g., $LaCo_4P_8Si_4$, $BaIr_4Sb_{10}Sn_2$) [34].

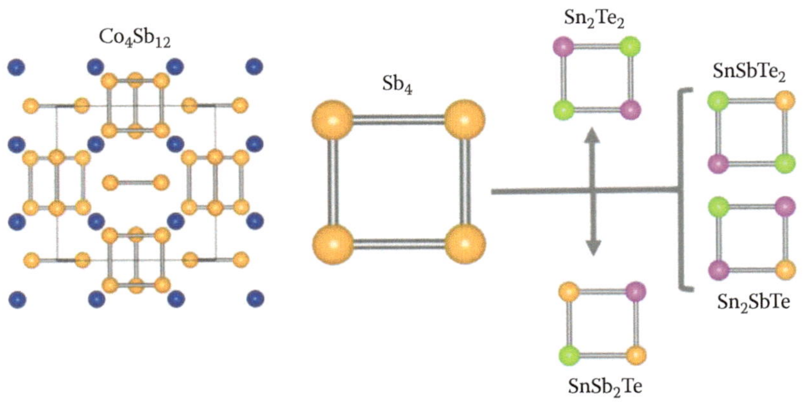

FIGURE 19.3
The Sb_4 rings can be fully replaced with Sn_2Te_2 rings (top) while maintaining charge balance. Lower Sn and Te concentrations may lead to charge imbalanced rings due to the geometric constraints of the rings (right, $SbSn_{1.5}Te_{1.5}$), phase separation, or ternary rings (bottom, $SnSb_2Te$) depending on stoichiometry.

19.3 Electronic Properties

In contrast to many other thermoelectric materials, the high *ZT* found in skutterudites can be largely attributed to the exceptional electronic properties rather than ultralow thermal conductivity (i.e., <0.5 W m^{-1} K^{-1}). Within the skutterudite family, the *n*- and *p*-type electronic transport is quite distinct and serves as the division of this section.

19.3.1 *n*-Type Skutterudites

n-Type filled $A_xCo_4Sb_{12}$ skutterudites have proven to be excellent thermoelectric materials due to their high electronic mobility, heavy effective mass, and low thermal conductivity associated with the addition of filler atoms into the void site. In the following, we explore the interplay between dopants and optimized carrier concentration and consider how *n*-type skutterudites simultaneously achieve high mobility and heavy effective mass.

19.3.1.1 Carrier Concentration Control

A broad range of dopants has been developed to render skutterudite *n*-type. Te and Ni are classic framework donors and have been the subject of extensive investigation. In recent work, the donor is often the filler species (A). Here, the filler nominally provides qx electrons per primitive cell $A_xCo_4Sb_{12}$ (with q the valence state of the filler and x the filling fraction). For skutterudites, the dopant effectiveness, as measured by the Hall effect, is typically low. Figure 19.4 shows that the measured carrier concentration is approximately one-third of the theoretical carrier concentration qx. This is likely due to a combination of compensating defects and a breakdown of assumptions embedded in the Hall effect. The optimum doping range for *n*-type thermoelectric performance is quite high, about 0.4–0.6 electrons per primitive cell, as seen in Figure 19.4 [29].

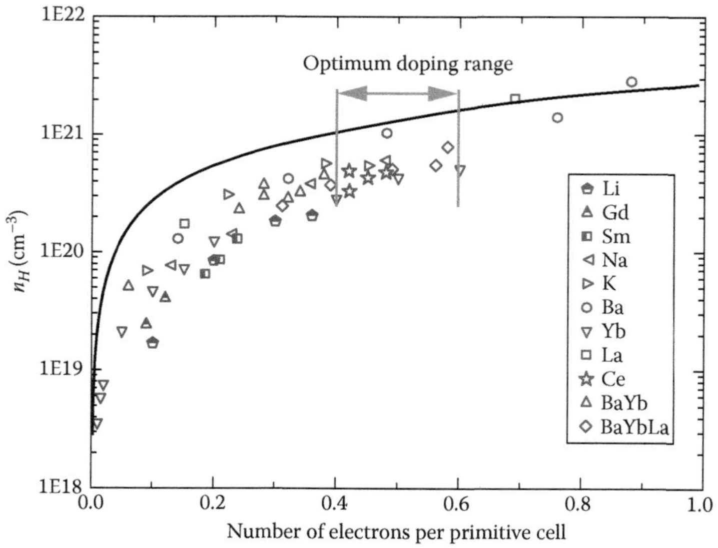

FIGURE 19.4
Many different filling atoms can be used to provide electrons to dope $CoSb_3$ *n*-type. The region of optimal doping ranges from 0.4–0.6 electrons per primitive cell.

Carrier concentration is particularly challenging to unravel when the dopant sits on several sites. For example, Ga and In have similar ionic radii to Sb and serve as acceptors when substituting for Sb. However, these elements will also occupy the interstitial site and serve as donors [40,41]. For Ga-doped $CoSb_3$, both defects form in charge-balanced concentrations, and Ga doping has little impact on charge carrier concentration [40]. In contrast, compound defects with In are not fully charge compensated, with the ratio of void-filling over Sb-substituting larger than two [41].

19.3.1.2 Electronic Sources of High ZT

Until recently, the favorable electronic band structure in *n*-type $CoSb_3$ had typically been attributed to threefold degeneracy at the conduction band minimum accompanied by linear band behavior at higher carrier concentrations. This had been related to the increase in effective mass as the doping level increases. Using combined experimental and computational studies, it was shown instead that a secondary conduction band with 12 conducting carrier pockets, as seen in Figure 19.5 (that probably converges with the primary band at high temperatures), is responsible for the extraordinary thermoelectric performance of *n*-type $CoSb_3$ skutterudites [42]. While the importance of a secondary heavy conduction band in enhancing the power factor is well acknowledged, the addition of doping atoms themselves may alter the band structure [43]. For example, Shi et al. [44] showed that Ga dual-site occupancy breaks the symmetry of the Sb–Sb network and causes high-energy conduction bands to descend within 100 meV of the conduction band edge.

19.3.2 *p*-Type Skutterudites

Historically, *p*-type systems have lower peak *ZT* values than their *n*-type counterparts, with few reports of *p*-type *ZT* above 1. This difference can largely be viewed in terms of

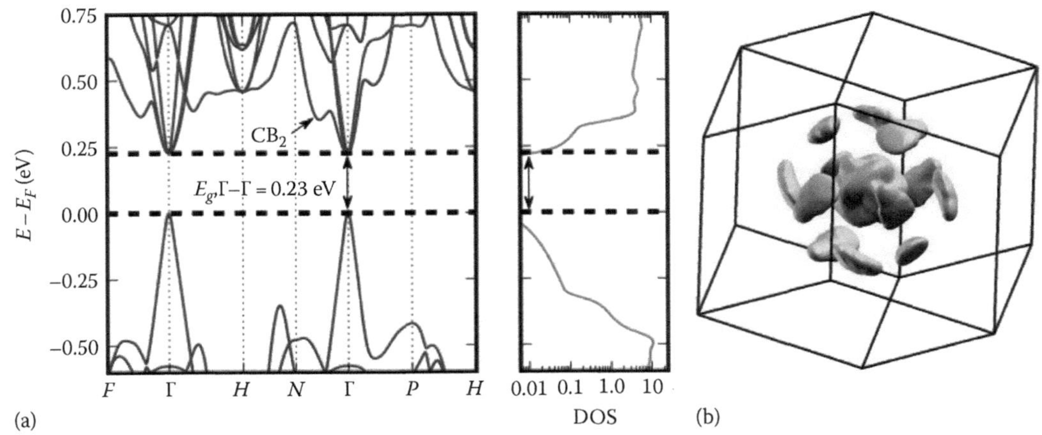

FIGURE 19.5
(a) Electronic band structure of $CoSb_3$ showing the band gap formed between the light bands at Gamma (Γ) and the second conduction band, CB_2, with high density of states that is essential for efficient thermoelectric performance. (b) Fermi surface of $CoSb_3$ with Fermi level for optimum *n*-type performance showing the 12 isolated Fermi pockets due to CB_2 in addition to the surface at Γ.

electronic transport; modifications to the valence band dispersion and charge carrier scattering thus represent critical opportunities that have begun to be explored.

19.3.2.1 Valence Band Structure

For *p*-type $CoSb_3$ skutterudites, the valence band edge has strong anion character associated with Sb lone pairs and their coordination with the adjacent Co. Figure 19.5 shows the valence band edge is defined by a nearly linear band with low effective mass and a band degeneracy of unity. Several other bands reside approximately 500 meV below the band edge, which may come into play at high temperatures or when $CoSb_3$ is sufficiently degenerately h^+ doped. Modifications to the valence band electronic structure can arise from either the incorporation of fillers to add in new bands or altering the framework chemistry to modify the curvature and position of existing bands. For example, recent calculations have shown that splitting the Sb site to Group IV–VI has real potential; secondary bands for the $Co(Sn/Te)_3$ system rise up and potentially enhance band degeneracy [39]. One can also consider "inductive" effects from the guest site on the valence band edge electronic structure. For example, Figure 19.6 highlights the interplay of the top two valence bands for various filler atoms in XFe_4Sb_{12} [45]. For Nb, Ca, Yb, and Na, an anticrossing occurs approximately 200 meV below the band edge that may reduce group velocity with an associated reduction in carrier mobility. In contrast, Sn and In show nested bands that avoid such deleterious effects. Additional hole pockets are also created at the H points in these systems. To date, a comprehensive understanding of how chemistry affects the valence band structure remains elusive due to the breadth of chemical space under consideration.

19.3.2.2 Scattering and Carrier Concentration Control

Single crystals of cobalt antimonide exhibit exceptionally high hole mobilities with values around 3000 $cm^2\ V^{-1}\ s^{-1}$ at room temperature [46]. This is consistent with a light band mass

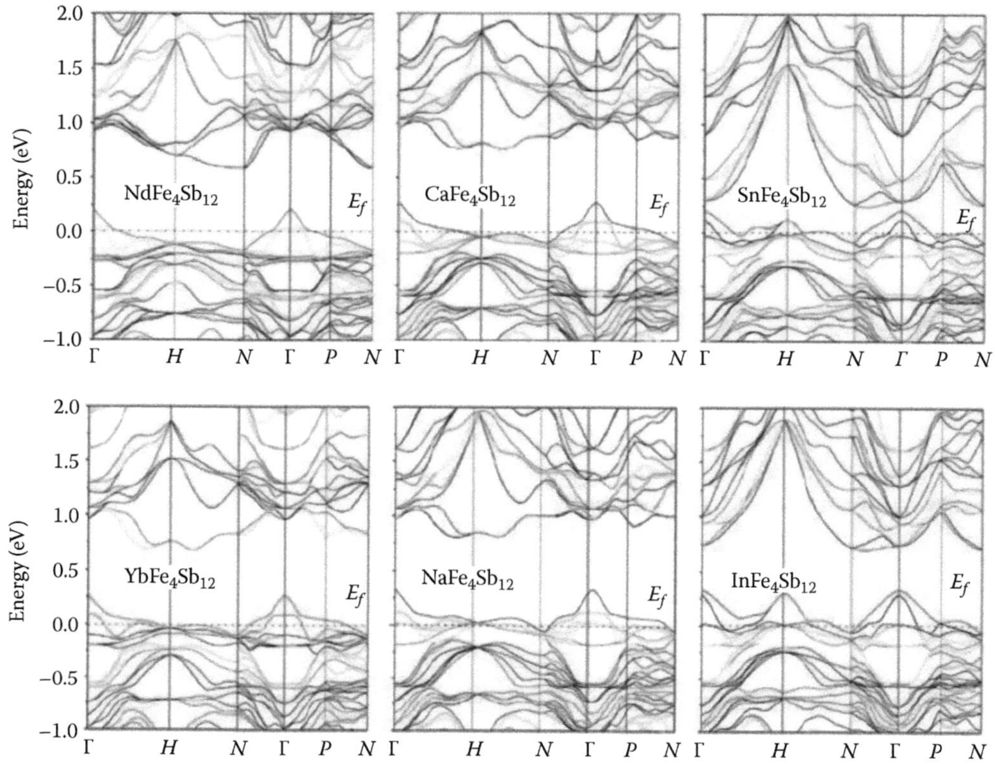

FIGURE 19.6
Choice of filler species can profoundly affect the band edges, as shown for AFe_4Sb_{12}. In these systems, the filler donors are overcompensated by the Fe acceptor, yielding degenerate *p*-type behavior. (Reproduced from Zhou, A., et al., *J. Electron. Mater.*, 40, 974, 2011. Copyright by the TMS Publications. With permission.)

(~0.25 m_e) and band degeneracy of one, which ultimately leads to high carrier velocities and little scattering due to low density of states. Regardless of the large mobility values that pure $CoSb_3$ exhibits, altering the framework chemistry of skutterudite to drive the Fermi level into the valence band can have a deleterious impact on mobility. Figure 19.7 shows that Sn on the Sb site has a minor impact on mobility, but the effect of Fe on the Co site is quite substantial. When electropositive fillers are present, back-doping on the Co–Sb framework with acceptors (e.g., $Ce_yCo_{3-x}Fe_xSb_{12}$ [47]) is necessary to achieve a *p*-type composition. Much like the filler-free discussion earlier, the addition of the back-dopant on the Co–Sb framework has a negative impact on the high hole mobility, μ_p. The low mobility found in $A_xCo_{4-x}Fe_xSb_{12}$ in Figure 19.7 may appear to be attributed to the Fe back-dopant, as seen in filler-free $Co_{4-x}Fe_xSb_{12}$ [15,48–52]. However, even in the absence of Fe, filled skutterudites ($A_xCo_4Sb_{12-x}Sn_x$) can exhibit reduced hole mobility [15,48–52].

While changing the framework leads to a decrease in mobility, as seen in Figure 19.7, it opens up the doors to exploring various flavors of *p*-type fillers in the $CoSb_3$ interstitial site. Recent works have placed Ce, La, Yb, In, and Ba in the site in hopes to reduce the thermal conductivity and correct for the detrimental decrease in hole mobility to preserve the figure of merit [50,53–55]. While the additions of these filler atoms have helped overcome the pinned *ZT*, several works are using a combination of filler species to dope the voids.

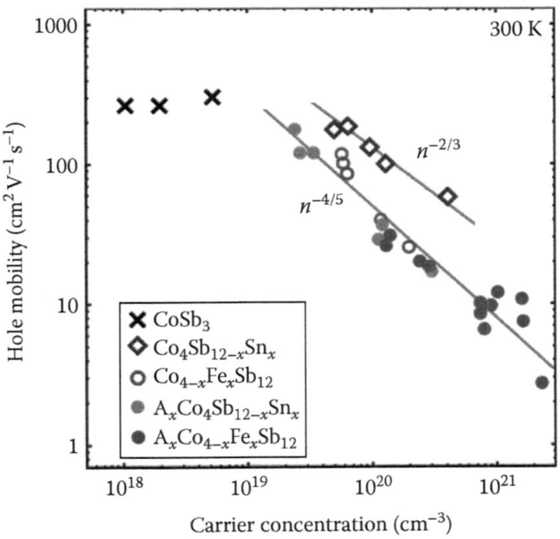

FIGURE 19.7
Room temperature mobility as a function of carrier concentration for various polycrystalline $CoSb_3$ compounds. The addition of electropositive fillers in $A_xCo_4Sb_{12}$ causes a significant drop in mobility, showing $n^{-4/5}$ dependence. Back-doping Sn on the Co–Sb framework has a less dramatic effect on the mobility values with increasing carrier concentration.

Double filling of p-type $A_xM_yFe_3CoSb_{12}$ was recently reported by Liu et al. by mixing various combinations of Ba, Ce, Nd, and Yb [56]. The use of multiple fillers could ultimately introduce additional phonon scattering for various frequencies and increase the point defect scattering mechanism within the system.

In attempt to find electronegative acceptor filler for skutterudites, recent work has placed iodine within the interstitial void in effort to drive cobalt antimonide p-type [20,57,58]. Sadly, these works show $I_xCo_4Sb_{12}$ is only stable at high pressures (5–12 GPa). Despite the metastability of the compound at ambient pressures, the iodine system shows a large decrease in the lattice thermal conductivity, pushing the κ_L to 0.7 W m^{-1} K^{-1} [20] as opposed to the normal values of base skutterudite (~6–8 W m^{-1} K^{-1}) [59]. The iodine acts as one of the first electronegative acceptors, pushing the Fermi level into the valence band without having to back dope on the Co–Sb framework. The interstitial void radius for $CoSb_3$ is reported to be 1.892 Å [4], while the ionic radius of iodine is 2.20 Å [60], lending itself to the metastability of the system and ultimately straining the structure [20]. Based on atomic radii, future works could include various flavors of halogens on the interstitial site, such as bromine [61], to create a more stable compound (in comparison iodine). Since the addition of electronegative acceptors does not require counterdopants on the framework to push the system p-type, the mobility may be preserved and higher ZT may be achievable.

19.4 Thermal Properties

Due to the broad spectrum of phonons that carry heat, the impact of modifications from atomic to microstructural must be considered. We begin by discussing the community's evolving understanding of vibrational dynamics at the atomic scale and how chemical modifications impact heat transfer. Nanostructured materials are then discussed in the context of scattering long wavelength phonons.

19.4.1 Filler and Framework Vibrational Dynamics

The narrative associated with why filler atom addition lowers the thermal conductivity continues to evolve as more advanced techniques are used to probe the skutterudite vibrational dynamics. While a single answer remains elusive, significant reductions in the lattice thermal conductivity have been realized by multiple filler species.

19.4.1.1 Localized Rattling versus Correlated Atomic Motion

The source of reducing lattice thermal conductivity upon incorporation from filler atoms has typically been attributed to their "rattling" effect. Here, "rattling" is often described as a localized motion of the filler atom independent of the host structure. This simple description can be misleading, however, as the filler atom clearly must interact with the host structure through electrostatic forces and the Pauli exclusion principle. Furthermore, such interaction is required if lattice phonons are to be scattered by these localized modes. This localized mode paradigm has been gradually challenged over the past decade [62]. Whether the filling atom reduces the lattice thermal conductivity by rattling or simply by introducing a low-frequency optical mode may be difficult to distinguish. Not only do both have a similar effect on κ (Figure 19.8) [63] but also real materials defy simple

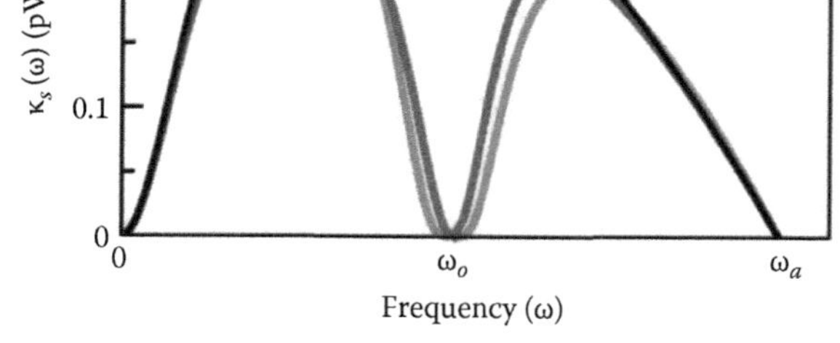

FIGURE 19.8
A simple Born–von Karman model allows the spectral thermal conductivity κ_s to be investigated with several different assumptions concerning phonon group velocity and scattering. Case A considers the acoustic branch of an unfilled material with boundary and Umklapp scattering present. In Case B, the dispersion is unaltered but a rattling (resonant scattering) term is included. The impact on transport is virtually indistinguishable from Case C, where an optical mode intersects the dispersion and lowers the group velocity. (From Toberer, E.S., Zevalkink, A., and Snyder, G.J., Phonon engineering through crystal chemistry, *J. Mater. Chem.*, 21, 15843, 2011. Reproduced by permission of The Royal Society of Chemistry.)

characterization and the debate assumes a false dichotomy. Additionally, different skutterudite compositions may have fundamentally different lattice dynamics due to differing filler chemistry and occupancy, preventing universal statements to be made.

Li and Mingo [64] found by *ab initio* calculations that the reduction of κ upon filling in $CoSb_3$ and $BaCo_4Sb_{12}$ is mainly due to increased phonon–phonon scattering. This is partially due to the filler modifying the phonon dispersion and partly due to the reduction of group velocities. By considering conventional anharmonic scattering alone, a significant reduction of κ upon filling is obtained without introducing any extra scattering mechanism. The overall reduction of phonon lifetimes further rules out resonant scattering as suggested by a rattling model, which claims that the increase of scattering should happen in a narrow range of frequencies. Consequently, the mechanism proposed is unrelated to the previous rattling hypothesis and does not highlight any particular resonant frequency.

Theoretical investigations in $LaFe_4Sb_{12}$ suggest the localized rattling mode and the acoustic modes hybridize and the resulting phonon modes have a large Grüneisen parameter. This is expected to result in increased phonon–phonon scattering [65]. High pressure offers an interesting strategy to change the interatomic interactions and test these theoretical investigations. A high-pressure study of the lattice dynamics in the filled skutterudite $Eu_{0.84}Fe_4Sb_{12}$ was carried out by means of x-ray powder diffraction and nuclear inelastic scattering. A large Grüneisen parameter was found for the rattling mode, which is hybridized with the acoustical phonons at ambient and moderate pressure. At a critical pressure of 12 GPa the Grüneisen parameter for the rattling mode is reduced and the phonon modes decouple. This result supports the hypothesis that anharmonic coupling between acoustic and optical phonon modes plays a central role in the reduced thermal conductivity in filled skutterudites. The existence of this transition in which essentially the "rattling" is switched off is reported for the first time [66]. The transferability of this conclusion to other skutterudite compositions will require additional investigations.

19.4.1.2 Multiple Fillers

Despite uncertainty in the exact vibrational dynamics underlying filler atoms, the last decade has seen significant progress in empirical efforts to screen fillers for reduced thermal conductivity. In particular, the advent of multiple filler species has offered even lower thermal conductivity. Heat capacity and Raman measurements show unequivocally that specific frequencies can be associated with each filler element; in concert, the presumption is that these different modes serve to scatter a broader range of phonon frequencies than a single filler species. Reconciling this resonant mode scattering description with the calculations on fully ordered $BaCo_4Sb_{12}$ discussed earlier remains an ongoing investigation.

19.4.2 Microstructural Effects

The lowest frequency phonons are challenging to scatter *via* point defects or other phonons. This is due to a combination of their long wavelengths and the difficulties of finding an allowed Umklapp transition. As such, there remains significant interest in developing nanostructured materials for phonon scattering. One way to quantify this opportunity is *via* phonon mean free path (MFP) analysis. The lattice thermal conductivity κ_L of $CoSb_3$ was calculated with first-principles methods with a comprehensive analysis on phonon mode contribution, relaxation time, and MFP distributions [67]. The MFP corresponding to 50% κ_L accumulation in $CoSb_3$ at 300 K was found to be ~135 nm, suggesting significant opportunities for nanostructured composites.

19.4.2.1 Skutterudite-Based Composites

Many skutterudite-based composites have been synthesized with improved thermoelectric performance, such as Ba–$CoSb_3$/Ag nanoparticles [68], $Yb_{0.2}Co_4Sb_{12}/AgSbTe_2$ nanoinclusions [69], $CoSb_3$/graphene [70], and (Ba, In) double-filled skutterudite/microsize pores [71]. Unraveling the impact on electronic and thermal transport can be difficult in such composites. For example, the addition of $AgSbTe_2$ nanoinclusions increased the Hall mobility, electrical conductivity, and absolute value of the Seebeck coefficient while decreasing the lattice thermal conductivity [69]. Such alteration in the electronic properties was attributed to the energy filtering effect at the matrix–nanoinclusion interfaces. Porous structures may also be beneficial to thermoelectric performance. Porous (Ba, In) double-filled skutterudite materials with pore diameter about 1–4 μm were prepared by Zn and Sb sublimation from ZnSb inclusions. Here again, improvements in electronic transport were attributed to electron filtering [71]. Definitively proving electron filtering effects are occurring is challenging in the face of changing carrier concentration and the inherent uncertainty of electronic transport measurements.

19.4.2.2 Grain Boundary Scattering via Spontaneous Segregation

Thermoelectric performance can be enhanced when grain boundary scattering of phonons is not overshadowed by a reduction in charge carrier mobility. Recent efforts to introduce such nanostructures in skutterudites have utilized spontaneous phase separations to produce nanocomposites. These transformations have yielded both incoherent and coherent grain boundaries. One example of incoherent grain boundaries is the $Yb_{0.6}Fe_2Co_2Sb_{12}/0.05FeSb_2$ composite [72], which forms spontaneously on cooling from a single phase. Here, the composite morphology can be controlled precisely *via* the parameters of the heat treatment. Coherent phase separation has also emerged due to differences in filler species and the associated lattice strain. For example, *p*-type $La_{0.8}Ti_{0.1}Ga_{0.1}Fe_3CoSb_{12}$ was found to possess a microstructure where each grain of the hot-pressed sample had phase separated La-poor and La-rich skutterudite phases. For $(Ca_xBa_{1-x})_yFe_4Sb_{12}$, phase separated samples show evidence of increased thermopower and decreased thermal conductivity [73,74]. Maintaining the framework translational symmetry in coherent composites may provide a route to high charge carrier mobility in nanostructured composites.

19.5 Generator-Related Activities

The rise of high ZT in skutterudites has inspired significant research into the design, construction, and testing of thermoelectric generators (TEGs). With a hot side temperature between 500°C and 600°C, such generators are attractive for mid-grade waste heat and cogeneration. Associated with generator construction are challenges involving thermomechanical stresses and device stability. Given the low power conversion efficiency of TEGs compared to mechanical heat engines, ensuring long operational lifetimes is critical if TEGs are to be commercially competitive.

19.5.1 Skutterudite *ZT* and Device Efficiencies

Before investing significant effort into understanding a thermoelectric material's mechanical properties and stability, the beginning of life performance should justify this investment. To that end, we consider both material *ZT* values achieved and reported skutterudite TEG efficiencies.

19.5.1.1 Skutterudite ZT

For *n*-type compositions, single fillers of various flavors (e.g., Na, Ba, In, Yb, Ce) [9,12,28,29,41] have yielded peak *ZT* values between 1.0 and 1.3. Double and triple filling has driven the lattice thermal conductivity down even further and yielded reported peak *ZT* values between 1.25 and 1.9 [1,3,24]. In contrast, *p*-type compositions have been significantly harder with which to achieve high *ZT* (peak values are reported from 1.0 to 1.3) [2,74,75]. The difference between *n*- and *p*-type performance is primarily due to differences in electronic structure between the lightly filled Co-rich (*n*-type) materials and the heavily filled Fe-rich (*p*-type) materials that have lower effective band gaps.

19.5.1.2 Device Efficiency

With new material design strategies to improve the *ZT* of skutterudites, there has been significant interest in developing the skutterudite-based TEGs. On the device level, the energy conversion efficiency has the potential to reach >10% [76–78]. Salvador et al. [76] reported the power output and conversion efficiency of a 32 couple, high-temperature skutterudite module at varying applied loading pressures and with different interface materials between the module and the heat source and sink of the test system. A 7% conversion efficiency at the module level was demonstrated when a temperature difference of 460°C was established. Extrapolated values indicate that 7.5% is achievable when proper thermal interfaces and loading pressures are used. Similar efficiency (about 8%) was obtained by Ochi et al. [77]. For a mid-temperature *n*-type doped skutterudite material, the efficiency of over 11% on a single-leg level at a temperature difference of 600°C could be achieved with the cold side maintained at a temperature close to 50°C [78]. To overcome the low *ZT* of skutterudite near ambient temperature, a segmented design using Bi_2Te_3 until ~250°C is expected to enable efficiencies of >12%.

19.5.2 Mechanical Properties and Device Stability

With the potential for high generator efficiency established, we consider recent studies of the mechanical properties and device stability of skutterudite-based TEGs.

19.5.2.1 Mechanical Properties

The mechanical properties of $CoSb_3$-based skutterudites are of critical concern for commercial application. Density functional theory calculations of the response along various shear and tensile deformations by Li et al. [79] reveal that the Co–Sb bond dominates the ideal strength. Because the Sb–Sb covalent bond is softer than the Co–Sb bond, the

Sb-rings are less rigid than the Co–Sb bonds, which lead to the Sb-rings stretching while the Co–Sb bonds break. It was also found that the filling of the $CoSb_3$ void site with typical fillers as Na, Ba, or Yb has little effect on the ideal strength and failure mode, consistent with the filler coupling only loosely to the framework.

The fracture toughness of undoped $CoSb_3$ and indium doped $In_{0.1}Co_4Sb_{12}$ skutterudites was studied by Eilertsen et al., who found no interstitial embrittlement due to indium void-site filling [80]. The significantly lower values of toughness for $CoSb_3$ skutterudites (~0.5 MPa $\sqrt{m}$) compared to previous results [81] indicate that there may be concern over the durability of skutterudite-based power-producing thermoelectric modules if care is not taken to ensure adequate toughness.

Thermoelectric materials must also be sufficiently mechanically robust to withstand the thermal fatigue while in service. The temperature dependence of the tensile fracture strength was measured on the *n*-type (La, Ba, Yb)-$CoSb_3$ skutterudite ($La_{0.50}Ba_{0.07}Yb_{0.08}Co_{4.00}Sb_{12.02}$) and *p*-type $Ce_{0.30}Co_{2.57}Fe_{1.43}Sb_{11.98}$ [82]. Both materials show tensile fracture strengths that are temperature independent up to 500°C and are ~140 MPa as measured by a three-point bend flexure test. The thermal expansion coefficients were measured to be similar (10% difference) between *n*- and *p*-type skutterudites (10.3 ppm for the *n*-type material and 11.5 ppm for the *p*-type material up to 450°C).

While skutterudite-based composites can have enhanced transport properties, there may also be a trade-off between thermoelectric and mechanical properties. For example, while inserting nanopores into thermoelectric materials [83] can remarkably reduce the thermal conductivity, the elastic modulus decreases with increasing porosity. The addition of carbon fibers in $CeFe_4Sb_{12}$ skutterudites [84] results in the enhancement of both the flexural strength and the fracture toughness. The optimal carbon fiber content was found to be 1 vol.%, which led to a 53% improvement of the flexural strength, yet did not change fracture toughness nor *ZT*. Above 1 vol.%, the thermoelectric properties start to deteriorate from the carbon fiber phase.

19.5.2.2 Thermal Stability

The thermal stability of skutterudite thermoelectric materials has been widely studied with the increasing potential for commercial application. Both *n*- and *p*-type skutterudites have reported thermal stability under vacuum, that is, *ZT* values remained unchanged or were slightly improved after annealing at 873 K for 5000 h [85]. However, two Sb-based oxides, $La_{0.9}Fe_3CoSb_{12}$ [86] and $Yb_yCo_4Sb_{12}$ [87], were preferably formed when aged at high temperatures in air.

To protect the materials from oxidation and sublimation during thermal cycles, there are two methods. One approach is through external coatings such as glass composites [88]. Composite glass coatings were fabricated on thermoelectric skutterudite materials using the dispersion system of modified nanosilica aerogel and micro glass powder. The glass coated skutterudite materials showed good oxidation resistance; no oxides were observed even after heating at 923 K in ambient air. The other method is to create internal barriers for the diffusion of oxygen ions. One study on the high-temperature stability of $Yb_yCo_4Sb_{12}$/Yb_2O_3 composites showed that there are a large number of dislocations in the grain interior. Dislocations are considered the major factor to effectively block the migration of the Yb^{2+} ions and O^{2-} ions, thus preventing the generation of oxide particles [89]. It was also observed that the oxidation could happen in a "pesting"-like way [90], which is related to the oxide layer periodically peeling off at ~800 K. Such behavior makes the $Ce_{0.9}Fe_3CoSb_{12}$

samples easy to be oxidized as the fresh substrate surface is exposed to a high concentration oxygen atmosphere. In this case, both coating and adding internal diffusion barriers (dislocations in the grain interior) could help reduce the oxidation.

19.6 Conclusions

Over the last few decades, the incredible diversity of skutterudites has begun to emerge, making these compounds the intermetallic analog to the perovskite oxides. Since 2010, major advances in skutterudite chemistry, transport, and devices have been achieved. The expansion of filler chemistry and the rise of multiple filler elements have enabled exceptionally low thermal conductivity and paths to alter the electronic structure. The development of an iodine-based filler at high pressure is particularly intriguing, as fillers to date have largely been electropositive. High throughput determination of hypothetical skutterudite compositions has also begun to reveal intriguing new opportunities well beyond the traditional $CoSb_3$ framework. Electronic transport has seen significant advances in terms of understanding the cause of excellent n-type performance and introducing modifications to enhance the electronic structure. For p-type compositions, performance remains reduced due to the single valence band edge. The incorporation of multiple filler elements has been effective at reducing the lattice component of the thermal conductivity. The enthusiasm for skutterudites and their thermoelectric performance is translating to a growth in device studies, including module construction.

Acknowledgments

G. J. S. acknowledges support by Solid-State Solar-Thermal Energy Conversion Center (S3TEC), an Energy Frontier Research Center funded by the U.S. Department of Energy, Office of Science, Basic Energy Sciences under Award # DE-SC0001299. Y. T. acknowledges support by the Swiss Federal Office of Energy under contract SI/501310-01. C. M. C. and E.S.T. acknowledges support from the NASA Science Missions Directorate's Radioisotope Power Systems Thermoelectric Technology Development Project.

References

1. Shi, X., J. Yang, J. R. Salvador et al. 2011. Multiple-filled skutterudites: High thermoelectric figure of merit through separately optimizing electrical and thermal transports. *J. Am. Chem. Soc.* 133: 7837.
2. Rogl, G., A. Grytsiv, P. Rogl et al. 2014. Nanostructuring of p- and n-type skutterudites reaching figures of merit of approximately 1.3 and 1.6, respectively. *Acta Mater.* 76: 434.
3. Rogl, G., A. Grytsiv, P. Rogl et al. 2014. N-type skutterudites (R, Ba, Yb)$_y$$Co_4Sb_{12}$ (R = Sr, La, Mm, Dd, Srmm, Srdd) approaching $ZT \approx 2.0$. *Acta Mater.* 63: 30.

4. Uher, C. 2001. Skutterudites: Prospective novel thermoelectrics. In *Semiconductors and Semimetals*, Academic Press, New York, Vol. 69, pp. 139–253.
5. Uher, C. 2005. Skutterudite-based thermoelectrics. In *Thermoelectrics Handbook: Macro to Nano*, ed. Rowe, D. M. CRC, Taylor & Francis Group, Boca Raton, FL, Ch. 34.
6. Zhang, Q. H., X. Y. Huang, S. Q. Bai et al. 2016. Thermoelectric devices for power generation: Recent progress and future challenges. *Adv. Eng. Mater.* 18: 194.
7. Rull-Bravo, M., A. Moure, J. F. Fernández, and M. Martín-González. 2015. Skutterudites as thermoelectric materials: Revisited. *RSC Adv.* 5: 41653.
8. Jeitschko, W. and D. Braun. 1977. $LaFe_4P_{12}$ with filled $CoAs_3$-type structure and Lsotypic lanthanoid-transition metal polyphosphides. *Acta Cryst. B* 33: 3401.
9. Pei, Y. Z., J. Yang, L. D. Chen et al. 2009. Improving thermoelectric performance of caged compounds through light-element filling. *Appl. Phys. Lett.* 95: 042101.
10. Pei, Y. Z., L. D. Chen, W. Zhang et al. 2006. Synthesis and thermoelectric properties of $K_yCo_4Sb_{12}$. *Appl. Phys. Lett.* 89: 221107.
11. Zhao, X. Y., X. Shi, L. D. Chen et al. 2006. Synthesis and thermoelectric properties of Sr-filled skutterudite $Sr_yCo_4Sb_{12}$. *J. Appl. Phys.* 99: 053711.
12. Chen, L. D., T. Kawahara, X. F. Tang et al. 2001. Anomalous barium filling fraction and *n*-type thermoelectric performance of $Ba_yCo_4Sb_{12}$. *J. Appl. Phys.* 90: 1864.
13. Nolas, G. S., M. Kaeser, R. T. Littleton, and T. M. Tritt. 2000. High figure of merit in partially filled ytterbium skutterudite materials. *Appl. Phys. Lett.* 77: 1855.
14. Yang, J., Q. Hao, H. Wang et al. 2009. Solubility study of Yb in *n*-type skutterudites $Yb_xCo_4Sb_{12}$ and their enhanced thermoelectric properties. *Phys. Rev. B* 80: 115329.
15. Nolas, G. S., J. L. Cohn, and G. A. Slack. 1998. Effect of partial void filling on the lattice thermal conductivity of skutterudites. *Phys. Rev. B* 58: 164.
16. Morelli, D. T., G. P. Meisner, B. Chen, S. Hu, and C. Uher. 1997. Cerium filling and doping of cobalt triantimonide. *Phys. Rev. B* 56: 7376.
17. Pei, Y. Z., S. Q. Bai, X. Y. Zhao, W. Zhang, and L. D. Chen. 2008. Thermoelectric properties of $Eu_yCo_4Sb_{12}$ filled skutterudites. *Solid State Sci.* 10: 1422.
18. Zhang, J., B. Xu, L.-M. Wang et al. 2012. High-pressure synthesis of phonon-glass electron-crystal featured thermoelectric $Li_xCo_4Sb_{12}$. *Acta Mater.* 60: 1246.
19. Yang, J., L. Zhang, Y. Liu et al. 2013. Investigation of skutterudite $Mg_yCo_4Sb_{12}$: High pressure synthesis and thermoelectric properties. *J. Appl. Phys.* 113: 113703.
20. Li, X., B. Xu, L. Zhang et al. 2014. Synthesis of iodine filled $CoSb_3$ with extremely low thermal conductivity. *J. Alloys Compd.* 615: 177.
21. Zhang, Q., C. Chen, Y. Kang et al. 2015. Structural and thermoelectric characterizations of samarium filled $CoSb_3$ skutterudites. *Mater. Lett.* 143: 41.
22. Liu, R., X. Chen, P. Qiu et al. 2011. Low thermal conductivity and enhanced thermoelectric performance of Gd-filled skutterudites. *J. Appl. Phys.* 109: 023719.
23. Yang, J., W. Zhang, S. Q. Bai, Z. Mei, and L. D. Chen. 2007. Dual-frequency resonant phonon scattering in $Ba_xR_yCo_4Sb_{12}$ (R = La, Ce, and Sr). *Appl. Phys. Lett.* 90: 192111.
24. Shi, X., H. Kong, C. P. Li et al. 2008. Low thermal conductivity and high thermoelectric figure of merit in *n*-type $Ba_xYb_yCo_4Sb_{12}$ double-filled skutterudites. *Appl. Phys. Lett.* 92: 182101.
25. Shi, X., W. Zhang, L. Chen, and J. Yang. 2005. Filling fraction limit for intrinsic voids in crystals: Doping in skutterudites. *Phys. Rev. Lett.* 95: 185503.
26. Mei, Z., J. Yang, Y. Pei et al. 2008. Alkali-metal-filled $CoSb_3$ skutterudites as thermoelectric materials: Theoretical study. *Phys. Rev. B* 77: 045202.
27. Miotto, F., C. A. Figueiredo, G. R. Ramos et al. 2011. Antimony desinsertion reaction from Sb_xCoSb_{3-x}. *J. Appl. Phys.* 110: 043529.
28. Tang, Y., S.-W. Chen, and G. J. Snyder. 2015. Temperature dependent solubility of Yb in Yb–$CoSb_3$ skutterudite and its effect on preparation, optimization and lifetime of thermoelectrics. *J. Materiomics* 1: 75.
29. Tang, Y., R. Hanus, S. W. Chen, and G. J. Snyder. 2015. Solubility design leading to high figure of merit in low-cost Ce- $CoSb_3$ skutterudites. *Nat. Commun.* 6: 7584.

30. Kim, H., M. Kaviany, J. C. Thomas et al. 2010. Structural order-disorder transitions and phonon conductivity of partially filled skutterudites. *Phys. Rev. Lett.* 105: 265901.
31. Zhang, T., K. Zhou, X. F. Li et al. 2015. Structural transition of partially Ba-filled thermoelectric $CoSb_3$ investigated by positron annihilation spectroscopy. *J. Appl. Phys.* 117: 055103.
32. Park, C.-H. and Y.-S. Kim. 2010. *Ab initio* study of native point-defects in $CoSb_3$: Understanding off-stoichiometric doping properties. *Phys. Rev. B* 81: 085206.
33. Grosvenor, A., R. Cavell, and A. Mar. 2006. X-ray photoelectron spectroscopy study of the skutterudites $LaFe_4Sb_{12}$, $CeFe_4Sb_{12}$, $CoSb_3$, and CoP_3. *Phys. Rev. B* 74: 125102.
34. Luo, H., J. W. Krizan, L. Muechler et al. 2015. A large family of filled skutterudites stabilized by electron count. *Nat. Commun.* 6: 6489.
35. Schmokel, M. S., L. Bjerg, J. Overgaard et al. 2013. Pushing x-ray electron densities to the limit: Thermoelectric $CoSb_3$. *Angew. Chem. Int. Ed.* 52: 1503.
36. Tan, G., S. Wang, and X. Tang. 2013. High thermoelectric figure of merit of *p*-type ternary unfilled skutterudite $FeSb_2Te$ via Ge doping. *Sci. Adv. Mater.* 5: 1974.
37. Rogl, G., A. Grytsiv, E. Royanian et al. 2013. New *p*- and *n*-type skutterudites with $ZT > 1$ and nearly identical thermal expansion and mechanical properties. *Acta Mater.* 61: 4066.
38. Kaltzoglou, A., P. Vaqueiro, K. S. Knight, and A. V. Powell. 2012. Synthesis, characterization and physical properties of the skutterudites $Yb_xFe_2Ni_2Sb_{12}$ ($0 \leq x \leq 0.4$). *J. Solid State Chem.* 193: 36.
39. Zevalkink, A., K. Star, U. Aydemir et al. 2015. Electronic structure and thermoelectric properties of pnictogen-substituted $ASn_{1.5}Te_{1.5}$ (A = Co, Rh, Ir) skutterudites. *J. Appl. Phys.* 118: 035107.
40. Qiu, Y., L. Xi, X. Shi et al. 2013. Charge-compensated compound defects in Ga-containing thermoelectric skutterudites. *Adv. Funct. Mater.* 23: 3194.
41. Tang, Y., Y. Qiu, L. Xi et al. 2014. Phase diagram of In–Co–Sb system and thermoelectric properties of In-containing skutterudites. *Energy Environ. Sci.* 7: 812.
42. Tang, Y., Z. M. Gibbs, L. A. Agapito et al. 2015. Convergence of multi-valley bands as the electronic origin of high thermoelectric performance in $CoSb_3$ skutterudites. *Nat. Mater.* 14: 1223.
43. Yang, J., L. Xi, W. Zhang, L. D. Chen, and J. Yang. 2009. Electrical transport properties of filled $CoSb_3$ skutterudites: A theoretical study. *J. Electron. Mater.* 38: 1397.
44. Shi, X., J. Yang, L. Wu et al. 2015. Band structure engineering and thermoelectric properties of charge-compensated filled skutterudites. *Sci. Rep.* 5: 14641.
45. Zhou, A., L.-S. Liu, C.-C. Shu et al. 2011. Electronic structures and transport properties of RFe_4Sb_{12} (R = Na, Ca, Nd, Yb, Sn, In). *J. Electron. Mater.* 40: 974.
46. Morelli, D., T. Caillat, J. P. Fleurial et al. 1995. Low-temperature transport properties of *p*-type $CoSb_3$. *Phys. Rev. B* 51: 9622.
47. Chen, B., J.-H. Xu, C. Uher et al. 1997. Low-temperature transport properties of the filled skutterudites $CeFe_{4-x}Co_xSb_{12}$. *Phys. Rev. B* 55: 1476.
48. Qiu, P., X. Shi, X. Chen et al. 2011. Effects of Sn-doping on the electrical and thermal transport properties of *p*-type cerium filled skutterudites. *J. Alloys Compd.* 509: 1101.
49. Park, K.-H., J. Y. Jung, J. I. Lee et al. 2010. Synthesis and electronic transport properties of Sn-doped $CoSb_3$. *Mater. Sci. Forum* 658: 21.
50. Park, K.-H., I.-H. Kim, S.-M. Choi et al. 2012. Preparation and thermoelectric properties of *p*-type Yb-filled skutterudites. *J. Electron. Mater.* 42: 1377.
51. Liu, R., P. Qiu, X. Chen, X. Huang, and L. Chen. 2011. Composition optimization of *p*-type skutterudites $Ce_yFe_xCo_{4-x}Sb_{12}$ and $Yb_yFe_xCo_{4-x}Sb_{12}$. *J. Mater. Res.* 26: 1813.
52. Kim, I.-H. and S.-C. Ur. 2007. Electronic transport properties of Fe-doped $CoSb_3$ prepared by encapsulated induction melting. *Mater. Lett.* 61: 2446.
53. Park, K.-H., S. Lee, W.-S. Seo, and I.-H. Kim. 2014. Synthesis and thermoelectric properties of $Ce_zFe_{4-x}Co_xSb_{12}$ skutterudites. *J. Korean Phys. Soc.* 64: 84.
54. Park, K.-H., S. Lee, W.-S. Seo, D.-K. Shin, and I.-H. Kim. 2014. The synthesis of $R_zFe_{4-x}Co_xSb_{12}$ (R: Yb, La, Ce) skutterudites and their thermoelectric properties. *J. Korean Phys. Soc.* 64: 863.
55. Park, K.-H., S.-W. You, S.-C. Ur et al. 2012. Thermoelectric properties of skutterudite in $YFe_xCo_{4-x}Sb_{12}$. *J. Korean Phys. Soc.* 60: 1485.

56. Liu, R., J. Yang, X. Chen et al. 2011. *P*-type skutterudites $R_xM_yFe_3CoSb_{12}$ (R, M = Ba, Ce, Nd, and Yb): Effectiveness of double-filling for the lattice thermal conductivity reduction. *Intermetallics* 19: 1747.
57. Fukuoka, H. and S. Yamanaka. 2010. High-pressure synthesis, structure, and electrical property of iodine-filled skutterudite $I_{0.9}Rh_4Sb_{12}$—First anion-filled skutterudite. *Chem. Mater.* 22: 47.
58. Zhang, L. B. Xu, X. Li et al. 2015. Iodine-filled $Fe_{4-x}Co_xSb_{12}$ polycrystals: Synthesis, structure, and thermoelectric properties. *Mater. Lett.* 139: 249.
59. Nolas, G. S., D. T. Morelli, and T. M. Tritt. 1999. Skutterudites: A phonon-glass-electron crystal approach to advanced thermoelectric energy conversion applications. *Annu. Rev. Mater. Sci.* 29: 89.
60. Shannon, R. D. and C. T. Prewitt. 1969. Effective ionic radii in oxides and fluorides. *Acta Cryst. B* 25: 925.
61. Ortiz, B. R., C. M. Crawford, R. W. McKinney, P. A. Parilla and E. S. Toberer. 2016. Thermoelectric properties of bromine filled $CoSb_3$ skutterudite. *J. Mater. Chem. A.* 4:8444.
62. Koza, M. M., M. R. Johnson, R. Viennois et al. 2008. Breakdown of phonon glass paradigm in La- and Ce-filled Fe_4Sb_{12} skutterudites. *Nat. Mater.* 7: 805.
63. Toberer, E. S., A. Zevalkink, and G. J. Snyder. 2011. Phonon engineering through crystal chemistry. *J. Mater. Chem.* 21: 15843.
64. Li W. and N. Mingo. 2014. Thermal conductivity of fully filled skutterudites: Role of the filler. *Phys. Rev. B* 89: 184304.
65. Bernstein, N., J. L. Feldman, and D. J. Singh. 2010. Calculations of dynamical properties of skutterudites: Thermal conductivity, thermal expansivity, and atomic mean-square displacement. *Phys. Rev. B* 81: 134301.
66. Sergueev, I., K. Glazyrin, I. Kantor et al. 2015. Quenching rattling modes in skutterudites with pressure. *Phys. Rev. B* 91: 224304.
67. Guo, R., X. Wang, and B. Huang. 2015. Thermal conductivity of skutterudite $CoSb_3$ from first principles: Substitution and nanoengineering effects. *Sci. Rep.* 5: 7806.
68. Schmidt, R. D., E. D. Case, Z. Lobo et al. 2014. Influence of silver nanoparticle addition, porosity, and processing technique on the mechanical properties of $Ba_{0.3}Co_4Sb_{12}$ skutterudites. *J. Mater. Sci.* 49: 7192.
69. Peng, J., L. Fu, Q. Liu et al. 2014. A study of $Yb_{0.2}Co_4Sb_{12}$–$AgSbTe_2$ nanocomposites: Simultaneous enhancement of all three thermoelectric properties. *J. Mater. Chem. A* 2: 73.
70. Feng, B., J. Xie, G. Cao, T. Zhu, and X. Zhao. 2013. Enhanced thermoelectric properties of *p*-type $CoSb_3$/graphene nanocomposite. *J. Mater. Chem. A* 1: 13111.
71. Yu, J., W.-Y. Zhao, P. Wei et al. 2014. Enhanced thermoelectric performance of (Ba,In) double-filled skutterudites via randomly arranged micropores. *Appl. Phys. Lett.* 104: 142104.
72. Zhou, C., J. Sakamoto, and D. Morelli. 2011. High-temperature thermoelectric properties of *p*-type Yb-filled skutterudite nanocomposites with $FeSb_2$ nanoinclusions. *J. Electron. Mater.* 41: 1030.
73. Rogl, G., A. Grytsiv, M. Falmbigl et al. 2012. Spinodal decomposition in $(Ca_xBa_{1-x})_yFe_4Sb_{12}$. *Acta Mater.* 60: 4487.
74. Meng, X., W. Cai, Z. Liu et al. 2015. Enhanced thermoelectric performance of *p*-type filled skutterudites via the coherency strain fields from spinodal decomposition. *Acta Mater.* 98: 405.
75. Rogl, G., A. Grytsiv, P. Heinrich et al. 2015. New bulk *p*-type skutterudites $Dd_{0.7}Fe_{2.7}Co_{1.3}Sb_{12-x}X_x$ (X = Ge, Sn) reaching $ZT > 1.3$. *Acta Mater.* 91: 227.
76. Salvador, J. R., J. Y. Cho, Z. Ye et al. 2014. Conversion efficiency of skutterudite-based thermoelectric modules. *Phys. Chem. Chem. Phys.* 16: 12510.
77. T. Ochi, G. Nie, S. Suzuki et al. 2014. Power-generation performance and durability of a skutterudite thermoelectric generator. *J. Electron. Mater.* 43: 2344.
78. George, T., G. Chen, D. Kraemer et al. 2011. Thermoelectric energy conversion using nanostructured materials. *Proceeding of the SPIE on Micro- and Nanotechnology Sensors, Systems, and Applications III*, Orlando, FL, Vol. 8031, pp. 80311J.

79. Li, G., Q. An, W. Li et al. 2015. Brittle failure mechanism in thermoelectric skutterudite $CoSb_3$. *Chem. Mater.* 27: 6329.
80. Eilertsen, J., M. A. Subramanian, and J. J. Kruzic. 2013. Fracture toughness of Co_4Sb_{12} and $In_{0.1}Co_4Sb_{12}$ thermoelectric skutterudites evaluated by three methods. *J. Alloys Compd.* 552: 492.
81. Ravi, V., S. Firdosy, T. Caillat et al. 2008. Mechanical properties of thermoelectric skutterudites. *Proceedings of the AIP Conference on Space Technology and Application International Forum, STAIF 2008*, Albuquerque, NM, Vol. 969, pp. 656–662.
82. Salvador, J. R., J. Yang, A. A. Wereszczak, H. Wang, and J. Y. Cho. 2011. Temperature dependent tensile fracture stress of *n*- and *p*-type filled skutterudite materials. *Sci. Adv. Mater.* 3: 577.
83. Li, W., G. Li, X. Yang, L. Liu, and P. Zhai. 2014. Influence of nanopores on the tensile/compressive mechanical behavior of crystalline $CoSb_3$: A molecular dynamics study. *J. Electron. Mater.* 44: 1477.
84. Wan, S., X. Huang, P. Qiu, S. Bai, and L. Chen. 2015. The effect of short carbon fibers on the thermoelectric and mechanical properties of *p*-type $CeFe_4Sb_{12}$ skutterudite composites. *Mater. Design* 67: 379.
85. Nie, T. G., Ochi, S. Suzuki et al. 2013. Stability of skutterudite thermoelectric materials. *J. Electron. Mater.* 43: 1752.
86. Shin, D.-K., I.-H. Kim, K.-H. Park, S. Lee, and W.-S. Seo. 2014. Thermal stability of $La_{0.9}Fe_3CoSb_{12}$ skutterudite. *J. Electron. Mater.* 44: 1858.
87. Xia, X., P. Qiu, X. Shi et al. 2012. High-temperature oxidation behavior of filled skutterudites $Yb_yCo_4Sb_{12}$. *J. Electron. Mater.* 41: 2225.
88. Dong, H., X. Li, Y. Tang et al. 2012. Fabrication and thermal aging behavior of skutterudites with silica-based composite protective coatings. *J. Alloys Compd.* 527: 247.
89. Ding, J., R. Liu, H. Gu, and L. D. Chen. 2014. Study on the high temperature stability of $Yb_yCo_4Sb_{12}/Yb_2O_3$ composite thermoelectric material. *J. Inorg. Mater.* 29: 209.
90. Qiu, P. X. Xia, X. Huang et al. 2014. "Pesting"-like oxidation phenomenon of *p*-type filled skutterudite $Ce_{0.9}Fe_3CoSb_{12}$. *J. Alloys Compd.* 612: 365.

20

SHS-Processed Thermoelectric Materials

Xinfeng Tang*, Xianli Su*, Qingjie Zhang*, and Ctirad Uher**
**Wuhan University of Technology*
***University of Michigan*

CONTENTS

20.1 Introduction and Background

Thermoelectric (TE) materials convert heat into electricity directly through the Seebeck effect or operate as heat pumps using the Peltier effect. TE energy conversion has drawn a great attention for applications in areas such as solar thermal conversion [1], industrial waste heat recovery [2], and spot cooling of electronic devices [3]. The efficiency of a TE material is strongly related to its dimensionless figure of merit ZT, defined as $ZT = \alpha^2\sigma T/\kappa$, where α, σ, κ, and T are the Seebeck coefficient, electrical conductivity, total thermal conductivity, and the absolute temperature, respectively. To achieve high efficiency, a large ZT is required. High electrical conductivity (corresponding to low Joule heating), large Seebeck coefficient (corresponding to a large potential difference), and low thermal conductivity (corresponding to a large temperature difference) are the necessary attributes of an efficient TE material. However, these three transport parameters are strongly interdependent, and thus, it is a challenging task to optimize the TE figure of merit. Most research efforts have focused on doping to optimize the carrier transport and on lowering the thermal conductivity [4–10]. According to the requirements for achieving high TE performance, the ideal TE material should have a glass-like thermal transport while also an excellent crystalline character assuring good electronic transport properties. This notion, first brought forward by Slack [11–13], is succinctly referred to as the phonon glass–electron crystal (PGEC) paradigm. Over the past dozen or so years, great efforts have been made in the following three areas to improve the TE performance:

1. The first strategy is to search for new materials possessing open crystalline environment with excellent electronic properties while poorly conducting heat, materials such as skutterudites [9–15,17,18] and clathrates [6–8,16]. Skutterudites are an excellent example in this respect as they have the cubic crystalline lattice with structural voids, which can be filled with guest species called fillers [17,18]. Skutterudites possess excellent electronic properties and therefore a high power factor $S^2\sigma$, and, by partially filling the voids, their intrinsically high thermal conductivity can be dramatically reduced [9–15,17,18]. As guest atoms enter the voids, their valence electrons are stripped off and go into the conduction band, dramatically enhancing the carrier concentration. Suitable fillers are a variety of species including rare-earth ions, alkaline-earth ions, and alkali metal ions. An ion in a cage is very loosely bonded to atoms of the framework and can vibrate with large amplitude; one often refers to this motion as "rattling." Such Einstein-like vibrations interact with the normal phonon modes of the structure in a resonant way and suppress their ability to conduct heat [14,19]. A broader range of phonon modes can be affected by filling the cages with species that have different Einstein frequencies; this is known as the multifilling concept. While a typical unfilled binary skutterudite such as $CoSb_3$ has the TE figure of merit at 850 K of less than 0.5, multiple-filled skutterudites are far more efficient, and their figure of merit reaches values in the range of 1.7–1.9 [14].
2. The second strategy to enhance TE properties consists of using nanostructure engineering, especially in its hierarchical form [20–31]. The concept of nanostructuring has drawn much attention not only because of its obvious relevance to phonon boundary and interface scattering, which is enhanced by the presence of numerous interfaces associated with nanoinclusions in the matrix or the entire sample being a nanocomposite structure, but also for a possibility of achieving quantum confinement of charge carriers [32] and carrier filtering [33] on account of potential barriers forming at the interface of nanoinclusions and the matrix. The essential point is that in order to obtain a high Seebeck coefficient in a semiconductor having a reasonably large carrier density, one needs to create sharp features such as spikes in the density of states and locate the Fermi level within such regions. While this is easily accomplished in lower-dimensional systems where the density of states is a sharply varying function of energy, in bulk materials, the density of states is a smooth function, and it is exceedingly difficult to perturb it and create sharp variations. This is the main reason why in most of the experiments, the benefit of nanostructuring does not come from the enhancement of the power factor but, rather, from a reduction of the lattice thermal conductivity. Of course, while the nanoinclusions will always scatter phonons to a greater or lesser extent, it is essential that the phonon mean-free path is reduced more than any degradation in the carrier mobility, which inevitably happens as a result of charge

carriers being scattered by nanoinclusions. If this is not so, the nanostructure will actually be detrimental to thermoelectricity. As a general rule, nanostructuring is likely to be effective only in structures where the phonon mean-free path is significantly larger than the mean-free path of charge carriers.

Among the first demonstrations of the benefit of nanoinclusions in enhancing the Seebeck coefficient was the work of Heremans et al. [34] with PbTe-containing nanoscale grains. A truly spectacular success was achieved by the Kanatzidis team [35] with modified PbTe structures of the form $AgPb_mSbTe_{2+m}$ referred to as the "LAST" family of materials where the presence of endotaxial nanoinclusions (nanodots) dramatically reduced the lattice thermal conductivity while the charge carrier transport was left essentially unaffected. This resulted in ZT values reaching 1.7 at 700 K. Eventually, by embedding nanoinclusions of varied sizes spanning from nano- to micrometers (hierarchical structures) to scatter a broad range of phonons, exceptionally large ZTs in excess of two were achieved [22,36], and the authors coined the term panascopic to describe this approach. While the panascopic approach is certainly effective in reducing the lattice thermal conductivity, it must be remembered that there is actually a limit how far one can mess up the structure and lower the thermal conductivity. This is given by the concept of minimum thermal conductivity first put forward by Slack [37], which basically states that the usual picture of lattice phonons loses its meaning when their mean-free path becomes less than the phonon wavelength λ. The range of phonons was later marginally extended by Cahill and Pohl [38] by requiring that their mean-free path cannot be smaller than one-half of the phonon wavelength. As the lattice thermal conductivity of some of the families of TE materials already approaches this limiting value, it is clear that any further dramatic progress in the performance of TE materials will have to come from improvements of the power factor.

3. The third important approach on how to enhance the TE properties is band structure engineering through appropriate doping, nonstoichiometry, or other chemical manipulations [39–53]. The effort in this challenging area is necessitated by a realization mentioned earlier that further reductions in the lattice thermal conductivity will, sooner or later, come to a halt and the only way how to improve the performance of TE materials will be to focus on the power factor, that is, on the electronic properties, which are ultimately determined by the band structure. Traditionally, doping shifts the Fermi level either up closer to the bottom of the conduction band or down closer to the valence band, in the process increasing the density of electrons or, respectively, holes. If this is all the doping does, one talks in terms of the rigid band model. However, some dopants may also modify the band structure. For instance, doping PbTe by Tl gives rise to a dramatically increased Seebeck coefficient as a consequence of strongly enhanced density of states due to the hybridization between the excited state of a Tl atom and the neighboring p-states of Te, yielding a doubling of the figure of merit ZT to above 1.5 at 773 K [39]. This phenomenon, referred to as the band resonant state, has also been observed in other systems, for example, in In-doped SnTe [40] and Ge-doped Zn_4Sb_3 [41], and Sn-doped Bi_2Te_3 [42]. While effective when present, there is no straightforward recipe predicting what dopant in which system should be used to achieve benefits of band resonant states [43].

Another approach how to enhance the density of states close to the Fermi level position is *via* band convergence [44–53]. It so happens that many systems have strongly temperature-dependent bands, which shift their position as the temperature is altered. A canonical example in this respect is PbTe where the valence band structure consists of two fairly closely lying bands in energy located at different *k*-points of the Brillouin zone: the upper and lighter valence band at the *L*-point (valley degeneracy of 4) and the lower lying and heavier valence band at the Σ-point with the valley degeneracy of 12. While the Σ band is a little temperature dependent, the *L*-point band moves down as the temperature increases, and at about 450 K, the band edges of the two valence bands coincide in energy, that is, the band convergence has been achieved, and the valley degeneracy has increased to 4 + 12 = 16. The enhanced valley degeneracy increases the density of states effective mass m^* given by $m^* = N_v^{2/3} m_b^*$, where m_b^* is the band effective mass without degrading the carrier mobility, which would have happened had the band effective mass been increased. Using the band convergence concept with Na-doped $PbTe_{0.85}Se_{0.15}$, Pei et al. [44] achieved an exceptionally high figure of merit of 1.8 at 850 K. The presence of Se in the structure, apart from providing enhanced phonon scattering, also effectively tuned the temperature where the valence band edges coincided in energy because it tended to increase the energy separation between the two bands and thus shifted the band convergence to higher temperatures. Band convergence can also be achieved by a chemical manipulation of the structure as demonstrated by Liu et al. [45] in the case of $Mg_2Si_{1-x}Sn_x$ solid solutions. It had been known that the band structure of both Mg_2Si and Mg_2Sn consists of two closely lying conduction bands: the lighter conduction band and the heavier conduction band [46]. While in Mg_2Si the band edge of the lighter conduction band lies lower in energy than the band edge of the heavier conduction band, in Mg_2Sn the order of the bands is reversed with the heavier band lying lower [47]. This means that for some particular value x of the Sn content, the two band edges must cross over and therefore coincide in energy. This has been confirmed theoretically by first-principles pseudopotential calculations [48], which suggested that the band convergence sets in when the Sn content is near $x = 0.625$. Indeed, at this solid solution composition, the Seebeck coefficient is maximized due to an enhanced density of states effective mass as the valley degeneracy N_v increases from the original 3 to 6 at the band convergence [49]. As a consequence, the dimensionless figure of merit attained the value of 1.3 at 750 K, a value by some 40% higher than previously reported figures of merit for $Mg_2Si_{1-x}Sn_x$ solid solutions. The chemically driven band convergence has also been observed in the $Sn_{1-x}Mn_xTe$ system yielding the optimized ZT value [50,51].

All the earlier strategies have contributed greatly to remarkable enhancements in the TE performance achieved in the past few years with both the classic TE materials and new families of TEs. In spite of these advances, thermoelectricity still remains a more-or-less peripheral and *niche* technology without a significant penetration of the mainstream industrial base. To make major inroads in the areas of power generation and cooling/air-conditioning, the performance of TE materials must still be improved further and the cost of manufacturing and assembling TE modules must be brought down considerably.

One of the significant factors in the overall cost structure of thermoelectricity is the synthesis of efficient TE materials. This includes the ready availability of elemental materials, the expense associated with processing the desired alloys and compounds, and scalability to a large-volume industrial operation. So far, TE materials have been synthesized mostly by one of the following methods: melting followed by slow cooling [4,6–8], melting followed by long time annealing [9–15], multistep solid state reactions [29,49], and mechanical alloying [40]. Each such synthesis process is time and energy consuming and not always easily scalable. Although some synthesis methods are faster, particularly the melt-spinning (MS) [18,20–21,25,26], where the entire processing can be completed in a couple of hours, such synthesis relies on the availability of specialized equipment. Moreover, it is often very difficult to control the desired stoichiometry and microstructure. Thus, there is a very strong demand to develop novel synthesis routes, which can be easily scalable and yield large quantities of high-performance TE materials at a low cost of manpower and energy in a very short time.

Recently, we reported on the adaptability of a well-known synthesis method used in the production of high-temperature refractory materials and known as the self-propagating high-temperature synthesis (SHS), which we applied in the previously unexplored and, in fact, considered *proscribed domain* of low temperatures [54–58]. Combining this technique with a subsequently applied plasma-activated sintering (PAS) process, we have demonstrated not only the rapid nature and scalability of this synthesis approach with several different TE materials (Bi_2Te_3-based alloys, skutterudite, SnTe, Cu_2Se, ZrNiSn, Cu_2SnSe_3, and $Mg_2Si_{1-x}Sn_x$, among others) but also the fact that the synthesis results in improved TE performance and enhanced mechanical properties of the resulting structures. Indeed, this synthesis process shows a great potential for large-scale synthesis of efficient TE materials. In this chapter, we discuss in details the essential elements of the SHS + PAS synthesis, and we illustrate on several examples the unique features of this processing technique. In addition, we also present a new empirically based criterion for the applicability of the SHS process that opens the door for using this method in the synthesis of low-melting-point compound semiconductors, the domain deemed previously inaccessible for the SHS synthesis.

20.2 Combustion Synthesis

Combustion synthesis is a method for synthesizing compounds by exothermic reactions, that is, by their own heat of fusion. Early research into this process and explorations of its feasibility for fabrication of materials started in the 1960s in what was then the Soviet Union. The technique was primarily targeting the synthesis of advanced structural materials such as refractories [59], high purity ceramics [60], and intermetallic compounds [61]. In general, the process can be subdivided according to two distinct modes of ignition, as shown in Figure 20.1. The first one is termed SHS in which an ingot of compacted

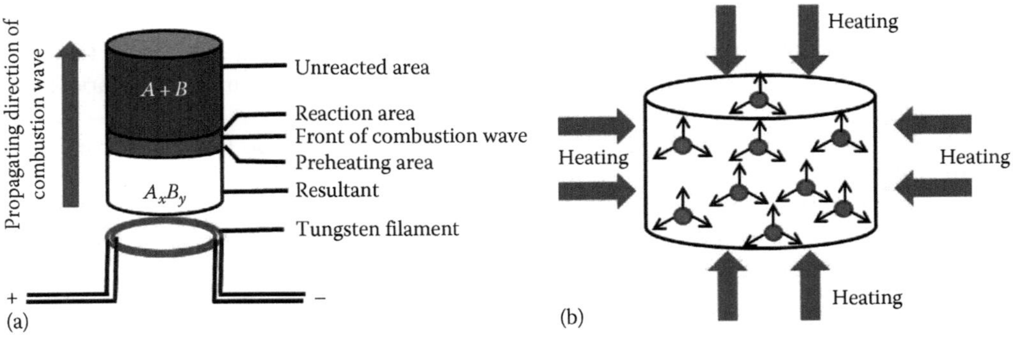

FIGURE 20.1
Schematic diagram of the two ignition modes: (a) self-propagating high-temperature synthesis and (b) thermal explosion.

elemental powders is ignited by point-heating its small section (usually the top or the bottom end of the ingot) using a heated tungsten coil. In this controlled combustion process, once the reaction is initiated, a combustion wave passes through the remaining material as the liberated heat of fusion in one section is sufficient to maintain the reaction in the neighboring section of the compact, leaving behind a fully reacted product. This type of ignition generates a temperature distribution along the length of the reacting ingot since the material reacted first will also start to cool first. The synthesis time is given by the time the combustion wave takes to pass through the entire length of the ingot. The second mode of ignition is termed thermal explosion, in which case the whole volume of the compacted powder is heated to the ignition temperature and the compacted powder fuses simultaneously to yield the reacted product. In this case, the synthesis time depends on how rapidly one can heat the whole ingot of compacted elemental powders to the ignition temperature. Whichever ignition mode is chosen, the combustion synthesis is exceptionally fast, typically on the time-scale of seconds. The substantially contactless nature of the combustion synthesis assures minimal contamination of the reacted product from the enclosure and the speed of the process mitigates evaporation of more volatile elements. A distinct advantage of the SHS process over the thermal explosion is a purification of the compact as the combustion wave passes through, resembling the process of zone leveling. In addition, due to the high heating and cooling rates, nonequilibrium structures can be achieved in the SHS-reacted products [59,62]. Since the combustion synthesis is a thermodynamic as well as kinetic processing, among the key parameters specifying the combustion process are the velocity of the combustion wave, the adiabatic temperature, and the combustion temperature. The velocity of the combustion wave is determined by the composition and the degree of compaction of the elemental powders and typically ranges from 1 to 250 mm s^{-1} [63–66]. Once the combustion is ignited, very high temperatures are reached in a very short time due to the highly exothermic nature of reactions. It is therefore reasonable to assume that one deals with a thermally isolated exothermic system because there is very little time for the heat to disperse to the surroundings. Therefore, the maximum temperature to which the product is raised can be assumed to be the adiabatic temperature, T_{ad}. As the enthalpy of the system is a state function,

$$-\Delta_f H_{298\,\mathrm{K}} = H_T^0 - H_{298\,\mathrm{K}}^0 = \int_{298\,\mathrm{K}}^{T_{ad}} C_p d_T \tag{20.1}$$

where $\Delta_f H_{298\,\mathrm{K}}$ is the standard enthalpy of formation of the product at 298 K, C_p is the molar-specific heat of the product, the integral includes latent heats of melting and vaporization if any phase transitions are present.

For most compounds, the required thermodynamic data can be obtained from the literature. Using Equation 20.1, the all-important adiabatic temperature T_{ad} can be calculated for any reaction. In practice, because the process is not fully adiabatic (some heat from the surface of the compacted ingot escapes to the surroundings), the experimentally determined combustion temperature is usually somewhat lower than the theoretically estimated T_{ad}.

Merzhanov [63–66] suggested that the combustion reaction will not be self-propagating unless T_{ad} reaches at least 1800 K. Of course, if this were really so, there would be no hope for using SHS in the synthesis of compound TE materials as the majority of such structures would either decompose, melt, or outright vaporize at such high temperatures. To our great surprise, when we attempted to use the SHS process to synthesize several different compound TE materials, the process worked and the reactions took place at considerably lower temperatures than 1800 K. Clearly, the Merzhanov's original criterion regarding the value of the adiabatic temperature necessary to attain in order for the combustion reaction to propagate is not universal. A new criterion, based on our understanding of the combustion process, is discussed in Section 20.7.

Since the relative density of the reacted compacts following the SHS process is low, it is necessary to densify the material further either by hot-pressing or, even better, by plasma-activated sintering (PAS). Thus, in all our studies of the combustion synthesis, we have used a combination of SHS–PAS processing [54–58].

In the following sections, we will illustrate the SHS–PAS synthesis process on four materials: classic Bi_2Te_3, $CoSb_3$-based skutterudite, recently intensively studied SnTe, and Cu_2Se possessing a phase transition within its operational range and having a substantial ionic conduction at high temperatures. In each case, we provide the relevant thermodynamic and kinetic parameters and briefly describe the resulting TE performance.

20.3 $Bi_2Te_{3-x}Se_x$ Synthesized by SHS–PAS

Bismuth telluride is a narrow-gap semiconductor and its solid solutions with Sb_2Te_3 and Bi_2Se_3 are the best *p*-type, respectively, *n*-type room temperature TE materials with the TE figure of merit ZT ~1 [67,68]. Commercial Bi_2Te_3-based materials are typically prepared by zone melting (ZM) [69,70]. However, the ZM process is time and energy intensive. Moreover, ingots prepared by ZM are near single-crystalline and cleave easily along the basal plane (perpendicular to the trigonal *z*-axis) due to weak van der Waals bonding between the neighboring $Te^{(1)}$–$Te^{(1)}$ layers. Therefore, it is of great interest to develop preparation methods for Bi_2Te_3-based TE compounds that are simple, inexpensive, rapid, and scalable and yield samples with high TE performance and robust mechanical properties. Although some new promising synthesis routes have been developed in the recent years, among them are ball-milling (BM), spark plasma sintering (SPS) [71], hot deformation (HP) [72], MS-SPS [20,25,26], and microwave-assisted synthesis [73]; these new technologies require specialized equipment, and the material often oxidizes and contains impurities [25,71–73]. Here, we illustrate advantages of the SHS–PAS synthesis in the preparation of high-performance Bi_2Te_3-based TE materials. We emphasize the single-phase nature of the resulting material and the superior TE and mechanical properties of such SHS–PAS processed material.

20.3.1 Phase Transformation Mechanism

We illustrate the phase transformation mechanism on *n*-type $Bi_2Te_{3-x}Se_x$ prepared directly from single elements and without any prior processing [55]. High-purity elemental powders of 99.99% Bi (under 200 mesh), 99.999% Se (under 200 mesh), and 99.99% Te (under 200 mesh) were weighed according to the stoichiometry of $Bi_2Te_{3-x}Se_x$, and then mixed uniformly in an agate mortar for 15 min. The mixed powders were then cold-pressed into a pellet with a steel die, and the pellet was placed and sealed in a quartz tube under vacuum. The SHS process was initiated by igniting the pellet at the bottom of the quartz tube. Since the combustion process of $Bi_2Te_{3-x}Se_x$ is almost the same as the combustion process of Bi_2Te_3 and Bi_2Se_3, we illustrate the process on the two-end members. Figure 20.2

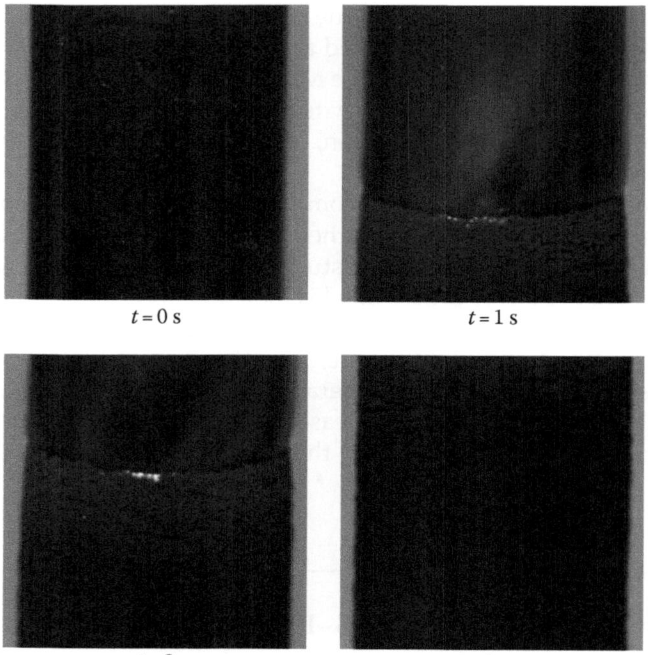

FIGURE 20.2
Typical images of a cold-pressed pellet of a mixture of Bi and Te powders at different times as the combustion wave passes through. (From Zheng, G., Su, X.-L., Liang, T. et al., High thermoelectric performance of mechanically robust *n*-type $Bi_2Te_{3-x}Se_x$ prepared by combustion synthesis, *J. Mater. Chem. A*, 3, 6603–6613, 2015. Reproduced by permission of The Royal Society of Chemistry.)

shows different stages of the SHS process of Bi_2Te_3. Following the ignition of the compacted powder of Bi and Te, the heat released in the exothermic reaction is high enough to ignite the neighboring layers of the pellet, the reaction becomes self-propagating, and the combustion wave passes through the entire pellet. Finally, the reacted pellet cools down. The measured propagation speed of the combustion wave in Bi_2Te_3 was 6 mm s^{-1}, which is in the range of reported values of 1–250 mm s^{-1}, demonstrating that the phenomenon is a typical SHS process [63–66]. The measured combustion temperature on the surface of Bi_2Te_3 was 722 K, while on the surface of Bi_2Se_3, it was a bit higher at about 740 K, as shown in the temperature profile of the SHS process in Figure 20.3a. Since the SHS synthesis is a nonequilibrium process, a question always arises regarding how the combustion reaction (fusion) proceeds and how the microstructure evolves during sintering. In order to gain more insight into the process, differential scanning calorimetry (DSC) in conjunction with x-ray diffraction (XRD) was employed. Figure 20.3b through d show the DSC curve with the heating rate of 10 K min^{-1} and XRD patterns of the products collected at 523, 573, 623, 673, 723, and 773 K. In Figure 20.3b, two endothermic peaks and two exothermic peaks are observed for Bi_2Te_3. The endothermic peak at 544 K corresponds to the melting process of Bi and the other endothermic peak at around 685 K depicts the formation of a eutectic phase between Te and Bi_2Te_3. The exothermic peak at 688 K corresponds to the formation process of Bi_2Te_3 while the exothermic peak at 630 K depicts the formation of intermediate phases of BiTe and Bi_4Te_3. All processes are confirmed by XRD patterns of the mixture heated to different temperatures.

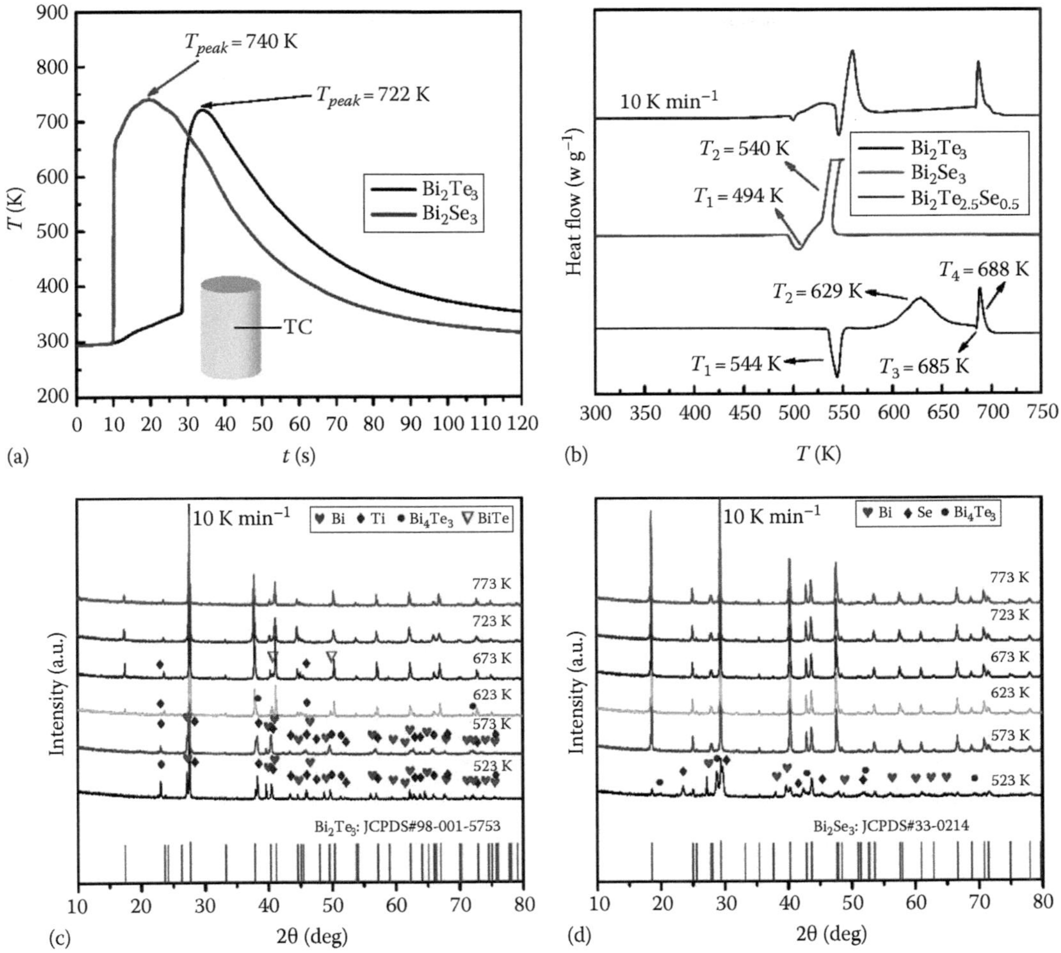

FIGURE 20.3
(a) Temperature versus time as recorded by a thermocouple attached to a compact during the self-propagating high-temperature synthesis processing of Bi_2Te_3 and Bi_2Se_3. (b) Differential scanning calorimetry curves with the heating rate of 10 K min^{-1} recorded from 300 to 750 K for Bi–Te, Bi–Se, and Bi–Te–Se mixed powders. (c) X-ray diffraction (XRD) patterns obtained on a Bi-Te mixed powder compact heated to 523, 573, 623, 673, 723, and 773 K, respectively, with the heating rate of 10 K min^{-1}. (d) XRD patterns obtained on a Bi–Se mixed powder compact heated to temperatures 523, 573, 623, 673, 723, and 773 K, respectively, with the heating rate of 10 K min^{-1}. (From Zheng, G., Su, X.-L., Liang, T. et al., High thermoelectric performance of mechanically robust *n*-type $Bi_2Te_{3-x}Se_x$ prepared by combustion synthesis, *J. Mater. Chem. A*, 3, 6603–6613, 2015. Reproduced by permission of The Royal Society of Chemistry.)

In contrast with the formation process of Bi_2Te_3, the fusion of Bi_2Se_3 seems more simple as just one endothermic peak (corresponding to the melting of Se), and one exothermic peak (corresponding to the formation of Bi_2Se_3) is detected; see Figure 20.3d. No intermediate compounds are observed during the fusion of Bi and Se forming the Bi_2Se_3 compound. After a short time of a few seconds, the time for the combustion wave to pass through the compacted powder of stoichiometric quantities of Bi, Te, and Se, a single-phase $Bi_2Te_{3-x}Se_x$ solid solution is obtained. This reacted product is ground into a powder

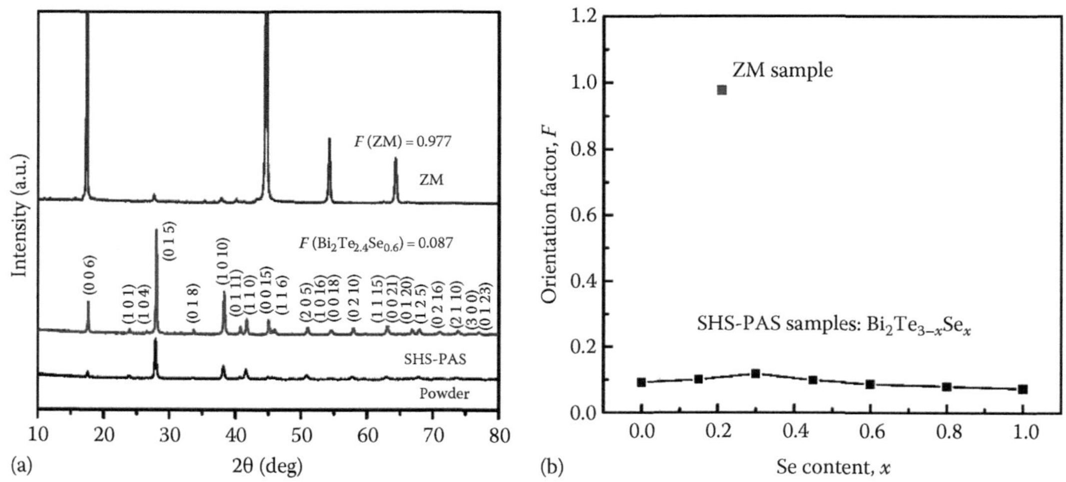

FIGURE 20.4
(a) X-ray diffraction peak assignment for $Bi_2Te_{2.4}Se_{0.6}$ and the zone melting (ZM) sample. Bottom trace, powder of $Bi_2Te_{2.4}Se_{0.6}$ after SHS. Middle trace, powder of $Bi_2Te_{2.4}Se_{0.6}$ after self-propagating high-temperature synthesis-plasma-activated sintering (SHS–PAS). Top trace, ZM sample, reflections taken. (From Zheng, G., Su, X.-L., Liang, T. et al., High thermoelectric performance of mechanically robust *n*-type $Bi_2Te_{3-x}Se_x$ prepared by combustion synthesis, *J. Mater. Chem. A*, 3, 6603–6613, 2015. Reproduced by permission of The Royal Society of Chemistry.)

and then sintered under pressure of 30 MPa at 753 K in vacuum using the PAS apparatus. The surface XRD of the pellet after the SHS–PAS process taken along the direction of applied pressure shows no preferential orientation while the zone-melted ingots show fully oriented structures with the orientation factor equal to 1; see Figure 20.4. Figure 20.5 shows the microstructure of $Bi_2Te_{3-x}Se_x$ after SHS–PAS. The grain structure has no preferred orientation and the elements are distributed homogeneously on the micron scale. In contrast, the grain size of the zone-melted ingot is very large and the grains display parallel sets of cleavage planes consistent with the large orientation factor. The essentially isotropic morphology of the SHS–PAS processed samples will be reflected in their transport properties.

20.3.2 Thermoelectric and Mechanical Properties of $Bi_2Te_{3-x}Se_x$

Figure 20.6 shows the temperature dependence of the electrical conductivity and the Seebeck coefficient for SHS–PAS processed $Bi_2Te_{3-x}Se_x$. The electrical conductivity decreases with the increasing temperature displaying the behavior of a highly degenerate semiconductor. As the content of Se increases, the electrical conductivity decreases. The Seebeck coefficient of all SHS–PAS processed samples is negative in the entire temperature range, indicating an *n*-type material. With the increasing content of Se, the absolute value of the Seebeck coefficient initially increases and then decreases. The highest Seebeck coefficient of −191 μV K^{-1} is obtained for $Bi_2Te_{2.4}Se_{0.6}$ at 448 K. Figure 20.7 shows the temperature dependence of the thermal conductivity for SHS–PAS synthesized $Bi_2Te_{3-x}Se_x$. With a greater fraction of Se in the compound, the thermal conductivity of $Bi_2Te_{3-x}Se_x$ prepared by the SHS–PAS process decreases. The lattice thermal conductivity initially decreases due to the dominance of Umklapp phonon scattering but, at elevated temperatures, it turns

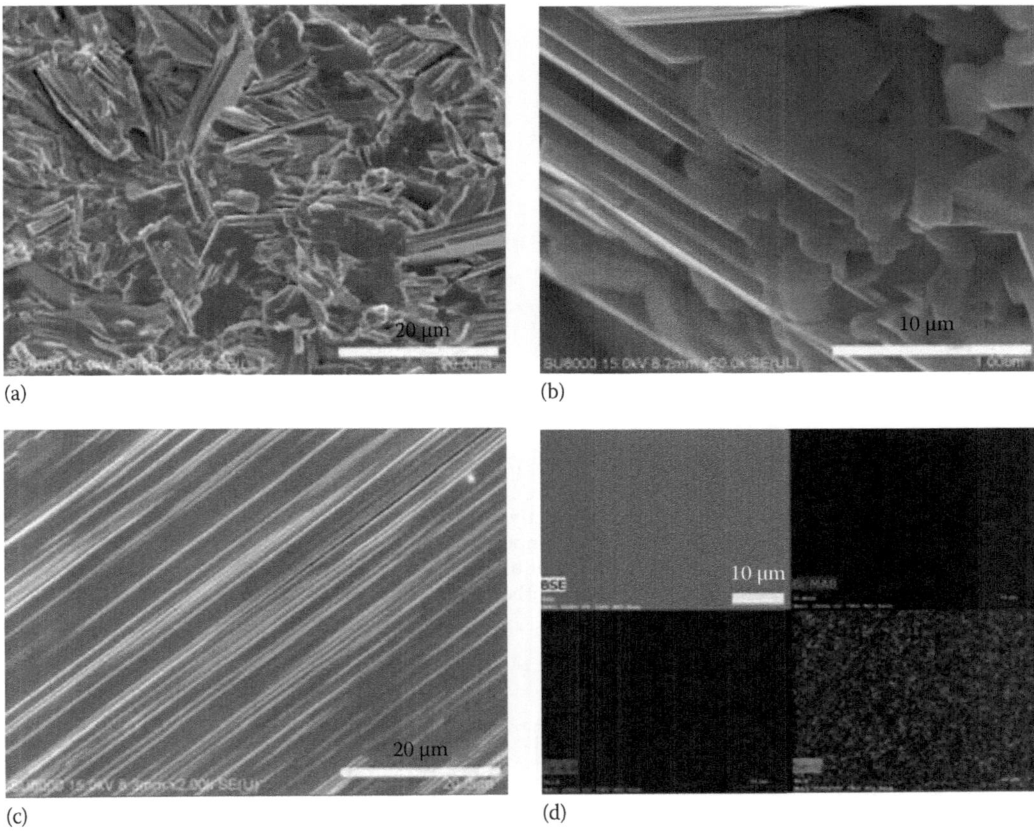

FIGURE 20.5
(a) Field emission scanning electron microscope (FESEM) image of the $Bi_2Te_{2.4}Se_{0.6}$ sample. (b) Higher magnification FESEM of the $Bi_2Te_{2.4}Se_{0.6}$ sample. (c) FESEM of the zone melting sample. (d) Element maps of the self-propagating high–temperature synthesis-plasma-activated sintering processed $Bi_2Te_{2.4}Se_{0.6}$. (From Zheng, G., Su, X.-L., Liang, T. et al., High thermoelectric performance of mechanically robust *n*-type $Bi_2Te_{3-x}Se_x$ prepared by combustion synthesis, *J. Mater. Chem. A*, 3, 6603–6613, 2015. Reproduced by permission of The Royal Society of Chemistry.)

around and increases with increasing temperature on account of a bipolar contribution associated with intrinsic excitations across the bandgap. Typically, one would expect the lattice thermal conductivity to decrease with the increasing content of Se due to more intense alloy scattering. While, indeed, the lattice thermal conductivity initially decreases, it then reverses the trend and actually increases as the content of Se increases. Thus, the sample with the Se content of 0.6 has the lowest lattice thermal conductivity instead of the sample with the Se content of $x = 1.5$ where the maximum disorder would be expected. This may be ascribed to changes in the interatomic coupling force as the Se content increases because the difference in electronegativities of Bi and Te is not the same as between Bi and Se. The highest ZT value of 0.95 is reached near 423 K with $Bi_2Te_{2.4}Se_{0.6}$ while the maximum ZT value obtained with zone-melted samples is 0.77 at 373 K, almost the same as the literature value [69,74]. We note that the ZT value of $Bi_2Te_{2.4}Se_{0.6}$ synthesized by the SHS–PAS process is superior compared to the ZT value of the zone-melted sample of the same composition over essentially the entire range of temperatures; see Figure 20.8a. The average

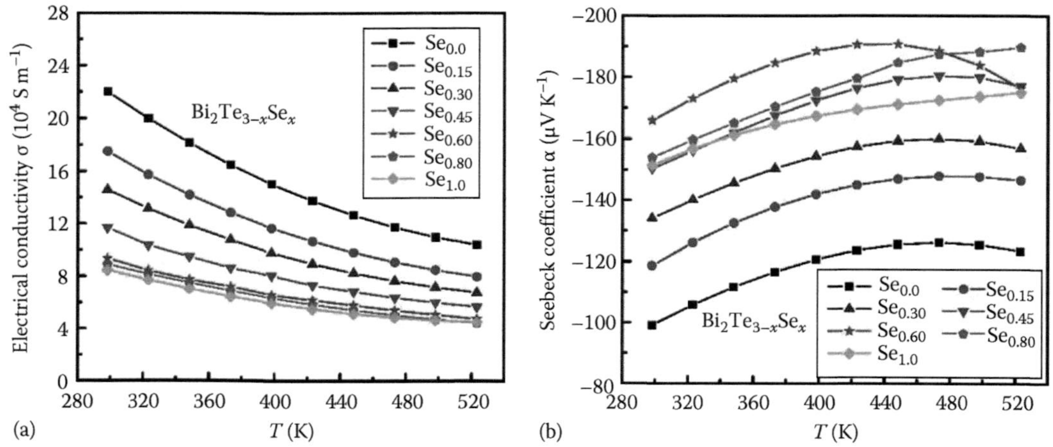

FIGURE 20.6
(a) Temperature dependence of the electrical conductivity for self-propagating high-temperature synthesis–plasma-activated sintering (SHS–PAS) samples. (b) Temperature dependence of the Seebeck coefficient for SHS–PAS samples. (From Zheng, G., Su, X.-L., Liang, T. et al., High thermoelectric performance of mechanically robust *n*-type $Bi_2Te_{3-x}Se_x$ prepared by combustion synthesis, *J. Mater. Chem. A*, 3, 6603–6613, 2015. Reproduced by permission of The Royal Society of Chemistry.)

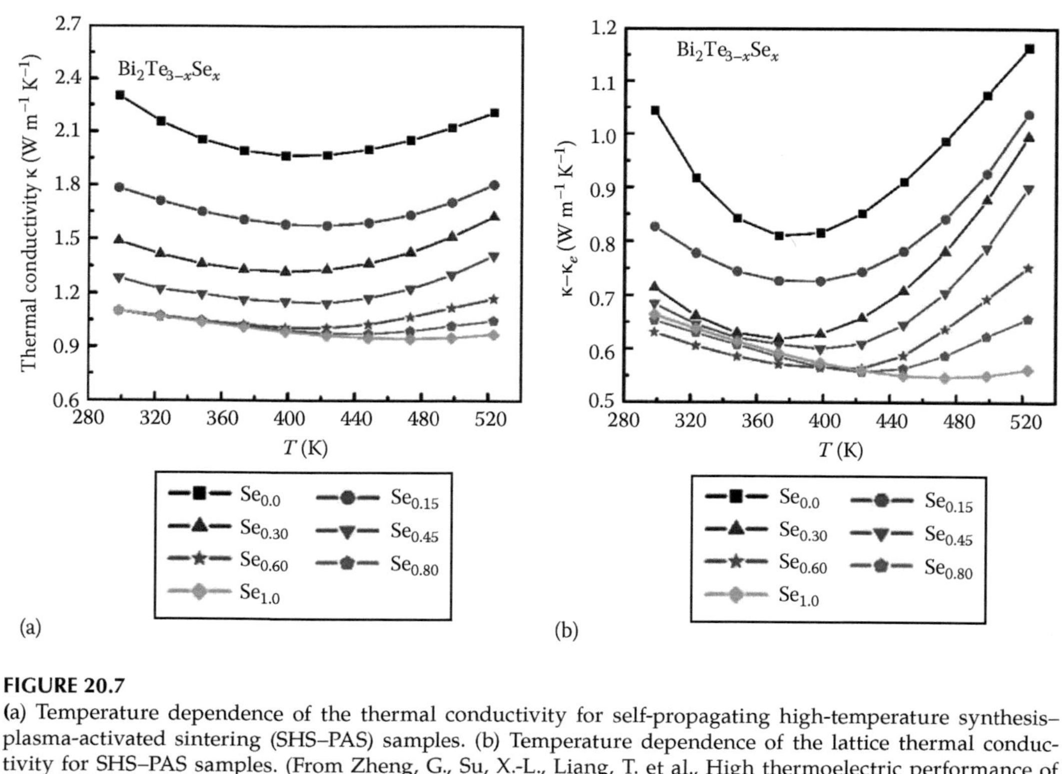

FIGURE 20.7
(a) Temperature dependence of the thermal conductivity for self-propagating high-temperature synthesis–plasma-activated sintering (SHS–PAS) samples. (b) Temperature dependence of the lattice thermal conductivity for SHS–PAS samples. (From Zheng, G., Su, X.-L., Liang, T. et al., High thermoelectric performance of mechanically robust *n*-type $Bi_2Te_{3-x}Se_x$ prepared by combustion synthesis, *J. Mater. Chem. A*, 3, 6603–6613, 2015. Reproduced by permission of The Royal Society of Chemistry.)

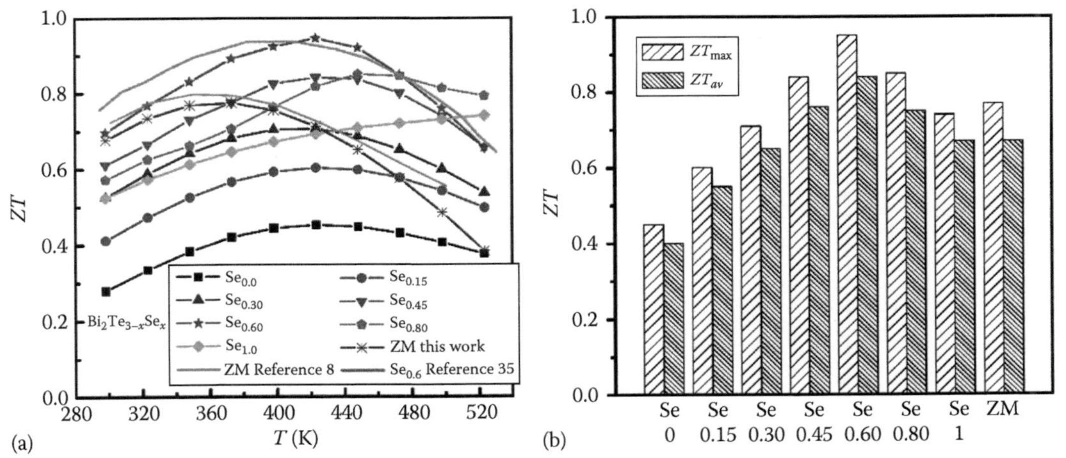

FIGURE 20.8
(a) Temperature dependence of ZT for self-propagating high-temperature synthesis–plasma-activated sintering (SHS–PAS) samples and the zone-melting (ZM) sample measured parallel to the growth direction, that is, its long axis being perpendicular to the *c*-axis. (b) The maximum and the average ZT over the entire temperature range from 298 to 523 K for SHS–PAS samples and the ZM sample. (From Zheng, G., Su, X.-L., Liang, T. et al., High thermoelectric performance of mechanically robust *n*-type $Bi_2Te_{3-x}Se_x$ prepared by combustion synthesis, *J. Mater. Chem. A*, 3, 6603–6613, 2015. Reproduced by permission of The Royal Society of Chemistry.)

ZT value (see Figure 20.8b), of samples in the temperature range studied, 298–523 K, was calculated through an integral. For the zone-melted sample, it yields $(ZT)_{av} = 0.67$, while the average figure of merit of $Bi_2Te_{2.4}Se_{0.6}$ synthesized by SHS–PAS is $(ZT)_{av} = 0.84$, about a 25% enhancement compared with the zone-melted sample. In addition, due to the grain size refinement and the nearly isotropic structure of SHS–PAS prepared samples, the compressive strength and the bending strength of SHS–PAS samples displayed in Figure 20.9 is 104 and 70 MPa, respectively, values much higher than the corresponding parameters of zone-melted samples, no matter in which direction measured. Because of its simple processing nature, short synthesis duration, simple equipment, and low energy needs, the SHS–PAS technology provides a new economical approach for large-scale fabrication of bismuth telluride-based TE compounds.

20.4 Cu_2Se Synthesized by SHS–PAS

Cu_2Se has two crystallographic forms: α-Cu_2Se, which is a low-temperature phase of the compound, and β-Cu_2Se, which is cubic, high-temperature phase. β-Cu_2Se is a superionic conductor and attracts considerable attention due to its excellent TE performance and earth abundant and low-cost elements as well as due to its electron-crystal phonon-liquid transport behavior [75,76]. The structure shows a great potential for power generation applications. However, the TE performance of β-Cu_2Se is very sensitive to its composition [75,76]. Controlling the composition precisely is the key to achieving the optimized TE performance. Meanwhile, developing the technology that enables the synthesis on a large scale and at low cost is very important from a commercial point of view. Cu_2Se has

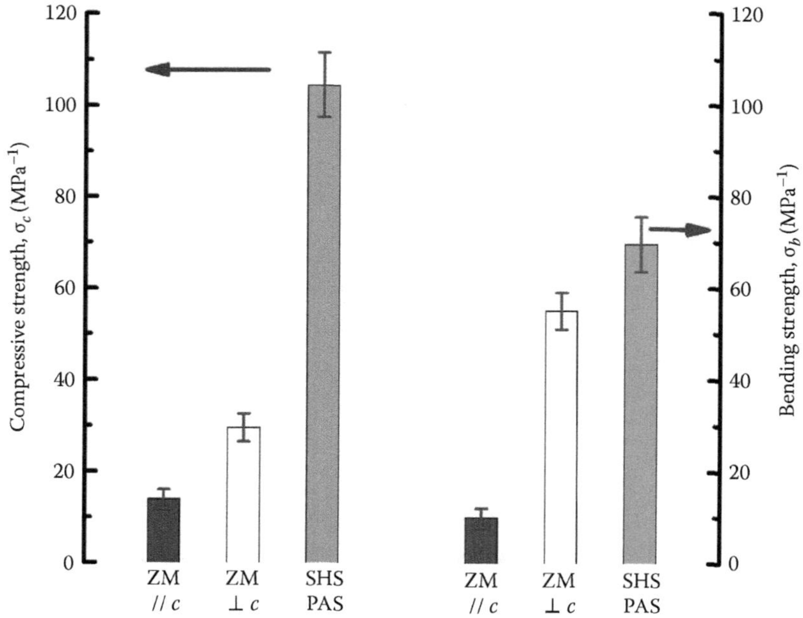

FIGURE 20.9
The average compressive strength and the bending strength of $Bi_2Te_{2.4}Se_{0.6}$ and zone melting samples. (From Zheng, G., Su, X.-L., Liang, T. et al., High thermoelectric performance of mechanically robust *n*-type $Bi_2Te_{3-x}Se_x$ prepared by combustion synthesis, *J. Mater. Chem. A*, 3, 6603–6613, 2015. Reproduced by permission of The Royal Society of Chemistry.)

primarily been synthesized by melting-quenching [76], melting-annealing [75], and by BM [77]. It is very difficult to control the composition precisely due to low melting (494 K) and boiling (958 K) points of Se, oxidation of Cu, and high–vapor pressure of Se. Moreover, all the synthesis approaches mentioned earlier take a long time, consume a lot of energy, and are not easy to scale up. In this section, we describe the SHS synthesis of Cu_2Se including the phase transformation mechanism, microstructure, and TE properties.

20.4.1 Phase Transformation Mechanism during the SHS Synthesis of Cu_2Se

Figure 20.10a shows infrared images of Cu_2Se in three different stages of the SHS process: (1) local ignition, (2) combustion wave passing through, and (3) product cooling. As already mentioned, the entire process takes a few seconds. Temperature changes as the combustion wave passes through at different positions along the length of a rectangular-shaped pellet are recorded by thermocouples TC1–TC3 and are shown in the inset of Figure 20.10b. The initial heating rate was more than 200 K s^{-1}. The maximum temperature reached in the center of the pellet was 835 K and the pellet cools down rapidly thereafter. The speed of the combustion front movement, 5.6 mm s^{-1}, was calculated by dividing the distance between the thermocouples by the time difference at which the thermocouples reached a certain temperature. The resulting speed is rather low, but it falls within the range of 1–250 mm s^{-1} reported in the literature for combustion fronts in other SHS-processed materials [63–66]. Figure 20.10c shows the DSC curves for the stoichiometric mixture of Cu and Se powders corresponding to two very different heating rates. When slowly heated

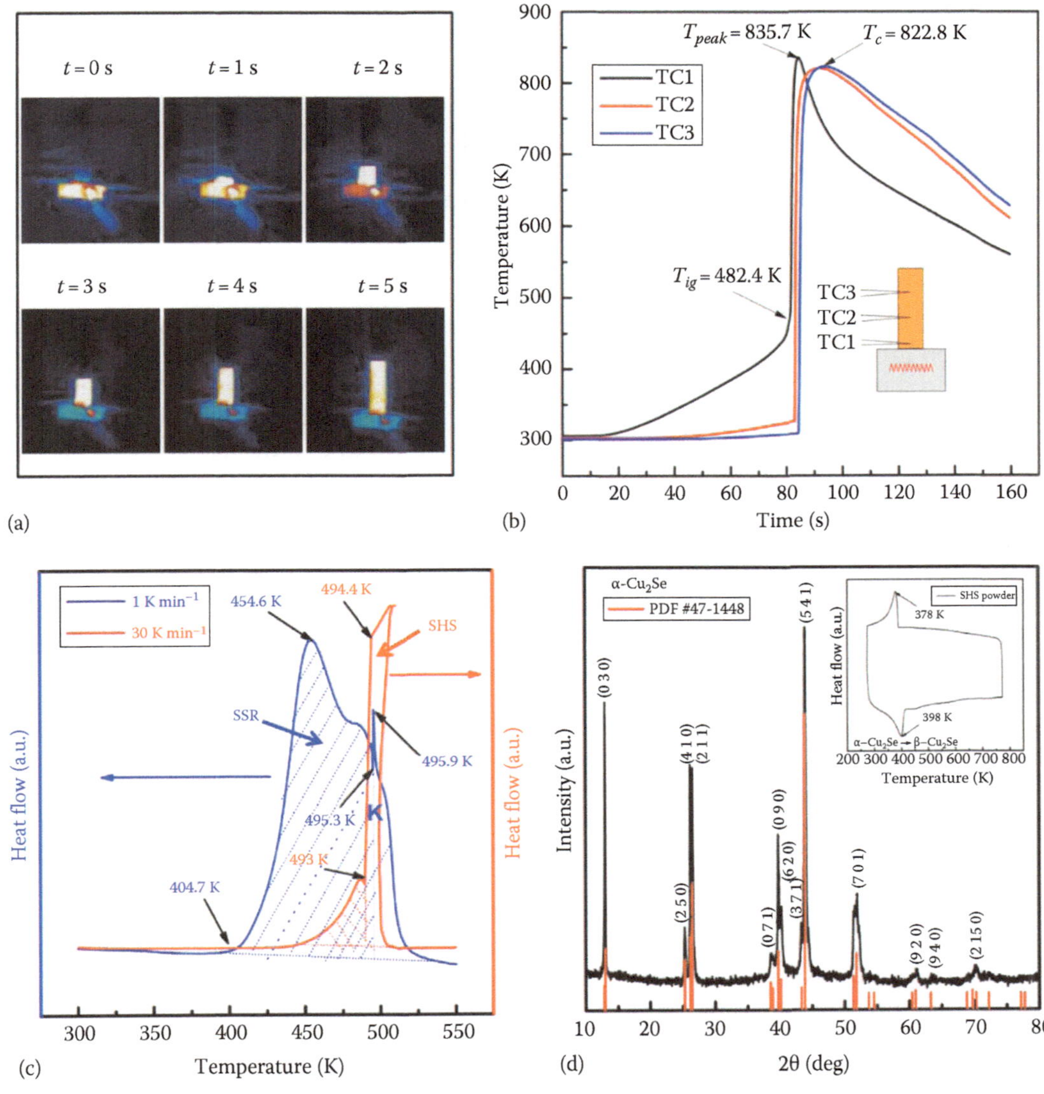

FIGURE 20.10
Characteristics of the self-propagating high-temperature synthesis (SHS). (a) Infrared photos at different stages of the SHS process, (b) temperature profile of thermocouples at different positions, (c) differential scanning calorimetry (DSC) curves of the SHS process, (d) x-ray diffraction pattern of the SHS product; the inset shows DSC cycling of the SHS product. (Reprinted by permission from Macmillan Publishers Ltd., *Nat. Commun.*, Su, X.-L., Fu, F., Yan, Y. et al., Self-propagating high-temperature synthesis for compound thermoelectrics and new criterion for combustion processing, 5, 4908, 2014, copyright 2014.)

(the rate of 1 K min^{-1}), the exothermic peak is much wider, and the heat flow rate is low, a situation corresponding to solid-state reaction processing. In contrast, when the heating rate increases to 30 K min^{-1}, the exothermic peak becomes sharp and huge. Figure 20.10c also shows that the ignition temperature is actually the melting point of Se. Figure 20.10d depicts the XRD pattern of the reacted material after SHS. The data demonstrate that the reacted material is a single-phase α-Cu_2Se structure in accord with JSPDS47-1448. The inset of Figure 20.10d shows the DSC curve of the material after SHS. The characteristic feature of Cu_2Se—its reversible α–β phase transition taking place at about 396 K—is clearly revealed, and it confirms that the Cu–Se ratio is, indeed, 2:1 since the phase transition temperature is a function of this ratio.

In order to determine how Cu_2Se forms from powders of Cu and Se during the SHS processing, a series of DSC traces and x-ray diffraction images were collected on 20 mg cold-pressed pellets heated to different temperatures, at different heating rates and different temperature resting stages. First, the pellet was cycled in turn to temperatures of 413, 433, 453, 473, 493, 513, 533, 553, and 573 K at a rate of 10 K min^{-1} and held at each temperature for 1 h. From each temperature it was cooled down to 273 K with the cooling rate of 30 K min^{-1}. All thermal cycling was done under the protection of an Ar gas flowing at a rate of 50 mL min^{-1}. Figure 20.11a shows XRD patterns of the pellet after each thermal cycle. For temperatures lower than 493 K, the Cu and Se started to react and form intermediate compounds, but no Cu_2Se was detected. Cycling to temperatures in the range of 513–573 K, XRDs of such powders were almost identical and indicated the presence of α-$Cu_{2-x}Se$ as the main phase. A small quantity of $Cu_{1.8}Se$ and Cu is also observed in the XRD pattern. DSC curves document that the α–β phase transition temperature increases with the increasing cycling temperature. The phase transition temperature settles at 396 K after the pellet was cycled to 573 K and was held there for 1 h, consistent with the literature value of the α–β phase transition in Cu_2Se. Using the heating rate of 10 K min^{-1}, single-phase Cu_2Se cannot be achieved except upon cycling to 573 K. However, since the real heating rate for the SHS process is much larger than 10 K min^{-1}, a question arises on what is the real ignition temperature. In order to determine it, the pellet was cycled to the same range of temperatures as before and under the same conditions except that a much faster heating rate of 30 K min^{-1} was used. As Figure 20.12 indicates, a single-phase Cu_2Se is obtained following cycling to merely 493 K and even without the need to park there for 1 h. We have also tried to cycle the pellet to 493 K with a heating rate of 20 K min^{-1}, but Cu_2Se did not form; see Figure 20.12b. Hence, according to the XRD and DSC results, the ignition temperature is 493 K, and for the process to be self-sustaining, the heating rate of at least 30 K min^{-1} is required. The temperature of 493 K is the melting point of Se, and thus, the ignition point coincides with melting of Se and dissolution of Cu in it. The data indicate that the reaction mechanism is determined chiefly by the rate with which the pellet is heated. When the heating rate is less than 30 K min^{-1}, the synthesis process is a solid state reaction resulting in intermediate products that require temperatures higher than 573 K to transform to Cu_2Se. In the SHS process, Cu_2Se is formed directly following the passage of the combustion wave and such synthesis effectively controls the composition and yields homogeneous Cu_2Se in a timespan of a few seconds.

20.4.2 Microstructure of Cu_2Se Synthesized by SHS

Since the SHS synthesis is a nonequilibrium process with high combustion temperature and high heating and cooling rates [63–66], a question arises as to what is its impact on the microstructure of Cu_2Se.

Figure 20.13a through f present the microstructure of Cu_2Se following the ultrafast SHS. Field emission scanning electron microscope (FESEM) photographs of the Cu_2Se powder after SHS are shown in Figure 20.13a and b. Two kinds of microstructure are present: a matrix of micron-scale size grains and a corn-like distributed nanostructure with the size of 5–10 nm. Figure 20.13c is a TEM image displaying several nanostructural features in the micron-scale matrix consistent with FESEM results. Figure 20.13d and e are High

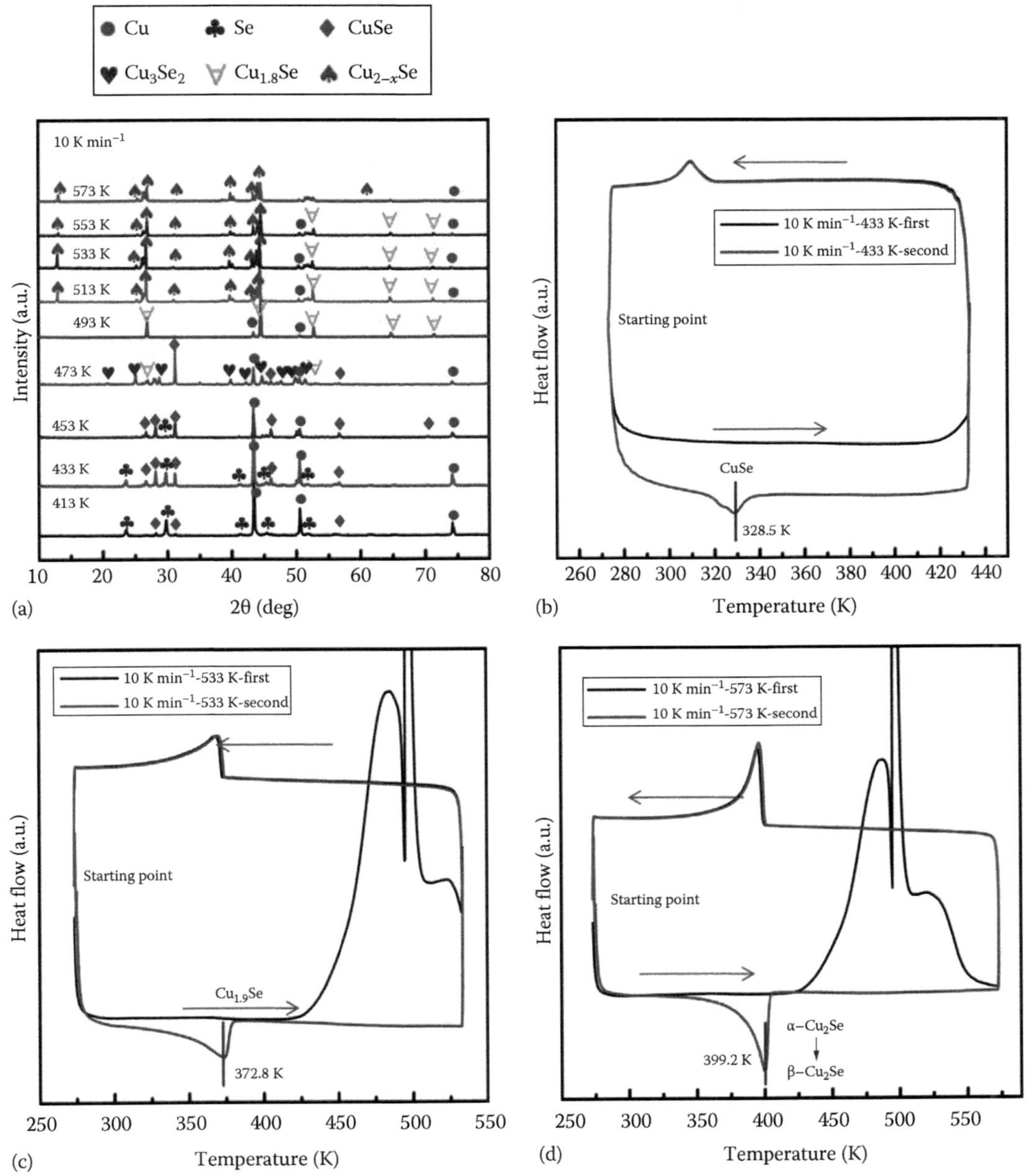

FIGURE 20.11
X-ray diffraction (XRD) patterns and differential scanning calorimetry (DSC) curves of Cu–Se pellets (with the stoichiometry of 2:1) cycled to different temperatures with the rate of 10 K min^{-1} and held there for 1 h. (a) XRD patterns, (b) DSC curves of pellets cycled once and twice to 433 K and held there for 1 h, (c) DSC curves of pellets cycled once and twice to 533 K and held there for 1 h, and (d) DSC curves of pellets cycled once and twice to 573 K and held there for 1 h. A small arrow on the curves indicates the direction of thermal cycling. (Reprinted by permission from Macmillan Publishers Ltd., *Nat. Commun.*, Su, X.-L., Fu, F., Yan, Y. et al., Self-propagating high-temperature synthesis for compound thermoelectrics and new criterion for combustion processing, 5, 4908, 2014, copyright 2014.)

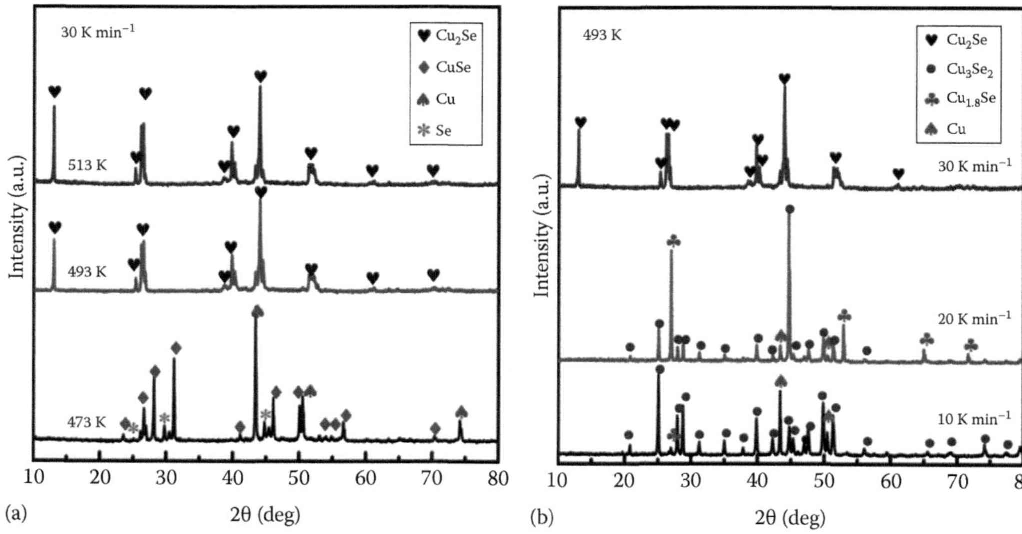

FIGURE 20.12
(a) X-ray diffraction (XRD) patterns of Cu-Se pellets (with the stoichiometry of 2:1) cycled to different temperatures with the rate of 30 K min^{-1} and without any temperature arrest. (b) XRD patterns of Cu-Se pellets (with the stoichiometry of 2:1) cycled to 493 K with different heating rates and no temperature arrest. (Reprinted by permission from Macmillan Publishers Ltd., *Nat. Commun.*, Su, X.-L., Fu, F., Yan, Y. et al., Self-propagating high-temperature synthesis for compound thermoelectrics and new criterion for combustion processing, 5, 4908, 2014, copyright 2014.)

resolution transmission electron microscopy (HRTEM) images showing interplanar spacing of 0.206 and 0.336 nm, respectively, corresponding to the (541) and (211) crystal faces of α-Cu_2Se according to JPDS47-1448. Semiquantitative energy dispersive spectrometer (EDS) results demonstrate that the corn-like nanostructure and the matrix are of the same composition normalized to Cu_2Se. Figure 20.13e shows that the interface between the nanostructure and the matrix is substantially coherent, which has implications for the charge carrier and phonon transport discussed later.

Just as other SHS processed compounds, Cu_2Se obtained after SHS does not have its full density and must be crushed and compacted by using PAS, resulting in ingots of better than 98% theoretical density. How the microstructure evolves following this brief PAS process is shown in Figure 20.13g through l. The grain size of the matrix remains at about 1–3 μm. While nanodots are observed on grain faces of the matrix with the size of about 30 nm, HRTEM images in Figure 20.13j through l indicate that at least some of the nanodots maintain their size of less than 10 nm and are essentially identical to nanodots prior to PAS. The interface between the nanostructure and the matrix remains substantially coherent. All these features should be beneficial for efficient TE properties.

20.4.3 Thermoelectric Properties of Cu_2Se Synthesized by SHS–PAS

The temperature dependence of TE properties of Cu_2Se is shown in Figure 20.14. To verify the measurements and their reproducibility, transport studies were carried out at

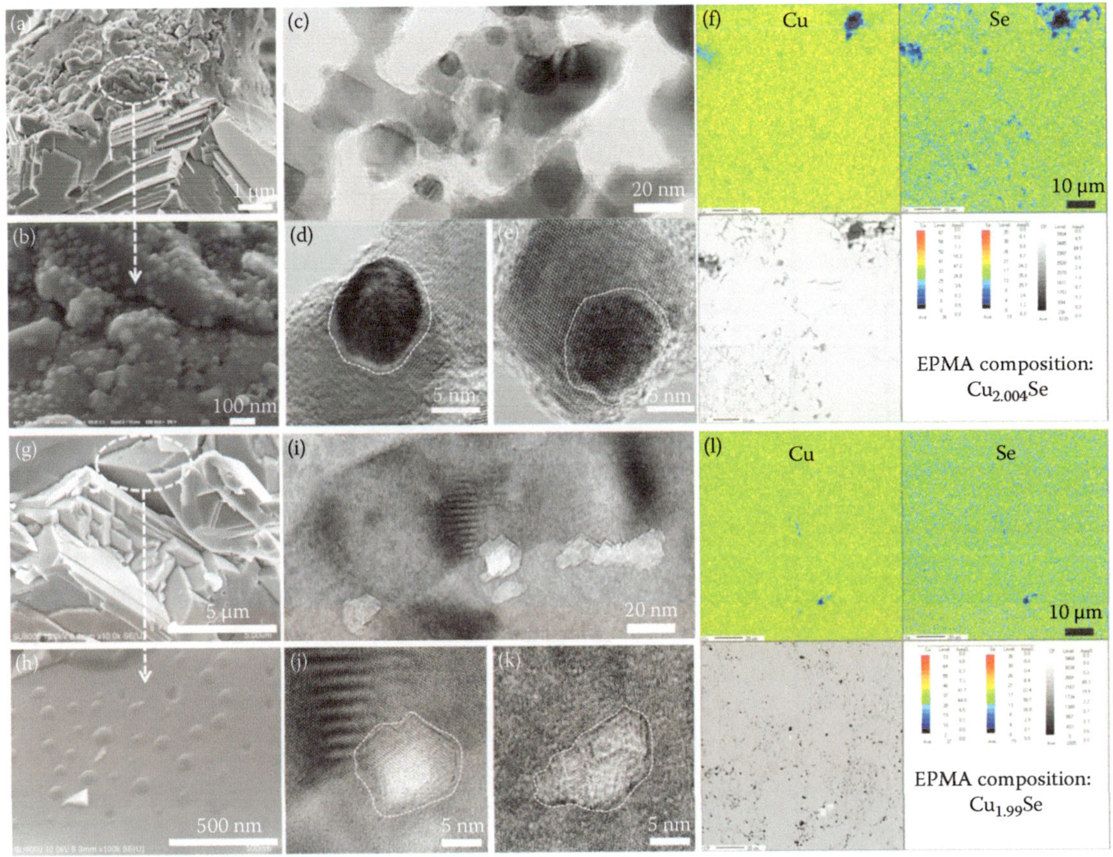

FIGURE 20.13
Microstructure of the Cu_2Se compound prepared by self-propagating high-temperature synthesis–plasma-activated sintering. (a, b) Field emission scanning electron microscope (FESEM) images of the self-propagating high-temperature synthesis (SHS) product, (c–e) HRTEM photos of the SHS product, (f) wavelength dispersive spectroscopy (WDS) of element distribution of the SHS product, (g–h) FESEM images of the self-propagating high-temperature synthesis–spark plasma sintering (SHS–SPS) bulk, (i–k) HRTEM photos of the SHS–SPS bulk, (l) WDS element distribution of the SHS–SPS bulk. (Reprinted by permission from Macmillan Publishers Ltd., *Nat. Commun.*, Su, X.-L., Fu, F., Yan, Y. et al., Self-propagating high-temperature synthesis for compound thermoelectrics and new criterion for combustion processing, 5, 4908, 2014, copyright 2014.)

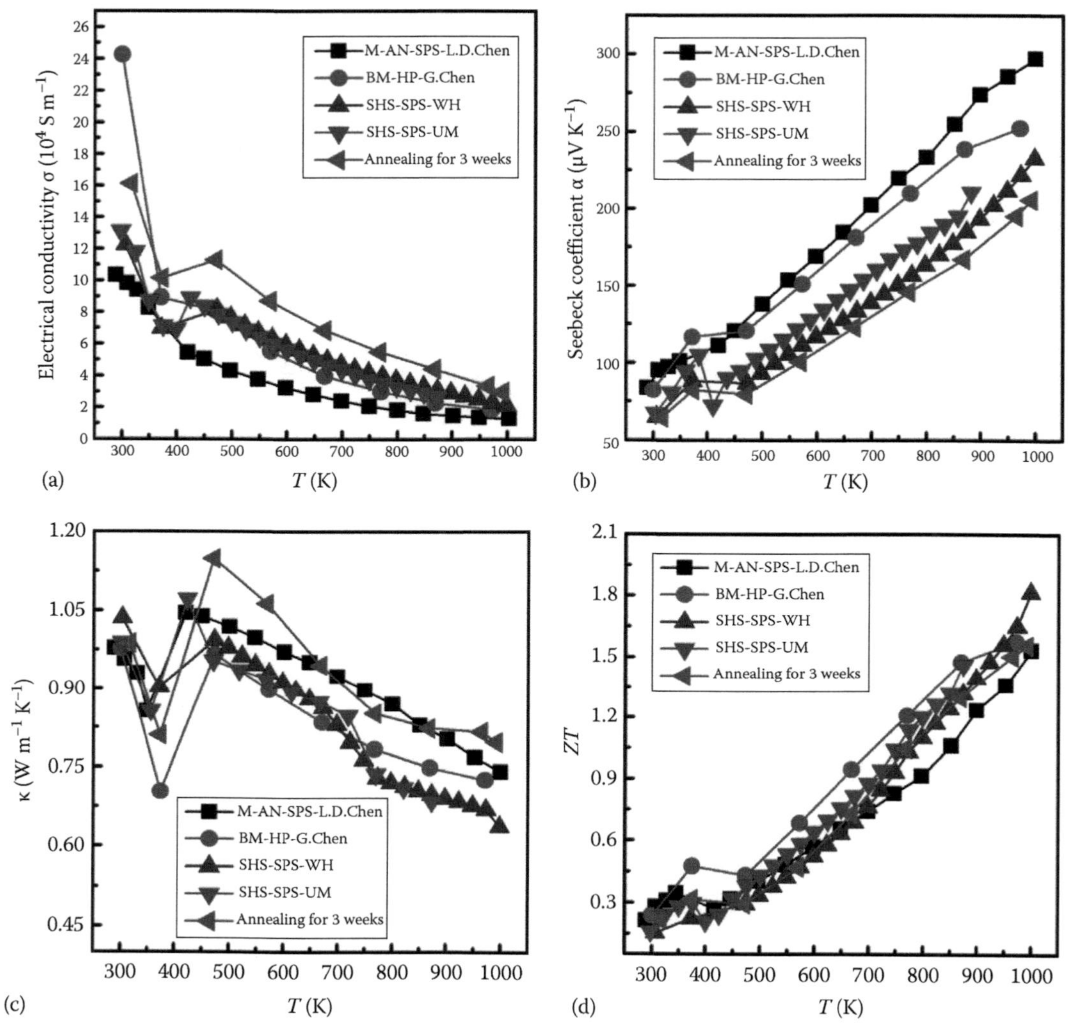

FIGURE 20.14
Temperature dependence of (a) electrical conductivity, (b) Seebeck coefficient, (c) thermal conductivity, and (d) *ZT* for Cu_2Se bulk prepared by self-propagating high-temperature synthesis–spark plasma sintering. (Reprinted by permission from Macmillan Publishers Ltd., *Nat. Commun.*, Su, X.-L., Fu, F., Yan, Y. et al., Self-propagating high-temperature synthesis for compound thermoelectrics and new criterion for combustion processing, 5, 4908, 2014, copyright 2014.)

both the Wuhan University of Technology and the University of Michigan on samples cut from the same ingot. Overall, the data agree to within 10%. The sample marked as SHS–PAS-Wuhan University of Technology (WH) was subsequently annealed at 800 K for 3 weeks and is marked as such. The temperature dependence of the electrical conductivity of Cu_2Se is shown in Figure 20.14a. The electrical conductivity decreases with increasing temperature, displaying a metallic behavior. The higher electrical conductivity of Cu_2Se synthesized by SHS–PAS in comparison to the M-AN-SPS sample is ascribed to its higher carrier concentration. All transport data show a discontinuity near 396 K associated with the α–β phase transition. The temperature dependence of the Seebeck coefficient of Cu_2Se is shown in Figure 20.14b. In the entire temperature range, the Seebeck coefficient is positive in accord with the behavior of the Hall effect. The Seebeck coefficient increases with the increasing temperature and reaches values larger than 230 $\mu V\ K^{-1}$ at 1000 K. Although the carrier concentration of the SHS–PAS sample is higher than that of M-AN-SPS, its Seebeck coefficient remains relatively high, resulting in the higher power factor. Apparently, the mostly coherent nature of the nanostructure in SHS–PAS samples exerts little impact on the carrier transport. The temperature dependence of the thermal conductivity of Cu_2Se is shown in Figure 20.14c. With the increasing temperature, the thermal conductivity decreases and reaches a low value of 0.65 $W\ m^{-1}\ K^{-1}$ near 1000 K. The lower thermal conductivity of the SHS–PAS sample is likely due to enhanced phonon boundary scattering on nanostructural features in this sample. As a result of low thermal conductivity and excellent electronic properties, the maximum *ZT* value as high as 1.8 was achieved at 1000 K. Moreover, Cu_2Se synthesized by SHS–PAS is thermally stable.

20.5 Te-Doped $CoSb_3$ Synthesized by SHS–PAS

Skutterudite is an intermediate temperature TE material that possesses a complicated crystal structure with structural voids, which can be filled by foreign species such as, ions of rare earths or alkaline earths. The resulting structure is called a filled skutterudite [9–15,17,18]. Due to weak bonding between the filler and frame atoms, the filler can undergo large amplitude oscillations in the void frequently referred to as rattling. These substantially Einstein-type of phonon modes interact with the normal phonon modes and significantly decrease the lattice thermal conductivity [17,18]. Moreover, the filler enters into the structure as a positively charged ion with its outer electrons stripped off and entering the conduction band. Void filling is thus an exceptionally effective *n*-type doping. Hence, skutterudites are often considered examples of the *PGEC* paradigm even though such a comparison must be taken with a grain of salt as the electronic properties are somewhat degraded upon void filling.

Although relatively simple binary cubic compounds, skutterudites such as $CoSb_3$ are not easy to synthesize due to slow kinetics associated with the peritectic nature of the reaction. To achieve a single-phase $CoSb_3$ structure requires long-time annealing. The synthesis process is therefore time and energy intensive and typically takes at least a week. We have shown recently that skutterudites, such as $CoSb_3$, can be conveniently and rapidly synthesized using the SHS process in combination with PAS. The resulting structure is a high-performance skutterudite and its synthesis is dramatically shortened from one week to one-half of an hour.

20.5.1 Phase Composition and Microstructure of Skutterudites after SHS

Figure 20.15 shows the XRD pattern of the product after SHS. The structure after SHS is a single-phase material that matches the standard PDF card for $CoSb_3$ very well (JCPDS# 01-083-0055). The inset in Figure 20.15 shows that the peak shifts to lower angles with Te doping, indicating that Te substitutes on the Sb site after SHS. Thus, a single-phase Te-doped skutterudite is obtained in a very short time on the order of seconds via SHS.

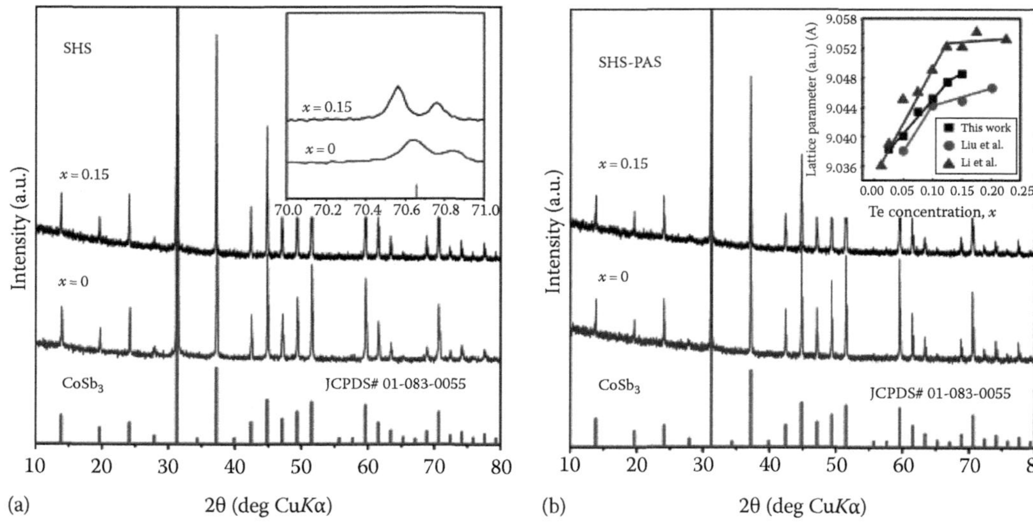

FIGURE 20.15
X-ray diffraction (XRD) patterns of $CoSb_{3-x}Te_x$ compounds after (a) self-propagating high-temperature synthesis and (b) after self-propagating high-temperature synthesis–plasma-activated sintering (SHS-PAS). The inset of (b) shows the lattice parameter of $CoSb_{3-x}Te_x$ compounds after SHS-PAS. (From Liang, T., Su, X.-L., Yan, Y. et al., Ultra-fast synthesis and thermoelectric properties of Te doped skutterudites, *J. Mater. Chem. A*, 2, 17914–17918, 2014. Reproduced by permission of The Royal Society of Chemistry.)

Figure 20.16 shows the microstructure and elemental distribution in the Te-doped $CoSb_{3-x}Te_x$ compound after SHS and after SHS–PAS. A spongelike structure with many holes is observed and the grain size is about 1–2 μm. After PAS, numerous nanopores with the size of about 20–100 nm are distributed homogeneously at grain boundaries with the matrix grain size slightly increased to above 2 μm. In spite of the presence of some nanopores, the relative density of all specimens after SHS–PAS is higher than 97%. All constituent elements (Co, Sb, and Te) are distributed homogeneously in bulk $CoSb_{2.85}Te_{0.15}$, indicating that highly homogeneous Te-doped $CoSb_{3-x}Te_x$ can be achieved in a very short time through the use of the SHS–PAS process.

20.5.2 Thermoelectric Properties of $CoSb_{3-x}Te_x$ after SHS–PAS

Figure 20.17 depicts the temperature dependence of transport properties of $CoSb_{3-x}Te_x$ synthesized by SHS–PAS. As shown in Figure 20.17a, the electrical conductivity increases with the increasing content of Te due to an increased charge carrier concentration and decreases as a function of increasing temperature indicating characteristics of a degenerate semiconductor. In contrast with skutterudites having the same carrier concentration [78,79] but synthesized by some other methods, the electrical conductivity of samples prepared by SHS–PAS is much higher. The inset of Figure 20.17a shows the temperature dependence of the electrical conductivity for pure skutterudite having no Te. At room temperature, the electrical conductivity is 6.2×10^4 S m^{-1}. Figure 20.17b shows the Seebeck coefficient of all Te-doped $CoSb_{3-x}Te_x$ as negative, indicting *n*-type transport properties. The Seebeck coefficient displays an inverse trend with the content of Te and its absolute magnitude decreases

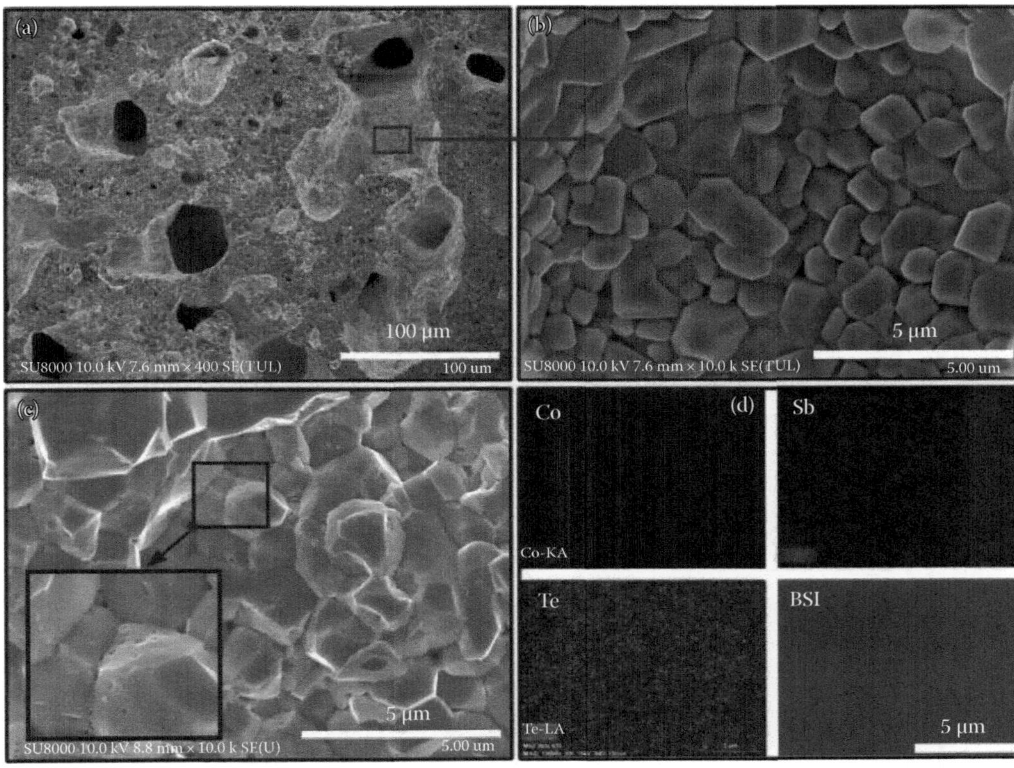

FIGURE 20.16
(a, b) Field emission scanning electron microscope (FESEM) images of bulk $CoSb_{3-x}Te_x$ compounds after self-propagating high-temperature synthesis process, (c) FESEM image of the bulk $CoSb_{3-x}Te_x$ compound after self-propagating high-temperature synthesis–plasma-activated sintering (SHS-PAS) processing, and (d) Energy dispersive X-ray detector (EDX) mapping of bulk sample after SHS-PAS. (From Liang, T., Su, X.-L., Yan, Y. et al., Ultra-fast synthesis and thermoelectric properties of Te doped skutterudites, *J. Mater. Chem. A*, 2, 17914–17918, 2014. Reproduced by permission of The Royal Society of Chemistry.)

from 163 to 130 μV K^{-1} at room temperature as the content of Te increases. However, the Seebeck coefficient of pure undoped skutterudite prepared by SHS–PAS is positive; see the inset in Figure 20.17b. The Seebeck coefficient of the Te-doped sample with $x = 0.15$ synthesized by SHS–PAS is almost the same as the sample of the same Te content synthesized by MA-SPS (dashed line), as reported by Liu et al. [79]. On the other hand, the Seebeck coefficient of the SHS–PAS synthesized sample with $x = 0.125$ is much higher than the sample (solid line) with the same Te content fabricated by the traditional long-term annealing method [78]. This may be ascribed to an energy-filtering effect, induced by the nanopores in SHS–PAS samples as indicated in Figure 20.16c, which can effectively scatter low-energy carriers.

Figure 20.17c shows the temperature-dependent thermal conductivity. The thermal conductivity decreases gradually with the increasing content of Te in spite of the increasing electronic part of the thermal conductivity. The likely reason is distortions on pnicogen rings created by Te substituting on sites of Sb. Since low-frequency spectrum of phonons is dominated by vibrations of the rings, the heat-conducting phonons are particularly

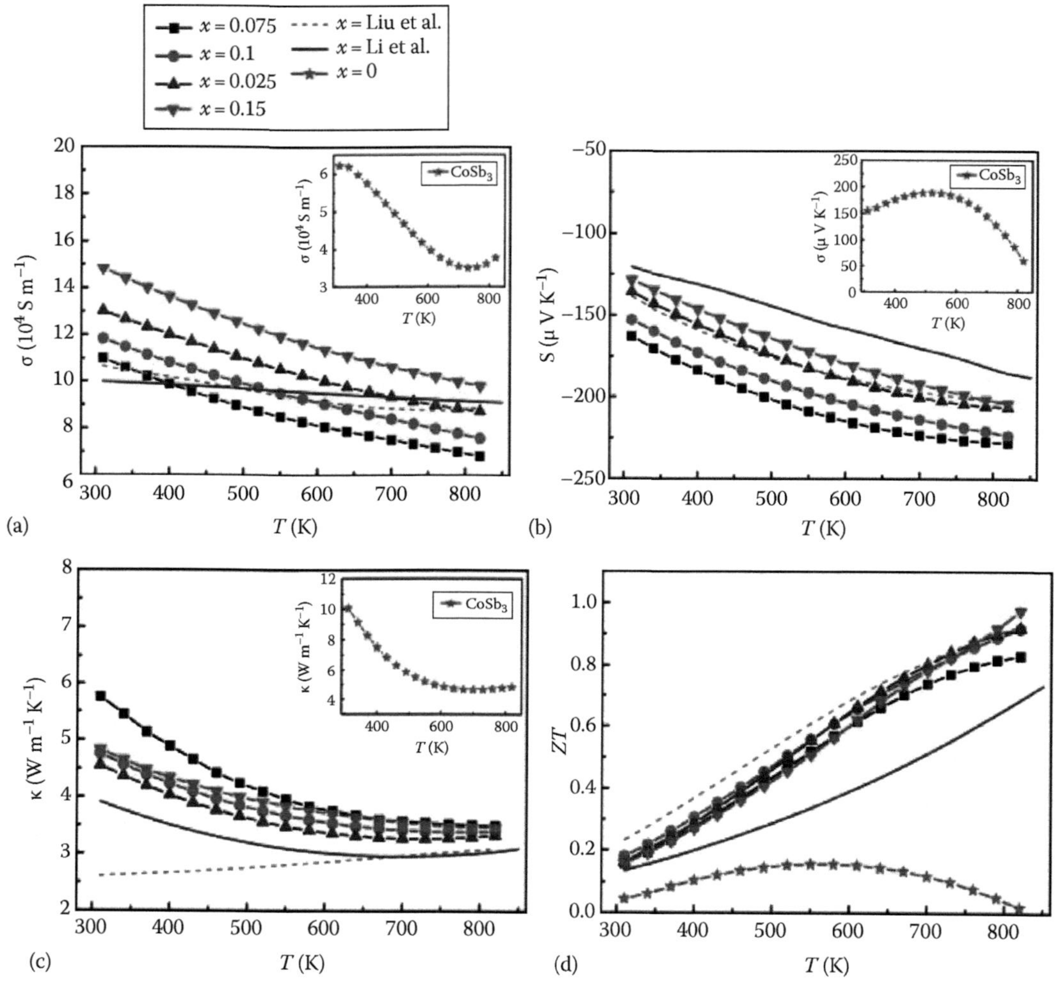

FIGURE 20.17
(a) Temperature dependence of the electrical conductivity and (b) of the Seebeck coefficient. (c) Temperature dependence of the thermal conductivity and (d) the dimensionless figure of merit ZT. Insets in (a), (b), and (c) show thermoelectric properties of pure $CoSb_3$ skutterudite. (From Liang, T., Su, X.-L., Tan, X. et al., Ultra-fast non-equilibrium synthesis and phase segregation in $In_xSn_{1-x}Te$ thermoelectrics by SHS-PAS processing, *J. Mater. Chem. C*, 3, 8550–8558, 2015. Reproduced by permission of The Royal Society of Chemistry.)

affected by such distortions, and the overall thermal conductivity is degraded. The thermal conductivity of pure $CoSb_3$ is 10.1 W m^{-1} K^{-1} at room temperature, while the thermal conductivity of $CoSb_{2.85}Te_{0.15}$ is only about 4.8 W m^{-1} K^{-1} at room temperature, decreased by some 50%.

Temperature dependence of the figure of merit ZT is presented in Figure 20.17d and shows a dramatically rising ZT with increasing temperature. All Te-doped samples have considerably higher figures of merit than pure $CoSb_3$. However, the dependence on the actual content of Te is only notable at high temperatures where the $CoSb_{2.85}Te_{0.15}$

compound achieves the highest ZT of 0.98 at 820 K. This value is superior to ZT values of binary skutterudites prepared by other methods, including the traditional long-term annealing and mechanical alloying [78,79]. The skutterudites prepared by the SHS–PAS technique also possess high purity and are well crystallized. The SHS–PAS synthesis of binary skutterudites results in an outstanding TE material obtained with a minimum use of energy and in a dramatically shortened time compared to any other synthesis process tried with skutterudites. As such, the SHS–PAS process offers a novel and economical method of fabricating skutterudites, which is also easily scalable to industrial-size production.

20.6 $In_xSn_{1-x}Te$ Synthesized by SHS–PAS

Chalcogenides and their solid solutions (PbQ, GeQ, SnQ; Q = S, Se, and Te) have long been considered promising high-temperature TE materials. The highest ZT above two has been achieved by using the band engineering concept and microstructure modification, especially in the case of PbTe [22,35,44]. SnTe is considered a viable alternative to PbTe on account of their similar structures and is also perceived as a more environmentally friendly material than PbTe because it does not contain toxic elements such as Pb, restricted all over the world. Thus, SnTe has recently attracted great attention in the TE community. SnTe is a narrow band semiconductor with high background carrier concentration due to the high density of intrinsic cation vacancies [80]. Moreover, the energy separation between the heavy hole band and the light hole band is about 0.30–0.40 eV, which is much larger than that in PbTe [50,51,80]. Thus, lowering the carrier concentration and increasing the band degeneracy are key issues on how to improve the TE performance. Attempts have been made to modify the band structure in order to enhance TE properties. For instance, Zhang et al. [40] found that indium atoms form the band resonance state in SnTe, which improves the Seebeck coefficient. Tan et al. [51] reported that the presence of Cd gives rise to band convergence in the SnTe system. However, the solubility limit of all such dopants in SnTe is very low. Since one dopes SnTe in order to lower the carrier concentration, the limited solubility effectively prevents doing so.

In this section, we report on a supersaturated solid solution of $In_xSn_{1-x}Te$, which we were able to synthesize by SHS because of the very fast and nonequilibrium nature of the process. The phase transformation mechanism and related microstructure were investigated. The phase segregation and nanostructure after annealing were achieved in this system for the first time.

20.6.1 Phase Composition and Microstructure of $In_xSn_{1-x}Te$ after SHS

Figure 20.18 displays the typical characteristic of SHS process and XRD pattern of SnTe after SHS. Different stages of the SHS process for a cold-pressed pellet of the powder mixture of pure Sn and Te with the stoichiometric SnTe as recorded by a high-speed camera are shown in Figure 20.18a. The bottom of the pellet was ignited by a hand torch. Once ignited, the combustion wave propagated releasing enough heat to ignite the neighboring areas and thus maintaining the self-propagating character of the combustion wave. For the SnTe pellet with dimensions of about $\phi 15 \times 30$ mm, the entire reaction was completed

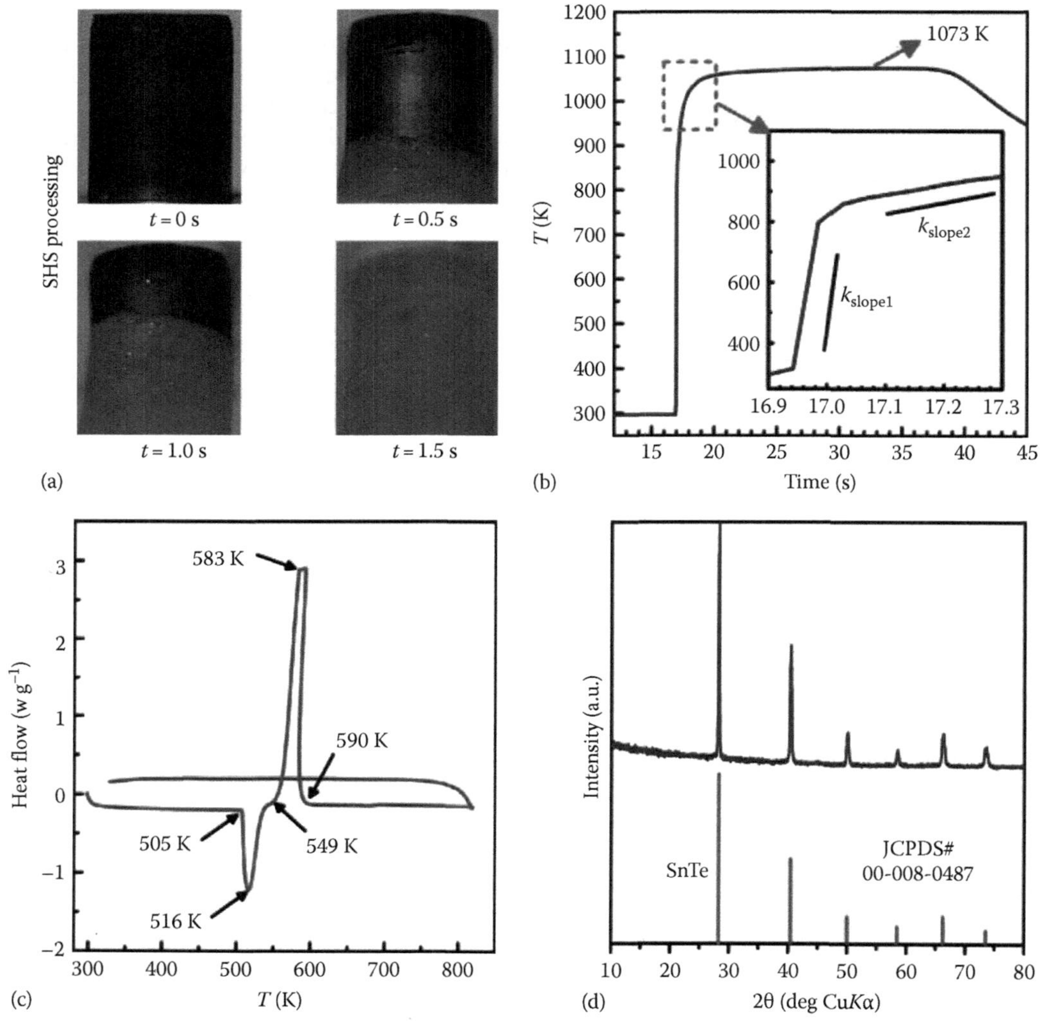

FIGURE 20.18
(a) Different stages of the self-propagating high-temperature synthesis (SHS) process in SnTe, (b) the pellet's temperature as the combustion wave passes by, (c) differential scanning calorimetry heat flow curves for Sn/Te = 1:1 mixed powder with a heating rate of 50 K min^{-1}, and (d) x-ray diffraction patterns of products synthesized by SHS. (From Liang, T., Su, X.-L., Tan, X. et al., Ultra-fast non-equilibrium synthesis and phase segregation in $In_xSn_{1-x}Te$ thermoelectrics by SHS-PAS processing, *J. Mater. Chem. C*, 3, 8550–8558, 2015. Reproduced by permission of The Royal Society of Chemistry.)

in only 1.5 s, leaving behind a fully reacted SnTe. The speed of the combustion wave front was thus about 20 mm s^{-1}, which is in the range of typical combustion wave speeds. Figure 20.18b shows the temperature profile on the SnTe pellet during the SHS process. The combustion temperature for SnTe, as measured by a thermocouple inserted into the pellet, was 1073 K, close to the melting point of SnTe (1079 K). As already noted, the adiabatic temperature is a critical parameter for the combustion synthesis and is defined as the highest temperature the sample reaches when the combustion wave passes through a thermally isolated pellet. The adiabatic temperature can be calculated using Equation 20.1

for the chemical reaction Sn + Te → SnTe. The calculated adiabatic temperature for SnTe turns out to be the melting point of SnTe (1079 K). Moreover, the calculation also indicates that, following the SHS reaction, the resulting SnTe is partially in a liquid and partially in a solid state. The mixed solid and liquid phases of SnTe are maintained as the sample reaches the adiabatic temperature. To gain more insight into the formation mechanism of SnTe during the SHS process, DSC cycling of the mixed Sn and Te powders was carried out with the heating rate of 50 K min^{-1}, as shown in Figure 20.18c. An endothermic peak, corresponding to melting of Sn, is observed at 505 K. With the increasing temperature, a sharp exothermic peak appears at 583 K, which corresponds to a fusion of Sn and Te, releasing an intense heat that maintains the reaction self-sustaining. Upon completion of the SHS process, a single-phase SnTe is obtained, as shown in Figure 20.18d. We conclude that a single-phase SnTe material can be synthesized using the SHS process in a timespan of a few seconds.

After the combustion wave has passed through, the mixture of liquid and solid phases of SnTe cools down rapidly to room temperature. What impact the nonequilibrium process has on the microstructure? As shown in Figure 20.19, the typical spongelike structure

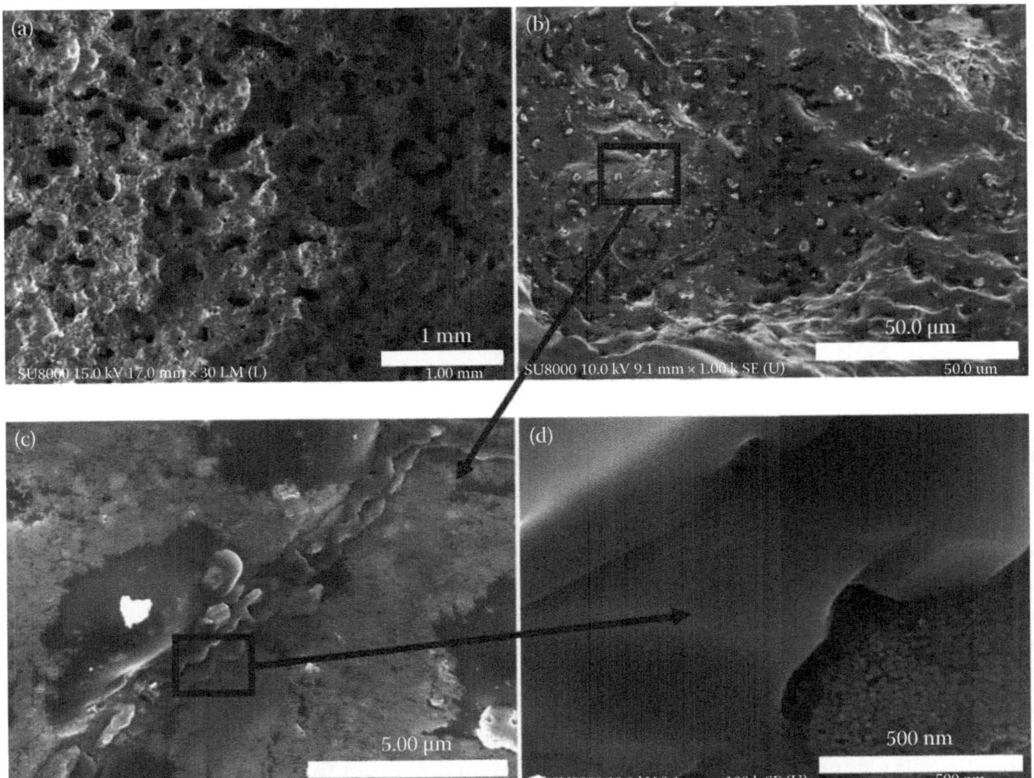

FIGURE 20.19
Morphology of SnTe prepared by self-propagating high-temperature synthesis (SHS) processing at different magnifications (a–d) field emission scanning electron microscope images of SnTe synthesized by SHS processing. (From Liang, T., Su, X.-L., Tan, X. et al., Ultra-fast non-equilibrium synthesis and phase segregation in $In_xSn_{1-x}Te$ thermoelectrics by SHS-PAS processing, *J. Mater. Chem. C*, 3, 8550–8558, 2015. Reproduced by permission of The Royal Society of Chemistry.)

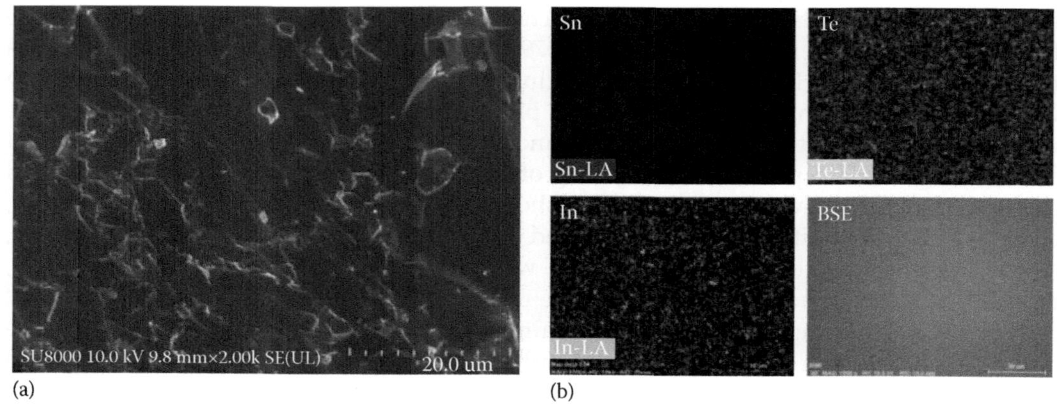

FIGURE 20.20
Morphology and elemental distribution in $In_xSn_{1-x}Te$ prepared by self-propagating high-temperature synthesis–plasma-activated sintering (SHS–PAS) processing. (a) Field emission scanning electron microscope images of $In_{0.015}Sn_{0.985}Te$ prepared by SHS–PAS processing and (b) EDX mapping of $In_{0.03}Sn_{0.97}Te$ prepared by SHS–PAS processing. (From Liang, T., Su, X.-L., Tan, X. et al., Ultra-fast non-equilibrium synthesis and phase segregation in $In_xSn_{1-x}Te$ thermoelectrics by SHS-PAS processing, *J. Mater. Chem. C*, 3, 8550–8558, 2015. Reproduced by permission of The Royal Society of Chemistry.)

with a number of holes on a micrometer scale forms after SHS. Numerous nanostructures with the size of about 10–50 nm are distributed homogeneously at grain boundaries. After applying PAS, a dense bulk material with the relative density higher than 98% is obtained. The grain size has increased dramatically to more than 10 μm and no nanoprecipitates are observed. A backscattering electron image in Figure 20.20 shows that the elements are distributed homogeneously on a scale of micrometers, demonstrating that In has entered the sites of Sn in the SnTe structure. Interestingly, following annealing of the bulk SnTe at 673 K for 2 days, high concentration of nanoprecipitates shows up with the size of about 10–50 nm distributed homogeneously at grain boundaries; see Figure 20.21. EDS results indicate that, in contrast with the composition of the matrix, the nanoprecipitates are rich in In and Te, as depicted in Figure 20.21d. This means that the matrix is somewhat deficient of Te, which leads to a decrease of Sn vacancies. Moreover, as the content of In increases, the concentration and the average grain size of nanoprecipitates increases; see Figure 20.22. According to a pseudobinary $Sn_{1-x}In_xTe$ phase diagram shown in Figure 20.23, the solubility limit of In in $Sn_{1-x}In_xTe$ is strongly dependent on the temperature. Considering the fast and nonequilibrium nature of the SHS process and taking into account the pseudobinary phase diagram, the high temperature state was quenched and maintained its high In solubility down to room temperature, and, hence, no nanoprecipitates were detected. However, upon annealing, the solubility limit of In in $Sn_{1-x}In_xTe$ at 673 K is smaller than at high temperatures. Thus, nanoprecipitates form after annealing. We conclude that nonequilibrium oversaturated $In_xSn_{1-x}Te$ solid solutions can be achieved through SHS processing. Moreover, the microstructure (morphology and concentration) can be modulated by the SHS–PAS process in conjunction with annealing. This extra tuning knob should be very beneficial for the development of outstanding TE properties.

20.6.2 Thermoelectric Properties of $In_xSn_{1-x}Te$ after SHS–PAS

Figure 20.24 shows the temperature dependence of TE properties for $In_xSn_{1-x}Te$ compounds. In order to assess the impact of nanoprecipitates on transport properties, $In_xSn_{1-x}Te$ samples were annealed at 673 K for 2 days. The electrical conductivity of all samples decreases with the increasing temperature indicating a highly degenerate semiconducting behavior. After annealing, the electrical conductivity of undoped SnTe is almost unchanged,

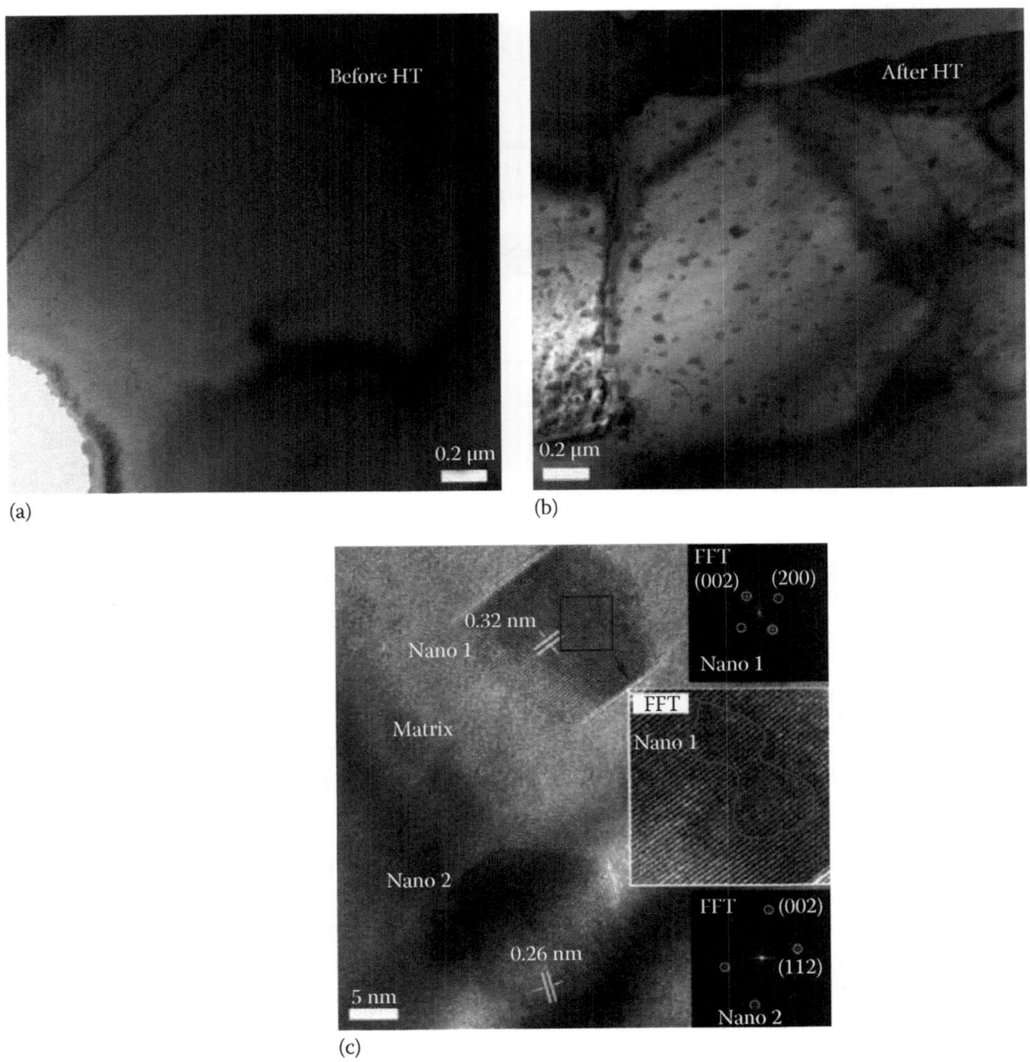

FIGURE 20.21
(a) Transmission electron microscope (TEM) image of $In_{0.015}Sn_{0.985}Te$ prepared by self-propagating high-temperature synthesis–plasma-activated sintering (SHS–PAS) before heat treatment (HT), (b) TEM image of $In_{0.015}Sn_{0.985}Te$ prepared by SHS–PAS after HT, (c) high-resolution TEM image of $In_{0.015}Sn_{0.985}Te$ prepared by SHS–PAS after HT. *(Continued)*

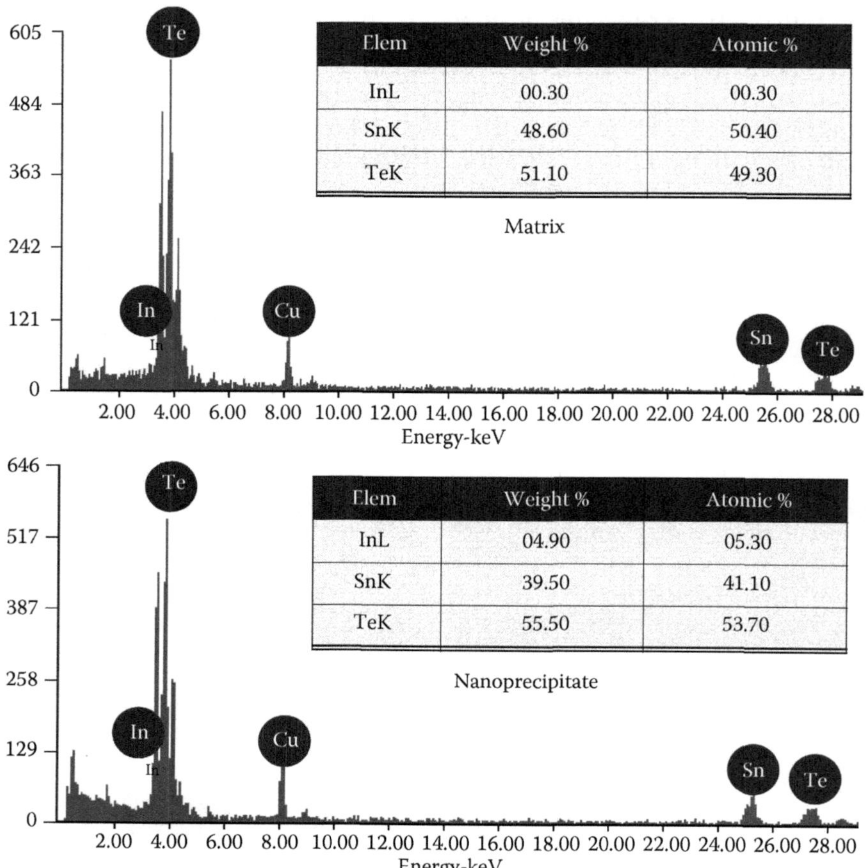

Elem	Weight %	Atomic %
InL	00.30	00.30
SnK	48.60	50.40
TeK	51.10	49.30

Matrix

Elem	Weight %	Atomic %
InL	04.90	05.30
SnK	39.50	41.10
TeK	55.50	53.70

Nanoprecipitate

(d)

FIGURE 20.21 ***(Continued)***
(d) EDS analysis of nanoprecipitates and the matrix of $In_{0.015}Sn_{0.985}Te$ prepared by SHS–PAS. (From Liang, T., Su, X.-L., Tan, X. et al., Ultra-fast non-equilibrium synthesis and phase segregation in $In_xSn_{1-x}Te$ thermoelectrics by SHS-PAS processing, *J. Mater. Chem. C*, 3, 8550–8558, 2015. Reproduced by permission of The Royal Society of Chemistry.)

especially at high temperatures. In sharp contrast, the electrical conductivity of In-doped samples decreases dramatically due to the decreasing carrier concentration as the number of Sn vacancies diminishes. An opposite trend is observed in the behavior of the Seebeck coefficient shown in Figure 20.24b. The temperature-dependent thermal conductivity of $In_xSn_{1-x}Te$ compounds is displayed in Figure 20.24c. As the temperature increases, the thermal conductivity of $In_xSn_{1-x}Te$ compounds decreases, ascribed to the influence of intense Umklapp scattering. As the In content increases, the thermal conductivity decreases for both unannealed and annealed samples due to the enhanced point defect scattering. In contrast to the unannealed sample, the thermal conductivity of annealed In-doped samples is further suppressed due to the presence of nanoprecipitations, which scatter low-frequency phonons. Making use of the Wiedemann–Franz law, we calculated the lattice thermal conductivity with the Lorenz number deduced by fitting the Seebeck coefficient and assuming a single parabolic band model dominated by acoustic phonon scattering. As

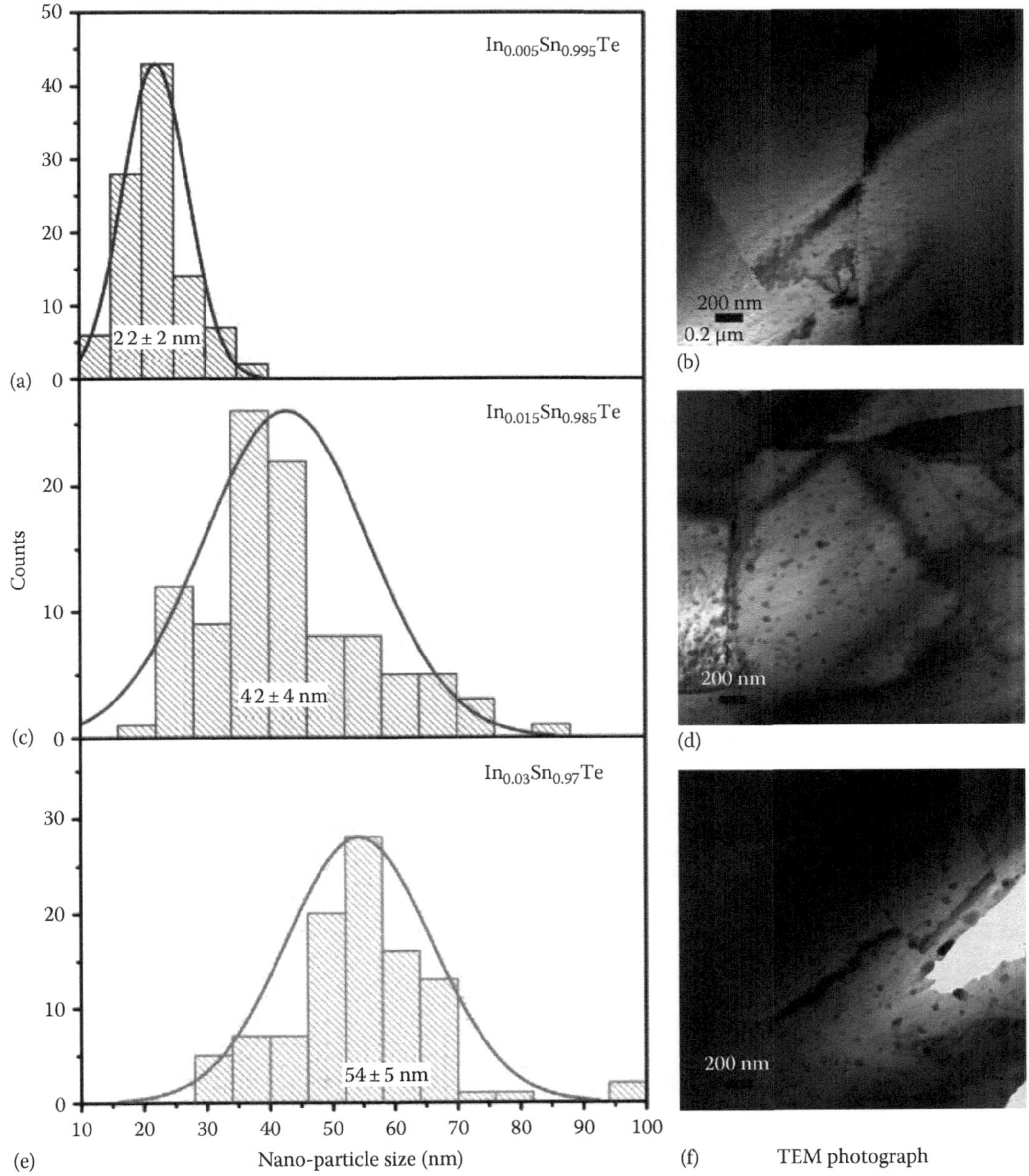

FIGURE 20.22
(a)–(f) Statistical analysis and TEM images of nanoparticle distributions at boundaries of $In_xSn_{1-x}Te$ for three in contents $x = 0.005$, 0.015, and 0.03 after the heat treatment. (From Liang, T., Su, X.-L., Tan, X. et al., Ultra-fast non-equilibrium synthesis and phase segregation in $In_xSn_{1-x}Te$ thermoelectrics by SHS-PAS processing, *J. Mater. Chem. C*, 3, 8550–8558, 2015. Reproduced by permission of The Royal Society of Chemistry.)

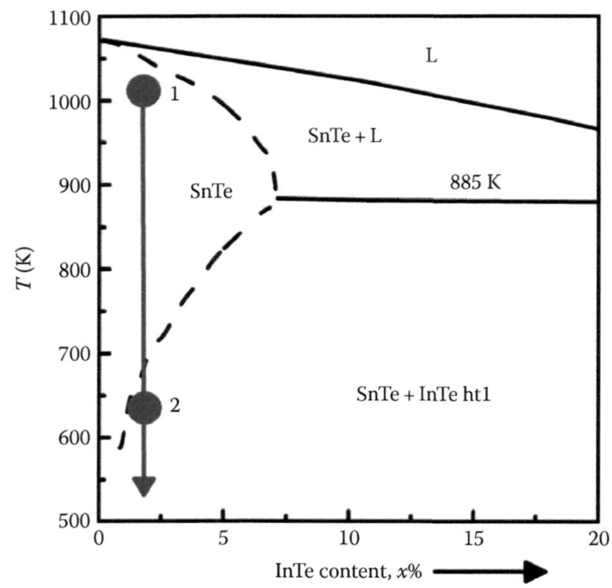

FIGURE 20.23
Estimated section of a pseudobinary phase diagram of the SnTe-InTe system indicating the significant temperature dependent solubility of InTe in SnTe. (Adapted from Tan, G. et al., *J. Am. Chem. Soc.*, 137, 11507, 2015.) Starting with a fast nonequilibrium self-propagating high-temperature synthesis processing at high temperature (point 1), the specimen is cooled down quickly and then phase separation is achieved by annealing at around 673 K (point 2). (From Liang, T., Su, X.-L., Tan, X. et al., Ultra-fast non-equilibrium synthesis and phase segregation in $In_xSn_{1-x}Te$ thermoelectrics by SHS-PAS processing, *J. Mater. Chem. C*, 3, 8550–8558, 2015. Reproduced by permission of The Royal Society of Chemistry.)

displayed in Figure 20.24d, the lattice thermal conductivity decreases with the increasing temperature and the content of In. The lowest lattice thermal conductivity approaching 0.5 W m^{-1} K^{-1} is obtained in $In_{0.005}Sn_{0.995}Te$ after annealing. Such a low value is a result of the synergistic effect of point defect and grain boundary phonon scattering processes. On account of the lower thermal conductivity and the enhanced Seebeck coefficient, the highest ZT value of about 0.92 was obtained at 920 K for the sample with $In_{0.005}Sn_{0.995}Te$ after annealing; see Figure 20.25.

20.7 New Thermodynamic Criterion for the SHS Process

Based on the initial successful combustion synthesis of single-phased Bi_2Te_3, $CoSb_3$, SnTe, and Cu_2Se, we subsequently synthesized numerous other TE compounds, including Cu_2SnSe_3, ZrNiSn, PbS, PbSe, Mg_2Sn, and Mg_2Si as single-phase structures; see Figure 20.26, typically in the timespan of a few seconds. Thermodynamic parameters and the respective fusion reactions of such compounds are presented in Table 20.1. Therefore, without any doubts, we have shown that the SHS process can be used for the synthesis of TE compounds. Moreover, examining the adiabatic temperatures in Table 20.1, one observes that they all are dramatically lower than 1800 K, the adiabatic temperature postulated by Merzhanov [63–66] as the

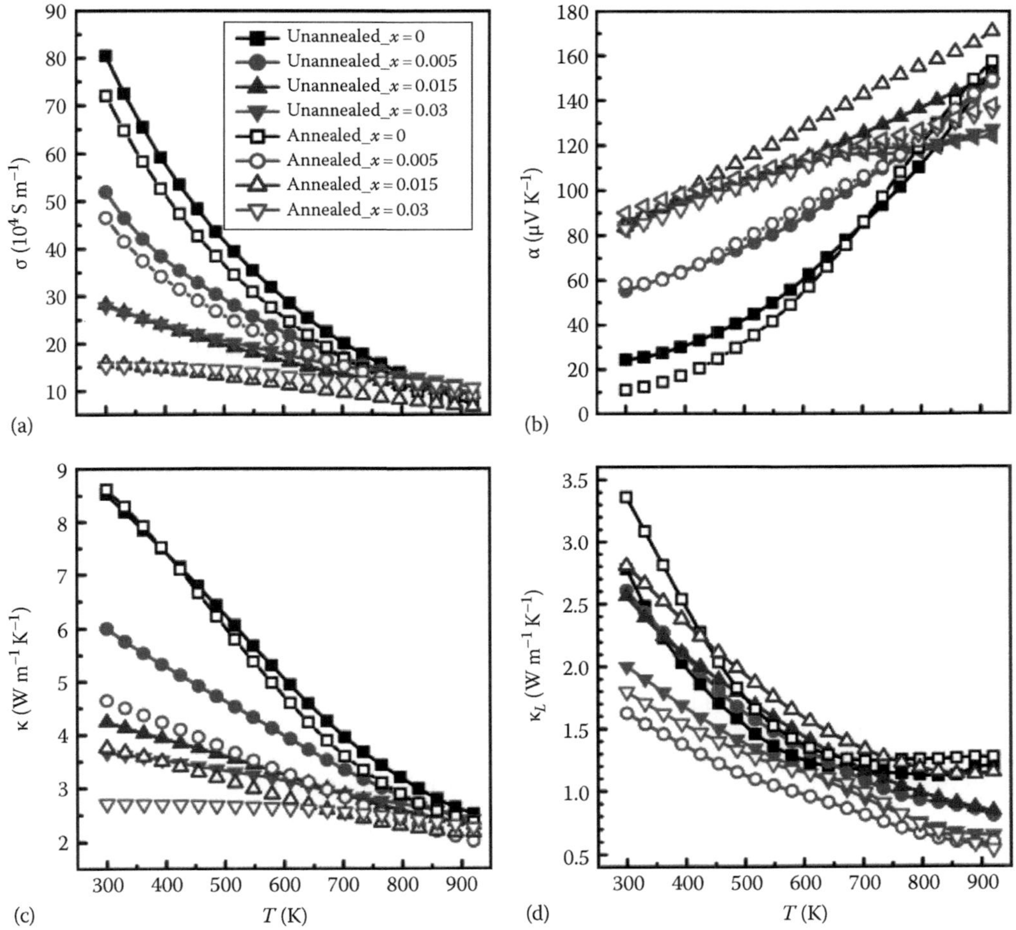

FIGURE 20.24
Temperature dependence of thermoelectric properties of $In_xSn_{1-x}Te$ samples prepared by self-propagating high-temperature synthesis–plasma-activated sintering before and after heat treatment. (a) Temperature dependence of the electrical conductivity and (b) the Seebeck coefficient. (c) Temperature dependence of the thermal conductivity and (d) temperature dependence of the lattice thermal conductivity. (From Liang, T., Su, X.-L., Tan, X. et al., Ultra-fast non-equilibrium synthesis and phase segregation in $In_xSn_{1-x}Te$ thermoelectrics by SHS-PAS processing, *J. Mater. Chem. C*, 3, 8550–8558, 2015. Reproduced by permission of The Royal Society of Chemistry.)

necessary condition for fusion reactions being self-sustaining and, therefore, the SHS process operational. The Merzhanov's empirical criterion was based on a vast collection of studies of high-temperature ceramics and intermetallics and for such materials the adiabatic temperature was, indeed, always in excess of 1800 K. Being able to use the SHS synthesis process with TE compounds where the adiabatic temperature is significantly below 1800 K, in some cases barely half of this temperature, clearly indicated that the Merzhanov's criterion is not universally valid.

A more vivid representation of this statement is depicted in Figure 20.27 where on the right-hand-side ordinate are presented values of the ratio between the calculated formation energy and the heat capacity, which, according to Munir [65,66], should have been larger than 2000 K, yet clearly failed in the case of TE compounds. Well, if that is the case,

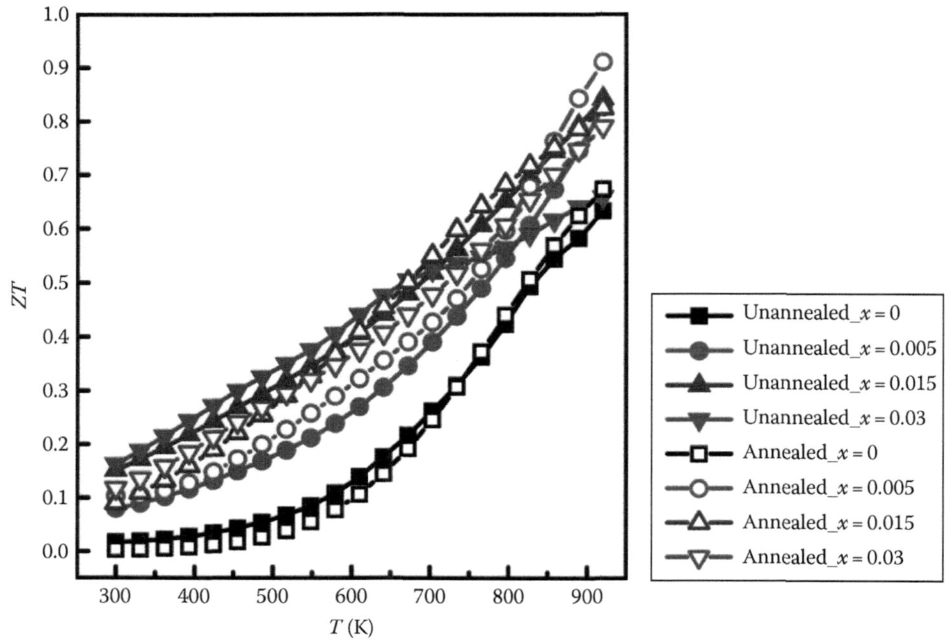

FIGURE 20.25
Temperature dependence of ZT of $In_xSn_{1-x}Te$ samples prepared by self-propagating high-temperature synthesis–plasma-activated sintering before and after heat treatment. (From Liang, T., Su, X.-L., Tan, X. et al., Ultra-fast non-equilibrium synthesis and phase segregation in $In_xSn_{1-x}Te$ thermoelectrics by SHS-PAS processing, *J. Mater. Chem. C*, 3, 8550–8558, 2015. Reproduced by permission of The Royal Society of Chemistry.)

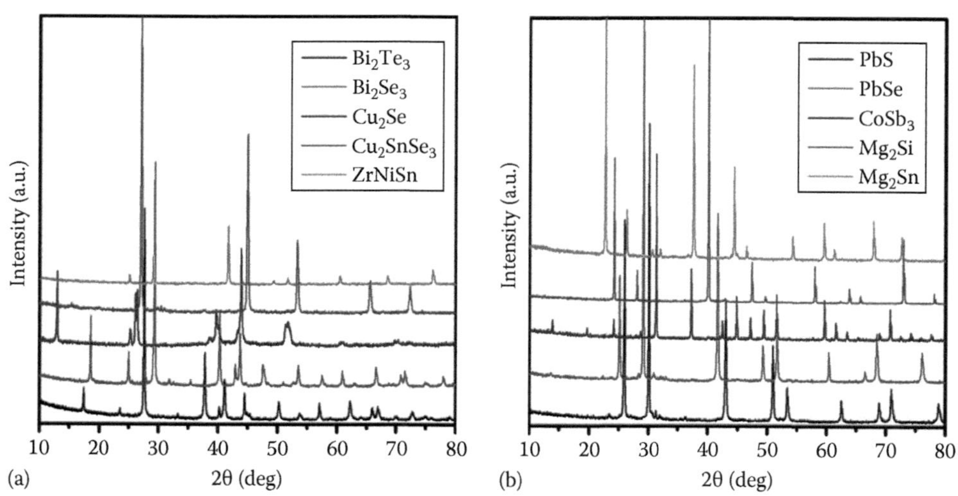

FIGURE 20.26
X-ray diffraction (XRD) pattern of thermoelectric compounds synthesized by the self-propagating high-temperature synthesis. (a) XRD pattern of Bi_2Te_3, Bi_2Se_3, Cu_2Se, Cu_2SnSe_3, and ZrNiSn compounds, (b) XRD pattern of PbS, PbSe, SnTe, $CoSb_3$, Mg_2Sn, and Mg_2Si compounds. (Reprinted by permission from Macmillan Publishers Ltd. *Nat. Commun.*, Su, X.-L., Fu, F., Yan, Y. et al., Self-propagating high-temperature synthesis for compound thermoelectrics and new criterion for combustion processing, 5, 4908, 2014, copyright 2014.)

TABLE 20.1
Parameters of the Self-Propagating High-Temperature Synthesis Process for Thermoelectric Materials

Material System	Reaction	Sample Dimension (mm)	Ignition Method	Adiabatic Temperature T_{ad} (K)	Combustion Wave Velocity (mm s^{-1})	Ignition Temperature (K)	Duration of Reaction (s)
Bi_2Te_3	$2Bi + 3Te \rightarrow Bi_2Te_3$	ϕ15 × 18	SHS	860	6	623	3
Bi_2Se_3	$2Bi + 3Se \rightarrow Bi_2Se_3$	ϕ15 × 18	SHS	995	6	543	3
Cu_2Se	$2Cu + Se \rightarrow Cu_2Se$	ϕ15 × 20	SHS	1001	5.6	493	3.6
Cu_2SnSe_3	$2Cu + Sn + 3Se \rightarrow Cu_2SnSe_3$	ϕ15 × 18	SHS	—	6	—	3
ZrNiSn	$Zr + Ni + Sn \rightarrow ZrNiSn$	ϕ15 × 20	SHS	—	23	—	0.87
PbS	$Pb + S \rightarrow PbS$	ϕ15 × 18	SHS	1427	4.5	—	4
PbSe	$Pb + Se \rightarrow PbSe$	ϕ15 × 18	SHS	1350	6	—	3
SnTe	$Sn + Te \rightarrow SnTe$	ϕ15 × 22	SHS	—	19.1	504	1.15
Mg_2Sn	$2Mg + Sn \rightarrow Mg_2Sn$	ϕ15 × 20	SHS	1053	4	841	5
Mg_2Si	$2Mg + Si \rightarrow Mg_2Si$	ϕ15 × 20	SHS	1282	3.9	850	5.1
Sb_2Te_3	$2Sb + 3Te \rightarrow Sb_2Te_3$	ϕ15 × 20	—	702	—	—	—
$MnSi_{1.70}$	$Mn + 1.70Si \rightarrow MnSi_{1.70}$	ϕ15 × 20	—	1314	—	—	—

Source: Reprinted by permission from Macmillan Publishers Ltd., *Nat. Commun.*, Su, X.-L., Fu, F., Yan, Y. et al., Self-propagating high-temperature synthesis for compound thermoelectrics and new criterion for combustion processing, 5, 4908, 2014, copyright 2014.

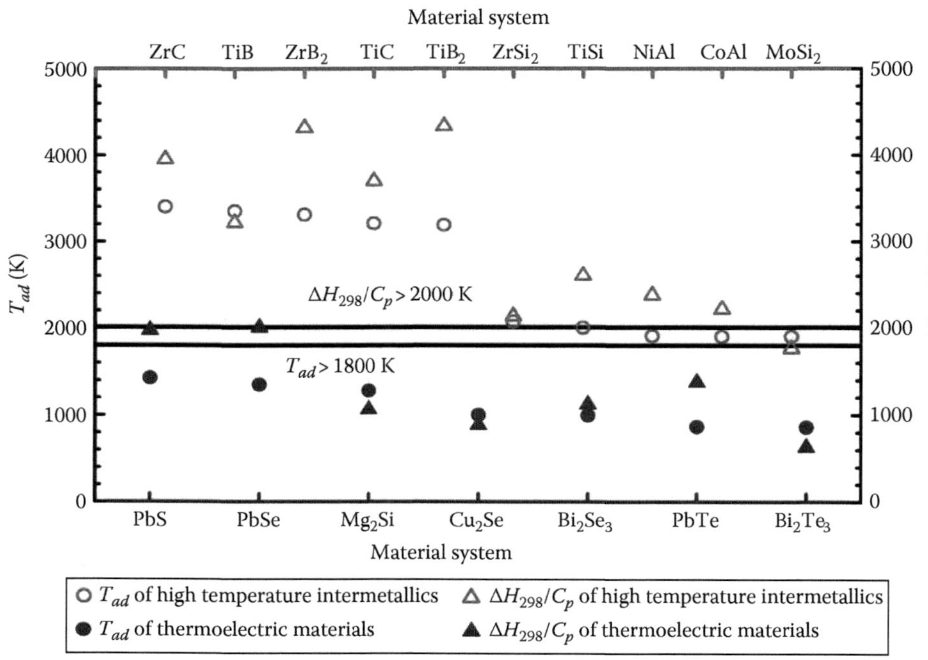

FIGURE 20.27
Adiabatic temperature T_{ad} and $\Delta H_{298}/C_{p(298)}$ of materials prepared by self-propagating high-temperature synthesis. (Reprinted by permission from Macmillan Publishers Ltd. *Nat. Commun.*, Su, X.-L., Fu, F., Yan, Y. et al., Self-propagating high-temperature synthesis for compound thermoelectrics and new criterion for combustion processing, 5, 4908, 2014, copyright 2014.)

what is then the appropriate criterion that determines whether the SHS synthesis will or will not be successful? In other words, what determines whether a fusion reaction will be self-propagating?

An important hint to answer this question was provided by our repeated failure to synthesize two important TE compounds, Sb_2Te_3 and $MnSi_{1.70}$, using the SHS process. Although we managed to ignite the bottom part of pellets of Sb_2Te_3 and $MnSi_{1.70}$, the combustion wave would simply not self-propagate. XRD patterns of the bottom and top parts of pellets of Sb_2Te_3 and $MnSi_{1.70}$ are shown in Figure 20.28. Sb_2Te_3 and $MnSi_{1.70}$ compounds were found at the bottom part of pellets, but the top part of pellets contained merely unreacted elemental species of Sb and Te and Mn and Si, respectively, confirming that the combustion wave has never passed through.

Why the SHS synthesis failed in the case of Sb_2Te_3 and $MnSi_{1.70}$? We have carefully examined the ratio between the adiabatic temperature and the melting point of the lower melting point component, $T_{ad}/T_{m,L}$, of compounds and plotted the result in Figure 20.29. The data also contain relevant ratio for high-temperature ceramics and intermetallics. It is clear from Figure 20.29 that a compound (regardless of whether a TE or high-temperature ceramic or intermetallic) can be synthesized by SHS with room temperature as the initial temperature, provided that the ratio $T_{ad}/T_{m,L}$ is larger than 1. When this happens, the higher melting point component rapidly dissolves in the liquid phase of the lower melting point component and generates heat at a rate high enough to sustain propagation of the combustion wave. Since this critical ratio falls below 1 in the case of Sb_2Te_3 and $MnSi_{1.70}$, the combustion wave front cannot self-propagate and the SHS synthesis will fail. The new

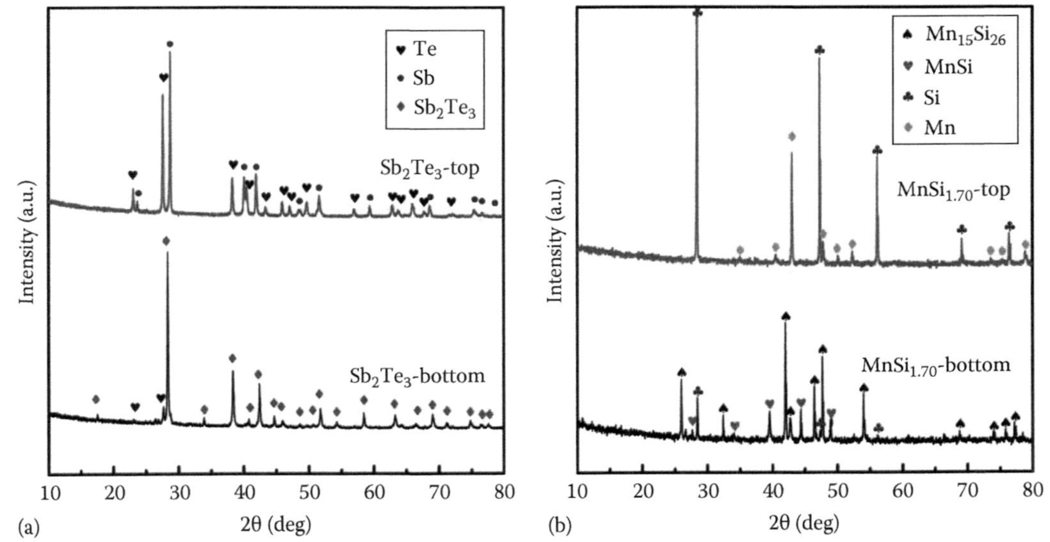

FIGURE 20.28
X-ray diffraction pattern of the bottom and top parts of pellets of (a) Sb_2Te_3 and (b) $MnSi_{1.70}$. (Reprinted by permission from Macmillan Publishers Ltd., *Nat. Commun.*, Su, X.-L., Fu, F., Yan, Y. et al., Self-propagating high-temperature synthesis for compound thermoelectrics and new criterion for combustion processing, 5, 4908, copyright 2014.)

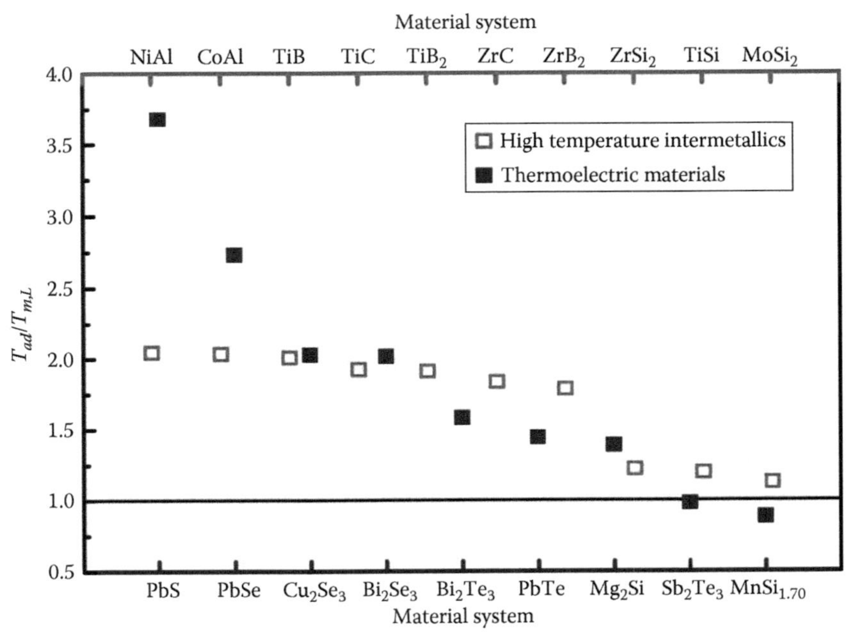

FIGURE 20.29
The ratio between the adiabatic temperature T_{ad} and the melting point of the lower melting point component $T_{m,L}$. (Reprinted by permission from Macmillan Publishers Ltd., *Nat. Commun.*, Su, X.-L., Fu, F., Yan, Y. et al., Self-propagating high-temperature synthesis for compound thermoelectrics and new criterion for combustion processing, 5, 4908, copyright 2014.)

criterion $T_{ad}/T_{m,L} \geq 1$ for the viability of the SHS process applies to all materials regardless of whether they are high-temperature compound ceramics or low-temperature compound TEs. Unlike the previous criterion for the self-sustaining nature of the combustion process, which was rather arbitrarily set, our criterion has a clear physical basis in terms of thermodynamic parameters of the system.

20.8 Conclusions and Outlook

High-performance TE compounds including $Bi_2Te_{3-x}Se_x$, $CoSb_{3-x}Te_x$, Cu_2Se, and $In_xSn_{1-x}Te$ have been successfully synthesized by SHS in conjunction with PAS. Thermodynamic and kinetic parameters for the SHS reaction such as the adiabatic temperature, the combustion temperature, and the speed of the combustion wave front were determined. The morphology of synthesized materials is modified due to the nonequilibrium nature of the process. The SHS process is extremely rapid and shortens the synthesis duration from several days to less than 20 min and does so with a minimum use of energy. For example, a single-phase, homogeneous $Bi_2Te_{3-x}Se_x$ with no preferred orientation synthesized by SHS–PAS possesses high TE performance with the peak ZT value of about 0.95 at 423 K and excellent mechanical properties. Moreover, the technologically relevant average ZT value from 298 to 523 K is 0.84, about a 25% improvement in comparison with the ZM sample. The compressive strength and the bending strength of $Bi_2Te_{2.4}Se_{0.6}$ prepared by SHS–PAS is 104 and 70 MPa, respectively, values much larger than those of the ZM sample measured perpendicular to the *c*-axis. In the case of Cu_2Se, a highly efficient, perfectly stoichiometric, and stable material is obtained when synthesized by SHS–PAS. The morphology of such Cu_2Se consists of 5–50 nm corn-like nanostructural features dispersed in the matrix with the grain size of 1–3 μm. Such microstructures minimize the thermal conductivity while excellent electronic properties are maintained. As a consequence, the ZT value of Cu_2Se reached 1.2 at 800 K, and values in excess of 1.8 were obtained near 1000 K. Te-doped skutterudites, $CoSb_{3-x}Te_x$, were also synthesized by the SHS–PAS process as single-phase structures. In this case, the synthesis period was shortened dramatically from times on the scale of a week or so necessary to complete the peritectic reaction in the traditional synthesis to less than 20 min upon application of SHS–PAS. The electrical conductivity increases with the rising content of Te, while the thermal conductivity decreases. The maximum ZT of 0.98 was reached for $CoSb_{2.85}Te_{0.15}$ at 820 K, which is superior to values obtained on comparably doped $CoSb_3$ prepared by the traditional synthesis methods. In the case of $In_xSn_{1-x}Te$, supersaturated InTe dissolves in SnTe at high temperatures as a consequence of the ultrafast and nonequilibrium nature of the combustion process and, upon annealing, nanometer scale inclusions precipitate in the SnTe matrix. The size of these nanoprecipitates increases with the increasing In content. Te vacancies, which formed during the phase separation process, reduce the concentration of holes. The formation of nanoprecipitates, coupled with the adjustment of the charge carrier concentration, alter transport parameters and give rise to a notable enhancement of the Seebeck coefficient and a degradation of the thermal conductivity. Doping SnTe with In enhances the TE figure of merit and the annealed sample with $x = 0.005$ recorded the highest $ZT = 0.92$ at about 800 K, which represents some 40% enhancement compared to the figure of merit of undoped SnTe.

In addition to demonstrating outstanding TE properties of SHS–PAS fabricated TE materials, we also developed a new criterion for applicability of the SHS synthesis process.

The traditional criterion of Merzhanov, which stated that the adiabatic temperature should exceed 1800 K for the fusion reaction to be self-propagating, clearly does not apply to the SHS synthesis of low-temperature compound TE materials. Based on our empirical data and, specifically, the input gained by failing to synthesize *via* SHS two important TE materials, Sb_2Te_3 and $MnSi_{1.70}$, we proposed a new criterion, which has a physically intuitive basis. Namely, the fusion reaction will be self-propagating only when the adiabatic temperature exceeds the melting point of the lower melting point component of the compound semiconductor, that is, $T_{ad}/T_{m,L} \geq 1$. This new criterion covers all materials successfully synthesized by SHS, including the high-temperature ceramics and intermetallics and low-temperature TE compound semiconductors. Moreover, the new criterion greatly expands the range of materials that can be successfully synthesized by this extraordinarily fast and inexpensive synthesis method. SHS–PAS opens the way for ultrafast, low-cost, mass production fabrication of efficient TE materials.

Acknowledgments

The authors acknowledge the support from the National Basic Research Program of China (973 program) under project 2013CB632502, the International Science & Technology Cooperation Program of China (Grant No. 2011DFB60150), the Natural Science Foundation of China (Grants No. 51402222, 51172174, 51521001), and the 111 Project of China (Grant No. B07040). C. U. and X. T. also acknowledge the support provided by CERC-CVC under the Award Number DE-PI0000012.

References

1. Tritt, T. M., H. Böttner, and L.-D. Chen. 2008. Thermoelectrics: Direct solar thermal energy conversion. *MRS Bull.* 33: 366–368.
2. Bell, L. E. 2008. Cooling, heating, generating power, and recovering waste heat with thermoelectric systems. *Science* 321: 1457–1461.
3. Chowdhury, I., R. Prasher, K. Lofgreen et al. 2009. On-chip cooling by superlattice-based thin-film thermoelectrics. *Nat. Nanotech.* 4: 235–238.
4. Su, X.-L., H. Li, and X.-F. Tang. 2010. Synthesis and thermoelectric properties of p-type Zn-doped $Zn_xIn_{1-x}Sb$ compounds. *J. Phys. D: Appl. Phys.* 43: 015403.
5. Liu, W., X.-F. Tang, and J. Sharp. 2010. Low-temperature solid state reaction synthesis and thermoelectric properties of high-performance and low-cost Sb-doped $Mg_2Si_{0.6}Sn_{0.4}$. *J. Phys. D: Appl. Phys.* 43: 085406.
6. Cao, W.-Q., Y.-G. Yan, X.-F. Tang, and X.-K. Deng. 2008. The effects of In isoelectronic substitution for Ga on the thermoelectric properties of $Sr_8Ga_{16-x}In_xGe_{30}$ type-I clathrates. *J. Phys. D: Appl. Phys.* 41: 215105.
7. Tang, X.-F., P. Li, S. Deng, and Q.-J. Zhang. 2008. High temperature thermoelectric transport properties of double-atom-filled clathrate compounds $Yb_xBa_{8-x}Ga_{16}Ge_{30}$. *J. Appl. Phys.* 104: 013706.
8. Deng, S., X.-F. Tang, P. Li, and Q.-J. Zhang. 2008. High temperature thermoelectric transport properties of p-type $Ba_8Ga_{16}Al_xGe_{30-x}$ type-I clathrates with high performance. *J. Appl. Phys.* 103: 073503.

9. Su, X.-L., X.-F. Tang, L. Han and S.-K. Deng. 2008. Structure and thermoelectric properties of n-type $Ga_xCo_4Sb_{12}$ skutterudite compounds. *Acta Phys. Sinica* 57: 6488–6493.
10. Shi, X., H. Kong, C.-P. Li et al. 2008. Low thermal conductivity and high thermoelectric figure of merit in n-type $Ba_xYb_yCo_4Sb_{12}$ double-filled skutterudites. *Appl. Phys. Lett.* 92: 182101.
11. Slack, G. A. 1995. New materials and performance limits for thermoelectric cooling. In *CRC Handbook of Thermoelectrics*, ed. D. M. Rowe, CRC Press, Boca Raton, FL, p. 407.
12. Nolas, G. S., G. A. Slack, D. T. Morelli, T. M. Tritt, and A. C. Ehrlich. 1996. The effect of rare-earth filling on the lattice thermal conductivity of skutterudites. *J. Appl. Phys.* 79: 4002–4008.
13. Tritt, T. M., G. S. Nolas, G. A. Slack, A. C. Ehrlich, D. J. Gillespie, and J. L. Cohn. 1996. Low-temperature transport properties of the filled and unfilled $IrSb_3$ skutterudite system. *J. Appl. Phys.* 79: 8412–8418.
14. Shi, X., J. Yang, J. R. Salvador et al. 2011. Multiple-filled skutterudites: High thermoelectric figure of merit through separately optimizing electrical and thermal transports. *J. Am. Chem. Soc.* 133: 7837–7846.
15. Su, X.-L., H. Li, Y. Yan et al. 2012. The role of Ga in $Ba_{0.30}Ga_xCo_4Sb_{12+x}$ filled skutterudites. *J. Mater. Chem.* 22: 15628–15634.
16. Nolas, G. S., J. L. Cohn, G. A. Slack, and S. B. Schujman. 1998. Semiconducting Ge clathrates: Promising candidates for thermoelectric applications. *Appl. Phys. Lett.* 73: 178–180.
17. Meisner, G. P., D. T. Morelli, S. Hu, J. Yang, and C. Uher. 1998. Structure and lattice thermal conductivity of fractionally filled skutterudites: Solid solutions of fully filled and unfilled end members. *Phys. Rev. Lett.* 80: 3551–3554.
18. Sales, B. C., D. Mandrus, and R. K. Williams. 1996. Filled skutterudite antimonides: A new class of thermoelectric materials. *Science* 272: 1325–1328.
19. Yang, J., W. Zhang, S. Q. Bai, Z. Mei, and L.-D. Chen. 2007. Dual-frequency resonant phonon scattering in $Ba_xR_yCo_4Sb_{12}$ (R = La, Ce, and Sr). *Appl. Phys. Lett.* 90: 192111.
20. Zheng, Y., Q. Zheng, Q. Zhang et al. 2015. Mechanically robust BiSbTe alloys with superior thermoelectric performance: A case study of stable hierarchical nanostructured thermoelectric materials. *Adv. Energy Mater.* 5: 1401391.
21. Tan, G., Y. Zheng, and X.-F. Tang. 2013. High thermoelectric performance of nonequilibrium synthesized $CeFe_4Sb_{12}$ composite with multi-scaled nanostructures. *Appl. Phys. Lett.* 103: 183904.
22. Biswas, K., J. He, I. D. Blum et al. 2012. High-performance bulk thermoelectrics with all-scale hierarchical architectures. *Nature* 489: 414–418.
23. Zhao, L.-D., S. Hao, S.-H. Lo et al. 2013. High thermoelectric performance via hierarchical compositionally alloyed nanostructures. *J. Am. Chem. Soc.* 135: 7364–7370.
24. Li, H., X.-F. Tang, Q.-J. Zhang, and C. Uher. 2009. High performance $In_xCe_yCo_4Sb_{12}$ thermoelectric materials with in situ forming nanostructured InSb phase. *Appl. Phys. Lett.* 94: 102114.
25. Xie, W.-J., X.-F. Tang, Y. Yan, Q.-J. Zhang, and T. M. Tritt. 2009. Unique nanostructures and enhanced thermoelectric performance of melt-spun BiSbTe alloys. *Appl. Phys. Lett.* 94: 102111.
26. Xie, W.-J., J. He, H. J. Kang et al. 2010. Identifying the specific nanostructures responsible for the high thermoelectric performance of $(Bi, Sb)_2Te_3$ nanocomposites. *Nano Lett.* 10: 3283–3289.
27. Su, X.-L., H. Li, G. Wang et al. 2011. Structure and transport properties of double-doped $CoSb_{2.75}Ge_{0.25-x}Te_x$ ($x = 0.125$–0.20) with in situ nanostructure. *Chem. Mater.* 23: 2948–2955.
28. Su, X.-L., H. Li, Y. Yan et al. 2012. Microstructure and thermoelectric properties of $CoSb_{2.75}Ge_{0.25-x}Te_x$ prepared by rapid solidification. *Acta Mater.* 60: 3536–3544.
29. Liu, W., X.-F. Tang, H. Li et al. 2012. Enhanced thermoelectric properties of n-type $Mg_{2.16}(Si_{0.4}Sn_{0.6})_{1-y}Sb_y$ due to nano-sized Sn-rich precipitates and an optimized electron concentration. *J. Mater. Chem.* 22: 13653–13661.
30. Wang, S., H. Li, D. Qi, W.-J. Xie, and X.-F. Tang. 2011. Enhancement of the thermoelectric performance of β-Zn_4Sb_3 by in situ nanostructures and minute Cd-doping. *Acta Mater.* 59: 4805–4817.
31. Xu, J.-J., H. Li, B. Du, X.-F. Tang, Q.-J. Zhang, and C. Uher. 2010. High thermoelectric figure of merit and nanostructuring in bulk $AgSbTe_2$. *J. Mater. Chem.* 20: 6138–6143.
32. Hicks, L. D. and M. S. Dresselhaus. 1993. Effect of quantum-well structures on the thermoelectric figure of merit. *Phys. Rev. B* 47: 12727.

33. Faleev, S. V. and F. Leonard. 2008. Theory of enhancement of thermoelectric properties of materials with nanoinclusiosn. *Phys. Rev. B* 77: 214304.
34. Heremans, J. P., C. M. Thrush, and D. T. Morelli. 2004. Thermopower enhancement in lead telluride nanostructures. *Phys. Rev. B* 70: 115334.
35. Hsu, K. F., S. Loo, F. Guo et al. 2004. Cubic $AgPb_mSbTe_{2+m}$: Bulk thermoelectric materials with high figure of merit. *Science* 303: 818–821.
36. Zhao, L.-D., V. P. Dravid, and M. G. Kanatzidis. 2014. The panascopic approach to high performance thermoelectrics. *Energy Environ. Sci.* 7: 251–268.
37. Slack, G. A. 1979. Thermal conductivity of nonmetallic crystals. In *Solid State Physics*, eds. F. Seitz and D. Turnbull, Academic Press, New York, Vol. 34, p. 1.
38. Cahill, D. G. and R. O. Pohl. 1987. Thermal conductivity of amorphous solids above the plateau. *Phys. Rev. B* 35: 4067–4073.
39. Heremans, J. P., V. Jovovic, E. S. Toberer et al. 2008. Enhancement of thermoelectric efficiency in PbTe by distortion of the electronic density of states. *Science* 321: 554–558.
40. Zhang, Q., B. Liao, Y. Lan et al. 2013. High thermoelectric performance by resonant dopant indium in nanostructured SnTe. *Proc. Natl. Acad. Sci. USA* 110: 13261–13266.
41. Wang, S., X. Tan, G. Tan et al. 2012. The realization of a high thermoelectric figure of merit in Ge-substituted β-Zn_4Sb_3 through band structure modification. *J. Mater. Chem.* 22: 13977–13985.
42. Jaworski, C. M., V. A. Kulbachinskii, and J. P. Heremans. 2009. Tin forms a resonant level in Bi_2Te_3 that enhances the room temperature thermoelectric power. *Phys. Rev. B* 80: 2333201.
43. Heremans, J. P. 2012. The effect of resonant energy levels on the thermoelectric power and the power factor. In *Materials, Preparation, and Characterization in Thermoelectrics*, ed. D. M. Rowe, CRC Press, Taylor & Francis, Boca Raton, FL, pp. 12-1–12-20.
44. Pei, Y., X. Shi, A. LaLonde, H. Wang, L. D. Chen, and G. J. Snyder. 2011. Convergence of electronic bands for high performance bulk thermoelectrics. *Nature* 473: 66–69.
45. Liu, W., X. Tan, K. Yin et al. 2012. Convergence of conduction bands as a means of enhancing thermoelectric performance of n-type $Mg_2Si_{1-x}Sn_x$ solid solutions. *Phys. Rev. Lett.* 108: 166601.
46. Zaitsev, V. K., M. I. Fedorov, I. S. Eremin, and E. A. Gurieva. 2006. Thermoelectrics on the base of solid solutions of Mg_2BIV compounds. In *Thermoelectrics Handbook*, ed. D. M. Rowe, CRC Press, Taylor & Francis, Boca Raton, FL, Ch. 29.
47. Zaitsev, V. K., M. I. Fedorov, E. A. Gurieva et al. 2006. Highly effective $Mg_2Si_{1-x}Sn_x$ thermoelectrics. *Phys. Rev. B* 74: 045207.
48. Tan, X. J., W. Liu, H. J. Liu, J. Shi, X.-F. Tang, and C. Uher. 2012. Multiscale calculations of thermoelectric properties of n-type $Mg_2Si_{1-x}Sn_x$ solid solutions. *Phys. Rev. B* 85: 205212.
49. Liu, W., H. Chi, H. Sun et al. 2014. Advanced thermoelectrics governed by a single parabolic band: $Mg_2Si_{0.3}Sn_{0.7}$, a canonical example. *Phys. Chem. Chem. Phys.* 16: 6893–6897.
50. Tan, G., F. Shi, S. Hao et al. 2015. Valence band modification and high thermoelectric performance in SnTe heavily alloyed with MnTe. *J. Am. Chem. Soc.* 137: 11507–11516.
51. Tan, G., L.-D. Zhao, F. Shi et al. 2014. High thermoelectric performance of p-type SnTe via a synergistic band engineering and nanostructuring approach. *J. Am. Chem. Soc.* 136: 7006–7017.
52. Hoang, K., S. D. Mahanti, and M. G. Kanatzidis. 2010. Impurity clustering and impurity-induced bands in PbTe-, SnTe-, and GeTe-based bulk thermoelectrics. *Phys. Rev. B* 81: 115106.
53. Heremans, J. P., B. Wiendlocha, and A. M. Chamoire. 2012. Resonant levels in bulk thermoelectric semiconductors. *Energy Environ. Sci.* 5: 5510–5530.
54. Su, X.-L., F. Fu, Y. Yan et al. 2014. Self-propagating high-temperature synthesis for compound thermoelectrics and new criterion for combustion processing. *Nat. Commun.* 5: 4908.
55. Zheng, G., X.-L. Su, T. Liang et al. 2015. High thermoelectric performance of mechanically robust n-type $Bi_2Te_{3-x}Se_x$ prepared by combustion synthesis. *J. Mater. Chem. A* 3: 6603–6613.
56. Liang, T., X.-L. Su, Y. Yan et al. 2014. Ultra-fast synthesis and thermoelectric properties of Te doped skutterudites. *J. Mater. Chem. A* 2: 17914–17918.
57. Liang, T., X.-L. Su, X. Tan et al. 2015. Ultra-fast non-equilibrium synthesis and phase segregation in $In_xSn_{1-x}Te$ thermoelectrics by SHS-PAS processing. *J. Mater. Chem. C* 3: 8550–8558.

58. She, X., X.-L. Su, H. Du et al. 2015. High thermoelectric performance of higher manganese silicides prepared by ultra-fast thermal explosion. *J. Mater. Chem. C* 3: 12116–12122.
59. Yi, H. C. and J. J. Moore. 1990. Self-propagating high-temperature (combustion) synthesis (SHS) of powder-compacted materials. *J. Mater. Sci.* 25: 1159–1168.
60. Barsoum, M. W. and T. El-Raghy. 1996. Synthesis and characterization of a remarkable ceramic: Ti_3SiC_2. *J. Am. Ceram. Soc.* 79: 1953–1956.
61. Li, B. Y., L. G. Rong, and V. E. Gjunter. 2000. Synthesis of porous Ni–Ti shape-memory alloys by self-propagating high-temperature synthesis: Reaction mechanism and anisotropy in pore structure. *Acta Mater.* 48: 3895–3904.
62. Merzhanov, A. G. 2004. The chemistry of self-propagating high-temperature synthesis. *J. Mater. Chem.* 14: 1779–1786.
63. Merzhanov A.G. 1975. Technological combustion problems. In: Protsessy Goreniya v Khemicheskoi Tekhnologii i Metallurgii (*Combustion Processes in Chemical Technology and Metallurgy*), Chernogolovka, Moscow, 5–28.
64. Novikov, P.N., P. I. Borovinskaya, and G. A. Merzhanov. *Combustion in chemical Engineering and Metallurgy* (in Russian), ed. by A. G. Merzhanov, Chernogolovka, Moscow, p. 174.
65. Munir, Z. A. and U. Anselmi-Tamburini. 1989. Self-propagating exothermic reactions: The synthesis of high-temperature materials by combustion. *Mater. Sci. Rep.* 3: 277–365.
66. Moore, J. J. and H. J. Feng. 1995. Combustion synthesis of advanced materials: Part I. Reaction parameters. *Progr. Mater. Sci.* 39: 243–273.
67. Drabble, J. R., and C. H. L. Goodman. 1958. Chemical bonding in bismuth telluride. *J. Phys. Chem. Solids* 5: 142–144.
68. Goldsmid, H. J. 1961. Recent studies of bismuth telluride and its alloys. *J. Appl. Phys.* 32: 2198–2202.
69. Jiang, J., L.-D. Chen, S. Q. Bai, Q. Yao, and Q. Wang. 2005. Fabrication and thermoelectric performance of textured n-type $Bi_2(Te, Se)_3$ by spark plasma sintering. *Mater. Sci. Eng. B* 117: 334–338.
70. Zheng, Y., H. Xie, S. Shu, Y. Yan, H. Li, and X. F. Tang. 2014. High-temperature mechanical and thermoelectric properties of p-type $Bi_{0.5}Sb_{1.5}Te_3$ commercial zone melting ingots. *J. Electron. Mater.* 43: 2017–2022.
71. Zhao, L.-D., B.-P. Zhang, J.-F. Li, M. Zhou, W.-S. Liu, and J. Liu. 2008. Thermoelectric and mechanical properties of nano-SiC-dispersed Bi_2Te_3 fabricated by mechanical alloying and spark plasma sintering. *J. Alloys Compd.* 455: 259–264.
72. Hu, L. P., X. H. Liu, H. H. Xie, J. J. Shen, T. J. Zhu, and X. B. Zhao. 2012. Improving thermoelectric properties of n-type bismuth–telluride-based alloys by deformation-induced lattice defects and texture enhancement. *Acta Mater.* 60: 4431–4437.
73. Dong, G.-H., Y.-J. Zhu, and L.-D. Chen. 2010. Microwave-assisted rapid synthesis of Sb_2Te_3 nanosheets and thermoelectric properties of bulk samples prepared by spark plasma sintering. *J. Mater. Chem.* 20: 1976–1981.
74. Prokofieva, L. V., D. A. Pshenay-Severin, P. P. Konstantinov, and A. A. Shabaldin. 2009. Optimum composition of a $Bi_2Te_{3-x}Se_x$ alloy for the n-type leg of a thermoelectric generator. *Semiconductors* 43: 973–976.
75. Liu, H., X. Shi, F. Xu et al. 2012. Copper ion liquid-like thermoelectrics. *Nat. Mater.* 11: 422–425.
76. Xiao, X.-X., W.-J. Xie, X.-F. Tang, and Q.-J. Zhang. 2011. Phase transition and high temperature thermoelectric properties of copper selenide $Cu_{2-x}Se$ ($0 \leq x \leq 0.25$). *Chin. Phys. B* 20: 087201.
77. Yu, B., W. Liu, S. Chen et al. 2012. Thermoelectric properties of copper selenide with ordered selenium layer and disordered copper layer. *Nano Energy* 1: 472–478.
78. Li, X.Y., L.-D. Chen, J.-F. Fan, W.-B. Zhang, T. Kawahara, and T. Hirai. 2005. Thermoelectric properties of Te-doped $CoSb_3$ by spark plasma sintering. *J. Appl. Phys.* 98: 083702.
79. Liu, W.-S., B.-P. Zhang, D.-F. Li, H.-L. Zhang, and L.-D. Zhao. 2007. Enhanced thermoelectric properties in $CoSb_{3-x}Te_x$ alloys prepared by mechanical alloying and spark plasma sintering. *J. Appl. Phys.* 102: 103717.
80. Santhanam, S. and A. K. Chaudhuri. 1981. Transport properties of SnTe interpreted by means of a two valence band model. *Mater. Res. Bull.* 16: 911–917.

21

Prospective Thermoelectrics among Topological Insulators

Jihui Yang
University of Washington

CONTENTS

21.1 Introduction

Since the theoretical prediction of the existence of topological insulators (TIs) [1–5], there have been widespread interests in these remarkable materials. Both two-dimensional (2D) and three-dimensional (3D) TIs have been experimentally observed and verified [6–11]. A distinctive feature of this class of new quantum state of materials is its electronic density of states, which consists of insulating bulk bands and metallic edge (2D) or surface (3D) bands. Figure 21.1 shows the electronic band structure characteristics of Bi_2Se_3 measured by angle-resolved photoemission spectroscopy [12]. Data shown are from References 9 and 13 and modified in Reference 12. The Dirac cone is the hallmark of TIs.

The formation of TIs necessitates strong spin–orbit coupling. The strong spin–orbit coupling induces electronic band structures that feature no spin degeneracy edge/surface states, which are protected by the time-reversal symmetry. Therefore, nonmagnetic impurities or any other time-reversal invariant local perturbations cannot cause elastic backscattering of the edge/surface states. This warrants the topological robustness of the edge/surface state conductance [1,6]. In addition, spin–orbit coupling brings about a sign change of parity eigenvalues in the Brillouin zone, hence leading to electronic band inversion [14–16]. This key feature of TIs has many ramifications. For example, it has been used to open a bandgap in semimetallic HgTe and turn HgTe into a TI, with applied strain [17], as well as for the findings of new TIs in ternary Heusler compounds [16].

The discovery of TIs has certainly opened the door for potential applications in spintronics [18], quantum computing [19], etc. Recently, there have been growing interests in

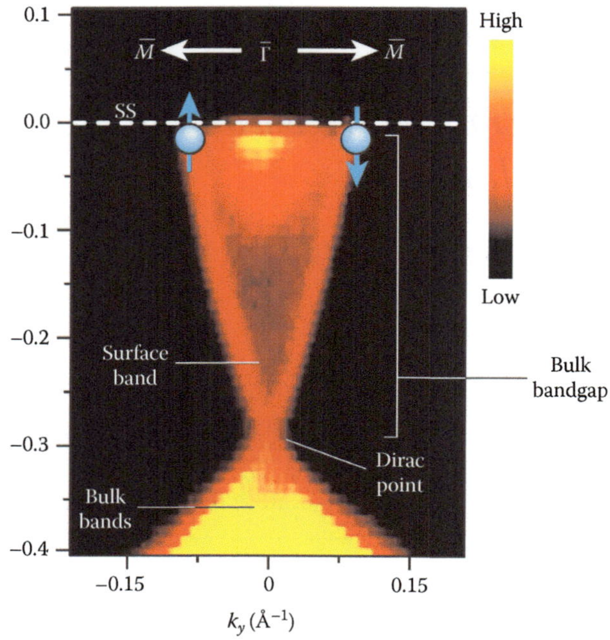

FIGURE 21.1
The electronic band structure features in topological insulator Bi_2Se_3 measured by angle-resolved photoemission spectroscopy. Γ and *M* denote the Brillouin-zone center and the center of the Brillouin-zone edge, respectively. Electron energy, E_B, is plotted against electron momentum, k_y. High intensity (red and yellow areas) indicates a nonzero electronic density of states. The surface bands crossing the bulk bandgap enclose a single Dirac point at the Γ point. The direction of electron spin is indicated by blue arrows. (Data shown are from Xia, Y. et al., *Nat. Phys.*, 5, 398, 2009; Hsieh, D. et al., *Nature*, 460, 1101, 2009; Modified in Moore, J.E., *Nature*, 464, 194, 2010.)

building a bridge between the TIs and high-performance thermoelectrics. This has been motivated by the following observation. First, strong spin–orbit coupling favors heavy elements for TIs, while most of the good thermoelectric materials have heavy elemental constituents in order to achieve low lattice thermal conductivity [20,21], with a few exceptions [22,23]. Second, TIs typically have the bulk energy gap comparable to the value of the spin–orbit coupling constant; hence, they are small gap semiconductors. The optimized thermoelectric materials are semiconductors with bandgap approximately equal to $10k_BT_{op}$ [24], where k_B is the Boltzmann constant and T_{op} the operating temperature of thermoelectric devices. So the best thermoelectric materials mostly possess bandgap values less than 1 eV. For room temperature thermoelectrics, this value is 0.25 eV. Therefore, most of the 3D TIs such as $Bi_{1-x}Sb_x$ [25], Bi_2Se_3, Bi_2Te_3, Sb_2Te_3 [10], band-inverted half-Heuslers [16,26], and skutterudites [26] are also the parent compounds of state-of-the-art thermoelectric materials. Despite these similarities, there exist major differences between TIs and thermoelectric materials. For TIs, the Fermi levels need to be within the energy bandgap, whereas thermoelectric materials are doped semiconductors, so their Fermi levels are in the conduction or valence bands. In addition, the need to have heavy constituents serves different purposes in TIs and thermoelectric materials. It helps to facilitate strong spin–orbit coupling in the former and low lattice thermal conductivity in the latter.

According to the conventional theory of thermoelectricity, a good thermoelectric material should simultaneously have low thermal conductivity κ (which has lattice κ_L and electronic κ_e components), high Seebeck coefficient S, and high electrical conductivity σ [20,21]. A dimensionless figure of merit ZT is used to characterize the thermoelectric performance of a material, where $ZT = S^2\sigma T/\kappa$ and T is the absolute temperature. It has also been well established that materials that have superior thermoelectric performance are typically heavily doped narrow gap semiconductors with low lattice thermal conductivity [20,21]. It seems, therefore, contradictory for insulating TIs to retain good thermoelectric properties. Recent theoretical and experimental efforts aimed to explore the *size effect* on the electronic transport properties ($S^2\sigma$) of TIs in order to increase the surface contribution to the overall electronic transport but had limited success. Moreover, strong spin–orbit coupling–induced band inversion has been investigated to reveal new TIs in known thermoelectric compounds, as well as for developing better thermoelectrics. Even though the lattice dynamics and thermal transport properties of the aforementioned TIs are well understood, weak TIs show unusual lattice dynamics and room temperature minimum lattice thermal conductivity, due to their structural and bonding characteristics. The influence of the topological states on the thermoelectric properties of materials is, hence, of considerable current interest. The goal of this chapter is to provide an overview of such stimulating activities.

21.2 Electronic Transport in Topological Insulators for Thermoelectricity

21.2.1 Theoretical Studies

Takahashi and Murakami studied the thermoelectric transport of both 2D and 3D TIs based on a linear response theory [27,28]. Because TIs are insulating in the bulk while metallic in the boundary (edge/surface) states that are stable against elastic scattering by nonmagnetic impurities, transport in the bulk and the boundary states were treated separately. They found that the electronic transport depends very sensitively on the inelastic scattering length. For the 2D TI, there is an enhancement of thermoelectric properties at low temperatures due to the long inelastic scattering length and, hence, the edge states dominate the electronic transport. At room temperature, the inelastic scattering length is significantly reduced, the bulk states dominate, and ZT is lowered. The crossover occurs when the chemical potential is at the band edge. In the case of 3D TIs, similar bulk-surface competition exists. However, the surface states no longer influence the electronic transport significantly due to the impurity effect, and ZT is less prominent, unless a gap is opened, in which case the problem falls into the general description of traditional narrow gap semiconductors [28].

In a more recent study, Xu et al. [29] uncovered an interesting *side effect* due to the unique energy dependence of the scattering time in the TIs, when they used intrinsic properties of the edge states instead of truncating them in the computations, as done in the previous study [28]. The calculations were performed for a general geometry with the cross-sectional area A and length L, while λ_l denoted the inelastic mean free path. Figure 21.2 shows the length dependence of the Seebeck coefficient from the edge states S_1. S_1 increases with increasing L/λ_l, from nearly zero in the ballistic limit (small L) to large saturation values in the diffusive limit ($L \sim \lambda_l$). Furthermore, this general behavior is independent of the choice of scattering time ratio $r_\tau = \tau_1/\tau_2$, where τ_1 and τ_2 are constant scattering times for the edge and the bulk states, respectively. The maximum value of S_1 can be drastically improved when varying the Fermi level E_F. Based on the fact that the sign of S_1 is different from the

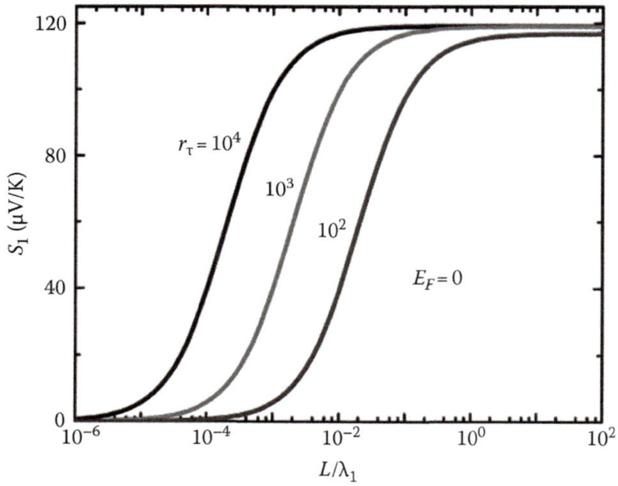

FIGURE 21.2
The length dependence of S_1 (S from the edge states) for the Fermi level $E_F = 0$ and various scattering time ratios $r_\tau = \tau_1/\tau_2$, where τ_1 and τ_2 are constant scattering times for the edge and the bulk states, respectively. L is the length of the sample and λ_l the inelastic mean-free path of the edge states. (Reprinted with permission from Xu, Y., Gan, Z., and Zhang, S.-C., *Phys. Rev. Lett.*, 112, 226801. Copyright 2014 by the American Physical Society.)

expected sign of the charge carriers (electrons, as E_F is near the conduction band minimum), the authors concluded that this anomalous S_1 originates from the unique energy dependence of the scattering time in the TIs and that there is no need to open the gap, as suggested by the previous study [28], in order to increase the Seebeck coefficient. The edge states can contribute to large S by optimizing E_F.

Figure 21.3 shows the calculated size dependence of ZT and S for the 2D TI-fluorinated stanene at 300 K [29], taking into account the Seebeck contribution from both the edge states as well as the bulk states. The results shown in Figure 21.3 were maximized by

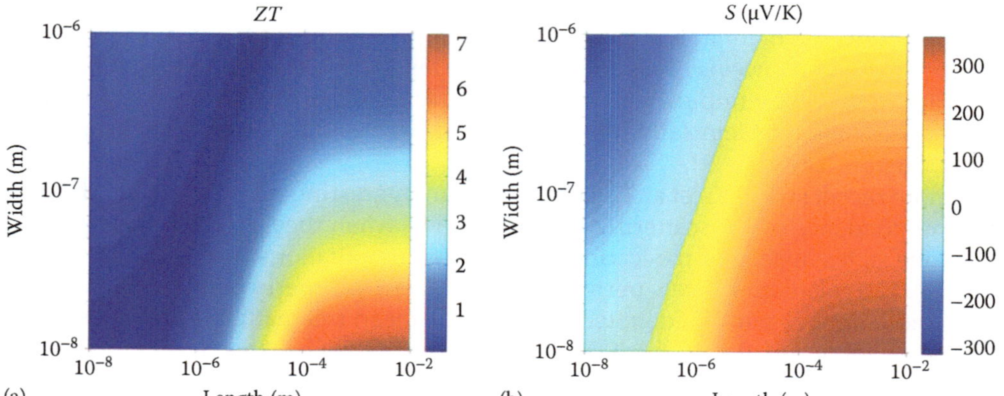

FIGURE 21.3
The size dependence of (a) ZT and (b) S for the 2D TI-fluorinated stanene at 300 K. (Reprinted with permission from Xu, Y., Gan, Z., and Zhang, S.-C., *Phys. Rev. Lett.*, 112, 226801. Copyright 2014 by the American Physical Society.)

choosing the optimal E_F. It is shown that bulk states dominate TE transport with small L and large W (the width of the sample), leading to small ZT and negative S. With increasing L and decreasing W, edge states start to increase their contribution to the transport processes and, eventually, become prevalent. Consequently, ZT can reach the maximum value of 7, with S changing sign from positive to negative. This effect is maximized by choosing $L \sim \lambda_1$ and $W \sim 2\xi$ (~10 nm), where ξ is the localization width. While this *size effect* on improving thermoelectric properties of TIs seems very encouraging, it certainly calls for direct experimental validation, which is very challenging both in sample fabrication and in measurements.

21.2.2 Experimental Studies

There have been many experimental attempts to investigate the potentially beneficial effects of the edge states in TIs, mostly with samples in the form of high-quality nanowires. The nanowires were chosen for the benefit of high surface-to-volume ratio in order to maximize the edge-to-bulk ratio. Typically, the nanowire measurements were carried out in microfabricated devices with some minor variations. One such device is shown here in Figure 21.4 [30]. It is understood that the thermal conductivity of nanowires would be significantly reduced from their bulk values [31]. So the majority of the recent studies have been focused on the electrical conductivity and Seebeck coefficient measurements.

The electrical transport properties have been reported in many TI nanowires including Bi_2Te_3, $(Bi_{1-x}Sb_x)_2Te_3$, and $Bi_{1.5}Sb_{0.5}Te_{1.7}Se_{1.3}$ [30,32–34]. None of the prior studies, however, has unambiguously demonstrated an improvement in the power factor ($S^2\sigma$) over the bulk phases. In some cases [30], the existence of edge states were verified by the magnetotransport measurements. The reasons for not measuring the power factor improvement over the bulk phases are a matter of an ongoing debate. Some argue that since the edge states and

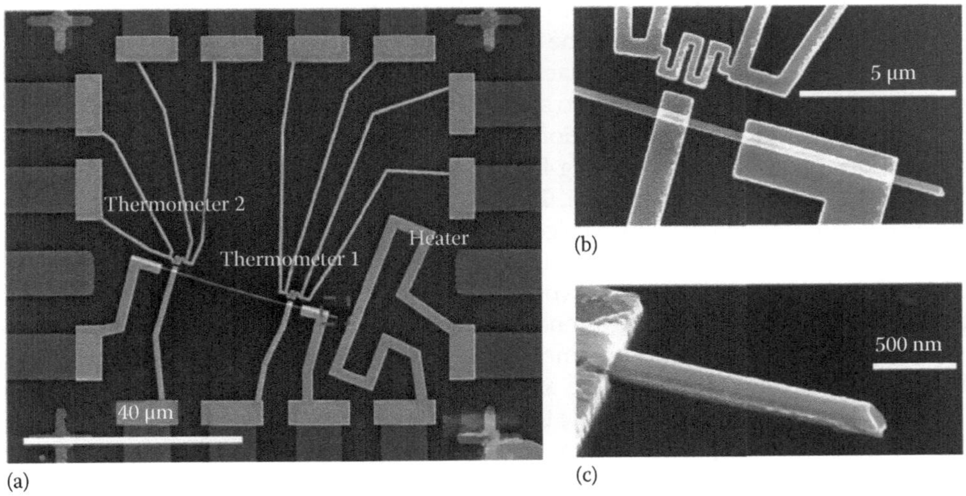

FIGURE 21.4
(a) Seebeck coefficient and electrical conductivity measurements device. (b) High-magnification image of thermometer 1. (c) Cross-section image of the dashed line area of (a). (From Hsiung, T.-C., Mou, C.-Y., Lee, T.-K., and Chen, Y.-Y., Surface-dominated transport and enhanced thermoelectric figure of merit in topological insulator $Bi_{1.5}Sb_{0.5}Te_{1.7}Se_{1.3}$, *Nanoscale*, 7, 518–523, 2015. Reproduced by permission of The Royal Society of Chemistry.)

the bulk states are supposed to be optimized at different Fermi levels, using nanowires to demonstrate the benefits of TIs may not be ideal [33]. It is interesting to note that the nanowires or stacks of Bi_2Te_3 quintuples used in the studies all have lateral dimensions (diameter or width) significantly higher than the optimal 10 nm (twice the localization length) estimated by Reference 29. Figure 21.3 suggests a rapid decrease of ZT and S upon increasing W. Experimental efforts to synthesize smaller diameter nanowires are therefore highly desirable to substantiate the theoretical predictions.

21.3 Connecting Topological and Thermoelectric Behavior

While it is still very challenging to demonstrate the benefits of the TI behavior for thermoelectricity, this topic continues to draw a wide-ranging interest and extensive effort. New insights and understanding are constantly evolving and being generated. Two recent examples of uncovering TIs in thermoelectric compounds and improving the thermoelectric performance of materials based on band inversion, the issue central to TIs, are reviewed here.

Ternary half-Heusler compounds have been studied as high efficiency thermoelectric materials in the past two decades [35–38], due to their heavy carrier effective mass, narrow bandgap, and modest lattice thermal conductivity. Many ternary half-Heusler compounds of XYZ composition (with X and Y as the transition or rare-earth metals and Z as the main-group element) are semiconductors with structural and electronic similarity to the binary semiconductors. Chadov et al. [16] showed that, structurally, the half-Heusler compounds are similar to the zinc-blende structure and can be viewed as consisting of an X^{n+} ion *stuffing* the zinc-blende YZ^{n-} sublattice. The study also showed that, electronically, the ternary half-Heuslers can also be readily compared to the binary semiconductors.

Figure 21.5 compares the calculated band structures of CdTe and HgTe with those of ScPtSb and ScPtBi. Both CdTe and ScPtSb are trivial semiconductors, with direct (CdTe) and indirect (ScPtSb) bandgaps between the conduction (blue) and the valence (red) bands of Γ_6 and Γ_8 symmetries, respectively. Replacing Cd with Hg closes the bandgap and induces a band inversion, Γ_6 now below Γ_8, due to stronger spin–orbit coupling (heavy Hg and Te) and the change in the degree of hybridization governed by the lattice parameter. This is a necessary condition for the TI state, making HgTe a topological system. The same phenomenon is seen in going from ScPtSb to ScPtBi. This analog between the ternary half-Heuslers and CdTe/HgTe prompted the study of a large phase space of ternary half-Heusler compositions.

The authors [16] found that all half-Heusler compounds (~50) with zero bandgap at the Fermi level under certain conditions would undergo the same type of band inversion as in the case of HgTe. Figure 21.6 plots the energy difference between Γ_6 and Γ_8, $E_{\Gamma_6} - E_{\Gamma_8}$, as functions of the experimental lattice parameter and the average nuclear charge over the atoms in the unit cell $\langle Z \rangle$, indicative of the average spin–orbit coupling. The compounds with $E_{\Gamma_6} - E_{\Gamma_8} > 0$ are trivial insulators, while those with $E_{\Gamma_6} - E_{\Gamma_8} < 0$ are TI candidates. It was also confirmed that increasing the lattice constant reduces the hybridization and closes the non-zero bandgap; coupled with sufficiently strong spin–orbit coupling, it would lead to a pronounced Γ_6–Γ_8 band inversion, which is the key to realize the TI state. Furthermore, the large variety of half-Heusler compounds available with different bandgaps and lattice parameters makes it possible to construct quantum wells with closely matched lattice parameters, such as in the case of HgTe/CdTe quantum wells [2]. The candidate pairs can be chosen near the same vertical line in Figure 21.6a to form a smooth transition from a trivial insulator to a TI.

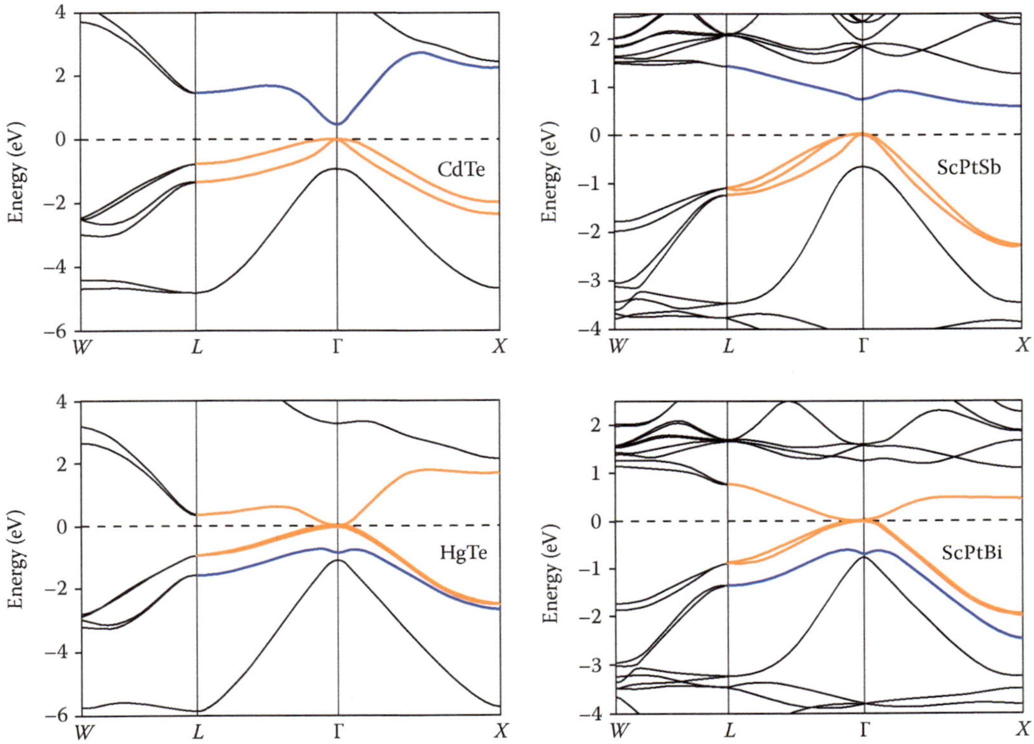

FIGURE 21.5
Band structures of CdTe and HgTe compared with ScPtSb and ScPtBi Heusler alloys. Red and blue mark the bands with Γ_8 and Γ_6 symmetry, respectively. Both CdTe and ScPtSb are trivial insulators with Γ_6 situated above Γ_8, which sits at the Fermi energy (set to zero). Both HgTe and ScPtBi are topological with inverted band order; Γ_6 is below Γ_8. (Reprinted by permission from Macmillan Publishers Ltd., *Nat. Mater.*, Chadov, S., Qi, X., Kübler, J., Fecher, G.H., Felser, C., and Zhang, S.-C., Tunable multifunctional topological insulators in ternary Heusler compounds, 9, 541–545, copyright 2010.)

The authors [16] concluded that half-Heusler compounds with band inversion are topologically nontrivial. They can be tuned from trivial insulators to TIs *via* changing the lattice parameter (using pressure, strain provided by a substrate) or *via* elemental substitutions to change the spin–orbit coupling. Finally, those compositions with *f*-electrons could offer opportunities to study exotic physical phenomena, such as superconductivity in TIs.

The topologically necessary band inversion in TIs not only changes the parity of wave functions in the gap but also brings about complex electronic structure modifications, which have recently been shown to benefit thermoelectricity [39]. Band structure calculations reveal that in two TIs—Bi_2Te_3 and Bi_2Te_2Se—spin–orbit coupling in these materials introduces changes in the bandgap, band symmetry, as well as, and more importantly for thermoelectrics, nonparabolicity.

Figure 21.7 shows the calculated band structures for two materials with varying spin–orbit coupling strength. Without spin–orbit, both materials are direct-gap semiconductors with band edges at the Γ point and comparatively parabolic bands. In both cases, increasing the spin–orbit interaction reduces the bandgap rapidly until in the 0.75 plot; the gap is very small and visibly nonparabolic. When the full strength of the spin–orbit is applied,

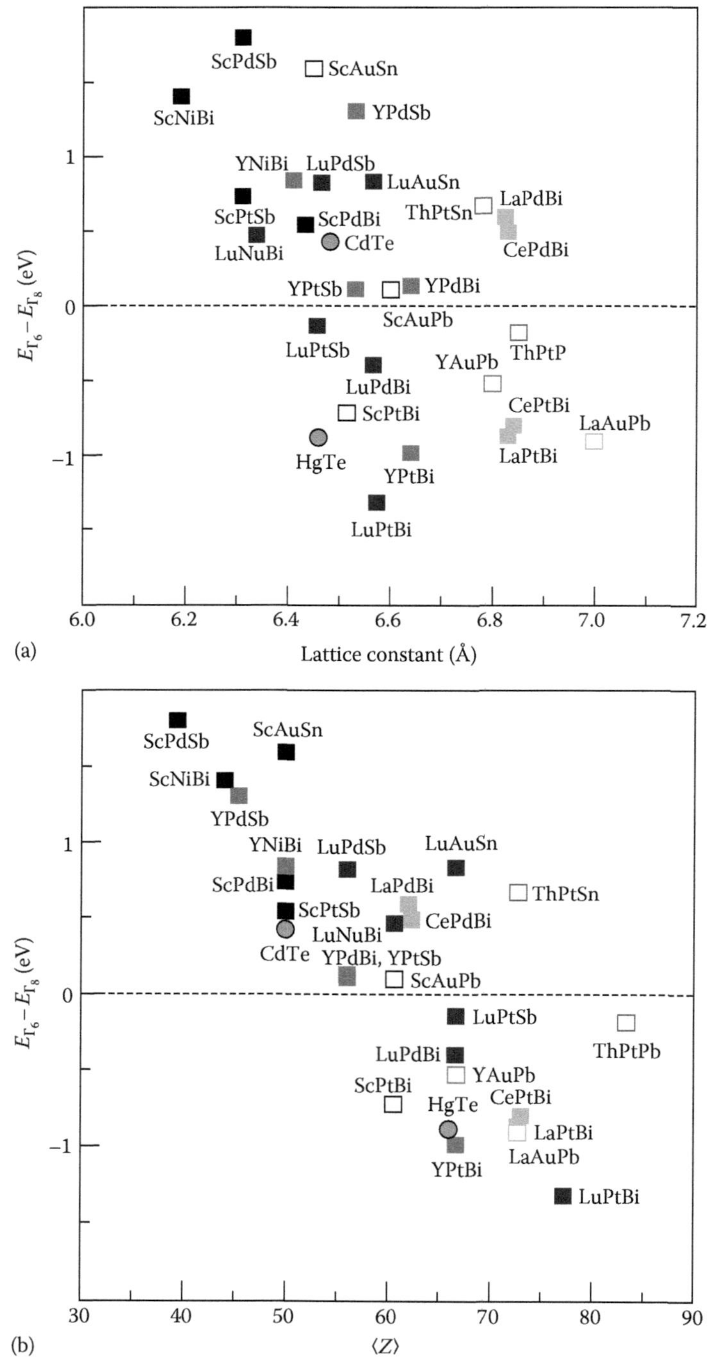

FIGURE 21.6
$E_{\Gamma_6} - E_{\Gamma_8}$ calculated for various half-Heusler alloys at their experimental lattice constants. (Reprinted by permission from Macmillan Publishers Ltd., *Nat. Mater.*, Chadov, S., Qi, X., Kübler, J., Fecher, G.H., Felser, C., and Zhang, S.-C., Tunable multifunctional topological insulators in ternary Heusler compounds, 9, 541–545, copyright 2010.)

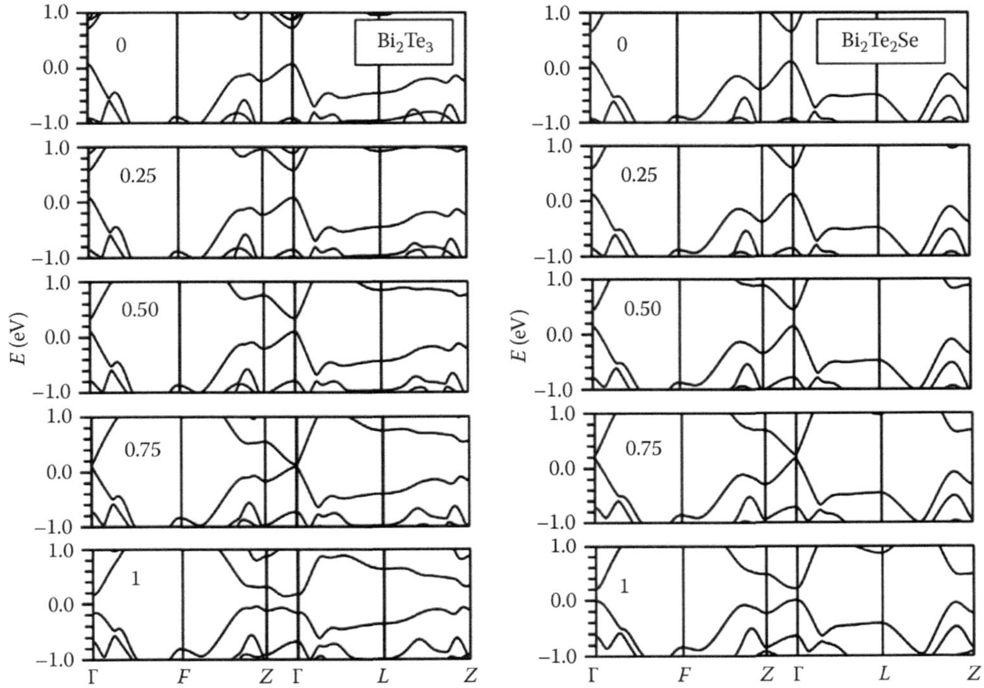

FIGURE 21.7
The calculated band structures of Bi_2Te_3 and Bi_2Te_2Se with the spin–orbit coupling included in strengths (relative to the actual physical value) of 0, 0.25, 0.5, 0.75, and unity. (Reprinted with permission from Shi, H., Parker, D., Du, M.-H., and Singh, D.J., *Phys. Rev. Appl.*, 3, 014004. Copyright 2015 by the American Physical Society.)

a new gap opens up between the Z and Γ points for Bi_2Te_3 but returns to the Γ point for Bi_2Te_2Se, and again, these band structures do appear to be comparatively nonparabolic. The importance of band nonparabolicity for thermoelectrics has been documented in the past [40]. This study clearly illustrates the effect of electronic band structure complexity, more specifically band nonparabolicity induced by the band inversion (or strong spin–orbit coupling), which is responsible for the superior thermoelectric performance in this class of materials.

21.4 Lattice Dynamics and Thermal Transport in Weak Topological Insulators

The TIs we have discussed so far are *strong* TIs, such as Bi_2Te_3, Bi_2Se_3, and Sb_2Te_3, or the half-Heuslers, whose lattice dynamics and thermal transport have been covered by a large body of literature [20,21,36]. Therefore, most of the recent studies of thermoelectric properties of TIs have been primarily focused on the electronic transport—the electrical conductivity and the Seebeck coefficient. A recent study by Wei et al. revealed that weak TIs, formed by an alternating stack of quantum spin Hall (QSH) layers and normal insulator

(NI) layers, possess a very interesting lattice dynamics phenomenon and a rather unusual room temperature lattice thermal conductivity approaching the minimum thermal conductivity [41].

As a weak TI, $Bi_{14}Rh_3I_9$ belongs to the Z2 (0; 001) class [42], implying that the nontrivial surface states with an even number of Dirac cones can be present at any of the crystal faces except those parallel to the (001) plane. Its crystal structure, shown in Figure 21.8, consists of an alternating stack of the topologically nontrivial graphene-analog QSH layers

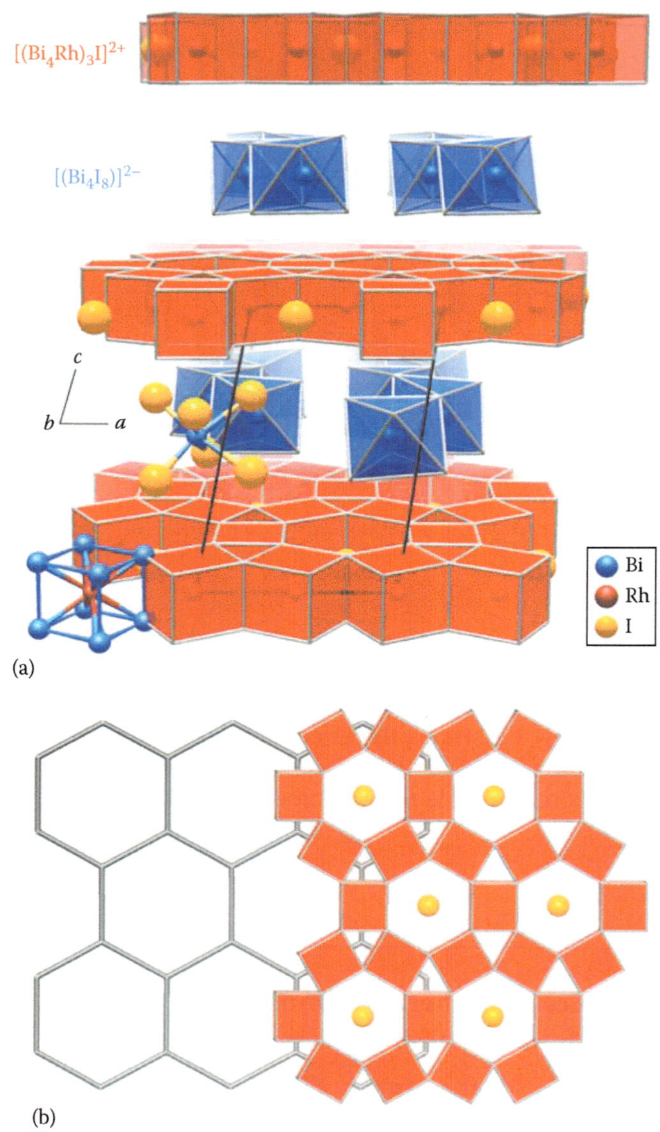

FIGURE 21.8
(a, b) Structure of $Bi_{14}Rh_3I_9$ and its relation to graphene. (Reprinted by permission from Macmillan Publishers Ltd., *Nat. Mater.*, Rasche, B., Isaeva, A., Ruck, M. et al., Stacked topological insulator built from bismuth-based graphene sheet analogues, 12, 422–425, copyright 2013.)

($[(RhBi_4)_3I]^{2+}$) and the topologically trivial NI layers ($[Bi_2I_8]^{2-}$) along the *c*-axis. The QSH layers crystallize in a ruby lattice where the iodine-centered honeycombs are surrounded by Rh-centered Bi_8 cubes, while the NI layers consist of parallel one-dimensional (1D) zigzag chains, containing edge-sharing BiI_6 octahedrons running along the *b*-axis. Weak bonding in the NI layers is predicted from the large interchain distance since the nearest interchain iodine–iodine distance is about 426 pm [42]. Lattice dynamics phonon calculations indicate the significantly low sound velocity in $Bi_{14}Rh_3I_9$. The average sound velocity is 1345 m s^{-1} along the in-plane ([00ξ]) direction and 1339 m s^{-1} along the cross-plane direction ([ξξ0]), lower than those in Bi_2Te_3 [43].

It is found that the topologically trivial NI layers also play an important role in lattice dynamics. Figure 21.9 shows the low-frequency phonon dispersion of $Bi_{14}Rh_3I_9$ projected onto the QSH layers and NI layers. There are two low-frequency optical branches near 18.7 cm^{-1} at the zone center (denoted by the circles and squares) that are due to the in-plane breathing and shear modes of the zigzag chains. The lowest zone-center frequency (16.7 cm^{-1}, denoted by triangles) acquires the character mainly from the QSH layers. This branch involves the bending mode of the QSH layers.

These low-frequency optical modes are attributable to the weak bonding in the NI layers and the heavy constituents in the QSH layers. The avoided crossing between the optical branches and the LA branch near the midpoint of the Brillouin zone (Figure 21.9) is a signature of strong acoustic–optical interactions and is caused by the localized vibrations of the NI layers together with a symmetry match between the interacting phonon modes. Because of the strong acoustic–optical interactions, the LA branch is remarkably flattened all the way to the zone boundary, resulting in both reduced group velocities and extremely low cutoff frequencies. Furthermore, the strong acoustic–optical interactions could introduce additional resonant phonon scattering [44]. In addition, the phonon calculations also reveal that the acoustic modes of this material have a very large Grüneisen parameter in excess of 2.5, indicative of strong lattice anharmonicity.

The low sound velocity, avoided crossing, and large lattice anharmonicity collectively lead to very low lattice thermal conductivity values approaching the minimum lattice thermal conductivity [45]. Figure 21.10 shows the temperature dependence of the lattice

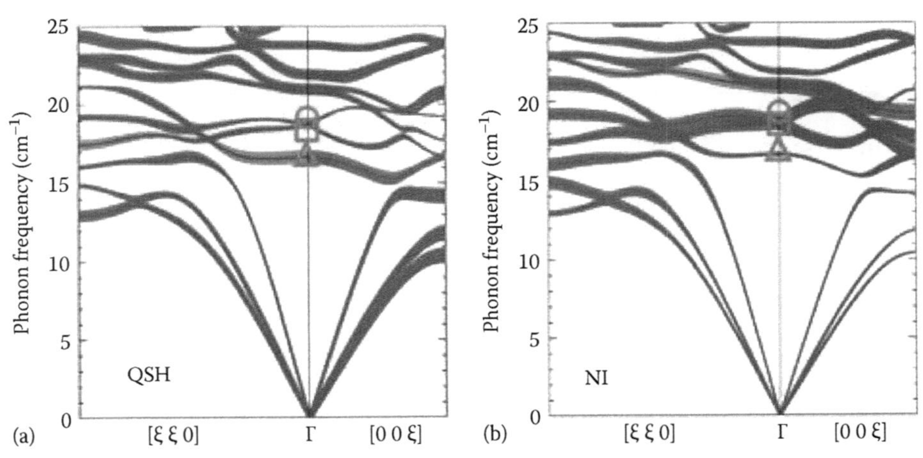

FIGURE 21.9
Phonon dispersion projected on the (a) QSH and (b) NI layers. (From Wei, P. et al., *Adv. Func. Mater.*, DOI: 10.1002/adfm 201600718, 2016.)

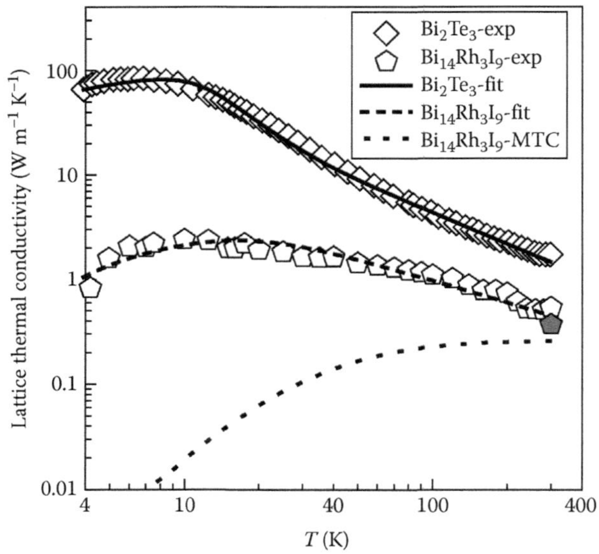

FIGURE 21.10
Low-temperature lattice thermal conductivity of $Bi_{14}Rh_3I_9$ and Bi_2Te_3 along the in-plane direction. The symbols represent the experimentally measured in-plane thermal conductivity of $Bi_{14}Rh_3I_9$ and Bi_2Te_3 ($Bi_{14}Rh_3I_9$-exp and Bi_2Te_3-exp). The solid and dashed lines are fits to the data. The dotted lines represent the calculated minimum thermal conductivity ($Bi_{14}Rh_3I_9$-MTC) for $Bi_{14}Rh_3I_9$. The filled symbol indicates data acquired based on the laser flash technique.

thermal conductivity for $Bi_{14}Rh_3I_9$ along the in-plane direction. Data for Bi_2Te_3 are plotted for comparison. The dotted line in Figure 21.10 represents the minimum thermal conductivity. It is clear that the lattice thermal conductivity of this layered material approaches the minimum thermal conductivity near room temperature, which is very unusual for a crystalline material [46]. This is attributed to the structural arrangement in this weak TI as well as the weak chemical bonds.

21.5 Conclusions

In this chapter, I have reviewed recent attempts to find superior thermoelectrics among a new class of quantum matter—the TIs. The theoretical studies aimed at distinguishing the contributions from the bulk and the boundary states in order to identify mechanisms and geometrical factors that could lead to high *ZT*. To utilize the remarkable boundary states promised by the theory, nanowires with high surface-to-volume ratios have been synthesized and measured. So far, there have yet to be any concrete data to experimentally demonstrate the benefits of the boundary states, which calls for additional future investigations. Very interestingly, the intrinsic band inversion in TIs caused by strong spin–orbit coupling has been used to discover new TIs in thermoelectric half-Heusler compounds, as well as for providing insight on the band structure origin of the superior thermoelectric properties in Bi_2Te_3-based materials. Finally, the minimum lattice thermal conductivity near room temperature has been found in weak TIs. The data presented here are just a small fraction of a very vibrantly evolving field, with promise for new scientific advances and technological innovations.

Acknowledgment

The author acknowledges the support by the US National Science Foundation under award No. 1235535.

References

1. Kane, C. and E. Z. Mele. 2005. Topological order and the quantum spin Hall effect. *Phys. Rev. Lett.* 95: 146802.
2. Bernevig, B. A., T. L. Hughes, and S.-C. Zhang. 2006. Quantum spin Hall effect and topological phase transition in HgTe quantum wells. *Science* 314: 1757–1761.
3. Fu, L., C. Kane, and E. Mele. 2007. Topological insulators in three dimensions. *Phys. Rev. Lett.* 98: 106803.
4. Moore, J. E. and L. Balents. 2007. Topological invariants of time-reversal-invariant band structures. *Phys. Rev. B* 75: 121306(R).
5. Roy, R. 2009. Topological phases and the quantum spin Hall effect in three dimensions. *Phys. Rev. B* 79: 195322.
6. König, M., S. Wiedmann, C. Brüne et al. 2007. Quantum spin Hall insulator state in HgTe quantum wells. *Science* 318: 766–770.
7. Hsieh, D., D. Qian, L. Wray et al. 2008. A topological Dirac insulator in a quantum spin Hall phase. *Nature* 452: 970–974.
8. Hsieh, D., Y. Xia, L. Wray et al. 2009. Observation of unconventional quantum spin textures in topological insulators. *Science* 323: 919–922.
9. Xia, Y., D. Qian, D. Hsieh et al. 2009. Observation of a large-gap topological-insulator class with a single Dirac cone on the surface. *Nat. Phys.* 5: 398–402.
10. Zhang, H., C.-X. Liu, X.-L. Qi, X. Dai, Z. Fang, and S.-C. Zhang. 2009. Topological insulators in Bi_2Se_3, Bi_2Te_3 and Sb_2Te_3 with a single Dirac cone on the surface. *Nat. Phys.* 5: 438–442.
11. Chen, Y. L. J. G. Analytis, J. H. Chu et al. 2009. Experimental realization of a three-dimensional topological insulator Bi_2Te_3. *Science* 325: 178–181.
12. Moore, J. E. 2010. The Birth of topological insulators. *Nature* 464: 194–198.
13. Hsieh, D., Y. Xia, D. Qian et al. 2009. A tunable topological insulator in the spin helical Dirac transport regime. *Nature* 460: 1101–1105.
14. Chadi, D. J., J. P. Walter, M. L. Cohen, Y. Petroff, and M. Balkanski. 1972. Reflectivities and electronic band structures of CdTe and HgTe. *Phys. Rev. B* 5: 3058–3064.
15. Dai, X., T. L. Hughes, X.-L. Qi, Z. Fang, and S.-C. Zhang. 2008. Helical edge and surface states in HgTe quantum wells and bulk insulators. *Phys. Rev. B* 77: 125319.
16. Chadov, S., X. Qi, J. Kübler, G. H. Fecher, C. Felser, and S.-C. Zhang. 2010. Tunable multifunctional topological insulators in ternary Heusler compounds. *Nat. Mater.* 9: 541–545.
17. Brüne, C., C. X. Liu, E. G. Novik et al. 2011. Quantum Hall effect from the topological surface states of strained bulk HgTe. *Phys. Rev. Lett.* 106: 126803.

18. Pan, Z.-H., A. V. Fedorov, D. Gardner, Y. S. Lee, S. Chu, and T. Valla. 2012. Measurement of an exceptionally weak electron-phonon coupling on the surface of the topological insulator Bi_2Se_3 using Angle-Resolved Photoemission Spectroscopy. *Phys. Rev. Lett.* 108: 187001.
19. Xu, Y., I. Miotkowski, C. Liu et al. 2014. Observation of topological surface state quantum Hall effect in an intrinsic three-dimensional topological insulator. *Nat. Phys.* 10: 956–963.
20. Goldsmid, J. H. 1986. *Thermoelectric Refrigeration*. London, U.K.: Pion Limited.
21. Ioffe, A. F. 1957. *Semiconductor Thermoelements and Thermoelectric Cooling*. London, U.K.: Infosearch Limited.
22. Terasaki, I., Y. Sasago, and K. Uchinokura. 1997. Large thermoelectric power in $NaCO_2O_4$ single crystals. *Phys. Rev. B* 56: R12685–R12687.
23. Funahashi, R., I. Matsubara, H. Ikuta, T. Takeuchi, U. Mizutani, and S. Sodeoka. 2000. An oxide single crystal with high thermoelectric performance in air. *Jpn. J. Appl. Phys.* 39: L1127–L1129.
24. Mahan, G. D. 1989. Figure of merit for thermoelectrics. *J. Appl. Phys.* 65: 1578–1583.
25. Fu, L. and C. L. Kane. 2007. Topological insulators with inversion symmetry. *Phys. Rev. B* 76: 045302.
26. Müchler, L., F. Casper, B. Yan, S. Chadov, and C. Felser. 2013. Topological insulators and thermoelectric materials. *Phys. Stat. Solidi RRL* 7: 91–100.
27. Takahashi, R. and S. Murakami. 2010. Thermoelectric transport in perfectly conducting channels in quantum spin Hall systems. *Phys. Rev. B* 81: 161302(R).
28. Takahashi, R. and S. Murakami. 2012. Thermoelectric transport in topological insulators. *Semicond. Sci. Technol.* 27: 24005.
29. Xu, Y., Z. Gan, and S.-C. Zhang. 2014. Enhanced thermoelectric performance and anomalous Seebeck effects in topological insulators. *Phys. Rev. Lett.* 112: 226801.
30. Hsiung, T.-C., C.-Y. Mou, T.-K. Lee, and Y.-Y. Chen. 2015. Surface-dominated transport and enhanced thermoelectric figure of merit in topological insulator $Bi_{1.5}Sb_{0.5}Te_{1.7}Se_{1.3}$. *Nanoscale* 7: 518–523.
31. Mavrokefalos, A., A. L. Moore, M. T. Pettes, L. Shi, W. Wang, and X. Li. 2009. Thermoelectric and structural characterizations of individual electrodeposited bismuth telluride nanowires. *J. Appl. Phys.* 105: 104318.
32. Hamdou, B., J. Kimling, A. Dorn et al. 2013. Thermoelectric characterization of bismuth telluride nanowires, synthesized via catalytic growth and post-annealing. *Adv. Mater.* 25: 239–244.
33. Hamdou, B., J. Gooth, T. Böhnert et al. 2015. Thermoelectric properties of band structure engineered topological insulator $(Bi_{1-x}Sb_x)_2Te_3$ nanowires. *Adv. Energy Mater.* 5: 1500280.
34. Goyal, V., D. Teweldebrhan, and A. A. Balandin. 2010. Mechanically-exfoliated stacks of thin films of Bi_2Te_3 topological insulators with enhanced thermoelectric performance. *Appl. Phys. Lett.* 97: 133117.
35. Aliev, F. G., N. B. Brandt, V. V. Moshchalkov, V. V. Kozyrkov, R. V. Skolozdra, and A. I. Belogorokhov. 1989. Gap at the Fermi level in the intermetallic vacancy system RNiSn (R = Ti, Zr, Hf). *Z. Phys. B: Condens. Matter* 75: 167–171.
36. Uher, C., J. Yang, S. Hu, D. T. Morelli, and G. P. Meisner. 1999. Transport properties of pure and doped MNiSn (M = Zr, Hf). *Phys. Rev. B* 59: 8615–8621.
37. Yang J., H. Li, T. Wu, W. Zhang, L. Chen, and J. Yang. 2008. Evaluation of half-Heusler compounds as thermoelectric materials based on the calculated electrical transport properties. *Adv. Funct. Mater.* 18: 2880–2888.
38. Zhu T., C. Fu, H. Xie, Y. Liu, and X. Zhao. 2015. High efficiency half-Heusler thermoelectric materials for energy harvesting. *Adv. Energy Mater.* 5: 1500588.
39. Shi, H., D. Parker, M.-H. Du, and D. J. Singh. 2015. Connecting thermoelectric performance and topological-insulator behavior: Bi_2Te_3 and Bi_2Te_2Se from first principles. *Phys. Rev. Appl.* 3: 014004.
40. Chen, X., D. Parker, and D. J. Singh. 2013. Importance of nonparabolic band effects in the thermoelectric properties of semiconductors. *Sci. Rep.* 3: 3168.

41. Wei, P., Jiong Yang, L. Guo, S. Wang, L. Wu, X. Xu, W. Zhao, Q. Zhang, W. Zhang, M. S. Dresselhaus, and J. Yang, 2016. Minimum thermal conductivity in weak topological insulators with bismuth-based stack structure. *Adv. Func. Mater.*, DOI: 10.1002/adfm 201600718.
42. Rasche, B., A. Isaeva, M. Ruck et al. 2013. Stacked topological insulator built from bismuth-based graphene sheet analogues. *Nat. Mater.* 12: 422–425.
43. Huang, B. L. and M. Kaviany. 2008. *Ab initio* and molecular dynamics predictions for electron and phonon transport in bismuth telluride. *Phys. Rev. B* 77: 125209.
44. Tse, J. S. and M. A. White. 1988. Origin of glassy crystalline behavior in the thermal properties of clathrate hydrates: A thermal conductivity study of tetrahydrofuran hydrate. *J. Phys. Chem.* 92: 5006–5011.
45. Slack, G. A. 1979. The thermal conductivity of nonmetallic crystals. In *Solid State Physics*, eds. H. Ehrenreich, F. Seitz, and D. Trunbull, pp. 1–69. New York: Academic Press.
46. Skoug, E. C. and D. T. Morelli. 2011. Role of lone-pair electrons in producing minimum thermal conductivity in nitrogen-group chalcogenide compounds. *Phys. Rev. Lett.* 107: 235901.

Index

T

An environmentally friendly book printed and bound in England by www.printondemand-worldwide.com

PEFC Certified

This product is from sustainably managed forests and controlled sources

www.pefc.org

PEFC

PEFC/16-33-415

This book is made of chain-of-custody materials; FSC materials for the cover and PEFC materials for the text pages.